Passive Margins: Tectonics, Sedimentation and Magmatism

The Geological Society of London Books Editorial Committee

Chief Editor
Rick Law (USA)

Society Books Editors
Jim Griffiths (UK)
Dan Le Heron (Austria)
Mads Huuse (UK)
Rob Knipe (UK)
Phil Leat (UK)
Teresa Sabato Ceraldi (UK)
Lauren Simkins (US)
Randell Stephenson (UK)
Gabor Tari (Austria)
Mark Whiteman (UK)

Society Books Advisors
Kathakali Bhattacharyya (India)
Anne-Christine Da Silva (Belgium)
Xiumian Hu (China)
Jasper Knight (South Africa)
Spencer Lucas (USA)
Dolores Pereira (Spain)
Virginia Toy (Germany)
Georg Zellmer (New Zealand)

Geological Society books refereeing procedures

The scientific and production quality of the Geological Society's books matches that of its journals. Since 1997, all book proposals are reviewed by two individual experts and the Society's Books Editorial Committee. Proposals are only accepted once any identified weaknesses are addressed.

The Geological Society of London is signed up to the Committee on Publication Ethics (COPE) and follows the highest standards of publication ethics. Once a book has been accepted, the volume editors agree to follow the Society's Code of Publication Ethics and facilitate a peer review process involving two independent reviewers. This is overseen by the Society Book Editors who ensure these standards are adhered to.

Geological Society books are timely volumes in topics of current interest. Proposals are often devised by editors around a specific theme or they may arise from meetings. Irrespective of origin, editors seek additional contributions throughout the editing process to ensure that the volume is balanced and representative of the current state of the field.

Submitting a book proposal
More information about submitting a proposal and producing a book for the Society can be found at https://www.geolsoc.org.uk/proposals.

It is recommended that reference to all or part of this book should be made in one of the following ways:

McClay, K. R. and Hammerstein, J. A. (eds) 2020. *Passive Margins: Tectonics, Sedimentation and Magmatism*. Geological Society, London, Special Publications, **476**, https://doi.org/10.1144/SP476.0

Tamara, J., McClay, K. R. and Hodgson, N. 2020. Crustal structure of the central sector of the NE Brazilian equatorial margin. Geological Society, London, Special Publications, **476**, 163–192, https://doi.org/10.1144/SP476-2019-54

Geological Society Special Publication No. 476

Passive Margins: Tectonics, Sedimentation and Magmatism

Edited by

K. R. McClay
Royal Holloway, University of London, UK
The University of Adelaide, Australia

and

J. A. Hammerstein
Royal Holloway, University of London, UK

2020
Published by
The Geological Society
London

The Geological Society of London

The Geological Society of London is a not-for-profit organisation, and a registered charity (no. 210161). Our aims are to improve knowledge and understanding of the Earth, to promote Earth science education and awareness, and to promote professional excellence and ethical standards in the work of Earth scientists, for the public good. Founded in 1807, we are the oldest geological society in the world. Today, we are a world-leading communicator of Earth science – through scholarly publishing, library and information services, cutting-edge scientific conferences, education activities and outreach to the general public. We also provide impartial scientific information and evidence to support policy-making and public debate about the challenges facing humanity. For more about the Society, please go to https://www.geolsoc.org.uk/

The Geological Society Publishing House (Bath, UK) produces the Society's international journals and books, and acts as European distributor for selected publications of the American Association of Petroleum Geologists (AAPG), the Geological Society of America (GSA), the Society for Sedimentary Geology (SEPM) and the Geologists' Association (GA). GSL Fellows may purchase these societies' publications at a discount. The Society's online bookshop is at https://www.geolsoc.org.uk/bookshop

To find out about joining the Society and benefiting from substantial discounts on publications of GSL and other Societies go to https://www.geolsoc.org.uk/membership or contact the Fellowship Department at: The Geological Society, Burlington House, Piccadilly, London W1J 0BG: Tel. +44 (0)20 7434 9944; Fax +44 (0)20 7439 8975; E-mail: enquiries@geolsoc.org.uk

For information about the Society's meetings, go to https://www.geolsoc.org.uk/events. To find out more about the Society's Corporate Patrons Scheme visit https://www.geolsoc.org.uk/patrons

Proposing a book
If you are interested in proposing a book then please visit: https://www.geolsoc.org.uk/proposals

Published by The Geological Society from:
The Geological Society Publishing House, Unit 7, Brassmill Enterprise Centre, Brassmill Lane, Bath BA1 3JN, UK

The Lyell Collection: www.lyellcollection.org
Online bookshop: www.geolsoc.org.uk/bookshop
Orders: Tel. +44 (0)1225 445046, Fax +44 (0)1225 442836

The publishers make no representation, express or implied, with regard to the accuracy of the information contained in this book and cannot accept any legal responsibility for any errors or omissions that may be made.

© The Geological Society 2020. Except as otherwise permitted under the Copyright, Designs and Patents Act, 1988, this publication may only be reproduced, stored or transmitted, in any form or by any other means, with the prior permission in writing of the publisher, or in the case of reprographic reproduction, in accordance with the terms of a licence issued by the Copyright Licensing Agency in the UK , or the Copyright Clearance Center in the USA. In particular, the Society permits the making of a single photocopy of an article from this issue (under Sections 29 and 38 of this Act) for an individual for the purposes of research or private study. Open access articles, which are published under a CC-BY licence, may be re-used without permission, but subject to acknowledgement.

Full information on the Society's permissions policy can be found at https://www.geolsoc.org.uk/permissions

British Library Cataloguing in Publication Data
A catalogue record for this book is available from the British Library.
ISBN 978-1-78620-385-4
ISSN 0305-8719

Distributors
For details of international agents and distributors see:
http://www.geolsoc.org.uk/agentsdistributors

Typeset by Nova Techset Private Limited, Bengaluru & Chennai, India
Printed and bound by CPI Group (UK) Ltd, Croydon CR0 4YY, UK

Contents

McClay, K. and Hammerstein, J. Introduction — 1

Gerdes, K. D. Professor David Gwyn Roberts – a Life in Geoscience — 11

Tugend, J., Gillard, M., Manatschal, G., Nirrengarten, M., Harkin, C., Epin, M.-E., Sauter, D., Autin, J., Kusznir, N. and McDermott, K. Reappraisal of the magma-rich versus magma-poor rifted margin archetypes — 23

Khalil, S. M. and McClay, K. R. Extensional fault-related folding in the northwestern Red Sea, Egypt: segmented fault growth, fault linkages, corner folds and basin evolution — 49

Kusznir, N. J., Roberts, A. M. and Alvey, A. D. Crustal structure of the conjugate Equatorial Atlantic Margins, derived by gravity anomaly inversion — 83

Scarselli, N., Duval, G., Martin, J., McClay, K. and Toothill, S. Insights into the Early Evolution of the Côte d'Ivoire Margin (West Africa) — 109

Pérez-Díaz, L. and Eagles, G. Estimating palaeobathymetry with quantified uncertainties: a workflow illustrated with South Atlantic data — 135

Tamara, J., McClay, K. R. and Hodgson, N. Crustal structure of the central sector of the NE Brazilian equatorial margin — 163

Restrepo-Pace, P. A. 'Ductile v. Brittle' – Alternative structural interpretations for the Niger Delta — 193

McCormack, K. D. and McClay, K. R. Orthorhombic faulting in the Beagle Sub-basin, North West Shelf, Australia — 205

Bilal, A., McClay, K. and Scarselli, N. Fault-scarp degradation in the central Exmouth Plateau, North West Shelf, Australia — 231

Deng, H. and McClay, K. Tectono-stratigraphy of the Dampier Sub-basin, North West Shelf of Australia — 259

Dooley, T. P., Hudec, M. R., Pichel, L. M. and Jackson, M. P. A. The impact of base-salt relief on salt flow and suprasalt deformation patterns at the autochthonous, paraautochthonous and allochthonous level: insights from physical models — 287

Pindell, J., Graham, R. and Horn, B. W. Role of outer marginal collapse on salt deposition in the eastern Gulf of Mexico, Campos and Santos basins — 317

Rowan, M. G. The South Atlantic and Gulf of Mexico salt basins: crustal thinning, subsidence and accommodation for salt and presalt strata — 333

Jagger, L. J., Bevan, T. G. and McClay, K. R. Tectono-stratigraphic evolution of the SE Mediterranean passive margin, offshore Egypt and Libya — 365

Llave, E., Hernández-Molina, F. J., García, M., Ercilla, G., Roque, C., Juan, C., Mena, A., Preu, B., Van Rooij, D., Rebesco, M., Brackenridge, R., Jané, G., Gómez-Ballesteros, M. and Stow, D. Contourites along the Iberian continental margins: conceptual and economic implications   403

Index   437

Introduction

KEN McCLAY[1,2]* & JAMES HAMMERSTEIN[1]

[1]Department of Earth Sciences, Royal Holloway, University of London, Egham, Surrey, TW20 0EX, UK

[2]Australian School of Petroleum, Energy and Resources, The University of Adelaide, Adelaide, South Australia 5000, Australia

KM, 0000-0002-4077-7645

*Correspondence: terratectonicsaustralia8115@yahoo.com

Passive margins

Passive margins evolve by rifting of continental plates, rupture and separation that forms new plate boundaries and new oceanic basins. Ancient and modern plate margins have long been a focus of discovery, research and exploration for resources.

Present day passive margins form the greater part of the world's continental margins. They extend on both sides of the Atlantic Ocean, surround much of the African, Australian and Antarctic continents and form both Indian margins as well as much of the circum Arctic margins (Fig. 1). In this volume key studies are presented from Central and South Atlantic margins, the eastern Indian margin and the NW Shelf of Australia as well as from the eastern Mediterranean (Fig. 1).

Many passive margins, such as the Gulf of Mexico, offshore Brazil (Fig. 1), as well as the Angolan and Congo margins, also contain either pre- and or post-rift salt that dramatically affects their structural and stratigraphic architectures. The formation of passive margins may be influenced by many factors, including plate geodynamics, patterns of underlying mantle flow, inherited continental basement fabrics and discontinuities, extensional fault geometries and evolution of initial rift architectures, magmatic processes and breakup kinematics and timings, as well as sediment loads and their distributions during margin evolution. Recent studies and some chapters in this volume have been based on the ION Span deep seismic profiles acquired over many passive margins as well as new broadband 2D and 3D seismic surveys, particularly over the deep-water sectors of margins such as offshore Brazil (**Rowan 2018**; Tugend *et al.* 2018) and offshore West Africa (Scarselli *et al.* 2018).

This volume of 16 papers is in honour of the late Professor David Roberts, who made significant contributions to understanding of the development of passive margins and their sedimentary basins.

The introduction paper by **Gerdes (2018)** celebrates the life and contributions of Professor David Roberts to the study of passive margins from his days at the Institute of Oceanographic Studies to leading the basins studies group at BP, and lastly as a consultant and Visiting Professor at Royal Holloway, University of London, Institut Français du Pétrole and other universities. This first paper also outlines some of the history of the study of passive margins from the mid to latter part of the twentieth century and to the present day. In particular one should acknowledge David's extremely fruitful collaborations with Professors Lucien Montadert and Pierre-Charles Graciansky as this partnership produced many of the first modern insights into passive margin systems (e.g. Graciansky *et al.* 2011; also see references in **Gerdes 2018**). These studies focused in particular on the structure of the Iberian margin and the importance of comparisons with field examples of the exposed Tethyan passive margins preserved in the western Alps. Many of the more recent studies of passive margins have built upon these earlier insights into passive margin evolution (e.g. Lavier & Manatschal 2006; Mohn *et al.* 2012; Péron-Pinvidic *et al.* 2013, 2015; Manatschal *et al.* 2015; Osmundsen & Péron-Pinvidic 2018).

Structure of passive margins

The broad classification of passive margins into magma-poor and magma-rich archetypes is reviewed by Tugend *et al.* (2018). These authors analyse two ION Span deep seismic reflection profiles – one offshore eastern India (magma poor) and the other offshore Uruguay (magma-rich), in order to study the nature and magmatism of the transition from stretched continental to proto-oceanic and to oceanic crust. They conclude that the mechanics of lithospheric breakup for both archetypes are essentially similar but differences in the timing of onset of decompression melting and magmatism may

From: McCLAY, K. R. & HAMMERSTEIN, J. A. (eds) 2020. *Passive Margins: Tectonics, Sedimentation and Magmatism*. Geological Society, London, Special Publications, **476**, 1–9.
First published online March 26, 2020, https://doi.org/10.1144/SP476-2019-246
© 2020 The Author(s). This is an Open Access article distributed under the terms of the Creative Commons Attribution License (http://creativecommons.org/licenses/by/4.0/). Published by The Geological Society of London.
Publishing disclaimer: www.geolsoc.org.uk/pub_ethics

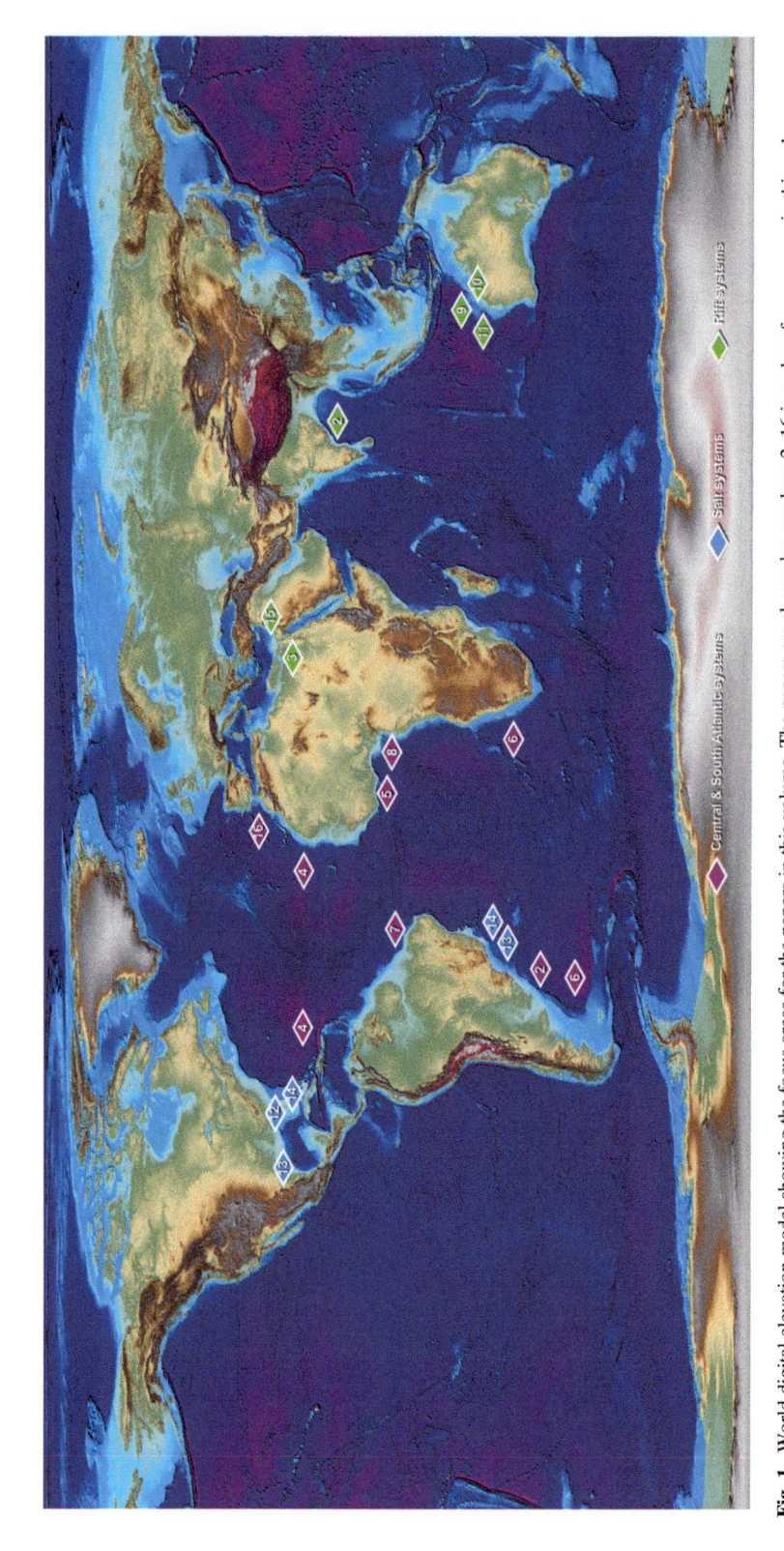

Fig. 1. World digital elevation model showing the focus areas for the papers in this volume. The papers are shown by numbers 2–16 in order of appearance in this volume. Regions of oceanic crust are shown in shades of blue to purple with the trenches of subduction systems highlighted by narrow elongate zones of intense purple colour. The colours for continental terranes vary from low elevation greens and yellows to greys and reds at the highest elevations (e.g. Himalayas). 2, Tugend *et al.*; 3, Khalil & McClay; 4, Kusznir *et al.*; 5, Scaselli *et al.*; 6, Perez-Diaz & Eagles; 7, Tamara *et al.*; 8, Restrepo-Pace; 9, McCormack & McClay; 10, Bilal & McClay; 11, Deng & McClay; 12, Dooley *et al.*; 13, Pindell *et al.*; 14, Rowan; 15, Jagger *et al.*; 16, Llave *et al.* Global DEM data courtesy of NASA.

generate different archetypal geometries and kinematics. Three possible interpretations and models for the architecture of lithospheric breakup are presented for each margin studied. Passive margin models from this research group (e.g. stylized in Fig. 2) are widely used as interpretational 2D templates for many recent passive margin studies (Péron-Pinvidic & Osmundsen 2018; Manatschal *et al.* 2015; Péron-Pinvidic *et al.* 2015; Osmundsen & Péron-Pinvidic 2018) as well as in papers presented in this volume (**Restrepo-Pace 2018**; **Rowan 2018**; **Scarselli *et al.* 2018**).

The second paper in this section (**Khalil & McClay 2018**) examines rift margin structures in the Northern Red Sea in Egypt as an example of the architecture of the inboard rift basin systems that occur on many passive margins. Passive margins commonly evolve from continental rift basins, and the structures formed at these early stages of extension prior to breakup are inherited as the passive margin evolves. The field examples illustrate the complex fault linkages between segmented and en-echelon rift faults (Fossen & Rotevatn 2016) as well as associated extensional folding in segmented rift half-grabens. These types of fault architectures are likely to also be found in other rifts (e.g. Henstra *et al.* 2015), as well as in the marginal basins of many passive margins (e.g. Northwest Shelf of Australia, **Deng & McClay 2019**).

Central and South Atlantic margins

The Atlantic margins have long been studied as type examples of passive margins, particularly in the context of the formation of conjugate margins such as those offshore Iberia–offshore Newfoundland (Péron-Pinvidic *et al.* 2013). Papers in this section focus mainly on the Equatorial Atlantic margins (**Kusznir *et al.* 2018**; **Scarselli *et al.* 2018**; **Tamara *et al.* 2020**), but also include the development of palaeobathymetry in the South Atlantic realm (**Pérez-Díaz & Eagles 2018**) and the shale tectonics of the Niger delta system (**Restrepo-Pace 2018**; see Fig. 1).

Kusznir *et al.* (2018) use regional-scale gravity inversion of the equatorial Atlantic region to map the 3D crustal thickness from the continental margins to the oceanic basins. In a series of regional maps and cross-sections they show depth to Moho and total crustal thicknesses including continental and oceanic, residual continental crustal thicknesses as well as continental lithosphere stretching and thinning factors. Using G Plate reconstructions they demonstrate that the Equatorial Atlantic opened using a series of stepped transform-rift segments that gave rise to variable crustal architectures along both the Brazilian and West African margins. The results from the gravity inversion and their crustal cross-sections are validated by strong agreement with published crustal cross-sections derived from wide-angle seismic studies. In particular this research highlights the anomalous crustal thicknesses encountered along transform margins. Gravity inversion results can be an important input into continental margin studies and plate reconstructions, for regional palaeogeographical reconstructions and palaeoheat flow predictions. The authors demonstrate that gravity inversion can be a valuable tool for deep-water basin analysis and may be applicable to other passive margins, particularly where there is little other data available.

Scarselli *et al.* (2018) analyse a modern depth migrated 3D broadband seismic volume offshore Côte d'Ivoire to determine the tectono-stratigraphic evolution and the deep structure underlying this part of the equatorial margin of West Africa (Fig. 1). Deep-rotated fault blocks with hanging-wall Aptian to Albian fanning growth strata, in places exceeding 3.5 km thickness, detach on a low-angle basal detachment. Aptian–Albian sill complexes are developed as well as prominent volcanic complexes that together form a widespread igneous province. Significant ridges occur perpendicular to the main extensional fault trends and these are interpreted to result from lateral movements owing to the proximity of the transform systems that occur in this region. The authors interpret the deep structure to be in the hyper-extensional to exhumed domains (similar to the model shown in Fig. 2) with transform elements and associated igneous intrusions and volcanic edifices. The results indicate that the development of the equatorial margin of West Africa may have initiated in the Aptian and the deep structures are not simply cylindrical but vary significantly along-strike and thus illustrate the necessity of deep 3D seismic data in order to fully characterize passive margins with oblique and transform elements.

Palaeobathymetric reconstructions from the Early Cretaceous of the South Atlantic basins (Fig. 1) are presented by **Pérez-Díaz & Eagles (2018)**. The methodology incorporates oceanic-age grids, satellite gravity data, dynamic topography and sediment thickness distributions and ages in order to calculate the depths of basement and oceanic crust at time slices for the South Atlantic evolution. Comparisons with measured palaeobathymetries from drill sites indicate that this method is accurate to within 700 m over large parts of the deep ocean to but less so near large igneous provinces and at earlier time periods. These reconstructions have significant implications for palaeo-oceanographic studies and the prediction of palaeodepostional environments on passive margins, and the methodology can be applied to other passive margins.

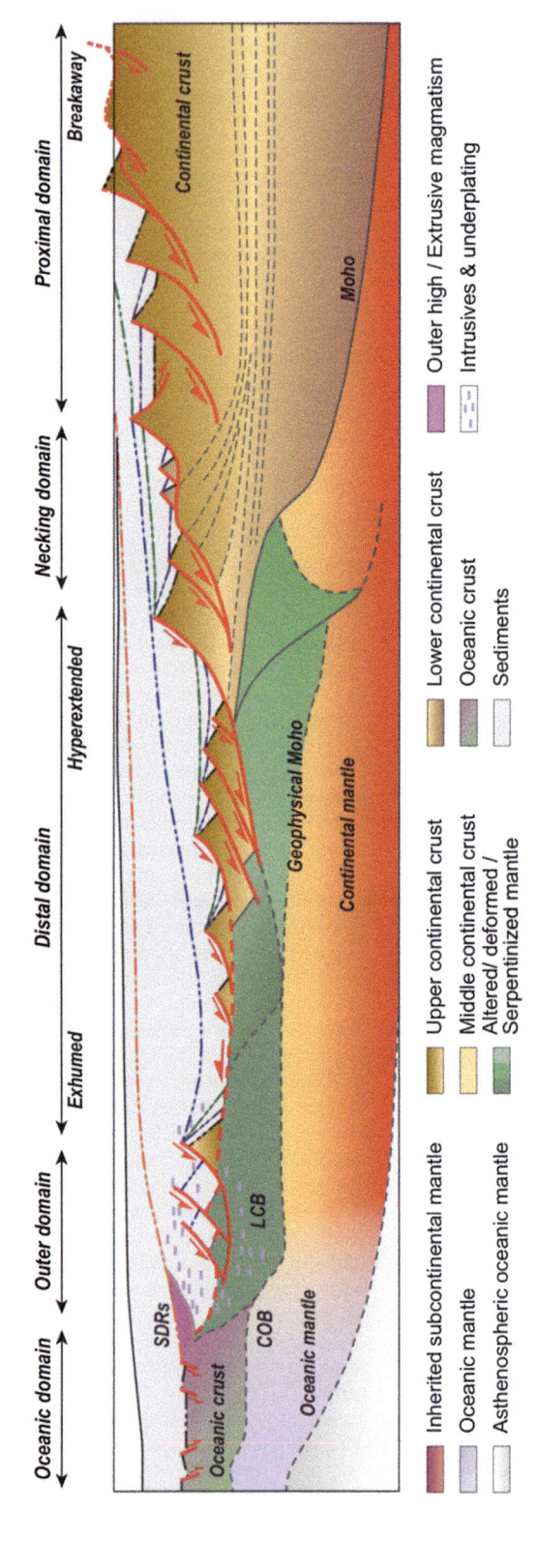

Fig. 2. Synoptic and schematic 2D lithospheric model of a passive margin based on Osmundsen & Péron-Pinvidic (2018) and Péron-Pinvidic & Osmundsen (2018). This figure summarizes some the main elements of many passive margins. Individual passive margins may vary greatly along-strike and not display all of these elements in every 2D section. (**a**) *Proximal domain* – the inboard rift fault systems with syn-rift sediments in hanging-wall basins; (**b**) *necking domain* – greater extension and rotation of crustal fault blocks – ductile stretching in the lower crust; (**c**) *distal – hyper-extension domain* – an outer domain of highly extended and rotated crustal blocks with very low angle to horizontal detachment faults. Large displacements and highly thinned lower crust; (**d**) *distal – exhumed domain* – an outer domain of highly extended boudinaged crust in places riding on exhumed continental mantle; (**e**) *outer domain – break up zone* with igneous intrusions (dykes and sills) as well as seaward-dipping reflectors formed by lava flows and volcanoclastic wedges that dip oceanward; continent–ocean boundary; (**f**) *Oceanic domain* – thin high-velocity oceanic crust commonly with a well-defined seismic Moho (e.g. figures in Tamara *et al.* 2020).

The crustal architecture of the central equatorial margin of northeastern Brazil is documented by **Tamara et al. (2020)** (Fig. 1). This oblique divergent margin is segmented by transform structures where the Saint Paul Fracture Zone impinges on the extensional sectors. Analysis of high-quality deep regional seismic lines has revealed distinct domains similar to those shown in Figure 2 but varying along the margin. In three dimensions along-trike there is an inboard rift domain, a necking domain passing to an outboard hyper-extension domain and exhumation domain transitioning to oceanic crust. The margin is steep with distinct gravity-driven slide complexes with linked updip listric extensional faults and downdip fold-and-thrust systems. Post-rift volcanic complexes are found in the exhumation and oceanic domains. The results of this study show that the architecture of this passive margin formed by the Cretaceous breakup of the Equatorial Atlantic varies significantly along-strike and evolutionary models need to be built on a number of seismic sections and not just on a single 'hero' line.

Progradational delta systems have formed on many passive margins such as those of the Mississippi delta on the Gulf of Mexico margin (Wu & Bally 2000), the Nile delta on the Mediterranean margin of Egypt and the Niger delta system offshore Nigeria, West Africa. They commonly contain significant hydrocarbon resources. **Restrepo-Pace (2018)** (Fig. 1) reviews the tectono-sedimentary evolution of the Niger delta and challenges the long-held model of ductile, overpressured 'mobile shale' in the deformed zones (cf. also Krueger & Grant 2013). In previous studies seismic blank zones have been usually interpreted as mobile ductile shale that formed 'shale ridges and shale diapirs' (Ajakaiye & Bally 2002; Weiner et al. 2010; Wood 2010). This paper demonstrates that, in modern long-offset seismic data, the apparently chaotic, 'seismic wipe-out' zones show coherent internal stratal geometries. This indicates that the 'overpressured shales' (e.g. Cobbold et al. 2009) did not deform like salt by continuous internal plastic flow but instead deformed by brittle processes – coherent folding, extension and imbrication on discrete fault surfaces facilitated by high porefluid pressures (cf. Krueger & Grant 2013). This 'brittle' model has important implications for understanding the structural styles in shale-detached deltas, for section balancing and for models of trap formation, particularly in deep-water fold-and-thrust belts on passive margins as well as in accretionary prisms.

Northwestern Australian passive margin

The three papers in this group focus on structural styles on the NW Shelf passive margin of northern Australia (Fig. 1). This is a Mesozoic–Cenozoic passive margin that underwent Permo-Carboniferous rifting followed by Late Triassic–Late Jurassic rifting with Late Cretaceous breakup and Cenozoic thermal subsidence (Baillie et al. 1994; Longley et al. 2002; Heine & Müller 2005). It is a major hydrocarbon province with significant gas and oil accumulations (Longley et al. 2002).

The paper by **McCormack & McClay (2018)** is a detailed structural analysis of a high-quality 3D survey on the outer part of the Beagle Sub-Basin – an inboard rift basin on the NW Shelf passive margin. The complex fault architectures display a 'pseudo conjugate' pattern of horsts and grabens but are interpreted to be an orthorhombic strain pattern in keeping with the regional WNW extension direction as the passive margin developed in the Late Triassic through to the Early Cretaceous. This analysis shows that different orientations of extensional fault systems do not necessarily imply that the regional driving stresses have rotated but rather that complex fault geometries and 'pseudo conjugate systems' may be developed in a single, uniform regional stress field.

Bilal et al. (2018) describe the extensional fault systems in the Exmouth Plateau of the NW Australian passive margin. Here domino-style extensional faults display significant footwall uplift and rotation together with fault scarp degradation and retreat – in places up to 1.8 km. Detailed fault displacement analyses show a clear correlation between the maximum uplift, maximum fault displacements and maximum scarp retreat/degradation. In other rift basins such as the Central and Northern North Sea (e.g. Fraser et al. 2002; Welbon et al. 2007), fault scarp degradation significantly changes fault and footwall geometries as well as potential reservoir architectures and volumes. Fault scarp degradation, whether by erosion or gravitational collapse, is likely to be found in the inboard sectors of many passive margins in the marginal rift blocks and in the necking zones (e.g. Fig. 2).

The paper by **Deng & McClay (2019)** focuses on the tectono-stratigraphic development of the Dampier Sub-Basin in the NW Australian passive margin. This is a Late Triassic to Late Jurassic inboard, marginal rift basin that was controlled by an earlier Permo-Carboniferous rift system along the NW margin of Australia. A new tectono-stratigraphic model is presented for this inboard rift basin that formed as a result of WNW extension superposed on an older Late Paleozoic NE–SW rift system. This tectonic history generated en-echelon overlapping, segmented and complexly linked Mesozoic extensional faults. Similar fault architectures have been documented in the northern Red Sea (e.g. Khalil & McClay 2016, 2018) and the Norwegian margin (e.g. Henstra et al. 2015) and illustrate the importance of understanding basement inheritance in both rifts and on passive margins.

Salt basins on passive margins

Many passive margins such as the northern Gulf of Mexico, the Campos and Santos basins offshore Brazil and the Angola and Congo basins of West Africa have significant salt basins that host major hydrocarbon resources (e.g. Heine et al. 2013; Hudec et al. 2013; Rowan 2014). Salt sections produce complex structures and strongly affect the stratal architectures and sedimentation within passive margin basins.

Scaled physical modelling of salt tectonics has become the principal tool in understanding the progressive evolution of salt systems in basins as copiously illustrated in the outstanding text book by Jackson & Hudec (2017). **Dooley et al. (2018)** describe some stunning analogue model experiments that demonstrate how salt may flow up over ridges and over complex topography that can be developed in extensional basins on passive margins. These experiments applied to the eastern Gulf of Mexico demonstrate that inherited topography controls salt dynamics and the types of salt accumulations formed by downslope flow on a subsiding passive margin. The models can be compared with the sections shown in the papers by **Pindell et al. (2018)** and **Rowan (2018)** in this volume and provide insights into the progressive evolution of these systems.

Pindell et al. (2018) use regional deep seismic lines to demonstrate 'outer margin collapse' (e.g. Pindell et al. 2014) and its effects on the evolution of the mega-salt basins on the Florida passive margin of the eastern Gulf of Mexico, as well as on the Santos and Campos basins of eastern Brazil. They propose a multistage extension model for the initial rift phase with deposition at or near sea-level followed by a widespread sag phase in response to thermal subsidence. This is followed by incipient 'breakup' with extension focused on outer margin fault systems followed by margin collapse and initial salt deposition. At the breakup stage depth-dependent stretching and subsidence of the outer margin produce regional tilting into the basin. The post-rift sag strata and base of salt dip ocean-ward together with down-slope salt flow towards the newly formed spreading centre. This model and variations of it may account for the salt distributions on the Florida passive margin of the eastern Gulf of Mexico as well as the salt basin architectures in the South Atlantic salt basins.

The third paper in this section by **Rowan (2018)** examines the similarities shared by salt basins in the Gulf of Mexico with those in the South Atlantic. Rowan analyses the salt geometries with respect to the basement architectures formed in the necking zone at the transition from highly extended continental crust to fully developed oceanic crust. He concludes that the salt formed in isolated basins at or near sea-level and that subsequent hyper-extension resulted in large, deep underfilled depocentres with basin-ward flow of salt producing inflation and allochthonous salt, in places on top of the oceanic crust. **Dooley et al. (2018)** simulated similar features in the analogue models of salt flow into the basin.

Other margins

Jagger et al. (2018) present new models for the tectono-stratigraphic evolution of the SE Mediterranean basin offshore Libya, Egypt and Lebanon. This passive margin is segmented into both rift- and transform-dominated segments. The extensional fault systems are strongly influenced by repeated reactivation of Pan-African basement fabrics and shear zones, as shown by the inheritance of Neo-tethyan rift systems and their control on the Santonian–Maastrichtian and Mid-Eocene to Oligocene inversion events. This study gives insights into the complex transform components of passive margins as also seen in the Ghana–Côte d' Ivoire margin on the equatorial margin of West Africa (cf. **Scarselli et al. 2018**), as well as on the Brazilian margin (**Tamara et al. 2020**).

The development of contourite systems around the Iberian passive margins is reviewed by **Llave et al. (2019)**. Contourites on passive margins are important potential new locations for hydrocarbon exploration. Using high-resolution seismic sections, this paper summarizes the key characteristics and architectures of contourites on the Atlantic margin of Iberia well as on the Mediterranean margin of Iberia. The models and examples of contourite depositional systems illustrated in this paper may be applied to other passive margins (e.g. South Atlantic), where they will be relevant to research as well as to exploration for new hydrocarbon reservoirs.

Summary

Passive margins are far from passive. They display a variety of dynamic structures developed during, as well as after, rifting, necking, thinning and breakup. These include hyper-extended crustal blocks, uplift and exposure of subcontinental mantle and stretching-related magmatism. Syn- and post-breakup loading by continental margin wedges, by progradational delta systems as well as by gravity-driven slide complexes from margin uplift and subsequent collapse substantially affect the architectures of many passive margins (cf. **Pindell et al. 2014, 2018**). It is noteworthy that most of the current models for passive margin evolution rely largely on 2D regional seismic lines and relatively little is known about the 3D architectures along-strike, and 3D seismic data is needed particularly for oblique and transform margins (e.g. Scarselli *et al.* 2018).

It is clear that in many margins inherited basement structures control the initial rift architectures (e.g. as in in the Gulf of Suez and Red Sea Bosworth *et al.* 2005; **Khalil & McClay** 2016, **2018**) as well as other rift basins (Tvedt *et al.* 2013; Henstra *et al.* 2015). These patterns are commonly reflected in the inboard structures of passive margins (e.g. Cobbold *et al.* 2001; Mohn *et al.* 2012). In particular the scales of rift margin uplift, syn- and post-rift exhumation and subsequent collapse into the evolving passive margin have not been widely studied. Invaluable insights have been gained from 2D deep reflection seismic profiles but future research on passive margins also needs to include 3D reflection seismic surveys. Results from such studies will need to be integrated with refraction, gravity and magnetic data in order to take the current 2D models (e.g. Fig. 2) into 3D, as recently shown for the distal Iberian margin by Lymer *et al.* (2019). Many passive margins are under-explored and new seismic and drilling data are much needed to improve current models for their evolution.

Acknowledgements The editors would like to thank the Department of Earth Sciences and the Fault Dynamics project at Royal Holloway, University of London for their support whilst editing this Special Publication. Gwen Peron-Pinvidic is thanked for permission to use her figures as a basis for our Figure 2. The numerous authors are thanked for their patience during the compilation of this volume. In particular the generosity of time and effort by the many referees is greatly appreciated in helping to fine tune the papers. The assistance and patience of the editorial and production staff at the Geological Society of London publication section has been fundamental to bringing this collection of papers to print – thank you all.

Funding This summary received no specific grant from any funding agency in the public, commercial, or not-for-profit sectors.

Author contributions KM: conceptualization (equal), writing – original draft (equal); JH: conceptualization (equal), writing – original draft (equal).

References

AJAKAIYE, D.E. & BALLY, A.W. 2002. Manual and atlas of structural styles. Niger Delta. *AAPG Continuing Education Course Notes Series*, **41**.

BAILLIE, P.W., POWELL, C.McA., LI, Z.X. & RYALL, A.M. 1994. The tectonic framework of Western Australia's Neoproterozoic to recent sedimentary basins. *In*: PURCELL, P.G. & PURCELL, R.R. (eds) The Sedimentary Basins of Western Australia: Proceedings of the Petroleum Exploration Society of Australia Symposium. Petroleum Exploration Society of Australia, Perth, 45–62.

BILAL, A., MCCLAY, K. & SCARSELLI, N. 2018. Fault-scarp degradation in the central Exmouth Plateau, North West Shelf, Australia. *In*: MCCLAY, K.R. & HAMMERSTEIN, J.A. (eds) 2020. *Passive Margins: Tectonics, Sedimentation and Magmatism*. Geological Society, London, Special Publications, **476**, https://doi.org/10.1144/SP476.11

BOSWORTH, W., HUCHON, P. & MCCLAY, K. 2005. The Red Sea and Gulf of Aden basins. *Journal of African Earth Sciences*, **43**, 334–378, https://doi.org/10.1016/j.jafrearsci.2005.07.020

COBBOLD, P.R., MEISLING, K.E. & MOUNT, V.S. 2001. Reactivation of an obliquely rifted margin, Campos and Santos basins, southeastern Brazil. *AAPG Bulletin*, **85**, 1925–1944.

COBBOLD, P.R., CLARKE, J.B. & LOSETH, H. 2009. Structural consequences of fluid overpressure and seepage forces in the outer thrust belt of the Niger Delta. *Petroleum Geoscience*, **15**, 3–15, https://doi.org/10.1144/1354-079309-784

DENG, H. & MCCLAY, K. 2019. Tectono-stratigraphy of the Dampier Sub-basin, North West Shelf of Australia. *In*: MCCLAY, K.R. & HAMMERSTEIN, J.A. (eds) 2020. *Passive Margins: Tectonics, Sedimentation and Magmatism*. Geological Society, London, Special Publications, **476**, https://doi.org/10.1144/SP476-2018-180

DOOLEY, T.P., HUDEC, M.R., PICHEL, L.M. & JACKSON, M.P.A. 2018. The impact of base-salt relief on salt flow and suprasalt deformation patterns at the autochthonous, paraautochthonous and allochthonous level: insights from physical models. *In*: MCCLAY, K.R. & HAMMERSTEIN, J.A. (eds) 2020. *Passive Margins: Tectonics, Sedimentation and Magmatism*. Geological Society, London, Special Publications, **476**, https://doi.org/10.1144/SP476.13

FOSSEN, H. & ROTEVATN, A. 2016. Fault linkages and relay structures in extensional settings – a review. *Earth Science Reviews*, **154**, 14–28, https://doi.org/10.1016/j.earscirev.2015.11.014

FRASER, S.I., ROBINSON, A.M. ET AL. 2002. Upper Jurassic. *In*: ARMOUR, A., EVANS, D. & HICKEY, C. (eds) *The Millennium Atlas: Petroleum Geology of the Central and Northern North Sea*. The Geological Society, London, 157–189.

GERDES, K.D. 2018. Professor David Gwyn Roberts – a Life in Geoscience. *In*: MCCLAY, K.R. & HAMMERSTEIN, J.A. (eds) 2020. *Passive Margins: Tectonics, Sedimentation and Magmatism*. Geological Society, London, Special Publications, **476**, https://doi.org/10.1144/SP476.7

GRACIANSKY, P.C., ROBERTS, D.G. & TRICART, P. 2011. The Western Alps, from Passive Margin to Orogenic Belt. *Developments in Earth Surface Processes*, **14**.

HEINE, C. & MÜLLER, R.D. 2005. Late Jurassic rifting along the Australian North West Shelf: margin geometry and spreading ridge configuration. *Australian Journal of Earth Sciences*, **52**, 27–39, https://doi.org/10.1080/08120090500100077

HEINE, C., ZOETHOUT, J. & MÜLLER, R.D. 2013. Kinematics of the South Atlantic rift. *Solid Earth*, **4**, 215–253, https://doi.org/10.5194/se-4-215-2013

HENSTRA, G.A., ROTEVATN, A., GAWTHORPE, R.L. & RAVNÅS, R. 2015. Evolution of a major segmented normal fault during multiphase rifting: the origin of plan-view zigzag geometry. *Journal of Structural Geology*, **74**, 45–63, https://doi.org/10.1016/j.jsg.2015.02.005

HUDEC, M.R., NORTON, I.O., JACKSON, M.P.A. & PEEL, F.J. 2013. Jurassic evolution of the Gulf of Mexico salt basin. *AAPG Bulletin*, **97**, 1683–1710, https://doi.org/10.1306/04011312073

JACKSON, M.P.A. & HUDEC, M.R. 2017. *Salt Tectonics: Principles and Practice*. Cambridge University Press, Cambridge.

JAGGER, L.J., BEVAN, T.G. & MCCLAY, K.R. 2018. Tectono-stratigraphic evolution of the SE Mediterranean passive margin, offshore Egypt and Libya. *In*: MCCLAY, K.R. & HAMMERSTEIN, J.A. (eds) 2020. *Passive Margins: Tectonics, Sedimentation and Magmatism*. Geological Society, London, Special Publications, **476**, https://doi.org/10.1144/SP476.10

KHALIL, S.M. & MCCLAY, K.R. 2016. 3D Geometry and kinematic evolution of extensional fault-related folds, northwestern Red Sea, Egypt. *In*: CHILDS, C., HOLDSWORTH, R.E., JACKSON, C.A.-L., MANZOCCHI, T., WALSH, J.J. & YIELDING, G. (eds) *The Geometry and Growth of Normal Faults*. Geological Society, London, Special Publications, **439**, 109–130, https://doi.org/10.1144/SP439.11

KHALIL, S.M. & MCCLAY, K.R. 2018. Extensional fault-related folding in the northwestern Red Sea, Egypt: segmented fault growth, fault linkages, corner folds and basin evolution. *In*: MCCLAY, K.R. & HAMMERSTEIN, J.A. (eds) 2020. *Passive Margins: Tectonics, Sedimentation and Magmatism*. Geological Society, London, Special Publications, **476**, https://doi.org/10.1144/SP476.12

KRUEGER, W.S. & GRANT, N.T. 2013. The growth history of toe thrusts of the Niger Delta and the role of pore pressure. *In*: MCCLAY, K., SHAW, J. & SUPPE, J. (eds) *Thrust Fault-Related Folding*. American Association of Petroleum Geologists, Memoirs, **94**, 357–390.

KUSZNIR, N.J., ROBERTS, A.M. & ALVEY, A.D. 2018. Crustal structure of the conjugate Equatorial Atlantic Margins, derived by gravity anomaly inversion. *In*: MCCLAY, K.R. & HAMMERSTEIN, J.A. (eds) 2020. *Passive Margins: Tectonics, Sedimentation and Magmatism*. Geological Society, London, Special Publications, **476**, https://doi.org/10.1144/SP476.5

LAVIER, L.Y. & MANATSCHAL, G. 2006. A mechanism to thin the continental lithosphere at magma-poor margins. *Nature*, **440**, 324–328; https://doi.org/10.1038/nature04608

LLAVE, E., JAVIER HERNÁNDEZ-MOLINA, F. ET AL. 2019. Contourites along the Iberian continental margins: conceptual and economic implications. *In*: MCCLAY, K.R. & HAMMERSTEIN, J.A. (eds) 2020. *Passive Margins: Tectonics, Sedimentation and Magmatism*. Geological Society, London, Special Publications, **476**, https://doi.org/10.1144/SP476-2017-46

LONGLEY, I. & BUESSENSCHUETT, C. ET AL. 2002. The North West Shelf of Australia – a Woodside perspective. *In*: KEEP, M. & MOSS, S. (eds) *The Sedimentary Basins of Western Australia 3: Proceedings of the Petroleum Exploration Society of Australia Symposium, Perth, Australia*. Petroleum Exploration Society of Australia, Perth, 27–88.

LYMER, G., CRESSWELL, D.J.F. ET AL. 2019. 3D development of detachment faulting during continental breakup. *Earth and Planetary Science Letters*, **515**, 90–99, https://doi.org/10.1016/j.epsl.2019.03.018

MANATSCHAL, G., LAVIER, L. & CHENIN, P. 2015. The role of inheritance in structuring hyperextended rift systems: some considerations based on observations and numerical modeling. *Gondwana Research*, **27**, 140–164, https://doi.org/10.1016/j.gr.2014.08.006

MCCORMACK, K.D. & MCCLAY, K.R. 2018. Orthorhombic faulting in the Beagle Sub-basin, North West Shelf, Australia. *In*: MCCLAY, K.R. & HAMMERSTEIN, J.A. (eds) 2020. *Passive Margins: Tectonics, Sedimentation and Magmatism*. Geological Society, London, Special Publications, **476**, https://doi.org/10.1144/SP476.3

MOHN, G., MANATSCHAL, G., BELTRANDO, M., MASINI, E. & KUZNIR, N. 2012. Necking of continental crust in magma-poor rifted margins: evidence from the fossil Alpine Tethys margins. *Tectonics*, **31**, TC1012, https://doi.org/10.1029/2011TCC002961

OSMUNDSEN, P.T. & PÉRON-PINVIDIC, G. 2018. Crustal-scale fault interaction at rifted margins and the formation of domain-bounding breakaway complexes: insights from offshore Norway. *Tectonics*, **37**, 935–964, https://doi.org/10.1002/2017TC004792

PÉREZ-DÍAZ, L. & EAGLES, G. 2018. Estimating palaeobathymetry with quantified uncertainties: a workflow illustrated with South Atlantic data. *In*: MCCLAY, K.R. & HAMMERSTEIN, J.A. (eds) 2020. *Passive Margins: Tectonics, Sedimentation and Magmatism*. Geological Society, London, Special Publications, **476**, https://doi.org/10.1144/SP476.1

PÉRON-PINVIDIC, G., MANATSCHAL, G. & OSMUNDSEN, P.T. 2013. Structural comparison of archetypal Atlantic rifted margins: a review of observations and concepts. *Marine and Petroleum Geology*, **43**, 21–47, https://doi.org/10.1016/j.marpetgeo.2013.02.002

PÉRON-PINVIDIC, G., MANATSCHAL, G., MASINI, E., SUTRA, E., FLAMENT, J.M., HAUPERT, I. & UNTERNEHR, P. 2015. Unravelling the along-strike variability of the Angola–Gabon rifted margin: a mapping approach. *In*: SABATO CERALDI, T., HODGKINSON, R.A. & BACKE, G. (eds) 2017. *Petroleum Geoscience of the West Africa Margin*. Geological Society, London, Special Publications, **438**, 49–76, https://doi.org/10.1144/SP438.1

PÉRON-PINVIDIC, G. & OSMUNDSEN, P.T. 2018. The Mid Norwegian–NE Greenland conjugate margins: rifting evolution, margin segmentation, and breakup. *Marine and Petroleum Geology*, **98**, 162–184, https://doi.org/10.1016/j.marpetgeo.2018.08.011

PINDELL, J., GRAHAM, R. & HORN, B.W. 2014. Rapid outer marginal collapse at the rift to drift transition of passive margin evolution, with a Gulf of Mexico case study. *Journal of Basin Research*, **26**, 1–25 https://doi.org/10.1111/bre.12055

PINDELL, J., GRAHAM, R. & HORN, B.W. 2018. Role of outer marginal collapse on salt deposition in the eastern Gulf of Mexico, Campos and Santos basins. *In*: MCCLAY, K.R. & HAMMERSTEIN, J.A. (eds) 2020. *Passive Margins: Tectonics, Sedimentation and Magmatism*.

Geological Society, London, Special Publications, **476**, https://doi.org/10.1144/SP476.4

RESTREPO-PACE, P.A. 2018. 'Ductile v. Brittle' – alternative structural interpretations for the Niger Delta. *In*: MCCLAY, K.R. & HAMMERSTEIN, J.A. (eds) 2020. *Passive Margins: Tectonics, Sedimentation and Magmatism*. Geological Society, London, Special Publications, **476**, https://doi.org/10.1144/SP476.2

ROWAN, M.G. 2014. Passive-margin salt basins: hyperextension, evaporite deposition, and salt tectonics. *Journal of Basin Research*, **26**, 154–182, https://doi.org/10.1111/bre.12043

ROWAN, M.G. 2018. The South Atlantic and Gulf of Mexico salt basins: crustal thinning, subsidence and accommodation for salt and presalt strata. *In*: MCCLAY, K.R. & HAMMERSTEIN, J.A. (eds) 2020. *Passive Margins: Tectonics, Sedimentation and Magmatism*. Geological Society, London, Special Publications, **476**, https://doi.org/10.1144/SP476.6

SCARSELLI, N., DUVAL, G., MARTIN, J., MCCLAY, K. & TOOTHILL, S. 2018. Insights into the Early Evolution of the Côte d'Ivoire Margin (West Africa). *In*: MCCLAY, K.R. & HAMMERSTEIN, J.A. (eds) 2020. *Passive Margins: Tectonics, Sedimentation and Magmatism*. Geological Society, London, Special Publications, **476**, https://doi.org/10.1144/SP476.8

TAMARA, J., MCCLAY, K. & HODGSON, N. 2020. Crustal structure of the central sector of the NE Brazilian equatorial margin. *In*: MCCLAY, K.R. & HAMMERSTEIN, J.A. (eds) 2020. *Passive Margins: Tectonics, Sedimentation and Magmatism*. Geological Society, London, Special Publications, **476**, https://doi.org/10.1144/SP476-2019-54

TUGEND, J., GILLARD, M. *ET AL.* 2018. Reappraisal of the magma-rich v. magma-poor rifted margin archetypes. *In*: MCCLAY, K.R. & HAMMERSTEIN, J.A. (eds) 2020. *Passive Margins: Tectonics, Sedimentation and Magmatism*. Geological Society, London, Special Publications, **476**, https://doi.org/10.1144/SP476.9

TVEDT, A.B.M., ROTEVATN, A., JACKSON, C.A.L., FOSSEN, H. & GAWTHORPE, R.L. 2013. Growth of normal faults in multilayer sequences: a 3D seismic case study from the Egersund Basin, Norwegian North Sea. *Journal of Structural Geology*, **55**, 1–20, https://doi.org/10.1016/j.jsg.2013.08.002

WEINER, R.W., MANN, M.G., ANGELICH, M.T. & MOLYNEUX, J.B. 2010. *Mobile Shale in the Niger Delta: Characteristics, Structure, and Evolution*. American Association of Petroleum Geologists, Memoirs, **95**, 145–161.

WOOD, L. (ed.) 2010. *Shale Tectonics*. American Association of Petroleum Geologists, Memoirs, **95**.

WELBON, A.I.F., BROCKBANK, P.J., BRUNSDEN, D. & OLSEN, T.S. 2007. Characterizing and producing from reservoirs in landslides: challenges and opportunities. *In*: JOLLEY, S.J., BARR, D., WALSH, J.J. & KNIPE, R.J. (eds) *Structurally Complex Reservoirs*. Geological Society, London, Special Publications, **292**, 49–74, https://doi.org/10.1144/SP292.3

WU, S. & BALLY, A.W. 2000. Slope tectonics – Comparison and contrasts in structural styles of salt and shale tectonics in the Gulf of Mexico with Shale Tectonics in the Niger Delta in the Gulf of Guinea. American Geophysical Union, Geophysical Monographs, **115**, 151–172.

Professor David Gwyn Roberts – a Life in Geoscience

KEITH D. GERDES

Shell Centre for Exploration Geoscience, Applied Geosciences Unit, School of Energy, Geoscience, Infrastructure & Society, Heriot–Watt University, Edinburgh EH14 4AS, UK

keithgerdes@hotmail.com

Abstract: This conference was held to celebrate the outstanding career of a remarkable man. Professor David Gwyn Roberts was one of the most creative, productive and influential geoscientists of his generation. The range of his achievements and the scope of his influence on the world of geoscience are remarkable. He had a profound impact on geological thinking and understanding that serves as his lasting legacy. His Life in Geoscience commenced as a marine geoscientist at the Institute of Oceanographic Sciences, where he rapidly developed a global reputation for innovative research; this was followed by a highly influential spell as an explorer, technical expert and mentor in BP, and culminated in a successful career as a globally respected consultant, professor, teacher and author. He also edited numerous journals, founded the publication *Marine and Petroleum Geology* and held high office in a number of professional organizations. Through his lectures, workshops and courses he served as an inspiration to hundreds of geoscientists around the world. Any phase of his career would identify David Gwyn Roberts as an outstanding geoscientist. The fact that these achievements were the work of one man mark Professor Roberts's life and contribution to the geosciences as truly remarkable.

This conference was held to celebrate the outstanding career in geosciences of a remarkable man. Professor David Gwyn Roberts was one of the most creative, productive and influential geoscientists of his generation (Fig. 1). Running like a red thread throughout David's career were his passion for geoscience; an infectious enthusiasm to better understand the Earth and its governing processes; and a desire to give back to the geoscience community which he served so well in so many ways. Professor Roberts was a force of nature, ahead of his time both intellectually and in the integrated manner in which he worked.

His Life in Geoscience comprised three phases, all of which overlapped to the benefit of the discipline. His initial career years as a marine geoscientist and research project leader at Institute of Oceanographic Sciences (IOS) were followed by his time in the energy industry as a technical expert, teacher and mentor with BP. During his corporate life he continued to edit numerous journals, founded the publication *Marine and Petroleum Geology* and held high office in a number of professional organizations. The third phase commenced on his retirement from BP during which David enjoyed one of the most productive phases of his career as a globally respected exploration consultant, visiting professor, teacher and author. Through his lectures, workshops and courses, he served as an inspiration to hundreds of geoscientists around the world.

In addition to his research papers, journal contributions and publications, one of the greatest gifts that Professor Roberts gave to the geoscience community was his time. Any one of his career phases would have identified David as an extraordinary geoscientist. The fact that these achievements were the work of one man marks David's life and contribution to the geosciences as truly outstanding.

Early years

David Gwyn Roberts was born in 1943 in Welshpool. When David was six years old, and speaking only Welsh, the family moved to Stockport, England, where he completed his schooling. A love of the outdoors had been a feature of his childhood and on leaving school he studied Geology at the University of Manchester, where he received his BSc.

After graduation, David began a postgraduate study, mapping recently active volcanoes in the Caribbean. The year prior to his graduation, however, had seen the game-changing interpretation of oceanic magnetic anomalies as evidence of seafloor spreading by Vine and Matthews (1963). The significance of this work was obvious to the major oceanographic centres and marine geological institutes around the world – and also to any geoscience student fascinated by the history of the world's continents and oceans. A major expansion of global oceanographic surveying and mapping followed and these were exciting times for geoscience. Marine geology and geophysics were developing rapidly and the integration of the results of the Deep-Sea Drilling Project (DSDP) and Joint Oceanographic Institutions for Deep Earth Sampling programmes with the first publicly available multichannel seismic

From: McCLAY, K. R. & HAMMERSTEIN, J. A. (eds) 2020. *Passive Margins: Tectonics, Sedimentation and Magmatism.* Geological Society, London, Special Publications, **476**, 11–21.
First published online March 1, 2018, https://doi.org/10.1144/SP476.7
© 2018 The Author(s). Published by The Geological Society of London. All rights reserved.
For permissions: http://www.geolsoc.org.uk/permissions. Publishing disclaimer: www.geolsoc.org.uk/pub_ethics

Fig. 1. Professor David Gwyn Roberts.

surveys were providing new insights into the nature and structure of the world's oceans and leading to major advances in our understanding of plate tectonics. David realized that marine geoscience was the place to be and when the opportunity arose to join the National Institute of Oceanography at Wormley, UK (now the IOS), David left his postgraduate studies to commence what was to become a lifetime of research into the structural and stratigraphic evolution of continental margins and their underlying basins.

Dave at IOS

At IOS, David initially worked on geophysical data from the Gulf of Aden (Roberts & Whitmarsh 1968) and then joined a team focussed on the study of the continental margins of the NE Atlantic. This was a particularly challenging area to survey, not least because of the metocean conditions. Some of David's earliest research cruises were to the Rockall area of the Atlantic, often on-board some of the smaller vessels in the IOS surveying fleet. The surveying results were also challenging – the conventional method of mapping the bathymetry of the oceans and assigning a geological age to the overlying magnetic anomalies by dating sediments cored from stratigraphic wells sited beneath them worked well for oceanic crust of Tertiary age (Scrutton & Roberts 1971). Extensive mapping along both margins of the Atlantic, however, revealed broad areas, up to 200 km wide, where the oceanic anomalies were of lower amplitude and broader wavelength, making their interpretation highly ambiguous (Laughton et al. 1975). The bathymetric profiles from these margins did not conform to the simple increase in water depth with distance from the spreading centre observed over areas of 'normal' oceanic crust. Geological mapping and shallow coring of the margins in the Rockall area also revealed that acoustic basement was far from uniform and included continental crust of Pre-Cambrian age and Mesozoic mafic intrusions (Roberts et al. 1972, 1973a, b; Miller et al. 1973; Flemming et al. 1975). In other areas of the continental margins increased sediment thicknesses precluded the direct sampling of the stratigraphy immediately above basement (Roberts & Kidd 1979). The presence of these broad transitional zones was at odds with crustal models of the day, which envisioned a near-vertical boundary between oceanic and continental crust of normal thicknesses at the continental margins. Geophysical surveying technology was in its infancy compared with the present day with 2D multichannel seismic surveys utilizing single 2 km-long streamers. These arrays acquired seismic data of limited resolution and penetration that could not resolve the complex deeper crustal structure. As a result, an array of terms was used to describe large areas of the margins – Cretaceous Magnetic Quiet Zones, Transition Zones, Oceanic Plateaux – and the method of identifying the continent – ocean boundary varied from one study group to another, often along the same continental margin.

Early continental margin research

David's early career was anchored in this area of transition between 'normal' oceanic and continental crust. After taking part in a series of research cruises in the NE Atlantic and on the UK continental shelf, he quickly became a recognized authority on the marine geology of both areas (Roberts 1974, 1975a, b; Roberts & Caston 1975; Roberts et al. 1979a). As was the practice at the time, David was also allowed to name the various undersea features that his studies of the Rockall area had revealed (Roberts 1975c, d). Being a devotee of the *Lord of the Rings* novels, he wrote to J.R.R. Tolkien asking for permission to name the structures on the Rockall Plateau after features in Middle Earth from the book trilogy. As a result, Rockall bathymetric maps contain structures such as the Eriador, Rohan and Gondor Seamounts, the Fangorn and Edoras Banks and the Isengard Ridge (Roberts 1975e).

Game-changing DSDP Leg 48

David's research interests, however, were not limited to just the NE Atlantic. He was fascinated by all forms of continental margin, the contrasting plate boundaries underlying them and the possible processes controlling the various structures being observed by geological and geophysical surveys over them (Laughton & Roberts 1978). During the 1970s, David and his colleagues at IOS joined forces with the Institut Francais du Petrole (IFP) in Paris to plan and operate DSDP Leg 48 (Roberts & Montadert 1979a; Roberts et al. 1979b; Montadert et al. 1979c). During Leg 48 the *Glomar Challenger* drilled a suite of shallow investigative wells in the Bay of Biscay and Western Approaches (Montadert et al. 1976). The integration of these well data with 2D regional seismic data led to a sequence of game-changing publications on the structure and stratigraphic evolution of continental margins. These papers also mark the beginning of a prolific collaboration and life-long friendship with his fellow chief scientist, Lucien Montadert. Wells were deliberately sited at locations where acoustic basement was thought to be at its shallowest depth (Fig. 2a). Many of the wells sampled stratigraphy characterized by condensed sequences and welded unconformities (Fig. 2b; Roberts & Montadert 1979b). The subsequent integration of these well results with the seismic data acquired over the DSDP drill sites provided the first evidence that these margins were underlain by rotated blocks of attenuated continental crust separated by shallow angle listric faults (Fig. 3; Roberts & Montadert 1979c). Some of these faults also acted as conduits for bodies of mantle material to ascend close to, and in some cases breach, the seabed (Montadert et al. 1979a,b; Roberts et al. 1979b). These results and interpretations were captured in more than 30 subsequent publications and the implication was clear. An entirely new multidisciplinary geoscience approach was required to fully understand the complexity of the Earth's continental margins (Montadert et al. 1977; De Charpal et al. 1978).

Decades of further study and technological advances in surveying techniques have revealed that these continental margin transition zones play host to a wide range of crustal types and structural morphologies – attenuated continental crust, extruded lava flows, thickened oceanic crust, salt domes, exhumed bodies of mantle material and isolated continental blocks. The significance of the publications derived from the data gathered during DSDP Leg 48, however, cannot be overstated

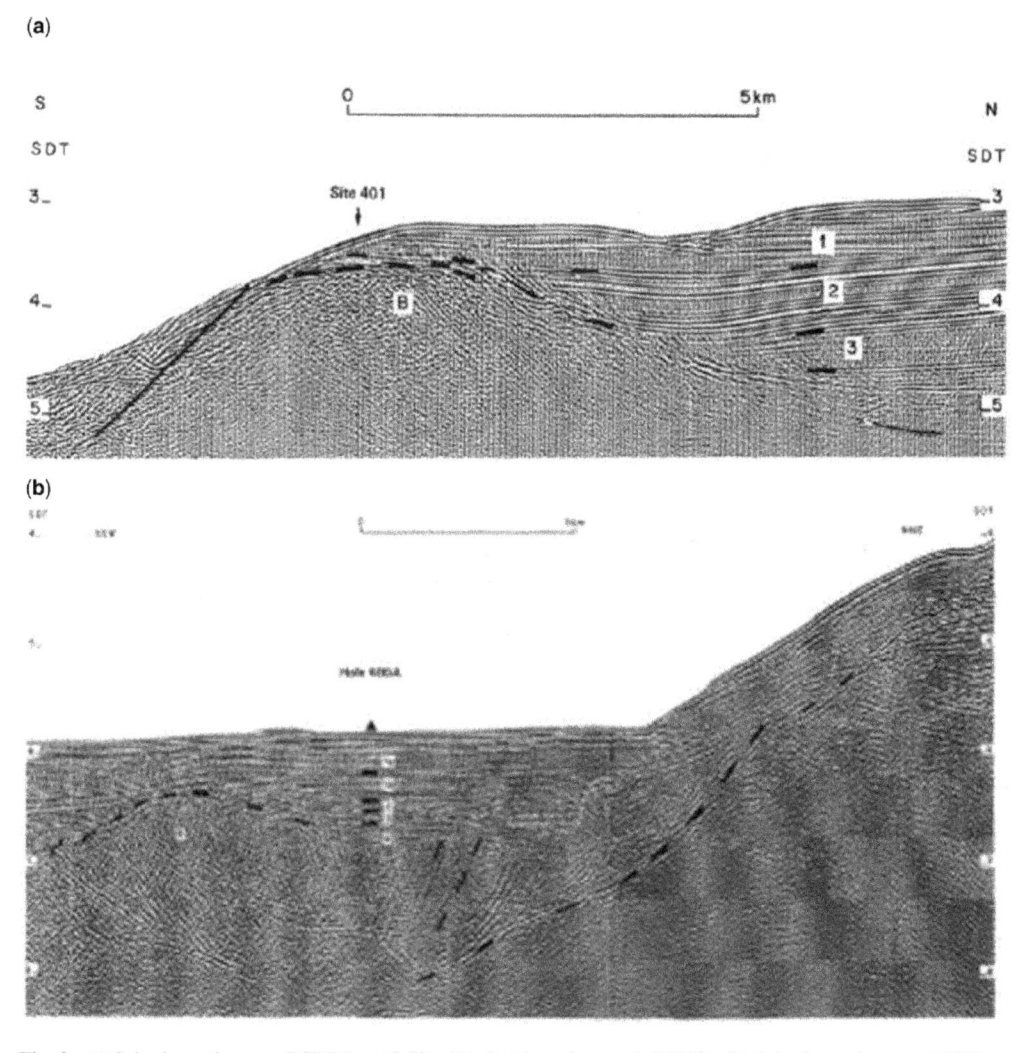

Fig. 2. (a) Seismic section over DSDP Leg 48 Site 401. In Montadert *et al.* (1979*b*). (b) Seismic section over DSDP Leg 48 Site 400A. In Montadert *et al.* (1979*b*).

(Roberts & Montadert 1979*c*; Montadert *et al.* 1979*a*). It is arguable that the geoscience concepts and ideas developed in the papers derived from the results of DSDP Leg 48 were as important for the future understanding of continental margins as those of Vine and Matthews were to the study of seafloor spreading and plate tectonics (Roberts & Montadert 1980).

The success of the publications that resulted from DSDP Leg 48 meant that David's group was well funded and his team received many offers of new datasets and surveying opportunities to further extend their research along the Atlantic margins and beyond. David continued to be a key contributor, often in the role of chief scientist, on research cruises throughout the NE Atlantic, extending as far south as the continental margin of NW Africa (Roberts *et al.* 1981, 1984*a*, *b*; Roberts & Ginzburg 1984). His rapidly developing reputation and influence as a government-funded researcher was instrumental in securing funding from the UK Department of Energy for a research cruise to the Indian Ocean to study the Seychelles microcontinent. This structure was considered at the time to be the best structural analogue for the Rockall Plateau, an area he knew well and which was of particular interest to the Department at the time. His insistence that his entire team accompany him on the research cruise was typical of David – an example of his loyalty to his co-workers which was a

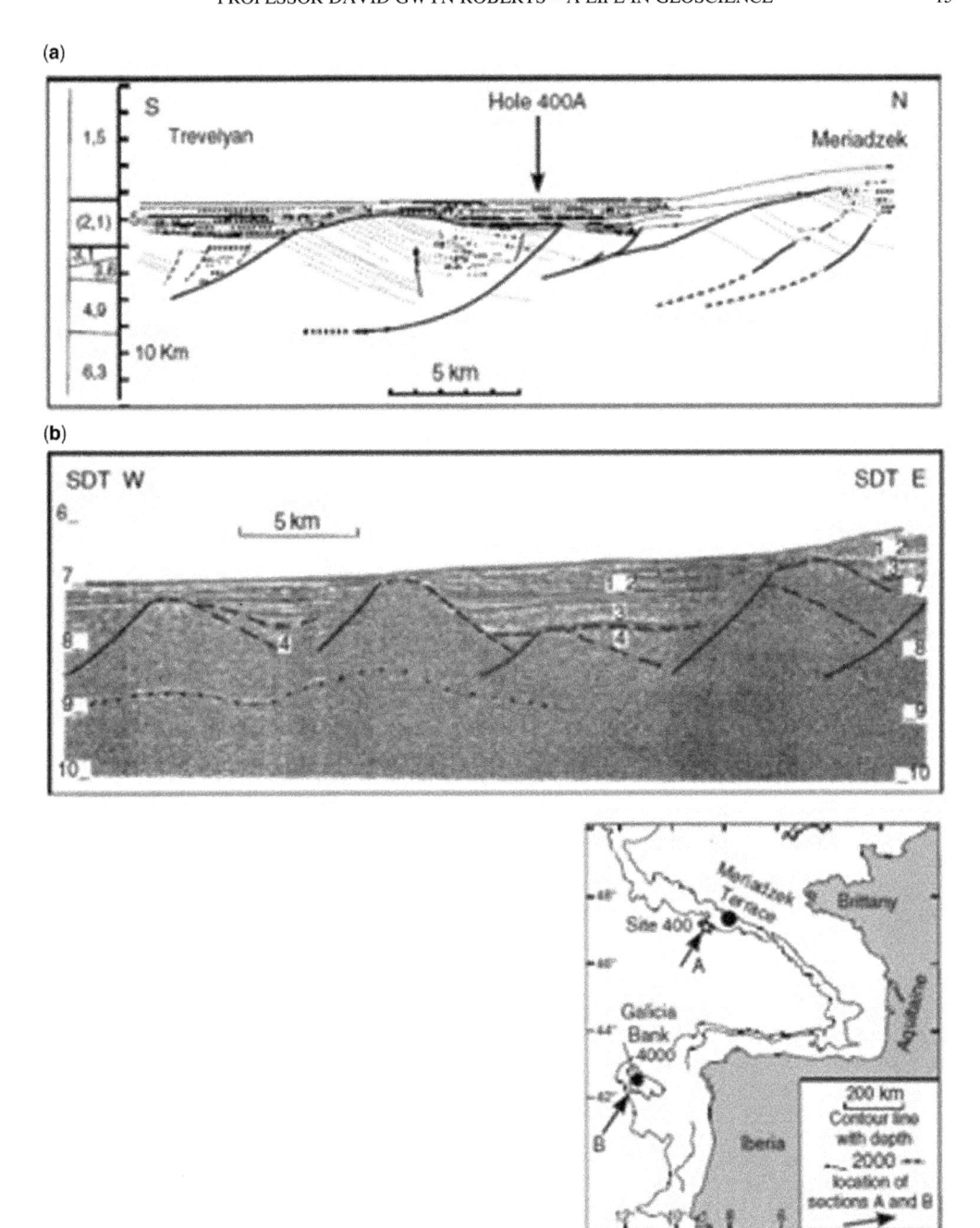

Fig. 3. Crustal structures in the Bay of Biscay and Galicia Bank. (**a**) geological cross-section through Site 400 with same horizontal and vertical scale, constructed from seismic reflection – Profile OC412 migrated. Note the listric faults bounding the tilted blocks. Near base of listric faults is a deeper horizontal reflector corresponding to interface between 4.9 and 6.3 km s^{-1} layers defined by seismic refraction data. (**b**) seismic reflection profile immediately west of Galicia Bank showing tilted blocks and listric faults and a deeper horizontal reflector as in the section above – seismic formations 1–4 are the same as defined in northern Biscay. Profile IFP-CNEXO-CEPM GP, processed. In Montadert *et al.* (1979*b*).

Fig. 4. Dave Roberts in the field in the Alps with P-C Graciansky (courtesy of Steve Matthews).

characteristic of his career. In 1979 the importance of his published research led to his Alma mater, the University of Manchester, awarding him a Doctorate of Science in recognition of his outstanding contribution to the study of continental margins.

Introduction to the Alps

David's collaboration with French oceanographers and geoscientists, which commenced during this time, was to become a feature of his career. During their studies of the margins of the north Atlantic, David's French colleagues encouraged him to visit and study the structures and geometries of the exhumed Tethyan margins exposed in the western Alps. These exposures are of a size and scale similar to those of the structures the team was observing on the regional seismic data acquired in their study areas. The use of seismic-scale field examples in the interpretation of geophysical datasets was something that David was to champion throughout his career. His interest in the Alpine examples facilitated a series of geoscience collaborations and life-long friendships with expert Alpine geologists such as Pierre-Charles de Graciansky and the late Marcel Lemoine, through whom David also met Pierre Tricart (Fig. 4). David developed a strong affection for the western Alps during his regular visits to the region. This love of the region eventually led to David and his family relocating to Comblaux in Haut Savoie.

Dave at BP

In 1981, and after 15 years of service with IOS, David left IOS to join the energy industry with BP. This was a time of great change for publicly funded research. Successive governments on both sides of the Atlantic enacted major reductions in the financial support for research institutes such as IOS and parts of the research fleet were sold off to the site survey industry. In contrast, technological advances in geophysical imaging and well operations led to rapid and substantial increases in the volume and resolution of data available to industry geoscientists. Improvements in resolution and the depth of penetration of 2D seismic data allied with the development of 3D seismic techniques provided seismic interpreters with clear images of stratigraphic sections, sequence boundaries and structures at previously unresolvable depths. The costs of acquiring these data, however, were beyond the restricted financial capacity of most research institutions and so much of this rich body of data remained inaccessible to publicly funded researchers. Marine geoscience surveying and research were largely led by the energy industry as it expanded global offshore activities in the search for additional resources. This was particularly the case with regard to the study of continental margins, which were the focus of much of the heightened industry activity and David's key area of research interest.

Breaking the mould

As the oil industry acquired data of ever-improving quality and resolution, it began to explore in progressively more complicated and challenging geological settings. The newly acquired datasets held the potential to elucidate the structure and stratigraphy of sedimentary basins targeted for exploration, and hence their hydrocarbon potential, like never before (Hubbard et al. 1985a). To maximize the value of these data required the geological interpretation of an integrated geoscience database with seismic data providing the areal framework. In David, BP saw the ideal man for the job – his multidisciplinary approach and reputation for excellence in the integrated interpretation of geoscience data was well established; his knowledge of continental margins was unsurpassed and his skills as a teacher and communicator in both office and field clear to all (Hubbard et al. 1985b).

David's first job title in BP was Head of the Basin Studies Group. It soon became very clear, however, that the needs for David's expertise were far reaching. David's role was quickly expanded to include worldwide technical assurance. He became the geoscience 'conscience' of the Exploration Department, often used as a metaphorical Trojan Horse in visits to offices where technical standards had fallen. Those fortunate enough to work with him on these trips soon got used to working after office hours in the hotel – often completely re-interpreting the exploration portfolio well into the night – before preparing a presentation of the results to the office leadership the following day. David had exacting standards, irrespective of time constraints. Everyone involved knew, however, that he would work all through the night to get the job done if necessary and his enthusiasm for doing the best job possible was infectious.

Raising the bar

During his site visits David recognized the requirement to further blur the boundaries between the geological and geophysical corporate cohorts and also the need to provide formal geoscience training to the technical community. In particular, he recognized the importance of using seismic scale field examples to enhance the geological interpretation of seismic datasets from his experience of comparing the Atlantic margin data with the structures of the western Alps. David proceeded to use his contacts in industry and academia to scour the globe for the best seismic and field examples to use in his courses. Members of his team were sent out to acquire data from a wide range of contacts and also plan the field excursions which were to be an essential part of each course. In the space of two years his team designed and led courses to Sicily, Sardinia, Mallorca, the Dolomites, the Gulf of Suez, East Greenland and (of course) the western Alps (Fig. 5). When delivering these courses David's ability to challenge, motivate, inspire and empower the BP geoscience community came to the fore. David's Sequence Stratigraphy course became a thing of legend with teams of geoscientists working into the early hours in the hope that their final presentations would survive his critical review. These courses quickly became mandatory for all BP geoscientists and within a few years Play Based Exploration, Basin Analysis and Sequence Stratigraphy were embedded into the exploration cohort and all of their work.

The impact of the exploration geoscience curriculum that David established in BP on the industry and global energy reserves cannot be underestimated. Alumni from these courses formed the backbone of the team of exploration geoscientists that transformed the fortunes of the company in the following decades. In addition, many alumni that formed part of the BP diaspora of the late twentieth century went on hold key leadership roles in the industry and have been associated with some of the largest worldwide exploration discoveries in the decades that followed.

Global exploration success

Corporate acquisition in the 1990s led David to Houston in the role of chief geologist. David immediately identified the need for the company to improve their understanding of the key controls on the petroleum systems of the Gulf of Mexico to maximize the value of the combined BP/Sohio portfolio. He melded together a team of experienced local knowledge holders and enthusiastic young explorers from Europe whose work formed the basis for BP's record of Gulf of Mexico exploration success during the 1990s, eclipsing the long-established competition, and leading the industry move into deep-water exploration.

As the BP global exploration portfolio expanded, David returned to the UK as leader of Frontier and International Exploration. During this period David's insistence on a consistency of approach, allied to the local knowledge of John Dolson and others, contributed significantly to a sequence of major discoveries in the Nile Delta. David then became BP's Global Exploration Advisor, responsible for geoscience assurance of all new ventures and exploration appraisal worldwide before becoming BP's Distinguished Exploration Advisor, the highest technical leadership role in the company, during his two years prior to retirement.

Service to the geoscience community and awards

David's posts in BP involved an enormous amount of international travel. He was always excellent

Fig. 5. Dave Roberts leading an industry field course in the western Alps.

company and many of his companions consider their time spent travelling and working with David as some of the happiest and most rewarding times of their careers. It is even more remarkable, therefore, that David continued to publish, promote research and serve the greater geoscience community so actively whilst in these exacting corporate roles. Within three years of joining BP he founded and served as editor-in-chief of *Marine and Petroleum Geology*. This journal quickly became well respected by the geoscience community and recently entered its fourth decade in print. During his time based in Houston David also co-edited AAPG Memoir 65, the first global collation of halokinetic observations, section balancing and modelling of salt bodies (Jackson *et al.* 1995). He continued to convene and chair conferences and meetings on the hot topics of the day for a whole range of professional organizations. In fact, the volume and scope of technical sessions, conferences and workshops that David convened or contributed to during his time with BP is remarkable. In addition to co-convening the AAPG Hedberg Conference on salt tectonics in 1993, he was also co-chairman of the 1995 AAPG International Conference in Nice; Chair of the technical program for the AAPG 1998 Vienna International Conference; General Chair of the AAPG International Conference in Birmingham in 1999; Chair of the 2003 AAPG Hedberg Research Conference on Palaeozoic and Triassic Petroleum Systems of North Africa in Algiers; and Chair of the Conference on the Hydrocarbon Habitat of Volcanic Passive Margins held in Stavanger in 2002. He was elected AAPG President for Europe between 2001 and 2003 and further served the organization on both its Corporate Liaison Committee (1991–04) and Distinguished Lecturer Committee (1992–95). He also served as chairman of the EAGE Petroleum Division 2000–01. His outstanding contributions to these professional organizations did not go unnoticed and his support for the AAPG in particular was repeatedly recognized; he received Certificates of Merit in 1995 and 1998; the Distinguished Service Award in 1997; the R.H. Dott Memorial award in 1998; and was made an Honorary Member of the AAPG in 2001 for 'outstanding service to petroleum geology'. In 1999, he also received the Petroleum Medal from the Geological Society of London.

Public service and academia

David never forgot his roots in marine geoscience, however, and he continued to support both academic and publicly funded research by serving on a host of government and industry/academia advisory panels. The list of panels, committees and boards that gratefully received his good counsel is extensive and included the NERC Earth Sciences Technology Board and Polar Science Expert Group; the ODP Industrial Liaison Panel and Co-ordinating Committee; the Governing Board of the Scott Polar Research Institute; the Governing Board of CASP; the Euromargins Steering Committee; the NERC/DTI Ocean Margins Board; and the Advisory Boards of Royal Holloway College, University of London and the Centre for Exploration IFP School, Paris.

All of these activities reflected his belief that oil and gas geoscience professionals had a responsibility to actively promote and enhance geoscience in its broadest sense. During this time he also reconnected with his earlier research links, becoming a visiting professor at the IFP, Paris and Senior Research Fellow at the UK National Oceanography Centre in Southampton. In 2001, his contribution to geoscience was recognized by the Royal Holloway College of the University of London, when he was made an Honorary Fellow and Visiting Professor.

Dave: the global consultant

On his retirement from BP David's career entered another highly influential and productive phase. By now he was recognized as a world expert in marine geology, tectonophysics, basin analysis and petroleum geology. His encyclopaedic knowledge and deep understanding of sedimentary basins, the processes driving their evolution and the implications for their hydrocarbon prospectivity put him in high demand as an independent consultant. His teaching and mentoring skills, first utilized during his time in BP and developed further during his teaching assignments at the Masters courses at Royal Holloway and IFP, were also in constant demand. His worldwide teaching activities enabled a continuous stream of young professionals and future geoscience professionals to be educated and inspired by his knowledge and enthusiasm for the subject. He also maintained strong links with the oil industry, holding board positions at companies such as Premier Oil, Getech and Medserv and consulting for a range of other companies.

International recognition

The significance of his contribution to geoscience continued to be recognized by both industry and academia. In 2005 the Geological Society of London awarded David one of their highest awards, the Coke Medal, for his 'outstanding contribution to geology and service to the geoscience community'. In 2006, he was made an Honorary Fellow of the Geological Society of America – an award granted to only two non-Americans each year – for his 'Internationally recognised contribution to geoscience through original research and scientific advances'.

Professor and author

David also continued to publish on many aspects of regional geology and the nature of continental margins. He re-united with friends and colleagues from earlier phases of his career on a series of projects. One such project was to revise, expand and translate into English the volume on the geology of the western Alps published in 2000 by his friends and co-workers Marcel Lemoine, Pierre-Charles de Graciansky and Pierre Tricart. The resultant publication was dedicated by David, Pierre and Pierre-Charles to their late mentor and inspiration Marcel Lemoine (De Graciansky *et al.* 2010). Perhaps the greatest challenge he set himself was to compile and edit a set of volumes covering the regional geology and tectonics of the world (Roberts & Bally 2012*a, b, c*). This was a labour of love as it enabled David to work again with colleagues from all the geoscience disciplines, many of whom had become firm friends. It also provided David with the ideal opportunity to work closely with his great friend from his early days with the DSDP, Bert Bally. The resultant publication comprised three volumes under the title, *Regional Geology and Tectonics* and received an award as one of the Outstanding Academic Titles of 2013 from the American Library Association.

The man and his legacy

Professor Roberts was above all a warm, humble and compassionate man with an enormous generosity of spirit and a great passion for life. His work ethic and exacting standards could be intimidating for some, but led many others to achieve well beyond their expectations. David would often work through the night to customize a course or workshop to the needs of a particular group or audience. Many recall with affection how the combination of his exhausting work schedule and the fine wine served at formal dinners or closing ceremonies led to him dozing off at the table! He was a natural integrator and networker before the latter term gained common parlance. A creator and nurturer of teams, he was always insistent that the voices of all the team were heard. Being in possession of one of the most creative and productive geoscience minds of his generation did not prevent him, however, from being able to show immense patience with those less gifted than himself, but who shared his passion for geoscience. These attributes made him such a valued teacher and mentor to so many, and a great friend and colleague to those fortunate enough to have worked and known him.

Professor David Gwyn Roberts was a geologist *par excellence*, respected worldwide for his innovative thought and encyclopaedic knowledge. He had a profound impact on geological thinking and understanding that continues to this day. The full range and volume of his career achievements are extraordinary. The ease with which he moved seamlessly between academia, research and industry was remarkable. His achievements in each of these settings and the respect in which he was held by differing cohorts was unique amongst modern day geoscientists. Any single aspect of his career would identify him as a great geoscientist. Any geoscience researcher would be considered outstanding having co-authored two seminal books, founded a major journal and with over 100 publications to their name. Any industry explorer with a career associated with so many exploration discoveries and which ascended to the highest echelons of the corporate world would be viewed as exceptional. Any geoscience teacher and mentor who has been the source of inspiration for so many successful careers in geoscience would be lauded as hugely influential. The fact that all of these achievements were the work of one man makes David Gwyn Roberts a true giant amongst geoscientists and the gift of his contribution to the geoscience community to be a true cause for celebration.

Acknowledgements I would like to acknowledge the considerable assistance I received from numerous past friends, colleagues and students of David during the compilation of this chapter. I am indebted to his family, Elizabeth (Robin) and daughter Nikki for their kind assistance and review of the final draft. This task benefitted significantly from the excellent memorials to David written by Mike Bowman for the AAPG and *The Times*. I would particularly like to thank Bob Stephenson, Rod Graham, Doug Masson, Bert Bally, Mark Thompson, Joe Pape and Steve Matthews for anecdotes and images submitted to either Mike or myself. Finally, I wish to thank Ken McClay, the organizing committee and all the contributors to the original conference and this volume which are such fitting tributes to the extraordinary contribution to geoscience made by Professor David Gwyn Roberts.

References

DE CHARPAL, O., GUENNOC, P, MONTADERT, L. & ROBERTS, D.G. 1978. Rifting, crustal attenuation and subsidence in the Bay of Biscay. *Nature*, **275**, 706–711.

DE GRACIANSKY, P.-C., ROBERTS, D.G. & TRICART, P. 2010. *The Western Alps, from Rift to Passive Margin to Orogenic Belt*. Developments in Earth Surface Processes. Elsevier, **14**.

FLEMMING, N.C., BINNS, P.E. & ROBERTS, D.G. 1975. *Helen's Reef: geology and bathymetry*. Institute of Geological Science Report **75/1**.

HUBBARD, R.J., PAPE, J. & ROBERTS, D.G. 1985a. Depositional sequence mapping as a technique to establish tectonic and stratigraphic framework and evaluate hydrocarbon potential on a passive continental margin. *In*: BERG, O.R. & WOOLVERTON, D.G. (eds) *Seismic Stratigraphy II: An Integrated Approach to Hydrocarbon Exploration*. AAPG Memoirs, AAPG, Tulsa, USA, **39**, 79–91.

HUBBARD, R.J., PAPE, J. & ROBERTS, D.G. 1985b. Depositional sequence mapping to illustrate the evolution of a passive continental margin. *In*: BERG, O.R. & WOOLVERTON, D.G. (eds) *Seismic Stratigraphy II: An Integrated Approach to Hydrocarbon Exploration*. AAPG Memoirs, **39**, 93–115.

JACKSON, P.A., ROBERTS, D.G. & SNELSON, S. (eds) 1995. *Salt Tectonics: A Global Perspective*. AAPG Memoir, **65**, AAPG, Tulsa, USA.

LAUGHTON, A.S. & ROBERTS, D.G. 1978. Morphology of continental margins. *Philosophical Transactions of the Royal Society, London*, A **290**, 75–85.

LAUGHTON, A.S., ROBERTS, D.G. & GRAVES, R. 1975. Bathymetry of the North East Atlantic: Mid-Atlantic Ridge to Europe. *Deep-Sea Research*, **22**, 791–810.

MILLER, J.A.M., ROBERTS, D.G. & MATTHEWS, D.H. 1973. Rocks of Grenville age from Rockall Bank. *Nature Physical Sciences*, **246**, 61.

MONTADERT, L., ROBERTS, D.G. ET AL. 1976. Glomar Challenger sails on Leg 48. *Geotimes*, **21**, 19–23.

MONTADERT, L., ROBERTS, D.G. ET AL. 1977. Rifting and subsidence on passive continental margins in the North-East Atlantic. *Nature*, **268**, 305–309.

MONTADERT, L., DE CHARPAL, O., ROBERTS, D., GUENNOC, P. & SIBUET, J.C. 1979a. Deep drilling results in the Atlantic Ocean: continental margins and palaeoenvironment. *In*: TALWANI, M., HAY, W. & RYAN, B.F. (eds) *Northeast Atlantic Passive Continental Margins: Rifting and Subsidence Processes*. Maurice Ewing Series, **3**, AGU, Washington, DC, 154–186.

MONTADERT, L., ROBERTS, D.G., DE CHARPAL, O. & GUENNOC, O. 1979b. Rifting and subsidence of the northern continental margin of the Bay of Biscay. *In*: MONTADERT, L., ROBERTS, D.G. ET AL. (eds) *Initial Reports DSDP*, 48. US Government Printing Office, Washington, DC, 1025–1060.

MONTADERT, L., ROBERTS, D.G. & THOMPSON, R.W.T. 1979c. Introduction and explanatory notes, leg 48, IPOD phase of the deep-sea drilling project. *In*: MONTADERT, L., ROBERTS, D.G. ET AL. (eds) *Initial Reports DSDP*, 48. US Government Printing Office, Washington, DC, 9–32.

ROBERTS, D.G. 1974. Structural development of the British Isles, continental margin and the Rockall Plateau. *In*: BURK, C.A. & DRAKE, C.L. (eds) *The Geology of Continental Margins*. Springer, New York, 343–359.

ROBERTS, D.G. 1975a. *The Solid Geology of the Rockall Plateau*. Institute of Geological Science Report **75/1**, 1–10.

ROBERTS, D.G. 1975b. Sediment distribution on the Rockall Bank, Rockall Plateau. *Marine Geology*, **19**, 239–257.

ROBERTS, D.G. 1975c. Tectonic and stratigraphic evolution of the Rockall Plateau and Trough. *In*: WOODLAND, A.W. (ed.) *Petroleum and the continental shelf of North West Europe, Vol. 1, Geology*. Applied Science Publishers, London, **1**, 71–91.

ROBERTS, D.G. 1975d. Marine Geology of the Rockall Plateau and Trough, Phil. *Transactions of the Royal Society, London*, **A 258**, 447–509.

ROBERTS, D.G. 1975e. Geology and tectonics of the area beyond the shelf west of the British Isles. Paper 0B-75202 in Offshore Europe 75, Kingston Upon Thames, Offshore Services Magazine, 8

ROBERTS, D.G. & BALLY, A.W. (eds) 2012a. *Regional Geology and Tectonics: Volume 1 A Principles of Geologic Analysis*. 1st edn. Elsevier, Amsterdam, The Netherlands.

ROBERTS, D.G. & BALLY, A.W. (eds) 2012b. *Regional Geology and Tectonics: Volume 1B. Phanerozoic Passive Margins, Cratonic Basins and Global Tectonic Maps*. 1st edn. Elsevier.

ROBERTS, D.G. & BALLY, A.W. (eds) 2012c. *Regional Geology and Tectonics: Volume 1C Phanerozoic Rift Systems and Sedimentary Basins*. 1st edn. Elsevier.

ROBERTS, D.G. & CASTON, V.N.D. 1975. Petroleum potential of the Deep Atlantic Ocean. *In*: *Proceedings of the 9th World Petroleum Congress*, Tokyo, **2**, 281–298.

ROBERTS, D.G. & GINZBURG, A. 1984. Deep crustal structure of southwest Rockall Plateau. *Nature*, **308**, 435–437.

ROBERTS, D.G. & KIDD, R.B. 1979. Abyssal sediment wave fields on Feni Ridge, Rockall Trough. Long range sonar studies. *Marine Geology*, **33**, 175–191.

ROBERTS, D.G. & MONTADERT, L. 1979a. Objectives of N.E. Atlantic Passive Margin drilling. *In*: MONTADERT, L., ROBERTS, D.G. ET AL. (eds) *Initial Reports DSDP*, 48. US Government Printing Office, Washington, DC, 3–5.

ROBERTS, D.G. & MONTADERT, L. 1979b. North East Atlantic margin palaeoenvironments. *In*: MONTADERT, L., ROBERTS, D.G. ET AL. (eds) *Initial Reports DSDP*, 48. US Government Printing Office, Washington, DC, 1099–1118.

ROBERTS, D.G. & MONTADERT, L. 1979c. Evolution of passive rifted margins – perspective and retrospective of DSDP Leg 48. *In*: MONTADERT, L., ROBERTS, D.G. ET AL. (eds) *Initial Reports DSDP*, 48. US Government Printing Office, Washington, DC, 1143–1153.

ROBERTS, D.G. & MONTADERT, L. 1980. Contrast in the structural style of the passive margins of Biscay and Rockall. *Philosophical Transactions of the Royal Society, London*, **A294**, 97–104.

ROBERTS, D.G. & WHITMARSH, R.B. 1968. A bathymetric and magnetic survey of the Gulf of Tadjura, Western Gulf of Aden. *Earth and Planetary Science Letters, J.*, 235–258.

ROBERTS, D.G., MATTHEWS, D.H. & EDEN, R.A. 1972. Metamorphic rocks from the southern end of Rockall Bank. *Journal of the Geological Society, London*, **128**, 501–506.

ROBERTS, D.G., ARDUS, D.A. & DEARNLEY, R. 1973a. Pre-Cambrian rocks drilled from the Rockall Bank. *Nature Physical Sciences*, **244**, 21–33.

ROBERTS, D.G., FLEMMING, N.C., HARRISON, R.K. & BINNS, P. 1973b. Helen's reef: a Cretaceous microgabbroic intrusion in the Rockall intrusive centre. *Marine Geology*, **16**, M21–M30.

ROBERTS, D.G., HUNTER, P.M. & LAUGHTON, A.S. 1979a. Bathymetry of the North East Atlantic. The Continental margin around the British Isles. *Deep Sea Research*, **26A**, 417–428.

ROBERTS, D.G., MONTADERT, L. & SEARLE, R.C. 1979b. The Western Rockall Plateau stratigraphy and structural evolution. *In*: MONTADERT, L., ROBERTS, D.G. ET AL. (eds) *Initial Reports DSDP*, 48. US Government Printing Office, Washington, DC, 1061–1088.

ROBERTS, D.G., MASSON, D.G. & MILES, P.R. 1981. Age and structure of the southern Rockall Trough: new evidence. *Earth and Planetary Science Letters*, **52**, 115–121.

ROBERTS, D.G., BACKMANN, J., MORTON, A.C., MURRAY, J.W. & KEENE, J.P. 1984a. Evolution of volcanic rifted margins: synthesis of Leg 81 results on the west margin of Rockall Platea. *In*: ROBERTS, D.G., BACKMAN, J., MORTON, A.C. & KEENE, J.B. (eds) *Initial Reports DSDP*, 81. US Government Printing Office, Washington, DC, 883–911.

ROBERTS, D.G., MORTON, A.C. & BACKMANN, J. 1984b. Late Palaeocene-Eocene volcanic events in the northern North Atlantic Ocean. *In*: ROBERTS, D.G., BACKMAN, J., MORTON, A.C. & KEENE, J.B. (eds) *Initial Reports DSDP*, 81. US Government Printing Office, Washington, DC, 913–923.

SCRUTTON, R.A. & ROBERTS, D.G. 1971. Structure of the Rockall Plateau and Trough, North East Atlantic. ICSU/SCORWG31 Symposium, Cambridge, 1970: The geology of the East Atlantic Continental Margin. *Institute of Geological Science Report*, **70/14**, 77–87.

VINE, F.J. & MATTHEWS, D.H. 1963. Magnetic anomalies over oceanic ridges. *Nature*, **199**, 947–949.

Reappraisal of the magma-rich versus magma-poor rifted margin archetypes

JULIE TUGEND[*,1,5], MORGANE GILLARD[1], GIANRETO MANATSCHAL[1], MICHAEL NIRRENGARTEN[2], CAROLINE HARKIN[3], MARIE-EVA EPIN[1], DANIEL SAUTER[1], JULIA AUTIN[1], NICK KUSZNIR[3] & KEN McDERMOTT[4]

[1]*Institut de Physique du Globe de Strasbourg, CNRS-UMR 7516, Université de Strasbourg, 1 rue Blessig, F-67084 Strasbourg Cedex, France*

[2]*Département Géosciences et Environnement, Université de Cergy-Pontoise, Neuville-sur-Oise, France*

[3]*Department of Earth, Ocean and Ecological Sciences, University of Liverpool, Liverpool L69 3GP, UK*

[4]*ION, 31 Windsor St, Chertsey, KT16 8AT, UK*

[5]*Current address: Sorbonne Université, CNRS-INSU, Institut des Sciences de la Terre Paris, ISTeP UMR 7193, F-75005 Paris, France / Total SA, R&D departement CSTJF, Pau, France*

J.T., 0000-0003-3724-7878
Correspondence: julie.tugend@sorbonne-universite.fr

Abstract: Rifted margins are commonly defined as magma-poor or magma-rich archetypes based on their morphology. We re-examine the prevailing model inferred from this classification that magma-rich margins have excess decompression melting at lithospheric breakup compared with steady-state seafloor spreading, while magma-poor margins have inhibited melting. We investigate the magmatic budget related to lithospheric breakup along two high-resolution long-offset deep reflection seismic profiles across the SE Indian (magma-poor) and Uruguayan (magma-rich) rifted margins.

Resolving the magmatic budget is difficult and several interpretations can explain our seismic observations, implying different mechanisms to achieve lithospheric breakup and melt production for each archetype. We show that the Uruguayan and other magma-rich margins may indeed involve excess decompression melting compared with steady-state seafloor spreading but could also be explained by a gradual increase with an early onset relative to crustal breakup. A late onset of decompression melting relative to crustal breakup enables mantle exhumation characteristic of magma-poor margin archetypes (e.g. SE India).

Despite different volumes of magmatism, the mechanisms suggested at lithospheric breakup are comparable between both archetypes. Considerations on the timing of decompression melting onset relative to crustal thinning may be more important than the magmatic budget to understand the evolution and variability of rifted margins.

Rifted margins used to be classified as 'volcanic' or 'non-volcanic' (e.g. Mutter *et al.* 1988; White & McKenzie 1989; Boillot & Coulon 1998). Used in the strictest sense, this classification quickly became somewhat binary and confusing (Mutter 1993), implying different mechanisms for lithospheric thinning and breakup. Because magmatism is observed even in settings initially considered as non-volcanic (e.g. Desmurs *et al.* 2001; Whitmarsh *et al.* 2001a, b), this terminology has been later adjusted to 'magma-poor' ('magma-starved') or 'magma-rich' ('magma-dominated') rifted margins (e.g. Sawyer *et al.* 2007; Reston 2009; Reston & Manatschal 2011; Doré & Lundin 2015). The definition of these end-member archetypes relies on the identification of a number of morphological features considered as characteristic for magma-poor or magma-rich rifted margins (e.g. Menzies *et al.* 2002; Reston 2009; Franke 2013; Doré & Lundin 2015). This terminology leads to assumptions on the magmatic budget: magma-rich rifted margins have a high magmatic budget during rifting and at lithospheric breakup while magma-poor margins have a very low magmatic budget. In particular, magma-rich margins are

From: McCLAY, K. R. & HAMMERSTEIN, J. A. (eds) 2020. *Passive Margins: Tectonics, Sedimentation and Magmatism.* Geological Society, London, Special Publications, **476**, 23–47.
First published online May 2, 2018, https://doi.org/10.1144/SP476.9
© 2018 The Author(s). Published by The Geological Society of London. All rights reserved.
For permissions: http://www.geolsoc.org.uk/permissions. Publishing disclaimer: www.geolsoc.org.uk/pub_ethics

thought to have excess decompression melting, often associated with elevated asthenosphere temperatures, compared with steady-state seafloor spreading. In contrast, magma poor margins are suggested to have inhibited decompression melting. However, this simplification based on the magmatic budget can be misleading. In this work, we re-examine this prevailing model. In fact, most rifted margins show complex and polyphase tectono-magmatic evolutions during rifting and at lithospheric breakup, preceding steady-state seafloor spreading onset, and can preserve characteristic features of both end-member archetypes. Magma-poor rifting can precede magma-rich lithospheric breakup (e.g. North-West shelf of Australia, Belgarde et al. 2015; Mid-Norwegian margin, e.g. Lundin & Doré 2011; Gernigon et al. 2015) and vice versa (e.g. India–Seychelles, Armitage et al. 2012). Deciphering the interaction between tectonic and magmatic processes at rifted margins is, therefore, important to understand the mechanisms controlling their rift-to-drift transition whether they are considered as magma poor or magma rich.

Magmatic processes occurring at the rift-to-drift transition, i.e. related to lithospheric breakup, are recorded continentward of the first unambiguous oceanic domains, in so-called 'transitional' (Welford et al. 2010; Sibuet & Tucholke 2013), 'embryonic' (Jagoutz et al. 2007), 'proto-oceanic' (Gillard et al. 2015) or 'outer domains' (Peron-Pinvidic et al. 2013; Peron-Pinvidic & Osmundsen 2016). At the rift-to-drift transition, melt production appears transient (Gladczenko et al. 1997; Nielsen & Hopper 2004; Pérez-Gussinyé et al. 2006) and tectonic deformation is not yet localized at a stable spreading centre (Gillard et al. 2015, 2016b). This domain replaces the classical continent–ocean boundary (Peron-Pinvidic & Osmundsen 2016), which is difficult to identify unambiguously (e.g. Eagles et al. 2015). The nature of the basement remains poorly constrained and the underlying lithosphere is often described as transitional or hybrid between continental and oceanic (Welford et al. 2010; Franke 2013; Sibuet & Tucholke 2013; Gillard et al. 2015, 2017; Peron-Pinvidic & Osmundsen 2016).

We describe and discuss observations from two high-resolution long-offset deep-reflection seismic profiles provided by ION Geophysical across the SE Indian and Uruguayan rifted margins. These examples are respectively considered as representative of magma-poor (e.g. Nemčok et al. 2013; Haupert et al. 2016; Sinha et al. 2016) and magma-rich rifted margins (e.g. Gladczenko et al. 1997; Blaich et al. 2011; Franke 2013). We apply the same seismic interpretation approach to describe and characterize their first-order architecture and magmatic budget. We focus on the location, timing and amount of magmatic additions emplaced at lithospheric breakup within ultra-distal rifted margins, in so-called proto-oceanic domains. The determination of the magmatic budget and the nature of the basement remains non-unique based on seismic reflection or from other indirect geophysical methods. For that reason, we present several hypotheses that can fit our observations for both examples. These alternative interpretations represent end-member scenarios for the magmatic budget at lithospheric breakup, resulting in different architectures of proto-oceanic domains. Based on these three 'end-member' interpretations, we suggest distinct mechanisms to achieve lithospheric breakup implying variable melt production, applicable to both rifted margin archetypes (magma poor or magma rich).

More generally, our work highlights the difficulty in determining a magmatic budget at rifted margins, showing the limitations and strong assumptions inherent to classifications based on this criterion. Despite different volumes of magmatism, the different mechanisms suggested at lithospheric breakup appear comparable between the magma-poor and magma-rich archetypes. Considerations on the onset of decompression melting relative to crustal thinning appear equally, if not more, important than the overall magmatic budget to understand the evolution and worldwide variability of rifted margins.

Dataset and interpretational approach

Reflection seismic data

We describe and interpret two industrial high-resolution long-offset deep-reflection seismic profiles acquired, processed and provided by ION Geophysical. Located across the SE Indian and Uruguayan rifted margins, these two profiles are part of the IndiaSPAN and UruguaySPAN projects (locations Figs 1a & 2a). These surveys are respectively composed of c. 27 700 km and 2800 km of seismic data acquired using powerful deep-penetrating sources. Some details on acquisition parameters of these two seismic surveys are available from ION Geophysical website (http://www.iongeo.com/Data_Library/India/ and http://www.iongeo.com/Data_Library/South_America/Uruguay/) and described in Nemčok et al. (2013) for the IndiaSPAN project. Kirchhoff prestack time and depth migrations (PSTM and PSDM) were performed on both seismic surveys following proprietary ION Geophysical processing workflow (example of processing workflow in Sauter et al. 2016). PSTM profiles were initially available with a 18 s record length. PSDM profiles image the crustal architecture down to 25 km for the IndiaSPAN and 40 km for the Uruguay SPAN.

Our interpretational work remained focused along the two seismic profiles (locations Figs 1b & 2b).

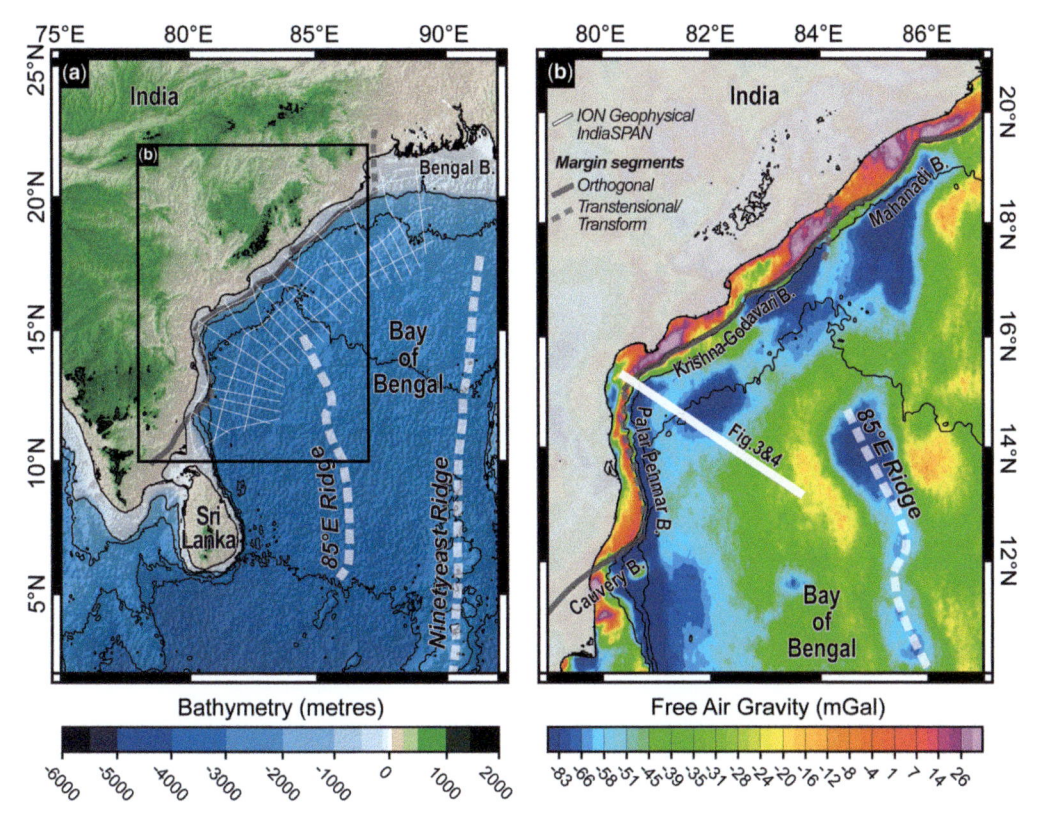

Fig. 1. (**a**) Topographic/bathymetric map of the East Indian rifted margin and Bay of Bengal (ETOPO1, Amante & Eakins 2009). (**b**) Free-air gravity anomaly map (Sandwell *et al.* 2014) showing the first-order morpho-tectonic features of the study area and location of the ION Geophysical IndiaSPAN (http://www.iongeo.com/Data_Library/India/, Radhakrishna *et al.* 2012; Nemčok *et al.* 2013). Topographic/bathymetric contours are given every 1000 m. Equidistant cylindrical projection, geographic coordinate system WGS 84.

Previous observations and interpretations are nevertheless available for the SE Indian margin, some also focused on the same seismic profile (e.g. Radhakrishna *et al.* 2012; Nemčok *et al.* 2013; Mangipudi *et al.* 2014; Pindell *et al.* 2014; Haupert *et al.* 2016). Seismic data and interpretations are available from other surveys offshore Uruguay or from adjacent lines also part of the Uruguay SPAN (e.g. Franke *et al.* 2007; Soto *et al.* 2011; Clerc *et al.* 2015).

Methodology

We applied on both case examples a seismic interpretation approach similar to the one described in Tugend *et al.* (2015) and summarized hereafter. We adapted the workflow to the interpretation of first-order characteristic features of both magma-poor and magma-rich rifted margins. First, we focused on the definition of first-order interfaces (where observable) on both PSTM and PSDM seismic lines, including the seafloor, top basement (i.e. base syn-tectonic sediments or base passive infill), base of Seaward Dipping Reflectors (SDRs)/extrusive and seismic Moho. Second, based on descriptions of the stratigraphic architecture and its relation to the underlying basement, we identified potential low- and high-β extensional settings (Wilson *et al.* 2001), this latter being associated with tectonically exhumed surfaces (Wilson *et al.* 2001; Tugend *et al.* 2015). Third, we identified and characterized different forms of magmatic additions (e.g. SDRs, sill intrusions, volcanic edifices; Planke *et al.* 2000; 2005; Calvès *et al.* 2011). The relation of these magmatic additions to key stratigraphic horizons (pre-, syn-, post-rift), where observable, can provide information on the timing of magma-emplacement relative to the evolution of the margin.

The identification of first-order interfaces on PSDM sections illustrates the evolution of total accommodation space (between sea-level and top

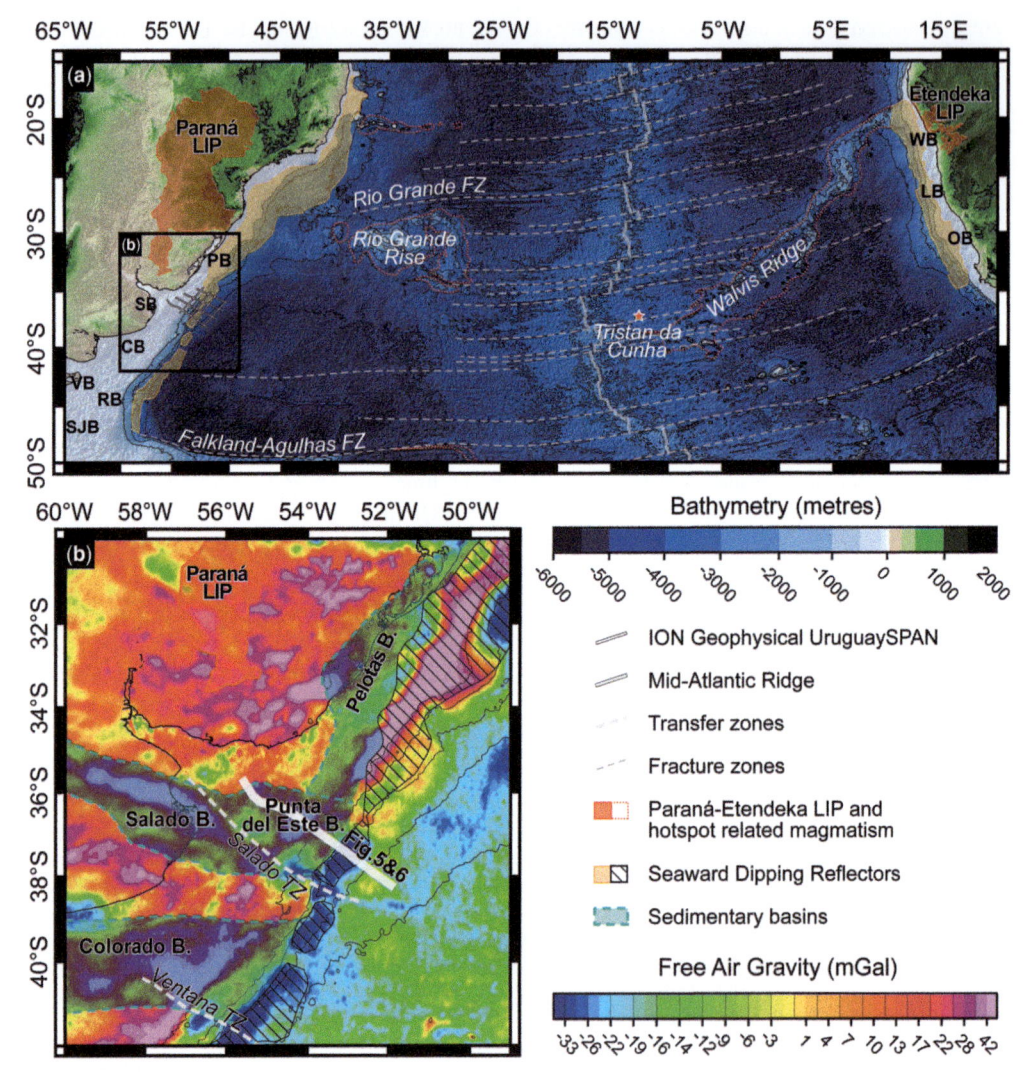

Fig. 2. (**a**) Topographic/bathymetric map of the South Atlantic rifted margins (ETOPO1, Amante & Eakins 2009). (**b**) Free-air gravity anomaly map of the Uruguayan segment (Sandwell *et al.* 2014). Approximate location of the UruguaySPAN as given on ION Geophysical website (http://www.iongeo.com/Data_Library/South_America/Uruguay/). First-order structures and magmatism compiled from the literature (Gladczenko *et al.* 1997, 1998; Franke *et al.* 2007; Koopmann *et al.* 2014; Stica *et al.* 2014; Clerc *et al.* 2015). Topographic/bathymetric contours are given every 1000 m. Equidistant cylindrical projection, geographic coordinate system WGS 84. SJB, San Jorge Basin; VB, Valdes Basin; RB, Rawson Basin; CB, Colorado Basin; SB, Salado Basin; WB, Walvis Basin LB, Lüderitz Basin; OB, Orange Basin.

basement; i.e. the present-day depth to top basement) and crustal thickness (between top basement and seismic Moho). In the case of magma-rich rifted margins, defining the interface between the base of SDRs/extrusive magmas is often difficult. Determining the accommodation space created during rifting thus remains challenging, as well as the relative proportion between magmatic additions and sediments.

Terminology

Based on this workflow, we define a set of first-order comparable architectural features that we consider as building blocks, i.e. corresponding to structural domains of rifted margins. From continent to ocean, we distinguish the proximal, necking, hyper-thinned, exhumed mantle, proto-oceanic and oceanic domains based on the terminology and definitions of Peron-Pinvidic et al. (2013), Sutra et al. (2013), Tugend et al. (2015), Peron-Pinvidic et al. (2017), Gillard et al. (2015). For the purpose of this contribution, we do not discriminate between the necking and hyperthinned domain and refer to the combination of both as 'thinned domain'. These structural domains are considered to correspond to genetic domains recording the interplay between successive extensional and/or magmatic processes (e.g. Lavier & Manatschal 2006; Péron-Pinvidic & Manatschal 2009; Sutra et al. 2013). Related processes are, however, also likely to interact and overlap in time and space during rifted margin evolution (e.g. Péron-Pinvidic & Manatschal 2009). As a result, structural domains are often not delimited by strict boundaries and the passage from one to the other is probably more complex and in some examples gradual (Peron-Pinvidic et al. 2013).

In this work, we distinguish 'crustal' and 'lithospheric' breakup. We consider that crustal breakup is achieved when the continental crust of two conjugate rifted margins is separated. Following Minshull et al. (2001), crustal breakup (referred to as 'continental breakup' in Minshull et al. 2001) corresponds to the seaward limit of stretched continental crust. We define lithospheric breakup as a tectono-magmatic process recording the rift-to-drift transition (Peron-Pinvidic & Osmundsen 2016) at proto-oceanic domains (Gillard et al. 2015) defined continentward of the first unambiguous oceanic domain. Following Gillard et al. (2015; 2016b), we consider that lithospheric breakup is achieved through the emplacement of a steady-state, self-sustaining, seafloor-spreading system, i.e. corresponding to stable and localized oceanic accretion. As emphasized by Minshull et al. (2001) and further discussed in this work, the location and timing of 'crustal breakup' may or may not correspond to 'lithospheric breakup'.

The SE India rifted margin case example

Geological setting and first-order tectono-magmatic context

The SE Indian rifted margin was once conjugate to East Antarctica through the Enderby Basin, Princess Elizabeth Trough and Davis Sea Basin (e.g. Powell et al. 1988; Ramana et al. 1994; Reeves & de Wit 2000; Lal et al. 2009; Radhakrishna et al. 2012; Sinha et al. 2016). The present-day structure of the SE Indian rifted margin results from a complex and polyphase breakup history involving India, Antarctica and Australia (e.g. Powell et al. 1988; Gaina et al. 2003; Subrahmanyam & Chand 2006; Lal et al. 2009). Between India and Antarctica, the occurrence of a complex breakup, related to the formation of the Elan Bank microcontinent, now preserved offshore Antarctica, is generally accepted (Gaina et al. 2003, 2007; Radhakrishna et al. 2012; Sinha et al. 2016; Talwani et al. 2016). Still, owing to uncertainties on the identification of magnetic anomalies, the exact fit of Elan Bank as well as the detailed timing of rifting and lithospheric breakup remain debated. Elan Bank is interpreted as conjugate either to the Krishna–Godavari (Radhakrishna et al. 2012; Sinha et al. 2016) or to the Mahanadi segment (Talwani et al. 2016) of the SE Indian margin (Fig. 1). Rifting seems to have started already during late Early Jurassic time (Nemčok et al. 2013; Sinha et al. 2016 and references therein), but the main rift event shaping the SE Indian margin probably occurred at the beginning of Early Cretaceous time (Powell et al. 1988; Lal et al. 2009; Sinha et al. 2016 and references therein). Rifting might not be synchronous along the entire margin (e.g. Sinha et al. 2016) that appears quite segmented. From SW to NE (Fig. 1), the Cauvery, Palar–Penmar, Krishna–Godavari, Mahanadi and Bengal basins are characterized by variable transtensional deformation and magmatic budget (e.g. Subrahmanyam & Chand 2006; Radhakrishna et al. 2012; Nemčok et al. 2013; Talwani et al. 2016 and references therein).

In the Bay of Bengal, two oceanic ridges are identified trending roughly north–south: the 85° E ridge terminating toward the Mahanadi segment (e.g. Curray & Munasinghe 1991; Choudhuri et al. 2014) and the Ninetyeast ridge (e.g. Coffin et al. 2002) further east (Fig. 1). The Ninetyeast ridge is commonly interpreted to mark the Kerguelen hotspot track. There is no general agreement on the nature of the 85° E ridge. It is interpreted as a fracture zone (Talwani et al. 2016 and references therein), a hotspot track or a hotspot track along a transform (Curray & Munasinghe 1991; Choudhuri et al. 2014), the associated plume being debated. The Rajmahal Traps (c. 118 Ma, (Coffin et al. 2002; Kent et al. 2002) cropping out onshore East India are interpreted as the Early Cretaceous magmatic record of the Kerguelen plume (e.g. Baksi et al. 1987; Coffin et al. 2002; Kent et al. 2002; Olierook et al. 2016) associated with either the Ninetyeast and/or the 85° E ridge.

We focus on a high-resolution reflection seismic profile provided by ION Geophysical, striking NW–SE across the Krishna–Godavari segment of the SE India rifted margin (Fig. 1). This area presents the characteristic features generally attributed to magma-poor hyperextended rifted margins, including extremely thinned continental crust and exhumed mantle (Radhakrishna et al. 2012; Nemčok et al. 2013; Pindell et al. 2014; Haupert et al. 2016; Sinha et al. 2016). The 85° E and 90° E ridges identified in the Bay of Bengal probably correspond to hotspot-transform tracks (Fig. 1), but they were formed only after rifting and lithospheric breakup occurred along the segment considered in this work, as suggested by Gaina et al. (2007) and Sinha et al. (2016).

Seismic observations

Definition of first-order interfaces. Seafloor delimits the present-day shelf break at about distance 15 km and deepens oceanward, reaching c. 3 km depth in the Bay of Bengal (Fig. 3). *Top basement*, where characterized by high-amplitude reflectors, is fairly well recognizable along the profile (Fig. 3). Continentward, from 0 to c. 80 km, top basement progressively deepens from c. 2 to 9 km depth and is characterized by sharp topographic variations. It corresponds to the interface between the base of syn-rift sediments and acoustic basement (possibly also including pre-rift sediments or corresponding to crystalline basement). From c. 80 to c. 150 km, we define top basement as the base of passive infill (as indicated by onlap/downlap geometry of overlaying sediments, Fig. 3c). Underneath, we observe a reflective layer, locally well stratified and organized. Discontinuous high-amplitude reflectors characterize its base, defined in Figure 3a as the *'base reflective layer'*, and showing small depth variations. From c. 150 to c. 210 km, top basement is slightly shallower and identified at c. 9 km depth. Only minor topographic variations are observed except for local highs at top basement (c. 0.5 s in height and 5–10 km in width; Fig. 3d). Oceanward, from c. 210 to 420 km, top basement is almost flat and characterized by discontinuous high-amplitude reflectors. *Seismic Moho* is observable discontinuously as deep reflectors, and appears more clearly on the PSTM profile (Fig. 3a) than on the PSDM one (Fig. 3b). From 0 to 80 km, we define seismic Moho at the base of a reflective package, merging from 80 to 140 km with the interface we identified as the base reflective layer. From c. 130 to 210 km, we define seismic Moho at the base of irregular packages of strong reflectors observed between 10 and 11 s (c. 15–17 km depth) and dipping continentward or oceanward (Fig. 3d). Further oceanward, from c. 220 to 420 km, seismic Moho is only locally visible corresponding to a succession of short discontinuous reflectors (Fig. 3a, b).

Stratigraphic and basin architecture. We only highlight observations on the first-order stratigraphic and basin architecture. Further detailed descriptions are available in Mangipudi et al. (2014) and Haupert et al. (2016). From distance 0 to c. 80 km, mainly low-β extensional settings are observed. Basement morphology delimits graben and half-graben-type basins and their associated wedge-shaped stratigraphic architecture (Nemčok et al. 2013; Haupert et al. 2016). From c. 80 to c. 150 km, the oceanward onlapping geometry of the overlaying sediments defines the typical passive infill of a post-tectonic sag-sequence (as defined in Masini et al. 2013). This sag-sequence is younger than the first sediments onlapping onto oceanic crust (Fig. 3), suggesting that it may still be part of the syn-rift record. As no tectonic deformation is observed in this sequence, it implies that it is post-tectonic. At the base of this post-tectonic sag-sequence, we observe an enigmatic reflective layer that is locally well stratified and characterized by continentward downlaps as observed on the PSTM section, suggesting it may partly correspond to sediments and magmatic flows (Fig. 3c). The apparent occurrence of magmatism at c. 110 km prevents further detailed stratigraphic descriptions within this layer. Still, the geometric relationships described within this sequence are compatible with the occurrence of a high-β extensional setting floored by large-offset normal faults (i.e. exhumation faults, Fig. 3c) dipping continentward as indicated by the downlapping sediments getting younger in the same direction. This reflective layer was deposited prior to the deposition of post-tectonic sag-sequences and possibly corresponds to syn-exhumation sequences recording the evolution of large offset normal faults. From c. 150 to c. 420 km, mainly onlaps and passive infill are observed.

Magmatic additions. Magmatic additions seem to be only evidenced in the most distal parts of the SE Indian margin (Fig. 3c, d). From distance c. 80 to c. 150 km, spatially delimited, high-amplitude reflectors are observed locally crosscutting the overlying stratigraphy, possibly corresponding to sills (planar or saucer shaped morphologies; Planke et al. 2005; Fig. 3c). Similar spatially delimited, high-amplitude reflectors are observed within the interpreted basement, some possibly corresponding to sills intrusive in the basement (Fig. 3c). Locally, they show an angular shape similar to the fault block facies unit defined by Planke et al. (2005). At about 110 km, the dome-shaped topography of the top basement and the associated symmetric downlaps on both sides suggest the occurrence of a presently buried volcanic edifice (c. 0.8 s in height and c. 30 km in width, Fig. 3c). Its internal structure is not difficult to observe on the seismic profile, but its overall morphology is similar to the 'hyaloclastite mounts' described by Calvès et al. (2011). Further oceanward from 150 to 210 km (Fig. 3d), the observed local highs at top basement are similar in length and height to the 'outer highs' features described by Calvès et al. (2011). The palaeobathymetry during the emplacement of this volcanic

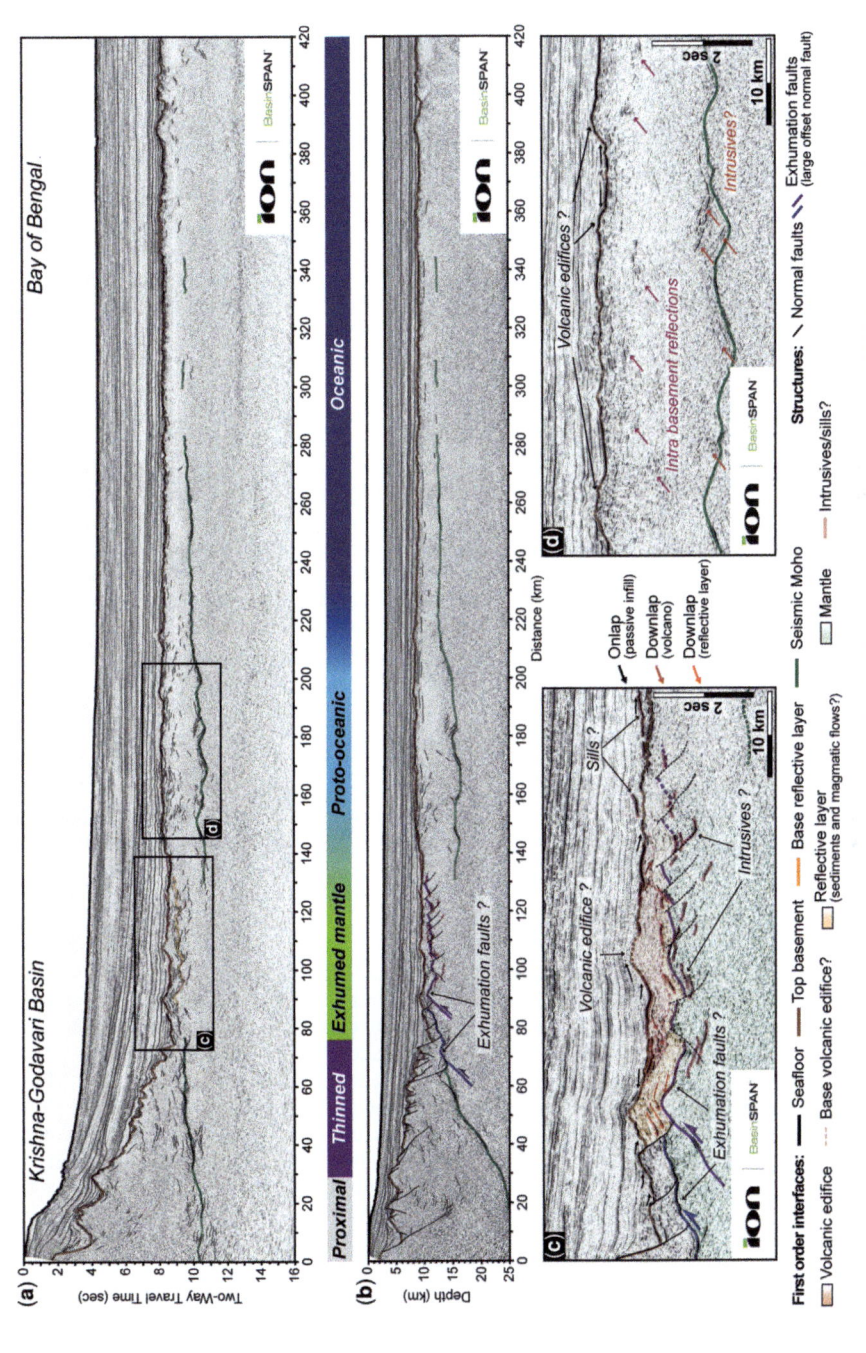

Fig. 3. Seismic observations from the SE Indian rifted margin case example (PSTM and PSDM seismic profiles, courtesy of ION Geophysical). (**a**) Line drawing of the PSTM seismic profile and interpretation of first-order interfaces. (**b**) Interpretation of first-order interfaces and tectonic structures of the corresponding PSDM seismic profile (vertical exaggeration ×2). Based on the evolution of accommodation space (between sea-level and top basement) and crustal thickness (between top basement and seismic Moho) along the PSDM profile, we define structural margin domains: the proximal, thinned, exhumed mantle, proto-oceanic and oceanic domains. (**c**) Zoom over the interpreted exhumed mantle domain showing hints for magmatic additions possibly syn- and post-exhumation. (**d**) Zoom over the interpreted proto-oceanic domain showing top basement, intra-basement reflectivity and pattern of seismic Moho.

edifice remains, however, difficult to constrain. As the interpreted sills locally crosscut the first sediments of the post-tectonic sag sequence, we suggest that part of the magmatic additions emplaced after the beginning of the passive infill.

Identification of structural domains

Continentward, from distance 0 to *c.* 30 km, accommodation space slightly increases (from *c.* 2 to *c.* 4 km), locally reaching >5 km within graben and half-graben basins. Continental basement shows little thickness variation (>25 to *c.* 22 km thick), representative of a *proximal domain* (Fig. 3b). From *c.* 30 to *c.* 80 km, accommodation space progressively increases (from *c.* 4 to *c.* 9 km). The associated deepening of the top basement and ascent of the seismic Moho delimit a progressive extreme thinning of the continental basement from 22 to less than 5 km thick, characteristic of the *thinned domain* (i.e. distal domain of Haupert *et al.* 2016). From *c.* 80 to *c.* 150 km, a large accommodation space is observed (locally >10 km) where we identified the occurrence of a potential high-β extensional setting. We define top basement as the base of the sag-sequence, but suggest the occurrence of syn-exhumation sediments (with a downlapping geometry) and magmatic flows underneath. The nature of the underlying basement cannot be directly constrained but the previously summarized observations are consistent with an *exhumed mantle domain* (see also Haupert *et al.* 2016). Potential field data support the exhumed mantle domain hypothesis as modelled by Nemčok *et al.* (2013). From *c.* 210 km to the end of the line, top basement and seismic Moho are almost parallel and define a *c.* 5 km-thick transparent basement characteristic of the *oceanic domain*. The observed thickness of oceanic crust is consistent with regional gravity inversion results in the Bay of Bengal (Radhakrishna *et al.* 2010).

From *c.* 150 to *c.* 210 km, accommodation space is reduced to *c.* 8 km and the top basement and seismic Moho define a 9–10 km-thick basement (see also Radhakrishna *et al.* 2010; Nemčok *et al.* 2013). Intra-basement reflectivity is frequent (Fig. 3d), dipping oceanward and continentward and often observed underneath the small volcanic edifices presented on Figure 3d. Therefore, this domain differs from the adjacent exhumed mantle and oceanic domains (see also Nemčok *et al.* 2013). The suggested increasing occurrence of magmatic additions towards the oceanward end of the exhumed mantle and the proximity of unambiguous oceanic crust (Fig. 3c, d) are characteristic features of *proto-oceanic domains* at magma-poor rifted margins (Iberia–Newfoundland: Welford *et al.* 2010; Peron-Pinvidic *et al.* 2013; Autralia–Antarctica: Gillard *et al.* 2015). The passage from the exhumed mantle to proto-oceanic domain is transitional highlighted by the progressive step-up morphology of the top basement (from *c.* 10 to *c.* 8 km depth). The passage from the proto-oceanic to oceanic domain appears equally transitional, here marked by an ascent of the Moho and slight deepening of the top basement.

Interpretations and scenarios for the nature of the proto-oceanic domain

General interpretation. Several interpretations have already been presented along this profile (e.g. Nemčok *et al.* 2013; Pindell *et al.* 2014; Haupert *et al.* 2016). The overall architecture presented in this work (Figs 3 & 4) shares many similarities with that of Haupert *et al.* (2016) except within the exhumed mantle domain and oceanward, where we defined the proto-oceanic domain.

The proximal domain is characterized by a weak thinning of the continental crust. We interpret a set of classical normal faults mainly dipping oceanward and delimiting half-graben basins, probably rooting at mid-crustal levels (possibly corresponding to some faint reflectivity observed at about 15 km depth, Fig. 3b). The beginning of the thinned domain coincides with the breakaway of a fault system corresponding to a major escarpment at about 30 km (R1 in Haupert *et al.* 2016), associated with a relatively large offset. Conjugate structures may occur at depth structuring the necking of the continental crust (Mohn *et al.* 2012). Further oceanward from 40 to 60 km, we interpret only small rift basins (a few kilometres wide). As the associated faults show a limited offset, they probably root within shallow crustal levels. Another important escarpment is observed at about 60 km (R2 in Haupert *et al.* 2016) where the crust is already thinned to less than 10 km thick. A set of oceanward-dipping faults possibly locally offsetting the Moho (Fig. 4) can be interpreted, suggesting that some of these faults can cut through the entire crust, and hence are probably embrittled. Such faults may allow the serpentinization of the underlying mantle (Pérez-Gussinyé & Reston 2001).

The exhumed mantle domain is characterized by the interpreted occurrence of exhumation faults on top of which an enigmatic reflective layer was identified and described (Fig. 3). The nature of this reflective layer is uncertain and may correspond to a volcano-sedimentary sequence (including both sediments and magmatic flows) consistent with the frequently suggested occurrence of magmatic additions (Fig. 3c). Similar sequences are notably described over the exhumed mantle of the Newfoundland (Peron-Pinvidic *et al.* 2010; Gillard *et al.* 2016b) and Australian–Antarctica rifted margins (Gillard *et al.* 2016b) recording the

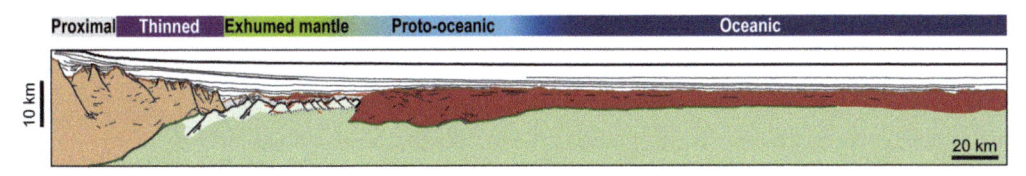

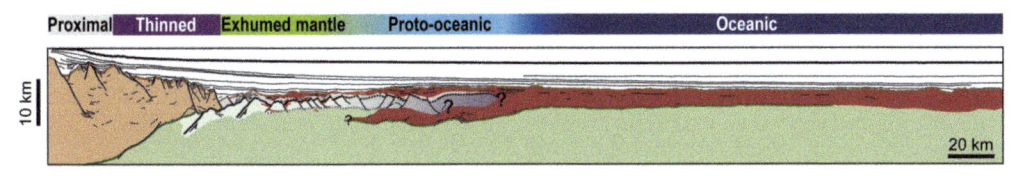

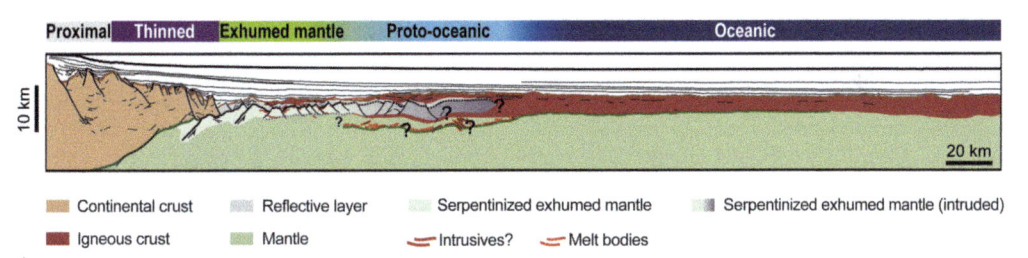

Fig. 4. Interpretations of the SE Indian rifted margin case example, illustrating different scenarios for the nature of the proto-oceanic domain. (**a**) Scenario 1: igneous crust. (**b**) Scenario 2: exhumed serpentinized mantle 'sandwiched' between extrusive and intrusive material. (**c**) Scenario 3: exhumed serpentinized mantle 'sandwiched' between extrusive and intrusive material and melt entrapment at depth.

progressive formation of new basement surfaces along exhumation faults and the associated magmatism. We interpret these exhumation faults to be dipping continentward, consistently with the interpretation of Haupert *et al.* (2016), based on the geometry observed in these syn-tectonic sequences. These exhumation faults are associated with topographic variations (Fig. 3a), possibly corresponding to normal faults crosscutting the previously exhumed basement as suggested from the fault block morphology (Planke *et al.* 2005) of some inferred intrusives (Figs 3b, c & 4). Magmatism is interpreted to occur within the exhumed mantle domain (Fig. 3c) and seems to become progressively more important toward the proto-oceanic domain as indicated by the increasing occurrence of magmatic intrusives at depth and in the overlying sediments (Fig. 3c). The first oceanic crust probably emplaced at a steady-state spreading system is relatively thin consistently with regional observations in the Bay of Bengal (Radhakrishna *et al.* 2010).

Scenarios of the nature of the proto-oceanic domain.
This domain is described in a few studies at magma-poor margins (e.g. Jagoutz *et al.* 2007; Welford *et al.* 2010; Bronner *et al.* 2011; Peron-Pinvidic *et al.* 2013; Gillard *et al.* 2015, 2016*b*, 2017). So far, two drill holes are publicly available in similar domains, at the most distal parts of the Iberia–Newfoundland rifted margins (Ocean Drilling Program (ODP), sites 1070 and 1277; Shipboard Scientific Party 1998, 2004). Potential analogues of proto-oceanic domains are identified in remnants of the Alpine Tethys rifted margins (e.g. Chenaillet ophiolite, Manatschal *et al.* 2011; Lower Platta nappe, Desmurs *et al.* 2002). Nevertheless, the nature of the basement, the architecture and magmatic budget of these domains is uncertain and likely to vary from one rifted margin to the other (Peron-Pinvidic *et al.* 2013, 2017). Therefore, we prefer presenting different interpretations involving variable magmatic budget rather than one solution.

The local highs observed in the proto-oceanic domain (Fig. 3d) are similar in shape to outer highs commonly interpreted as volcanic edifices near oceanic domains in settings considered as magma rich (e.g. Planke *et al.* 2000; Calvès *et al.* 2011). This analogy straightforwardly suggests that the proto-oceanic domain could be made of igneous crust only (scenario 1, Fig. 4a), locally c. 10 km thick (Fig. 3b; Nemčok et al. 2013). The intra-basement reflectivity observed underneath could reasonably be interpreted as corresponding to the deep structure of the volcanic edifices and the reflective patterns above seimic Moho as magmatic intrusives (Fig. 3d). Thick igneous crust and volcanic edifices are common in magma-rich rifted margin contexts adjacent to continental crust (Menzies et al. 2002; Nielsen & Hopper 2004) as observed, for example, at the West Indian rifted margin: (Calvès et al. 2011), Hatton Bank (White et al. 2008) or SE Greenland (Larsen et al. 1998; Hopper et al. 2003). If this interpretation is indeed possible, it is quite surprising for a rifted margin where adjacent mantle exhumation is inferred and considered to be magma poor (Radhakrishna et al. 2012; Nemčok et al. 2013; Pindell et al. 2014; Haupert et al. 2016; Sinha et al. 2016).

Various forms of magmatic additions seem to occur in the interpreted exhumed mantle domain (Fig. 3c). The oceanward limit of potentially exhumed mantle appears gradual and magmatic additions seem to become more important oceanward. Hence, an alternative interpretation for this domain (scenario 2, Fig. 4b) could be that the basement is composed of exhumed serpentinized mantle progressively 'sandwiched' between magmatic extrusive (basalts?) and intrusive material (gabbroic underplates?). Intra-basement reflectivity could correspond to the top of faulted exhumed mantle, variably intruded (by feeder dykes?), on top of which extrusives and local volcanic edifices can be emplaced. The additional presence of continental crust fragments cannot be excluded (e.g. Nemčok et al. 2013). The locally thick reflective packages observed above the interpreted seismic Moho could correspond to sill-like intrusives (gabbroic?) forming a mafic underplated body at the base of serpentinized exhumed mantle. Bronner et al. (2011) suggested a similar interpretation at the Iberia–Newfoundland rifted margins based on refraction and reflection seismic data and observations from the ODP Sites 1277 that penetrated exhumed mantle and recovered intrusives and extrusive mafic material (Jagoutz et al. 2007). The Chenaillet ophiolite preserved in the Alps (Manatschal et al. 2011) can be considered as an analogue of this interpretation of the proto-oceanic domain (Gillard et al. 2015, 2016b). There, basaltic rocks deposited on top of exhumed serpentinized peridotites are exposed (Manatschal et al. 2011). These volcanic sequences appear to seal normal faults that developed in the previously exhumed serpentinized mantle (Manatschal et al. 2011).

In our last alternative (scenario 3, Fig. 4c), we suggest that the reflective packages observed above the interpreted seismic Moho could in fact be within the mantle, possibly corresponding to a layer of magma entrapment. The overall architecture interpreted for the proto-oceanic domain is similar to the scenario 2 except for a thinner underplated magmatic layer and the suggested presence of melt entrapment within the mantle. The occurrence of melt impregnation and stagnation within lithospheric mantle is documented at the most distal parts of present-day rifted margins based on drilling results (Iberia–Newfoundland, Müntener & Manatschal 2006). Similar observations are made in onshore fossil analogues of exhumed mantle and embryonic oceanic domains preserved in the Alps (Muntener & Piccardo 2003; Müntener et al. 2004; Muntener et al. 2010; Picazo et al. 2016).

Uruguay rifted margin case example

Geological setting and first-order tectono-magmatic context

The Brazilian, Uruguayan and Argentinian rifted margins of South America (including the Pelotas, Salado and Colorado basins; Fig. 2) were initially conjugate to the Namibian and South African rifted margins through the Walvis, Lüderitz and Orange Basins (e.g. Rabinowitz & LaBrecque 1979; Gladczenko et al. 1997; Torsvik et al. 2009; Moulin et al. 2010; Blaich et al. 2011; Heine et al. 2013). The South Atlantic rifted margins result from the Late Jurassic–Early Cretaceous breakup of West Gondwana (e.g. Rabinowitz & LaBrecque 1979; Gladczenko et al. 1997; Torsvik et al. 2009; Moulin et al. 2010; Heine et al. 2013; Frizon De Lamotte et al. 2015). In between the Rio Grande and Falkland–Agulhas fracture zones, onset of rifting occurred in the latest Jurassic (e.g. Heine et al. 2013 and references therein). A first rift event is associated with the formation of several rift basins trending NW–SE, obliquely to the final margin structure, such as the Salado/Punta del Este (e.g. Stoakes et al. 1991; Soto et al. 2011) and Colorado (Autin et al. 2013) basins (Fig. 2). Then, the formation of the South Atlantic and onset of oceanic spreading occurred diachronously related to a progressive and segmented propagation from south to north (e.g. Franke et al. 2007; Blaich et al. 2013; Franke 2013; Heine et al. 2013; Koopmann et al. 2014; Stica et al. 2014 and references therein) between c. 137 and 126 Ma.

In the South Atlantic Ocean, the Rio Grande Rise and Walvis Ridge are generally interpreted to mark the passage of the Tristan Da Cunha hotspot (Fig. 2) responsible for the eruption of the Paraná–Etendeka Large Igneous Province (LIP) (Gibson et al. 2006) between 138 and 129 Ma (Turner et al. 1994; Stewart et al. 1996; Peate 1997). The relationship between LIP emplacement and rifting is complex and the detailed spatial and temporal relationship remains unclear (Franke et al. 2007; Franke 2013; Stica et al. 2014; Frizon De Lamotte et al. 2015). This complexity may partially be explained by the progressive and segmented northward propagation of the South Atlantic prior to and during the emplacement of the Paraná–Etendeka LIP (Franke et al. 2007; Koopmann et al. 2014). As a result, the Paraná–Etendeka LIP can be considered as pre-, syn- or post-rift depending on the margin segment considered (Stica et al. 2014).

We focus on a high-resolution reflection seismic profile provided by ION Geophysical, striking NW–SE offshore Uruguay across the Salado/Punta del Este basin and terminating in the South Atlantic Ocean (Fig. 2). The Uruguay rifted margin, as most margins of the southern South Atlantic, shows thick SDR sequences (Franke et al. 2007; Soto et al. 2011; Clerc et al. 2015) considered as characteristic of magma-rich rifted margins (Hinz 1981; Mutter 1985; Planke et al. 2000; Menzies et al. 2002; Doré & Lundin 2015).

Seismic observations

First order interfaces. Seafloor progressively deepens from less than 500 m continentward to more than 4 km at the oceanward end of the profile in the South Atlantic Ocean (Fig. 5). From distance 0 to *c.* 140 km, *top basement* is defined at the top of a reflective package, locally showing evidence of erosional truncations (e.g. near 40 km; Fig. 5). It corresponds to the interface between the base passive infill (from 0 to *c.* 80 km) or base syn-rift (from *c.* 80 to *c.* 120 km) and acoustic basement (including either pre-rift and magmatic sequences, or crystalline basement; Stoakes et al. 1991). It progressively deepens from *c.* 1.5 to *c.* 4 km and is characterized by local topographic variations (between *c.* 80 and *c.* 120 km). From *c.* 140 to *c.* 240 km, top basement is only characterized by faint local reflections. We define it at the base of syn-rift sediments where observable (Fig. 5c). From *c.* 240 km to the end of the profile, high-amplitude reflectors at the top of SDRs and at the base of passive infill characterize the top basement oceanward (Fig. 5d), corresponding to an almost flat interface. From *c.* 260 to *c.* 340 km, we tentatively define the *base of SDRs*. From *c.* 260 to *c.* 280 km, it corresponds to a relatively well-defined high-amplitude reflector at the base of the SDR package. From *c.* 290 to *c.* 340 km, we define it at the downward termination of SDRs (Fig. 5d). Along the profile, deep reflectors are commonly observed and interpreted as *seismic Moho*. They are notably well imaged on the PSTM profile (Fig. 5a). From 0 to *c.* 80 km, from *c.* 150 to *c.* 210 km and from *c.* 260 to *c.* 310 km, we define seismic Moho at the base of parallel discontinuous high-amplitude reflectors commonly forming packages locally more than 1 s thick (*c.* 5 km thick). Further oceanward, from *c.* 320 km to the end, seismic Moho corresponds to a succession of short parallel discontinuous reflectors.

Stratigraphic and basin architecture. From distance 0 to *c.* 260 km, basement morphology defines graben and half-graben-type basins corresponding to low-β extensional settings. Still, these basins are locally quite deep (*c.* 12–13 km depth), associated with the relatively thick rift sequences of the Punta del Este basin (locally more than 6–7 km thick, Fig. 5, Stoakes et al. 1991). Symmetric onlaps of sedimentary sequences are more commonly observed than typical wedge-shaped geometries (Fig. 5a). From 260 to 380 km, we observe continentward onlaps onto top basement marking a progressive post-SDRs emplacement continentward passive infill (Fig. 5a, d). From 380 km to the end of the line, oceanward downlaps can be observed (Fig. 5a).

Magmatic additions. Continentward, from distance *c.* 120 to *c.* 260 km, magmatic additions are suggested to occur mainly as sill-like intrusives into sedimentary sequences of graben and half-graben-type basins (Fig. 5). Sills appear as spatially limited high-amplitude reflectors either parallel to or cross-cutting the stratigraphy corresponding to 'planar', 'planar transgressive' and 'saucer-shaped' morphologies (Planke et al. 2005; e.g. at 150 km). Evidence of magmatism is interpreted at *c.* 220 km (Fig. 5c). The overall dome-shaped morphology with roughly symmetric flanks is characteristic of a volcanic edifice, possibly also associated with sill intrusions. Interestingly, this volcanic edifice appears to be sitting on top of syn-rift and possible early post-rift sequences, suggesting that magmatic activity mainly occurred after the formation of graben-type basins. The exact onset and timing remain nevertheless difficult to constrain in more detail. From *c.* 260 to *c.* 380 km, we observe SDRs characterized by high-amplitude reflectors terminating rather abruptly at depth. SDRs are classically interpreted as volcanic flows (Hinz 1981; Mutter et al. 1982) emplaced in sub-aerial to shallow conditions based on drilling results, e.g. off Norway (Eldholm et al. 1987; 1989) and SE Greenland (Larsen et al. 1994; Duncan et al. 1996; Larsen & Saunders 1998). To first order, the most continentward SDR sequence has a wedge-shaped geometry. Further oceanward (Fig. 5d) the SDR sequences correspond to a succession of nearly parallel reflections, whereas further oceanward, they show again a clear wedge-shaped geometry. From *c.* 320 to *c.* 340 km, we observe a progressive decrease in the length of these SDRs. From *c.* 340 to *c.* 360 km, we observe local high-amplitude reflectors (mainly observable on the PSTM section)

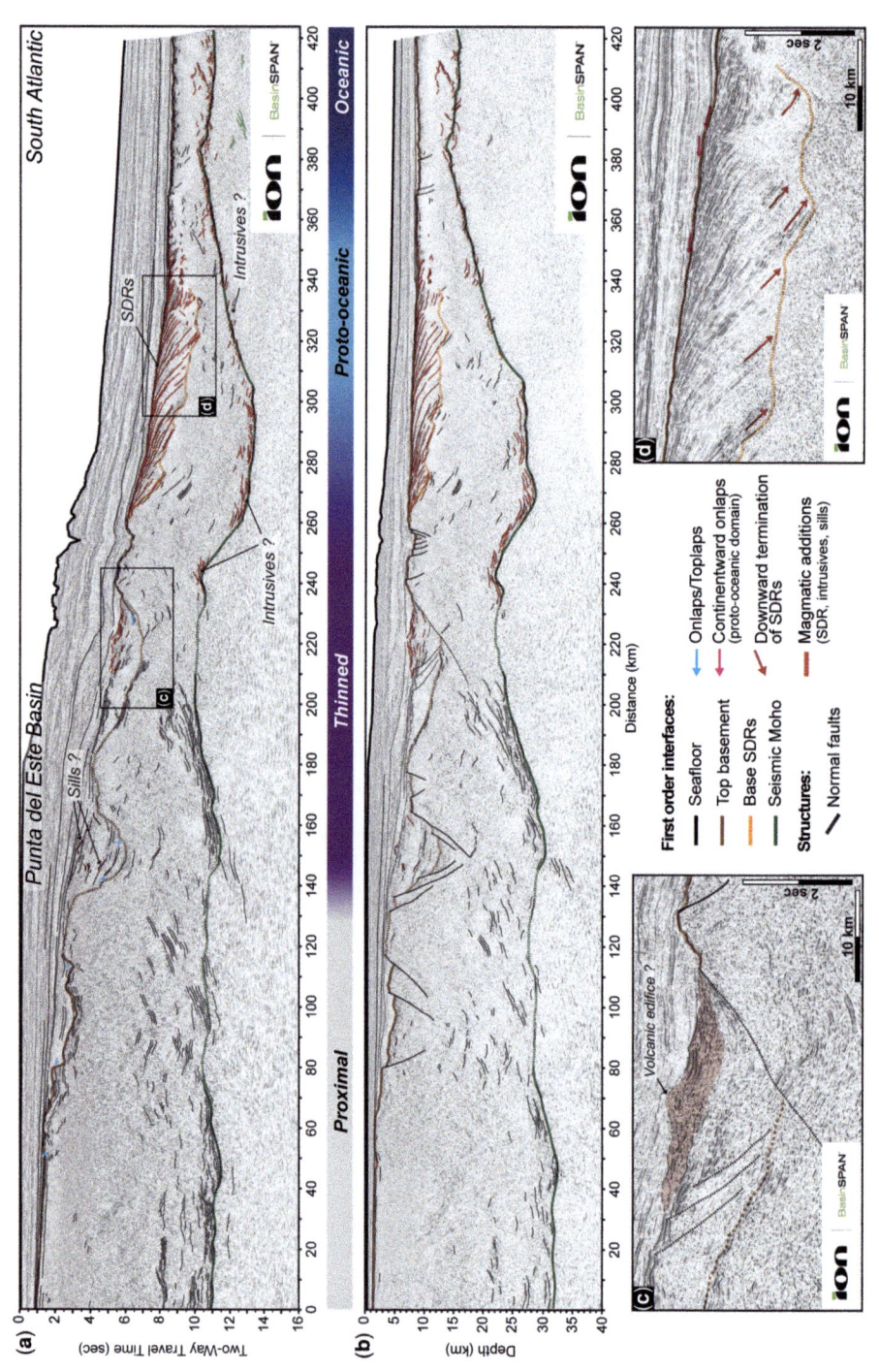

Fig. 5. Seismic observations from the Uruguayan rifted margin case example (PSTM and PSDM seismic profiles, courtesy of ION Geophysical). (**a**) Line drawing of the PSTM seismic profile and interpretation of first-order interfaces. (**b**) Interpretation of first-order interfaces and structures of the PSDM of the same seismic profile (vertical exaggeration × 2). Based on the evolution of accommodation space (between sea-level and top basement) and crustal thickness (between top basement/base SDRs and seismic Moho) along the PSDM profile, we define structural margin domains: the proximal, thinned, proto-oceanic and oceanic domains. (**c**) Zoom over a possible volcanic edifice in the interpreted thinned domain. (**d**) Zoom over the interpreted proto-oceanic domain showing top basement, SDRs, base SDRs and continentward onlaps.

dipping oceanward, still possibly corresponding to short SDR sequences. At their base, we identify weak horizontal discontinuous reflectors.

Magmatic additions at the base or within the basement are also likely but remain difficult to identify unambiguously. We note that, where we identified potential sill intrusions and volcanic buildups into rift basins, the base of the crust defining seismic Moho is often ill defined. This observation contrasts with the relative ubiquitous occurrence of high-amplitude discontinuous parallel reflectors at the base of the crust along the profile. Some of these deep reflective packages are observed oceanward (c. 240 to c. 370 km; Fig. 5). They may partly correspond to intrusive magmatic bodies emplaced at the base of the crust synchronously with the SDR sequences.

Identification of structural domains

Continentward, from distance 0 to c. 140 km, accommodation space slightly increases from c. 1.5 to c. 3.5 km, locally reaching a maximum of 5 km within a graben-type basin (Fig. 5). Top basement and seismic Moho are almost parallel and flat, defining a c. 25–30 km-thick continental basement, consistent with the occurrence of a *proximal domain*. From c. 140 to c. 260 km, evidence for locally deep graben and half-graben basins is identified, associated with a progressive important increase in accommodation space oceanward (from c. 3.5 to c. 7.5 km, locally >12 km). The progressive deepening of top basement and Moho depth variations reflects crustal thickness variations from 25 km to possibly locally less than 15 km (at about 220 km), characteristic for a *thinned domain*. Magmatism is evidenced in this domain possibly occurring during early post-rift time. From 380 km to the end of the line, top basement and Moho are roughly parallel, defining a 6–7 km-thick transparent basement, except for the local occurrence of internal reflectors, characteristic of the *oceanic domain*.

From 260 to 320 km, accommodation space shows only little increase oceanward (from 7 to 8 km), as the top basement remains relatively flat. In contrast, at depth, seismic Moho variations define a 'crustal keel' locally delimiting a 20 km-thick basement continentward, getting progressively thinner oceanward (from c. 310 to c. 380 km). SDRs are observed at the base of passive infill, below top basement (as defined in this work) and becoming thicker oceanward, as suggested by the base of SDR geometry (Fig. 5d). Associated intrusions possibly also occur at the base of the crust (Fig. 5). The occurrence of SDRs and proximity of standard oceanic crust (c. 7 km thick, White et al. 1992) are characteristic features of 'continent–ocean transitions' (COT), 'transitional crust' or 'outer domains' at magma-rich rifted margins, (Menzies et al. 2002; Franke 2013; Peron-Pinvidic et al. 2013), referred to in this work as *proto-oceanic domain*. A deepening of Moho reflections (from c. 240 to c. 270) marks the transition from the thinned to proto-oceanic domain, while top basement remains flat. The transition from our defined proto-oceanic domain to a standard oceanic domain possibly occurs where an inflection in seismic Moho topography is observed.

Interpretation and scenarios for the nature of the proto-oceanic domain

General interpretation. Interpretations of the overall rifted margin architecture were previously published based on adjacent lines and other seismic surveys (e.g. Franke et al. 2007; Soto et al. 2011; Clerc et al. 2015) sharing some similarities with the architecture interpreted in this work (Fig. 6). The proximal domain shows a weak thinning of the continental crust consistently with the occurrence of shallow graben-type basins. The transition to the thinned domain is defined where we interpret the breakaway of an important fault delimiting one side of a roughly symmetric deep graben. The overall architecture of our interpreted thinned domain is atypical. This is possibly related to the fact that the profile crosses the Punta del Este basin oriented obliquely (NW–SE) to the final rifted margin segmentation (Soto et al. 2011), explaining why this domain is not observed on adjacent profiles located further north (Clerc et al. 2015). Local evidence of magmatism is interpreted within this domain corresponding to sill complexes, a volcanic edifice and possible intrusions at depth (Fig. 5). The transition from the thinned to proto-oceanic domain is interpreted as gradual, approximately at the continentward end of the first SDR sequences. Similarly, the passage from proto-oceanic to oceanic appears transitional. The first oceanic crust likely emplaced at a steady-state spreading system is 6–7 km thick, consistent with global average (c. 7 ± 1 km; White et al. 1992).

Scenarios of the nature of the proto-oceanic domain. Most studies at magma-rich rifted margins place the COT, i.e. our proto-oceanic domain, where SDRs occur, either at their landward/seaward edge, or in the centre (Franke 2013). Several legs of the Deep Sea Drilling Project and ODP drilled SDR sequences off the British Isles (leg 81, Roberts and Schnitker 1984), offshore Norway (leg 104; Eldholm et al. 1987) and SE Greenland (legs 152, Larsen et al. 1994; and 163, Duncan et al. 1996). Drilling results confirmed the volcanic nature of SDRs, interweaved with some sediments and the geochemical signatures of the lava flows showed a decrease in continental contamination oceanward (Larsen & Saunders 1998; Saunders et al. 1998). Even if they may not

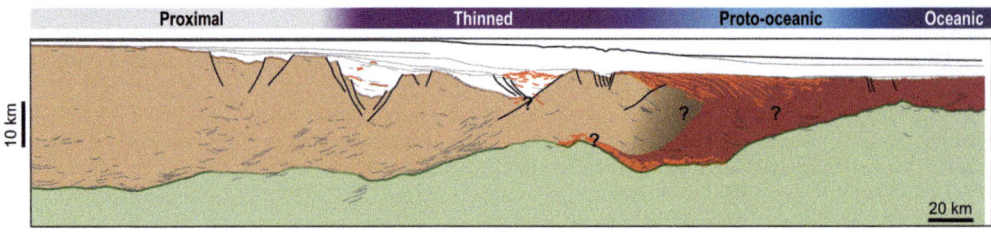

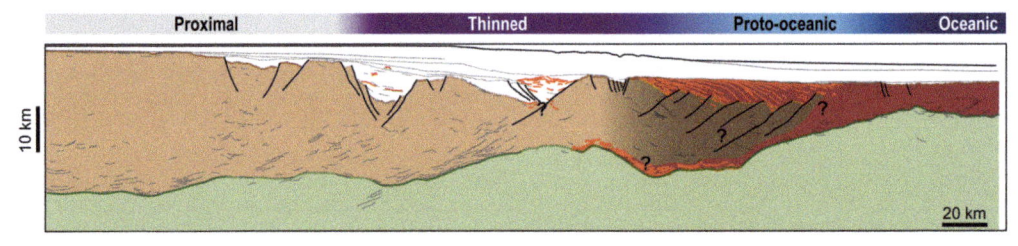

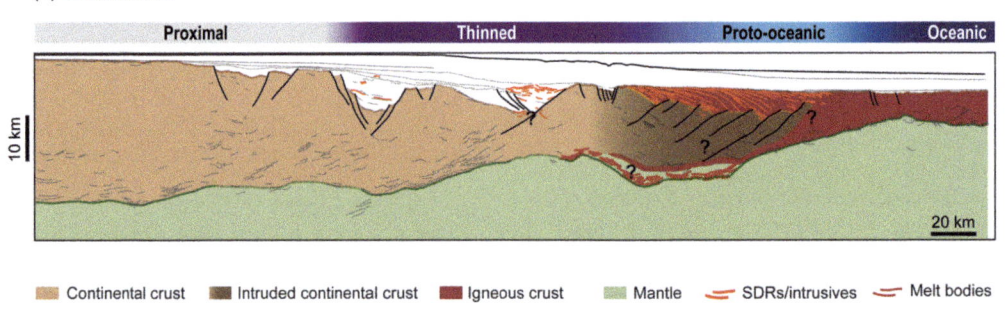

Fig. 6. Interpretations of the Uruguayan rifted margin case example, illustrating different scenarios for the nature of the proto-oceanic domain. (**a**) Scenario 1: igneous crust. (**b**) Scenario 2: intruded continental crust 'sandwiched' between extrusives (SDRs) and underplated material. (**c**) Scenario 3: intruded continental crust 'sandwiched' between extrusives (SDRs) and underplated material and melt entrapment at depth.

use the same terminology as the one used in this work, many studies focused on this domain (e.g. Hinz 1981; Mutter 1985; Gladczenko et al. 1997; Planke et al. 2000; Franke et al. 2010). Most debates are related to the emplacement mechanism related to the formation of SDRs (e.g. Larsen & Saunders 1998; Franke et al. 2010; Buck 2017; Paton et al. 2017) and on the nature of the underlying basement (e.g. Larsen et al. 1998; Hopper et al. 2003; Geoffroy 2005; Geoffroy et al. 2015). In the following, we present different interpretations for the nature and architecture of the proto-oceanic domain involving variable magmatic budget that are compatible with our seismic observations.

The first SDR package observed shows a wedge-shaped geometry, suggesting a fault-controlled geometry. Over this first sequence, SDRs appear sub-parallel 'prograding' oceanward, suggesting a minor role of faulting, consistent with a proto-oceanic domain made from igneous crust only (scenario 1; Fig. 6a), locally c. 20 km thick (considering a vertical section between top basement and seismic Moho). In this interpretation, we consider that the continental crust terminates abruptly at the downward termination of the first sub-parallel SDRs (Fig. 5d). The abrupt termination of the continental crust related to the emplacement of thick igneous crust along the proto-breakup axis is inspired from studies offshore SE Greenland (Larsen & Saunders 1998; Larsen et al. 1998; Hopper et al. 2003), Hatton Bank (White et al. 2008) and Norway (Eldholm et al. 1987), where SDRs were drilled. The deep reflective packages observed above the interpreted seismic Moho could then be interpreted as the intrusive equivalent of SDRs, possibly corresponding to mafic underplates (Skogseid *et al.* 2000; White *et al.* 2008).

The wedge-shaped geometries of the most continentward and oceanward SDRs suggest a possible syn-magmatic fault activity consistent with a proto-oceanic domain including thin continental crust getting more and more intruded oceanward (scenario 2, Fig. 6b). Intruded continental crust remnants are interpreted as sandwiched between SDRs and magmatic underplates at depth. This intruded basement is commonly referred to as 'transitional crust' (e.g. Geoffroy 2005; Franke *et al.* 2010; Franke 2013; Abdelmalak *et al.* 2015; Geoffroy *et al.* 2015). The base of the SDR sequences (Fig. 5) could then indicate the top of the intruded continental crust. The deep reflective packages at depth (Fig. 5) could correspond to intruded lower crust (White *et al.* 2008; Geoffroy *et al.* 2015) or to mafic underplates referred to as Lower Crustal Body (LCB; White & McKenzie 1989; Mjelde *et al.* 2008). Field observations from suggested fossil analogues preserved in the Scandinavian Caledonides (Abdelmalak *et al.* 2015) show complex dyke generations intruding a continental basement. Evidence for complex and polyphase syn-magmatic fault activity is documented in Afar (Geoffroy *et al.* 2014; Stab *et al.* 2016). The hypothesis of fault controlled emplacement of SDRs is commonly suggested at present-day magma-rich rifted margins of the South Atlantic (e.g. Franke *et al.* 2007, 2010; Stica *et al.* 2014; Geoffroy *et al.* 2015; Becker *et al.* 2016).

In our last alternative interpretation, we suggest that the deep reflective packages observed at depth could be within the mantle (scenario 3, Fig. 6c). The overall architecture of the proto-oceanic domain is interpreted to be similar to scenario 2 except for a thinner layer of underplated material and the suggested occurrence of melt bodies trapped within the mantle. Geochemical studies from the Main Ethiopian Rift considered as a tectonically active analogue of a magma-rich setting show a complex and protracted magmatic evolution associated with melt stagnation levels within the lithospheric mantle (Rooney *et al.* 2014, 2017).

Identifying the timing and amount of magmatic additions

Interpretations derived from seismic reflection data are non-unique and therefore imply significant uncertainties when it comes to suggesting geological interpretations. On the one hand, determining the precise timing of magma emplacement is problematic and can only be done for extrusives, based on relationships with the stratigraphy (Figs 3c & 5c). On the other hand, the identification of magmatic additions and in particular magmatic intrusives within crystalline basement is difficult to constrain as both rock types often share similar petrophysical properties (densities/velocities). In the absence of drill-hole data, based on seismic reflection data only, resolving the precise timing and exact volume of magmatic additions remains challenging.

Indirect determination of the timing of magma emplacement

On the Uruguay rifted margin, our observations suggest that most of the magmatism was emplaced early after the rifting phase related to the formation of the Salado/Punta del Este Basin (Fig. 5c), consistent with observations reported by regional studies (e.g. Heine *et al.* 2013). As SDR sequences, most likely corresponding to lava flows, progressively develop into unambiguous standard oceanic crust (7 ± 1 km, White *et al.* 1992), we can suggest that they were emplaced at lithospheric breakup. Intrusives probably occur at the base of the crust (Fig. 5). The timing of emplacement cannot be ascertained but the proximity of SDRs suggest that they may be similar to the distal magmatic intrusions (LCB) interpreted as related to lithospheric breakup in the Colorado Basin further south (Autin *et al.* 2016).

On the SE Indian rifted margin, magmatic additions are suggested to occur only in the ultra-distal parts and in continuity with the first unequivocal oceanic domain (Fig. 3). Part of the magmatic additions possibly emplaced during or early after mantle exhumation as indicated by the inferred occurrence of a volcanic edifice on top of possible syn-tectonic sequences (reflective layer, Fig. 3c). Part of the magmatism probably emplaced after the deposition of the post-tectonic sag-sequences, i.e. after the onset of the passive infill, as indicated by the likely presence of sill-like intrusions crosscutting the overlaying stratigraphy (Fig. 3c). Based on these deductions, we believe that most of the magmatism observed is likely to be part of lithospheric breakup processes.

Uncertainties in determining the magmatic budget

The identification of magmatic extrusives (volcanic edifices, SDRs) can reasonably be done based on our high-resolution PSTM and PSDM seismic reflection data. However, determining the overall magmatic budget and notably the amount of distal magmatic additions intrusive at the base or within basement remains challenging. High-resolution refraction data may help distinguish between pre-rift lower crust and intrusives emplaced at lithospheric breakup as shown from the Hatton Bank example (White *et al.* 2008). Integrated quantitative approaches can be used to examine the shape (Skogseid *et al.* 2000; Autin *et al.* 2013) and nature (Nirrengarten *et al.* 2014; Autin *et al.* 2016) of distal LCB characteristic of magma-rich rifted margins. Potential field modelling represents a useful tool to test different interpretations and to provide quantitative verifications that could narrow down the number of possible solutions. Nevertheless, they cannot provide unique solutions as different lithologies may present similar geophysical properties (Christensen & Mooney 1995). In addition, the intense tectonic and/or magmatic activity affecting the distal domains of rifted margins strongly alter the initial petrophysical proprieties of the rocks that form these domains. As a result, the average density and velocity used as input for forward modelling would in fact be very similar between our different scenarios (Peron-Pinvidic *et al.* 2016).

Because of these uncertainties, we decided to present and discuss three alternative interpretations for each of our two case examples (Figs 4 & 6). The hypotheses for the architectures of proto-oceanic domains result in different scenarios for the magmatic budget at lithospheric breakup. As these interpretations are compatible with the limited drilling results in offshore analogues (where available) and/or with onshore field observations previously described, we believe that all of them can be considered as geologically coherent and plausible. These interpretations represent non-unique 'end-member' scenarios based on which we aim to discuss fundamental processes related to lithospheric breakup at rifted margins whether they are considered as representative of magma-poor or magma-rich archetypes.

Magmatic budget at rifted margins: discussion

Architecture of proto-oceanic domains and implications for lithospheric breakup processes

Based on the interpretations of the proto-oceanic domain architecture previously suggested for the SE Indian and Uruguayan rifted margins (Figs 4 & 6), we first examine and tentatively estimate the magmatic budget at lithospheric breakup for each scenario (Fig. 7). Secondly, based on the inferred evolution of melt production in the proto-oceanic domain, we present different potential mechanisms for lithospheric breakup involving different tectono-magmatic interactions (Figs 7 & 8).

In scenario 1, whether we consider the SE Indian or the Uruguayan rifted margin, the proto-oceanic domain is suggested to be dominantly made from igneous crust, respectively reaching *c.* 10 and *c.* 20 km thick. In both cases, the thickness of magmatic additions exceeds the 7 ± 1 km standard thickness (Fig. 7; White *et al.* 1992; Bown & White 1994) predicted from decompression melting models (White & McKenzie 1989). The volume of magmatic additions is reduced to standard thicknesses in the oceanic domain (Fig. 7), even less in the case of SE India. The clear increase in magmatic additions suggested in the proto-oceanic domain of the scenario 1 is consistent with a relatively fast melt production at lithospheric breakup that would appear rather 'instantaneous' at a geological scale (Fig. 8). This excess magmatic event/pulse is transient and often advocated to occur at the rift to drift transition of magma-rich rifted margin (Nielsen & Hopper 2004). Nevertheless, the possibility of an excess magmatic event/magmatic pulse at lithospheric breakup has also been considered in the case of the Iberia–Newfoundland rifted margins, archetype of magma-poor settings (Bronner *et al.* 2011).

In scenario 2, the proto-oceanic domain corresponds to a complex basement respectively composed of intruded exhumed serpentinized mantle (SE India) or intruded continental crust (Uruguay) sandwiched in-between extrusive and intrusive material. As a result, in both cases, the apparent total crustal thickness (i.e. between top basement and seismic Moho, Figs 3 & 5) is due to the cumulative effect of magmatic additions and continental crust/exhumed serpentinized mantle thicknesses (Fig. 7). Interestingly, magmatic addition thickness is strongly reduced compared with scenario 1 and does not necessarily exceed the 7 ± 1 km standard thickness (White *et al.* 1992; Bown & White 1994) derived from decompression melting model predictions (White & McKenzie 1989). As a result, in both cases, a relative progressive increase in melt production can be suggested at lithospheric breakup that may appear 'gradual' (Fig. 8; Whitmarsh *et al.* 2001*a*), possibly recorded within wide areas (e.g. Gillard *et al.* 2015). Extensional tectonic processes (mechanical thinning) are probably dominant in the initial stages of lithospheric breakup either related to polyphase extensional deformation within exhumed mantle domain at magma-poor rifted margins (Gillard *et al.* 2016*a*, *b*) or to explain the formation of SDRs at magma-rich rifted margins (e.g. Franke *et al.* 2007, 2010). Magmatic processes become more important only towards the end of lithospheric breakup (Gillard *et al.* 2015, 2017; Peron-Pinvidic & Osmundsen 2016).

In scenario 3, the architectures suggested for the proto-oceanic domain are similar to the ones presented in scenario 2 except for the presence of melt stagnation levels within the mantle. The amount of remnants of continental crust/exhumed serpentinized mantle in between igneous material and the

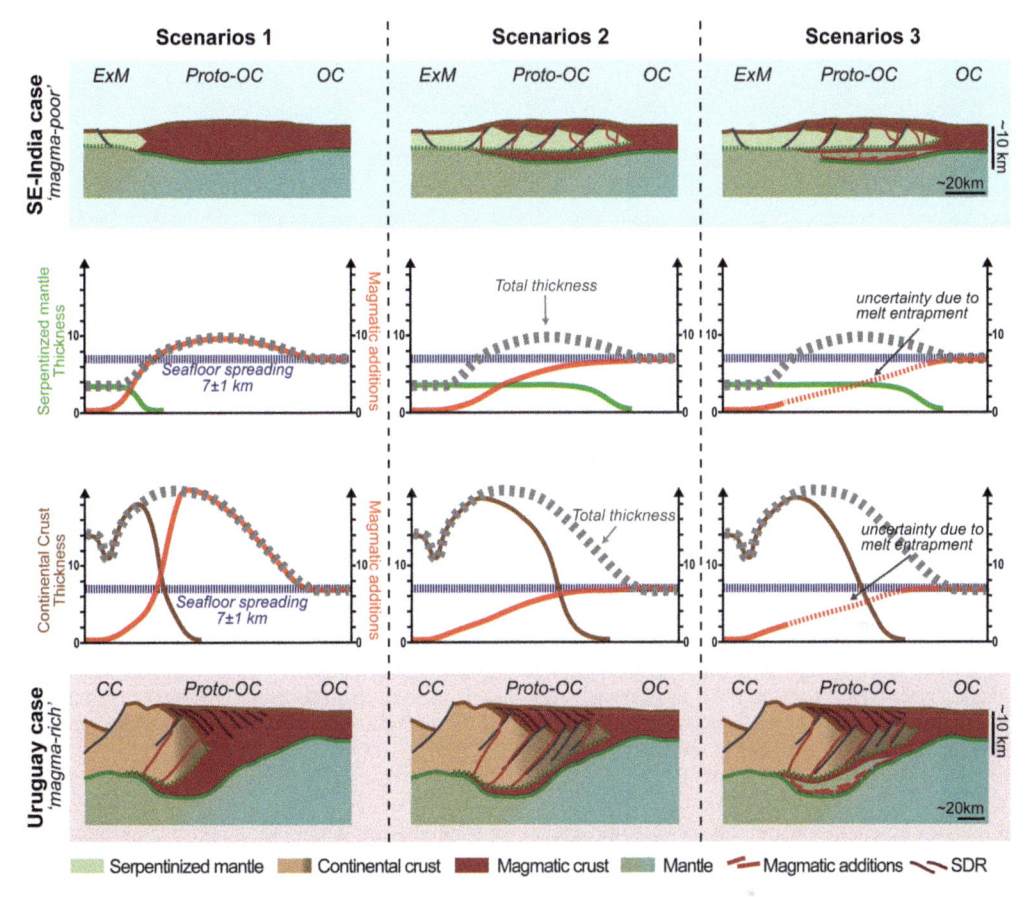

Fig. 7. Estimates of the magmatic budget at lithospheric breakup inferred from the different scenarios (1–3) proposed for the proto-oceanic domains at the SE Indian (upper part) and Uruguayan examples (lower part). The evolution of magmatic addition thickness, representing the magmatic budget is indicated by the red line. The green and brown curves respectively represent the thickness of exhumed serpentinized mantle (SE India) and continental crust (Uruguay). The dashed grey line represents the apparent total thickness of the proto-oceanic domain (ie. between top basement and seismic Moho). The thick dashed blue line represents the 7 ± 1 km-thick reference for oceanic crust thickness (White *et al.* 1992; Bown & White 1994) inferred from decompression melting models (White & McKenzie 1989). CC, Continental crust; ExM, exhumed serpentinized mantle; Proto-OC, proto-oceanic crust; OC, oceanic crust.

volume of melt entrapped in the mantle is difficult to estimate (Fig. 7). However, as suggested for scenario 2, magmatic addition thickness does not necessarily exceed the 7 ± 1 km standard thickness (White *et al.* 1992; Bown & White 1994) derived from decompression melting model predictions (White & McKenzie 1989), even for the Uruguayan case example. In both cases, the interpreted occurrence of melt entrapment implies an inefficient/incomplete extraction of melt out of the mantle, possibly suggesting variable melt production resulting in a polyphase or 'stuttering' (Jagoutz *et al.* 2007) lithospheric breakup (Fig. 8). Such lithospheric breakup processes are likely to be associated with important local variations in the magmatic budget comparable with what is observed in present-day ultra-slow spreading systems (Cannat *et al.* 2008; Sauter *et al.* 2016), often used as analogues to understand COT (Pérez-Gussinyé *et al.* 2006; Cannat *et al.* 2009). The Main Ethiopian Rift may correspond to a nascent analogue, where the transition from mechanical/tectonic-dominated to magmatic-dominated processes appears as largely spatially distributed and temporally protracted (Rooney *et al.* 2014).

Some implications for the reappraisal of magma-poor v. magma-rich rifted margin archetypes

Despite different volumes of magmatism, we presented several potential mechanisms for lithospheric

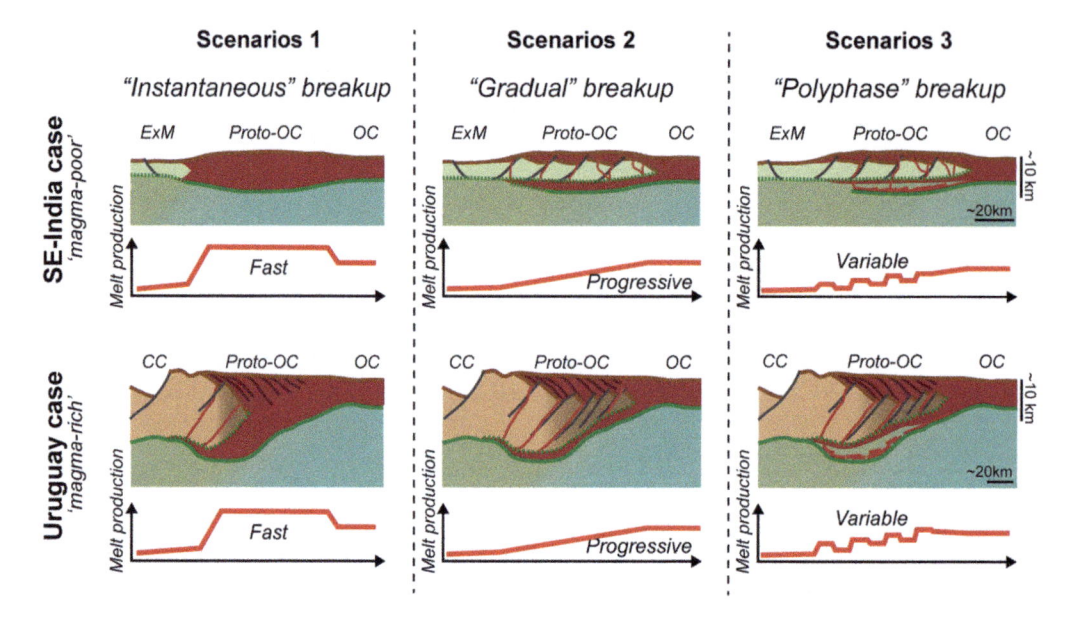

Fig. 8. Interpretations of lithospheric breakup mechanisms for each of the scenarios (1–3) proposed for the proto-oceanic domains at the SE Indian (upper part) and Uruguayan examples (lower part). The diagrams presented associated with each scenario show the inferred evolution of melt production at lithospheric breakup and recorded within the proto-oceanic domain. We distinguish the 'instantaneous', 'gradual' and 'polyphase' lithospheric breakup, respectively associated with fast, progressive and variable melt production. CC, Continental crust; ExM, exhumed serpentinized mantle; Proto-OC, proto-oceanic crust; OC, oceanic crust.

breakup applicable to both magma-poor and magma-rich archetypes related to different melt production and tectono-magmatic interplays (Figs 7 & 8). The Uruguayan and other rifted margins showing magma-rich morphologies may be explained by excess decompression melting compared with steady-state seafloor spreading (scenario 1, Figs 7 & 8) but could also involve a monotonic (scenarios 2 and 3, Figs 7 & 8) increase in decompression melting with an early onset relative to crustal breakup. The converse, where the onset of decompression melting occurs later relative to crustal breakup, allows for mantle exhumation characterizing magma-poor rifted margin. The transition from exhumed mantle to oceanic crust at the SE Indian and other rifted margins showing magma-poor characteristics could result from an excess decompression melting event compared with steady-state seafloor spreading (scenario 1, Figs 7 & 8). This contrasts with the progressive (scenario 2, Figs 7 & 8) or 'stuttering' (scenario 3, Figs 7 & 8) onset of decompression melting (Whitmarsh *et al.* 2001*a*; Jagoutz *et al.* 2007) more classically inferred. As a result, we highlight that the formation of each archetype (magma-poor or magma-rich) could result from different tectono-magmatic interactions and melt production at lithospheric breakup. Davis & Lavier (2017) draw a similar conclusion based on numerical simulations, showing that several variables can lead to the formation of an end-member archetype morphology.

To account for the uncertainty in determining the magmatic budget at rifted margins and notably the amount of underplated material, we presented three interpretations for each case study. A notable difference for all interpretations between the SE Indian/magma-poor and Uruguayan/magma-rich case example is related to the onset of decompression melting relative to the amount of crustal thinning (rift evolution). The timing of decompression melting onset relative to crustal thinning appears to be an important parameter to consider, equal to, if not more important than, the magmatic budget to understand the processes occurring at the rift-to-drift transition and the worldwide variability of rifted margins.

Parameters controlling melt production and onset of decompression melting: area for further research

Several studies have focused on the parameters controlling the onset of decompression melting and the amount of melt production at rifted margins (e.g. Minshull *et al.* 2001; Nielsen & Hopper 2004; Pérez-Gussinyé *et al.* 2006; Fletcher *et al.* 2009; Armitage et al. 2010; Lundin et al. 2014; Davis & Lavier 2017). They notably revealed the importance of mantle temperature, extension rates, mantle composition, preceding rift history (inheritance) and the absence or occurrence of active upwelling of the asthenosphere. Mantle temperature is classically considered to represent one of the main factors controlling the onset of decompression melting and the magmatic budget (e.g. White & McKenzie 1989). Elevated mantle temperatures enhance melt supply and are often considered as the main parameter controlling the magmatic budget at magma-rich rifted margins (e.g. Skogseid et al. 2000). In contrast, lower mantle temperatures inhibit and delay decompression melting onset. Extension rates at lithospheric breakup are also considered to have a significant effect on magma supply (Lundin et al. 2014). Lundin et al. (2014) notably suggested that magma-poor/magma-rich settings are mainly determined by the opening rate of regional tectonic plates and distance to the associated Euler pole. Mantle composition is also known to control melt production: the more primitive and volatile rich the mantle is, the more melt it may produce and vice versa, if the mantle is depleted (Cannat et al. 2009). Melt extraction efficiency (Muntener et al. 2010), rift-induced processes such as melt infiltration and stagnation resulting from melt–rock reactions within lithospheric mantle (Müntener et al. 2004; Picazo et al. 2016) also appear important.

In detail, the magmatic budget and formation of end-member archetypes is probably controlled by a complex interaction between these parameters (e.g. Pérez-Gussinyé et al. 2006; Fletcher et al. 2009; Armitage et al. 2010; Brown & Lesher 2014; Davis & Lavier 2017). Based on numerical simulations, Pérez-Gussinyé et al. (2006) showed that a decrease in melt production cannot solely be a function of extension rates requesting an additional key role of mantle temperature or composition. Some studies reveal the role on the magmatic budget of the timing of a mantle thermal anomaly emplacement relative to the rift evolution (Skogseid 2001; Armitage et al. 2010). Further work is required to better unravel the interplay of parameters controlling the timing and amount of melt production as well as to determine more precisely its volume in seismic sections.

Conclusions

Based on a number of morphological features, rifted margins are commonly defined as either 'magma poor' or 'magma rich' (e.g. Sawyer et al. 2007; Reston 2009; Reston & Manatschal 2011; Franke 2013; Doré & Lundin 2015). This terminology/classification results in assumptions on the magmatic budget of rifted margins during rifting and at lithospheric breakup. In this work, we re-appraised and questioned a presently prevailing model that magma-rich margins necessarily have excess decompression melting during lithospheric breakup compared with steady-state seafloor spreading and that magma-poor margins have inhibited melting.

We first highlighted the difficulty in resolving the magmatic budget at rifted margins based on seismic reflection data only. Quantitative analyses could be used to narrow down the number of potential hypotheses but would still provide non-unique solutions. To account for this uncertainty, we presented several interpretations, each supported by onshore field analogues and drilling results in similar settings, where available. As a result, we suggested several mechanisms to achieve lithospheric breakup for each end-member archetype, implying different tectono-magmatic interactions and melt production (scenarios 1–3). We showed that the Uruguayan and other magma-rich rifted margins could result from excess decompression melting compared with steady-state seafloor spreading but could also be explained by a gradual or stuttering increase with an early onset relative to crustal breakup (i.e. rupture and separation of continental crust). The converse, where the onset of decompression melting is late relative to crustal breakup, allows for mantle exhumation, characteristic of the magma-poor rifted margin archetype such as the SE Indian rifted margin.

Eventually, we show that different tectono-magmatic interactions and melt production can lead to the formation of magma-poor or magma-rich morphologies. In spite of different volumes of magmatism, the lithospheric breakup mechanisms suggested are comparable between magma-poor and magma-rich archetypes. Considerations on the timing of decompression melting onset relative to crustal thinning may be more important than the overall magmatic budget to unravel the processes occurring at the rift-to-drift transition and the worldwide variability of rifted margins.

Acknowledgments The authors acknowledge ION Geophysical for providing the two seismic profiles presented in this work. This paper largely benefited from the critical and constructive reviews of Erik Lundin and Tony Doré and additional comments from the editor James A. Hammerstein. We thank the developers of the free software QGis. We are grateful to Jakob Skogseid and Philippe Werner for fruitful discussions.

Funding The MM4 consortium (BP, Conoco Phillips, Statoil, Petrobras, Total, Shell, BHP-Billiton, and BG) financially supported this project.

References

Abdelmalak, M.M., Andersen, T.B. *et al.* 2015. The ocean–continent transition in the mid-Norwegian margin: insight from seismic data and an onshore Caledonian field analogue. *Geology*, **43**, G37086.1, https://doi.org/10.1130/G37086.1

Amante, C. & Eakins, B.W. 2009. ETOPO1 1 arc-minute global relief model: procedures, data sources and analysis. *NOAA Technical Memorandum NESDIS NGDC-24* National Geophysical Data Center, NOAA, https://doi.org/10.7289/V5C8276M

Armitage, J.J., Collier, J.S. & Minshull, T.A. 2010. The importance of rift history for volcanic margin formation. *Nature*, **465**, 913–917, https://doi.org/10.1038/nature09063

Armitage, J.J., Collier, J.S., Minshull, T.A. & Henstock, T.J. 2012. Thin oceanic crust and flood basalts: India-Seychelles breakup. *Geochemistry, Geophysics, Geosystems*, **12**, 1–25, https://doi.org/10.1029/2010GC003316

Autin, J., Scheck-Wenderoth, M. *et al.* 2013. Colorado Basin 3D structure and evolution, Argentine passive margin. *Tectonophysics*, **604**, 264–279, https://doi.org/10.1016/j.tecto.2013.05.019

Autin, J., Scheck-Wenderoth, M., Götze, H.J., Reichert, C. & Marchal, D. 2016. Deep structure of the Argentine margin inferred from 3D gravity and temperature modelling, Colorado Basin. *Tectonophysics*, **676**, 198–210, https://doi.org/10.1016/j.tecto.2015.11.023

Baksi, A.K., Barman, T.R., Paul, D.K. & Farrar, E. 1987. Widespread early Cretaceous flood basalt volcanism in eastern India: geochemical data from the Rajmahal–Bengal–Sylhet Traps. *Chemical Geology*, **63**, 133–141, https://doi.org/10.1016/0009-2541(87)90080-5

Becker, K., Tanner, D.C., Franke, D. & Krawczyk, C.M. 2016. Fault-controlled lithospheric detachment of the volcanic southern South Atlantic rift. *Geochemistry, Geophysics, Geosystems*, **17**, 887–894, https://doi.org/10.1002/2015GC006081

Belgarde, C., Manatschal, G., Kusznir, N., Scarselli, S. & Ruder, M. 2015. Rift processes in the Westralian Superbasin, North West Shelf, Australia: insights from 2D deep reflection seismic interpretation and potential fields modeling. *In: APPEA Conference 2015*. Australian Petroleum Production & Exploration Association, Melbourne, 1–8.

Blaich, O.A., Faleide, J.I. & Tsikalas, F. 2011. Crustal breakup and continent–ocean transition at South Atlantic conjugate margins. *Journal of Geophysical Research: Solid Earth*, **116**, 1–38, https://doi.org/10.1029/2010JB007686

Blaich, O.A., Faleide, J.I., Tsikalas, F., Gordon, A.C. & Mohriak, W. 2013. Crustal-scale architecture and segmentation of the South Atlantic volcanic margin. *In*: Mohriak, W.U., Danforth, A., Post, P.J., Brown, D.E., Tari, G.C., Nemčok, M. & Sinha, S.T. (eds) *Conjugate Divergent Margins*. Geological Society, London, Special Publications, **369**, 167–183, https://doi.org/10.1144/SP369.22

Boillot, G. & Coulon, C. 1998. *La Déchirure Continentale et L'ouverture Océanique: Géologie Des Marges Passives*. Gordon and Breach, Amsterdam.

Bown, J.W. & White, R.S. 1994. Variation with spreading rate of oceanic crustal thickness and geochemistry. *Earth and Planetary Science Letters*, **121**, 435–449, https://doi.org/10.1016/0012-821X(94)90082-5

Bronner, A., Sauter, D., Manatschal, G., Péron-Pinvidic, G. & Munschy, M. 2011. Magmatic breakup as an explanation for magnetic anomalies at magma-poor rifted margins. *Nature Geoscience*, **4**, 549–553, https://doi.org/10.1038/nphys1201

Brown, E.L. & Lesher, C.E. 2014. North Atlantic magmatism controlled by temperature, mantle composition and buoyancy. *Nature Geoscience*, **7**, 820–824, https://doi.org/10.1038/ngeo2264

Buck, W.R. 2017. The role of magmatic loads and rift jumps in generating seaward dipping reflectors on volcanic rifted margins. *Earth and Planetary Science Letters*, **466**, 62–69, https://doi.org/10.1016/j.epsl.2017.02.041

Calvès, G., Schwab, A.M. *et al.* 2011. Seismic volcanostratigraphy of the western Indian rifted margin: the pre-Deccan igneous province. *Journal of Geophysical Research: Solid Earth*, **116**, https://doi.org/10.1029/2010JB000862

Cannat, M., Sauter, D., Bezos, A., Meyzen, C., Humler, E. & Le Rigoleur, M. 2008. Spreading rate, spreading obliquity, and melt supply at the ultraslow spreading Southwest Indian Ridge. *Geochemistry, Geophysics, Geosystems*, **9**, https://doi.org/10.1029/2007GC001676

Cannat, M., Manatschal, G., Sauter, D. & Péron-Pinvidic, G. 2009. Assessing the conditions of continental breakup at magma-poor rifted margins: what can we learn from slow spreading mid-ocean ridges? *Comptes Rendus Geoscience*, **341**, 406–427, https://doi.org/10.1016/j.crte.2009.01.005

Choudhuri, M., Nemčok, M., Stuart, C., Welker, C., Sinha, S.T. & Bird, D. 2014. 85°E Ridge, India: constraints on its development and architecture. *Journal of the Geological Society of India*, **84**, 513–530, https://doi.org/10.1007/s12594-014-0160-9

Christensen, N.I. & Mooney, W.D. 1995. Seismic velocity structure and composition of the continental crust: a global view. *Journal of Geophysical Research*, **100**, 9761–9788, https://doi.org/10.1029/95JB00259

Clerc, C., Jolivet, L. & Ringenbach, J.-C. 2015. Ductile extensional shear zones in the lower crust of a passive margin. *Earth and Planetary Science Letters*, **431**, 1–7, https://doi.org/10.1016/j.epsl.2015.08.038

Coffin, M.F., Pringle, M.S., Duncan, R.A., Gladczenko, T.P., Storey, M., Müller, R.D. & Gahagan, L.A. 2002. Kerguelen hotspot magma output since 130 Ma. *Journal of Petrology*, **43**, 1121–1137, https://doi.org/10.1093/petrology/43.7.1121

Curray, J.R. & Munasinghe, T. 1991. Origin of the Rajmahal Traps and the 85°E Ridge: preliminary reconstructions of the trace of the Crozet hotspot. *Geology*, **19**, 1237, https://doi.org/10.1130/0091-7613(1991)01<1237:OOTRTA>92.3.CO;2

Davis, J.K. & Lavier, L.L. 2017. Influences on the development of volcanic and magma-poor morphologies during passive continental rifting. *Geosphere*, **13**, 1524–1540, https://doi.org/10.1130/GES01538.1

Desmurs, L., Manatschal, G. & Bernoulli, D. 2001. The Steinmann Trinity revisited: mantle exhumation and magmatism along an ocean-continent transition: the

Platta nappe, eastern Switzerland. *In*: WILSON, R.C.L., WHITMARSH, R.B., TAYLOR, B. & FROITZHEIM, N. (eds) *Non-Volcanic Rifting of Continental Margins*. Geological Society, London, Special Publications, **187**, 235–266, https://doi.org/10.1144/GSL.SP.2001.187.01.12

DESMURS, L., MÜNTENER, O. & MANATSCHAL, G. 2002. Onset of magmatic accretion within a magma-poor rifted margin: a case study from the Platta ocean–continent transition, eastern Switzerland. *Contributions to Mineralogy and Petrology*, **144**, 365–382, https://doi.org/10.1007/s00410-002-0403-4

DORÉ, T. & LUNDIN, E. 2015. Hyperextended continental margins: Knowns and unknowns. *Geology*, **43**, 95–96, https://doi.org/doi:10.1130/focus012015.1

DUNCAN, R.A., LARSEN, H.C. & ALLAN, J.F. (eds). 1996. *Proceedings of the Ocean Drilling Program. Initial Reports*, **163**. Ocean Drilling Program, Texas A&M University, College Station, TX, https://doi.org/10.2973/odp.proc.ir.163.1996

EAGLES, G., PÉREZ-DIAZ, L. & SCARSELLI, N. 2015. Getting over continent ocean boundaries. *Earth Science Reviews*, **151**, 244–265, https://doi.org/10.1016/j.earscirev.2015.10.009

ELDHOLM, O., THIEDE, J. & TAYLOR, E. (eds). 1987. *Proceedings of the Ocean Drilling Program. Initial Reports*, **104**. Ocean Drilling Program, Texas A&M University, College Station, TX, https://doi.org/10.2973/odp.proc.ir.104.1987

ELDHOLM, O., THIEDE, J. & TAYLOR, E. 1989. Evolution of the vøring volcanic margin. *In: Proceedings of the Ocean Drilling Program. Scientific Results*, **104**. Ocean Drilling Program, Texas A&M University, College Station, TX, https://doi.org/10.2973/odp.proc.sr.104.191.1989

FLETCHER, R., KUSZNIR, N. & CHEADLE, M. 2009. Melt initiation and mantle exhumation at the Iberian rifted margin: comparison of pure-shear and upwelling-divergent flow models of continental breakup. *Comptes Rendus Geoscience*, **341**, 394–405, https://doi.org/10.1016/j.crte.2008.12.008

FRANKE, D. 2013. Rifting, lithosphere breakup and volcanism: comparison of magma-poor and volcanic rifted margins. *Marine and Petroleum Geology*, **43**, 63–87, https://doi.org/10.1016/j.marpetgeo.2012.11.003

FRANKE, D., NEBEN, S., LADAGE, S., SCHRECKENBERGER, B. & HINZ, K. 2007. Margin segmentation and volcanotectonic architecture along the volcanic margin off Argentina/Uruguay, South Atlantic. *Marine Geology*, **244**, 46–67, https://doi.org/10.1016/j.margeo.2007.06.009

FRANKE, D., LADAGE, S. *ET AL.* 2010. Birth of a volcanic margin off Argentina, South Atlantic. *Geochemistry, Geophysics, Geosystems*, **11**, https://doi.org/10.1029/2009GC002715

FRIZON DE LAMOTTE, D., FOURDAN, B., LELEU, S., LEPARMENTIER, F. & CLARENS, P. 2015. Style of rifting and the stages of Pangea breakup. *Tectonics*, **34**, 1009–1029, https://doi.org/10.1002/2014TC003760

GAINA, C., MÜLLER, R.D., BROWN, B.J. & ISHIHARA, T. 2003. Microcontinent formation around Australia. *In: Evolution and Dynamics of the Australian Plate*. Geological Society of America, Boulder, CO, Special Papers, **372**, 405–416, https://doi.org/10.1130/0-8137-2372-8.405

GAINA, C., MÜLLER, R.D., BROWN, B., ISHIHARA, T. & IVANOV, S. 2007. Breakup and early seafloor spreading between India and Antarctica. *Geophysical Journal International*, **170**, 151–169, https://doi.org/10.1111/j.1365-246X.2007.03450.x

GEOFFROY, L. 2005. Volcanic passive margins. *Comptes Rendus Geoscience*, **337**, 1395–1408, https://doi.org/10.1016/j.crte.2005.10.006

GEOFFROY, L., LE GALL, B., DAOUD, M.A. & JALLUDIN, M. 2014. Flip-flop detachment tectonics at nascent passive margins in SE Afar. *Journal of the Geological Society*, **171**, 689–694, https://doi.org/10.1144/jgs2013-135

GEOFFROY, L., BUROV, E.B. & WERNER, P. 2015. Volcanic passive margins: another way to break up continents. *Scientific Reports*, **5**, 14828, https://doi.org/10.1038/srep14828

GERNIGON, L., BLISCHKE, A., NASUTI, A. & SAND, M. 2015. Conjugate volcanic rifted margins, seafloor spreading, and microcontinent: insights from new high-resolution aeromagnetic surveys in the Norway Basin. *Tectonics*, **34**, 907–933, https://doi.org/10.1002/2014TC003717

GIBSON, S.A., THOMPSON, R.N. & DAY, J.A. 2006. Timescales and mechanisms of plume–lithosphere interactions: $^{40}Ar/^{39}Ar$ geochronology and geochemistry of alkaline igneous rocks from the Paraná–Etendeka large igneous province. *Earth and Planetary Science Letters*, **251**, 1–17, https://doi.org/10.1016/j.epsl.2006.08.004

GILLARD, M., AUTIN, J., MANATSCHAL, G., SAUTER, D., MUNSCHY, M. & SCHAMING, M. 2015. Tectonomagmatic evolution of the final stages of rifting along the deep conjugate Australian–Antarctic magma-poor rifted margins: constraints from seismic observations. *Tectonics*, **34**, 753–783, https://doi.org/10.1002/2015TC003850

GILLARD, M., AUTIN, J. & MANATSCHAL, G. 2016*a*. Fault systems at hyper-extended rifted margins and embryonic oceanic crust: structural style, evolution and relation to magma. *Marine and Petroleum Geology*, **76**, 51–67, https://doi.org/10.1016/j.marpetgeo.2016.05.013

GILLARD, M., MANATSCHAL, G. & AUTIN, J. 2016*b*. How can asymmetric detachment faults generate symmetric ocean continent transitions? *Terra Nova*, **28**, 27–34, https://doi.org/10.1111/ter.12183

GILLARD, M., SAUTER, D., TUGEND, J., TOMASI, S., EPIN, M.-E. & MANATSCHAL, G. 2017. Birth of an oceanic spreading center at a magma-poor rift system. *Scientific Reports*, **7**, 15 072, https://doi.org/10.1038/s41598-017-15522-2

GLADCZENKO, T.P., HINZ, K., ELDHOM, O., MEYER, H., NEBEN, S. & SKOGSEID, J. 1997. South Atlantic volcanic margins. *Journal of the Geological Society*, **154**, 465–470, https://doi.org/10.1144/gsjgs.154.3.0465

GLADCZENKO, T.P., SKOGSEID, J. & ELDHOM, O. 1998. Namibia volcanic margin. *Marine Geophysical Researches*, **20**, 313–341, https://doi.org/10.1023/A:1004746101320

HAUPERT, I., MANATSCHAL, G., DECARLIS, A. & UNTERNEHR, P. 2016. Upper-plate magma-poor rifted margins: stratigraphic architecture and structural evolution. *Marine and Petroleum Geology*, **69**, 241–261, https://doi.org/10.1016/j.marpetgeo.2015.10.020

HEINE, C., ZOETHOUT, J. & MÜLLER, R.D. 2013. Kinematics of the South Atlantic rift. *Solid Earth*, **4**, 215–253, https://doi.org/10.5194/se-4-215-2013

HINZ, K. 1981. A hypothesis on terrestrial catastrophes: wedges of very thick oceanward dipping layers beneath passive margins; their origin and paleoenvironment significance. *Geologiches Jahrbuch*, **E22**, 345–363.

HOPPER, J.R., DAHL-JENSEN, T. ET AL. 2003. Structure of the SE Greenland margin from seismic reflection and refraction data: implications for nascent spreading center subsidence and asymmetric crustal accretion during North Atlantic opening. *Journal of Geophysical Research: Solid Earth*, **108**, 61–64, https://doi.org/10.1029/2002JB001996

JAGOUTZ, O., MÜNTENER, O., MANATSCHAL, G., RUBATTO, D., PÉRON-PINVIDIC, G., TURRIN, B.D. & VILLA, I.M. 2007. The rift-to-drift transition in the North Atlantic: a stuttering start of the MORB machine? *Geology*, **35**, 1087, https://doi.org/10.1130/G23613A.1

KENT, R.W., PRINGLE, M.S., MÜLLER, R.D., SAUNDERS, A.D. & GHOSE, N.C. 2002. ^{40}Ar/^{39}Ar geochronology of the Rajmahal basalts, India, and their relationship to the Kerguelen Plateau. *Journal of Petrology*, **43**, 1141–1153, https://doi.org/10.1093/petrology/43.7.1141

KOOPMANN, H., FRANKE, D., SCHRECKENBERGER, B., SCHULZ, H., HARTWIG, A. & STOLLHOFEN, H. 2014. Segmentation and volcano-tectonic characteristics along the SW African continental margin, South Atlantic, as derived from multichannel seismic and potential field data. *Marine and Petroleum Geology*, **50**, 22–39, https://doi.org/10.1016/j.marpetgeo.2013.10.016

LAL, N.K., SIAWAL, A. & KAUL, A.K. 2009. Evolution of east coast of India – A plate tectonic reconstruction. *Journal of the Geological Society of India*, **73**, 249–260.

LARSEN, H.C. & SAUNDERS, A.D. 1998. Tectonism and volcanism at the southeast Greenland rifted margin: a record of plume impact and later continental rupture. In: *Proceedings of the Ocean Drilling Program. Scientific Results*, **152**. Ocean Drilling Program, Texas A&M University, College Station, TX, https://doi.org/10.2973/odp.proc.sr.152.240.1998

LARSEN, H.C., SAUNDERS, A.D. & CLIFT, P.D. (eds). 1994. *Proceedings of the Ocean Drilling Program. Initial Reports*, **152**. Ocean Drilling Program, Texas A&M University, College Station, TX, https://doi.org/10.2973/odp.proc.ir.152.1994

LARSEN, H.C., DAHL-JENSEN, T. & HOPPER, J.R. 1998. Crustal structure along the Leg 152 drilling transect. In: *Proceedings of the Ocean Drilling Program. Scientific Results*, **152**. Ocean Drilling Program, Texas A&M University, College Station, TX, 463–475, https://doi.org/10.2973/odp.proc.sr.152.245.1998

LAVIER, L.L. & MANATSCHAL, G. 2006. A mechanism to thin the continental lithosphere at magma-poor margins. *Nature*, **440**, 324–328, https://doi.org/10.1038/nature04608

LUNDIN, E.R. & DORÉ, A.G. 2011. Hyperextension, serpentinization, and weakening: a new paradigm for rifted margin compressional deformation. *Geology*, **39**, 347–350, https://doi.org/10.1130/G31499.1

LUNDIN, E.R., REDFIELD, T.F. & PERON-PINDIVIC, G. 2014. Rifted continental margins: geometric influence on crustal architecture and melting. In: PINDELL, J., HORN, B. ET AL. *Sedimentary Basins: Origin, Depositional Histories, and Petroleum Systems. 33rd Annual Gulf Coast Section of the Society for Sedimentary Geology (SEPM) Foundation, Bob F. Perkins Conference*, 26–28 January, Houston, TX, 18–53.

MANATSCHAL, G., SAUTER, D., KARPOFF, A.M., MASINI, E., MOHN, G. & LAGABRIELLE, Y. 2011. The Chenaillet Ophiolite in the French/Italian Alps: an ancient analogue for an oceanic core complex? *Lithos*, **124**, 169–184, https://doi.org/10.1016/j.lithos.2010.10.017

MANGIPUDI, V.R., GOLI, A., DESA, M., TAMMISETTI, R. & DEWANGAN, P. 2014. Synthesis of deep multichannel seismic and high resolution sparker data: implications for the geological environment of the Krishna–Godavari offshore, Eastern Continental Margin of India. *Marine and Petroleum Geology*, **58**, 339–355, https://doi.org/10.1016/j.marpetgeo.2014.08.006

MASINI, E., MANATSCHAL, G. & MOHN, G. 2013. The Alpine Tethys rifted margins: reconciling old and new ideas to understand the stratigraphic architecture of magma-poor rifted margins. *Sedimentology*, **60**, 174–196, https://doi.org/10.1111/sed.12017

MENZIES, M.A., KLEMPERER, S.L., EBINGER, C.J. & BAKER, J. 2002. Characteristics of volcanic rifted margins. In: *Volcanic Rifted Margins*. Geological Society of America, Boulder, CO, Special Paper, **362**, 1–14, https://doi.org/10.1130/0-8137-2362-0.1

MINSHULL, T.A., DEAN, S.M., WHITE, R.S. & WHITMARSH, R.B. 2001. Anomalous melt production after continental break-up in the southern Iberia Abyssal Plain. In: WILSON, R.C.L., WHITMARSH, R.B., TAYLOR, B. & FROITZHEIM, N. (eds) *Non-volcanic Rifting of Continental Margins*. Geological Society, London, Special Publications, **187**, 537–550, https://doi.org/10.1144/GSL.SP.2001.187.01.26

MJELDE, R., RAUM, T., BREIVIK, A.J. & FALEIDE, J.I. 2008. Crustal transect across the North Atlantic. *Marine Geophysical Researches*, **29**, 73–87, https://doi.org/10.1007/s11001-008-9046-9

MOHN, G., MANATSCHAL, G., BELTRANDO, M., MASINI, E. & KUSZNIR, N. 2012. Necking of continental crust in magma-poor rifted margins: evidence from the fossil Alpine Tethys margins. *Tectonics*, **31**, https://doi.org/10.1029/2011TC002961

MOULIN, M., ASLANIAN, D. & UNTERNEHR, P. 2010. A new starting point for the South and Equatorial Atlantic Ocean. *Earth-Science Reviews*, **98**, 1–37, https://doi.org/10.1016/j.earscirev.2009.08.001

MÜNTENER, O. & MANATSCHAL, G. 2006. High degrees of melt extraction recorded by spinel harzburgite of the Newfoundland margin: the role of inheritance and consequences for the evolution of the southern North Atlantic. *Earth and Planetary Science Letters*, **252**, 437–452, https://doi.org/10.1016/j.epsl.2006.10.009

MUNTENER, O. & PICCARDO, G.B. 2003. Melt migration in ophiolitic peridotites: the message from Alpine–Apennine peridotites and implications for embryonic ocean basins. In: DILEK, Y. & ROBINSON, P.T. (eds) *Ophiolites in Earth History*. Geological Society, London, Special Publications, **218**, 69–89, https://doi.org/10.1144/GSL.SP.2003.218.01.05

MÜNTENER, O., PETTKE, T., DESMURS, L., MEIER, M. & SCHALTEGGER, U. 2004. Refertilization of mantle peridotite in embryonic ocean basins: trace element and Nd isotopic evidence and implications for crust–mantle relationships. *Earth and Planetary Science Letters*, **221**, 293–308, https://doi.org/10.1016/S0012-821X(04)00073-1

MUNTENER, O., MANATSCHAL, G., DESMURS, L. & PETTKE, T. 2010. Plagioclase peridotites in ocean–continent transitions: refertilized mantle domains generated by melt stagnation in the shallow mantle lithosphere. *Journal of Petrology*, **51**, 255–294, https://doi.org/10.1093/petrology/egp087

MUTTER, J.C. 1985. Seaward dipping reflectors and the continent–ocean boundary at passive continental margins. *Tectonophysics*, **114**, 117–131, https://doi.org/10.1016/0040-1951(85)90009-5

MUTTER, J.C. 1993. Margins declassified. *Nature*, **364**, 393–394, https://doi.org/10.1038/364393a0

MUTTER, J.C., TALWANI, M. & STOFFA, P.L. 1982. Origin of seaward-dipping reflectors in oceanic crust off the Norwegian margin by 'subaerial sea-floor spreading'. *Geology*, **10**, 353, https://doi.org/10.1130/0091-7613(1982)10<353:OOSRIO>2.0.CO;2

MUTTER, J.C., BUCK, W.R. & ZEHNDER, C.M. 1988. Convective partial melting: 1. A model for the formation of thick basaltic sequences during the initiation of spreading. *Journal of Geophysical Research*, **93**, 1031, https://doi.org/10.1029/JB093iB02p01031

NEMČOK, M., SINHA, S.T. ET AL. 2013. East Indian margin evolution and crustal architecture: integration of deep reflection seismic interpretation and gravity modelling. In: MOHRIAK, W.U., DANFORTH, A., POST, P.J., BROWN, D.E., TARI, G.C., NEMČOK, M. & SINHA, S.T. (eds) *Conjugate Divergent Margins*. Geological Society, London, Special Publications, **369**, 477–496, https://doi.org/10.1144/SP369.6

NIELSEN, T.K. & HOPPER, J.R. 2004. From rift to drift: mantle melting during continental breakup. *Geochemistry, Geophysics, Geosystems*, **5**, https://doi.org/10.1029/2003GC000662

NIRRENGARTEN, M., GERNIGON, L. & MANATSCHAL, G. 2014. Nature, structure and age of lower crustal bodies in the Møre volcanic rifted margin: facts and uncertainties. *Tectonophysics*, **59**, https://doi.org/10.1016/j.tecto.2014.08.004

OLIEROOK, H.K.H., JOURDAN, F., MERLE, R.E., TIMMS, N.E., KUSZNIR, N. & MUHLING, J.R. 2016. Bunbury Basalt: Gondwana breakup products or earliest vestiges of the Kerguelen mantle plume? *Earth and Planetary Science Letters*, **440**, 20–32, https://doi.org/10.1016/j.epsl.2016.02.008

PATON, D.A., PINDELL, J., MCDERMOTT, K., BELLINGHAM, P. & HORN, B. 2017. Evolution of seaward-dipping reflectors at the onset of oceanic crust formation at volcanic passive margins: insights from the South Atlantic. *Geology*, **45**, 439–442, https://doi.org/10.1130/G38706.1

PEATE, D.W. 1997. The Paraná–Etendeka province. In: MAHONEY, J.J. & COFFIN, M.F. (eds) *Large Igneous Provinces: Continental, Oceanic, and Planetary Flood Volcanism*, American Geophysical Union, Washington, DC, **100**, 217–245.

PÉREZ-GUSSINYÉ, M. & RESTON, T.J. 2001. Rheological evolution during extension at nonvolcanic rifted margins: onset of serpentinization and development of detachments leading to continental breakup. *Journal of Geophysical Research: Solid Earth*, **106**, 3961–3975, https://doi.org/10.1029/2000JB900325

PÉREZ-GUSSINYÉ, M., MORGAN, J.P., RESTON, T.J. & RANERO, C.R. 2006. The rift to drift transition at non-volcanic margins: insights from numerical modelling. *Earth and Planetary Science Letters*, **244**, 458–473, https://doi.org/10.1016/j.epsl.2006.01.059

PÉRON-PINVIDIC, G. & MANATSCHAL, G. 2009. The final rifting evolution at deep magma-poor passive margins from Iberia-Newfoundland: a new point of view. *International Journal of Earth Sciences*, **98**, 1581–1597, https://doi.org/10.1007/s00531-008-0337-9

PERON-PINVIDIC, G. & OSMUNDSEN, P.T. 2016. Architecture of the distal and outer domains of the Mid-Norwegian rifted margin: insights from the Rån–Gjallar ridges system. *Marine and Petroleum Geology*, **77**, 280–299, https://doi.org/10.1016/j.marpetgeo.2016.06.014

PERON-PINVIDIC, G., SHILLINGTON, D.J. & TUCHOLKE, B.E. 2010. Characterization of sills associated with the U reflection on the Newfoundland margin: evidence for widespread early post-rift magmatism on a magma-poor rifted margin. *Geophysical Journal International*, **182**, 113–136, https://doi.org/10.1111/j.1365-246X.2010.04635.x

PERON-PINVIDIC, G., MANATSCHAL, G. & OSMUNDSEN, P.T. 2013. Structural comparison of archetypal Atlantic rifted margins: a review of observations and concepts. *Marine and Petroleum Geology*, **43**, 21–47, https://doi.org/10.1016/j.marpetgeo.2013.02.002

PERON-PINVIDIC, G., OSMUNDSEN, P.T. & EBBING, J. 2016. Mismatch of geophysical datasets in distal rifted margin studies. *Terra Nova*, **28**, 340–347, https://doi.org/10.1111/ter.12226

PERON-PINVIDIC, G., MANATSCHAL, G., MASINI, E., SUTRA, E., FLAMENT, J.M., HAUPERT, I. & UNTERNEHR, P. 2017. Unravelling the along-strike variability of the Angola-Gabon rifted margin: a mapping approach. In: SABATO CERALDI, T., HODGKINSON, R.A. & BACKE, G. (eds) *Petroleum Geoscience of the West Africa Margin*. Geological Society, London, Special Publications, **438**, 49–76, https://doi.org/10.1144/SP438.1

PICAZO, S., MÜNTENER, O., MANATSCHAL, G., BAUVILLE, A., KARNER, G. & JOHNSON, C. 2016. Mapping the nature of mantle domains in Western and Central Europe based on clinopyroxene and spinel chemistry: evidence for mantle modification during an extensional cycle. *Lithos*, **266–267**, 233–263, https://doi.org/10.1016/j.lithos.2016.08.029

PINDELL, J., GRAHAM, R. & HORN, B. 2014. Rapid outer marginal collapse at the rift to drift transition of passive margin evolution, with a Gulf of Mexico case study. *Basin Research*, **26**, 701–725, https://doi.org/10.1111/bre.12059

PLANKE, S., SYMONDS, P.A., ALVESTAD, E. & SKOGSEID, J. 2000. Seismic volcanostratigraphy of large-volume basaltic extrusive complexes on rifted margins. *Journal of Geophysical Research*, **105**, 19 335–19 351, https://doi.org/10.1029/1999JB900005

PLANKE, S., RASMUSSEN, T., REY, S.S. & MYKLEBUST, R. 2005. Seismic characteristics and distribution of volcanic intrusions and hydrothermal vent complexes in the Vøring and Møre basins. In: *Petroleum Geology: North-West Europe and Global Perspectives*, Geological Society, London, Petroleum Geology Conference Series, **6**, 833–844, https://doi.org/10.1144/0060833

POWELL, C.M., ROOTS, S.R. & VEEVERS, J.J. 1988. Pre-breakup continental extension in East Gondwanaland and the early opening of the eastern Indian Ocean.

Tectonophysics, **155**, 261–283, https://doi.org/10.1016/0040-1951(88)90269-7

RABINOWITZ, P.D. & LABRECQUE, J. 1979. The Mesozoic South Atlantic Ocean and evolution of its continental margins. *Journal of Geophysical Research*, **84**, 5973, https://doi.org/10.1029/JB084iB11p05973

RADHAKRISHNA, M., SUBRAHMANYAM, C. & DAMODHARAN, T. 2010. Thin oceanic crust below Bay of Bengal inferred from 3-D gravity interpretation. *Tectonophysics*, **493**, 93–105, https://doi.org/10.1016/j.tecto.2010.07.004

RADHAKRISHNA, M., TWINKLE, D., NAYAK, S., BASTIA, R. & RAO, G.S. 2012. Crustal structure and rift architecture across the Krishna-Godavari basin in the central Eastern Continental Margin of India based on analysis of gravity and seismic data. *Marine and Petroleum Geology*, **37**, 129–146, https://doi.org/10.1016/j.marpetgeo.2012.05.005

RAMANA, M.V., NAIR, R.R. ET AL. 1994. Mesozoic anomalies in the Bay of Bengal. *Earth and Planetary Science Letters*, **121**, 469–475, https://doi.org/10.1016/0012-821X(94)90084-1

REEVES, C. & DE WIT, M. 2000. Making ends meet in Gondwana: retracing the transforms of the Indian Ocean and reconnecting continental shear zones. *Terra Nova*, **12**, 272–280, https://doi.org/10.1046/j.1365-3121.2000.00309.x

RESTON, T.J. 2009. The structure, evolution and symmetry of the magma-poor rifted margins of the North and Central Atlantic: a synthesis. *Tectonophysics*, **468**, 6–27, https://doi.org/10.1016/j.tecto.2008.09.002

RESTON, T. & MANATSCHAL, G. 2011. *Rifted Margins: Building Blocks of Later Collision*. In: Arc-Continent Collision, Springer, Berlin, Heidelberg, 3–21.

ROBERTS, D.G. & SCHNITKER, D. 1984. *Initial Reports of the Deep Sea Drilling Project, 81.* US Government Printing Office, Initial Reports of the Deep Sea Drilling Project https://doi.org/10.2973/dsdp.proc.81.1984

ROONEY, T.O., NELSON, W.R., DOSSO, L., FURMAN, T. & HANAN, B. 2014. The role of continental lithosphere metasomes in the production of HIMU-like magmatism on the northeast African and Arabian plates. *Geology*, **42**, 419–422, https://doi.org/10.1130/G35216.1

ROONEY, T.O., NELSON, W.R., AYALEW, D., HANAN, B., YIRGU, G. & KAPPELMAN, J. 2017. Melting the lithosphere: metasomes as a source for mantle-derived magmas. *Earth and Planetary Science Letters*, **461**, 105–118, https://doi.org/10.1016/j.epsl.2016.12.010

SANDWELL, D.T., MÜLLER, R.D., SMITH, W.H.F., GARCIA, E. & FRANCIS, R. 2014. New global marine gravity model from CryoSat-2 and Jason-1 reveals buried tectonic structure. *Science (New York)*, **346**, 65–67, https://doi.org/10.1126/science.1258213

SAUNDERS, A.D., LARSEN, H.C. & FITTON, J.G. 1998. Magmatic development of the southeast Greenland Margin and evolution of the Iceland Plume: geochemical constraints from Leg 152. In: *Proceedings of the Ocean Drilling Program. Scientific Results*, **152**. Ocean Drilling Program, Texas A&M University, College Station, TX, 479–501, https://doi.org/10.2973/odp.proc.sr.152.239.1998

SAUTER, D., UNTERNEHR, P. ET AL. 2016. Evidence for magma entrapment below oceanic crust from deep seismic reflections in the Western Somali Basin. *Geology*, **44**, G37747.1, https://doi.org/10.1130/G37747.1

SAWYER, D.S., COFFIN, M.F., RESTON, T.J., STOCK, J.M. & HOPPER, J.R. 2007. COBBOOM: the Continental Breakup and Birth of Oceans Mission. *Scientific Drilling*, **5**, 13–25, https://doi.org/10.2204/iodp.sd.5.02.2007

SHIPBOARD SCIENTIFIC PARTY 1998. Site 1070. In: WHITMARSH, R.B., BESLIER, M.-O. ET AL. (eds) *Proceedings of the Ocean Drilling Program. Initial Reports*, **173**. Ocean Drilling Program, Texas A&M University, College Station, TX, 265–294, https://doi.org/10.2973/odp.proc.ir.173.108.1998

SHIPBOARD SCIENTIFIC PARTY 2004. Site 1277. In: TUCHOLKE, B.E., SIBUET, J.-C., KLAUS, A. ET AL. (eds) *Proceedings of the Ocean Drilling Program. Initial Reports*, **210**. Ocean Drilling Program, Texas A&M University, College Station, TX, 1–39, https://doi.org/10.2973/odp.proc.ir.210.104.2004

SIBUET, J. & TUCHOLKE, B.E. 2013. The geodynamic province of transitional lithosphere adjacent to magma-poor continental margins. In: MOHRIAK, W.U., DANFORTH, A., POST, P.J., BROWN, D.E., TARI, G.C., NEMČOK, M. & SINHA, S.T. (eds) *Conjugate Divergent Margins*. Geological Society, London, Special Publications, **369**, 429–452, https://doi.org/10.1144/SP369.15

SINHA, S.T., NEMČOK, M., CHOUDHURI, M., SINHA, N. & RAO, D.P. 2016. The role of break-up localization in microcontinent separation along a strike-slip margin: the East India-Elan Bank case study. In: NEMČOK, M., RYBÁR, S., SINHA, S.T., HERMESTON, S.A. & LEDVÉNYIOVÁ, L. (eds) *Transform Margins: Development, Controls and Petroleum Systems*. Geological Society, London, Special Publications, https://doi.org/10.1144/SP431.5

SKOGSEID, J. 2001. Volcanic margins: geodynamic and exploration aspects. *Marine and Petroleum Geology*, **18**, 457–461, https://doi.org/10.1016/S0264-8172(00)00070-2

SKOGSEID, J., PLANKE, S., FALEIDE, J.I., PEDERSEN, T., ELDHOLM, O. & NEVERDAL, F. 2000. NE Atlantic continental rifting and volcanic margin formation. In: NØTTVEDT, A. (ed.) *Dynamics of the Norwegian Margin*. Geological Society, London, Special Publications, **167**, 295–326, https://doi.org/10.1144/GSL.SP.2000.167.01.12

SOTO, M., MORALES, E., VEROSLAVSKY, G., DE SANTA ANA, H., UCHA, N. & RODRÍGUEZ, P. 2011. The continental margin of Uruguay: crustal architecture and segmentation. *Marine and Petroleum Geology*, **28**, 1676–1689, https://doi.org/10.1016/j.marpetgeo.2011.07.001

STAB, M., BELLAHSEN, N., PIK, R., QUIDELLEUR, X., AYALEW, D. & LEROY, S. 2016. Modes of rifting in magma-rich settings: tectono-magmatic evolution of Central Afar. *Tectonics*, **35**, 2–38, https://doi.org/10.1002/2015TC003893

STEWART, K., TURNER, S., KELLEY, S., HAWKESWORTH, C., KIRSTEIN, L. & MANTOVANI, M. 1996. 3-D, $^{40}Ar/^{39}Ar$ geochronology in the Paraná continental flood basalt province. *Earth and Planetary Science Letters*, **143**, 95–109, https://doi.org/10.1016/0012-821X(96)00132-X

STICA, J.M., ZALÁN, P.V. & FERRARI, A.L. 2014. The evolution of rifting on the volcanic margin of the Pelotas Basin and the contextualization of the Paraná–Etendeka LIP in the separation of Gondwana in the South Atlantic. *Marine and Petroleum Geology*, **50**, 1–21, https://doi.org/10.1016/j.marpetgeo.2013.10.015

STOAKES, F.A., CAMPBELL, C.V., CASS, R. & UCHA, N. 1991. Seismic stratigraphic analysis of the Punta Del Este Basin, Offshore Uruguay, South America. *AAPG Bulletin*, **75**, 219–240.

SUBRAHMANYAM, C. & CHAND, S. 2006. Evolution of the passive continental margins of India-a geophysical appraisal. *Gondwana Research*, **10**, 167–178, https://doi.org/10.1016/j.gr.2005.11.024

SUTRA, E., MANATSCHAL, G., MOHN, G. & UNTERNEHR, P. 2013. Quantification and restoration of extensional deformation along the Western Iberia and Newfoundland rifted margins. *Geochemistry, Geophysics, Geosystems*, **14**, 2575–2597, https://doi.org/10.1002/ggge.20135

TALWANI, M., DESA, M.A., ISMAIEL, M. & SREE KRISHNA, K. 2016. The Tectonic origin of the Bay of Bengal and Bangladesh. *Journal of Geophysical Research: Solid Earth*, **121**, 4836–4851, https://doi.org/10.1002/2015JB012734

TORSVIK, T.H., ROUSSE, S., LABAILS, C. & SMETHURST, M.A. 2009. A new scheme for the opening of the South Atlantic Ocean and the dissection of an Aptian salt basin. *Geophysical Journal International*, **177**, 1315–1333, https://doi.org/10.1111/j.1365-246X.2009.04137.x

TUGEND, J., MANATSCHAL, G., KUSZNIR, N.J. & MASINI, E. 2015. Characterizing and identifying structural domains at rifted continental margins: application to the Bay of Biscay margins and its Western Pyrenean fossil remnants. *In*: GIBSON, G.M., ROURE, F. & MANATSCHAL, G. (eds) *Sedimentary Basins and Crustal Processes at Continental Margins: From Modern Hyper-extended Margins to Deformed Ancient Analogues*. Geological Society, London, Special Publications, **413**, 171–203, https://doi.org/10.1144/SP413.3

TURNER, S., REGELOUS, M., KELLEY, S., HAWKESWORTH, C. & MANTOVANI, M. 1994. Magmatism and continental break-up in the South Atlantic: high precision ^{40}Ar–^{39}Ar geochronology. *Earth and Planetary Science Letters*, **121**, 333–348, https://doi.org/10.1016/0012-821X(94)90076-0

WELFORD, J.K., SMITH, J.A., HALL, J., DEEMER, S., SRIVASTAVA, S.P. & SIBUET, J.-C. 2010. Structure and rifting evolution of the northern Newfoundland Basin from Erable multichannel seismic reflection profiles across the southeastern margin of Flemish Cap. *Geophysical Journal International*, **180**, 976–998, https://doi.org/10.1111/j.1365-246X.2009.04477.x

WHITE, R. & MCKENZIE, D. 1989. Magmatism at rift zones: the generation of volcanic continental margins and flood basalts. *Journal of Geophysical Research*, **94**, 7685, https://doi.org/10.1029/JB094iB06p07685

WHITE, R.S., MCKENZIE, D. & O'NIONS, R.K. 1992. Oceanic crustal thickness from seismic measurements and rare earth element inversions. *Journal of Geophysical Research*, **97**, 19683, https://doi.org/10.1029/92JB01749

WHITE, R.S., SMITH, L.K., ROBERTS, A.W., CHRISTIE, P.A.F. & KUSZNIR, N.J. 2008. Lower-crustal intrusion on the North Atlantic continental margin. *Nature*, **452**, 460–464, https://doi.org/10.1038/nature06687

WHITMARSH, R.B., MANATSCHAL, G. & MINSHULL, T.A. 2001a. Evolution of magma-poor continental margins from rifting to seafloor spreading. *Nature*, **413**, 150–154, https://doi.org/10.1038/35093085

WHITMARSH, R.B., MINSHULL, T.A., RUSSELL, S.M., DEAN, S.M., LOUDEN, K.E. & CHIAN, D. 2001b. The role of syn-rift magmatism in the rift-to-drift evolution of the West Iberia continental margin: geophysical observations. *In*: WILSON, R.C.L., WHITMARSH, R.B., TAYLOR, B. & FROITZHEIM, N. (eds) *Non-volcanic Rifting of Continental Margins*. Geological Society, London, Special Publications, **187**, 107–124, https://doi.org/10.1144/GSL.SP.2001.187.01.06

WILSON, R.C.L., MANATSCHAL, G. & WISE, S. 2001. Rifting along non-volcanic passive margins: stratigraphic and seismic evidence from the Mesozoic successions of the Alps and western Iberia. *In*: WILSON, R.C.L., WHITMARSH, R.B., TAYLOR, B. & FROITZHEIM, N. (eds) *Non-volcanic Rifting of Continental Margins*. Geological Society, London, Special Publications, **187**, 429–452, https://doi.org/10.1144/GSL.SP.2001.187.01.21

Extensional fault-related folding in the northwestern Red Sea, Egypt: segmented fault growth, fault linkages, corner folds and basin evolution

SAMIR M. KHALIL[1,2] & KEN R. McCLAY[1]*

[1]*Fault Dynamics Research Group, Department of Earth Sciences, Royal Holloway University of London, Egham, Surrey TW20 0EX, UK*

[2]*Department of Geology, Faculty of Science, Suez Canal University, Ismailia 41522, Egypt*

K.R.M., 0000-0002-4077-7645
Correspondence: k.mcclay@es.rhul.ac.uk

Abstract: Segmented, planar, domino-style extensional fault arrays and their associated hanging wall fault-related folds form complex linked basins along the onshore margin of the northwestern Red Sea, Egypt. The extensional fault systems form half-graben basins with kilometre-scale, asymmetrical, doubly plunging longitudinal synclines and narrow, plunging transverse anticlines and synclines. The axial traces of the hanging wall longitudinal folds are curvilinear, sub-parallel to the half-graben Border faults, and bend or are offset at relay ramps and at fault linkage points. Transverse corner fold systems occur at the fault linkage points and fault jogs. The fold geometries, variations in fault displacement, and fault slip indicators indicate that the fold and fault systems are kinematically related and formed during the Late Oligocene–Miocene rifting of the northern Red Sea. The folds were controlled by vertical and lateral fault propagation and by the mechanical anisotropy of the pre-rift strata. The proposed model for these extensional folds is the initial formation of monoclinal flexures above reactivated blind basement faults. Increased displacement, propagation and segment linkage formed hanging wall longitudinal folds and transverse corner folds. The longitudinal folds grew progressively at the expense of the transverse folds and merged along-strike into long hanging wall synclinal basins.

Extensional fault-related folds are now widely recognized in many rift basins and on passive margins. They are formed by variations in displacement along major fault systems, by vertical or lateral extensional fault propagation folding (forced folds), or by changes in shape or displacement along the extensional fault system.

Extensional fault-related folds have been described from the Newark basin, USA (Schlische 1993, 1995), the Basin and Range Province, USA (Janecke *et al.* 1998), the Gulf of Suez, Egypt (e.g. Patton 1984; Moustafa 1987; Moustafa & El Raey 1993; Gawthorpe *et al.* 1997; Gupta *et al.* 1999; Sharp *et al.* 2000; Khalil & McClay 2006; Lewis *et al.* 2015), the Rhine Graben (Maurin & Niviere 1999), offshore Norway (Withjack *et al.* 1990; Pascoe *et al.* 1999; Corfield & Sharp 2000; Richardson *et al.* 2005; Marsh *et al.* 2010) and the Red Sea rift (Khalil & McClay 2002, 2016). These studies generally highlighted the importance of extensional faulting and related folding in the development of rift basins and on continental margins. Some of these papers also discussed the role of extensional fault-related folds for sediment dispersal patterns and basin fill models as well as the formation of structural traps for petroleum accumulations.

Extensional fault-related folds

Classifications and descriptions of extensional fault-related folds have been summarized by Schlische (1995), Janecke *et al.* (1998), Rykkelid & Fossen (2003) and Lăpădat *et al.* (2016) and include the following types.

(1) Fault-bend folds, which are produced by changes in the dip of the fault surface, include hanging wall rollover anticlines (Dula 1991) and ramp synclines (e.g. Suppe 1985; McClay 1990).

(2) Normal and reverse drag folds are formed by frictional variations along the fault surfaces (Hamblin 1965; Barnett *et al.* 1987; Twiss & Moores 1992; Davis & Reynolds 1996; Peacock *et al.* 2000).

(3) Longitudinal and transverse folds are formed by along-strike variations in fault displacements (Schlische 1995; Janecke *et al.* 1998;

From: McClay, K. R. & Hammerstein, J. A. (eds) 2020. *Passive Margins: Tectonics, Sedimentation and Magmatism.*
Geological Society, London, Special Publications, **476**, 49–81.
First published online January 1, 2018, https://doi.org/10.1144/SP476.12
© 2018 The Author(s). Published by The Geological Society of London. All rights reserved.
For permissions: http://www.geolsoc.org.uk/permissions. Publishing disclaimer: www.geolsoc.org.uk/pub_ethics

Khalil & McClay 2002, 2016; Young et al. 2003; Wilson et al. 2009). Doubly plunging longitudinal footwall anticlines and hanging wall synclines are formed parallel to the fault surface, whereas transverse hanging wall synclines and transverse footwall anticlines form perpendicular to the fault surface (Schlische 1995).

(4) Longitudinal fault-propagation folds are caused by folding at the tip line of a propagating fault (e.g. Allmendinger 1998; Hardy & McClay 1999; Corfield & Sharp 2000; Ferrill et al. 2007, 2012) and are superimposed on the longitudinal hanging wall synclines formed by variations in along-strike displacement. In some papers, the term forced folds (Stearns 1978) is also used for fault-propagation folds (e.g. Withjack et al. 1990; Schlische 1995; Maurin & Niviere 1999; Finch et al. 2004; Hardy & Finch 2006; Ford et al. 2007).

(5) Compactional drape folds are generated by the differential compaction of sediments over a pre-existing, high-standing horst block bounded by extensional faults.

Mechanical stratigraphy

The variable mechanical properties of multi-layer stratigraphic packages have been widely recognized as key parameters in the development of extensional fault-propagation folds (forced folds) (e.g. Withjack et al. 1990; Sharp et al. 2000; Ferrill et al. 2007, 2012; Ferrill & Morris 2008; Jackson & Rotevatn 2013; Tvedt et al. 2013). Analogue model experiments (Horsfield 1977; Vendeville 1988; McClay 1990; Withjack et al. 1990; Patton & Fletcher 1995) and numerical and kinematic models (Groshong 1989; Dula 1991; Saltzer & Pollard 1992; Allmendinger 1998; Hardy & McClay 1999; Mandl 2000; Finch et al. 2004; Jin & Groshong 2006; Hardy 2011) have shown that changes in fault dip, strain rate and the thickness of the incompetent layer also control the development of extensional fault-related folds.

Despite the fact that field and model studies have highlighted the complex nature of extensional fault-related folds, details of their 3D geometries, origins and kinematic evolution are incompletely understood (Schlische et al. 2002; Kane et al. 2010; Brandes & Tanner 2016; Khalil & McClay 2016; Lăpădat et al. 2016). Previous research has mainly focused on folds developed by individual extensional faults, with relatively few studies on the development of folds associated with segmented extensional fault arrays and linked fault systems (Young et al. 2003; Wilson et al. 2009; Fossen & Rotevatn 2016; Lăpădat et al. 2016).

We analysed three well-exposed examples of extensional fault-related hanging wall folds on the NW margin of the Red Sea. Our aims were to determine the inter-relationships between the complex hanging wall fold geometries as well as the growth and along-strike linkage of their bounding extensional faults. New kinematic evolutionary models are proposed for the development of these hanging wall structures and, in particular, for the corner folds formed at fault segment linkage points and at transfer zones or breached relay ramps. Our research involved detailed field mapping and section construction combined with analysis of Landsat Thematic Mapper images and aerial photographs. The NW Red Sea folds have previously been interpreted as contraction-related folds, formed in response to a regional compressive stress or strike-slip movement along the rift faults (e.g. Jarrige et al. 1986, 1990; Montenat et al. 1988, 1998). Our kinematic models challenge this view because our extensive field experience in the Red Sea and the Gulf of Suez has found no evidence of regional contraction or regional strike-slip deformation (Khalil & McClay 2001, 2002, 2016; Bosworth et al. 2005; Bosworth 2015).

Geological setting of the NW Red Sea

The Red Sea is a Cenozoic NW-trending continental rift system that formed in response to the divergence of the Arabian and African continental plates. It is a modern example of an embryonic and young ocean basin formed by continental break-up (McKenzie et al. 1970; Girdler & Styles 1974; Le Pichon & Francheteau 1978; Hempton 1987; Meshref 1990; Coleman 1993). The onset of rifting in the Red Sea region occurred in the Late Oligocene–Early Miocene (McKenzie et al. 1970; Cochran 1983; Steckler et al. 1988; Coleman 1993; Bosworth et al. 2005; Bosworth 2015) and the rift extended to the NW into the Gulf of Suez (Fig. 1a). Extension and rift propagation dramatically decreased in the Gulf of Suez region during the Late Mid Miocene, whereas continued opening of the Red Sea rift basin was accommodated by major sinistral strike-slip faulting along the Aqaba–Dead Sea transform (Freund 1970; Ben-Menahem et al. 1976; Steckler et al. 1988; Bosworth et al. 2005; Bosworth 2015). At this stage, extension focused along the axial zone of the Red Sea resulted in uplift and exhumation of the rift margins.

The structure of the NW margin of the Red Sea is dominated by two oppositely dipping fault domains separated by the complex Duwi accommodation zone (Fig. 1a; Moustafa 1997; Khalil & McClay 2001, 2009; Younes & McClay 2001). The northern domain consists of a series of extensional faults that dip to the NE, whereas the southern domain is formed by a series of faults that dip to the SW (Fig. 1a). The area analysed in this paper is an 80 km long segment of the rift margin located in the southern dip domain (Fig. 1a). It is characterized by NW–SE- and north–south-striking, linked and generally SW-dipping fault systems (Fig. 1b, c). These extensional fault systems form a western Border fault system and an eastern Coastal fault system (Fig. 1b, c). The hanging walls of these faults are a number of kilometre-scale, doubly plunging synclines that deform the pre-rift and earliest synrift strata (Fig. 1b, c).

Stratigraphic and structural framework of the NW Red Sea margin

Stratigraphy

The stratigraphy of the NW margin of the Red Sea can be divided into pre-rift and synrift megasequences, which are well exposed in the rotated, basement-cored hanging wall blocks of the northwestern Red Sea margin (Figs 1 & 2).

Basement. The Precambrian basement consists of meta-volcanics and metasediments intruded by Precambrian to Eo-Cambrian syn- and post-tectonic granites and granodiorites (Akaad & Noweir 1980; Stoeser & Camp 1985; Said 1990; Stern 1994). These units are affected by a strong fabric of Precambrian structures, including faults, dykes and Pan-African shear zones and fractures, oriented north–south, NNW–SSE, NW/WNW–SE/ESE and NE–SW (Fig. 1b). The Precambrian structures were reactivated during the Late Oligocene–Miocene rifting of the Red Sea (e.g. Khalil & McClay 2001; Younes & McClay 2001) and control the present day rhomboidal and zigzag fault pattern of this part of the rift margin (Fig. 1b).

Pre-rift stratigraphy. The Precambrian basement is unconformably overlain by 500–700 m of Upper Cretaceous to Middle Eocene pre-rift strata (Fig. 2). These pre-rift units occur in a number of isolated, fault-bounded hanging wall sub-basins (the Duwi, Hamadat and Zug El Bahar sub-basins) and in small outcrops in the coastal area (Fig. 1b). The lowermost part of the pre-rift section consists of up to 130 m of massive and thick-bedded, fluvial to shallow marine siliciclastic sediments of the Nubia Formation (Fig. 2). The Nubia sandstones are overlain by a 215–385 m thick anisotropic sequence of thin-bedded shales, sandstones and limestones with phosphate units and thin oyster beds of the Upper Cretaceous to Paleocene Quseir, Duwi, Dakhla and Esna formations (Fig. 2) (Youssef 1957; Abd El-Razik 1967; Issawi et al. 1969; Said 1990). The uppermost pre-rift strata consist of a 120–280 m thick section of competent, thick-bedded chalky and cherty limestones of the Lower to Middle Eocene Thebes Formation (Fig. 2).

The anisotropic units of the pre-rift succession are sandwiched between the competent crystalline basement and Nubia sandstones below and the competent, thick-bedded limestones of the Thebes Formation above (Fig. 2). The anisotropic shale-rich units in this section form the internal detachments indicated in Figure 2. These strongly controlled the development of extensional fault-propagation folds in the research area during rift evolution as well as elsewhere in the Red Sea–Gulf of Suez rift system (Moustafa 1987, 1992a, b; Withjack et al. 1990; Patton et al. 1994; Sharp et al. 2000; Khalil & McClay 2002, 2006; Jackson et al. 2006; Wilson et al. 2009).

Synrift stratigraphy. Synrift strata in the NW margin of the Red Sea range in age from Late Oligocene to Recent (Fig. 2). The lowermost synrift strata consist of coarse-grained clastic sediments of the Nakheil and Ranga formations (Fig. 2). The Nakheil Formation (Late Oligocene; Akkad & Dardir 1966; Said 1990) is locally preserved in the hanging wall Nakheil and Hamadat synclines (Fig. 1b) and unconformably overlies the Eocene Thebes Formation with an c. 10° angular discordance. It consists of 60–120 m of lacustrine red mudstones and sandstones, limestones, chert breccias and conglomerates derived from the underlying Thebes Formation. The Ranga Formation is Aquitanian–Burdigalian in age (El Bassyony 1982) and consists of 100–186 m of continental to shallow marine sandstones and polymict conglomerates containing clasts of basement sediments, limestones and cherts. The Ranga Formation also unconformably overlies the pre-rift units and, in places, overlies the early synrift Nakheil strata (Fig. 2).

The Ranga Formation is unconformably overlain by a 20–60 m thick section of reef limestones and shallow marine fine-grained clastic sediments of the Late Burdigalian–Langhian Um Mahara Formation (Fig. 2) (El Bassyony 1982; Said 1990). In the area west of Quseir (Fig. 1b), the Um Mahara Formation develops reef build-ups and reef talus on the eastern slopes of the footwall basement blocks (Fig. 1b). The Um Mahara Formation is unconformably overlain by 90–400 m of massive to poorly bedded evaporites of the Middle–Late Miocene Abu Dabbab Formation. These are, in turn, overlain by 200–300 m of Upper Miocene to Quaternary coarse clastic sediments, mudstones and carbonates of the Mersa Alam, Shagara and Samadi formations (Fig. 2), which form the main exposures in the coastal area (Fig. 1). Detailed descriptions of the rift sequences along the NW Red Sea margin have been published by Purser et al. (1987), Purser & Philobbos (1993), Montenat et al. (1988, 1998) and Orszag-Sperber et al. (1998).

Regional structure

The structure of the research area is dominated by two main fault systems: the Coastal fault system in the east and the Border fault system in the west (Fig. 1b). The Duwi accommodation zone is a complex zone of tilted fault blocks and NW-trending synclines (Fig. 1b). This accommodation zone was probably controlled by the Precambrian Hamrawin shear zone, which was reactivated during the latest Oligocene to Miocene extension (Moustafa 1997; Younes & McClay 2001). The Coastal fault system generally strikes NW, sub-parallel to the present day Red Sea coastline, and consists of NW–SE-, NNW–SSE- and north–south-striking fault segments (Fig. 1b). North of the Hamrawin shear zone the Anz fault segment (Fig. 1b) dips to the NE, but to the south of this zone the Zug El Bahar fault segment

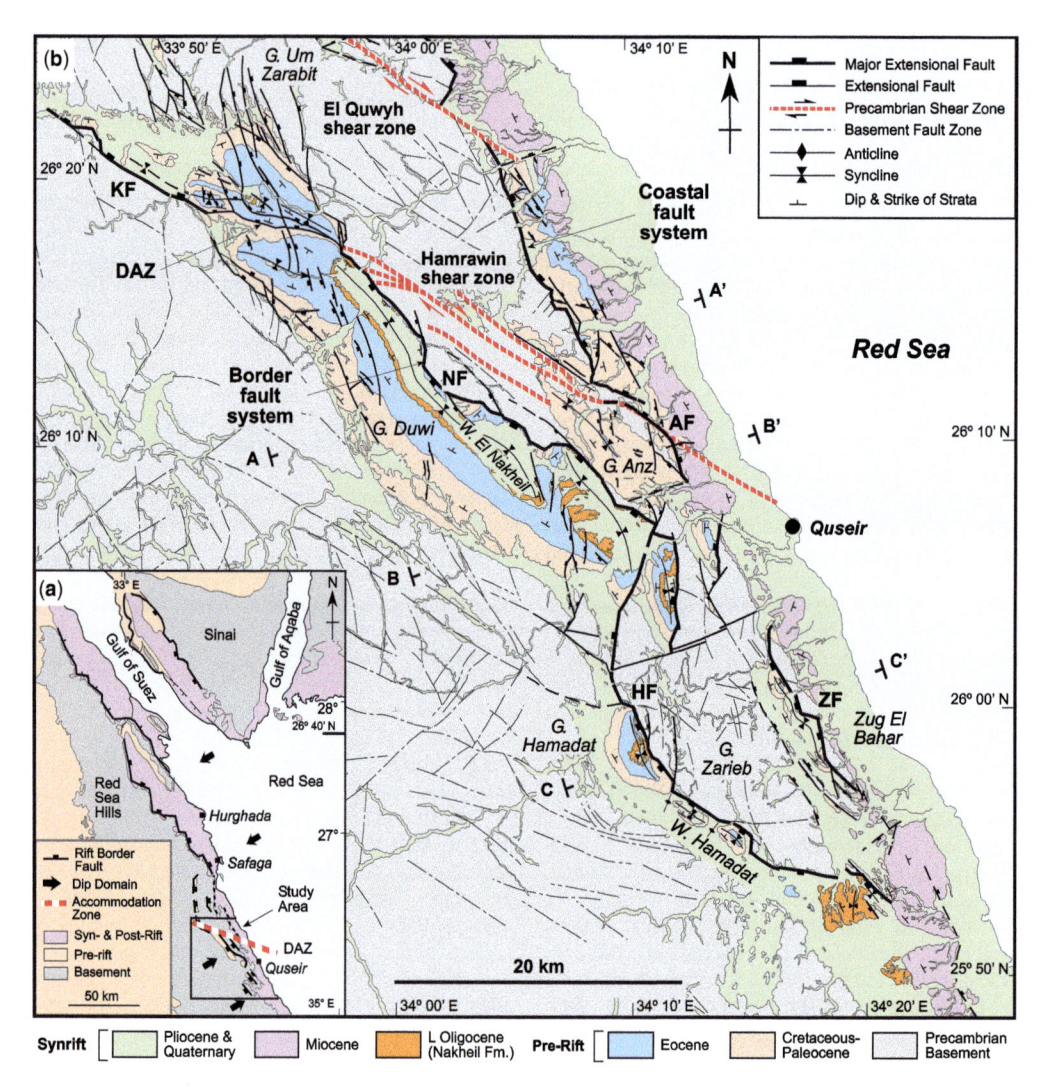

Fig. 1. (a) Location map of the study area showing the principal structural elements of the NW Red Sea. Bold arrows indicate the dominant stratal dip direction within the half-graben sub-basins. DAZ, Duwi accommodation zone. (b) Surface regional geological map of the study area showing the major fault systems and main fault blocks (after Khalil & McClay 2002). KF, NF and HF indicate the Kallahin, Nakheil and Hamadat segments of the Border fault system, respectively; AF and ZF indicate the Anz and Zug El Bahar segments of the Coastal fault system, respectively. G, Gebel (mountain); W, Wadi (dry valley).

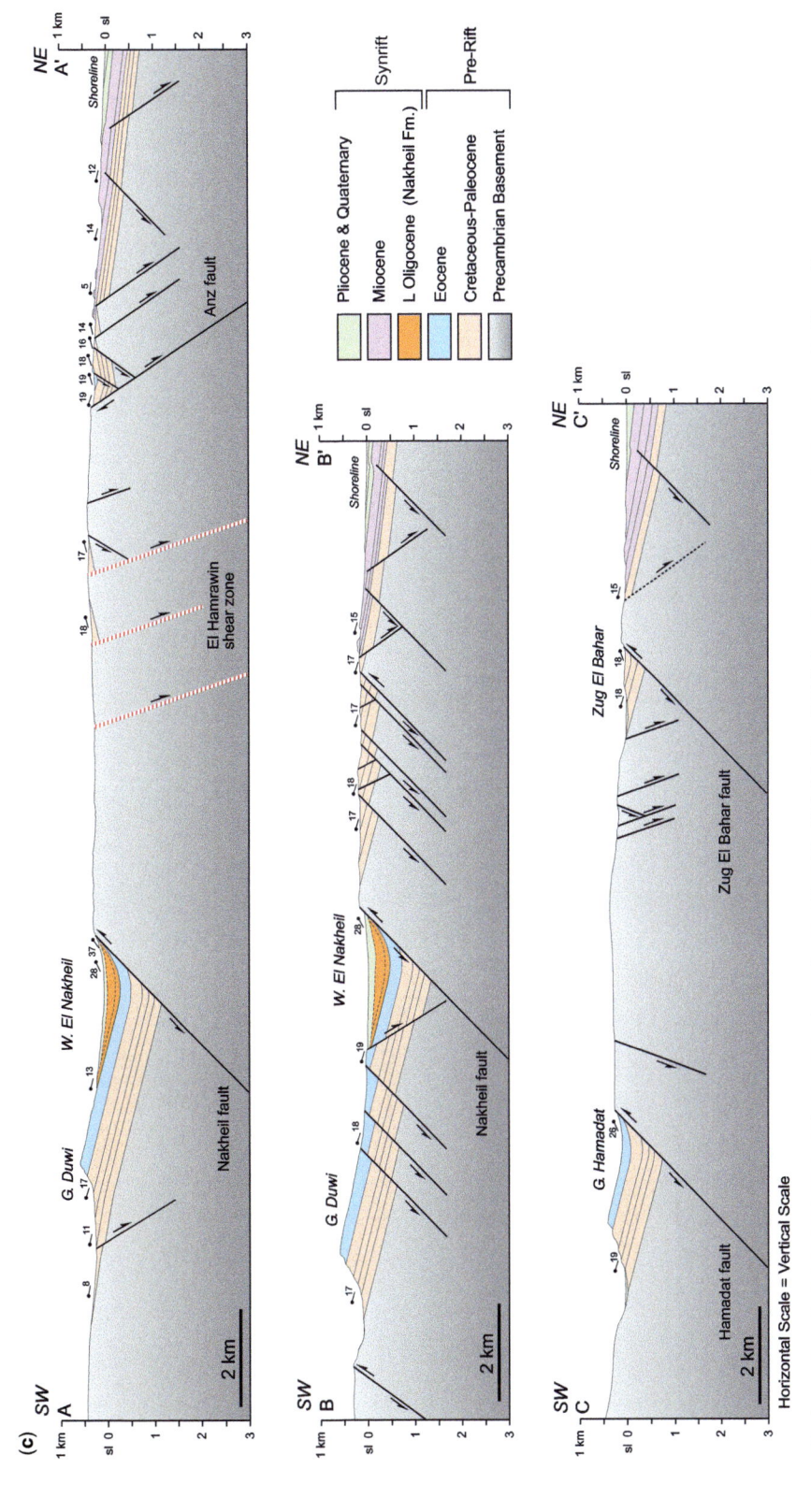

Fig. 1. *Continued.* (c) Regional cross-sections across the study area (after Khalil & McClay 2002). Locations of cross-sections are shown in Figure 1b.

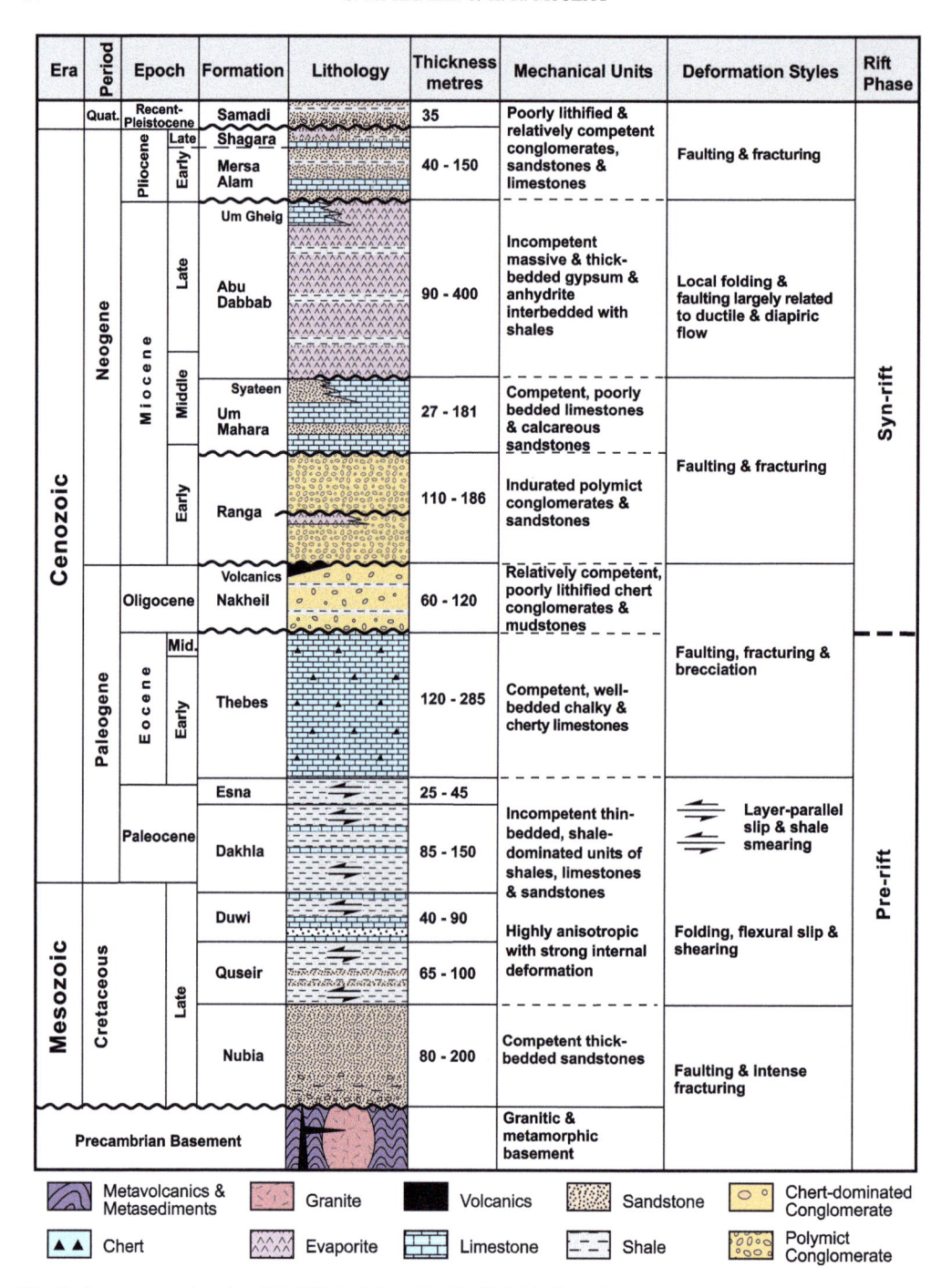

Fig. 2. Summary stratigraphy of the NW Red Sea (after Khalil & McClay 2016). The mechanical stratigraphy, deformation styles and internal detachment units within the pre-rift sequence are indicated.

dips SW (Fig. 1b). The rotated hanging wall of this system exposes pre- and synrift strata dipping to the NE and are locally preserved in half-graben (e.g. the Duwi block; Fig. 1b). The throw along the NE-dipping Coastal fault system varies from 0.5 to >2 km (Figs 1b, c & 3).

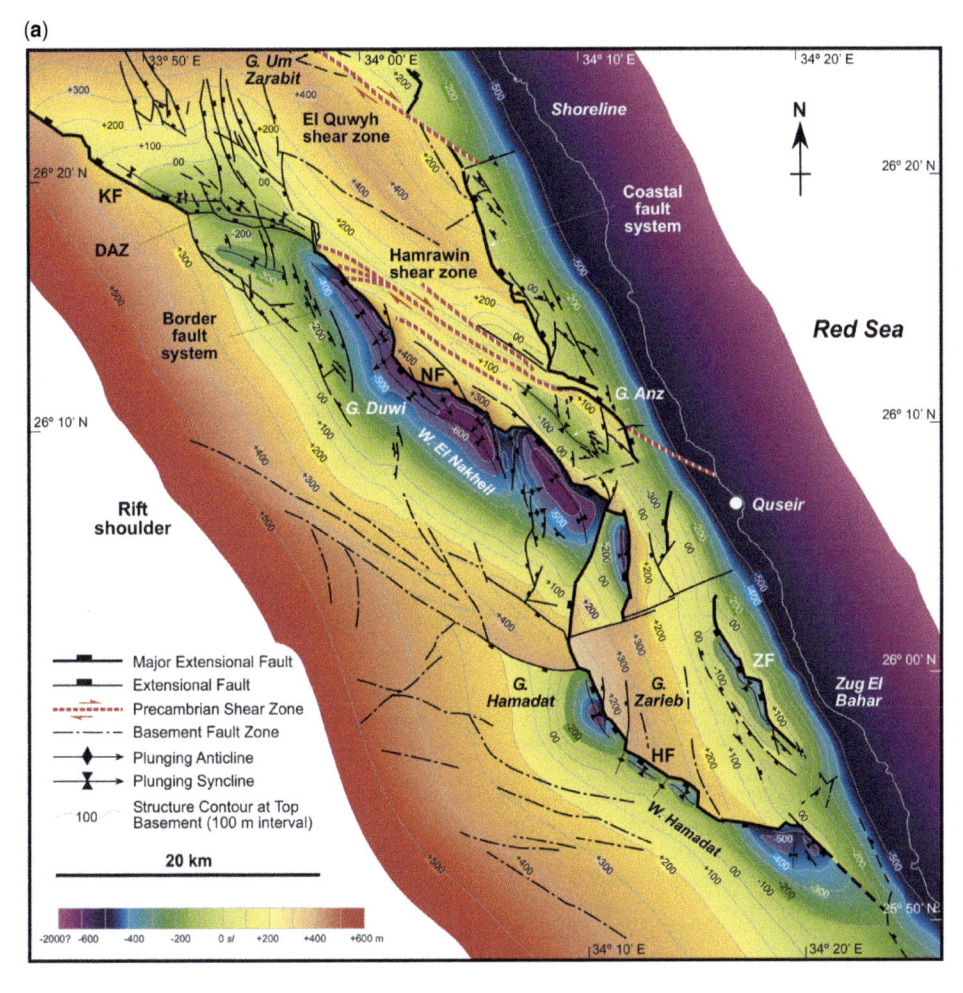

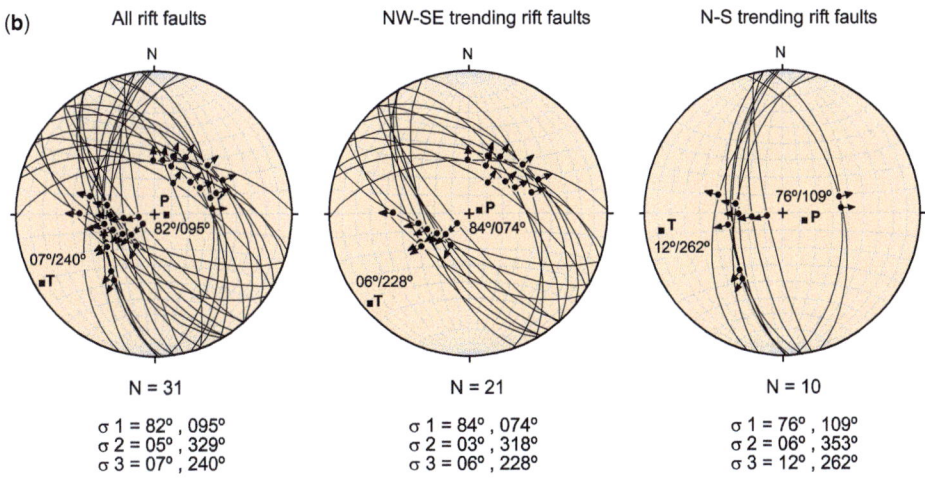

Fig. 3. (a) Structure–contour map for the study area showing the fault systems, main fault blocks and half-graben basins. See Figure 1b for fault segment abbreviations. (b) Lower hemisphere equal-area stereographic projections of faults and striae showing NE–SW extension direction. *T* and *P* indicate the extension and shortening axes, respectively.

The Border fault system trends mainly WNW–ESE, oblique to the Red Sea shoreline (Fig. 1b). It is characterized by a distinct zigzag pattern formed by linked WNW-, NW- and NNE-striking fault segments with throws varying from 0.5 to >2.5 km (based on topographic offsets against basement and offsets of the pre-rift strata; Fig. 3). The dip polarity of the Border fault system changes across the Duwi accommodation zone (Fig. 1b). NW of this zone the Kallahin fault segment (Fig. 1a) dips to the NE. South of the Hamrawin shear zone, the Nakheil and Hamadat fault segments (Fig. 1b) dip to the SW. The hanging wall of the Border fault system is dominated by several doubly plunging, asymmetrical, longitudinal hanging wall synclines 4–10 km wide and 10 to >40 km in length (Duwi and Hamadat synclines; Figs 1b & 3). The axial traces of these synclines are offset and/or bend along-strike and trend nearly parallel to the Border fault systems. In the immediate hanging wall of the Border fault, pre-rift strata occur in steeply dipping panels sub-parallel to the fault (Figs 1b, c). The Upper Oligocene synrift Nakheil conglomerates occur in small isolated outcrops in the cores of these hanging wall synclines.

Regional NE–SW-oriented cross-sections through the research area (Fig. 1c) show that the main segments of the Coastal and Border fault systems are planar and dip 55–65° NE at the surface. Smaller faults break-up the half-graben into 1–3 km wide, domino-style fault blocks. The throws of these smaller faults range from tens to a few hundreds of metres and their mean dip is $c.$ 60° (Fig. 1c).

North of Quseir, the Coastal and Border fault systems define a NW-trending, 40 km long, gently SE-plunging horst block extending from Gebel Um Zarabit in the north to Gebel Anz in the south (Fig. 3). The changes in structural relief and plunge directions of the footwall blocks are also obvious in the area south of Quseir, in the basement footwalls of the Hamadat and Zug El Bahar faults (Fig. 3). Khalil & McClay (2009) noted that the plunge direction of the tilted fault blocks, together with the relay ramps between the fault segments, exerted a fundamental control on the dispersal of the Miocene synrift clastic sediments and localization of the fan delta systems in the NW Red Sea.

Measurements of slicken-lineations on the exposed fault surfaces indicate that pure dip-slip displacements dominate, with only a few north–south- to NNE-trending faults showing minor sinistral oblique-slip movement and NW-trending faults showing minor dextral oblique-slip movement (Fig. 3b). Palaeostrain analyses of fault slip data indicate that the faults were developed in a regime of near-horizontal extensional (T) stress with an axis oriented (07°/240°) and near-vertical compressive (P) stress with an axis oriented (82°/095°) (Fig. 3b). This stress regime is interpreted to be related to the regional NE–SW-oriented extension that operated throughout most of the Oligocene–Miocene rifting of the Red Sea and Gulf of Suez (Khalil & McClay 2001, 2002, 2016; Bosworth et al. 2005; Bosworth 2015). The oblique-slip component found on some faults is minor and it is most likely related to the obliquity of the reactivated basement fault zones with respect to the regional NE–SW extension, rather than to NW–SE compression or an early stage of strike-slip faulting as proposed by Jarrige et al. (1986, 1990) and Montenat et al. (1988, 1998).

Extensional fault-related folding

This section describes the hanging wall folds of the Nakheil fault system for the main outcrops south of the Duwi accommodation zone, where they are well-developed in the Duwi half-graben (Figs 4 & 5).

Duwi half-graben

The Duwi half-graben shown in Figure 4a is $c.$ 40 km long and, on average, $c.$ 8 km wide. The NW-striking Nakheil fault forms the Border fault system with largely Precambrian basement units in the footwall to the east (Fig. 4a, b). The fault is highly segmented with WNW-, NW- and north–south-striking sectors, with distinct longitudinal synclines and complex breached relays linking these different segments (Fig. 4b). Sections across the main Duwi half-graben show that the bounding faults dip 58–66° SW with minimum stratigraphic offsets of 1.5–2.3 km (Fig. 4c).

Longitudinal synclines. The pre-rift Cretaceous to Eocene outcrops in the Duwi half-graben define four prominent, broadly NW-trending, hanging wall longitudinal synclines (L1–L4, Figs 4 & 5). The SE-plunging syncline L1 at the northern end of the main Duwi half-graben occurs where the Border fault polarity switches from the SW-dipping Nakheil fault to the NE-dipping Kallahin fault across the Duwi accommodation zone (Figs 1b & 4b). This northernmost hanging wall syncline merges to the south with the main 25 km long Nakheil syncline L2, which is characterized by a curvilinear axial surface trace sub-parallel to the Nakheil fault system with sharp bends at relay structures R1, R2 and R3 (Figs 4 & 5). It terminates to the SE in a zone of en echelon normal faults (Figs 4b & 5b). The southeastern Nakheil longitudinal syncline L3 is formed in the hanging wall of the Nakheil fault system, where it is offset to the east at breached relay R3 and where there is a series of north–south-striking en echelon extensional faults at the southern end of Gebel Duwi (Figs 5b & 6–8). The southernmost longitudinal syncline L4 occurs where the southernmost segment of the Nakheil fault system is offset to the

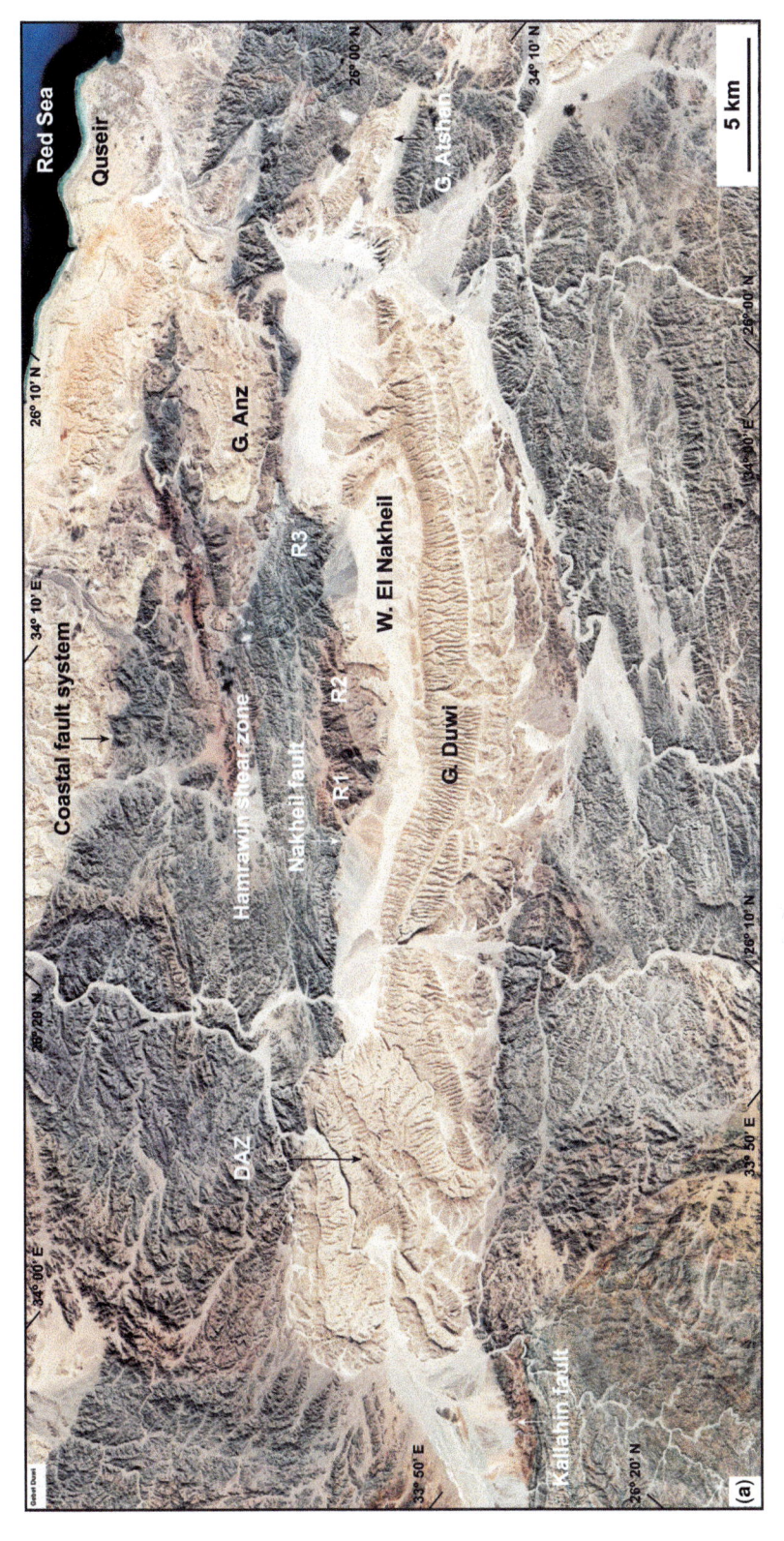

Fig. 4. (**a**) Google Earth image (©2016 DigitalGlobe) showing the Duwi half-graben and the main fault and fold systems. R1, R2 and R3 indicate the locations of breached relay ramps. (**b**) Detailed geological map of the Duwi half-graben showing the linked segments of the Nakheil fault system and the Duwi accommodation (DAZ) zone as well as the Nakheil hanging wall synclines (L1–L4). (**c**) NE–SW cross-sections through the northern and central Nakheil hanging wall synclines. Note the asymmetry of the synclines and the greatest structural relief in the central cross-section B–B′. Locations of cross-sections are shown in Figure 4b. (**d**) Lower hemisphere equal-area stereographic projections of poles to bedding showing the asymmetrical and doubly plunging geometry of the Nakheil hanging wall synclines.

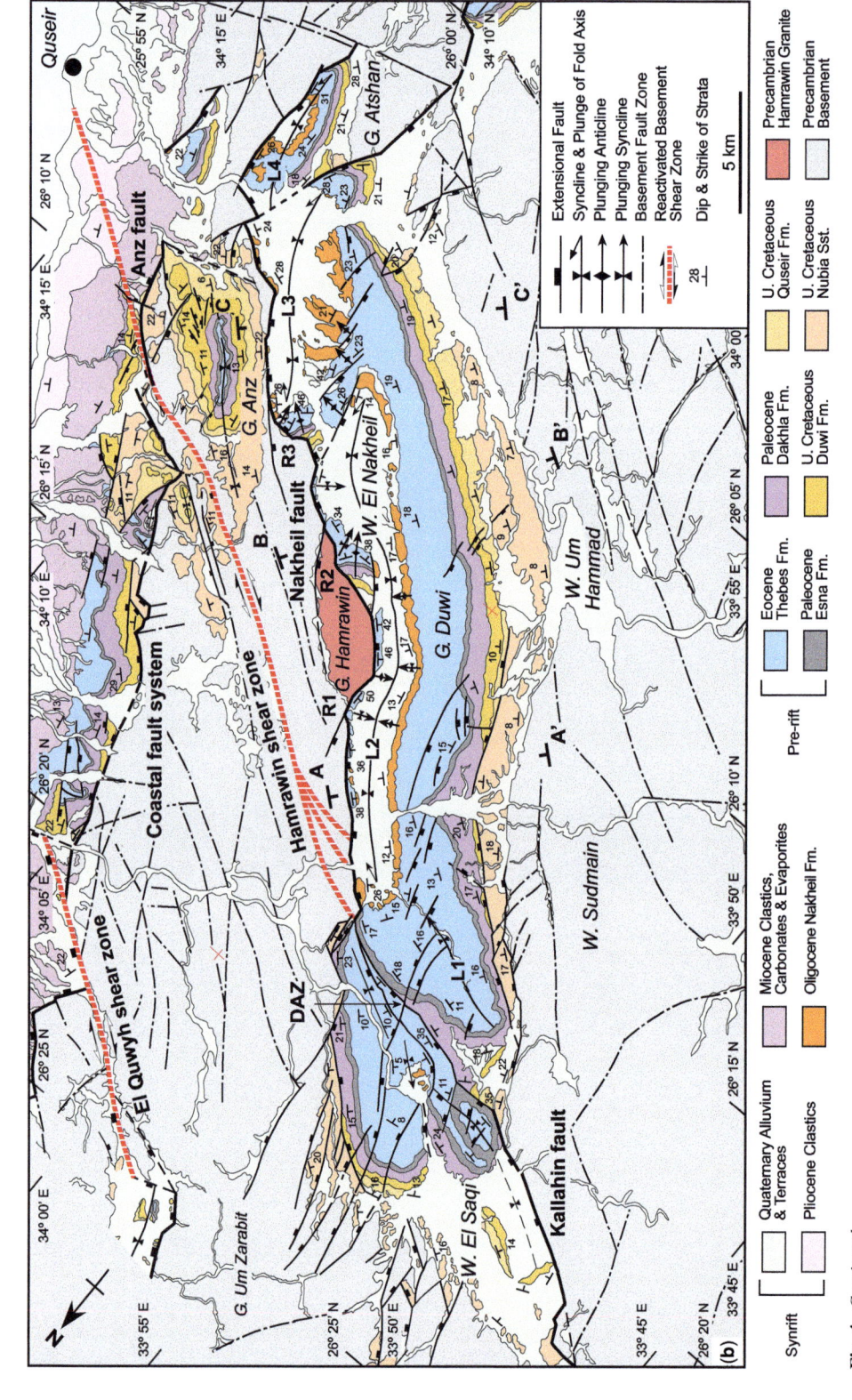

Fig. 4. *Continued.*

EXTENSIONAL FAULT-RELATED FOLDING, RED SEA

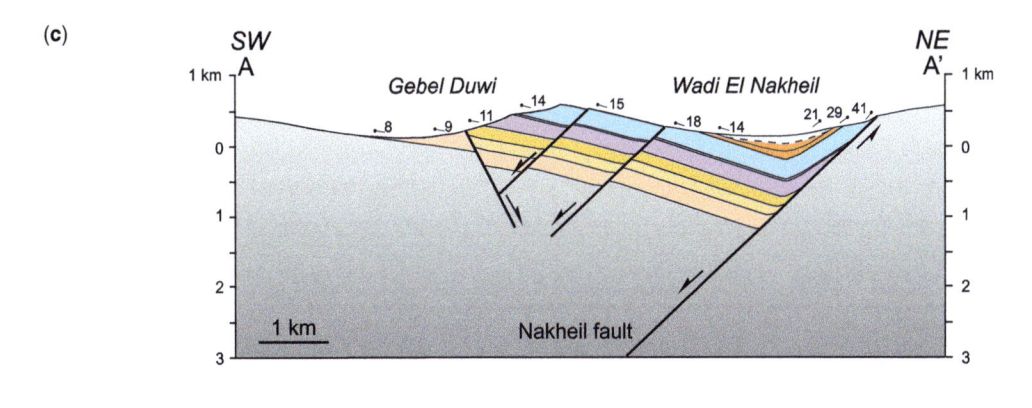

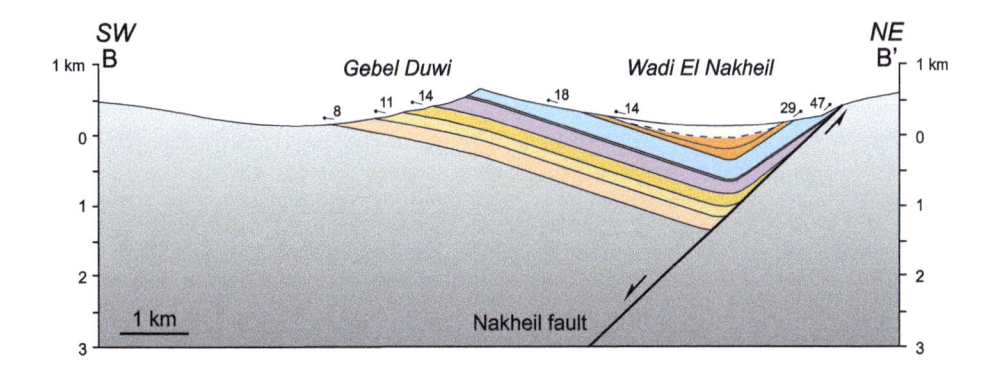

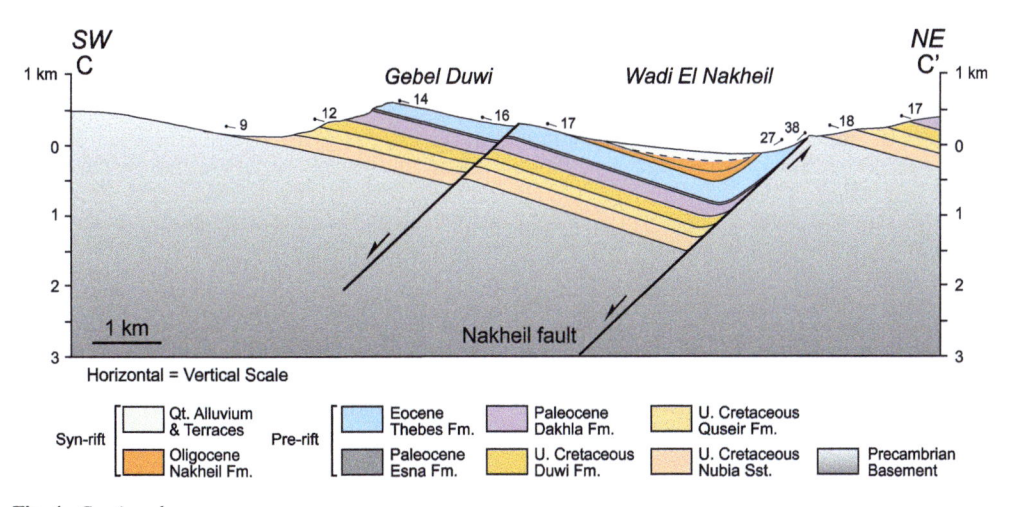

Fig. 4. *Continued.*

east by the north–south-trending fault-bounding Gebel Atshan (Figs 4b & 5b).

The main and southeastern Nakheil hanging wall longitudinal synclines L2 and L3 are doubly plunging and distinctly asymmetrical, with relatively narrow, steep (40° on average) west–SW-dipping limbs and wide, gentle (15° on average) east–NE-dipping limbs (Fig. 4c). The structural relief of the these synclines is *c.* 2 km and the wave lengths are 5–6 km (Fig. 4c). Outcrop observations reveal that the base of the pre-rift Upper Cretaceous Nubia sandstones (Fig. 2) are intensely fractured and lack significant

(d)

Main Nakheil Syncline L2 — N-Plunging Fold Axis 10°–007°; SSE-Plunging Fold Axis 11°–153°; N = 37

Southeastern Nakheil Syncline L3 — N-Plunging Fold Axis 16°–012°; SSE-Plunging Fold Axis 30°–174°; N = 15

Southernmost Nakheil Syncline L4 — NNE-Plunging Fold Axis 16°–009°; SSE-Plunging Fold Axis 06°–177°; N = 13

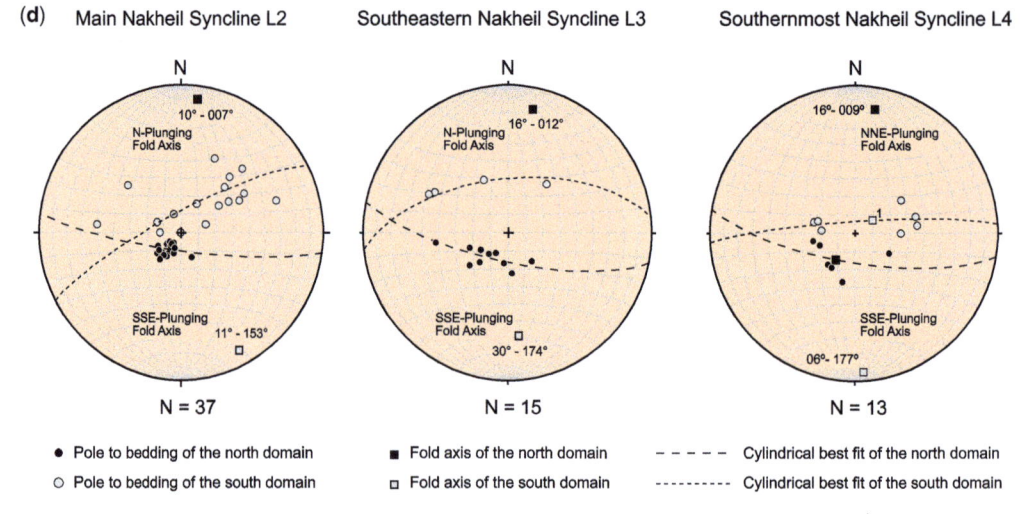

- • Pole to bedding of the north domain
- ○ Pole to bedding of the south domain
- ■ Fold axis of the north domain
- □ Fold axis of the south domain
- – – – – Cylindrical best fit of the north domain
- ---------- Cylindrical best fit of the south domain

Fig. 4. *Continued.*

folding adjacent to the fault. By contrast, the highly anisotropic units (interbedded sandstones, siltstones, shales and limestones; Fig. 2) of the Upper Cretaceous–Paleocene strata are strongly folded and, in places along the steep SW-dipping limbs, form narrow panels that dip up to 50°, sub-parallel to the main Nakheil fault system. In particular, the shale units of the Dakhla and Esna formations (Fig. 2) are commonly highly sheared and attenuated along the fault zone. By contrast, the thick-bedded competent limestones of the overlying Eocene Thebes Formation are highly fractured and brecciated in the immediate hanging wall of the faults. Stereographic projections of poles to bedding measured along the limbs and noses of the Nakheil hanging wall synclines show that they are strongly scattered (Fig. 4d), reflecting the non-cylindrical form and doubly plunging nature of the synclines.

Breached relay ramps and corner folds. Three major breached relay systems (R1, R2 and R3) occur where there are abrupt changes in strike along the Nakheil fault in the main sector of the Duwi half-graben (Fig. 5). Relay R1 occurs at the northwestern margin of the Hamrawin granite body (Fig. 5b). The Hamrawin relay (R2) formed along the southern margin of the Gebel Hamrawin granite body (Fig. 6a); the Nakheil relay (R3) links the southern segments of the Nakheil fault system (Figs 5b & 6b).

Small corner fold anticlines and synclines are formed in the SW-dipping hanging wall next to the Nakheil fault system, with hinge lines and axial surfaces at high angles to the main Nakheil fault surface (Figs 4b, 5b & 6a, b). These corner folds formed at the sharp bends and kinks in the main trace of the Nakheil fault at the breached (or linked) relay systems R1, R2 and R3 (Fig. 5b). The convex kinks formed anticlines, whereas the concave kinks formed synclines. In the immediate hanging wall of the Nakheil fault, these corner folds plunge inwards towards the axis of the longitudinal syncline along the Duwi half-graben (Figs 5b, 6a, b & 7).

The main long outcrop of the Eocene Thebes limestone forms the western limb of the main Duwi syncline L2 (Fig. 5a). Opposite relays R1 and R2, gentle, inward-plunging anticlines and synclines, can be mapped in the Eocene limestones opposite the corner folds. These developed on the more steeply dipping eastern limb (Fig. 5b).

A series of SE-plunging, en echelon anticlines formed in the Eocene limestones on the western limb of longitudinal syncline L3 opposite relay R3 (Figs 5a & 6a). These occur in the footwall of north–south-trending, en echelon extensional faults (Figs 6c & 7) with anticline axial traces perpendicular to the central sectors of the segmented extensional faults (Figs 6c, 7 & 8). These are interpreted as footwall transverse anticlines formed in response to the along-strike change in fault displacement at breached relay R3.

Hamadat half-graben

The Hamadat half-graben consists of three hanging wall longitudinal synclinal basins of Upper Cretaceous to Eocene pre-rift units (Fig. 9) and is *c.* 20 km long and 3–4 km wide. Isolated outcrops of the early synrift Oligocene Nakheil strata occur in the core of the northern syncline (Fig. 9). The bounding Hamadat fault system consists of linked, NW- and north-trending fault segments that dip 55–60° SW and offset the hanging wall pre-rift strata

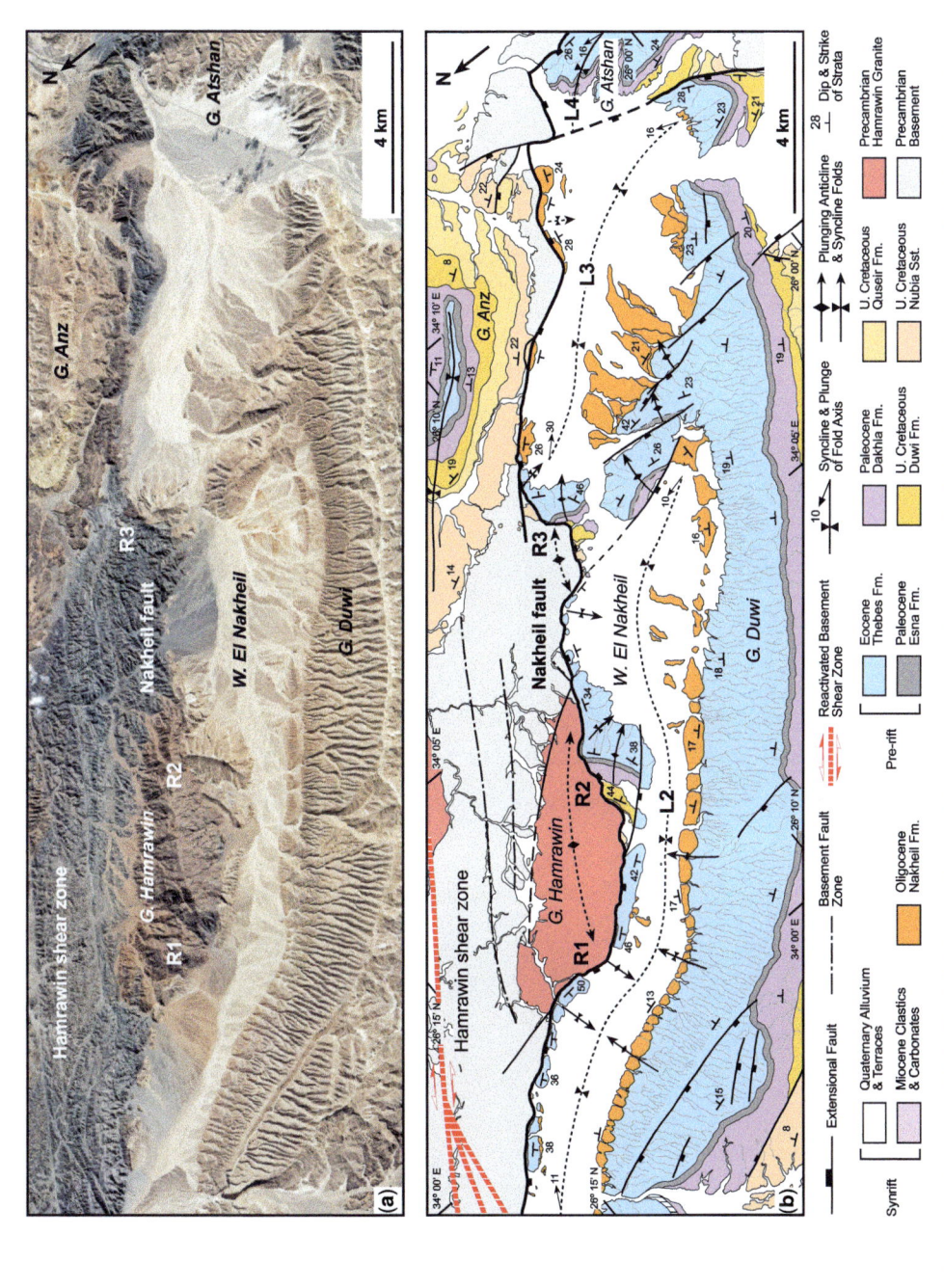

Fig. 5. (**a**) Google Earth image (©2016 DigitalGlobe) for the Gebel Duwi area showing the details of the Nakheil fault and fold systems. R1, R2 and R3 indicate the locations of the relay ramps. (**b**) Detailed map of the Gebel Duwi area showing the complex structure of the Nakheil doubly plunging, hanging wall synclines.

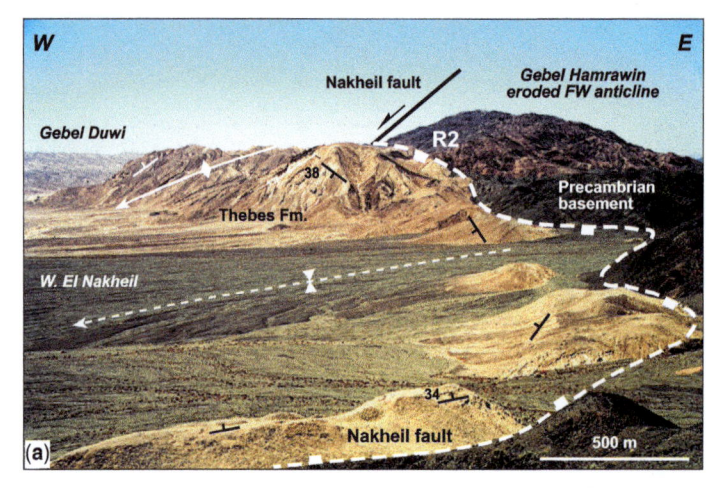

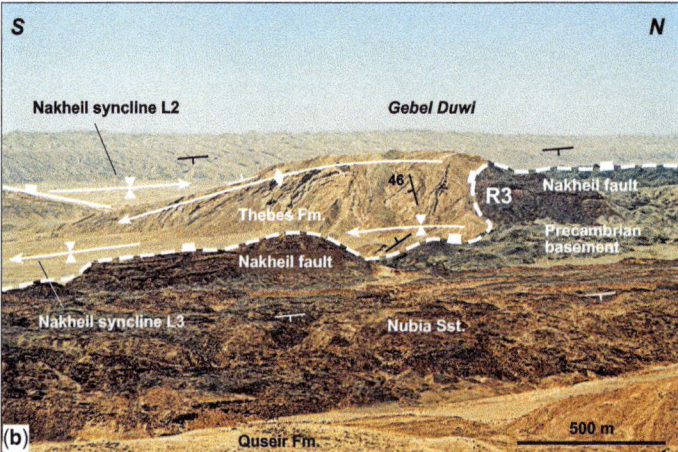

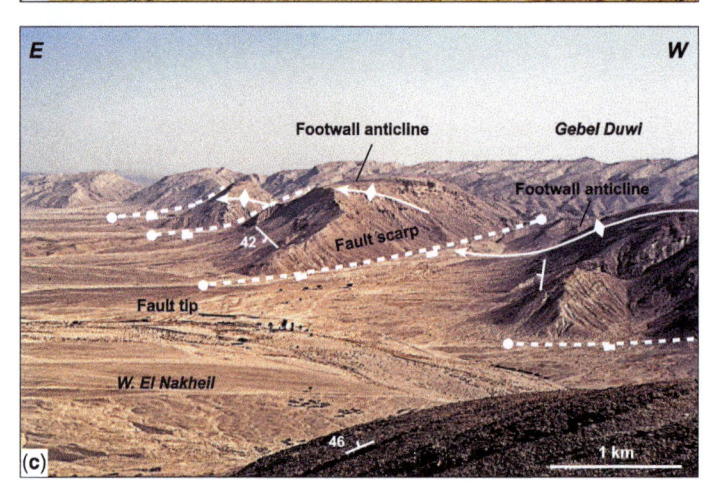

Fig. 6. Outcrop examples of structures in the Duwi half-graben. (**a**) View looking north showing the Hamrawin relay (R2) and the corner folds at the immediate hanging wall of the Nakheil fault. (**b**) View looking NW showing the Nakheil relay (R3) and the corner folds developed at the segment linkage points along the Nakheil fault. (**c**) View looking south showing an en echelon fault system together with footwall anticlines deforming the southern part of Gebel Duwi. This en echelon fault system forms the southern termination of the main Nakheil syncline L2.

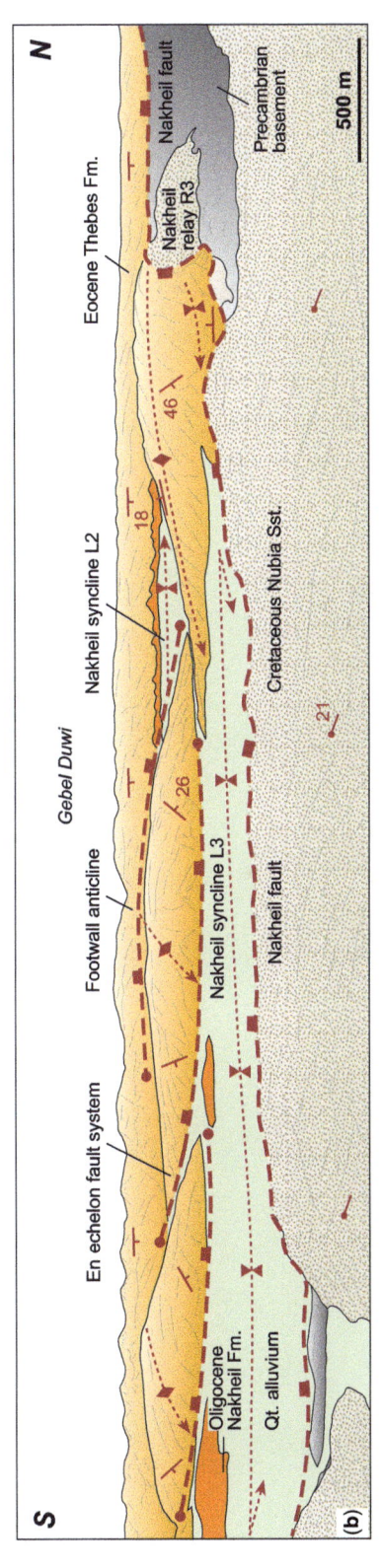

Fig. 7. (a) View looking west and (b) perspective sketch showing the corner and transverse folds and the north–south-trending en echelon fault system in the Duwi half-graben. The corner fold occurs at the breach position of the Nakheil relay (R3) and transverse footwall anticlines form on the footwalls of the en echelon fault systems.

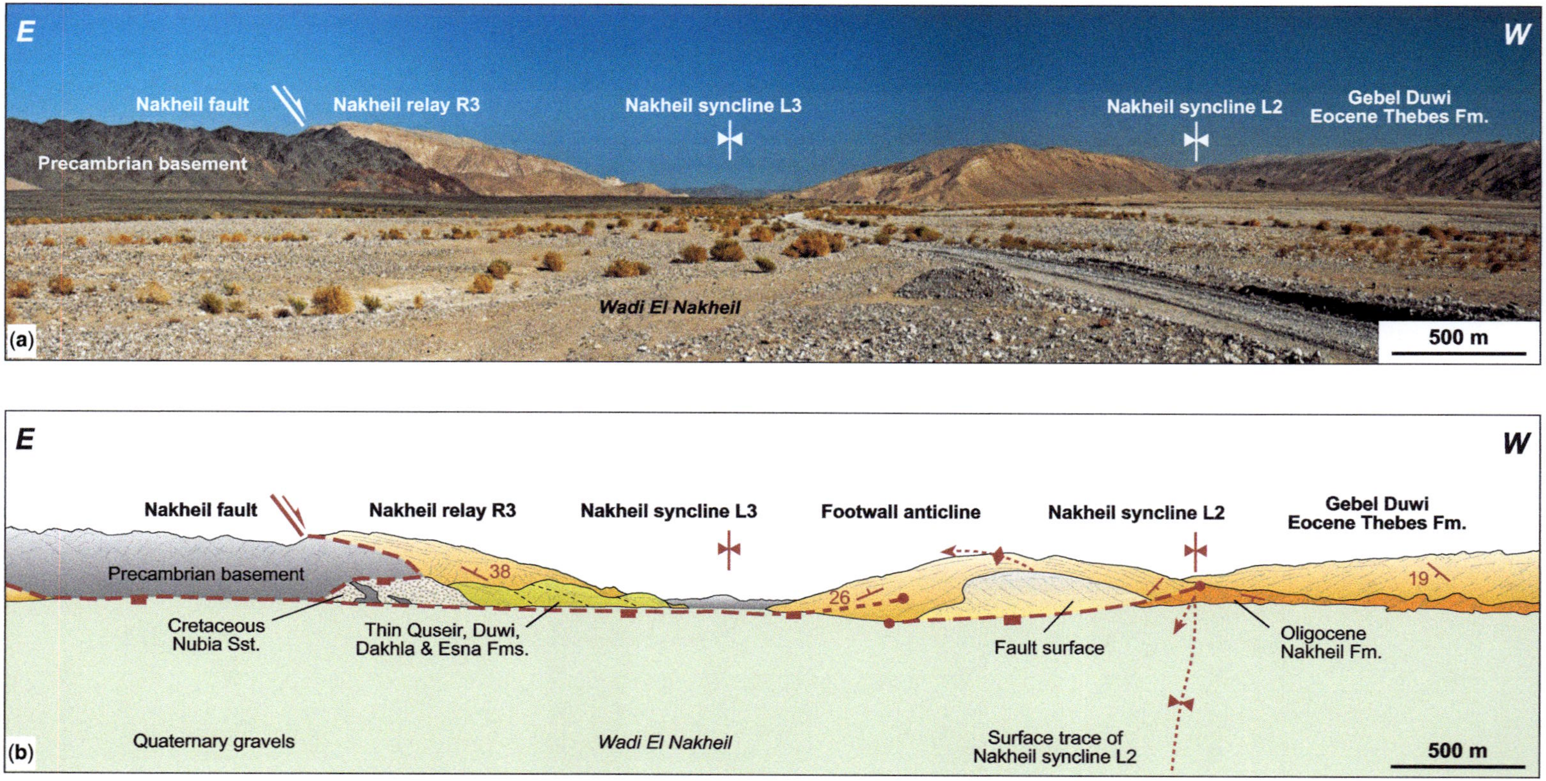

Fig. 8. (a) View looking south and (b) perspective sketch showing the offset of the axial traces of the main and southeastern Nakheil synclines L2 and L3, respectively. The juxtaposition of the basal pre-rift Nubia sandstone against the Precambrian basement at the position of the Nakheil relay R3 indicates a decreased offset of the Nakheil fault in this relay zone.

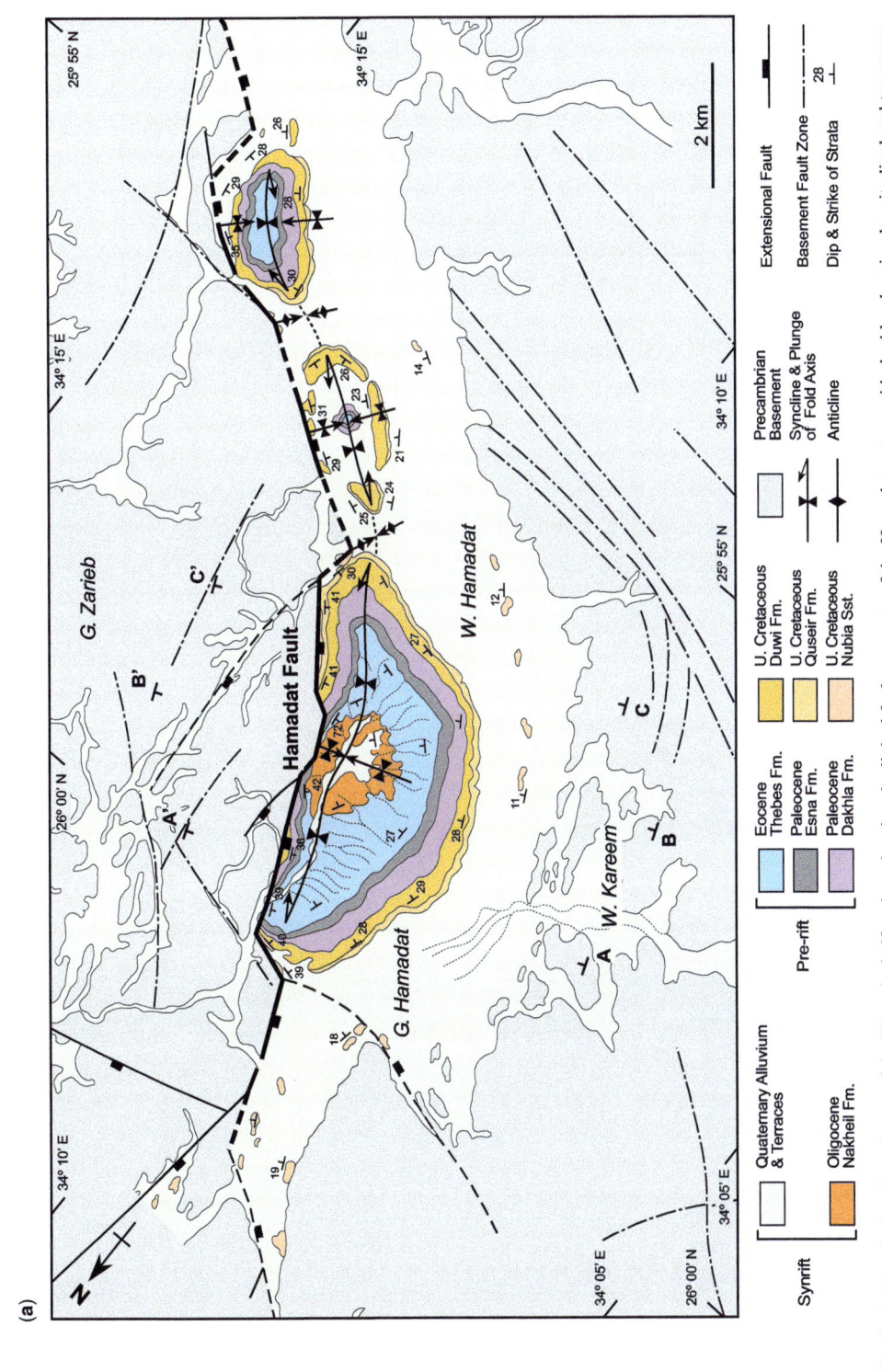

Fig. 9. (a) Detailed geological map of the Hamadat half-graben showing the linked fault segments of the Hamadat system with doubly plunging longitudinal and transverse hanging wall folds (modified after Khalil & McClay 2002). Inward-plunging anticlines separate the longitudinal synclines at the kink/linkage position of the Hamadat fault.

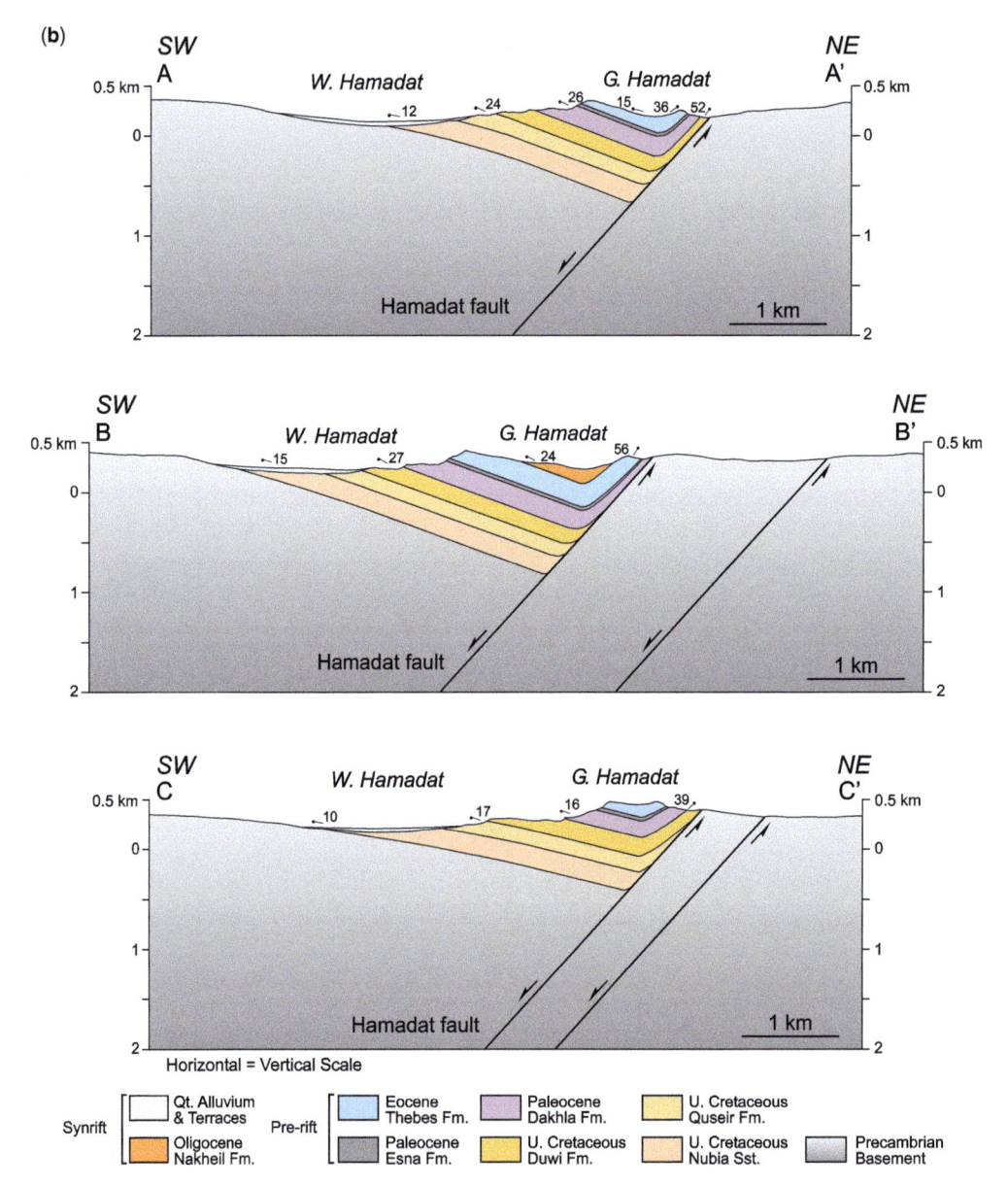

Fig. 9. *Continued.* (**b**) NE–SW cross-sections through the Hamadat half-graben display the asymmetry and change in structural relief of the north Hamadat syncline with the largest fault offset and structural relief shown in the central cross-section B-B'.

against the footwall Precambrian basement (Fig. 9). Stratigraphic offsets vary significantly along the strike of individual fault segments, giving rise to doubly plunging, hanging wall longitudinal synclines with their axial surface traces sub-parallel to the Hamadat fault system (Fig. 9). The northern Hamadat syncline has a wavelength of 4.5 km and the amplitude varies from 1 km near the northern and southern noses and to up to 1.5 km at the central section of the fold (Fig. 9b). The central and southern synclines have wavelengths from 2 to 3 km and fold amplitudes of *c.* 0.8 km (Fig. 9a). The NW-trending longitudinal synclines are superimposed onto smaller NE-trending, hanging wall transverse synclines (terminology of Schlische 1995). These transverse, inward-plunging folds are perpendicular

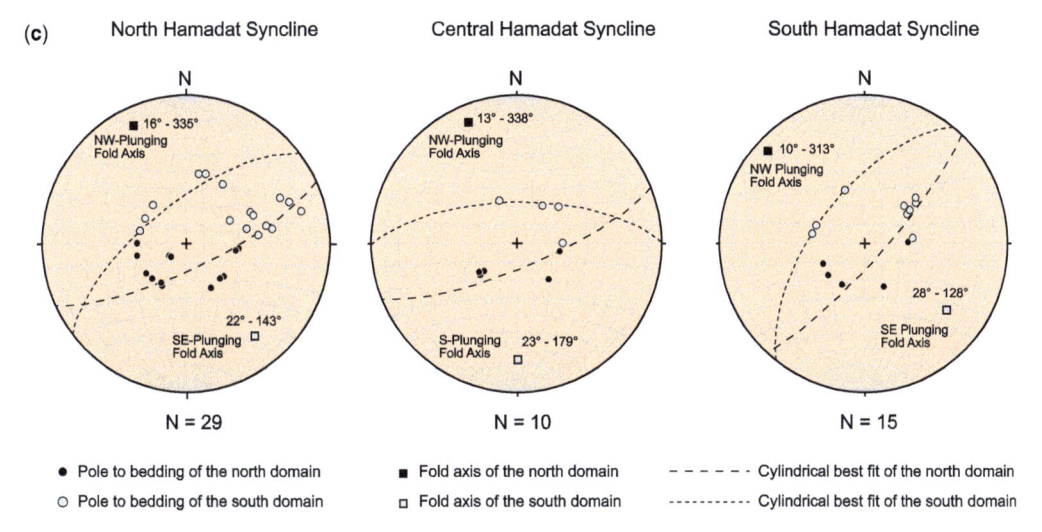

Fig. 9. *Continued.* (**c**) Lower hemisphere equal-area stereographic projections of poles to bedding showing the asymmetric and doubly plunging geometries of the Hamadat hanging wall synclines.

to the longitudinal folds (Fig. 9a). Between the three longitudinal synclines there are two pairs of narrow, NE-trending, inward-plunging anticlines (corner folds) located at kinks formed at the segment linkage points of the Hamadat fault system (Fig. 9a). These are points of minimum throw, as indicated by offset of the basal pre-rift Nubia sandstone in the immediate fault hanging wall against the footwall Precambrian basement (Fig. 9a).

The Hamadat hanging wall longitudinal synclines are noticeably asymmetrical in cross-section, with long, gentle (12–29°) NE-dipping limbs and short, steep (35–56°) SW-dipping limbs (Fig. 9b). In the steeply dipping limb, the shale units of the Esna and Dakhla formations are markedly thinned from 150 m to <60 m thick in the immediate hanging wall of the Hamadat fault. Stereographic projections of poles to bedding for these Hamadat synclines show a marked scatter, which indicates that the folds are non-cylindrical, with a gentle (10–16°) plunge towards the NW and a moderate (22–28°) plunge towards the south and SE directions (Fig. 9c).

Zug El Bahar-half-graben

The Zug El Bahar half-graben has a general NW–SE trend and is *c.* 8 km long and 2.5 km wide (Fig. 10a). It is bounded by the main Zug El Bahar fault in the east and a small displacement fault along the edge of the basement in the west. The Zug El Bahar fault offsets the pre-rift Upper Creteacous strata in the hanging wall against the footwall Precambrian basement and consists of NW- and north-trending segments linked in a zigzag pattern (Fig. 10a). The fault throw varies along-strike from 200 m at the fault terminations, where it offsets the basal pre-rift Nubia sandstones against the Precambrian basement, to *c.* 700 m near the centres of the main fault segments, where the Duwi and Dakhla formations are juxtaposed against the footwall Precambrian basement (Fig. 10a, b).

The outcrop pattern of pre-rift strata in the hanging wall of the Zug El Bahar fault outlines a NW-trending, asymmetrical, doubly plunging, hanging wall longitudinal syncline (Fig. 10). Two local, SW-plunging transverse corner folds on the eastern, SW-dipping limb of the Zug El Bahar syncline occur at the interpreted fault segment linkage points (Fig. 10a). NE–SW cross-sections show that the Zug El Bahar longitudinal syncline is characterized by a narrow, moderate to steep (26–34°) SW-dipping limb and a wider, gentle (14–22°) NE-dipping limb (Fig. 10). The structural relief of the syncline is 500–700 m and its average wavelength is 1.5 km (Fig. 10b). The stereographic plots of poles to bedding of the Zug El Bahar syncline display a non-cylindrical scatter and double-plunging geometry (Fig. 10c).

Fault displacement profiles

Plots of fault displacements along the strike of an extensional fault system provide valuable insights into fault growth, evolution and linkage, particularly for segmented systems (e.g. Muraoka & Kamata 1983; Walsh & Watterson 1987, 1988; Peacock & Sanderson 1991; Cowie & Scholz 1992a, b; Dawers *et al.* 1993; Cartwright *et al.* 1995, 1996; Childs *et al.* 1995; Dawers & Anders 1995; Peacock & Sanderson 1996; Willemse *et al.* 1996; Schultz 2000; Wilkins & Gross 2002; Kim & Sanderson 2005; Fossen &

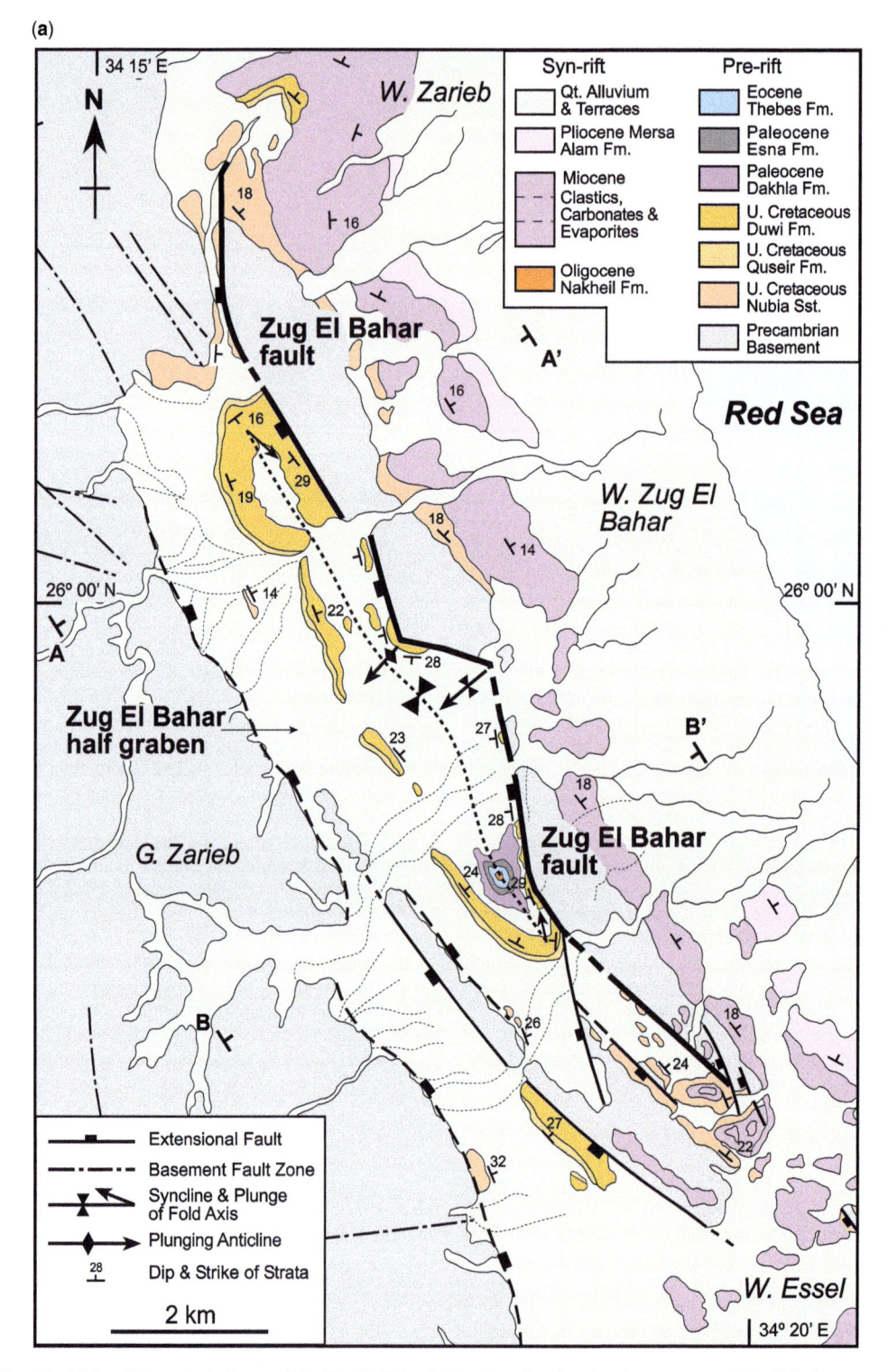

Fig. 10. (a) Detailed geological map of the Zug El Bahar half-graben showing the zigzag pattern of the linked segments of the Zug El Bahar fault system, together with the curved axial trace of the Zug El Bahar syncline.

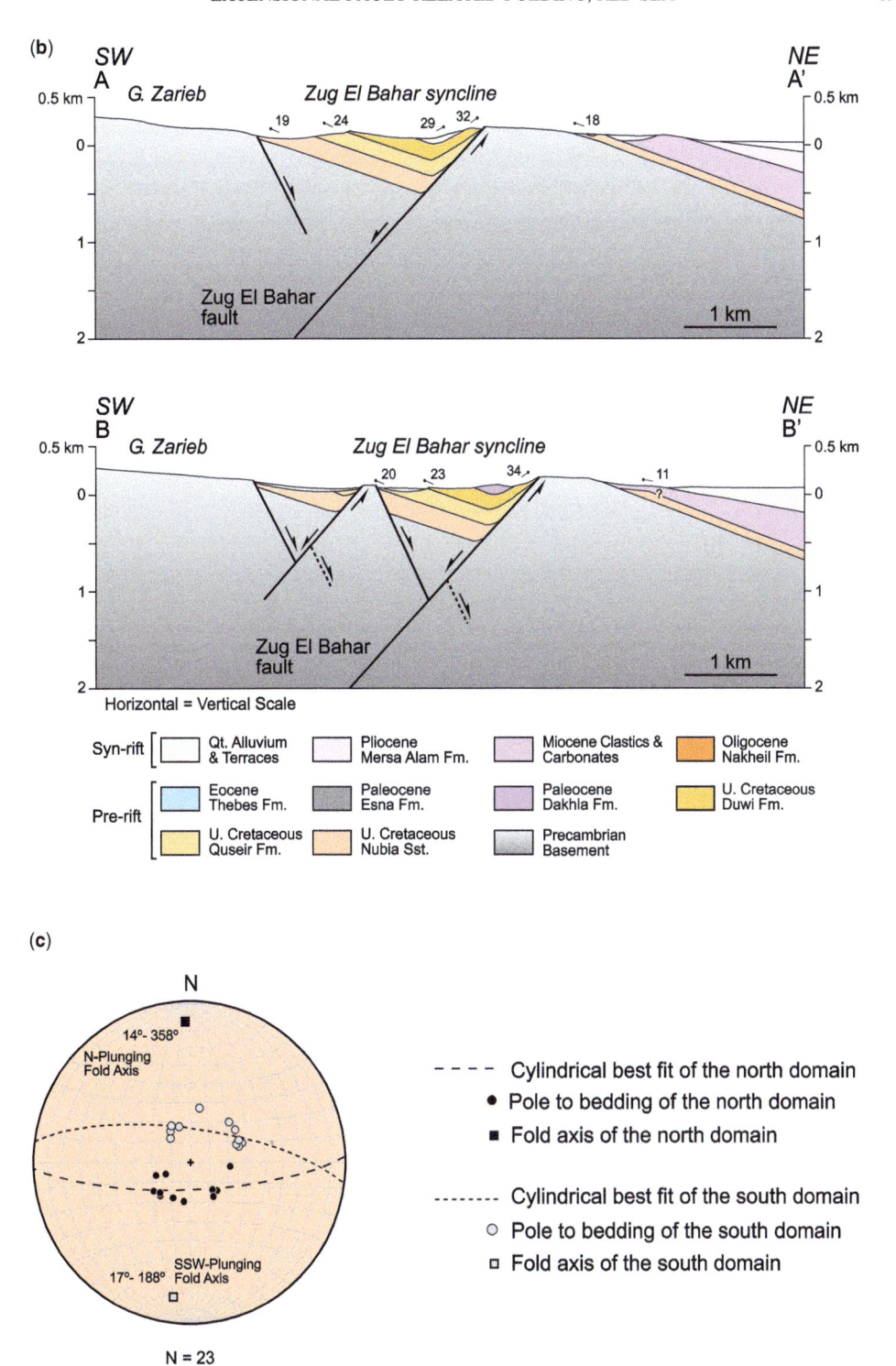

Fig. 10. *Continued.* (**b**) NE–SW cross-sections through the Zug El Bahar half-graben showing the asymmetry of the hanging wall syncline. (**c**) Lower hemisphere equal-area stereographic projections of poles to bedding showing the asymmetrical and doubly plunging geometry of the Zug El Bahar hanging wall syncline.

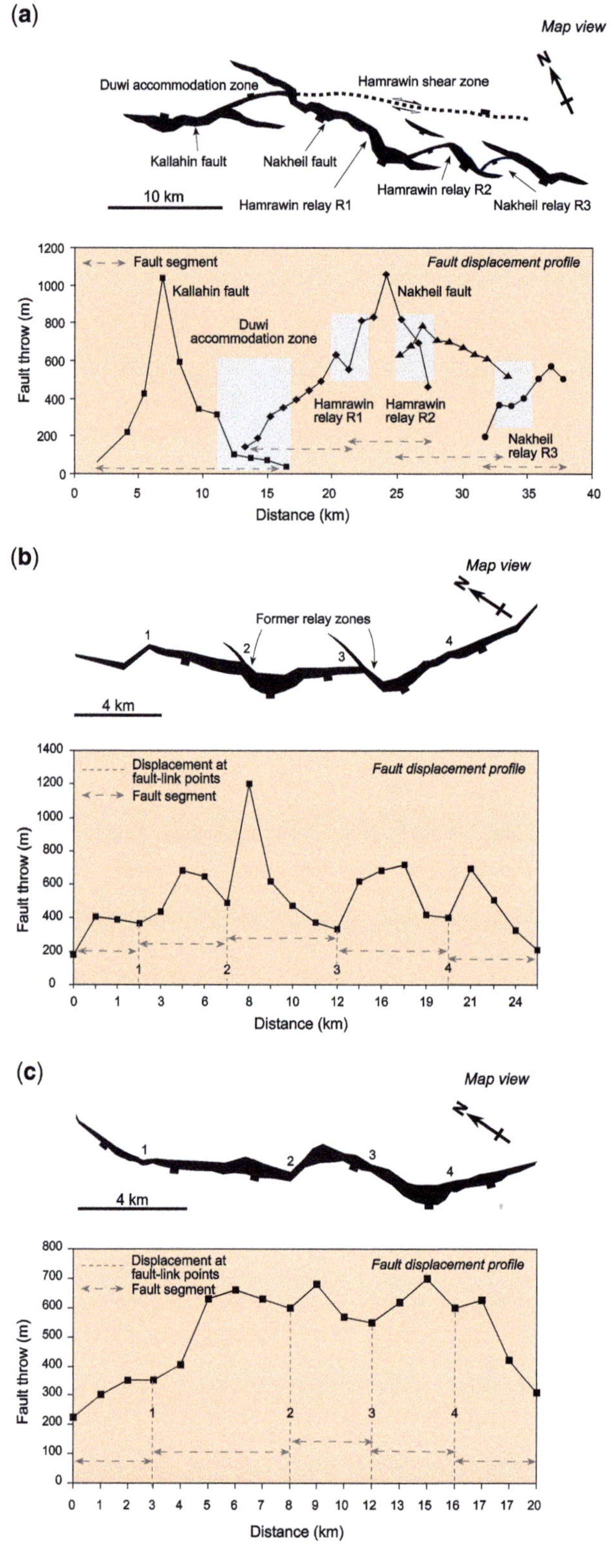
(a)
Map view
Duwi accommodation zone
Hamrawin shear zone
Kallahin fault
Nakheil fault
Hamrawin relay R1
Hamrawin relay R2
Nakheil relay R3
10 km
Fault segment
Fault displacement profile
Kallahin fault
Nakheil fault
Duwi accommodation zone
Hamrawin relay R1
Hamrawin relay R2
Nakheil relay R3
Fault throw (m)
Distance (km)
(b)
Map view
Former relay zones
1
2
3
4
4 km
Displacement at fault-link points
Fault segment
Fault displacement profile
Fault throw (m)
Distance (km)
(c)
Map view
1
2
3
4
4 km
Displacement at fault-link points
Fault segment
Fault displacement profile
Fault throw (m)
Distance (km)

Rotevatn 2016). The displacement profiles in Figure 11 were constructed from outcrop relationships, cross-sections and reconstructions of the basement topography shown in Figure 3a. The calculated displacements are a minimum as there are no pre-rift or synrift strata preserved on the uplifted and eroded footwall blocks of Precambrian basement (Figs 4, 5, 9 & 10).

The fault displacement profiles of the Nakheil and Kallahin fault systems indicate that the relays R1, R2 and R3 as well as the Duwi accommodation zone were developed at the fault displacement minima on the fault profiles (Fig. 11a).

The displacement profile of the Hamadat fault system (Fig. 11b) shows the highly segmented nature of this fault system. The profile is asymmetrical, strongly jagged and composed of several peaks and troughs that can be interpreted to indicate the linkage of initially isolated fault segments (e.g. Peacock & Sanderson 1991; Cartwright et al. 1995, 1996; Childs et al. 1995; Morley & Wonganan 2000). The displacement minima occur at the locations of fault linkage points, where the transverse anticlines and corner anticline folds are developed (Fig. 9a).

The fault displacement profile of the Zug El Bahar fault (Fig. 11c) is irregular, with multiple maxima and minima, reflecting the highly segmented nature of the fault system. The locations of minimum throws are the locations where the fault jogs or linkage points of differently oriented fault segments occur.

Discussion

The extensional faults and fault-related folds from the northern margin of the Red Sea described in this paper illustrate some of the 3D complexities of hanging wall geometries generated by regional extension in rift basins. These provide excellent outcrop analogues for subsurface structures where segmented and linked extensional fault geometries are found in intra-continental rift basins and on passive margins.

Extensional fault segmentation and linkage

Our understanding of the evolution of extensional fault systems in rift basins has been developed from many analyses of fault displacement relationships and fault linkage patterns, from the formation and growth of isolated faults to overlapping fault systems (relay ramps) and to fault linkages by the breaching or faulting of relay systems.

Isolated fault growth model for extensional faults. In the isolated fault growth model (e.g. Walsh & Watterson 1987, 1988, 1989; Peacock & Sanderson 1991; Walsh et al. 2003; Kim & Sanderson 2005; Nicol et al. 2005), individual separate faults nucleate and amplify by radial propagation and an increase in both fault lengths and displacements. However, in many extensional basins like-dipping fault systems are not isolated, but form segmented systems that overlap with adjacent faults forming relay ramps (e.g. Larsen 1988; Peacock & Sanderson 1994; Fossen & Rotevatn 2016).

Segment linkage models for extensional fault evolution. Segment linkage models involve the breaching or faulting of relay structures, which leads to increased effective fault lengths that can accommodate greater displacements (e.g. Trudgill & Cartwright 1994; Cartwright et al. 1995, 1996; Childs et al. 1995, 2003; Dawers & Anders 1995; Huggins et al. 1995; Baudon & Cartwright 2008; Fossen & Rotevatn 2016). As described in this paper, the northwestern margin of the Red Sea rift basin exhibits characteristic patterns of breached relay ramps that link early-formed fault segments at sharp bends and kinks along the fault traces of the main Nakeil fault system as well as those of the Hamadat and Zug El Bahar faults (Figs 4b, 5b, 9a & 10a). Their displacement profiles show minima where breaching and linkage occurred (Fig. 11). Similar fault linkage and displacement patterns are found elsewhere along the northern Red Sea (Khalil & McClay 2001, 2009, 2016) as well as in the Gulf of Suez (Gawthorpe & Hurst 1993; Gawthorpe et al. 1997, 2000; McClay et al. 1998; McClay & Khalil 1998; Gupta et al. 1999; Young et al. 2000; Jackson et al. 2002; Wilson et al. 2009; Moustafa & Khalil 2017). The North Sea rift basins and other extensional terranes (e.g. Muraoka & Kamata 1983; Walsh & Watterson 1987; Dawers et al. 1993; Fossen et al. 2010; Elliott et al. 2011; Wilson et al. 2013; Fossen & Rotevatn 2016) are also characterized by similarly linked segmented extensional fault systems.

Coherent fault model and reactivated basement faults. Walsh et al. (2002) introduced an alternative model for extensional fault growth whereby, upon nucleation, some extensional faults rapidly developed along-strike to establish a stable maximum fault length before accumulating significant displacements (Walsh et al. 2003; Schöpfer et al. 2007; Giba et al. 2012; Jackson & Rotevatn 2013). This

Fig. 11. Hanging wall displacement profiles for the major fault systems showing displacement variations along fault strike, together with the locations of transfer zones and fault linkage points. (**a**) Nakheil and Kallahine faults; (**b**) Hamadat fault; and (**c**) Zug El Bahar fault. Note that the displacements are a minimum as the footwall blocks have been eroded down into the basement and no footwall cut-off positions are preserved.

model appears to be particularly applicable to reactivated basement faults that splay vertically or laterally from a single fault surface at depth to form en echelon segmented faults higher in the overlying stratigraphy (Walsh et al. 2002, 2003; Giba et al. 2012; Jackson & Rotevatn 2013; Fossen & Rotevatn 2016). This coherent fault growth model is most likely applicable to the segmented and linked extensional fault systems of the northwestern Red Sea margin as well as to those of the Gulf of Suez rift, where there are strong controls by reactivated Precambrian basement fabrics (e.g. Moustafa 1997; McClay & Khalil 1998; Khalil & McClay 2001, 2009, 2016; Bosworth et al. 2005; Bosworth 2015; Moustafa & Khalil 2017).

Longitudinal synclines and transverse folds

The map patterns and the images of the Duwi, Hamadat and Zug El Bahar fault systems display prominent, segmented, longitudinal doubly plunging synclines with broad transverse synclines (Schlische 1995) focused at positions of maximum displacement along the linked fault systems (Figs 1b, 4, 5, 9 & 10). The segmented and complex hanging wall structures of the Duwi half-graben are the result of the along-strike linkage and breaching of the relay structures R1–R3 of the Nakheil Border fault system (Fig. 5).

Similar segmented, doubly plunging hanging wall synclines have been described elsewhere along the Red Sea margin (Khalil & McClay 2002, 2016) as well as from outcrops on the eastern margin of the Gulf of Suez (McClay et al. 1998; Khalil & McClay 2001; Wilson et al. 2009). Similar hanging wall syncline architectures have been recognized in other extensional terranes (e.g. Chapman et al. 1978; Muraoka & Kamata 1983; Walsh & Watterson 1987; Dawers et al. 1993; Lewis et al. 2015).

Comparative examples of extensional hanging wall synclines are shown in Figure 12. The 3D seismic analysis of the October fault zone in the Gulf of Suez (Jackson & Rotevatn 2013) shows segmented hanging wall fault-related folds with configurations similar to those described in this study (Fig. 12a). Similarly, North Sea examples, such as the southern Bremstein fault complex offshore mid-Norway (Wilson et al. 2013) and the Brage fault zone in the Oseberg East area of the Horda Platform (Lewis et al. 2015), display linked fault segments with hanging wall, fault-parallel longitudinal synclines together with fault-perpendicular, transverse anticlines and synclines (Fig. 12b, c).

Extensional fault-propagation folding

Maps and cross-sections of the Duwi, Hamadat and Zug El Bahar fault systems all show longitudinal extensional fault-propagation folding with panels of strata in the immediate fault hanging walls generally moderately to steeply dipping in the same direction as the adjacent fault surface (Figs 4c, 9c & 10c).

Extensional fault-propagation folding (sometimes called forced folding) normally occurs where extensional faults propagate up-section through mechanically weak, ductile and anisotropic units (Withjack et al. 1990; Schlische 1995; Cosgrove & Ameen 1999; Hardy & McClay 1999; Sharp et al. 2000; Jackson et al. 2006; Lewis et al. 2015). This usually produces hanging wall synclines tens of metres to kilometres wide, such as the northwestern Red Sea margin. These extensional fault-propagation folds were accommodated by flexural slip within the highly anisotropic shale-dominated Upper Cretaceous to Paleocene sequences (Fig. 2) to produce longitudinal folds parallel to the driving fault, with terminations in tip line monoclines along-strike (Khalil & McClay 2016). The large-scale folds in the northwestern Red Sea margin are unlikely to have been formed by frictional drag folding (Ferrill et al. 2012). Drag folds normally result from frictional drag along the fault surface (Hatcher 1994) and are generally restricted to only a few tens of metres adjacent to the fault (Twiss & Moores 1992; Davis & Reynolds 1996; Peacock et al. 2000).

Similar styles of extensional fault-propagation folding have been documented in many rift basins (e.g. Patton et al. 1994; Schlische 1995; Sharp et al. 2000; Jackson et al. 2006; Ferrill et al. 2007, 2011, 2012; Ford et al. 2007; Kane et al. 2010; Lewis et al. 2013; Lăpădat et al. 2016) as well as having been simulated in analogue and numerical models (e.g. Withjack et al. 1990; Hardy & McClay 1999; Withjack & Schlische 2006).

Corner folds in extensional fault systems

This paper proposes the term corner folds for the type of transverse and oblique fold systems that develop specifically at fault linkage positions (breached and faulted relay ramps) in segmented extensional fault arrays. Corner folds have been recognized in thrust fault systems, particularly where the thrust surfaces have complex 3D geometries with corners and bends in the thrust surfaces (Alvarez-Marron 1995), but they have not previously been recognized in extensional fault systems.

In the northwestern Red Sea outcrops described in this paper, the extensional fault systems display well-developed kilometre-scale corner folds, particularly along the main central Nakheil fault system (Figs 4b, 5b & 6). These formed either at the breached structures or faulted relay ramps where the main fault segments linked along-strike and formed a sharp curve or kink in the fault trace. The axial

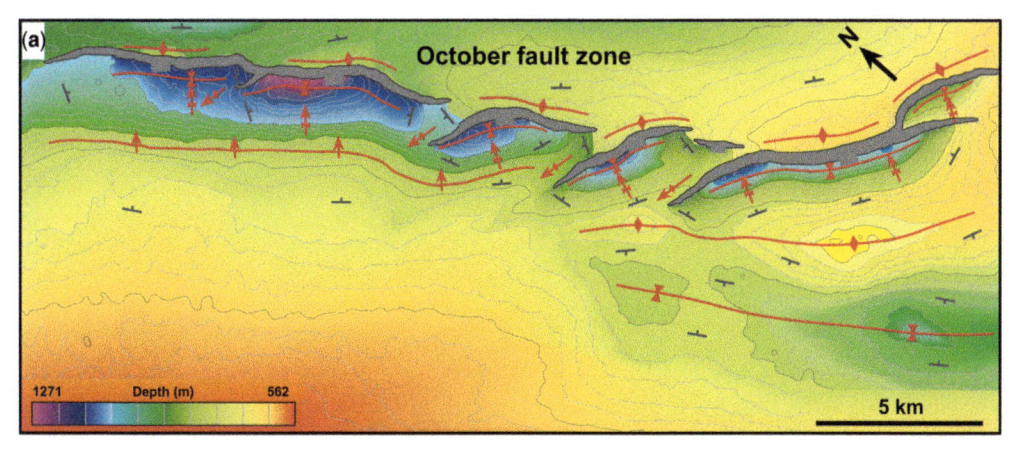

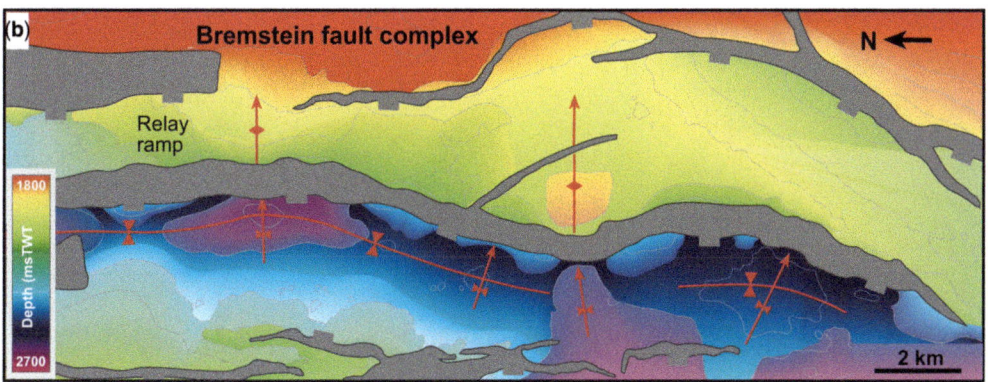

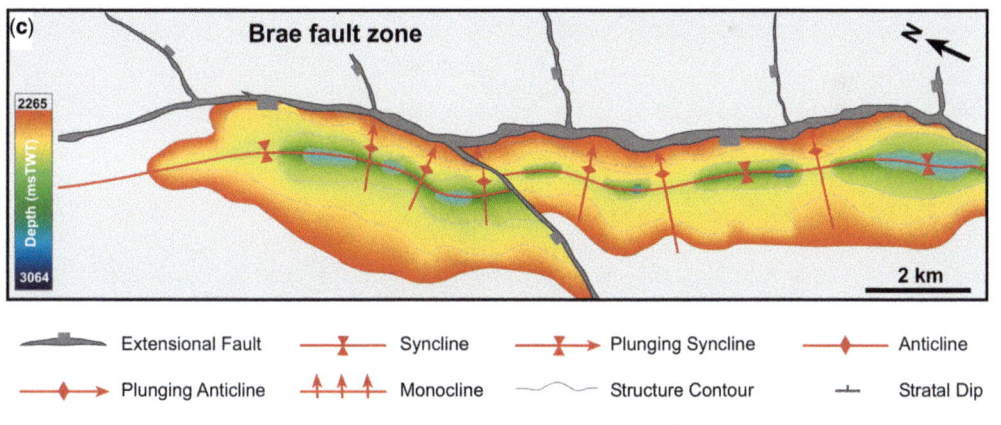

Fig. 12. 3D subsurface examples of segmented and linked extensional fault systems showing hanging wall synclines similar to those in the northern Red Sea margin. (**a**) Depth–structure map showing the October fault system and supra-salt structure at the top of the Wardan Formation in the Gulf of Suez rift (after Jackson & Rotevatn 2013). (**b**) Two-way travel time (TWT)–structure map of the Garn Formation showing a portion of the Bremstein fault complex offshore mid-Norway (after Wilson *et al.* 2013). (**c**) TWT–structure map of the base of synrift unit of the Bragge fault zone in the northern North Sea (after Lewis *et al.* 2015).

traces of the corner folds trend generally oblique to the fault strike and plunge inwards towards the axial trace of the larger, longitudinal hanging wall synclines (Figs 4b & 5). The dimensions of the corner folds directly relate to the tightness of the kinks and bends of the linked fault segments and are

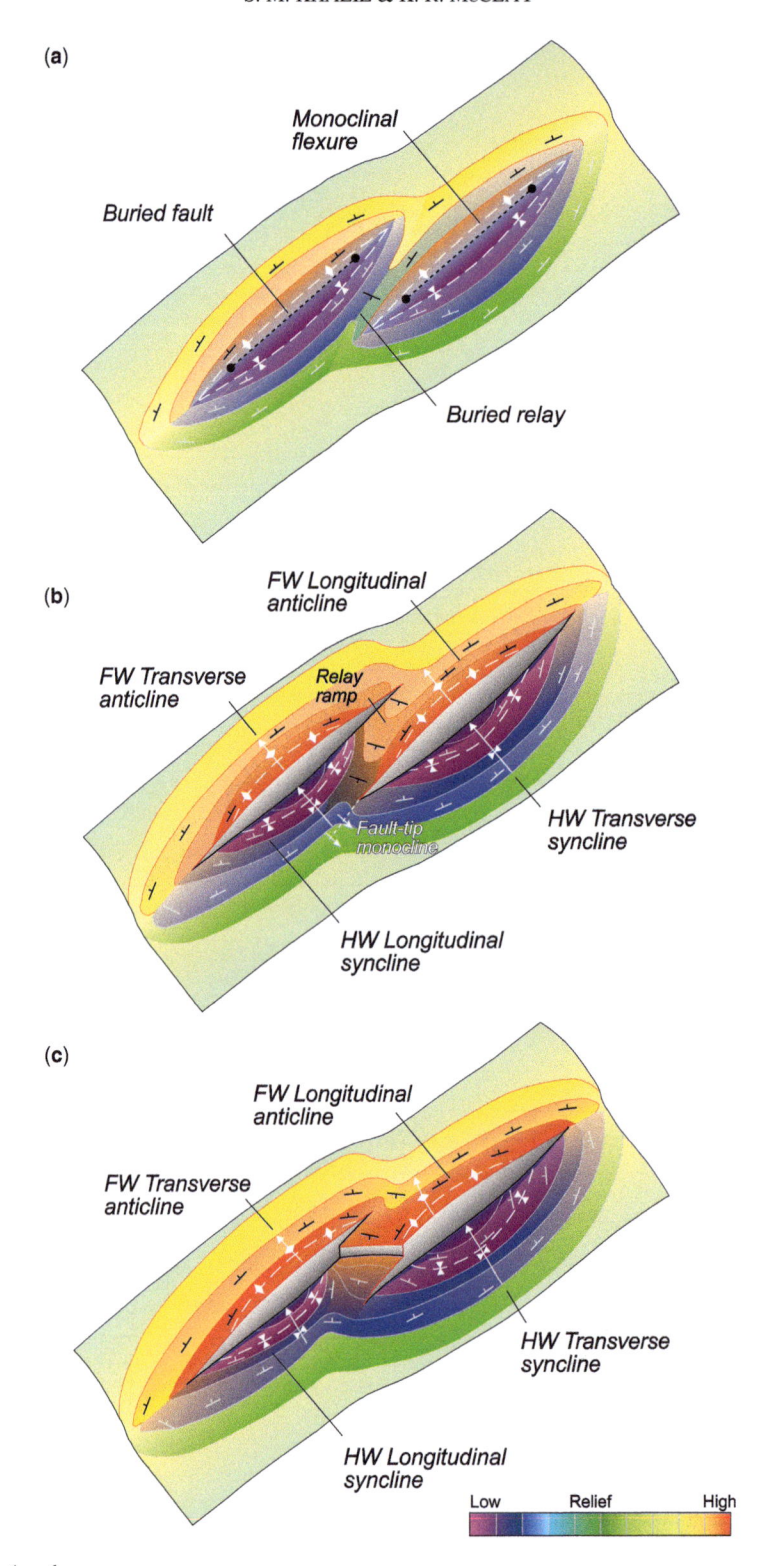

Fig. 13. *Continued.*

(d)

FW Longitudinal anticline

HW Longitudinal syncline

(e)

Outward plunging folds

Corner synclines

Corner anticlines

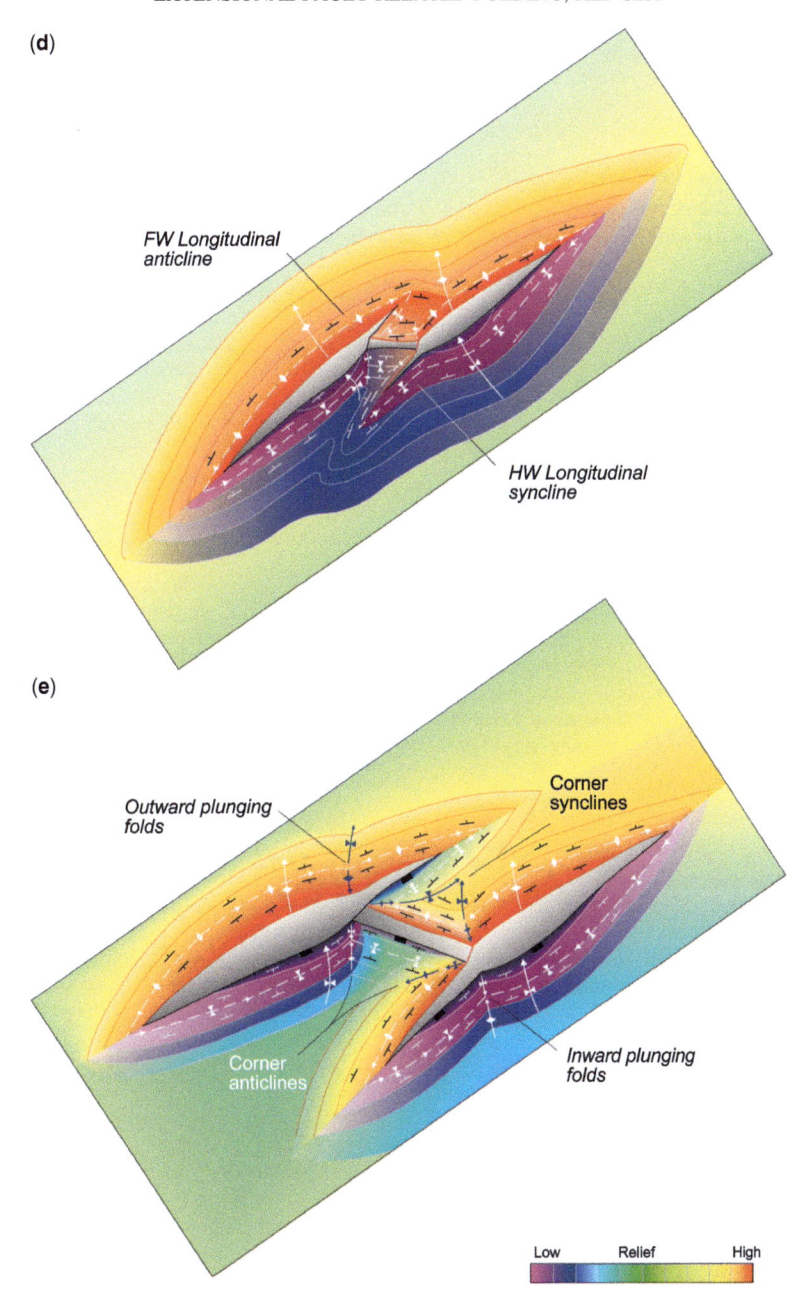

Inward plunging folds

Low Relief High

Fig. 13. Conceptual model illustrating the evolution of the northern Red Sea margin extensional fault-related folds as described in this paper. (**a**) Monoclinal synclinal flexures above blind (buried) low displacement offset extensional faults at depth. (**b**) Relay ramp system formed by upwards propagation of the offset extensional faults to the surface. Transverse folds have axial surface traces perpendicular to the fault system and are located at the positions of maximum displacement along the fault traces.(**c**) Increased fault displacement and along-strike propagation with an oblique transfer fault linking the two fault segments forming a breached relay. Longitudinal fault-propagation folds are superimposed onto the transverse folds and the displacement maxima migrate towards the transfer zone. (**d**) Initiation of corner folds at the breached relay (transfer zone) and at the fault linkage points; at this stage the larger longitudinal folds dominate at the expense of transverse folds. (**e**) Large displacement on the linked fault system with well-developed corner folds located at the linkage points between the breached relay ramp system and the main fault segments.

interpreted as accommodation structures formed to relieve the space problems generated by the vertical displacement across differently oriented fault segments at the fault kinks and bends. The strongly segmented Hamadat fault system (Fig. 9a) has three well-developed longitudinal synclines associated with the individual segments, together with distinct corner anticlines at the kinks/linkage points in the main Hamadat fault system (Fig. 9a).

Evolutionary model

The following section proposes an evolutionary model for the progressive development of these extensional fault-related fold systems on this part of the northwestern Red Sea rift margin (Fig. 13).

Stage 1: monoclinal flexures above buried faults. This shows the development of monoclinal synclinal flexures above blind extensional faults formed by reactivation of the Precambrian fabrics in the basement (Fig. 13a).

Stage 2: overlapping extensional faults–relay ramp. Figure 13b shows the overlapping, like-dipping, en echelon extensional faults at the surface forming a relay ramp. At this stage, the amount of offset is small, but the gentle longitudinal and transverse folds have formed.

Stage 3: breached relay ramp. With greater extension, fault propagation and increased displacement, the relay system became breached, either by lateral fault propagation and linkage or by the development of a new, discrete, hard-linked transfer fault breaking the relay ramp (Fig. 13c). At this stage, the increased displacements on the main fault systems would have increased the amplitude of the associated longitudinal folds.

Stage 4: breached relay ramp with extensional fault-propagation folds. Figure 13d shows the development of longitudinal hanging wall synclines and longitudinal footwall anticlines in response to the lateral propagation of the linked fault system, together with longitudinal hanging wall and footwall extensional fault-propagation folds. Oblique corner anticlines and corner synclines formed at the fault corners and fault linkage positions and the longitudinal folds increased in size, such that they tended to dominate the transverse folds formed during the earlier stages. This stage can be directly compared with the structural pattern of the Nakheil fault–fold system (Figs 4b & 5b), where the Nakheil hanging wall longitudinal synclines are offset and bend at the relay zones which, in places, are dominated by narrow corner folds trending oblique to the Nakheil fault system.

Stage 5: corner folds, breached relay ramps & fault linkage points. Figure 13e shows greater displacement on the linked fault system where the dimension of the longitudinal folds increases at the expense of the earlier formed transverse folds. At this stage, corner anticlines are well developed and focused at the linkage points, forming local structural highs within the larger, longitudinal synclines – for example, the Nakheil longitudinal synclines (Fig. 4b) – or separate different longitudinal syncline folds such as the Hamadat synclines (Fig. 9a).

As extension focused into the centre of the northern Red Sea rift, flexural uplift of the margin would have allowed the erosion of the footwalls of these extensional fault systems, leaving only the synclinal basins of the hanging walls to be preserved.

Conclusions

(1) Late Oligocene–Miocene, domino-style extensional fault systems in the northwestern Red Sea rift margin bound a series of half-graben basins dominated by doubly plunging, longitudinal hanging wall synclines and kilometre-scale fault-propagation folds. The linkage patterns of originally segmented and offset rift faults strongly controlled the basin architectures on this part of the northwestern Red Sea margin.

(2) Oblique and transverse corner folds formed at breached and faulted relay ramps as well as at linkage positions where originally offset segmented extensional faults joined to form longer, through-going linked fault arrays.

(3) An evolutionary model is proposed for the initiation and evolution of these complex fault systems on the northwestern Red Sea margin.
  (a) Initial development of monoclinal synclines above blind segmented extensional faults formed by the reactivation of pre-existing fabrics in the Precambrian basement (Fig. 13a).
  (b) Increased extension, lateral and vertical propagation of the en echelon fault segments formed relay ramps. Longitudinal folds with footwall and hanging wall transverse folds were developed, associated with individual en echelon fault segments (Fig. 13b).
  (c) Continued extension led to increased longitudinal folding, together with extensional fault-propagation folding and breaching of the relay ramps (Fig. 13c). Corner folds formed at the breached relay ramps and linkage positions of the originally segmented faults. The dimensions and amplitudes of the longitudinal

folds progressively increased in response to the increased displacement on the linked fault systems, eventually becoming the dominant hanging wall structures.
(4) This study shows that extremely complex hanging wall fold and fault geometries can be developed within a single, regional extensional stress regime with no need to invoke more complex interpretations involving strike-slip or changes in the regional stress field.
(5) The examples illustrated here show strong similarities to extensional fault-related fold systems elsewhere in the Gulf of Suez rift, in the rift basins of the North Sea and on the Norwegian margin.

The outcrops and models presented in this paper may provide analogues for understanding the complexities of extensional fault systems and associated hanging wall structures in intra-continental rifts and on passive margins, particularly in rift systems where basin margin structures are not well imaged in 2D and 3D seismic surveys.

Acknowledgements The authors gratefully acknowledge support from the Fault Dynamics Project, Royal Holloway University of London and Suez Canal University, Egypt. Bill Bosworth (Apache, Egypt), Maher Ayyad (BG Egypt) and Amgad Younes (Shell) are thanked for many fruitful discussions on the geology of the NW Red Sea. This paper has benefited greatly from the constructive comments of reviewers Martin Insley, Ian Sharp and associate editor James Hammerstein.

Funding Partially supported by the Fault Dynamics Research Goup and the STAR project funded by BG Group, BHPBilliton, ConocoPhillips, ENI, MarathonOil, Nexen, Shell, Talisman Energy, YPF.

References

ABD EL-RAZIK, T.M. 1967. Stratigraphy of the sedimentary cover of the Anz-Atshan-south Duwi district. *Bulletin of the Faculty of Science, Cairo University*, **431**, 135–179.

AKAAD, M.K. & NOWEIR, A.M. 1980. Geology and lithostratigraphy of the Arabian desert orogenic belt of Egypt between latitudes 25° 35′ and 26° 30′. In: COORAY, P.G. & TAHOUN, S.A. (eds) *Evolution and Mineralization of the Arabian-Nubian Shield*. Pergamon Press, New York, 127–135.

AKKAD, S. & DARDIR, A.A. 1966. *Geology and Phosphate Deposits of Wasif, Safaga Area*. Geological Survey of Egypt, Papers, **36**.

ALLMENDINGER, R.W. 1998. Inverse and forward numerical modelling of trishear fault-propagation folds. *Tectonics*, **17**, 640–656.

ALVAREZ-MARRON, J. 1995. Three-dimensional geometry and interference of fault-bend folds: examples from Ponga Unit, Variscan Belt, NW Spain. *Journal of Structural Geology*, **17**, 549–560.

BARNETT, J.A.M., MORTIMER, J., RIPPON, J.H., WALSH, J.J. & WATTERSON, J. 1987. Displacement geometry in the volume containing a single normal fault. *American Association of Petroleum Geologists Bulletin*, **71**, 925–937.

BAUDON, C. & CARTWRIGHT, J.A. 2008. 3D seismic characterisation of an array of blind normal faults in the Levant Basin, Eastern Mediterranean. *Journal of Structural Geology*, **30**, 746–760.

BEN-MENAHEM, A., NUR, A. & VERED, M. 1976. Tectonics, seismicity and structure of the Afro-Eurasian junction – the breaking of an incoherent plate. *Physics of the Earth and Planetary Interiors*, **12**, 1–50.

BOSWORTH, W. 2015. Geological evolution of the Red Sea: historical background, review and synthesis. In: RASAUL, N.M.A. & STEWART, I.C.F. (eds) *The Red Sea*. Springer, Berlin, 45–78.

BOSWORTH, W., HUCHON, P. & MCCLAY, K. 2005. The Red Sea and Gulf of Aden basins. *Journal of African Earth Sciences*, **43**, 334–378.

BRANDES, C. & TANNER, D.C. 2016. Fault-related folding: a review of kinematic models and their application. *Earth Science Reviews*, **138**, 352–370.

CARTWRIGHT, J.A., TRUDGILL, B. & MANSFIELD, C. 1995. Fault growth by segment linkage: an explanation for scatter in maximum displacement and trace length data from Canyonlands Grabens of S.E. Utah. *Journal of Structural Geology*, **17**, 1319–1326.

CARTWRIGHT, J.A., MANSFIELD, C. & TRUDGILL, B. 1996. The growth of normal faults by segment linkage. In: BUCHANAN, P.G. & NIEUWLAND, D.A. (eds) *Modern Development in Structural Interpretation, Validation and Modelling*. Geological Society, London, Special Publications, **99**, 163–177, https://doi.org/10.1144/GSL.SP.1996.099.01.13

CHAPMAN, G.R., LIPPARD, S.J. & MARTYN, J.E. 1978. The stratigraphy and structure of the Kamasia Range, Kenya rift valley. *Journal of the Geological Society, London*, **135**, 265–281, https://doi.org/10.1144/gsjgs.135.3.0265

CHILDS, C., WATTERSON, J. & WALSH, J.J. 1995. Fault overlap zones within developing normal fault systems. *Journal of the Geological Society, London*, **152**, 353–459, https://doi.org/10.1144/gsjgs.152.3.0535

CHILDS, C., NICOL, A., WALSH, J.J. & WATTERSON, J. 2003. The growth and propagation of synsedimentary faults. *Journal of Structural Geology*, **25**, 633–648.

COCHRAN, J.R. 1983. A model for development of the Red Sea. *American Association of Petroleum Geologists Bulletin*, **67**, 41–69.

COLEMAN, R.G. 1993. *Geologic Evolution of the Red Sea*. Oxford Monographs on Geology and Geophysics, **24**. Oxford University Press, Oxford.

CORFIELD, S. & SHARP, I.R. 2000. Structural style and stratigraphic architecture of fault propagation folding in extensional settings: a seismic example from the Smørbukk area, Halten Terrace, Mid-Norway. *Basin Research*, **12**, 329–341.

COSGROVE, J.W. & AMEEN, M.S. 1999. A comparison of the geometry, spatial organization and fracture patterns associated with forced folds. In: COSGROVE, J.W. & AMEEN, M.S. (eds) *Forced Folds and Fractures*. Geological Society, London, Special Publications, **169**, 7–21, https://doi.org/10.1144/GSL.SP.2000.169.01.02

Cowie, P.A. & Scholz, H. 1992a. Physical explanation for the displacement-length relationship of faults using a post-yield fracture mechanics model. *Journal of Structural Geology*, **14**, 1133–1148.

Cowie, P.A. & Scholz, C.H. 1992b. Growth of faults by accumulation of seismic slip. *Journal of Geophysical Research*, **97**, 11085–11095.

Davis, G.H. & Reynolds, S.J. 1996. *Structural Geology of Rocks and Regions*. 2nd edn. Wiley, Chichester.

Dawers, N.H. & Anders, M.H. 1995. Displacement-length scaling and fault linkage. *Journal of Structural Geology*, **17**, 607–614.

Dawers, N.H., Anders, M.H. & Scholz, C.H. 1993. Growth of normal faults; displacement-length scaling. *Geology*, **21**, 1107–1110.

Dula, W.D. 1991. Geometric models of listric normal faults and rollover folds. *American Association of Petroleum Geologists Bulletin*, **75**, 1609–1625.

El Bassyony, A.A. 1982. *Stratigraphical studies on Miocene and younger exposures between Quseir and Berenice, Red Sea coast, Egypt*. PhD thesis, Ain Shams University.

Elliott, G.M., Wilson, P., Jackson, C.A.-L., Gawthorpe, R.L., Michelsen, L. & Sharp, I.R. 2011. The linkage between fault throw and footwall scarp erosion patterns: an example from the Bremstein Fault Complex, offshore Mid-Norway. *Basin Research*, **23**, 1–18.

Ferrill, D.A. & Morris, A.P. 2008. Fault zone deformation controlled by mechanical stratigraphy, Balcones fault system, Texas. *American Association of Petroleum Geologists Bulletin*, **92**, 359–380.

Ferrill, D.A., Morris, A.P. & Smart, K.J. 2007. Stratigraphic control on extensional fault propagation folding: Big Brushy Canyon Monocline, Sierra Del Carmen, Texas. *In*: Jolley, S.J., Barr, D., Walsh, J.J. & Knipe, R.J. (eds) *Structurally Complex Reservoirs*. Geological Society, London, Special Publications, **292**, 203–217, https://doi.org/10.1144/SP292.12

Ferrill, D.A., Morris, A.P., McGinnis, R.N., Smart, K.J. & Ward, W.C. 2011. Fault zone deformation and displacement partitioning in mechanically layered carbonates: the Hidden Valley fault, central Texas. *American Association of Petroleum Geologists Bulletin*, **95**, 1383–1397.

Ferrill, D.A., Morris, A.P. & McGinnis, R.N. 2012. Extensional fault-propagation folding in mechanically layered rocks: the case against the frictional drag mechanism. *Tectonophysics*, **576–577**, 78–85.

Finch, E., Hardy, S. & Gawthorpe, R. 2004. Discrete-element modelling of extensional fault-propagation folding above rigid basement fault blocks. *Basin Research*, **16**, 489–506.

Ford, M., Le Carlier De Veslud, C. & Bourgeois, O. 2007. Kinematic and geometric analysis of fault-related folds in a rift setting: the Dannemarie Basin, Upper Rhine Graben, France. *Journal of Structural Geology*, **29**, 1811–1830.

Fossen, H. & Rotevatn, A. 2016. Fault linkages and relay structures in extensional settings – a review. *Earth Science Reviews*, **154**, 14–28.

Fossen, H., Schultz, R.A., Rundhovde, E., Rotevatn, A. & Buckley, S.J. 2010. Fault linkage and graben stepovers in the Canyonlands (Utah) and the Viking Graben, with implications for hydrocarbon migration and accumulation. *American Association of Petroleum Geologists Bulletin*, **94**, 597–611.

Freund, R. 1970. Plate tectonics of the Red Sea and Africa. *Nature*, **228**, 453.

Gawthorpe, R.L. & Hurst, J.M. 1993. Transfer zones in extensional basins: their structural style and influence on drainage development and stratigraphy. *Journal of the Geological Society, London*, **150**, 1137–1152, https://doi.org/10.1144/gsjgs.150.6.1137

Gawthorpe, R.L., Sharp, I., Underhill, J.R. & Gupta, S. 1997. Linked sequence stratigraphic and structural evolution of propagating normal faults. *Geology*, **25**, 795–798.

Gawthorpe, R.L., Jackson, C.A.-L., Young, M.J., Sharp, I.R. & Moustafa, A.R. & Leppard, C.W. 2000. Normal fault growth, displacement localisation and the evolution of normal fault populations: the Hammam Faraun fault block, Suez rift, Egypt. *Journal of Structural Geology*, **25**, 883–895.

Giba, M., Walsh, J.J. & Nicol, A. 2012. Segmentation and growth of an obliquely reactivated normal fault. *Journal of Structural Geology*, **39**, 253–267.

Girdler, R.W. & Styles, P. 1974. Two stage Red Sea floor spreading. *Nature*, **247**, 1–11.

Groshong, R.H. 1989. Half-graben structures: balanced models of extensional fault-bend folds. *Geological Society of America Bulletin*, **101**, 96–105.

Gupta, S., Underhill, J.R., Sharp, I. & Gawthorpe, R.L. 1999. Role of fault interactions in controlling syn-rift dispersal patterns: Miocene Abu Alaqa Group, Suez Rift, Sinai, Egypt. *Basin Research*, **11**, 167–189.

Hamblin, W.K. 1965. Origin of 'reverse drag' on the downthrown side of normal faults. *Geological Society of America Bulletin*, **76**, 1145–1164.

Hardy, S. 2011. Cover deformation above steep, basement normal faults: insights from 2D discrete element modeling. *Marine and Petroleum Geology*, **28**, 966–972.

Hardy, S. & Finch, E. 2006. Discrete element modelling of the influence of cover strength on basement-involved fault-propagation folding. *Tectonophysics*, **415**, 225–238.

Hardy, S. & McClay, K.R. 1999. Kinematic modelling of extensional fault-propagation folding. *Journal of Structural Geology*, **21**, 695–702.

Hatcher, R.D., Jr 1994. *Structural Geology: Principles, Concepts, Problems*. Prentice Hall, Englewood Cliffs, NJ.

Hempton, M. 1987. Constraints on Arabian plate motion and extensional history of the Red Sea. *Tectonics*, **6**, 687–705.

Horsfield, W.T. 1977. An experimental approach to basement-controlled faulting. *Geologie en Mijnbouw*, **56**, 363–370.

Huggins, P., Watterson, J., Walsh, J.J. & Childs, C. 1995. Relay zone geometry and displacement transfer between normal faults recorded in coal mine plans. *Journal of Structural Geology*, **17**, 1741–1755.

Issawi, B., Francis, M., El-Hinnawi, M. & Mehanna, A. 1969. *Contribution to the Structure and Phosphate Deposits of Quseir Area*. Geological Survey of Egypt, Papers, **50**.

Jackson, C.A.-L. & Rotevatn, A. 2013. 3D seismic analysis of the structure and evolution of a salt-influenced normal fault zone: a test of competing fault growth models. *Journal of Structural Geology*, **54**, 215–234.

JACKSON, C.A.-L., GAWTHORPE, R.L. & SHARP, I.R. 2002. Growth and linkage of the East Tanka fault zone, Suez rift: structural style and syn-rift stratigraphic response. *Journal of the Geological Society, London*, **159**, 175–187, https://doi.org/10.1144/0016-76490 1-100

JACKSON, C.A.-L., GAWTHORPE, R.L. & SHARP, I.R. 2006. Style and sequence of deformation during extensional fault-propagation folding: examples from the Hammam Faraun and El-Qaa fault blocks, Suez Rift, Egypt. *Journal of Structural Geology*, **28**, 519–535.

JANECKE, S.U., VANDENBURG, C.J. & BLANKENAU, J.J. 1998. Geometry, mechanisms and significance of extensional folds from examples in the Rocky Mountain Basin and Range Province, USA. *Journal of Structural Geology*, **20**, 841–856.

JARRIGE, J.J., OTT D'ESTEVOU, P. *ET AL*. 1986. Inherited discontinuities and Neogene structure: the Gulf of Suez and the northwestern edge of the Red Sea. *Philosophical Transactions of the Royal Society of London, Series A*, **317**, 129–139.

JARRIGE, J.J., OTT D'ESTEVOU, P., BUROLLET, P.F., MONTENAT, C., RICHET, J.P. & THIRIET, J.P. 1990. The multistage tectonic evolution of the Gulf of Suez and northern Red Sea continental rift from field observations. *Tectonics*, **9**, 441–465.

JIN, G. & GROSHONG, R.H., JR 2006. Trishear kinematic modeling of extensional fault propagation folding. *Journal of Structural Geology*, **28**, 170–183.

KANE, K.E., JACKSON, C.A.L. & LARSEN, E. 2010. Normal fault growth and fault-related folding in a salt-influenced rift basin: South Viking Graben, offshore Norway. *Journal of Structural Geology*, **32**, 490–506.

KHALIL, S.M. & McCLAY, K.R. 2001. Tectonic evolution of the northwestern Red Sea–Gulf of Suez rift system. *In*: WILSON, R.C.L., WHITMARSH, R.B., TAYLOR, B. & FROITZHEIM, N. (eds) *Non-Volcanic Rifting of Continental Margins: a Comparison of Evidence from Land and Sea*. Geological Society, London, Special Publications, **187**, 453–473, https://doi.org/10.1144/GSL.SP.2001. 187.01.22

KHALIL, S.M. & McCLAY, K.R. 2002. Extensional fault-related folding, northwestern Red Sea, Egypt. *Journal of Structural Geology*, **24**, 743–762.

KHALIL, S.M. & McCLAY, K.R. 2006. Extensional fault-related folding, Gulf of Suez, Egypt. *Middle East Research Centre, Ain Shams University, Earth Science Series*, **20**, 1–16.

KHALIL, S.M. & McCLAY, K.R. 2009. Structural control on syn-rift sedimentation, northwestern Red Sea margin, Egypt. *Marine and Petroleum Geology*, **26**, 1018–1034.

KHALIL, S.M. & McCLAY, K.R. 2016. 3D Geometry and kinematic evolution of extensional fault-related folds, northwestern Red Sea, Egypt. *In*: CHILDS, C., HOLDSWORTH, R.E., JACKSON, C.A.-L., MANZOCCHI, T., WALSH, J.J. & YIELDING, G. (eds) *The Geometry and Growth of Normal Faults*. Geological Society, London, Special Publications, **439**, 109–130, https://doi.org/ 10.1144/SP439.11

KIM, Y.S. & SANDERSON, D.J. 2005. The relationship between displacement and length of faults: a review. *Earth Science Reviews*, **68**, 317–334.

LĂPĂDAT, A., IMBER, J., YIELDING, G., IACOPINI, D., McCAFFREY, K.J.W., LONG, J.J. & JONES, R.R. 2016. Occurrence and development of folding related to normal faulting within a mechanically heterogeneous sedimentary sequence: a case study from Inner Moray Firth, UK. *In*: CHILDS, C., HOLDSWORTH, R.E., JACKSON, C.A.-L., MANZOCCHI, T., WALSH, J.J. & YIELDING, G. (eds) *The Geometry and Growth of Normal Faults*. Geological Society, London, Special Publications, **439**, 373–394, https://doi.org/10.1144/SP439.18

LARSEN, P.H. 1988. Relay structures in a lower Permian basement-involved extension system, East Greenland. *Journal of Structural Geology*, **10**, 3–8.

LE PICHON, X. & FRANCHETEAU, J. 1978. A plate tectonic analysis of the Red Sea–Gulf of Aden area. *Tectonophysics*, **46**, 369–406.

LEWIS, M.M., JACKSON, C.A.L. & GAWTHORPE, R.L. 2013. Salt-influenced normal fault growth and forced folding: the Stavanger fault system, North Sea. *Journal of Structural Geology*, **54**, 156–173.

LEWIS, M.M., JACKSON, C.A.L., GAWTHORPE, R.L. & WHIPP, P.S. 2015. Early synrift reservoir development on the flanks of extensional forced folds: a seismic-scale outcrop analog from the Hadahid fault system, Suez rift, Egypt. *American Association of Petroleum Geologists Bulletin*, **99**, 985–1012.

MANDL, G. 2000. *Faulting in Brittle Rocks: an Introduction to the Mechanics of Tectonic Faults*. Springer, Berlin.

MARSH, N., IMBER, J., HOLDSWORTH, R.E., BROCKBANK, P. & RINGROSE, P. 2010. The structural evolution of the Halten Terrace, offshore Mid-Norway: extensional fault growth and strain localisation in a multi-layer brittle–ductile system. *Basin Research*, **22**, 195–214.

MAURIN, J.-C. & NIVIERE, B. 1999. Extensional forced folding and décollement of the pre-rift series along the Rhine graben and their influence on the geometry of the syn-rift sequences. *In*: COSGROVE, J.W. & AMEEN, M.S. (eds) *Forced Folds and Fractures*. Geological Society, London, Special Publications, **169**, 73–86, https://doi.org/10.1144/GSL.SP.2000.169.01.06

McCLAY, K.R. 1990. Extensional fault systems in sedimentary basins: a review of analogue model studies. *Marine and Petroleum Geology*, **7**, 206–233.

McCLAY, K.R. & KHALIL, S.M. 1998. Extensional hard linkages, eastern Gulf of Suez, Egypt. *Geology*, **26**, 563–566.

McCLAY, K.R., NICHOLS, G.J., KHALIL, S.M., DARWISH, M. & BOSWORTH, W. 1998. Extensional tectonics and sedimentation, Eastern Gulf of Suez, Egypt. *In*: PURSER, B.H. & BOSENCE, D.W.J. (eds) *Sedimentation and Tectonics of the Gulf of Aden–Red Sea Rift System*. Chapman & Hall, London, 223–238.

McKENZIE, D.P., DAVIES, D. & MOLNAR, P. 1970. Plate tectonics of the Red Sea and east Africa. *Nature*, **226**, 243–248.

MESHREF, W.M. 1990. Tectonic framework. *In*: SAID, R. (ed.) *The Geology of Egypt*. A.A. Balkema, Rotterdam, 113–155.

MONTENAT, C., OTT D'ESTEVOU, P. *ET AL*. 1988. Tectonic and sedimentary evolution of the Gulf of Suez and the northern western Red Sea. *Tectonophysics*, **153**, 161–177.

MONTENAT, C., OTT D'ESTEVOU, P., JARRIGE, J.-J. & RICHERT, J.P. 1998. Rift development in the Gulf of Suez and the north-western Red Sea: structural aspects and related sedimentary processes. *In*: PURSER, B.H. & BOSENCE, D.W.J. (eds) *Sedimentation and Tectonics of Rift Basins: Red Sea–Gulf of Aden*. Chapman and Hall, London, 97–116.

MORLEY, C.K. & WONGANAN, N. 2000. Normal fault displacement characteristics, with particular reference to synthetic transfer zones, Mae Moh mine, northern Thailand. *Basin Research*, **12**, 307–327.

MOUSTAFA, A.R. 1987. Drape folding in the Baba-Sidri area, eastern side of the Suez rift. *Egyptian Journal of Geology*, **31**, 15–27.

MOUSTAFA, A.R. 1992a. Structural setting of the Sidri-Feiran area, eastern side of the Suez rift. *Middle East Research Center, Ain Shams University, Earth Science Series*, **6**, 44–54.

MOUSTAFA, A.R. 1992b. The Feiran tilted blocks: an example of a synthetic transfer zone, eastern side of Suez rift. *Annales Tectonicæ*, **2**, 193–201.

MOUSTAFA, A.R. 1997. Controls on the development and evolution of transfer zones: the influence of basement structure and sedimentary thickness in the Suez rift and Red Sea. *Journal of Structural Geology*, **19**, 755–768.

MOUSTAFA, A.R. & EL RAEY, A.K. 1993. Structural characteristics of the Suez rift margin. *Geologische Rundschau*, **82**, 101–109.

MOUSTAFA, A.R. & KHALIL, S.M. 2017. Control of extensional transfer zones on syntectonic and post-tectonic sedimentation: implications for hydrocarbon exploration. *Journal of the Geological Society*, **174**, 318–335. https://doi.org/10.1144/jgs2015-138

MURAOKA, H. & KAMATA, H. 1983. Displacement distribution along minor fault traces. *Journal of Structural Geology*, **5**, 483–495.

NICOL, A., WALSH, J., BERRYMAN, K. & NODDER, S. 2005. Growth of a normal fault by the accumulation of slip over millions of years. *Journal of Structural Geology*, **27**, 327–342.

ORSZAG-SPERBER, F., PURSER, B.H., RIOUAL, M. & PLAZIAT, J.-C. 1998. Post Miocene sedimentation and rift dynamics in the southern Gulf of Suez and northern Red Sea. *In*: PURSER, B.H. & BOSENCE, D.W.J. (eds) *Sedimentation and Tectonics of Rift Basins: Red Sea–Gulf of Aden*. Chapman and Hall, London, 427–447.

PASCOE, R., HOOPER, R., STORHAUG, K. & HARPER, H. 1999. Evolution of extensional styles at the southern termination of the Nordland Ridge, Mid-Norway: a response to variations in coupling above Triassic salt. *In*: FLEET, A.J. & BOLDY, S.A.R. (eds) *Petroleum Geology of Northwest Europe: Proceedings of the 5th Conference*. Geological Society, London, 83–90.

PATTON, T.L. 1984. Surface studies of normal-fault geometries in the pre-Miocene stratigraphy, west central Sinai Peninsula. *Egyptian General Petroleum Corporation, 6th Exploration Seminar,1982*. Cairo, 437–452.

PATTON, T.L. & FLETCHER, R.C. 1995. Mathematical block-motion model for deformation of a layer above a buried fault of arbitrary dip and sense of slip. *Journal of Structural Geology*, **17**, 1455–1472.

PATTON, T.L., MOUSTAFA, A.R., NELSON, R.A. & ABDINE, S.A. 1994. Tectonic evolution and structural setting of the Suez rift. *In*: LANDON, S.M. (ed.) *Interior Rift Basins*. American Association of Petroleum Geologists, Memoirs, **59**, 7–55.

PEACOCK, D.C.P. & SANDERSON, D.J. 1991. Displacement, segment linkage and relay ramps in normal fault zones. *Journal of Structural Geology*, **13**, 721–733.

PEACOCK, D.C.P. & SANDERSON, D.J. 1994. Geometry and development of relay ramps in normal fault systems. *American Association of Petroleum Geologists Bulletin*, **78**, 147–165.

PEACOCK, D.C.P. & SANDERSON, D.J. 1996. Effects of propagation rate on displacement variations along faults. *Journal of Structural Geology*, **18**, 311–320.

PEACOCK, D.C.P., KNIPE, R.J. & SANDERSON, D.J. 2000. Glossary of normal faults. *Journal of Structural Geology*, **22**, 291–306.

PURSER, B.H. & PHILOBBOS, E.R. 1993. The sedimentary expressions of rifting in the NW Red Sea, Egypt. *In*: PHILOBBOS, E.R. & PURSER, B.H. (eds) *Geodynamics and Sedimentation of the Red Sea–Gulf of Aden Rift System*. Geological Society of Egypt, Special Publications, **1**, 1–45.

PURSER, B.H., ORSZAG-SPERBER, F. & PLAZIAT, J.C. 1987. Sedimentation et rifting: les series Neogenes de la marge nord-occidental de la Mer Rouge (Egypte). *Notes et Memoires, Compagnie Francaise de Petroles, Paris*, **21**, 111–114.

RICHARDSON, N.J., UNDERHILL, J.R. & LEWIS, G. 2005. The role of evaporite mobility in modifying subsidence patterns during normal fault growth and linkage, Halten terrace, mid-Norway. *Basin Research*, **17**, 203–223.

RYKKELID, E. & FOSSEN, H. 2003. Layer rotation around vertical fault overlap zones: observations from seismic data, field examples, and physical experiments. *Marine and Petroleum Geology*, **19**, 181–192.

SAID, R. (ed.) 1990. *The Geology of Egypt*. A.A. Balkema, Rotterdam.

SALTZER, S.D. & POLLARD, D.D. 1992. Distinct element modeling of structures formed in sedimentary overburden by extensional reactivation of basement normal faults. *Tectonics*, **11**, 165–174.

SCHLISCHE, R.W. 1993. Anatomy and evolution of the Triassic–Jurassic continental rift system, eastern North America. *Tectonics*, **12**, 1026–1042.

SCHLISCHE, R.W. 1995. Geometry and origin of fault-related folds in extensional settings. *American Association of Petroleum Geologists Bulletin*, **79**, 1661–1678.

SCHLISCHE, R.W., WITHJACK, M.O. & EISENSTADT, G. 2002. An experimental study of the secondary deformation produced by oblique-slip normal faulting. *American Association of Petroleum Geologists Bulletin*, **86**, 885–906.

SCHÖPFER, M.P.J., CHILDS, C., WALSH, J.J., MANZOCCHI, T. & KOYI, H. 2007. Geometrical analysis of the refraction and segmentation of normal faults in periodically layered sequences. *Journal of Structural Geology*, **29**, 318–335.

SCHULTZ, R.A. 2000. Understanding the process of faulting: selected challenges and opportunities at the edge of the 21st century. *Journal of Structural Geology*, **21**, 985–993.

SHARP, I.R., GAWTHORPE, R.L., UNDERHILL, J.R. & GUPTA, S. 2000. Fault propagation folding in extensional settings: examples of structural style and synrift sedimentary response from the Suez rift, Sinai,

Egypt. *Geological Society of America Bulletin*, **112**, 1877–1899.

STEARNS, D.W. 1978. Faulting and forced folding in the Rocky Mountain foreland. *In*: MATTHEWS, V., III (ed.) *Laramide Folding Associated with Basement Block Faulting in the Western United States*. Geological Society of America, Memoirs, **151**, 1–38.

STECKLER, M.S., BERTHELOT, F., LYBERIS, N. & LE PICHON, X. 1988. Subsidence in the Gulf of Suez: implications for rifting and plate kinematics. *Tectonophysics*, **153**, 249–270.

STERN, R.J. 1994. Arc assembly and continental collision in the Neoproterozoic East African orogen: implications for the consolidation of Gondwanaland. *Annual Reviews of Earth and Planetary Sciences*, **22**, 319–351.

STOESER, D.B. & CAMP, V.E. 1985. Pan African microplate accretion of the Arabian shield. *Geological Society of America Bulletin*, **96**, 817–826.

SUPPE, J. 1985. *Principles of Structural Geology*. Prentice Hall, Englewood Cliffs, NJ.

TRUDGILL, B. & CARTWRIGHT, J. 1994. Relay-ramp forms and normal-fault linkages. *Geological Society of America Bulletin*, **106**, 1143–1157.

TVEDT, A.B.M., ROTEVATN, A., JACKSON, C.A.-L., FOSSEN, H. & GAWTHORPE, R.L. 2013. Growth of normal faults in multilayer sequences: a 3D seismic case study from the Egersund Basin, Norwegian North Sea. *Journal of Structural Geology*, **55**, 1–20.

TWISS, R.J. & MOORES, E.M. 1992. *Structural Geology*. Freeman, New York.

VENDEVILLE, B.C. 1988. Scale-models of basement induced-extension. *Comptes Rendus de l'Académie des Sciences*, **307**, 1013–1019.

WALSH, J.J. & WATTERSON, J. 1987. Distributions of cumulative displacement and seismic slip on a single normal fault surface. *Journal of Structural Geology*, **9**, 1039–1046.

WALSH, J.J. & WATTERSON, J. 1988. Analysis of the relationship between displacements and dimensions of faults. *Journal of Structural Geology*, **10**, 239–247.

WALSH, J.J. & WATTERSON, J. 1989. Displacement gradients on fault surfaces. *Journal of Structural Geology*, **11**, 307–316.

WALSH, J.J., NICOL, A. & CHILDS, C. 2002. An alternative model for the growth of faults. *Journal of Structural Geology*, **24**, 1669–1675.

WALSH, J.J., BAILEY, W.R., CHILDS, C., NICOL, A. & BONSON, C.G. 2003. Formation of segmented normal faults: a 3-D perspective. *Journal of Structural Geology*, **25**, 1251–1262.

WILKINS, S.J. & GROSS, M.R. 2002. Normal fault growth in layered rocks at Split Mountain, Utah: influence of mechanical stratigraphy on dip linkage, fault restriction and fault scaling. *Journal of Structural Geology*, **24**, 1413–1429.

WILLEMSE, E.J.M., POLLARD, D.D. & AYDIN, A. 1996. Three-dimensional analyses of slip distributions on normal fault arrays with consequences for fault scaling. *Journal of Structural Geology*, **18**, 295–309.

WILSON, P., GAWTHORPE, R.L., HODGETTS, D., RARITY, F. & SHARP, I.R. 2009. Geometry and architecture of faults in a syn-rift normal fault array: the Nukhul half-graben, Suez rift, Egypt. *Journal of Structural Geology*, **31**, 759–775.

WILSON, P., ELLIOTT, G.M., GAWTHORPE, R.L., JACKSON, C.A.-L., MICHELSEN, L. & SHARP, I.R. 2013. Geometry and segmentation of an evaporite-detached normal fault array: 3D seismic analysis of the southern Bremstein fault complex, offshore mid-Norway. *Journal of Structural Geology*, **51**, 74–91.

WITHJACK, M.O. & SCHLISCHE, R.W. 2006. Geometric and experimental models of extensional fault-bend folds. *In*: BUITER, S.J.H. & SCHREURS, G. (eds) *Analogue and Numerical Modelling of Crustal-Scale Processes*. Geological Society, London, Special Publications, **253**, 285–305, https://doi.org/10.1144/GSL.SP.2006.253.01.15

WITHJACK, M.O., OLSEN, J. & PETERSON, E. 1990. Experimental models of extensional forced folds. *American Association of Petroleum Geologists Bulletin*, **74**, 1038–1054.

YOUNES, A. & MCCLAY, K.R. 2001. Role of basement fabric on rift architecture: Gulf of Suez–Red Sea, Egypt. *American Association of Petroleum Geologists Bulletin*, **86**, 1003–1026.

YOUNG, M.J., GAWTHORPE, R.L. & SHARP, I.R. 2000. Sedimentology and sequence stratigraphy of a transfer zone coarse-grained delta, Miocene Suez Rift, Egypt. *Sedimentology*, **47**, 1081–1104.

YOUNG, M.J., GAWTHORPE, R.L. & SHARP, I. 2003. Normal fault growth and early syn-rift sedimentology and sequence stratigraphy: Thal Fault, Suez Rift, Egypt. *Basin Research*, **15**, 479–502.

YOUSSEF, M.I. 1957. Upper Cretaceous rocks in Kosseir area. *Bulletin de l'Institute du Desert d' Egypt*, **7**, 35–53.

Crustal structure of the conjugate Equatorial Atlantic Margins, derived by gravity anomaly inversion

NICK J. KUSZNIR[1,2], ALAN M. ROBERTS[2]* & ANDREW D. ALVEY[2]

[1]Department of Earth & Ocean Sciences, University of Liverpool, Liverpool L69 3BX, UK

[2]Badley Geoscience Ltd, North Beck House, North Beck Lane, Spilsby, Lincolnshire, PE23 5NB, UK

A.M.R., 0000-0003-4839-0741
*Correspondence: alan@badleys.co.uk

Abstract: The crustal structure of the Equatorial Atlantic conjugate margins (South America and West Africa) has been investigated using 3D gravity anomaly inversion, which allows for (1) the elevated geothermal gradient of the lithosphere following rifting and break-up and (2) magmatic addition to the crust during rifting and break-up. It is therefore particularly suitable for the analysis of rifted margins and their associated ocean basins. Maps of crustal thickness and conjugate-margin stretching, derived from gravity anomaly inversion, are used to illustrate how the Equatorial Atlantic opened as a set of stepped rift-transform segments, rather than as a simple orthogonal rifted margin. The influence of the transform faults and associated oceanic fracture zones is particularly clear when the results of the gravity anomaly inversion are combined with a shaded-relief display of the free-air gravity anomaly. A set of crustal cross-sections has been extracted from the results of the gravity inversion along both equatorial margins. These illustrate the crustal structure of both rifted-margin segments and transform-margin segments. The maps and cross-sections are used to delineate crustal type on the margins as (1) inboard, entirely continental, (2) outboard, entirely oceanic and (3) the ocean–continent transition in between where mixed continental and magmatic crust is likely to be present. For a given parameterization of melt generation the amount of magmatic addition within the ocean–continent transition is predicted by the gravity inversion. One of the strengths of the gravity-inversion technique is that these predictions can be made in the absence of any other directly acquired data. On both margins anomalously thick crust is resolved close to a number of oceanic fracture zones. On the South American margin we believe that this thick crust is probably the result of post-break-up magmatism within what was originally normal-thickness oceanic crust. On the West African margin, however, three possible origins are discussed: (1) continental crust extended oceanwards along the fracture zones; (2) oceanic crust magmatically thickened at the fracture zones; and (3) oceanic crust thickened by transpression along the fracture zones. Gravity inversion alone cannot discriminate between these possibilities. The cross-sections also show that, while 'normal thickness' oceanic crust (c. 7 km) predominates regionally, local areas of thinner (c. 5 km) and thicker (c. 10 km) oceanic crust are also present along both margins. Finally, using maps of crustal thickness and thinning factor as input to plate reconstructions, the regional palaeogeography of the Equatorial Atlantic during and after break-up is displayed at 10 Ma increments.

Supplementary material: Detailed illustrations of the crustal-thickness mapping, the crustal cross-sections and the plate reconstructions are available at: https://doi.org/10.6084/m9.figshare.c.4031266.v1

Gravity anomaly inversion is an excellent entry point into the analysis of rifted continental margins at the regional scale. Much, often all, of the required input information is available in the public domain, enabling geologically consistent analysis of large areas to be performed. Other techniques may provide more detailed information on a more local scale (e.g. Fletcher et al. 2013; Roberts et al. 2013; Cowie et al. 2015, 2016), but at the regional scale gravity anomaly inversion is the best starting point, not least because full data coverage is available for oceanic areas as well as for the continental margins themselves.

In this paper we take advantage of complete data coverage across the Atlantic to investigate the crustal structure of the conjugate margins of equatorial South America and West Africa, together with their linkage across the Atlantic ocean. The results were originally compiled as input to an industry workshop discussing the exploration potential of the Equatorial Atlantic (PESGB 2016). Significant hydrocarbon discoveries exist on both margins but their potential is not yet thought to be exhausted. Analysis of crustal basement structure, with its associated implications for basement heat flow (e.g. Cowie & Kusznir 2012a), is likely to be an

From: McClay, K. R. & Hammerstein, J. A. (eds) 2020. *Passive Margins: Tectonics, Sedimentation and Magmatism.* Geological Society, London, Special Publications, **476**, 83–107.
First published online March 19, 2018, https://doi.org/10.1144/SP476.5
© 2018 The Author(s). This is an Open Access article distributed under the terms of the Creative Commons Attribution License (http://creativecommons.org/licenses/by/3.0/). Published by The Geological Society of London.
Publishing disclaimer: www.geolsoc.org.uk/pub_ethics

important part of any future exploration screening strategy in these areas.

The gravity inversion technique applied to derive crustal structure has been described in several previous publications (e.g. Greenhalgh & Kusznir 2007; Alvey et al. 2008; Chappell & Kusznir 2008; Cowie & Kusznir 2012b; Roberts et al. 2013; Cowie et al. 2015) and so is recapped here only briefly. In Roberts et al. (2013) and Cowie et al. (2015, 2016) gravity inversion has been used alongside other techniques, such as subsidence analysis and analysis of residual depth anomalies, to provide a multifaceted view of rifted margin structure. Here, however, we wish to focus on what gravity inversion alone can provide, while recognizing that its use alongside other techniques can, of course, provide yet more information.

3D gravity inversion method

The gravity inversion method and workflow are summarized in Figure 1a (adapted from Alvey 2010; Roberts et al. 2013). The three principal sets of input data are maps/grids of:

- satellite free-air gravity anomaly data (Sandwell & Smith 2009 and subsequent updates at http://topex.ucsd.edu/WWW_html/mar_grav.html);
- bathymetric/topographic data (Smith & Sandwell 1997 and subsequent updates at http://topex.ucsd.edu/WWW_html/mar_topo.html);
- sediment thickness data (e.g. Laske & Masters 1997 and subsequent updates at http://igppweb.ucsd.edu/~gabi/sediment.html; and Divins 2003 and subsequent updates at http://www.ngdc.noaa.gov/mgg/sedthick/sedthick.html).

Each of these cited datasets is freely and publicly available. Public-domain sediment thickness information can be replaced by proprietary information for a particular area, should such information exist. Use of such proprietary sediment thickness information will almost certainly improve the reliability of the results for a study with local focus. For

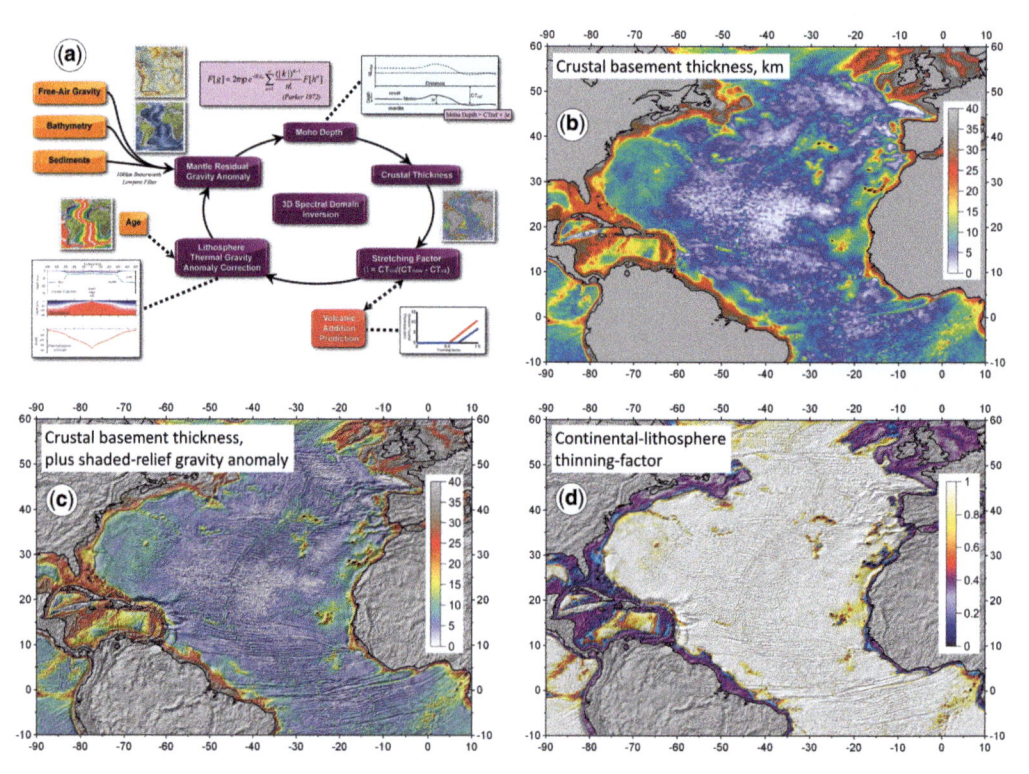

Fig. 1. (a) Schematic outline of the gravity inversion methodology to determine Moho depth, crustal basement thickness and lithosphere thinning factor, using gravity anomaly inversion incorporating a lithosphere thermal correction and decompression-melt prediction. Adapted from Alvey (2010) and Roberts et al. (2013). (b) Example map of total crustal basement thickness (continental and oceanic) produced by gravity inversion of the offshore central Atlantic region. Scale in kilometres. (c) Map of total crustal basement thickness (as in b) overlain by a display of the shaded-relief free-air gravity anomaly. The overlay helps to delineate major tectonic features within the results. Scale in kilometres. (d) Example map of continental-lithosphere thinning factor $(1 - 1/\beta)$ for the central Atlantic, overlain by the shaded-relief free-air gravity anomaly.

regional-scale analysis, however, and for coverage into the oceans, the global public sources are generally required.

The principal output from the gravity inversion comprises maps of:

- present-day depth to Moho, the primary output on which all other results are based (Fig. 1a);
- total crustal basement thickness (base sediment to Moho, no distinction between continental and oceanic crust);
- residual thickness of the continental crust (total crustal basement thickness minus predicted volcanic addition, see below); and
- lithosphere stretching factor (β) and thinning factor (γ), where $\gamma = 1 - 1/\beta$.

Where public-domain sediment thickness information can be calibrated against estimates of sediment thickness from good-quality seismic-reflection data we find, both in the Equatorial Atlantic area and elsewhere, that the public-domain data tend to underestimate the sediment thickness. The consequence of this in a regional gravity inversion study, as described here, is that estimates of crustal thickness will be a probable maximum and estimates of stretching/thinning will be a probable minimum.

Key to the success of the gravity inversion method employed in this paper is a correction for the gravity anomaly associated with the elevated geotherm within both continental-margin and oceanic lithosphere which results from rifting/break-up of the margin and the formation of an ocean basin. The lithosphere thermal gravity anomaly is negative and very large (c. -350 mgal at a young ocean ridge). Failure to include a correction for the lithosphere thermal gravity anomaly leads to a substantial over-estimate of Moho depth and crustal basement thickness and an under-estimate of continental-lithosphere thinning. The magnitude of the gravity anomaly decreases with time as the thermal anomaly cools following rifting/break-up but for a mid-Cretaceous break-up age, as in the case of the Equatorial Atlantic, it is essential to include the lithosphere thermal gravity anomaly correction. The methodology by which the thermal gravity anomaly correction is included in the gravity inversion is described in detail in Chappell & Kusznir (2008, see also Fig. 1a).

The gravity inversion method that we use determines Moho depth and crustal basement thickness but cannot itself distinguish between continental and oceanic crust. In order to differentiate oceanic crust from continental-basement crust we use a parameterization of decompression melting. Following McKenzie & Bickle (1988) and White & McKenzie (1989) it is assumed that decompression melting of the lithosphere occurs at high continental-lithosphere thinning and stretching factors, resulting in magmatic addition which contributes to the total thickness of the crust (Chappell & Kusznir 2008, fig. 3, see also Fig. 1a). As stretching proceeds beyond a given critical value the original continental crust will continue to stretch and thin, but the total thickness of the crust will be buffered by the addition of new magmatic material. Thus, following magmatic addition, the location of the base of the crust (Moho) is controlled both by the magnitude of stretching/thinning of the continental crust and by the amount of newly added magmatic material.

Using the central Atlantic as an illustration, Figure 1b–d shows how the results of the gravity inversion are compiled to reveal geological information in map form. The specific parameters constraining the inversion are not described at this point, but will be returned to later in the paper when detail of the equatorial region is discussed. Figure 1b shows a map of crustal thickness (Moho to base sediment) for the central Atlantic (offshore regions only). There is no distinction in this map between areas of predicted continental crust and oceanic crust. Thick crust is seen at the continental margins and within internal basins beyond the margins. Thin crust is seen in the centre of the Atlantic.

While Figure 1b illustrates the basic result of the gravity inversion, the information conveyed by this map can be greatly enhanced by adding an overlay of the shaded-relief free-air gravity anomaly (Fig. 1c), which is the gravity anomaly data used as input to the inversion. This additional information allows us to pick out features such as the mid-ocean ridge, transform faults and tectonic flowlines within the oceanic area and immediately aids the understanding of margin conjugacy. Within the paper all future results in map form are shown with an overlay of shaded-relief free-air gravity anomaly.

Figure 1d shows the map of continental-lithosphere thinning factor which results from the map of crustal thickness once the correction for volcanic addition is taken into account. A thinning factor value of 0 is no continental thinning. A value of 1 is complete thinning and removal of the continental crust and continental lithosphere. As expected, gradations in thinning factor occur across the continental margins on both sides of the Atlantic, while the large white area in the centre of the map corresponds to a thinning factor of 1 and the presence of oceanic crust. There are other maps from the gravity inversion which can be displayed, some of which are shown later, but the specific objective of Figure 1 is to introduce the technique. The technique itself has acquired the acronym of OCTek Gravity Inversion, derived from its focus on the tectonics of the ocean–continent transition (OCT).

A global compilation of crustal thickness

Figure 1, in common with all previous published maps derived from the OCTek gravity inversion technique (see references above), shows results only in offshore areas. This is not a technical restriction of the technique itself, but has come about because public-domain sediment-thickness information in offshore areas tends to be of better quality and more reliable than it is onshore. In particular the Divins (2003) data is higher resolution than the older Laske & Masters (1997) data, but it is only available offshore, whereas the Laske & Masters data has global offshore and onshore coverage at a coarse resolution of 1 × 1°. If the coarseness of this coverage can be accepted then the gravity inversion technique can be extended onshore.

Figure 2 shows, in a series of four 90° rotations, a global map of crustal basement thickness, derived from the OCTek gravity inversion technique. We believe that the first-order results, in terms of identifying areas of thick crust, thin crust and transition in between, will be correct in both onshore and offshore areas. We would advise caution, however, in taking the onshore crustal-thickness values too literally against the scale provided because of (1) the variability in the quality of the underlying sediment-thickness information and (2) uncertainties in parameterizing both break-up/rift ages (where appropriate) and magmatic addition from decompression melting.

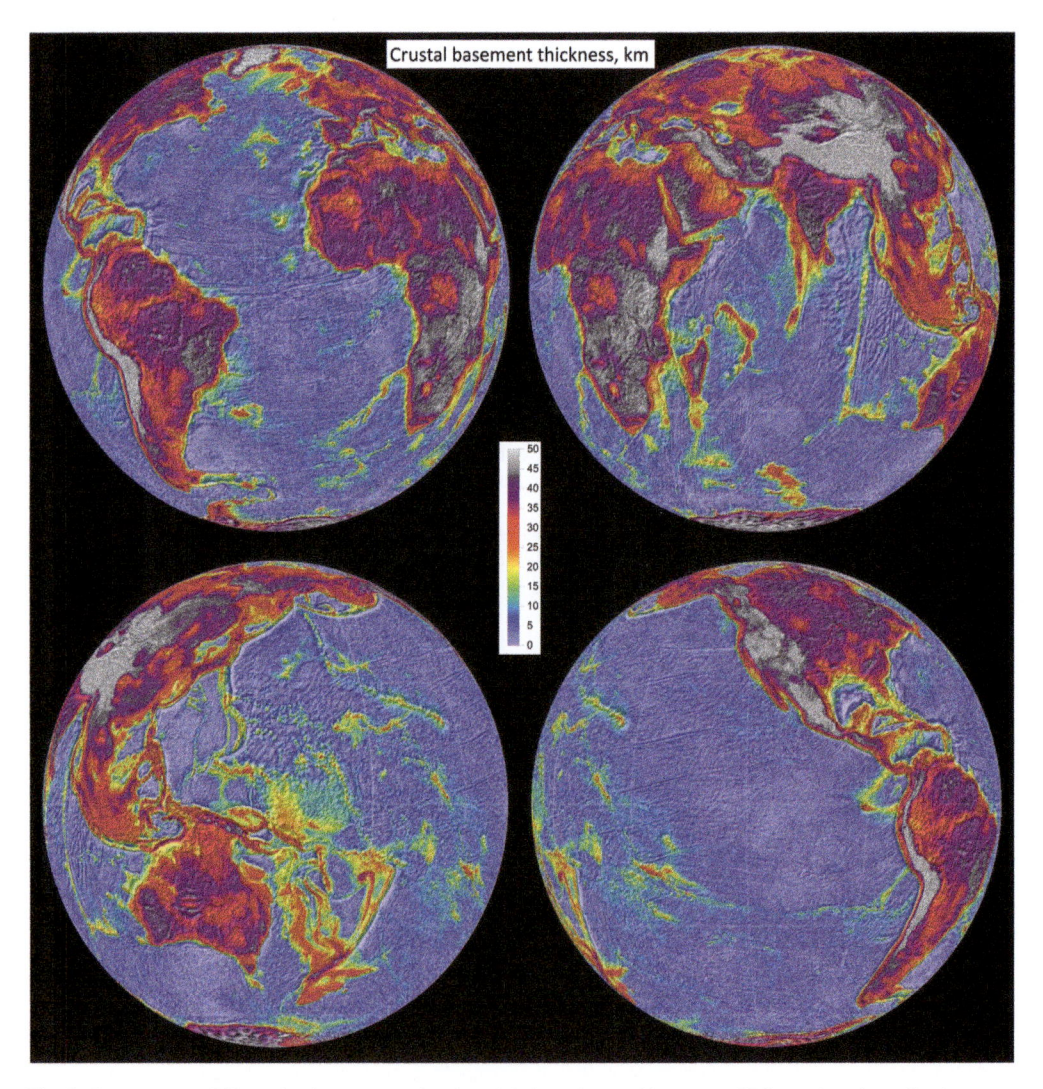

Fig. 2. Four maps, at 90° rotation increments, showing global, total crustal basement thickness (continental and oceanic) derived by OCTek gravity inversion. Scale in kilometres. Both onshore and offshore regions are included for the first time in such results.

The importance of being able to map crustal thickness at any location and overlay the results with tectonic information from the shaded-relief gravity anomaly is that this provides us with a window into areas in which little or no direct geological information has been acquired, specifically offshore deep-water areas with no seismic reflection data, seismic refraction data or drilling information. The thickness of continental crustal basement controls the crustal radiogenic heat input into a basin (e.g. Cowie & Kusznir 2012a). An understanding of crustal-thickness distribution and crustal composition therefore provides information which is essential for the prediction of basement heat flow, an important input to petroleum-systems analysis.

In the rest of the paper we will focus on a smaller area within Figure 2 but a large area in its own right, the Equatorial Atlantic.

Crustal thickness across the Equatorial Atlantic

Figure 3 shows a map of crustal basement thickness for the Equatorial Atlantic, extending north into the Central Atlantic and south into the South Atlantic. Crustal thickness for both onshore and offshore areas is included and the map itself is an extraction from Figure 2. There is no differentiation between continental and oceanic crust in this map, it is simply predicted crustal basement thickness. The gravity inversion results shown in Figure 3 are tuned to a break-up age of 110 Ma and are applicable to the Equatorial Atlantic. Other more specific input data and parameters are described below in the context of Figures 4 and 5.

In the onshore areas of South America and Central/North Africa, Figure 3 picks out areas of localized thinner crust (20–30 km) within the

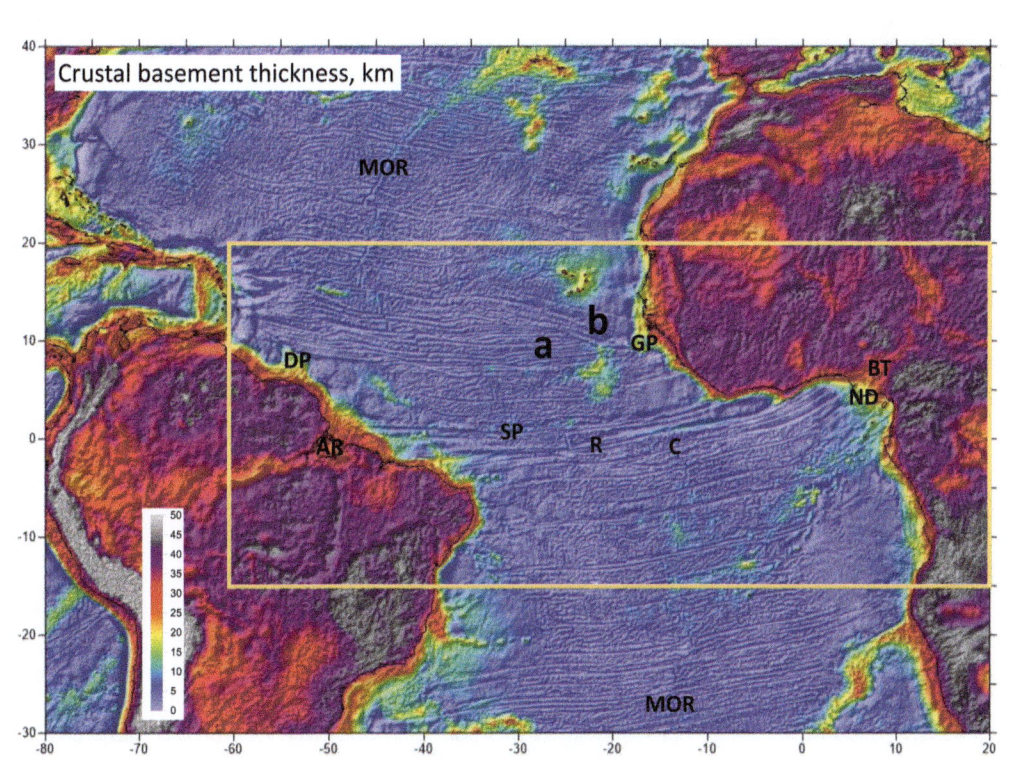

Fig. 3. Map of total crustal basement thickness (continental and oceanic) from gravity inversion of the Equatorial Atlantic, overlain by a display of the shaded-relief free-air gravity anomaly. Scale in kilometres. For regional clarity the map also extends north into the Central Atlantic and south into the South Atlantic. The focus of the main equatorial study is shown by the outline box. **a** identifies an area of prominent c. east–west oceanic fracture-zone flowlines related to Cretaceous opening of the Equatorial Atlantic. **b** identifies an area of clockwise-oblique fracture-zone flowlines related to the earlier Jurassic opening of the Central Atlantic. MOR, mid-ocean-ridge; DP, Demerara Plateau; GP, Guinea Plateau; AR, Amazon Rift; BT, Benue Trough; ND, Niger Delta; SP, St Paul's fracture zone; R, Romanche fracture zone; C, Chain fracture zone. (A higher-resolution display of the same map is available within Supplementary Material, Figure S1.)

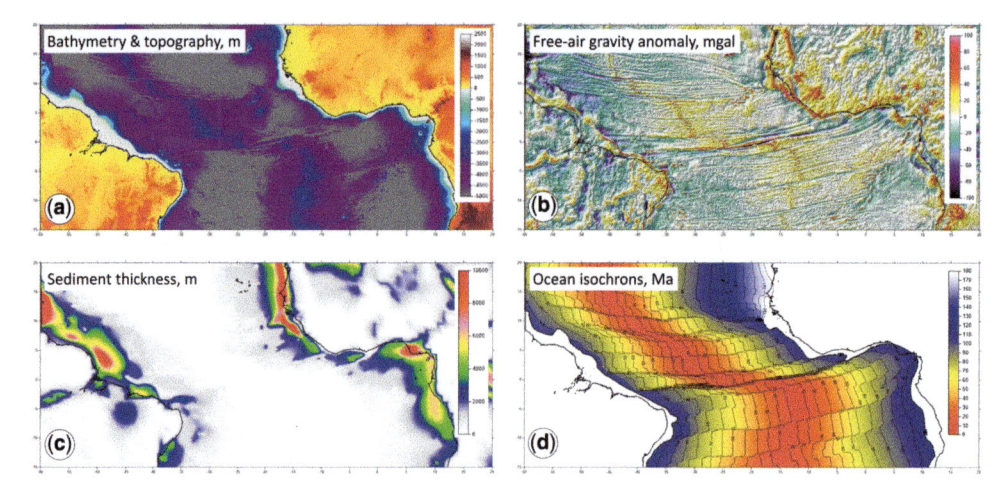

Fig. 4. Primary input data for gravity inversion of the Equatorial Atlantic. (**a**) Bathymetry and topography, scale in metres (Smith & Sandwell 1997 and updates). (**b**) Satellite free-air gravity anomaly, scale in mgal (Sandwell & Smith 2009 and updates), overlain by a shaded-relief display of itself. (**c**) Sediment thickness, scale in metres (Divins 2003 offshore; Laske & Masters 1997 onshore). (**d**) Ocean isochrons, scale in millions of years (Müller *et al.* 2008), used to determine the age of the lithosphere thermal-gravity anomaly in oceanic areas.

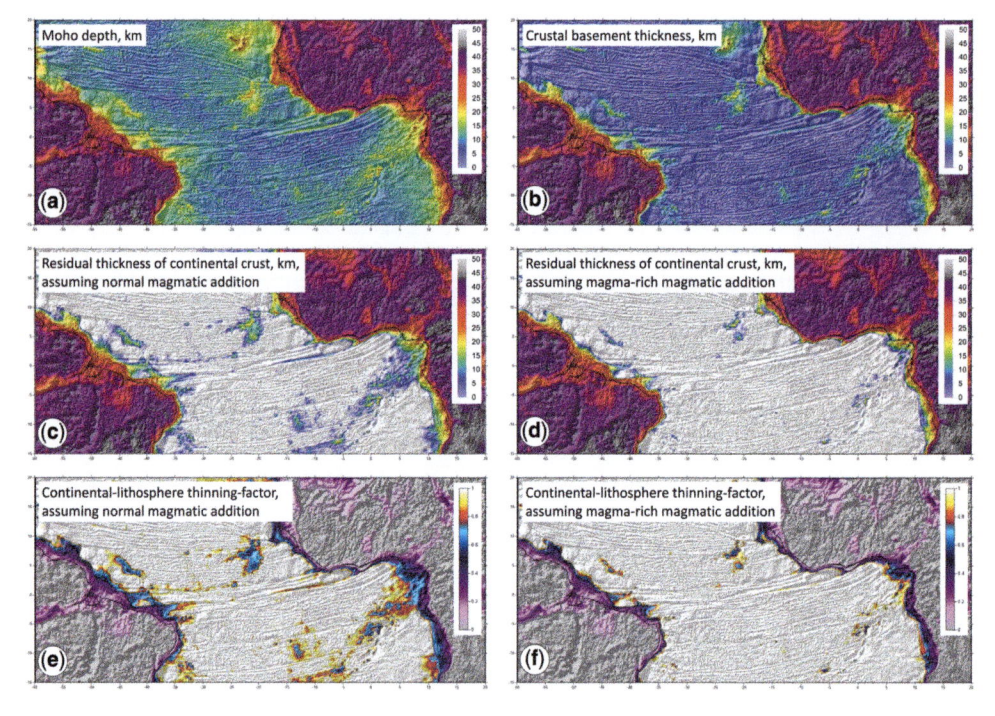

Fig. 5. Results from two gravity inversion models for the Equatorial Atlantic, one assuming 'normal' magmatic addition (decompression melting), the other assuming 'magma-rich' magmatic addition. Both models use a break-up age of 110 Ma and a Reference Moho Depth of 37.5 km. (**a**) Moho depth, common to both models. (**b**) Total crustal basement thickness (continental and oceanic), common to both models. (**c**) Residual thickness of the continental crust, assuming normal magmatic addition. (**d**) Residual thickness of the continental crust, assuming magma-rich magmatic addition. (**e**) Continental-lithosphere thinning factor $(1 - 1/\beta)$, assuming normal magmatic addition. Thinning factor 1 defines oceanic crust. (**f**) Continental-lithosphere thinning factor, assuming magma-rich magmatic addition. Scale in kilometres for Moho depth and crustal basement thickness.

broader cratonic areas where the crust is 40 km or more in thickness. The thinner areas correspond to known onshore rifts, perhaps the most obvious of which are the Amazon rift and the Benue Trough (Fig. 3). Some of these onshore rifts in the equatorial region are older than the Cretaceous break-up of the Atlantic margins and can be matched across the two continents by plate restoration (see later in this paper). We reiterate the point made above not to take the onshore crustal-thickness values too literally against the scale provided, but rather consider the first-order results in terms of mapped crustal thickness variations.

At the scale of Figure 3 the gradation from thick crust onshore to thin crust in the offshore oceanic areas is rapid. These are the areas of the stretched and thinned continental margins, details of which are investigated in the subsequent discussion which follows.

Within the broad oceanic area itself the crust has a thickness of $c.$ 7 km (normal thickness oceanic crust), with the exception of localized volcanic seamounts and seamount chains, which map as areas of thicker crust (10–20 km). The shaded-relief gravity adds detail to the oceanic areas which would not otherwise be apparent. The Atlantic mid-ocean ridge is clearly picked out. So too is its far-from-linear structure; it is offset by many small and large fracture zones/transform faults. This is particularly apparent across the prominent fracture zones of the equatorial area.

The fracture zones also allow us to identify the flowlines which define plate-separation direction over time. In the area labelled (a) the broadly east–west flowlines of separation in the Cretaceous and younger (110 Ma onwards) oceanic crust can be seen as a strong imprint across most of the ocean width (see also isochrons in Fig. 4). Just to the north of (a), however, in the area labelled (b), the flowlines strike clockwise of east–west with a clear angular discordance where the two sets meet. The flowlines at (b) relate to the earlier Jurassic opening of the Central Atlantic between North Africa and North America (see isochrons in Fig. 4). This initial Jurassic opening of the Central Atlantic occurred in a direction $c.$ 10° clockwise to the younger Cretaceous opening further south. The adjustment within the older Jurassic oceanic segment to the younger opening direction can be seen in the curve of the flowlines west of (b).

Conditioning the Equatorial Atlantic gravity inversion

Figures 4 and 5 illustrate the data input and model parameterization used for gravity inversion of the Equatorial Atlantic. Figure 4a shows the bathymetry and topography data for the equatorial region (Smith & Sandwell 1997 & updates). Figure 4b shows the free-air gravity anomaly data (Sandwell & Smith 2009 & updates) overlain by a shaded-relief display of itself. Figure 4c shows the sediment thickness data used in the gravity inversion, which is a merge of Divins (2003) offshore and Laske & Masters (1997) onshore. As expected, the thickest sediments are concentrated along the continental margins. As mentioned above, this compilation of public-domain data is likely to provide a minimum estimate of sediment thickness at the regional scale. Figure 4d shows the ocean isochrons (Müller et al. 2008) for the Equatorial Atlantic.

An area for which the public-domain sediment-thickness information (Fig. 4c) is known to be an under-estimate is the Niger Delta (Fig. 3). Here the gravity inversion shows the delta lying on thinned continental crust (Figs 3, 5 & 8a), in an area where the basement is known to be oceanic (see the ocean isochrons in Fig. 4d).

The ocean isochrons are required in order for the gravity inversion to work correctly in areas of known oceanic crust. In the Equatorial and South Atlantic the oceanic crust spans the age range 110 Ma to present. To the north, in the Central Atlantic, the age range is 180 Ma to present. The age of inception of the lithosphere thermal-gravity anomaly in the oceanic areas varies with the age of the ocean crust. This results in a spatially varying lithosphere thermal-gravity anomaly at the present day: very large at the mid-ocean ridge ($c.$ 350 mgal) and much lower ($c.$ 50 mgal or less) at the oldest oceanic lithosphere and at the rifted margins. Ocean isochrons are used to give the thermal re-equilibration time (cooling time) of the lithosphere thermal anomaly within the oceanic areas and thus produce gravity inversion results (e.g. Moho depth, crustal thickness, Fig. 3) that fully compensate for the underlying oceanic thermal structure.

In order to condition the lithosphere thermal-gravity anomaly across the rifted continental margins of West Africa and South America, a fixed break-up age is used for the lithosphere thermal re-equilibration time within the continental region and for the region of uncertain crustal affinity across the OCT. In the Equatorial Atlantic region (Fig. 4) this age is 110 Ma. Further to the north in the Central Atlantic (Fig. 3) it is 170 Ma.

Figure 5 shows a set of results from two gravity inversion models of the Equatorial Atlantic. Both models have used all of the input data in Figure 4, a break-up age of 110 Ma and a Reference Moho Depth of 37.5 km. Reference Moho Depth is a geophysical/geodetic parameter that represents the reference datum to which Moho relief determined by gravity inversion is applied in order to determine Moho depth. It is controlled by the long-wavelength component of the Earth's gravity field which results from deep (sublithosphere) mantle processes and structure (see Cowie & Kusznir 2012b; Cowie et al. 2015 for more detailed discussions).

Where the two inversion models of Fig. 5 differ is in their parameterization of magmatic addition. The model results shown in Figure 5c and e correspond to decompression melting assuming 'normal' magmatic addition, in which melting begins at thinning factor 0.7 and produces 7 km of oceanic crust when thinning factor reaches 1. This is the parameterization of melting for normal temperature asthenosphere (McKenzie & Bickle 1988; White & McKenzie 1989) first applied by Chappell & Kusznir (2008). The results shown in Figure 5d and f use decompression melting assuming 'magma-rich' magmatic addition, in which melting begins at thinning factor 0.5 and produces 10 km of oceanic crust when thinning factor reaches 1 (Fig. 1a, Chappell & Kusznir 2008 fig. 3; Roberts et al. 2013 fig. 1b). The gravity inversion method can parameterize any combination of critical thinning factor and magmatic addition, but here we concentrate on the base-case 'normal' model and an enhanced 'magma-rich' case.

The results of the gravity inversion for both Moho depth and total crustal basement thickness are largely insensitive to the parameterization of magmatic addition and thus the maps in Figure 5a and b are common to both models. Residual thickness of the continental crustal basement (Fig. 5c, d) and lithosphere thinning factor (Fig. 5e, f) are, however, both sensitive to magmatic addition and thus Figure 5c and e are different to 5d and f.

Maps of the residual thickness of the continental crust (Fig. 5c, d) are produced by subtracting the magmatic addition predicted within the gravity inversion from the calculated total crustal thickness. In oceanic areas this therefore results in a prediction of zero thickness for remaining continental crust (white in Fig. 5c, d). When comparing Figure 5c and d the difference is most apparent within the oceanic area, where the allowance for greater magmatic addition in Figure 5d 'cleans up' some of the areas of oceanic crust over-thickened by seamounts. There is, however, also a more subtle difference (at the scale of the current maps) across the area of the outer continental margin where the prediction of greater magmatic addition in Figure 5d results in thinner continental crust and a shift inboard of the continent–ocean boundary (COB; edge of oceanic white zone). This can be made clearer when the maps for a particular area are enlarged.

Maps of thinning factor $(1 - 1/\beta$, Fig. 5e, f) are produced by comparing the thickness of the residual continental crust (Fig. 5c, d) with the assumed initial thickness of the continental crust, which in the equatorial area is 37.5 km. Where the predicted thickness of continental crust is 37.5 km or greater the thinning factor is 0. Where the thickness of continental crust is zero (oceanic), the thinning factor is 1. Figures 5e and f differ from each other in the same areas and for the same reasons as Figures 5c and d, with oceanic areas again being displayed as white.

In the absence of any information to the contrary, 'normal' magmatic addition should generally be considered the base case in any particular area. The most straightforward calibration of magmatic addition can be obtained using Moho depth from deep, long-offset seismic reflection data which extends on to known oceanic crust, in which case magmatic addition in the gravity inversion can be calibrated by the thickness of the oldest oceanic crust. Under 'normal' conditions this will be c. 7 km and as calibration examples later in the paper show, this is probably the general case for the Equatorial Atlantic. The North American side of the Central Atlantic to the immediate north (Fig. 3), however, is known to show characteristics of a 'magma-rich' margin (Eldholm et al. 2000) and thus 'normal' and 'magma-rich' solutions are presented in Figure 5.

Following from the discussion of magmatic addition, Figure 6 (adapted from Manatschal et al. 2015) reminds us that rifted continental margins should not simply be considered as the two magmatic cases covered by the models in Figure 5. At one end of the 'magmatic scale' lie 'magma-poor' margins with little or no magmatic addition at the time of break-up (Fig. 6a). Magma-poor margins result in mantle exhumation rather than the formation of new oceanic crust. At the other end of the scale lie 'magma-rich' margins, with enhanced volcanic addition producing extrusive lavas and seaward-dipping reflectors, in addition to thick (c. 10 km) oceanic crust (Fig. 6b). In between these two end members lies a complete range which may be encountered in natural examples. It is within this range that margins approaching the 'normal' case will be the most common, but even within relatively local areas some variation in magmatic addition can be expected.

Crustal structure of the South American equatorial margin

Figure 7a shows a map of total crustal basement thickness for the South American equatorial margin, enlarged from Figures 3 and 5a. This result is independent of magmatic addition, because it does not differentiate between continental and oceanic crust. The regional strike of both the Atlantic ocean margin and the South American coast in this area is NW–SE, at c. 45° to the opening direction of the Atlantic as defined by the east–west-aligned fracture zones. The coast and margin are not orthogonal to the opening direction because this is not a simple, linear rifted margin. Rather the margin is subdivided into a

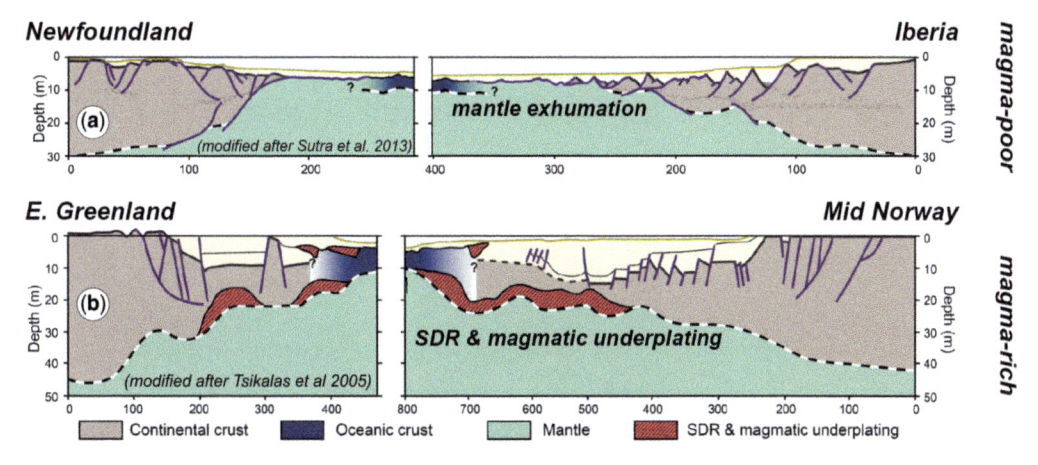

Fig. 6. Example cross-sections of end-member margin types showing: (**a**) the conjugate magma-poor margins of Newfoundland and Iberia; (**b**) the conjugate magma-rich margins of East Greenland and Norway. (Adapted from Manatschal *et al.* 2015 fig. 2 by Gianreto Manatschal and used with permission. Original sources Tsikalas *et al.* 2005; Sutra *et al.* 2013.)

number of relatively short rift segments (margin strike north–south) and transform segments (margin strike east–west) by the numerous oceanic fracture zones, which give the margin a stepped, oblique regional geometry. Figure 3 shows that the fracture zones are easily correlatable across to the African margin, which has a similar stepped rift/transform geometry (Fig. 8a). The well-known St Paul's, Romanche and Chain fracture zones are each identified in Figures 3, 7a and 8a.

Figure 7b comprises 15 crustal-scale cross-sections extracted from the results of the gravity inversion (Fig. 7a). The cross-sections are constructed from four key geological surfaces, two of which are input to the gravity inversion and two of which are results. The input surfaces are seabed/bathymetry (plus topography in onshore areas) and the base sediment (Divins 2003 offshore; Laske & Masters 1997 onshore). Together these two surfaces define the sediment thickness. The output surfaces are Moho depth (Fig. 5a) and the top of magmatic addition (see below).

The area between the Moho and base sediment defines the total crustal basement thickness in both continental and oceanic areas (Figs 3 & 7a). The area between the top of magmatic addition and the Moho defines the amount of magmatic addition predicted by the gravity model. The distribution of magmatic addition is then used in turn to define the three main crustal zones of the margin:

(1) inboard areas with no magmatic addition, where the crustal basement is entirely continental – the Moho and top of magmatic addition are here coincident in the cross-sections;

(2) outboard areas where the continental crust has been entirely replaced by magmatic addition and crustal basement is entirely oceanic – the base sediment and top of magmatic addition are here coincident in the cross-sections;

(3) the area between 1 and 2 comprising the OCT, where both highly thinned continental crust and new magmatic addition are likely to be present – in this area the top of magmatic addition is a distinct interface in the cross-sections.

Within the area of the OCT on the cross-sections, where continental and magmatic crust are both present, the new magmatic addition is displayed as underlying the thinned continental crust with an 'underplating' geometry. This is simply a graphical construction within the cross-sections and no specific volcanic mechanism or location is implied. The magmatic addition in each case could be a combination of underplating, intrusion and extrusion, but it is represented for simplicity on the cross-sections as a new layer underlying the thinned continental crust. This construction produces a typical 'feather-edge' geometry to the continental crust as it thins to zero and is replaced by oceanic crust.

The 15 cross-sections have been constructed so that some are orientated *c.* east–west, lying along the opening direction of the Equatorial Atlantic (i.e. they are dip lines). These sections define the crustal geometries of the rift segments to the margin. Others are orientated *c.* north–south and define the crustal geometries of the transform segments. Possible exceptions to this are cross-sections 1 and 2, in the NW corner of the map (Fig. 7a) and lying to the NW of the Demerara Plateau. These two sections

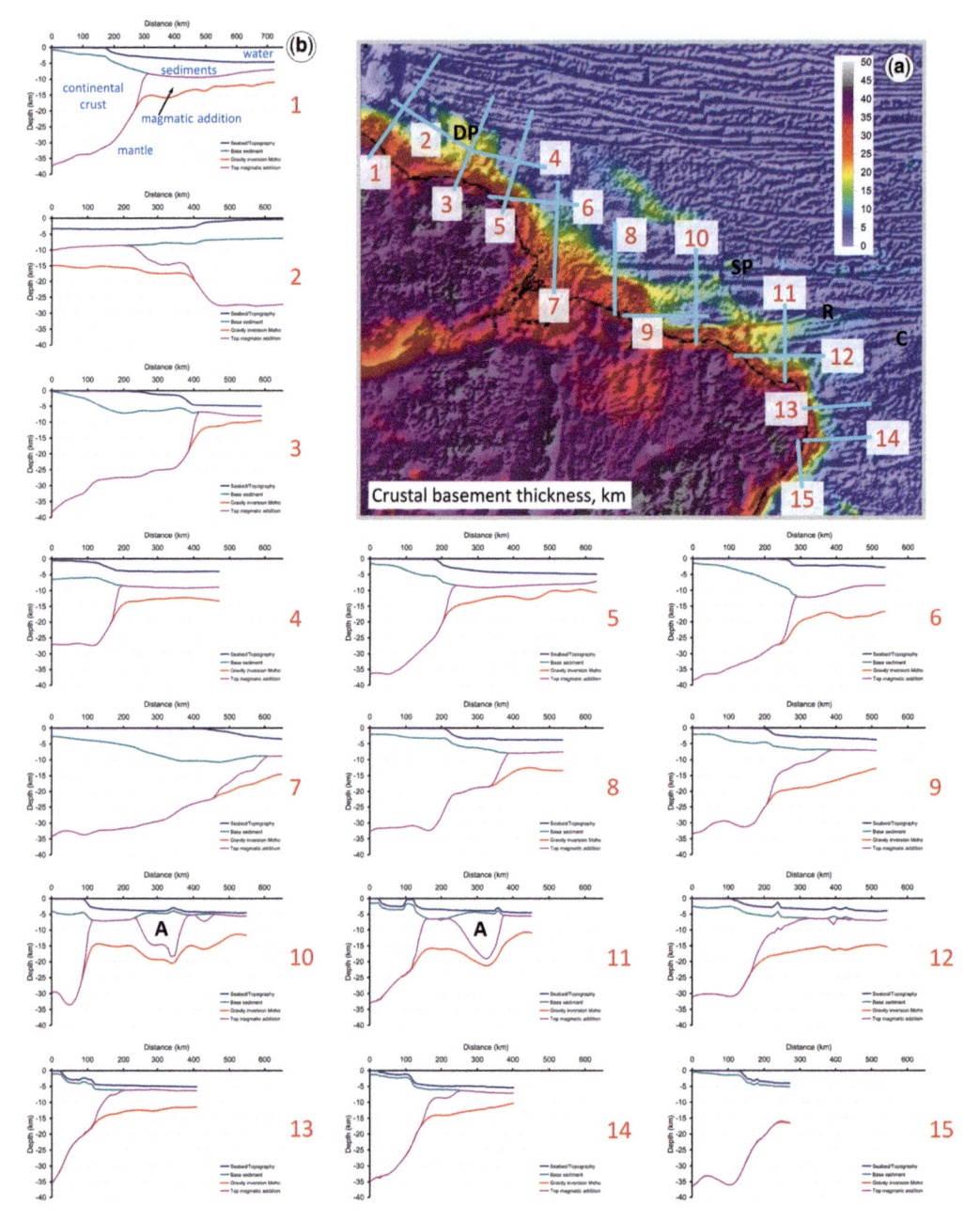

Fig. 7. (**a**) Map of total crustal basement thickness for the South American equatorial margin, enlarged from Figure 5a. Scale in kilometres. DP, Demerara Plateau; SP, St Paul's fracture zone; R, Romanche fracture zone; C, Chain fracture zone. The locations of the 15 cross-sections comprising (b) are shown. (**b**) Fifteen crustal-scale cross-sections extracted from the results of the gravity inversion. Sections 1 and 2 use a Jurassic break-up age of 170 Ma. Sections 3–15 use a Cretaceous break-up age of 110 Ma. Sections 1–5, 7, 8 and 13–15 were determined assuming 'normal' magmatic addition. Sections 6, 9–12 were determined assuming 'magma-rich' magmatic addition. Sections 1, 4, 5, 9 and 12–14 cross rifted margin segments. Sections 2, 3, 6–8, 10, 11 and 15 cross transform margin segments. On sections 10 and 11 A indicates areas of anomalously thick crust outboard of the COB (see text). (A higher-resolution display of the same cross-sections is available within Supplementary Material, Figure S1.)

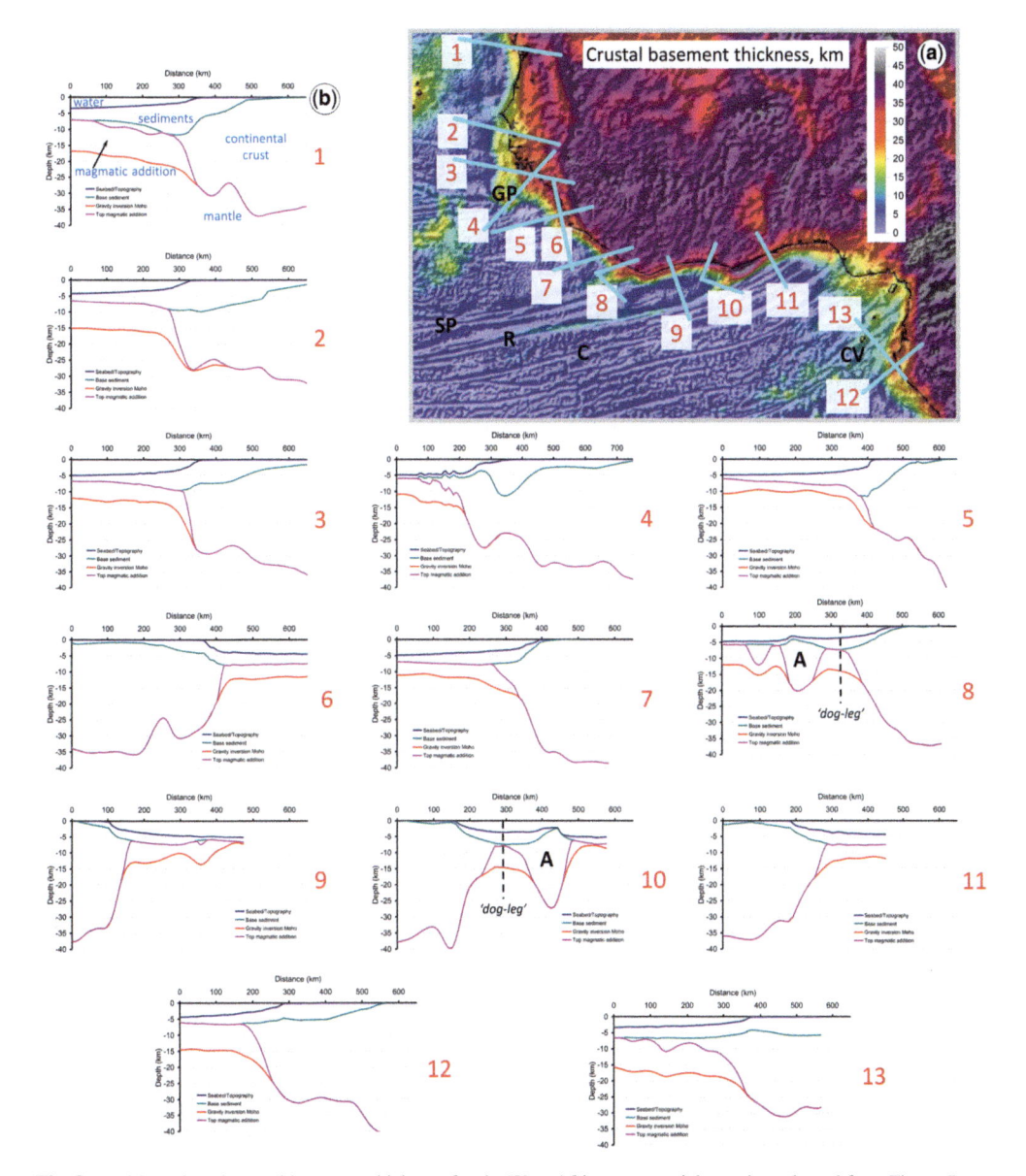

Fig. 8. (a) Map of total crustal basement thickness for the West African equatorial margin, enlarged from Figure 5a. Scale in kilometres. GP, Guinea Plateau; SP, St Paul's fracture zone; R, Romanche fracture zone; C, Chain fracture zone; CV, Cameroon volcanic line. The locations of the 13 cross-sections comprising (b) are shown. (b) Thirteen crustal-scale cross-sections extracted from the results of the gravity inversion. Sections 1–3 use a Jurassic break-up age of 170 Ma. Sections 4–13 use a Cretaceous break-up age of 110 Ma. Sections 3–11 determined assuming 'normal' magmatic addition. Sections 1, 2, 12 and 13 were determined assuming 'magma-rich' magmatic addition. Sections 1–3, 5, 7 and 12 cross rifted margin segments. Sections 4, 6, 9, 11 and 13 cross transform margin segments. Sections 8 and 10 are 'dog-leg' sections which cross both a rifted margin and an oceanic transform fault. On sections 8 and 10 A indicates areas of anomalously thick crust outboard of the COB (see text). (A higher-resolution display of the same cross-sections is available within Supplementary Material, Figure S1.)

extend on to oceanic crust which may not be related to the east–west Cretaceous opening of the Atlantic, but may instead be a fragment of oceanic crust related to earlier Jurassic opening of the Central Atlantic and Gulf of Mexico (e.g. Pindell & Kennan 2009). Sections 1 and 2 have been produced with a break-up age of 170 Ma (acknowledging Jurassic oceanic crust). Sections 3–15 have been produced with the standard Cretaceous break-up age for the Equatorial Atlantic of 110 Ma.

Ten of the 15 cross-sections have been produced from a gravity inversion model parameterized for 'normal' magmatic addition (Fig. 5c, e). The other five cross-sections (6, 9, 10, 11, 12) have been produced using decompression melting parameterized for 'magma-rich' magmatic addition (Fig. 5d, f).

With the exception of cross-section 15, all of the sections begin inboard on continental crust with no magmatic addition and end outboard on oceanic crust with no remaining continental crust. These 14 sections therefore all cross the OCT, either at a rifted margin or at a transform margin, depending on orientation. Cross-section 15 does not extend far enough south to reach predicted oceanic crust and in fact shows no magmatic addition because the trigger thinning factor of 0.7 (for normal decompression melting) has not been reached within the length of the section.

Five cross-sections have been produced using the parameterization for 'magma-rich' magmatic addition. The crustal thickness map (Fig. 7a) shows that all five sections (6, 9, 10, 11, 12) cross areas of crustal thickness at their outboard end which is thicker than the standard oceanic thickness of 7 km. For these lines the magmatic addition has therefore been increased to the magma-rich maximum of 10 km in order to test whether this will resolve thick oceanic crust (up to 10 km). This has worked for sections 6, 9 and 12. Caution should be used, however, in concluding that the break-up at these locations was necessarily magma-rich. Ocean drilling and deep-seismic-reflection data suggest that the oceanic crust in these regions experienced post-formation intra-plate magmatism of Late Cretaceous and Early Tertiary age, as observed on the Ceara Rise (Kumar & Embley 1977; Hekinian et al. 1978), which thickened what may have been normal (or even thin) oceanic crust.

On sections 10 and 11 even 'magma-rich' magmatic addition does not resolve continuous oceanic crust within the outboard domain and instead these two cross-sections show isolated crustal blocks greater than 10 km thickness (labelled A on the sections, Fig. 7b). These blocks almost certainly correspond to oceanic crust plus younger volcanic addition, resulting in magmatic thicknesses in excess of 10 km, rather than being isolated slivers of continental crust. This conclusion highlights that, wherever possible, regional geological knowledge should be used to assist interpretation of the gravity inversion results.

As with all predictive models, the reliability of the predictions is greater if the models can be independently validated. Figures 10–12 are used (below) to compare three of the South American cross-sections (3, 5, 6) with published cross-sections constructed from pre-existing data. Away from areas of validation, however, the power of extracting cross-sections from the gravity inversion models is that the sections provide an insight into rifted margin geometry in areas where this would not otherwise be possible. In this context such cross-sections may be used to help position long regional seismic lines across this particular margin, thus reducing uncertainty associated with an expensive commercial process.

Crustal structure of the West African equatorial margin

Figure 8 provides the equivalent display for the West African equatorial margin as that shown by Figure 7 for the South American margin. Figure 8a is a map of total crustal basement thickness, enlarged from Figures 3 and 5a. Figure 8b comprises 13 crustal-scale cross-sections extracted from the gravity inversion results.

The margin is again strikingly non-linear and partitioned into rift and transform segments by the same oceanic fracture zones which segment the South American margin, particularly so the very prominent St Paul's, Romanche and Chain fracture zones (Figs 3 & 8a). St Paul's and Romanche delineate longer transform-margin segments along the northern Gulf of Guinea than the width of the intervening rift segments. The 13 cross-sections have again been constructed so that they illustrate either rift segments within the current WSW–ENE opening direction, or transform segments orthogonal to this. Two of the sections (8 and 10) have right-angle bends in them so that they illustrate both rift and transform geometry, across the St Paul's and Romanche fracture zones respectively.

The northern three sections (1–3) lie north of the Cretaceous equatorial margin *sensu stricto* and extend on to Jurassic oceanic crust of the Central Atlantic and North Africa. These sections have therefore been produced with a break-up age of 170 Ma. Sections 4–13 have been produced with the standard Cretaceous break-up age for the Equatorial Atlantic of 110 Ma.

Nine of the 13 cross-sections have been produced from a gravity inversion model parameterized for 'normal' magmatic addition (Fig. 5c, e). The other four cross-sections (1, 2, 12 and 13) have been produced from a model parameterized for 'magma-rich' magmatic addition (Fig. 5d, f). All of the sections begin inboard on continental crust with no magmatic addition and end outboard on oceanic crust with no remaining continental crust. They therefore all cross the OCT, either at a rifted margin or at a transform margin, depending on orientation.

Some explanation of the 'magma-rich' sections is again required. The crustal thickness map (Fig. 8a) shows that all four sections (1, 2, 12, 13) extend outboard on to areas of crust which are thicker than the standard oceanic thickness of 7 km (c. 10 km thick). In these areas the magmatic addition within the gravity inversion has therefore been increased to the magma-rich maximum of 10 km in order to see if this will resolve thick oceanic crust (up to 10 km). This works for all four sections, producing oceanic crust thicker than 7 km at the outboard end of each section. For sections 1 and 2 (Jurassic oceanic crust) this result is not a surprise as the conjugate North American Jurassic margin is commonly considered to be a 'magma-rich' margin (Eldholm et al. 2000). Further south, section 13 crosses the Cameroon volcanic line (post-break-up seamounts), while section 12 lies close to it. Thus, while magmatic crust thicker than 7 km on these two sections may have been produced during break-up, it could also have been enhanced in thickness by the more recent post-break-up Cameroon Line volcanics (Gallacher & Bastow 2012).

Within the 13 West African sections the most complex crustal geometries are seen on sections 8 and 10, which are the two 'dog-leg' sections crossing major transform faults. On both sections the inboard c. 300 km crosses a rifted margin segment, with the continental crust thinning from c. 35 km to 0, at which point it is replaced by 7 km of oceanic crust. For the next (outboard) c. 300 km both sections dog-leg to the south, crossing the St Paul's and Romanche fracture zones, respectively. As the fracture zones are crossed the crustal basement thickens again to c. 15 km on section 8 and c. 20 km on section 10 (labelled A on the sections). On the South American margin, blocks of anomalously thick crust close to fracture zones (Fig. 7, sections 10 and 11) were interpreted as oceanic crust thickened by post-break-up magmatism. Along the St Paul's and Romanche fracture zones, however, there is no indication of post-break-up magmatism (it is encountered further to the south along the Cameroon volcanic line; Fig. 8). We therefore believe that other possible explanations are required for the presence of the thick crust adjacent to and immediately north of the two major fracture zones on the African side. These possibilities include:

(1) they are blocks of continental crust which have been extended oceanwards along the fracture zones during the break-up process;
(2) they are areas of oceanic crust magmatically thickened along the fracture zones by the early post-break-up passage of the spreading centre to the south;
(3) they are areas of oceanic crust tectonically thickened by mild transpression (during the passage of the spreading centre) along the inside arc of the curved fracture zones.

The gravity inversion alone cannot distinguish between these three possibilities, but given the prediction of oceanic crust on both cross-sections 8 and 10 north of the fracture zones (where the 'dog-leg' bends are located; Fig. 8), we consider option 1 the least likely and favour a magmatic composition for the areas of thick crust. These possibilities are discussed again later in the paper.

West African section 4 also shows a crustal geometry which is more complex than a 'simple' progressive thinning of continent into ocean. An enlargement of this cross-section, together with the associated profile of thinning factor $(1 - 1/\beta$; Fig. 5e) is shown in Figure 9. The section has its inboard (NE) end located c. 200 km onshore and across a distance of c. 400 km the continental basement thins steadily from c. 36 to c. 12 km, at which point the thinning factor has reached c. 0.7. Over the next 100 km, however, the continental basement thickens again to c. 24 km (thinning factor <0.4), before finally thinning once more, this time to zero (thinning factor 1) and being replaced by oceanic crust. There is thus an anomalously thick crustal block towards the outboard end of the section, which corresponds to the bathymetric feature known as the Guinea Plateau (Figs 3, 8 & 9). In this particular case we believe that the crustal block is indeed likely to be continental crust and we interpret the area of thin crust (c. 12 km) immediately inboard to be a 'failed break-up basin' (FBB on Fig. 9, e.g. Scotchman et al. 2010; Fletcher et al. 2013). Failed break-up basins occur at continental margins as rift basins which were the locus of an initial attempt at break-up, but which were subsequently abandoned as final break-up occurred further outboard. This is probably the consequence of two en echelon rift segments propagating towards each other but failing to connect directly (Scotchman et al. 2010, fig. 8; Fletcher et al. 2013, figs 12 & 13). Section 4 lies at the boundary between Jurassic Central Atlantic oceanic crust to the north and Cretaceous Equatorial Atlantic oceanic crust to the south. It is quite likely that the 'failed break-up basin' in this location results from the attempt (ultimately successful) to link the developing Cretaceous margin into the pre-existing Jurassic margin. If the Jurassic margin and the propagating Cretaceous margin were initially laterally offset from each other, the crustal configuration following final break-up could have been a failed break-up basin at the northern end of the initial Cretaceous rift. The pre-break-up conjugacy of the Guinea Plateau and 'failed break-up basin' (Fig. 9) with the Demerara Plateau on the South American margin

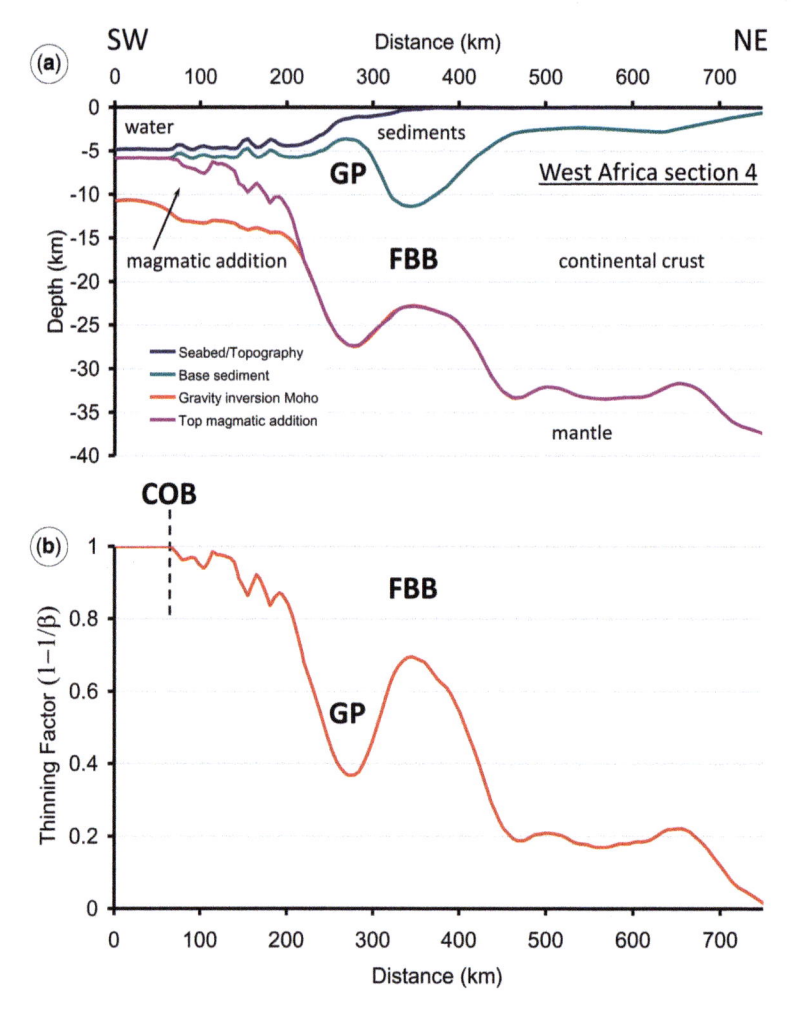

Fig. 9. (**a**) Enlargement of West African section 4, see Figure 8a for location. FBB indicates the location of thin continental crust and a possible 'failed break-up basin' inboard of the COB. GP indicates the location of thick continental crust below the Guinea Plateau, between the COB and the FBB. (**b**) Profile of thinning factor for section 4. Thinning factor rises to *c.* 0.7 within the 'failed break-up basin', before dropping to 0.4 to the west at the Guinea Plateau where the continental crust thickens once more. Thinning factor reaches 1 in the oceanic area at the western end of the section.

(Fig. 3) is illustrated later in the context of the plate reconstructions.

Validation of the cross-sections by comparison with pre-existing data

The power of the crustal cross-sections derived from gravity inversion is that they can be used to make predictions about crustal type and crustal structure for areas in which there has been no local acquisition of seismic data. As with all models, if the results can in some way be validated by comparison with pre-existing information then confidence in their predictions is increased. Figures 10–14 show five of the cross-sections produced by gravity inversion compared with five previously published cross-sections, three on the South American margin (Figs 10–12) and two on the West African margin (Figs 13 & 14).

South American section 3

Figure 10a shows an enlargement of South American cross-section 3 (Fig. 7), together with a crustal cross-section from nearby to the east constructed by Greenroyd *et al.* (2008) from wide-angle and reflection seismic data plus supporting gravity modelling

CRUSTAL STRUCTURE OF THE CONJUGATE EQUATORIAL ATLANTIC MARGINS

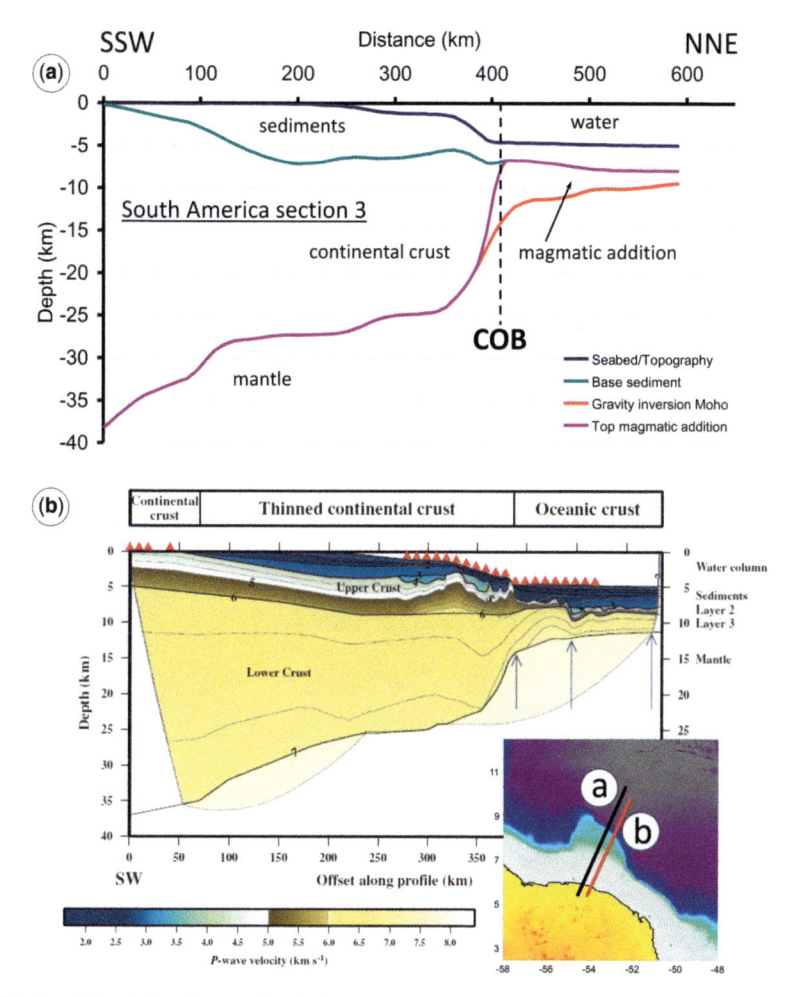

Fig. 10. Validation of South American section 3 determined using OCTek gravity inversion. (**a**) Enlargement of South American section 3, see Figure 7a for location. (**b**) Crustal cross-section produced from wide-angle and reflection seismic data by Greenroyd *et al.* (2008) (see inset map for location). The similarity in illustrated crustal geometries and crustal types is striking and thus (**b**) provides a good validation of the predictions from gravity inversion.

(Fig. 10b). The sections strike slightly east of north, crossing the Demerara Plateau (offshore Suriname and French Guyana) and its northern margin with the Atlantic (Fig. 7a). This margin is interpreted by both ourselves and Greenroyd *et al.* to be a Cretaceous transform margin. In the regional crustal-thickness map (Fig. 3) the northern margin of the Demerara Plateau is clearly bounded by a major oceanic fracture zone extending across the full width of the Equatorial Atlantic from South America to West Africa.

The crustal geometries and the crustal types picked out by both sections are very similar. Both sections show the same 'gentle' tapering of the continental crust northwards towards the transform margin. The transform margin then shows a very abrupt step from continental crust (*c.* 15 km thick) on to relatively thin oceanic crust (*c.* 5 km thick). The Greenroyd *et al.* section, derived primarily from acquired seismic data, provides a very good validation of the likely accuracy of the nearby cross-section from the regional gravity inversion.

South American section 5

Figure 11a shows an enlargement of South American cross-section 5 (Fig. 7), together with an intersecting but clockwise-oblique crustal cross-section constructed by Greenroyd *et al.* (2007) from wide-angle and reflection seismic data (Fig. 11b). The sections

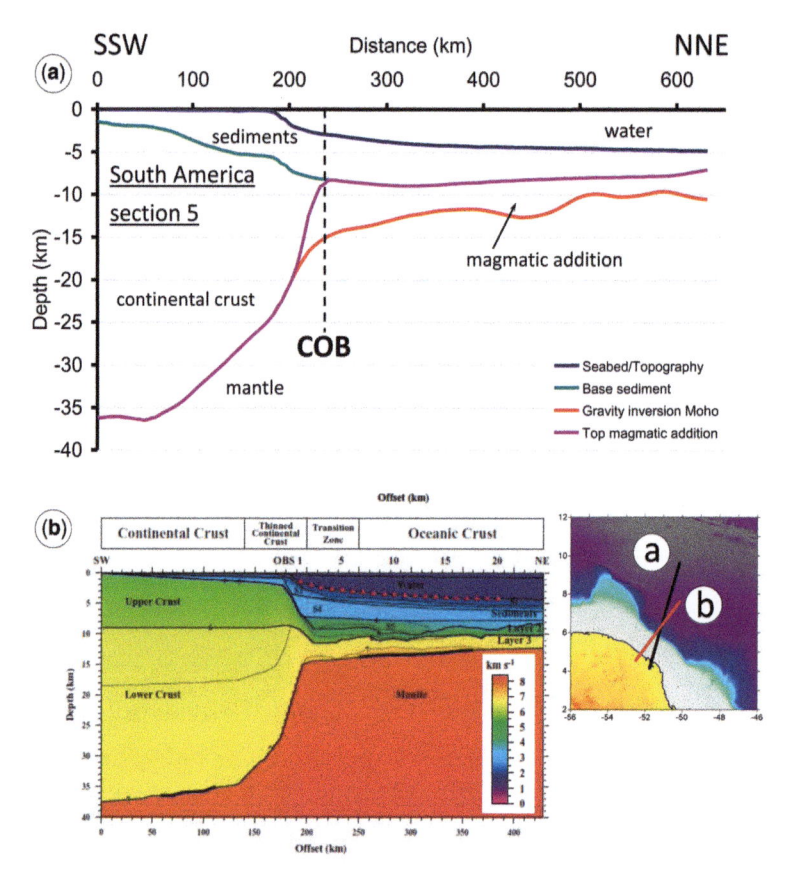

Fig. 11. Validation of South American section 5 determined using OCTek gravity inversion. (**a**) Enlargement of South American section 5, see Figure 7a for location. (**b**) Crustal cross-section produced from wide-angle and reflection seismic data by Greenroyd *et al.* (2007) (see inset map for location). The similarity in illustrated crustal geometries and crustal types is striking and thus (**b**) provides a good validation of the predictions made from gravity inversion.

both strike slightly east of north (offshore French Guyana) and are broadly parallel to section 3 (Fig. 10, *c.* 300 km to the west). Section 5 does not cross the offshore Demerara Plateau but rather passes relatively rapidly (<200 km) from onshore continent on to Cretaceous oceanic crust. Both sections show a sharp continental margin and abrupt thinning of continental crust which is replaced to the north by relatively thin (*c.* 5 km) oceanic crust.

Greenroyd *et al.* (2007) describe the line location as a 'rift-type setting', which may simply refer to the presence of an attenuated continental margin, but we believe that in detail the two sections quite clearly cross another sharp transform margin, delineated by a clear oceanic fracture zone at the COB (Figs 3 & 7).

Regardless of any potential kinematic interpretation, the similarities between crustal geometry and crustal type on both sections provide another very good validation of the predictive cross-section from the regional gravity inversion.

South American section 6

Figure 12a shows an enlargement of South American cross-section 6 (Fig. 7), together with a crustal cross-section from a nearby location (but with a different orientation) constructed by Watts *et al.* (2009) from wide-angle and reflection seismic data (Fig. 12b). Section 6 extends eastwards from the coast of French Guyana (Fig. 7a) and crosses what we interpret to be a rifted continental margin, with the section lying along the direction of oceanic spreading. At its eastern end, on oceanic crust, section 6 intersects the Watts *et al.* section, which itself extends SW from here on to the Brazilian coast, at *c.* 45° to section 6 (see inset map in Fig. 12b).

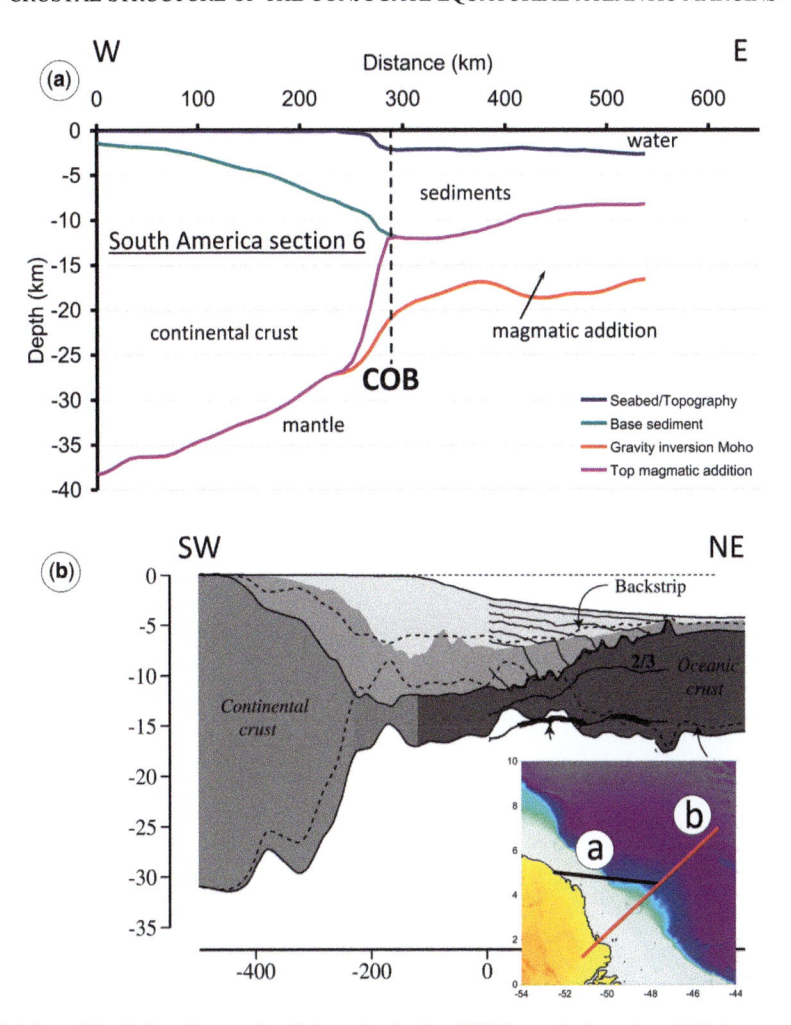

Fig. 12. Validation of South American section 6 determined using OCTek gravity inversion. (**a**) Enlargement of South American section 6, see Figure 7a for location. (**b**) Crustal cross-section orientated at *c.* 45° to (a) but intersecting it at its outboard, eastern oceanic end (see inset map for location). The section was produced from wide-angle and reflection seismic data by Watts *et al.* (2009). The similarity in illustrated crustal geometries and crustal types is striking and thus (b) provides a good validation of the predictions made from gravity inversion.

Despite their geographical divergence away from the oceanic intersection, the two cross-sections show very similar crustal geometries and predicted crustal types, providing further corroboration of the results from the gravity inversion. Both sections begin in the west on thick continental crust, which thins across *c.* 250 km (the rifted margin) into a layer of very thin crust (<5 km thick). Both our gravity inversion and the seismic study of Watts *et al.* interpret this thin crust to be thin oceanic crust (thinning factor 1 in the gravity inversion). The geometry of this thin crust is similar to that reported by Funck *et al.* (2003) and Hopper *et al.* (2007) for the SCREECH 1 line on the NE Newfoundland margin, which is a magma-poor margin on which mantle exhumation has been identified. Outboard of the thin crust, the crustal basement thickens to normal, or slightly greater than normal, oceanic values (Figs 3 & 7). Both ourselves and Watts *et al.* interpret this to be thicker oceanic crust, younger than the break-up event. In order to resolve this area of thick crust as oceanic (rather than thinned continental), the gravity inversion must be parameterized for magma-rich magmatic addition (see discussion above and Fig. 5).

The successful comparison and validation presented here for three cross-sections in the area of Suriname, French Guyana and northernmost Brazil gives some confidence in the interpretation of results from the gravity inversion and associated cross-sections further to the south along the South

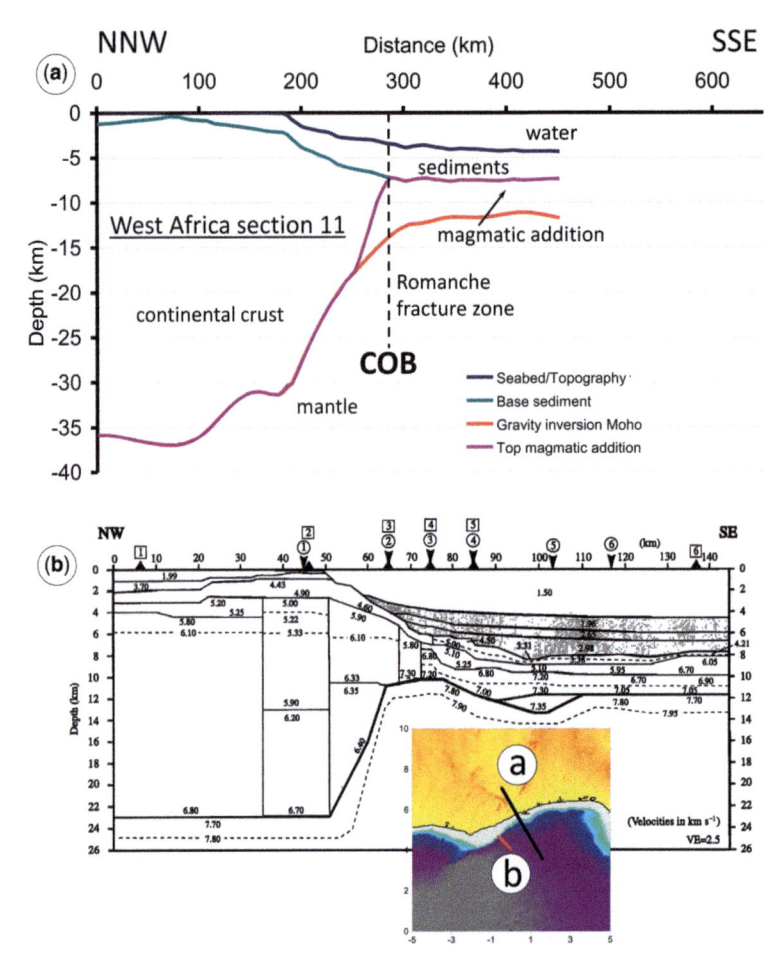

Fig. 13. Partial validation of West African section 11 determined using OCTek gravity inversion. (**a**) Enlargement of West African section 11, see Figure 8a for location. (**b**) Crustal cross-section produced from wide-angle seismic data by Edwards *et al.* (1997) (see inset map for location). The crustal geometries and crustal types illustrated in both sections are similar, but the continental Moho is predicted to be deeper on (a) from gravity inversion (*c.* 35 km) than it is on (b) from seismic (*c.* 23 km). The validation is therefore not complete, but the seismic Moho in (b) at *c.* 23 km does seem surprisingly shallow for onshore Africa.

American equatorial margin, in areas where there are no published cross-sections available for comparative purposes.

West African section 11

Figure 13a shows an enlargement of West African cross-section 11 (Fig. 8), together with a (shorter) crustal cross-section from further to the SW along the same margin segment, constructed by Edwards *et al.* (1997) from wide-angle seismic data (Fig. 13b). Both sections cross the transform margin delineated by the Romanche fracture zone, where it lies close to the coastline of Ghana. This is thus a COB in both geological and geographical terms.

Both sections illustrate the sharpness of the COB across the fracture zone, thick continental crust to the north being replaced by oceanic crust to the south across a few tens of kilometres. The continental Moho of Edwards *et al.* is, however, surprisingly shallow, essentially flat at 23 km. This is difficult to explain in the light of the regional gravity inversion (and also isostatic arguments), which consistently places the West African Moho at 35 km or deeper (Fig. 5a).

There is better agreement in crustal thickness south of the fracture zone where both sections show the oceanic crust to be relatively thin (*c.* 5 km). We have also made this observation about the oldest oceanic crust on the three validated South American

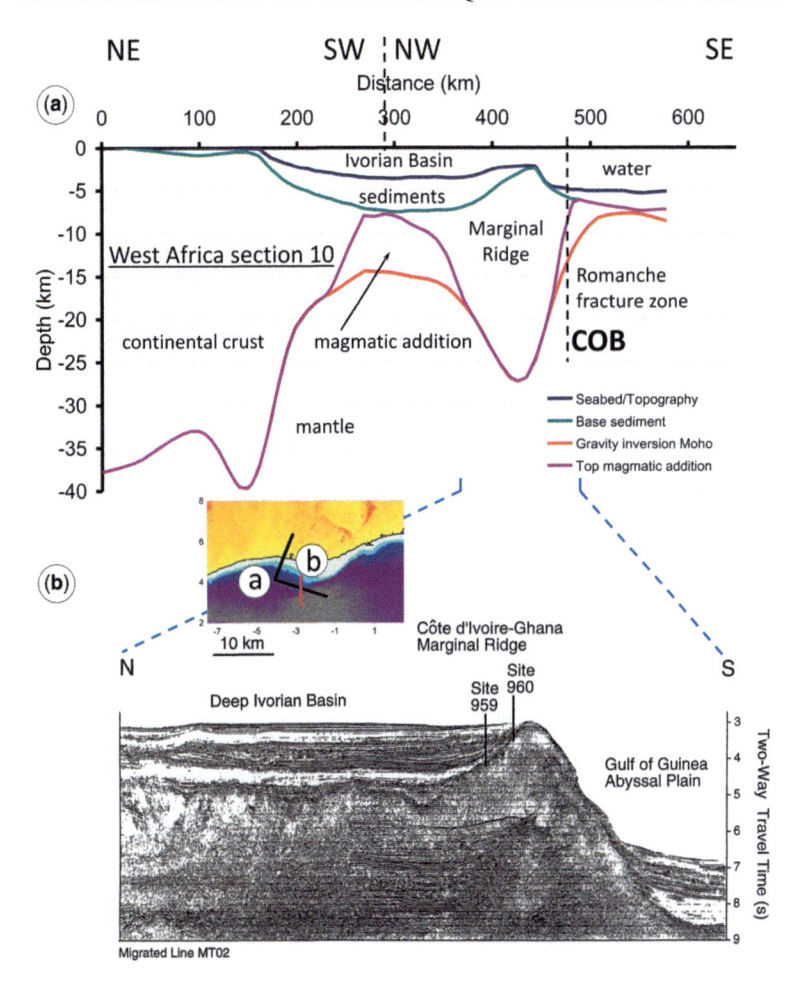

Fig. 14. (**a**) Enlargement of West African (dog-leg) section 10 determined using OCTek gravity inversion, see Figure 8a for location. (**b**) Seismic section from Clift *et al.* (1997), *c.* 100 km in length, crossing the Romanche fracture zone at an angle to section 10 (see inset map for location). The prominent Marginal Ridge immediately north of the Romanche fracture zone is clear on both sections. (a) shows the Marginal Ridge to be underlain by thick crust, *c.* 20 km. While this could be a continental block (Clift *et al.*), it could also a be block of thickened magmatic crust (see text).

profiles (Figs 10–12), a result which possibly indicates slow spreading on both margins of the early Equatorial Atlantic.

The crustal geometries and crustal types on the two sections match well, but the thickness of the continental crust in the two models is different. We therefore have a partial, but not complete, comparison and validation of the gravity inversion results. We suggest that, in the light of the regional gravity inversion results, the continental Moho depth from the wide-angle experiment could be re-examined.

West African section 10

Figure 14a shows an enlargement of West African cross-section 10 (Fig. 8), together with a seismic line (from Clift *et al.* 1997; also Basile *et al.* 1993; Sage *et al.* 2000) which intersects the southern part of section 10 (Fig. 14b). Both cross the Romanche fracture zone, offshore Côte d'Ivoire. In detail the seismic line strikes clockwise of section 10 (Fig. 14 inset map) and is considerably shorter (*c.* 100 km in length). North of the Romanche fracture zone the seismic line lies NE of section 10 and further inboard with respect to the continental margin. With no wide-angle seismic data available this is not a validation of the Moho and crustal-thickness predictions of the gravity inversion, but rather an illustration of the crustal geometries across the Romanche fracture zone resolved by the gravity inversion.

The central focus of the seismic line is the 'Marginal Ridge', along and to the immediate north of the

Romanche fracture zone. Both Clift *et al.* (1997) and Sage *et al.* (2000) considered the Marginal Ridge, at the location imaged by the seismic line, to be underlain by continental crust and thus part of the continental margin, which is anomalously thick along the fracture zone. Clift *et al.* (fig. 1) also showed that to the north of the Marginal Ridge the seismic line is interpreted to lie inboard (to the east) of the COB, which bounds the rifted margin of the Deep Ivorian Basin. Section 10 extends NW to a more distal location north of the Romanche fracture zone (Fig. 14 inset map) and makes predictions of possibly different crustal-types in this area.

The Ivorian Basin on section 10 (Fig. 14a) is predicted by the gravity inversion (parameterized for normal magmatic addition) to be underlain by magmatic crust, which in the centre of the basin has completely replaced the continental crust, making this an oceanic setting. Section 10 is therefore interpreted to lie outboard of the COB within the Ivorian Basin. The crustal affinity of the Marginal Ridge on section 10 (which is where it intersects the seismic line) is less definitive. The gravity inversion resolves a crustal block *c.* 20 km thick, which by default is identified as continental as it is too thick even for a parameterization with magma-rich magmatic addition. As outlined above, in discussion of West African sections 8 and 10 (Fig. 8), we believe there are three possible explanations for the presence of thick crust on the Marginal Ridge:

(1) continental crust extended along the fracture zone (the interpretation of Clift *et al.* 1997);
(2) oceanic crust magmatically thickened by the passage of the spreading centre;
(3) oceanic crust thickened by mild transpression along the fracture zone.

The latter two possibilities were recognized by Clift *et al.* as potential contributors to the geometry of the Marginal Ridge, although they favoured option 1. The gravity inversion alone cannot distinguish between these three possibilities on section 10, but given the prediction of oceanic crust within the Ivorian Basin north of the fracture zone we would favour a magmatic affinity for the Marginal Ridge at the location of section 10. It is notable that the Romanche, St Paul and Chain fracture zones all show a significant anti-clockwise change in their orientation to the NE, close to the African coast (Figs 3 & 8a). This indicates that a change in the divergence vector occurred between Africa and South America during early seafloor spreading, while the spreading centre migrated west along the fracture zones, providing a possible mechanism for transpressional thickening of pre-existing oceanic crust along the inside arc of the fracture zones (option 3 above).

In a recent paper Nemčok *et al.* (2016) have suggested that the Marginal Ridge (their Ghana Ridge) is a detached microcontinental fragment which has been translated 133 km WSW along the Romanche fracture zone. They equate it to two other microcontinental blocks which have been dredged further west along the fracture zone. We believe that it is unlikely that the Marginal Ridge is an allochthonous block because neither our maps (Figs 3 & 8a) nor the maps of Nemčok *et al.* (figs 1 & 2) show a continuous fracture zone or major fault along the northern margin of the Marginal Ridge. Such a fault is required in order to accommodate the 133 km displacement.

Finally, while there is no validation or comparison data available for the sections we have produced in the northern part of equatorial West Africa, three of which extend on to Jurassic oceanic crust (sections 1–3, Fig. 8), we would draw attention to the results of the DAKHLA wide-angle seismic experiment (Klingelhoefer *et al.* 2009) north of our AOI, offshore Western Sahara. The DAKHLA experiment provides crustal profiles extending on to the Central Atlantic Jurassic oceanic crust and shows a similar crustal structure in the OCT to that determined by OCTek gravity anomaly inversion in this area.

Equatorial Atlantic plate reconstructions

In the discussion so far we have attempted to show how gravity inversion of publicly available datasets can be used to provide information about 3D crustal geometries along deep-water continental margins. In this final discussion we will attempt to show how the results of the gravity inversion are not only relevant to our understanding of present-day margin structure but can also be used as input to quantitative models of margin history by constraining plate reconstructions.

Plate reconstructions commonly use ephemeral geomorphic features such as present-day bathymetry and coastlines to constrain the final 'closed' geometry of a restoration sequence. At the time of break-up neither the present-day coast nor shelf-breaks existed and so the restorations could potentially be better constrained by using geological features present at this time in the past.

Figure 15 shows a sequential plate reconstruction of the area covered by the crustal-thickness map in Figure 3. This focuses on the Equatorial Atlantic but also extends north into the Central Atlantic and south into the South Atlantic. The restorations have been performed in GPlates 1.5 (http://www.gplates.org), which is publicly available software. The geological property being restored by the restorations is total crustal basement thickness, which is the product of both the break-up process and subsequent seafloor spreading. Total crustal thickness

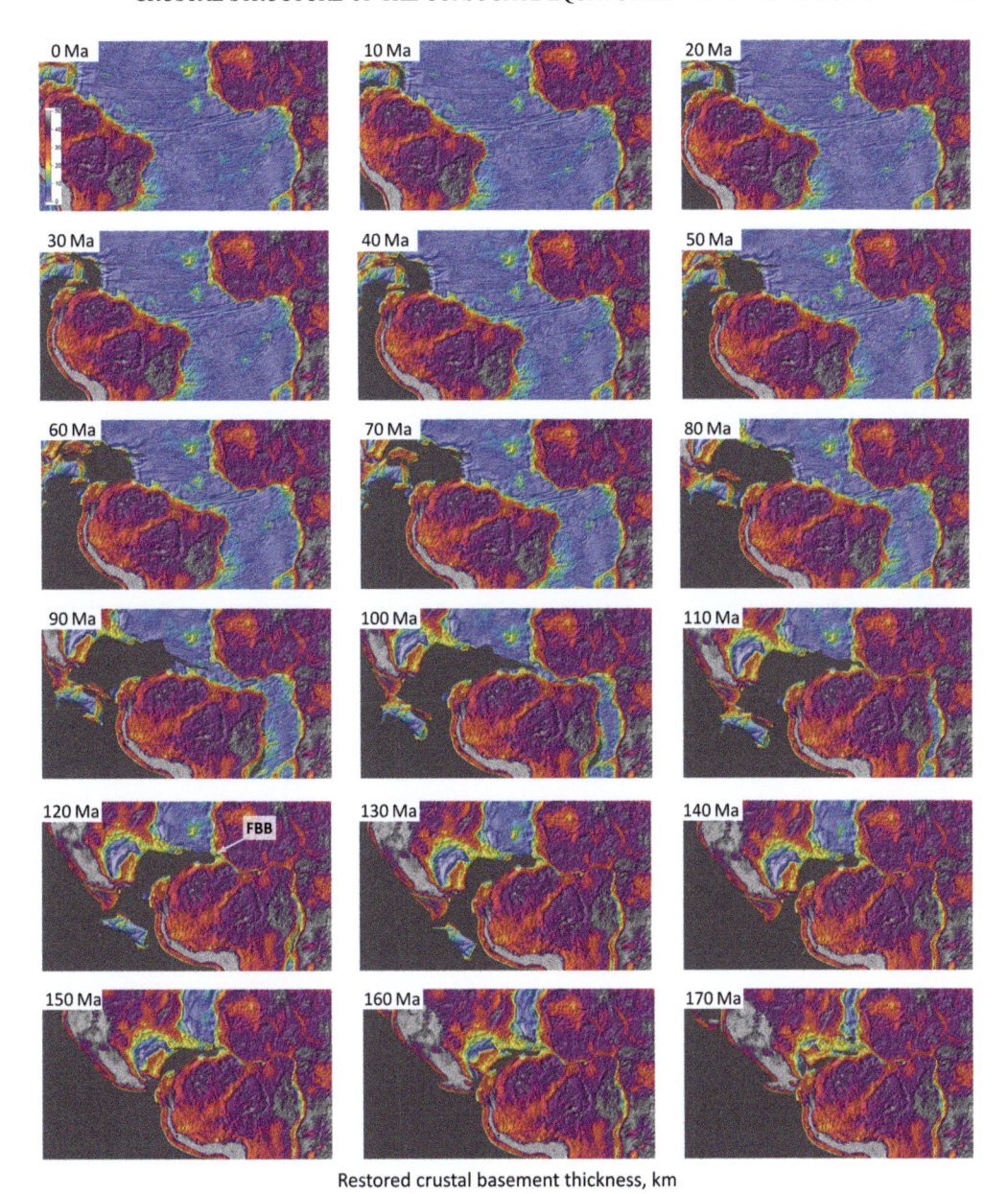

Fig. 15. Plate reconstruction, using GPlates 1.5, of the Central-Equatorial–South Atlantic. The plate reconstructions are populated with the map of total crustal basement thickness (continental and oceanic) from gravity inversion (Fig. 3). The reconstruction is illustrated at increments of 10 Ma, back to 170 Ma. First oceanic crust in the Equatorial Atlantic forms between 110 and 100 Ma. Use of the crustal thickness map allows long-term geological features to be restored, rather than the more typical use of present-day bathymetric features and coastlines. FBB on the restoration at 120 Ma indicates the location of a possible 'failed break-up basin' lying east of the restored and now adjacent Guinea Plateau and Demerara Plateau (see also Figs 3, 8 & 9). (A higher-resolution display of the same plate reconstructions is available within Supplementary Material, Figure S2a.)

does not distinguish crustal type (continental or oceanic), but Figure 16 shows the same set of restorations restoring the corresponding map of thinning factor (Fig. 5), which accounts for predicted magmatic addition and in which oceanic areas are defined by thinning factor = 1 (coloured white).

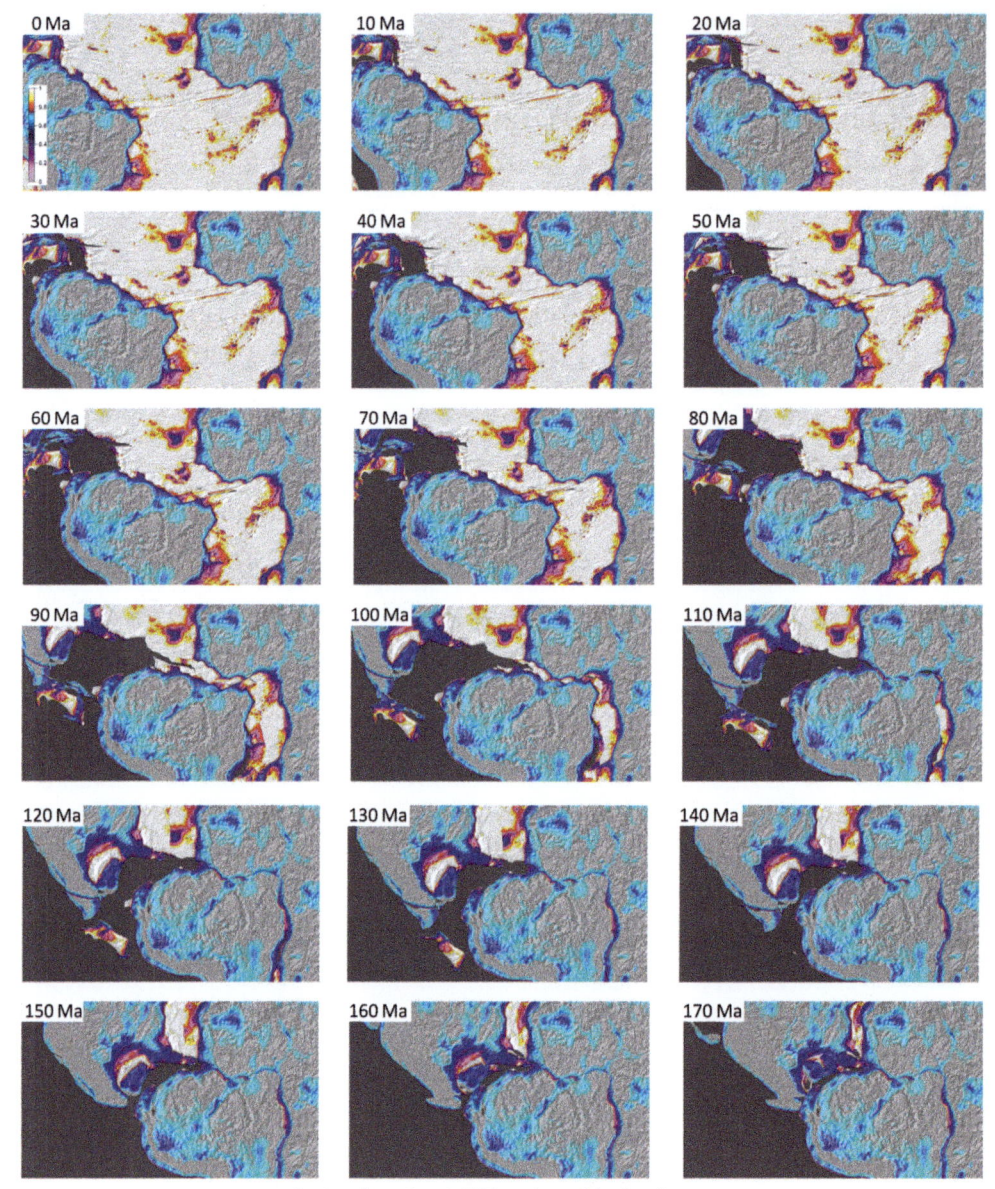

Restored continental-lithosphere thinning-factor

Fig. 16. The same plate restoration sequence as in Figure 15 but with the plate reconstructions populated with the map of thinning factor from gravity inversion (Fig. 5). Use of thinning factor allows oceanic areas to be identified as white (thinning factor 1) in each stage of the reconstruction. (A higher-resolution display of the same plate reconstructions is available within Supplementary Material, Figure S2b.)

The plate reconstructions (Figs 15 & 16) are shown at time increments of 10 Ma and use the default GPlates 1.5 rotation poles, plate polygons and ocean isochrons to constrain plate motions (Seton *et al.* 2012). The restorations extend from the present day back to 170 Ma, thus, not only do they illustrate restoration of the (Cretaceous) Equatorial Atlantic, but they also illustrate restoration of the older (Jurassic) Central Atlantic. Rather than describe each step of the restoration, we will simply highlight the key points of the full sequence.

Prior to 110 Ma restorations show that the Equatorial Atlantic was closed. Continental break-up for the Equatorial Atlantic, defined as the generation of the first oceanic crust, occurred between 110 and 100 Ma. By 100 Ma a series of discrete, isolated oceanic basins with a trapezoidal geometry had formed but did not form a through-going, deep-water, ocean–basin system. These trapezoidal basins were bounded north and south by transform faults and east and west by oblique rifted continental margins. The continental crust of South America and Africa was still connected in some places at this time. By 90 Ma, Africa and South America were separated by oceanic crust and we expect there to have been deep-water oceanic connectivity of the Equatorial Atlantic with both the Central and South Atlantic.

The restorations from 170 to 110 Ma show the development of the opening of the Central Atlantic prior to the opening of the Equatorial Atlantic. They show the plate-tectonic context of the development of the South American margin to the west of the Demerara Plateau (Fig. 3) and its relationship to the formation of the Central Atlantic and Gulf of Mexico.

To the south, the plate restorations show that, while the southern South Atlantic started to open at 130 Ma, significant oceanic connectivity of the Equatorial Atlantic to the south did not occur until Albian times (110 Ma or younger). The restorations to 110 Ma and older show the relationship of the (onshore) South American and West African basins formed prior to Equatorial Atlantic break-up.

The restoration at 120 Ma allows us to add more insight into the 'failed break-up basin' inboard of the Guinea Plateau on the West African margin (Figs 3 & 9). This restoration shows that the 'failed break-up basin' inboard of the Guinea Plateau was initially directly along strike to the north of (and linked to) the rifted margin defining the eastern flank of the Demerara Plateau on the conjugate South American margin. The Guinea Plateau and the Demerara Plateau probably began to extend together at c. 120 Ma or earlier, prior to break-up. By 110 Ma the eastern flank of the Demerara Plateau had successfully separated from West Africa, but the 'failed-break-up basin' and the Guinea Plateau remained attached to West Africa as the final break-up transferred displacement along the northern flank of the Demerara Plateau and propagated northwards along the western flank of the Guinea High (Figs 9 & 15).

The plate reconstructions in Figures 15 and 16 are not new in the sense of providing new kinematic information about plate motions across the Atlantic; they use the prior information about plate motions contained within GPlates. What is new, however, is that the restorations are the first to incorporate a full crustal model (continental and oceanic) within the properties of the restored plates and thus illustrate the restored plates in a new and informative way. As with most of the work presented in this paper, we have focused on the plate restorations at a very large scale, but at a more focused scale across the Equatorial Atlantic the same restorations have been used to help to constrain palaeogeographical models for exploration scoping studies.

Concluding summary

The principal objective of this paper has been to show how regional-scale gravity inversion can be used to provide a 3D crustal model of rifted margin geometry, in circumstances where little or no other data (particularly seismic data) may be available. The technique provides important geological information about any area to which it is applied, but its particular strength lies in application to petroleum exploration scoping studies, where it provides a cost-effective entry point into basin analysis.

In other papers (e.g. Roberts et al. 2013; Cowie et al. 2015, 2016) it has been shown how the gravity inversion technique can be used as one of a suite of tools for deep-water basin analysis. Such additional tools might include subsidence analysis and analysis of residual-depth-anomalies, both of which require supporting seismic-reflection data. Our purpose in this paper, however, has been to demonstrate the use of the gravity inversion technique on its own. The gravity inversion technique therefore provides a good starting point in a frontier exploration setting where little or no seismic reflection data is available.

The primary output from the gravity inversion is a suite of maps which show:

- Moho depth;
- total crustal basement thickness (continental or oceanic);
- residual thickness of the continental crust;
- continental lithosphere stretching factor (β) and thinning factor (γ), where $\gamma = 1 - 1/\beta$.

These maps capture a 3D crustal model for continental areas, oceanic areas and the rifted margins in between. Our understanding of the results is helped by extracting crustal cross-sections. The cross-sections themselves are a powerful predictive tool, but may themselves be compared with and validated against other data, such as wide-angle and reflection seismic data. Conversely the results of the gravity inversion can be used to validate or constrain interpretation of deep-seismic reflection data, by providing realistic bounds for crustal thickness and Moho position.

In our investigation of the Equatorial Atlantic, maps of crustal-thickness and conjugate-margin stretching have been used to illustrate how the Equatorial Atlantic opened as a set of stepped rift-transform segments, rather than as a simple orthogonal rifted margin. This has resulted in complex crustal geometries within the basins along the margins, which have been illustrated with a series of cross-sections. On both margins anomalously thick crust is resolved along a number of oceanic fracture zones and we have discussed the possible origins for this. The cross-sections also show that while 'normal thickness' oceanic crust (c. 7 km) predominates in the equatorial region, areas of thinner (c. 5 km) and thicker (c. 10 km) oceanic crust are also present on both margins.

By using the results of the gravity inversion as input to plate reconstructions, the regional palaeogeography of the Equatorial Atlantic during and after break-up has been displayed, in the context of the diachronous opening of the Central and South Atlantic on either side.

Although not covered specifically in this paper, the results of the gravity inversion can also be used as the input for further analysis. In particular, by quantifying (1) the thickness of the continental crust and (2) the magnitude of lithosphere stretching, two of the main uncertainties for the prediction of basement heat-flow are addressed and maps of top basement heat flow can be produced which draw directly on the results of the gravity inversion (e.g. Cowie & Kusznir 2012a).

Acknowledgements We would like to thank Ross Garden, Neil Frewin and PESGB colleagues for the original invitation to present this work at a PESGB workshop on the Equatorial Atlantic. We thank Gianreto Manatschal for his permission to use the redraughted cross-sections in Figure 6 and we thank Ken McClay for the invitation to contribute to this volume and thus write up the work. The original manuscript was improved by reviews from Garry Karner, Jonathan Turner and Tony Doré, who we thank for their helpful comments.

References

ALVEY, A. 2010. *Using crustal thickness and continental lithosphere thinning factors from gravity inversion to refine plate reconstruction models for the Arctic & North Atlantic*. PhD thesis, University of Liverpool.

ALVEY, A., GAINA, C., KUSZNIR, N.J. & TORSVIK, T.H. 2008. Integrated crustal thickness mapping and plate reconstructions for the high Arctic. *Earth and Planetary Science Letters*, **274**, 310–321, https://doi.org/10.1016/j.epsl.2008.07.036

BASILE, C., MASCLE, J., POPOFF, M., BOUILLIN, J.-P. & MASCLE, G. 1993. The Côte d'Ivoire–Ghana transform margin: a marginal ridge structure deduced from seismic data. *Tectonophysics*, **222**, 1–19.

CHAPPELL, A.R. & KUSZNIR, N.J. 2008. Three-dimensional gravity inversion for Moho depth at rifted continental margins incorporating a lithosphere thermal gravity anomaly correction. *Geophysical Journal International*, **174**, 1–13.

CLIFT, P.D., LORENZO, J., CARTER, A., HURFORD, A.J. & ODP 159 SCIENTIFIC PARTY 1997. Transform tectonics and thermal rejuvenation on the Côte d'Ivoire–Ghana margin, west Africa. *Journal of the Geological Society, London*, **154**, 483–489, https://doi.org/10.1144/gsjgs.154.3.0483

COWIE, L. & KUSZNIR, N.J. 2012a. Gravity inversion mapping of crustal thickness and lithosphere thinning for the eastern Mediterranean. *The Leading Edge*, July **2012**, 810–814.

COWIE, L. & KUSZNIR, N.J. 2012b. Mapping crustal thickness and oceanic lithosphere distribution in the Eastern Mediterranean using gravity inversion. *Petroleum Geoscience*, **18**, 373–380, https://doi.org/10.1144/petgeo2011-071

COWIE, L., KUSZNIR, N.J. & MANATSCHAL, G. 2015. Determining the COB location along the Iberian margin and Galicia Bank from gravity anomaly inversion, residual depth anomaly and subsidence analysis. *Geophysical Journal International*, **203**, 1355–1372, https://doi.org/10.1093/gji/ggv367

COWIE, L., ANGELO, R.M., KUSZNIR, N.J., MANATSCHAL, G. & HORN, B. 2016. Structure of the ocean–continent transition, location of the continent–ocean boundary and magmatic type of the northern Angolan margin from integrated quantitative analysis of deep seismic reflection and gravity anomaly data. *In*: SABATO CERALDI, T., HODGKINSON, R.A. & BACKE, G. (eds) *Petroleum Geoscience of the West Africa Margin*. Geological Society, London, Special Publications, **438**, 159–176, https://doi.org/10.1144/SP438.6

DIVINS, D.L. 2003. *Total Sediment Thickness of the World's Oceans & Marginal Seas*. NOAA National Geophysical Data Center, Boulder, CO. Updates: http://www.ngdc.noaa.gov/mgg/sedthick/sedthick.html

EDWARDS, R.A., WHITMARSH, R.B. & SCRUTTON, R.A. 1997. The crustal structure across the transform continental margin off Ghana, eastern equatorial Atlantic. *Journal of Geophysical Research*, **102**, 747–772, https://doi.org/10.1029/96JB02098

ELDHOLM, O., GLADCZENKO, T.P., SKOGSEID, J. & PLANKE, S. 2000. Atlantic volcanic margins: a comparative study. *In*: NØTTVEDT, A. (ed) *Dynamics of the Norwegian Margin*. Geological Society, London, Special Publications, **167**, 411–428, https://doi.org/10.1144/GSL.SP.2000.167.01.16

FLETCHER, R.F., KUSZNIR, N.J., ROBERTS, A.M. & HUNSDALE, R. 2013. The formation of a failed continental breakup basin: the Cenozoic development of the Faroe–Shetland Basin. *Basin Research*, **25**, 532–553, https://doi.org/10.1111/bre.12015

FUNCK, T.J., HOPPER, J.R., LARSEN, H.C., LOUDEN, K.E., TUCHOLKE, B.E. & HOLBROOK, W.S. 2003. Crustal structure of the ocean–continent transition at Flemish Cap: seismic refraction results. *Journal of Geophysical Research*, **108**, 2531, https://doi.org/10.1029/2003JB002434

GALLACHER, R.J. & BASTOW, I.D. 2012. The development of magmatism along the Cameroon Volcanic Line: evidence from teleseismic receiver functions. *Tectonics*, **31**, TC3018, https://doi.org/10.1029/2011TC003028

GREENHALGH, E.E. & KUSZNIR, N.J. 2007. Evidence for thin oceanic crust on the extinct Aegir Ridge, Norwegian Basin, NE Atlantic derived from satellite gravity inversion. *Geophysical Research Letters*, **34**, L06305, https://doi.org/10.1029/2007GL029440

GREENROYD, C.J., PEIRCE, C., RODGER, M., WATTS, A.B. & HOBBS, R.W. 2007. Crustal structure of the French Guiana margin, West Equatorial Atlantic. *Geophysical Journal International*, **169**, 964–987, https://doi.org/10.1111/j.1365-246X.2007.03372.x

GREENROYD, C.J., PEIRCE, C., RODGER, M., WATTS, A.B. & HOBBS, R.W. 2008. Demerara Plateau – the structure and evolution of a transform passive margin. *Geophysical Journal International*, **172**, 549–564, https://doi.org/10.1111/j.1365-246X.2007.03662.x

HEKINIAN, R., BONTE, P., DUDLEY, W., BLANC, P.L., JEHANC, C., LABEYRIE, L. & DUPLESSEY, J.C. 1978. Volcanics from the Sierra Leone Rise. *Nature*, **275**, 536–538.

HOPPER, J.R., FUNCK, T. & TUCHOLKE, B.E. 2007. Structure of the Flemish Cap margin, Newfoundland: insights into mantle and crustal processes during continental breakup. *In*: KARNER, G.D., MANATSCHAL, G. & PINHEIRO, L.M. (eds) *Imaging, Mapping, and Modelling Continental Lithosphere Extension and Breakup*. Geological Society, London, Special Publications, **282**, 47–61, https://doi.org/10.1144/SP282.3

KLINGELHOEFER, F., LABAILS, C. ET AL. 2009. Crustal structure of the SW-Moroccan margin from wide-angle and reflection seismic data (the DAKHLA experiment) Part A: wide-angle seismic models. *Tectonophysics*, **468**, 63–82, https://doi.org/10.1016/j.tecto.2008.07.022

KUMAR, N. & EMBLEY, R.W. 1977. Evolution and origin of the Ceara Rise: an aseismic rise in the western equatorial Atlantic. *Geological Society of America Bulletin*, **88**, 683–694.

LASKE, G. & MASTERS, G. 1997. A Global Digital Map of Sediment Thickness. *EOS Transactions, AGU*, **78**, F483. Updates: http://igppweb.ucsd.edu/~gabi/sediment.html

MANATSCHAL, G., LAVIER, L. & CHENIN, P. 2015. The role of inheritance in structuring hyperextended rift systems: some considerations based on observations and numerical modelling. *Gondwana Research*, **27**, 140–164, https://doi.org/10.1016/j.gr.2014.08.006

MCKENZIE, D.P. & BICKLE, M.J. 1988. The volume and composition of melt generated by extension of the lithosphere. *Journal of Petrology*, **29**, 625–679.

MÜLLER, R.D., SDROLIAS, M., GAINA, C. & ROEST, R.W. 2008. Age, spreading rates and spreading symmetry of the world's ocean crust. *Geochemistry Geophysics Geosystems*, **9**, Q04006, https://doi.org/10.1029/2007GC001743, https://www.ngdc.noaa.gov/mgg/ocean_age/ocean_age_2008.html

NEMČOK, M., SINHA, S.T., DORÉ, A.G., LUNDIN, E.R., MASCLE, J. & RYBÁR, S. 2016. Mechanisms of microcontinent release associated with wrenching-involved continental break-up; a review. *In*: NEMČOK, M., RYBÁR, S., SINHA, S.T., HERMESTON, S.A. & LEDVENYIOVA, L. (eds) *Transform Margins: Development, Controls and Petroleum Systems*. Geological Society, London, Special Publications, **431**, 323–359, https://doi.org/10.1144/SP431.14

PINDELL, J.L. & KENNAN, L. 2009. Tectonic evolution of the Gulf of Mexico, Caribbean and northern South America in the mantle reference frame: an update. *In*: JAMES, K.H., LORENTE, M.A. & PINDELL, J.L. (eds) *The Origin and Evolution of the Caribbean Plate*. Geological Society, London, Special Publications, **328**, 1–55, https://doi.org/10.1144/SP328.1

ROBERTS, A.M., KUSZNIR, N.J., CORFIELD, R.I., THOMPSON, M. & WOODFINE, R. 2013. Integrated tectonic basin modelling as an aid to understanding deep-water rifted continental margin structure and location. *Petroleum Geoscience*, **19**, 65–88, https://doi.org/10.1144/petgeo2011-046

SAGE, F., BASILE, Ch., MASCLE, J., PONTOISE, B. & WHITMARSH, R.B. 2000. Crustal structure of the continent–ocean transition off the Côte d'Ivoire–Ghana transform margin: implications for thermal exchanges across the palaeotransfrom boundary. *Geophysical Journal International*, **143**, 662–678.

SANDWELL, D.T. & SMITH, W.H.F. 2009. Global marine gravity from retracked Geosat and ERS-1 altimetry: ridge segmentation v. spreading rate. *Journal of Geophysical Research*, **114**, B01411, https://doi.org/10.1029/2008JB006008, updates http://topex.ucsd.edu/WWW_html/mar_grav.html

SCOTCHMAN, I.C., GILCHRIST, G., KUSZNIR, N.J., ROBERTS, A.M. & FLETCHER, R. 2010. The breakup of the South Atlantic Ocean: formation of failed spreading axes and blocks of thinned continental crust in the Santos basin, Brazil and its consequences for petroleum system development. *In*: VINING, B. & PICKERING, S.C. (eds) *Petroleum Geology: From Mature Basins to New Frontiers – Proceedings of the 7th Petroleum Geology Conference*. Geological Society, London, 855–866, https://doi.org/10.1144/0070855

SETON, M., MÜLLER, R.D. ET AL. 2012. Global continental and ocean basin reconstructions since 200 Ma. *Earth-Science Reviews*, **113**, 212–270, https://doi.org/10.1016/j.earscirev.2012.03.002

SMITH, W.H.F. & SANDWELL, D.T. 1997. Global seafloor topography from satellite altimetry and ship depth soundings. *Science*, **277**, 1957–1196. Updates: http://topex.ucsd.edu/marine_topo/

SUTRA, E., MANATSCHAL, G., MOHN, G. & UNTERNEHR, P. 2013. Quantification and restoration of extensional deformation along the Western Iberia and Newfoundland rifted margins. *Geochemistry, Geophysics, Geosystems*, **14**, 2575–2597.

TSIKALAS, F., FALEIDE, J.I., ELDHOLM, O. & WILSON, J. 2005. Late Mesozoic-Cenozoic structural and stratigraphic correlations between the conjugate mid-Norway and NE Greenland continental margins. *In*: DORÉ, A.G. & VINING, B.A. (eds) *Petroleum Geology: North-West Europe and Global Perspectives – Proceedings of the 6th Petroleum Geology Conference*. Geological Society, London, 785–801, https://doi.org/10.1144/0060785

WATTS, A.B., RODGER, M., PEIRCE, C., GREENROYD, C.J. & HOBBS, R.W. 2009. Seismic structure, gravity anomalies, and flexure of the Amazon continental margin, NE Brazil. *Journal of Geophysical Research*, **114**, B07103, https://doi.org/10.1029/2008JB006259

WHITE, R.S. & MCKENZIE, D.P. 1989. Magmatism at rift zones: the generation of volcanic continental margins and flood basalts. *Journal of Geophysical Research*, **94**, 7685–7729.

Insights into the Early Evolution of the Côte d'Ivoire Margin (West Africa)

NICOLA SCARSELLI[1]*, GREGOR DUVAL[2], JAVIER MARTIN[2], KEN McCLAY[1] & STEVE TOOTHILL[2]

[1]Department of Earth Sciences, Royal Holloway University of London, Egham, Surrey TW20 0EX, UK

[2]CGG, Crompton Way, Crawley, West Sussex RH10 9QN, UK

*Correspondence: nicola.scarselli@rhul.ac.uk

Abstract: A tectono-stratigraphic analysis of a broadband 3D seismic survey over the outer slope of Côte d'Ivoire margin, west Africa, revealed that Cenomanian and younger strata seal well-developed rift fault blocks up to 15 km across. Growth strata indicate that these were formed during rifting that culminated in seafloor spreading in the late Albian, challenging existing plate reconstructions for the opening of the equatorial Atlantic ocean. A previously unrecognized system of volcanic edifices linked at depth to a network of sill complexes has also been identified. These are aligned along a NE–SW trend, concordant with kilometre-wide ridges, interpreted as folds formed by steep, crustal faults with an oblique-slip component. These trends are similar to those of fracture zones in the region and indicate that the Côte d'Ivoire was a transform margin in the late Albian. These results highlight the potential of offshore Côte d'Ivoire for deep-water rift plays with large traps formed by extensional fault blocks together with prospective Albian reservoirs ponded in their hanging walls. In addition, the volcanoes and ridges generated seabed relief along the newly created transform margin, forming confined basins for potential deposition of Turonian and younger turbidites and the generation of stratigraphic traps.

Recent exploration success in the mid-slope offshore Ghana and Côte d'Ivoire (Fig. 1; Dailly et al. 2013; Martin et al. 2015) has prompted interest in the tectono-stratigraphic evolution and hydrocarbon potential of the equatorial African margin (Dailly et al. 2013; Nemčok et al. 2013, 2016; Coole et al. 2015; Davison et al. 2015; Martin et al. 2015; Sabato Ceraldi et al. 2016; Ye et al. 2017). The basins along this margin, especially those located offshore Côte d'Ivoire and Ghana, and in particular their deep-water sectors, are regarded as being poorly understood owing to limited drilling beyond the shelf break (Wells et al. 2012). Previous research was based on analyses of a coarse grid of 2D seismic lines mainly acquired across the Côte d'Ivoire–Ghana ridge, leaving the adjacent basins largely unexplored (Mascle & Blarez 1987; Basile et al. 1993; Sage et al. 2000; Attoh et al. 2004; Antobreh et al. 2009). Since the work of Antobreh et al. (2009) no research based on the analysis of new geophysical data acquired in the deep water of Côte d'Ivoire has been published.

A number of plate tectonic models for the opening of the South Atlantic ocean, including the equatorial African margin, have been proposed over the last 25 years (e.g. Nürnberg & Müller 1991; Jones et al. 1995; Eagles 2007; Moulin et al. 2010; Heine et al. 2013; Pérez-Díaz & Eagles 2014; Granot & Dyment 2015). These models are mainly driven by ages of seafloor spreading based on the recognition of magnetic anomalies. As a result, different models have been proposed for the equatorial African margin owing to the lack of magnetic anomalies in the Equator–magnetic 'quiet zone' (Rabinowitz & LaBrecque 1979; Clift et al. 1997; Antobreh et al. 2009; Granot & Dyment 2015), adding to uncertainties related to the timing of rifting and seafloor spreading along the margin.

This paper presents the interpretation of a high-quality, depth-migrated, 3D broadband seismic survey over a large, virtually unexplored area, in the deep-water Ivorian Tano Basin offshore Côte d'Ivoire (Fig. 1; Martin et al. 2015). Broadband technology has allowed deep penetration and provides high-quality images of features at depths in excess of 10 km below sea-level. The research provides an understanding of the early geological history offshore Côte d'Ivoire margin, with new evidence that constrains previously unrecognized tectonic phases. The wider implications of this research on the evolution of nearby basins along the equatorial African margin are also discussed.

Geological setting

The Côte d'Ivoire and Ghana shelf is regarded as a world-class example of a transform margin (Attoh

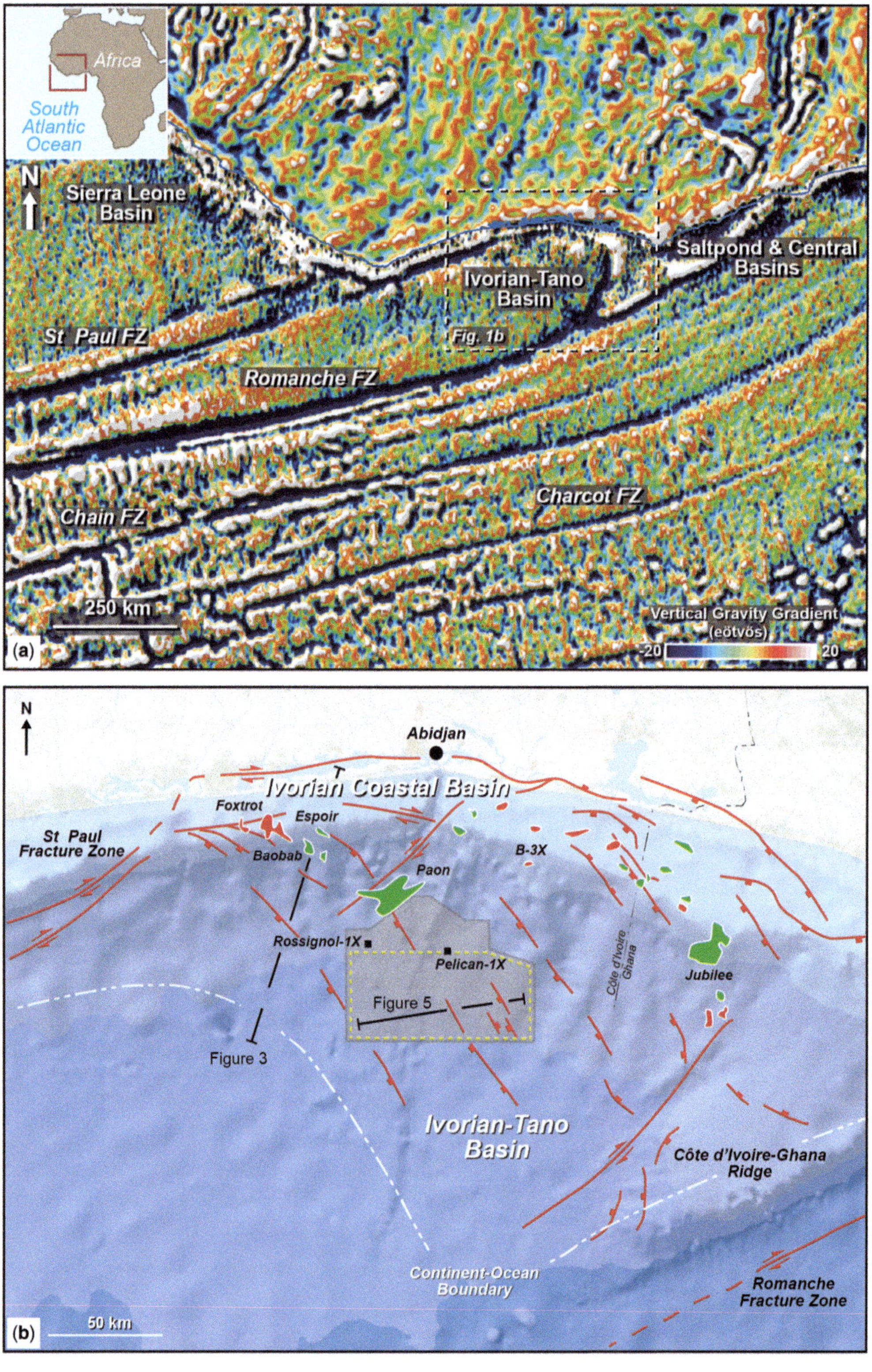
Africa
South Atlantic Ocean
N
Sierra Leone Basin
Saltpond & Central Basins
Ivorian-Tano Basin
Fig. 1b
St Paul FZ
Romanche FZ
Charcot FZ
Chain FZ
250 km
Vertical Gravity Gradient (eötvös)
-20
20
(a)
N
Abidjan
Ivorian Coastal Basin
Foxtrot
Espoir
St Paul Fracture Zone
Baobab
B-3X
Paon
Rossignol-1X
Pelican-1X
Côte d'Ivoire Ghana
Jubilee
Figure 5
Figure 3
Ivorian-Tano Basin
Côte d'Ivoire-Ghana Ridge
Continent-Ocean Boundary
Romanche Fracture Zone
50 km
(b)

et al. 2004; Antobreh et al. 2009; Nemčok et al. 2013; Basile 2015; Mercier de Lépinay et al. 2016). Seminal papers on transform margins were based on the analysis of the Côte d'Ivoire–Ghana ridge (e.g. Mascle & Blarez 1987), a prominent bathymetric feature that is thought to be related to the activation of the Romanche fracture zone, the largest transform fault and fracture system known with maximum lateral offset of the Mid-Atlantic ridge of c. 800 km (e.g. Davison et al. 2015).

The shallow water of the Côte d'Ivoire margin is relatively well explored with a number of wells drilled and several seismic acquisition campaigns since the 1970s (Attoh et al. 2004; Brownfield & Charpentier 2006; Antobreh et al. 2009; Martin et al. 2015). These investigations indicated the presence of Cretaceous strata in a series of rift depocentres commonly referred to as the Ivorian Coastal Basin (Figs 1–3; Attoh et al. 2004; Antobreh et al. 2009). Within these basins, terrestrial to shallow marine sediments of Aptian–Albian age have been drilled and found to contain reservoirs and source rocks of good quality (Fig. 2; Dailly et al. 2013; Martin et al. 2015). These units are offset by extensional faults that are sealed by Upper Cretaceous and younger sediments, and therefore the Aptian–Albian section has been regarded as a syn-rift sequence (Mascle & Blarez 1987; Basile et al. 1993; Antobreh et al. 2009). However, the previous research did not investigate the geometries of growth strata around these faults in order to elucidate timing of rifting and 'break-up'.

The timing of seafloor spreading and onset of post-rift conditions along the Côte d'Ivoire margin is contentious. The age proposed by plate tectonic models varies from late Aptian (Heine et al. 2013) to mid Albian (Antobreh et al. 2009; Torsvik et al. 2009; Mercier de Lépinay et al. 2016) to late Albian (Eagles 2007; Pérez-Díaz & Eagles 2014). Well penetrations in the shelf sector of the margin sampled post-rift Cenomanian marine siliciclastics and carbonates sealing fault-block crests (Fig. 2; Dumestre 1985; Chierici 1996; Brownfield & Charpentier 2006). Within this Upper Cretaceous post-rift section, a successful mid-slope play was proven by the Mahogany-1 discovery in Ghana (Jubilee field) and followed up recently in the Côte d'Ivoire with Anadarko's Paon, Pelican and Rossignol discoveries (Figs 1–3; Dailly et al. 2013; Martin et al. 2015).

In the sequence of events from rift to drift, it is thought that shear stresses owing to progressive development of the transform margin may have played a major role in the development of the Ivorian Tano Basin (Mascle & Blarez 1987; Antobreh et al. 2009). However, geological evidence is scant, primarily owing to the uncertainties related to the interpretation of structures using widely spaced 2D lines.

Seismic data

The dataset available for this study was a 3000 km^2 subset of a larger 3D seismic volume which covers 4400 km^2 of the Côte d'Ivoire outer slope. The volume is located c. 25 km south of the recent Paon discovery, at a present-day water depth of 2 km (Figs 1 & 3).

The dataset was acquired in 2014 by CGG using the latest broadband technology which allows enhanced low-frequency content for a deeper penetration, and in turn provides for clearer imaging at depth (Hill et al. 2006; Firth et al. 2014). The dataset was processed with 3D Kirchhoff PSDM (Pre-Stack Depth Migration) to a depth of 12 km. Processing included advanced de-multiple and de-ghosting techniques such as 3D SRME (surface related multiple elimination) and GWE (ghost wavefield elimination) to obtain the sharpest images possible. The velocity model used to drive the PSDM was obtained from ray-based tomography. The main velocity boundaries were built using a classic layer-stripping top-to-bottom approach encompassing a total of 12 iterations and making use of the following structural markers: water bottom, Eocene, Campanian, Turonian and Albian boundaries. Anisotropy was accounted for through a TTI approach (tilted transverse isotropy) in which the velocity parameters parallel to the structural horizon planes are isotropic but may differ perpendicularly to the structural trends. This research, which focuses on deep structures at depth of 5–12 km (e.g. Fig. 4), has greatly benefited from this advanced seismic acquisition and processing techniques.

The dataset is zero-phased and processed following the Society of Exploration Geophysicists amplitude polarity convention. 'Hard kicks' (an increase in acoustic impedance with depth) correspond to a positive reflection that appears grey to black in the

Fig. 1. Location maps of the study area. (**a**) Vertical gravity gradient map derived from Bouguer gravity data showing the location of the Ivorian Tano Basin along the West Africa equatorial margin. Note that the basin is bounded by two large fracture zones, the St Paul to the north and the Romanche to the south. Data from Sandwell et al. (2014). (**b**) Bathymetric map with major faults highlighted and showing the location of the 3D seismic data analysed in this study. The data is a subset (yellow dashed line) of a larger volume (grey shade) acquired by CGG in 2014. Main basin features outlined from this work, from Dailly et al. (2013) and from Basile et al. (1993).

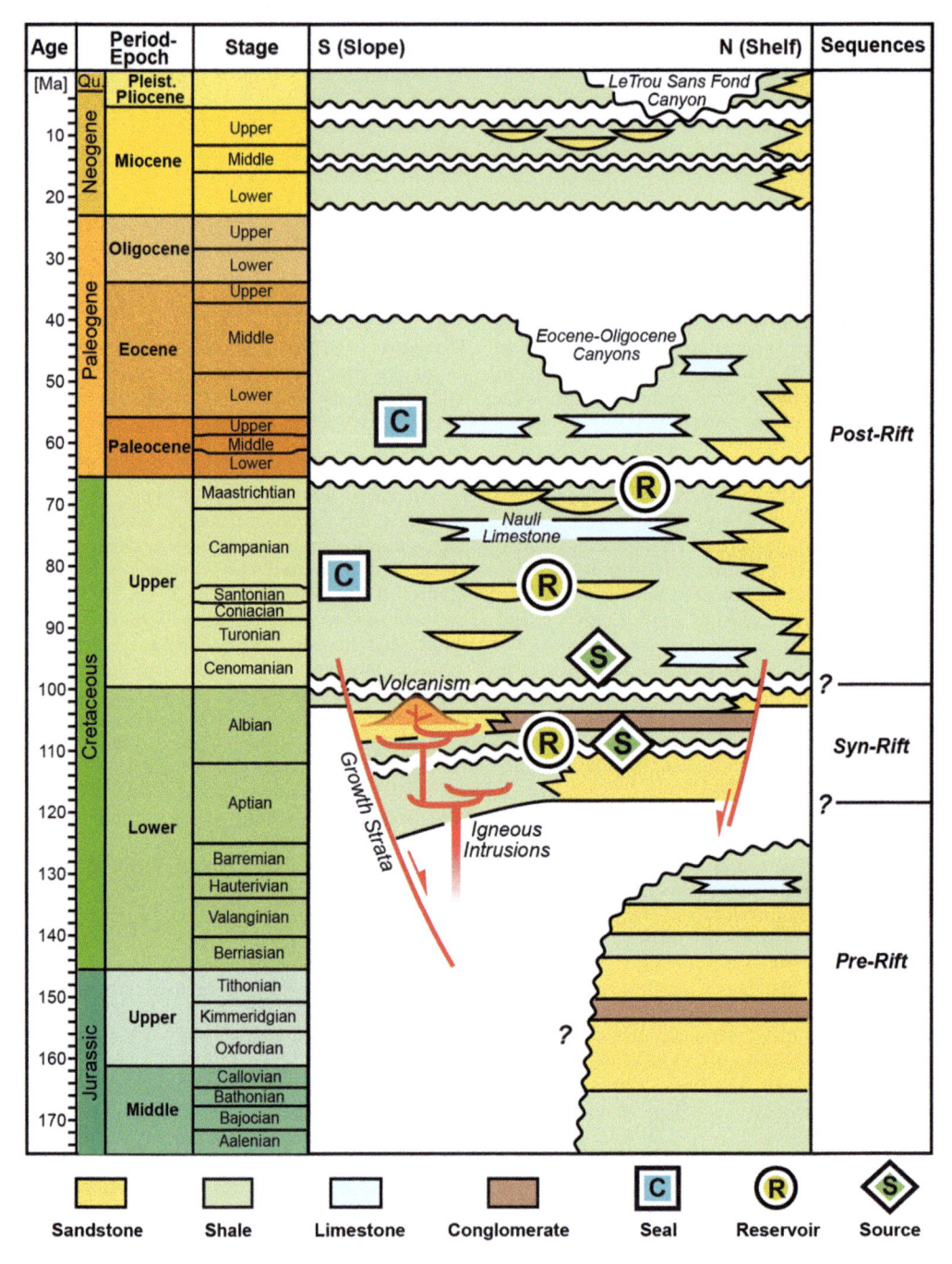

Fig. 2. Tectono-stratigraphic chart of the Ivorian Tano Basin compiled from results of this work and from Kjemperud *et al.* (1991), MacGregor *et al.* (2003) Brownfield & Charpentier (2006) and Wells *et al.* (2012).

seismic displays shown in this paper. Conversely, 'soft kicks' correspond to red–yellow reflections.

There was no well data directly available within the area covered by the survey. Seismic stratigraphy was calibrated by correlation of picks from wells outside the survey area, along regional 2D seismic lines intersecting the 3D survey. The correlation of Top Albian, Top Cenomanian and Top Turonian

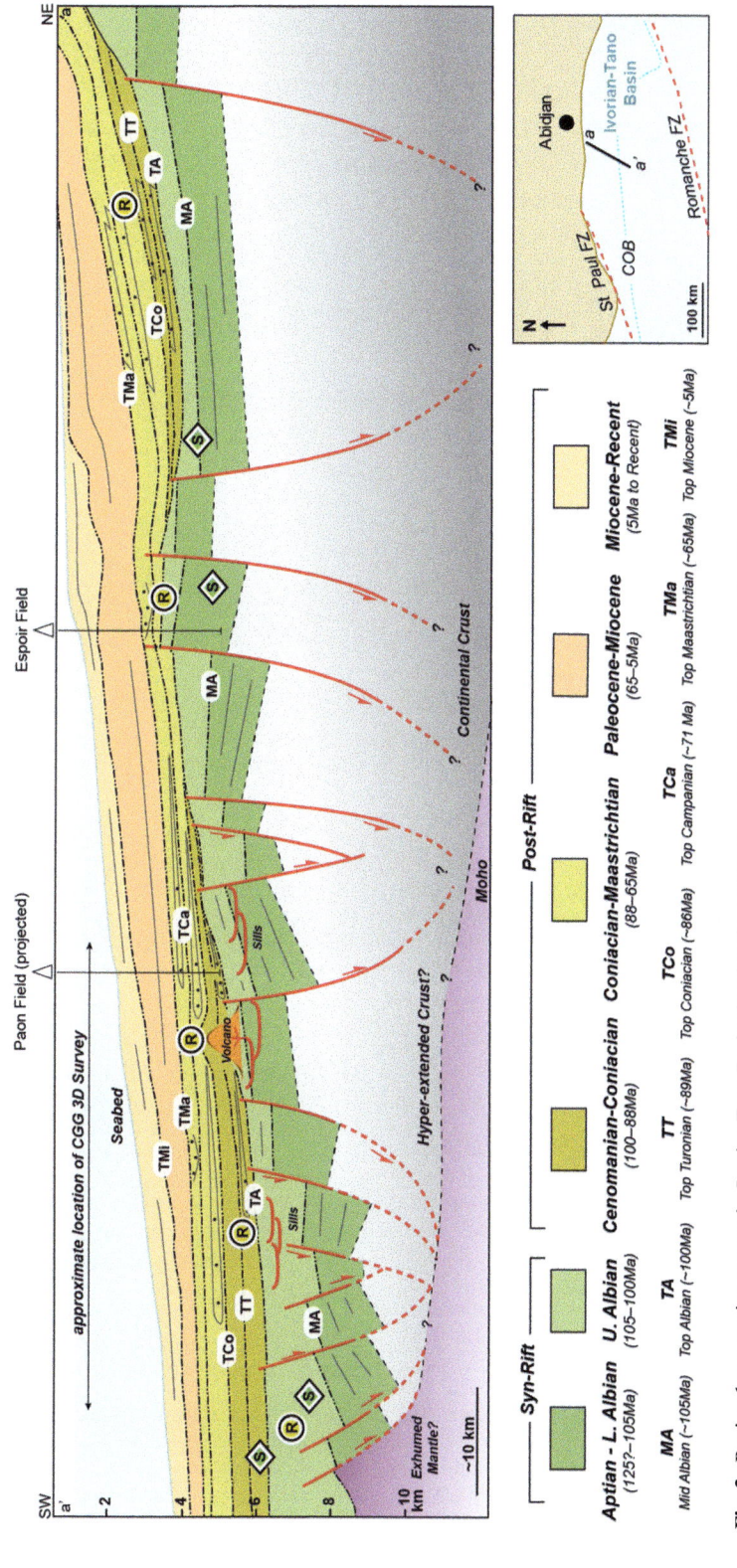

Fig. 3. Regional geo-section across the Ivorian Tano Basin showing the architecture of the Côte d'Ivoire margin. The section is compiled from results of this research and from Brownfield & Charpentier (2006).

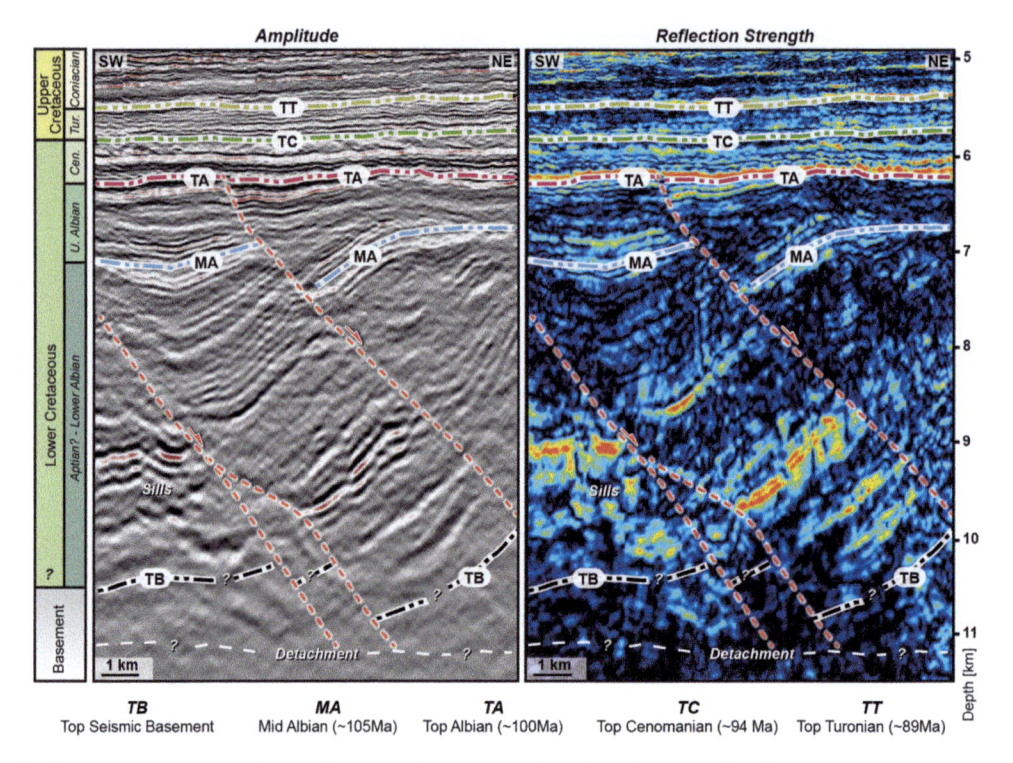

Fig. 4. Seismo-stratigraphic diagram illustrating the main seismic horizons interpreted in this study. This is focused at depths of 5–11 km where deep stratigraphic units offset by faults are clearly imaged on amplitude and reflection strength displays.

is considered to be good as these horizons were not affected by major unconformities or large faults (Fig. 4). The correlation of the Mid Albian horizon has a degree of uncertainty as this is offset by prominent faults. However, this stratigraphic pick was tied to a high-amplitude continuous reflection which is easily identifiable in the data (Fig. 4).

Interpretation workflow

Seismic analysis was carried out using Halliburton Geoprobe and DecisionSpace Desktop tools using established 3D seismic interpretation techniques (e.g. Cartwright & Huuse 2005; Posamentier et al. 2007). Standard workflows have been enhanced through the utilization of seismic attributes such as dip-steered coherency and reflection strength. Coherency was used to highlight faults (Bahorich & Farmer 1995; Marfurt et al. 1998; Chopra & Marfurt 2008; Iacopini & Butler 2011). The dip-steering variant of the attribute was implemented so as to achieve improved fault detection (de Groot & Bril 2005; Marfurt & Alves 2015). Reflection strength, also known as envelope or instantaneous amplitude (Barnes 2007; Chopra & Marfurt 2007), was used to accentuate amplitude variations and to highlight high-amplitude features such as igneous intrusions.

Key seismic horizons and depth slices have been interpreted to gather an understanding of the geology of the area of interest (Figs 4 & 5). Interpretation of the Mid Albian horizon was key in revealing a well-developed fault system at depth (Fig. 6). The Top Albian was also an important horizon as its interpretation showed a series of prominent ridges and younger faults (e.g. Figs 9 & 10). Where observed, growth strata geometries were studied to constrain the timing of fault activation (McClay & Ellis 1987; McClay 1990). The Top Cenomanian and a depth slice at 7.5 km were also interpreted to map a magmatic system of volcanoes and sills (Fig. 12).

Seismic stratigraphy

Figure 4 is a seismic stratigraphic diagram that focuses on the lower half of the survey where deep structures and Lower to Upper Cretaceous strata are clearly imaged. At the base of the survey (depth c. 11 km), the seismic basement horizon

EARLY EVOLUTION OF THE CÔTE D'IVOIRE MARGIN

Fig. 5. Key NE–SW section across the 3D survey showing the main structural and stratigraphic features observed in the study area. Note to the NE the presence of a well-developed extensional fault system affecting the Aptian–Albian sequences. Aptian–Albian faulted strata together with Cenomanian and younger units are folded by prominent ridges. Also note that to the SW an extensive igneous complex and an associated volcanic system are clearly imaged. Section location is shown in Figure 1.

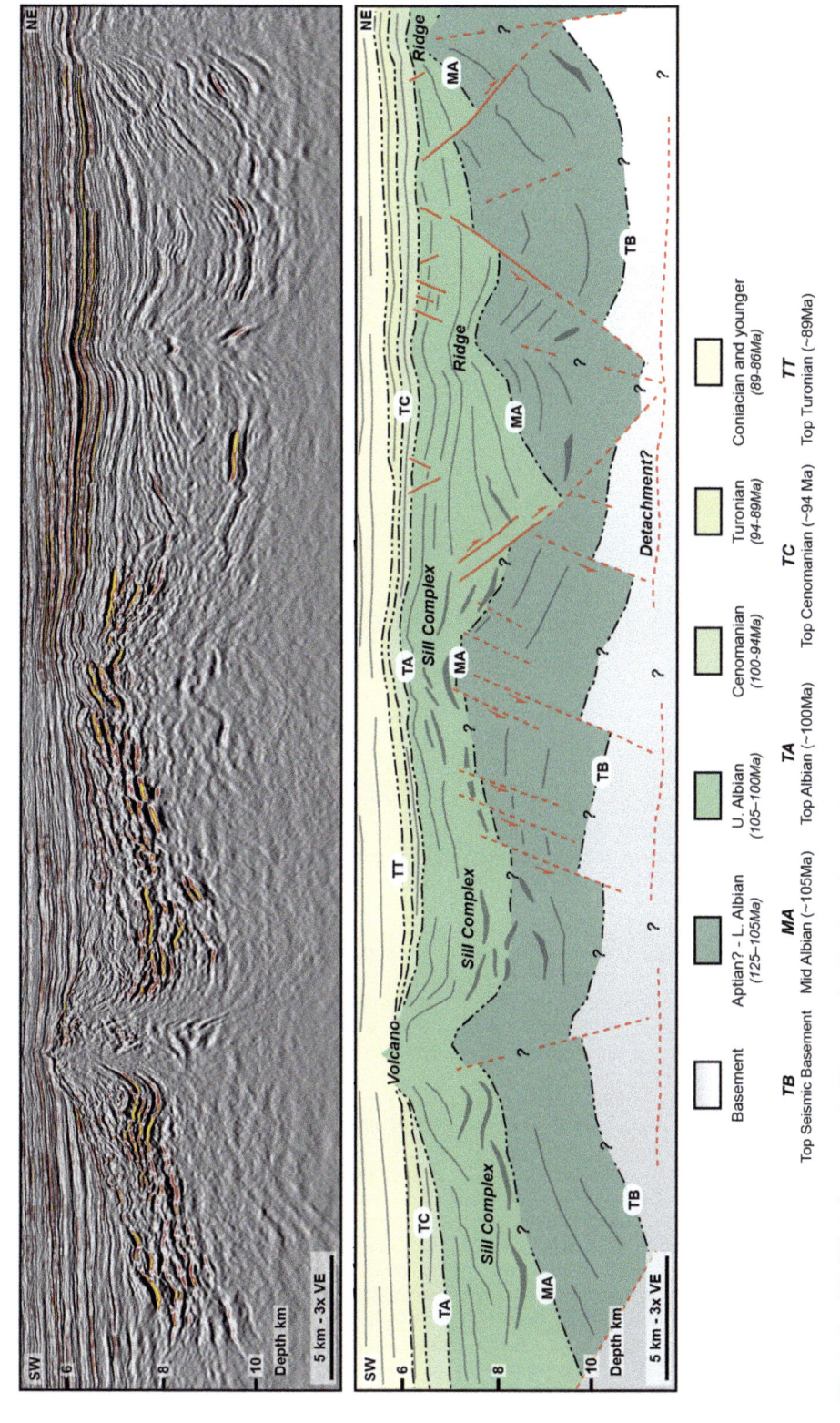

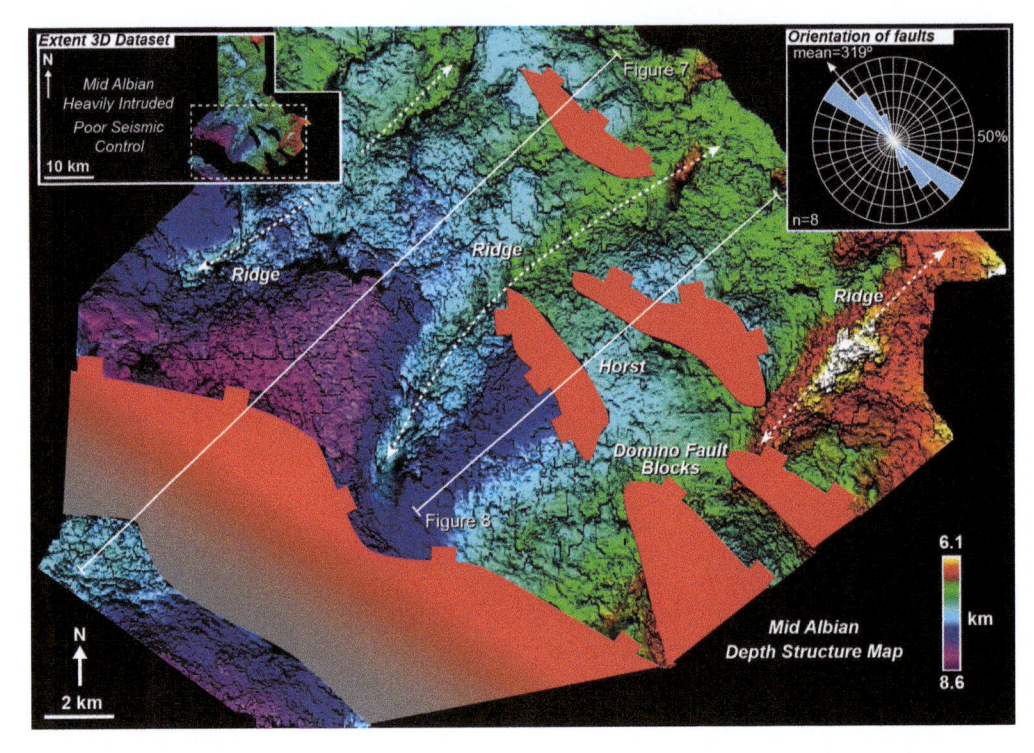

Fig. 6. Depth structure map of the Mid Albian horizon. The map shows that the mid Albian strata are affected by a series of NW–SE-trending extensional faults. The faults exhibit various structural styles, including half-graben structures and horst and graben structures, as well as domino-style fault blocks. Interpretation of the horizon was achieved in the eastern part of the dataset, where imaging of this deep section was not affected by magmatic intrusions, as shown in Figures 5 and 12b.

divides a low-amplitude, chaotic to reflection-free unit from packages with well-developed reflectivity above. This vertical arrangement of seismic facies suggests a transition from basement to a sedimentary cover above.

The age of the strata at the base of the sedimentary pile cannot be constrained. However, given that there are over 3 km of sediments below the Mid Albian horizon, it is proposed that the sediments immediately above the basement are lower Albian, possibly Aptian in age. These strata, together with the upper Albian, form a section up to 4 km thick, seismically characterized by prominent reflections with variable amplitudes. This package is offset by large normal faults which appear to terminate at depths of 11–12 km, possibly indicating the presence of a low-angle detachment (Figs 4, 5, 7 & 8). The Aptian to Albian units exhibit evidence of incisions several kilometres across (Figs 7 & 8) and, in places, are characterized by the presence of high-amplitude events, interpreted to be sills and dykes (Figs 4 & 5). These intrusions are elements of a large magmatic system imaged in the western-central part of the survey (see following sections and Fig. 12).

The Top Albian is a prominent, high-amplitude, continuous reflection that can be easily traced over a large extent of the 3D seismic volume as well as in nearby 2D seismic lines. The mapped horizon divides deformed Albian strata from packages of flat, mainly undeformed, continuous reflections of the Cenomanian, Turonian and Coniacian units (Fig. 4). The Top Cenomanian and Top Turonian seismic picks provide further subdivision within the upper Lower Cretaceous and Upper Cretaceous strata imaged within the survey.

Fault styles and timings

First-generation faults

The depth structure map of the Mid Albian horizon documents the presence of a well-developed, NW–SE-trending (mean strike 319°, $n = 8$), extensional fault system (Fig. 6). The presence of a large intrusive system meant that reliable interpretation of the Mid Albian horizon was possible only in restricted areas of the eastern part of the seismic survey (Figs 5 & 6). Here, individual faults were observed to

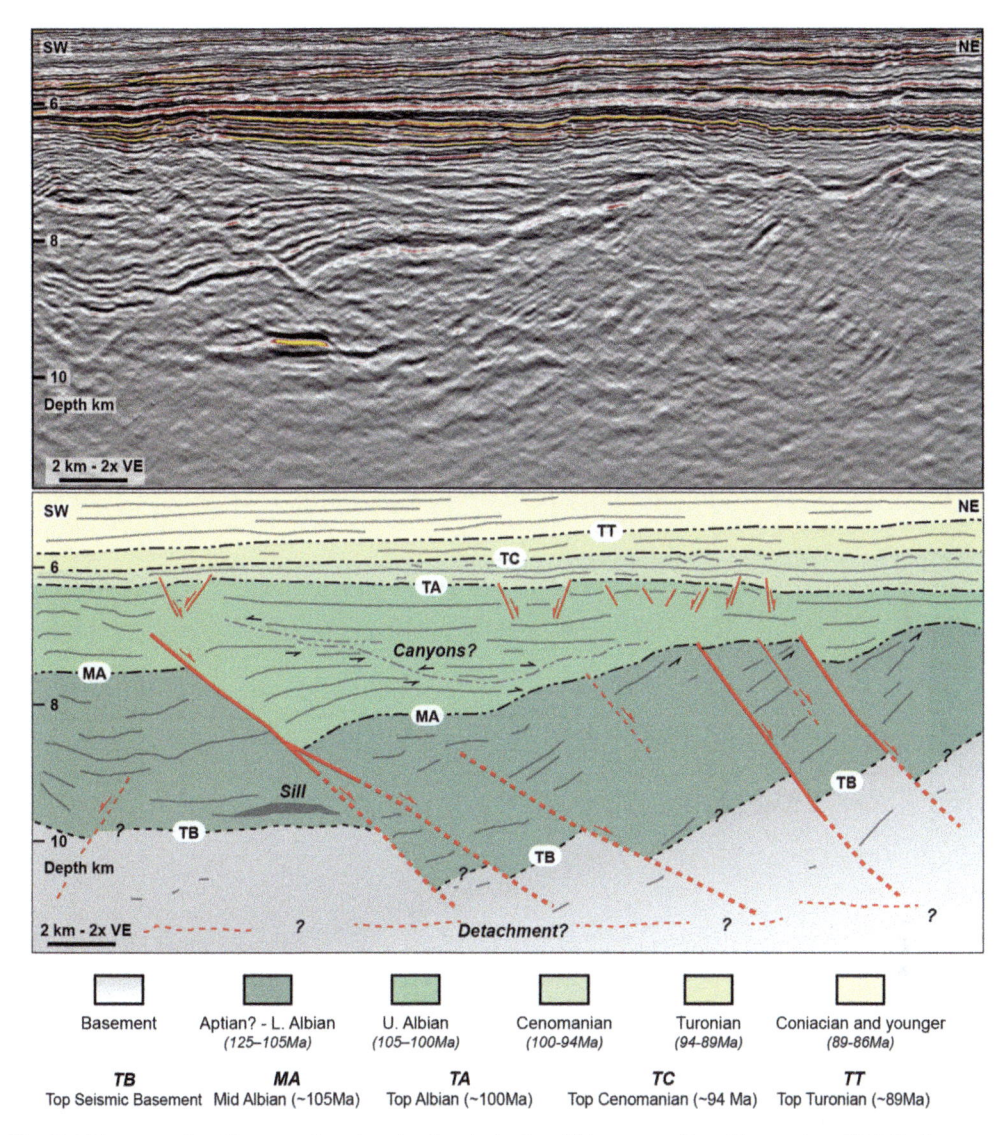

Fig. 7. Uninterpreted and interpreted section showing Aptian? to Albian strata affected by a series of NE-dipping extensional faults. The fault dips are shallowly dipping at 20–35° and delineate highly rotated fault blocks up to 10 km across (cf. Fig. 8). Shallow fault dips and marked rotation of faulted strata indicate a high level of extension above a low-angle, weak detachment surface.

extend laterally for 4–6 km, with large faults continuing well beyond the survey boundaries, possibly reaching lengths in excess of 10–20 km (Fig. 6).

Dip-oriented profiles in Figures 7 and 8 show that the fault system developed a variety of extensional geometries including half-grabens, horst and grabens as well as domino fault blocks. Individual fault blocks are commonly 3–15 km across and exhibit pronounced rotation, reaching values in excess of 35° along the larger faults. The block-bounding faults are planar and are characterized by shallow fault dips of 20–35°. Where seismic imaging allows clear mapping of the faults at depth, it can be seen that these progressively become listric as they sole on to the deep detachment level (Figs 7 & 8). Large faults extend vertically for up to 2.5 km and can attain maximum displacements in excess of 1.5 km at the mid Albian level (Fig. 7).

The hanging walls of the faults have fanning growth strata with geometries that thicken against the faults, indicating progressive rotation as extension developed (Fig. 8). In places these growth strata

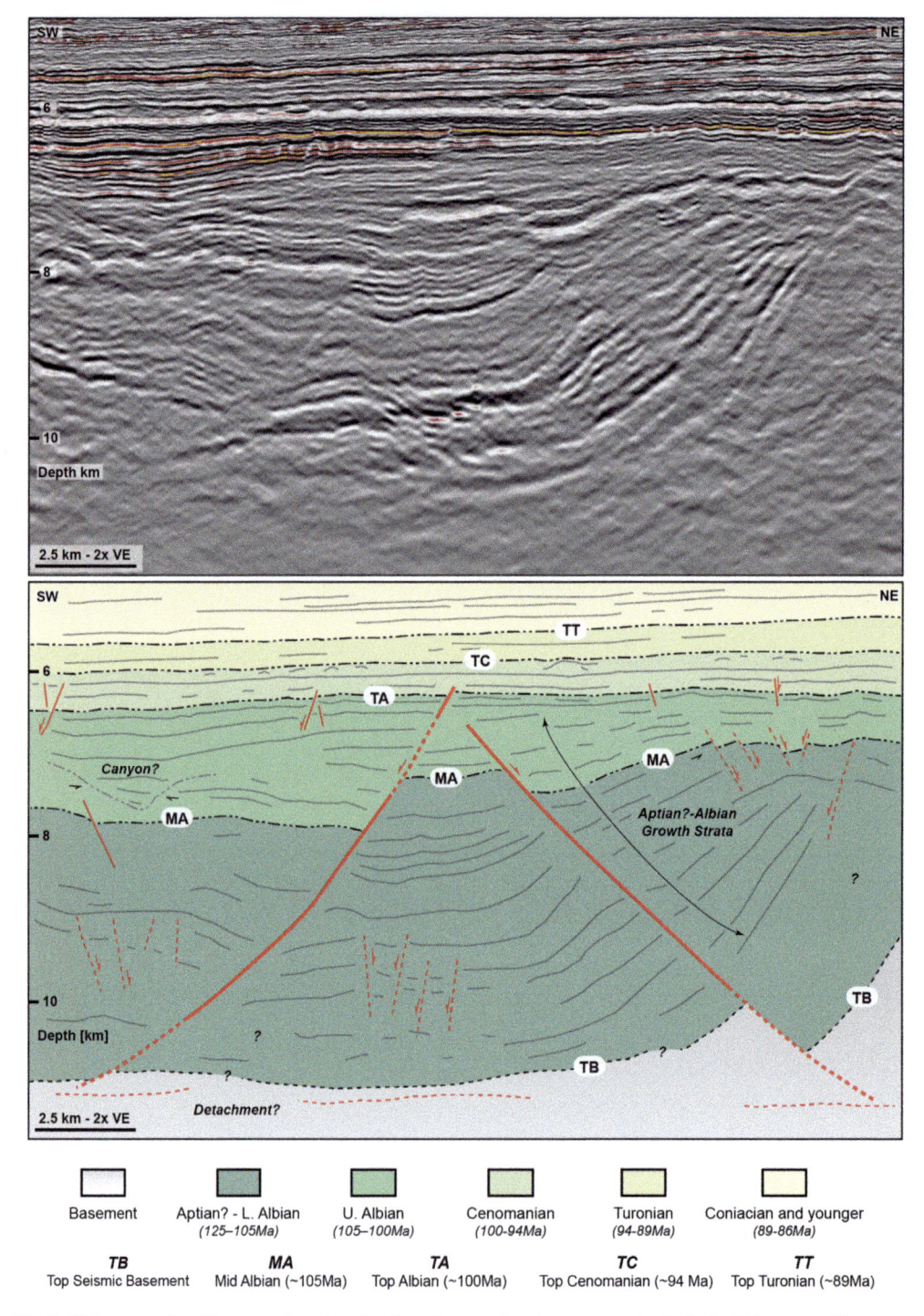

Fig. 8. Uninterpreted and interpreted section showing a horst and graben structure in Aptian?–Albian strata. The horst bounding faults are listric and are interpreted to terminate at depth on to a common low-angle detachment unit. Note that hanging wall strata to the NE clearly exhibit up to 2 km of fanning growth strata of possible Aptian-Albian age. These indicate long-lived extension until the end of the Albian. Evidence of erosional truncations below the Mid Albian horizon (cf. Fig. 7) may indicate two distinct phases of extension.

can exceed thicknesses of 3.5 km within the Aptian to Albian succession, suggesting a long-lived phase of extension at that time. Investigation of stratal terminations reveals truncations below the Mid Albian horizon (Figs 7 & 8), which could possibly indicate two phases of extension on these faults.

Second generation faults

Detailed analysis of the depth structure map and coherency extractions along the Top Albian horizon reveals numerous extensional faults affecting the uppermost part of the Albian strata (Figs 9, 10a & 11). The faults are abundant in the eastern part of the survey (Fig. 9a, c) and have an overall NW–SE trend (mean strike trend 328°; $n = 42$; Fig. 9). Attribute extractions and 3D views show that these faults are straight or slightly curvilinear with fault lengths of 1.5–8 km (Figs 9b & 10a).

Vertical profiles perpendicular to the faults show that these are planar with fault dips of c. 50° (Fig. 11). Large faults can attain offsets of c. 150 m and extend vertically for c. 2 km, offsetting the upper Albian and lower Cenomanian strata. In places, these structures seem to be linked to the larger 'first-generation' faults at depth (Fig. 8).

Given the limited displacement of these faults, growth strata are not readily apparent, limiting the determination of their exact age. However, the faults are sealed by the Top Cenomanian horizon but offset the Top Albian horizon (Fig. 11). Therefore, these relationships possibly indicate fault activity during the Cenomanian.

Ridges

Depth structure maps of the Mid Albian horizon (Fig. 6) and Top Albian horizon (Figs 9 & 10a) show a series of prominent, NE–SW-trending ridges (mean strike trend 49°, $n = 26$; Fig. 9c). The ridges range in length from 1.5 to 5 km with widths ranging from 500 m to 2 km (Figs 9 & 10a). Large ridges with widths in excess of 5 km seem to be formed by a number of coalesced structures (Fig. 10a). Three-dimensional displays of the Top Albian horizon show clear examples of ridges arranged in 'en echelon' fashion as well as examples of the Cenomanian faults offsetting, hence post-dating, the ridges. Vertical profiles through the ridges reveal that these features pertain to upper Albian and Cenomanian strata and are asymmetric, with steep limbs of 25–30° and shallow limbs of 15–20° (Fig. 10b). It is noted that asymmetry of the limbs changes along the same trend of ridges (Fig. 10a).

The 'en echelon' array of the ridges and their marked asymmetry may indicate that these features are fault-propagation folds controlled by steep faults at depth with a component of oblique slip. Lateral variations in their asymmetry also suggest an along-strike variation in the direction of dip of the faults at depth (Fig. 10b). The marked asymmetry of the ridges makes an alternative explanation of a volcanic origin less preferable. Thickness variations across the ridges as well as onlap terminations towards their crests can be clearly observed within the section bounded by Top Cenomanian and the Mid Albian horizons. These indicate a long-lived phase of growth of the ridges during the late Albian and Cenomanian.

Magmatic system

Volcanoes

Three prominent, subcircular to elliptical mounds are shown in the depth-structure maps of the Top Albian and Top Cenomanian (Figs 9 & 12a). The mounds are aligned in a NE–SW direction, which is similar to the strike trend of the ridges described above (Fig. 9). The mounds are spatially related to high-amplitude features at depths that are interpreted to be igneous sills forming an extensive magmatic system observed in the eastern part of the survey (Fig. 12a). For this reason, the mounds are ascribed to be the effusive part of such magmatic system.

Detailed interpretation of vertical profiles shows a clear volcanic stratigraphy embedded between two key surfaces, the top and the base of volcanoes (Fig. 13). The top volcano is a composite surface defined by a series of clear onlap terminations of Lower Cretaceous strata on to the volcanic edifices. The oldest onlapping strata are the uppermost Albian, capped by the Top Albian horizon that also terminates against the flanks of the volcanoes. These features suggest that extrusion and therefore formation of the magmatic system ended in the late Albian.

Below the top volcano surface, a series of well-imaged reflections dip away from the volcanic crest. These units, which are inferred to be mid Albian in age, mimic the shape of the flanks of the volcano and are therefore interpreted to be a series of volcanoclastic deposits forming the main body of the volcanic structure (Figs 12 & 13). These strata terminate downwards in a series of well-imaged downlaps that define the base of the volcano – i.e. the palaeodepositional surface at the time of extrusion (Fig. 13). It is noted that the base volcano horizon is stratigraphically located well above the Mid Albian horizon, suggesting that volcanic extrusion probably initiated in the mid to late Albian. Interpretation of the top and base volcano horizons shows that the volcanoes have basal widths from 7 to 10 km, with summit heights in excess of 1.5 km and flank dips of c. 25°. These large geomorphic

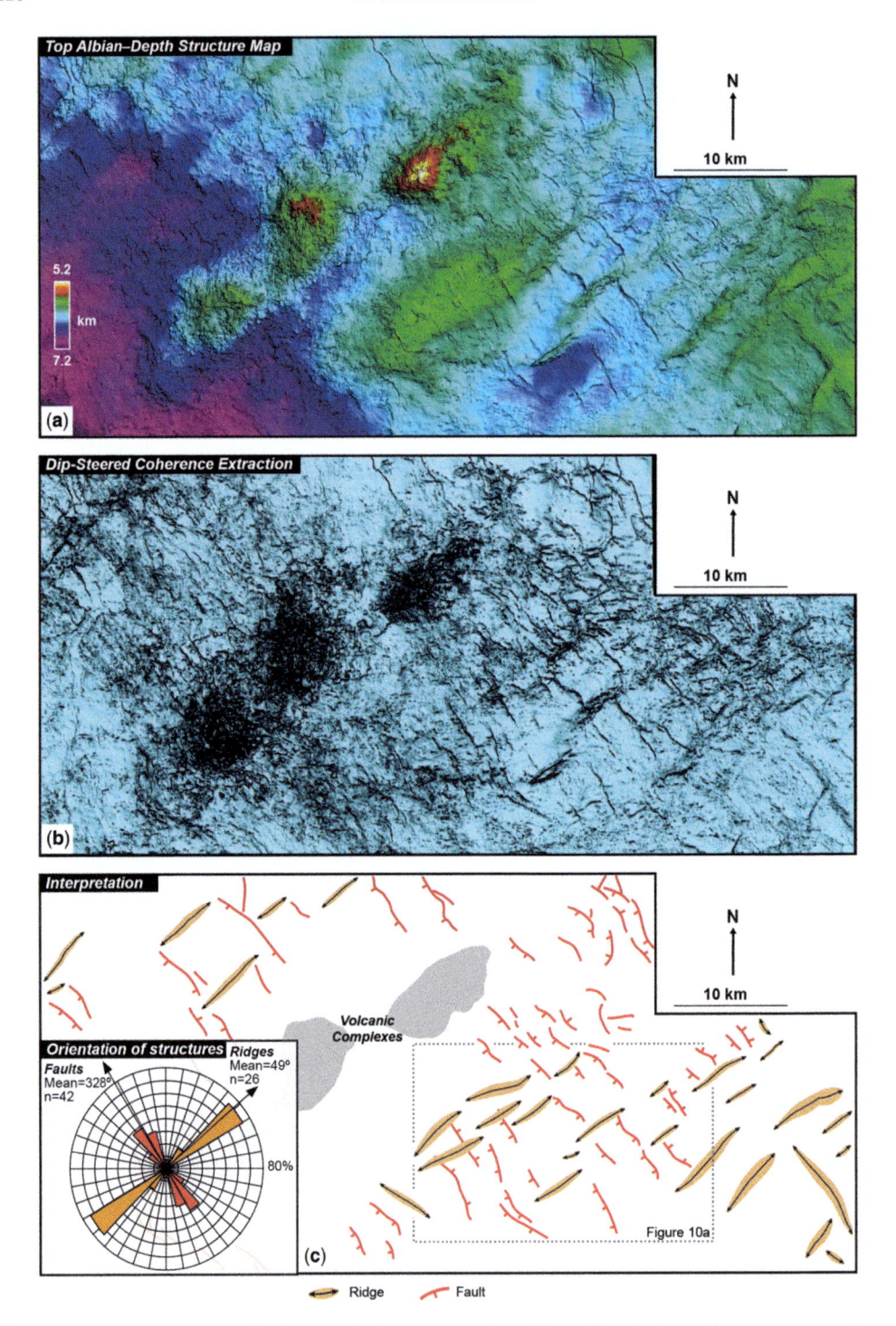

Fig. 9. (a) Depth structure map; (b) dip-steered coherency extraction of Top Albian horizon; (c) structure map with orientation of the main features described shown in the inset rose diagram. Note in (a) a series of prominent SW–NE-trending folds. In the same panel, three volcanic complexes also occur. A set of well-developed normal faults with a NNW–SSE orientation are clearly imaged in the coherency extraction shown in (b). Details of these two sets of structures are presented in Figure 10a.

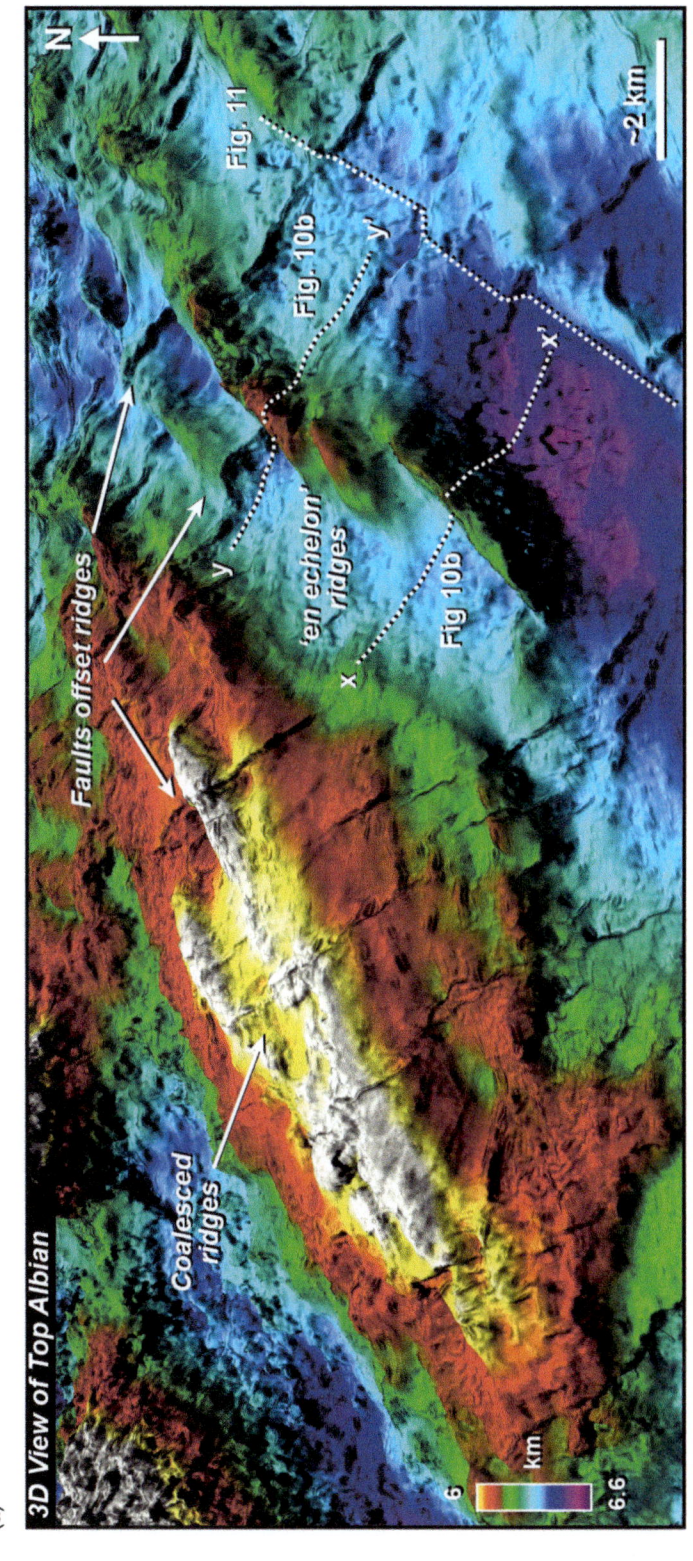

Fig. 10. (**a**) Detailed 3D view of Top Albian horizon showing folds and faults affecting this unit; (**b**) uninterpreted and interpreted detailed sections through key folds. The locations of these sections are shown in (a). Note in (a) the presence of numerous faults that offset and therefore post-date the ridges. The sections in (b) show that the ridges are markedly asymmetric, an indication that these may be associated with faults at depth. Stratal terminations and thickness variations across the fold in the Albian–Cenomanian intervals indicate active folding occurring at that time. Parallel packages of Turonian and younger stratigraphy form drape units that seal the folds. Also, note the 'en echelon' arrangement of the ridges observed in (a), possible evidence of oblique slip along the faults present in the core of these structures.

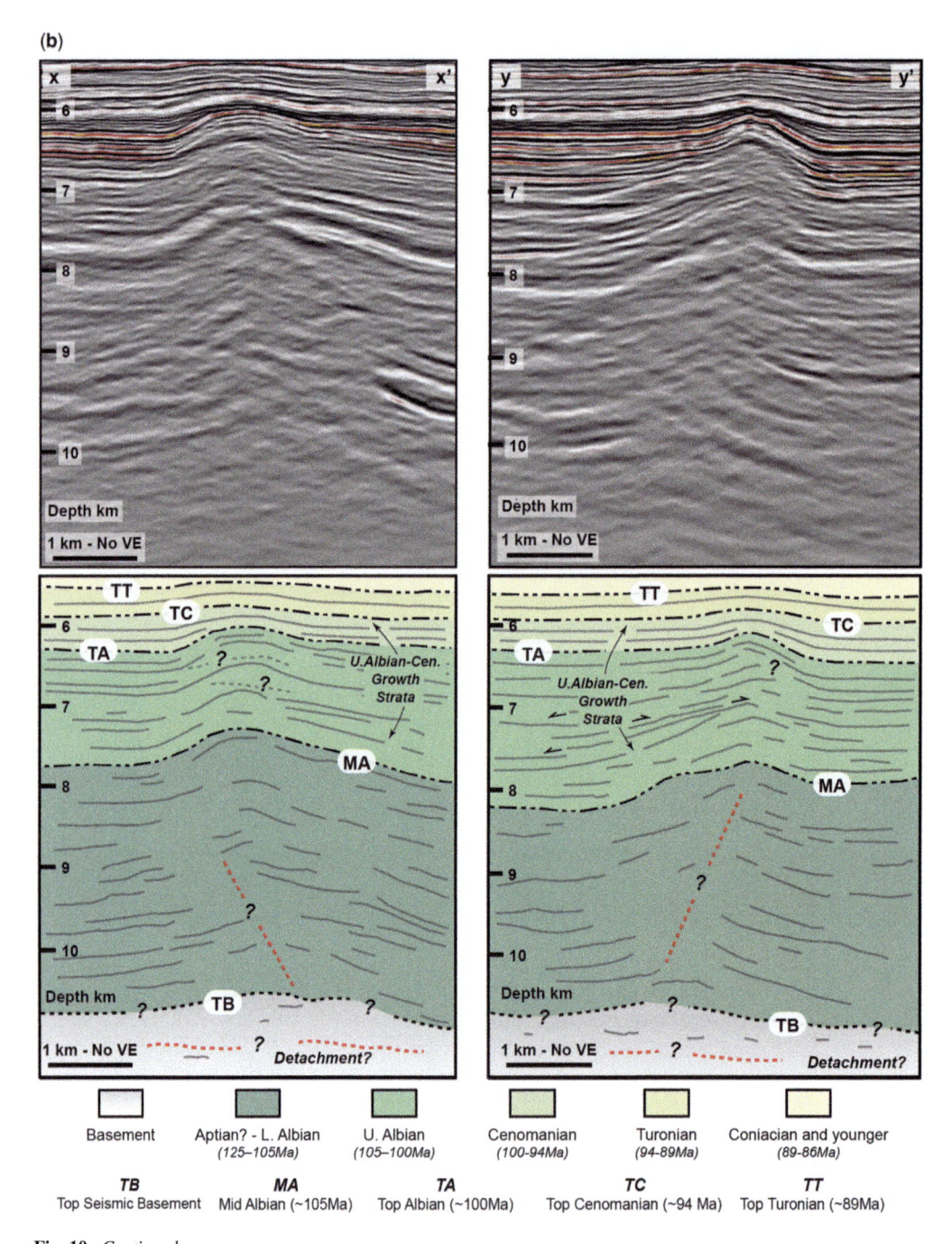

Fig. 10. *Continued.*

features undoubtedly had significant impact on the stratal architectures of the Upper Cretaceous stratigraphy as demonstrated by onlap terminations of Cenomanian and Turonian units as well as drape structures in Coniacian and younger strata.

Sills

Reflection strength extractions at a depth of 7.5 km reveal a subcircular area with a diameter of c. 70 km characterized by high-amplitude responses

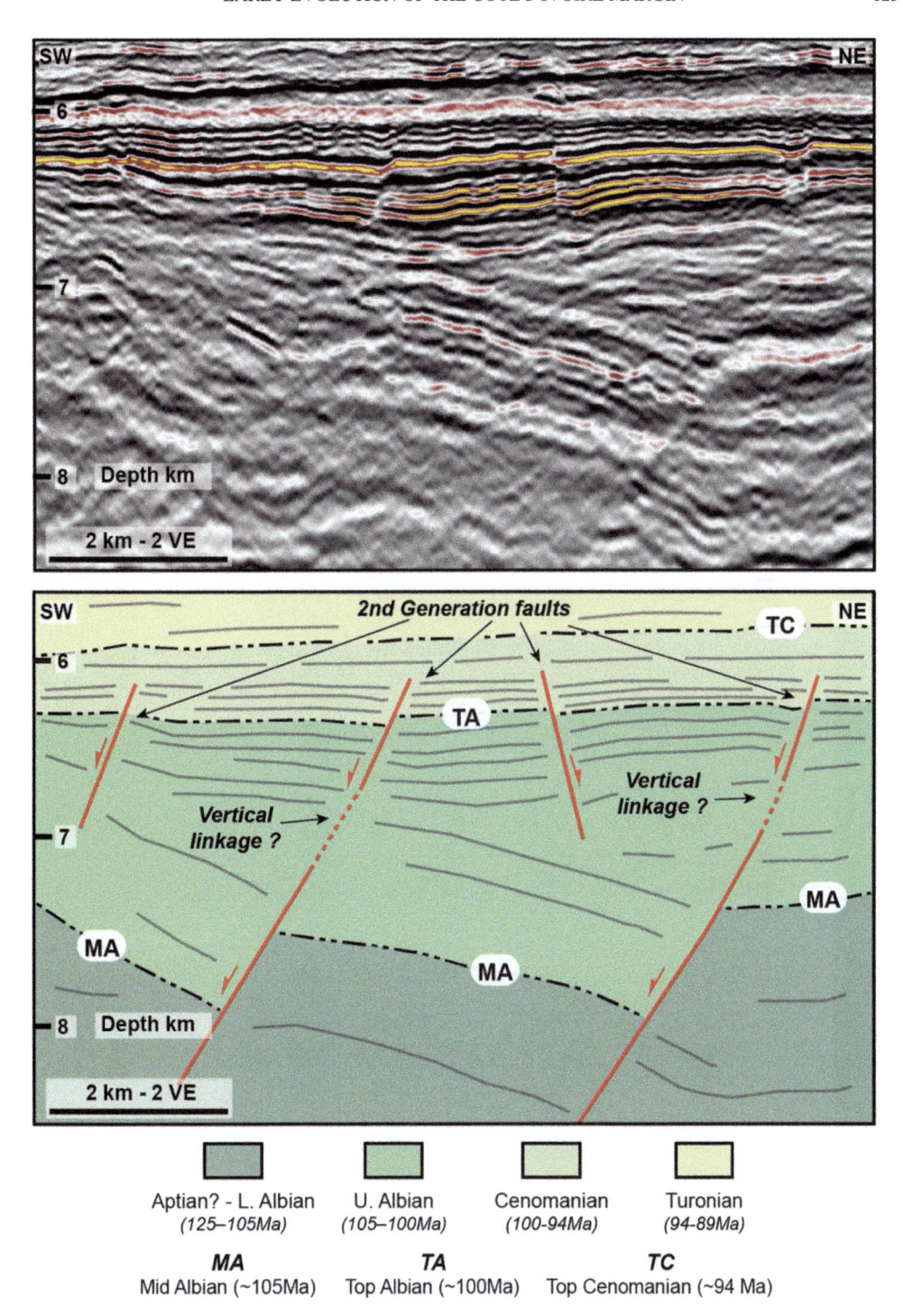

Fig. 11. Uninterpreted and interpreted detailed section showing planar faults affecting Albian–Cenomanian strata. These show limited displacement which hinders the recognition of clear growth strata. The upper tips of the faults clearly offset the Top Albian horizon, whereas the Top Cenomanian appears unaffected. This indicates that the faults were short lived, with active faulting at the transition between the Albian and Cenomanian. Some sections also show that these faults may have formed because of reactivation of older faults as observed in Figure 8.

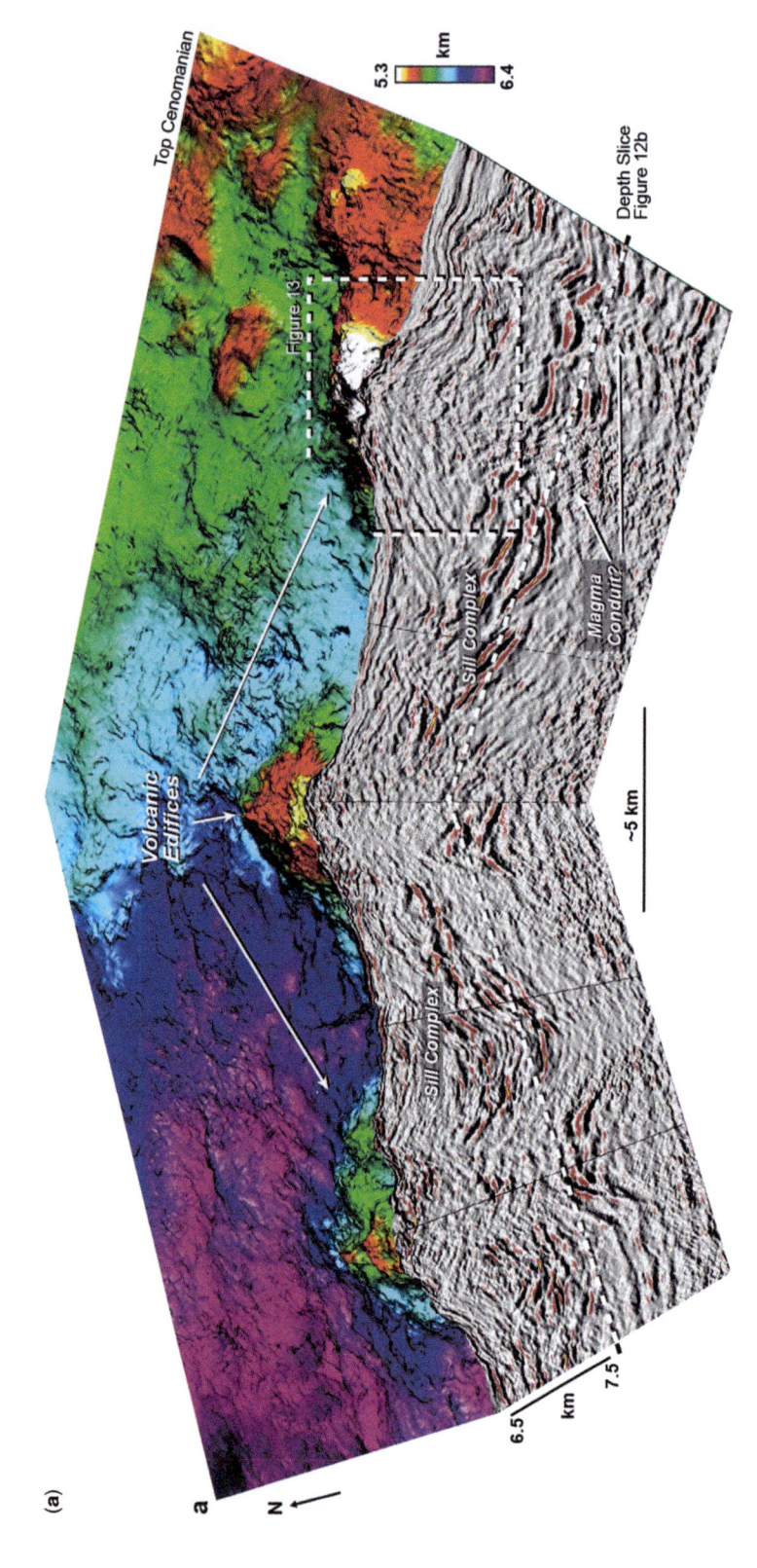

Fig. 12. (a) Seismic block diagram showing a well-developed magmatic plumbing system in the study area. The block is cropped at the Top Cenomanian horizon and shows three volcanic edifices with a series of sills at depth.

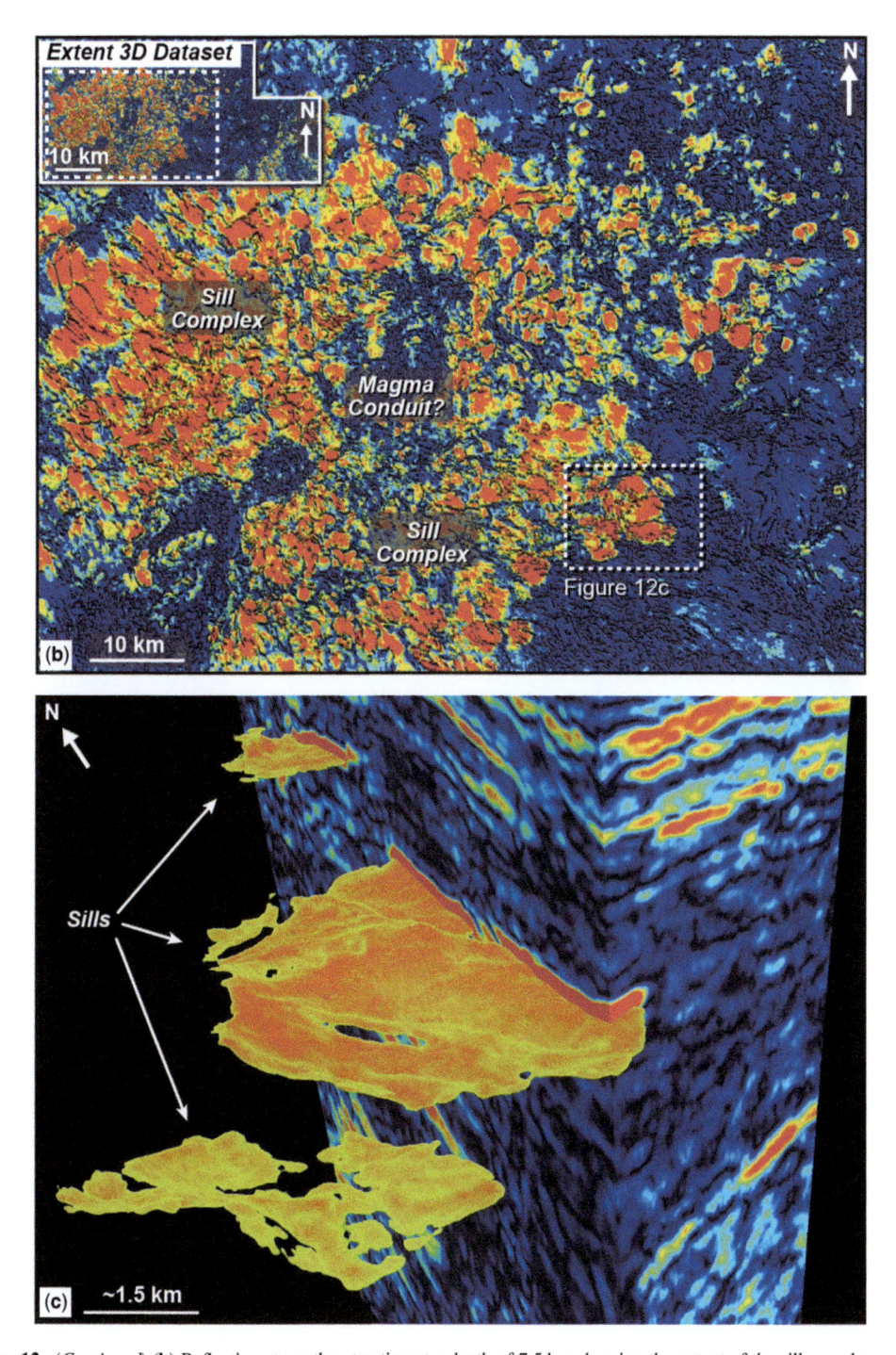

Fig. 12. (*Continued*) (**b**) Reflection strength extraction at a depth of 7.5 km showing the extent of the sill complexes. These extend over distances in excess of 70 km and developed around the volcanic edifices – in this extraction the magma conduits are imaged as circular features with a low strength response (see also Fig. 13). (**c**) Co-displayed geobodies and reflection strength in vertical section. The geobodies were constructed to map the high amplitudes associated with the sill complexes as seen on the reflection strength displays. Individual sills vary in shape and dimension from strata-concordant to saucer-shaped intrusions which can extend laterally for up to 8 km.

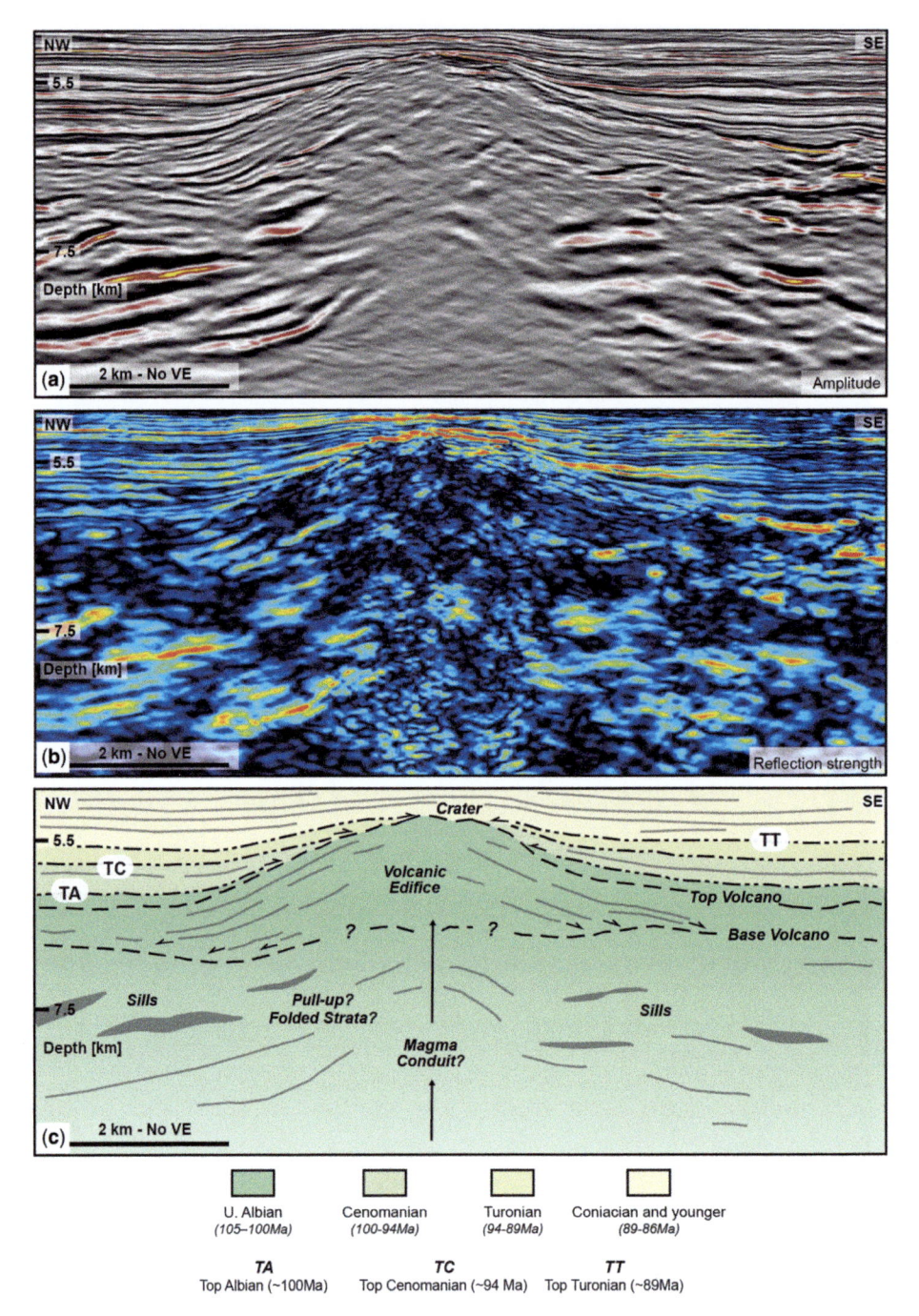

Fig. 13. Uninterpreted (**a**) amplitude and (**b**) reflection strength section showing a key example of a volcanic edifice in the study area. (**c**) Interpreted section. The surface delineating the top of the volcano (Top Volcano) is defined by a series of onlap terminations of Cenomanian and younger strata. Below this surface, a well-developed volcano stratigraphy can be recognized in the form of packages of continuous reflections of variable amplitudes dipping away from the summit and downlapping on to the palaeodepositional surface at the time of extrusion (Base Volcano). These stratal relationships indicate that volcanism mainly occurred in the late Albian. The reflection strength highlights the presence of a number of sill intrusions at depth as well as a disrupted, low-amplitude columnar zone below the volcanic edifice which may indicate the main magma conduit.

in the western part of the dataset (Fig. 12b). Detailed inspection shows numerous high-amplitude bodies 2–8 km wide (Fig. 12b, c). On vertical seismic sections these have hard reflections cutting through stratigraphy (e.g. Figs 7, 12 & 13), the typical diagnostic characteristics of igneous sills (e.g. Hansen *et al.* 2004). The sill complexes extend over an area of 2900 km^2 centred around the volcanic edifices (Fig. 12a, b). For this reason, it is interpreted that volcanoes and sills are part of the same magma system (Fig. 12). The lack of definitive age-diagnostic features, such as onlap terminations against sill-inflation anticlines, means that the timing of emplacement of the sill complexes is not constrained. However, it is assumed that sills are genetically related to the volcanics, forming the intrusive counterpart of the same magmatic system. This implies that the sills may have been emplaced slightly earlier or at the same time as the volcanoes formed, during the mid to late Albian.

Discussion

Rift-drift evolution of the Côte d'Ivoire margin

The results of this research show clear evidence of long-lived extension on the Côte d'Ivoire margin along NW–SE-trending faults that started in the Aptian and ended in the late Albian (Fig. 14). The resultant structures, with shallow, somewhat listric faults detached on to a low-angle detachment with highly rotated fault blocks (Figs 7 & 8). The detachment is probably located within the basement unit, and this together with the orientations of the faults along expected regional rift trends suggest that the extensional faults in this survey are related to the rift processes as the South Atlantic opened in the Early Cretaceous.

Hyper-extended rifted basins commonly display faults that detach above weak exhumed, serpentinized mantle (Doré & Lundin 2015). Seismic profiles from the west Iberia margin have shown low-angle, listric faults with fault dips of 20–30° that define fault blocks with rotations of 10–30° (Reston & Pérez-Gussinyé 2007; Ranero & Pérez-Gussinyé 2010; Bayrakci *et al.* 2016). A similar hyper-extension model may be invoked to explain the structural styles of the extensional faults observed offshore Côte d'Ivoire (Fig. 3). This may also be supported by analyses of refraction surveys and potential field data that indicate the presence of anomalously thin oceanic crust and serpentinized lower crustal bodies near the COB on the margin (Peirce *et al.* 1996; Antobreh *et al.* 2009).

Margin break-up

The timing of break-up from this study is constrained by the Top Albian horizon that clearly seals

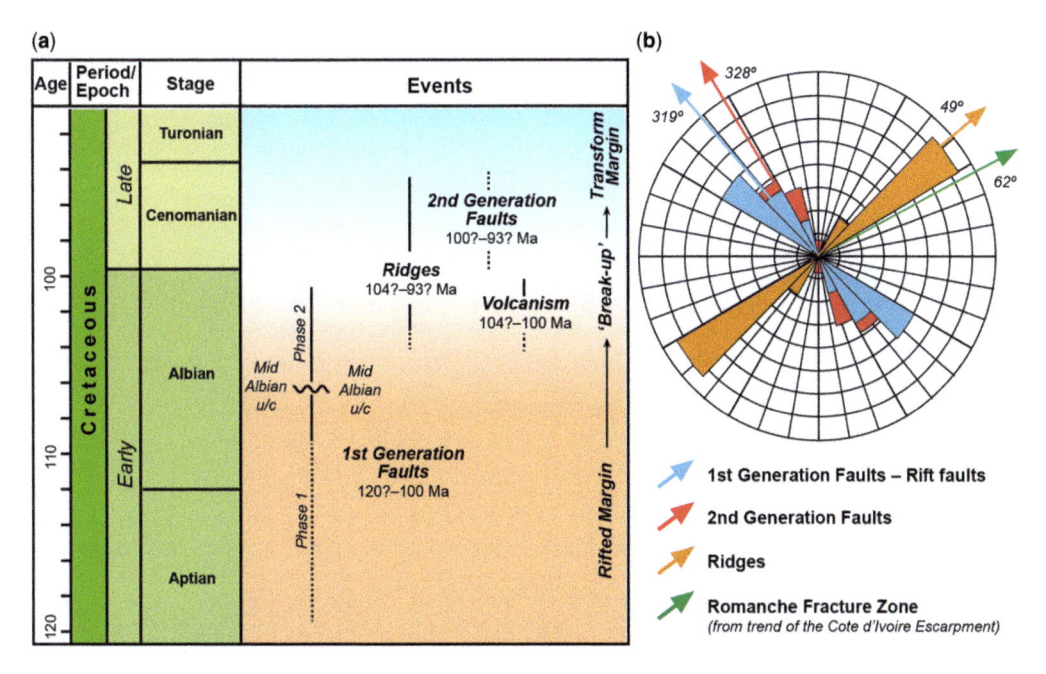

Fig. 14. (**a**) An event chart compiled from the seismic analysis presented in this study. (**b**) Summary rose diagram showing main orientations of structures identified in this research.

underlying syn-rift growth strata (Figs 6 & 7). This would imply seafloor spreading and a full marine environment offshore Côte d'Ivoire since Cenomanian time (100.5 Ma; Fig. 14a). This timing of break-up challenges some well-established plate reconstructions of the South Atlantic (Antobreh *et al.* 2009; Torsvik *et al.* 2009; Heine *et al.* 2013; Mercier de Lépinay *et al.* 2016) that indicate seafloor spreading in this segment of the Equatorial African Margin during Aptian–mid Albian. Other models from Pérez-Díaz & Eagles (2014), Granot & Dyment (2015) and Eagles (2007) predict a later break-up in the late Albian–Cenomanian, more in line with the findings of this research. Analysis of sedimentary facies (Pletsch *et al.* 2001) and stable isotope analyses from exceptionally well-preserved foraminifera from a number of ODP sites along the equatorial African margin (Friedrich *et al.* 2012) also supports a late break-up, as indicated by evidence that fully developed Atlantic circulation did not occur earlier than the mid Cenomanian (Granot & Dyment 2015).

Extension along the margin was complex, as shown by two distinct phases of stretching separated by an unconformity of mid Albian age, and the partially coeval development of ridges and volcanism (Fig. 14a). The first phase of rifting in the Aptian? to early Albian corresponds to the time that continental stretching initiated the formation of the Ivorian Tano Basin (Fig. 15a). It is envisaged that the mid Albian unconformity probably represents a time of thermally driven uplift and erosion in response to the development of a hot spreading centre to the south of the basin as stretching in the Ghana sector (to the east) of the Equatorial African Margin culminated in seafloor spreading and formation of a mid-oceanic ridge (Figs 14a & 15b). As the unconformity is embedded within the syn-rift growth sequence (Figs 4, 6 & 7), it implies that the mid Albian phase of uplift and erosion happened when rift conditions persisted in the Ivorian Tano Basin (Fig. 15b). The occurrence of this basin-wide event is also supported by studies on thermal diagenesis of clay minerals carried out on samples recovered from ODP sites along the Côte d'Ivoire–Ghana ridge, which have identified the onset in the mid–late Albian of an abnormal and short-lived palaeogeothermal gradient of *c.* 350°C/km (Holmes 1998). Similarly, Holmes (1998) and Antobreh *et al.* (2009) have indicated that this thermal event may have been induced by the passage of a seafloor-spreading centre south of the Ivorian Tano Basin which promoted and amplified the uplift of the Côte d'Ivoire–Ghana ridge. The accretion of new oceanic crust to the south of the Ivorian Tano Basin would imply the birth of an active transform margin and of the Romanche fracture zone in the mid Albian (Fig. 15b).

Magmatism

High rates of melt supply at the base of mid-oceanic ridges are thought to be responsible for formation of steady-state magma chambers along the length of ridge axes (Sinton & Detrick 1992; Kelley *et al.* 2002). Here it is proposed that the magmatic system found in the Ivorian Tano Basin (Figs 11 & 12) was linked to the magmatism associated with the mid-oceanic ridge that was supposedly located at the southern end of the basin during the mid to late Albian (Fig. 15b). This is supported by the fact that magmatism ceased in the Cenomanian (Fig. 14a), a time when, once rifting had ceased, seafloor spreading and accretion of oceanic crust progressively moved the Ivorian Tano Basin to the east, away from the mid-oceanic ridge to the south (Fig. 15c). Rift-related magmatism as an alternative interpretation for the genesis of the Ivorian magmatic system seems less plausible as volcanic activity was short-lived and only occurred at the latest stage of rifting (Fig. 14a).

Transform margin

The onset of a fully developed transform margin was also accompanied by the formation of the numerous mid Albian–Cenomanian ridges observed in the study area (Figs 9 & 10). With a strike trend similar to that of the Romanche fracture zone (Fig. 14b), the lack of laterally consistent vergence of these structures, together with their observed 'en echelon' arrangement (Fig. 10), suggest that the ridges may have developed as result of oblique slip along crustal faults that accommodated shear stresses ensued from the development of the transform margin. Evidence of folding as a mechanism of formation of the ridges may suggest that these crustal fractures are likely to be laterally complex with stepovers and bends. Such complexities are known to partition strands of strike-slip faults with distinct styles including folds and 'narrow push-up ridges' parallel to associated fault segments (Christie-Blick & Biddle 1985; Cunningham & Mann 2007). Other ridges of similar age have been reported to exist along the equatorial African margin and have been ascribed to be inversion anticlines (Tari 2006; Antobreh *et al.* 2009; Davison *et al.* 2015). In this study, however, no 'harpoon' structures (*sensu* Badley *et al.* 1989) were observed to occur associated with the ridges, and these are clearly oriented perpendicularly to the extensional fault systems (Fig. 6). This makes tectonic inversion a less preferable mechanism for the formation of the ridges in the Ivorian Tano Basin.

The fact that the volcanoes are aligned in the same direction as the strike trends of the ridges suggests that these deep structures may have developed as 'weak zones' allowing magma passage to the

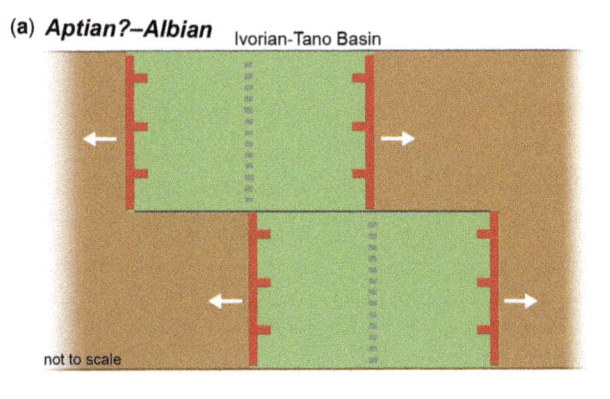

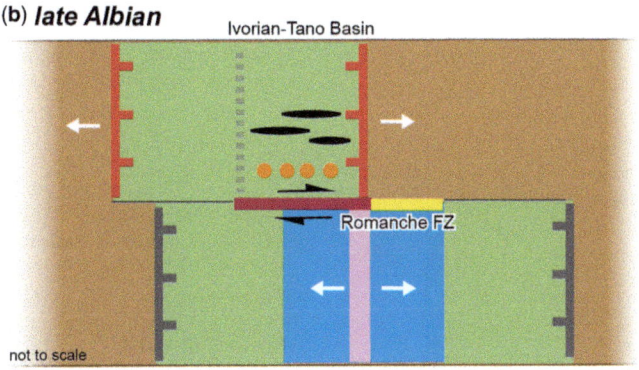

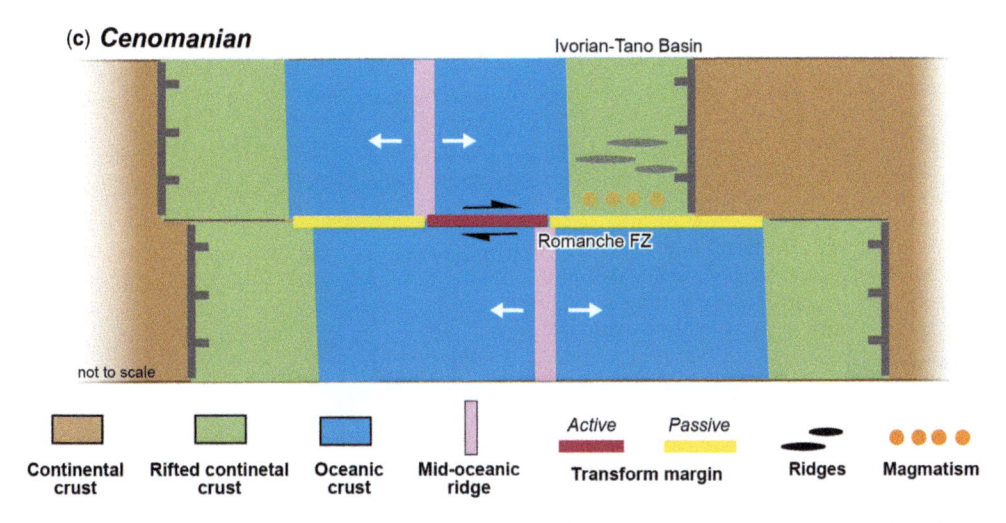

Fig. 15. Schematic evolutionary model of the Ivorian Tano Basin during the Aptian–Cenomanian. (**a**) Aptian–Albian rifting. (**b**) Continued rifting through the Albian, with a spreading centre formed to the south. Shearing along the newly formed Romanche Fracture Zone produced deformation and associated magmatism in the basin.
(**c**) Cenomanian, initiation of seafloor spreading contiguous to the Ivorian Tano Basin induced eastward migration of the basin. The basin slid past the southern spreading centre and became part of a tectonically inactive passive transform margin.

surface. The fact that the ridges became inactive in the late Cenomanian indicates the end of tectonic shearing in the Ivorian Tano Basin at that time. It is proposed that this coincided with when the basin finally slid past the southern spreading centre and entered the so-called 'passive transform margin stage' (Fig. 15c; *sensu* Mascle & Blarez 1987; Basile 2015; Mercier de Lépinay *et al.* 2016).

Cenomanian extensional faults

The last structural event documented in the basin is marked by the formation of short-lived, normal faults during the Cenomanian (Figs 11 & 14). Given their small size and limited displacements, these have not been described in previous papers that relied mainly on the analysis of widely spaced 2D seismic lines (Attoh *et al.* 2004; Antobreh *et al.* 2009).

The Cenomanian faults have a similar orientation to the rift faults to which they seem to be linked at depth (Fig. 14b). This suggests that the Cenomanian faults formed as a result of a phase of reactivation of the older and deeper rift structures. Given the evidence that rifting terminated in the late Albian (Fig. 14b), the extensional phase responsible for the formation of these late faults may have been triggered by processes of 'post-rift relaxation' (Morgan 1983; Burov & Cloetingh 1997; Burov & Poliakov 2003). The lack of known regional tectonic events in the margin during the Late Cretaceous makes it difficult to suggest alternative scenarios for the formation of the Cenomanian faults.

Implications for hydrocarbon prospectivity

The presence of a well-developed rift architecture (Figs 7 & 8) in the deep-water Ivorian Tano Basin provides additional opportunities for structural trapping in a basin thus far mostly reliant on subtle, post-rift, stratigraphic traps (MacGregor *et al.* 2003; Dailly *et al.* 2013). Within such a deep-water rift play, key controls on reservoir distribution would be in hanging wall basins where shallow marine and terrestrial drainage is likely to have developed as indicated by incisions in the hanging wall of rift structures mapped in the study area (Figs 2, 3, 7 & 8). Mid to upper Albian reservoirs have been found to be of good quality in well penetrations on the shelf of the Côte d'Ivoire margin (MacGregor *et al.* 2003). Lowermost Cretaceous, lacustrine, rift source rocks documented in the western Niger Delta (Haack *et al.* 2000) are thought to be present in similar basins along the equatorial African margin, including in the Ivorian Tano Basin (Elvsborg & Dalode 1985; MacGregor *et al.* 2003; Brownfield & Charpentier 2006). Proven rift source rocks offshore Côte d'Ivoire are found in the mid and upper Albian (Morrison *et al.* 2000; MacGregor *et al.* 2003). Therefore, contingent upon a favourable timing of migration and charge, the thick rift section documented in this study is likely to contain some of these source rocks able to charge rift reservoirs in large, fault-controlled traps (Fig. 3). An additional trapping mechanism in the rift play may come from the ridges (Figs 9 & 10). These can be up to 5 km long and 1.5 km wide with over 100 m of vertical relief and hence may offer considerable trap potential. The trapping potential is much higher for coalesced ridges that may be five times longer than individual ridges (Fig. 10a). As the Cenomanian post-rift faults have offset part of the upper Albian rift section (Fig. 11), they may pose a risk for seal integrity for shallower rift reservoirs. Migration along these faults, however, would have the potential to allow charging of deep-water post-rift reservoirs from rift source rocks. Additional post-rift charge may come from well-known Cenomanian source rocks on the equatorial African margin (Morrison *et al.* 2000; MacGregor *et al.* 2003; Dailly *et al.* 2013). Recent exploration effort has documented the presence of high-quality Upper Cretaceous turbidite reservoirs in the inner slope offshore Ghana (Jubilee field) and Côte d'Ivoire (Paon discovery; Fig. 1; Dailly *et al.* 2013; Coole *et al.* 2015; Martin *et al.* 2015). Detailed seismic analysis and drilling have indicated that these reservoirs extend into the deep water of the Ivorian Tano Basin (Martin *et al.* 2015). Here the main control on distribution of these sands would be the post-rift seabed geomorphology, that this research has demonstrated to likely be affected by the presence of prominent volcanoes and ridges (Figs 10, 12 & 13). These features, which may be ultimately related to deep crustal transforms, could develop confined basins where turbidite flows of Turonian to Campanian age would possibly pond, whereas highs would form barriers against which sands would onlap and pinch out, creating potential for accumulation of reservoirs and opportunity for stratigraphic traps to develop. The recent discoveries of Pelican and Rossignol (Fig. 1b) confirm that such an Upper Cretaceous turbidite play in the deep-water Ivorian Tano Basin is very prolific.

Conclusions

The main findings of this research are:

- Aptian? to late Albian Atlantic continental rifting produced a well-developed rift system in the Ivorian Tano Basin. Rift styles include rotated fault blocks and horst and graben structures up to 10 km across.
- The presence of a 'weak' detachment for the rift faults together with their low angle dips and the occurrence of highly rotated fault blocks provide evidence of hyper-extension during rifting.
- Rifting was punctuated by a phase of thermally driven uplift that brought about a basin-wide phase of erosion of the rift morphology. This thermal event is attributed to the development of

the mid-oceanic ridge south of the Ivorian Tano Basin during the mid Albian.
- The transition of the Côte d'Ivoire margin to a full transform margin occurred in the late Albian to Cenomanian with the formation of crustal, oblique-slip faults with 'en echelon' folds at their upper tips.
- Crustal weakening along these large faults controlled the emplacement of the mid to late Albian volcanic centres and subvolcanic sill complexes.
- This paper shows the first evidence of Albian magmatism in the region as indicated by the well-imaged volcanoes and associated sills at depth.
- In the Cenomanian, a phase of extension formed small-scale normal faults which, in places, are linked to rift faults at depth. The tectonic control on this late extensional phase is unclear and may include 'post-rift relaxation' processes or a previously unreported Late Cretaceous regional tectonic event that promoted reactivation of rift faults.
- The variety of stratigraphic and structural features developed through the Cretaceous history of the basin offers high-trapping potentials for a number of plays in the rift as well as in the post-rift sequences.

Acknowledgments The authors would like to thank CGG Multi-Client and New Ventures for providing the seismic data and permission to publish this paper. Resources and equipment for this research were kindly provided by the Fault Dynamics Research Group. Halliburton are thanked for kindly providing Landmark seismic interpretation software as a University Research Grant. The authors are indebted to Gabor Tari, Ian Davison and Teresa Sabato Ceraldi for their critical review of the manuscript.

References

ANTOBREH, A.A., FALEIDE, J.I., TSIKALAS, F. & PLANKE, S. 2009. Rift–shear architecture and tectonic development of the Ghana margin deduced from multichannel seismic reflection and potential field data. *Marine Petroleum Geology*, **26**, 345–368, https://doi.org/10.1016/j.marpetgeo.2008.04.005

ATTOH, K., BROWN, L., GUO, J. & HEANLEIN, J. 2004. Seismic stratigraphic record of transpression and uplift on the Romanche transform margin, offshore Ghana. *Tectonophysics*, **378**, 1–16, https://doi.org/10.1016/j.tecto.2003.09.026

BADLEY, M.E., PRICE, J.D. & BACKSHALL, L.C. 1989. Inversion, reactivated faults and related structures: seismic examples from the southern North Sea. *In*: COOPER, M.A. & WILLIAMS, G.D. (eds) *Inversion Tectonics*. Geological Society, London, Special Publications, **44**, 201–219, https://doi.org/10.1144/GSL.SP.1989.044.01.12

BAHORICH, M. & FARMER, S. 1995. The coherence cube. *Leading Edge*, **14**, 1053–1058.

BARNES, A. 2007. A tutorial on complex seismic trace analysis. *Geophysics*, **72**, W33–W43, https://doi.org/10.1190/1.2785048

BASILE, C. 2015. Transform continental margins – part 1: concepts and models. *Tectonophysics*, **661**, 1–10, https://doi.org/10.1016/j.tecto.2015.08.034

BASILE, C., MASCLE, J., POPOFF, M., BOUILLIN, J.P. & MASCLE, G. 1993. The Ivory Coast–Ghana transform margin: a marginal ridge structure deduced from seismic data. *Tectonophysics*, **222**, 1–19, https://doi.org/10.1016/0040-1951(93)90186-N

BAYRAKCI, G., MINSHULL, T.A. ET AL. 2016. Fault-controlled hydration of the upper mantle during continental rifting. *Nature Geoscience*, **9**, 384–388, https://doi.org/10.1038/ngeo2671

BROWNFIELD, M.E. & CHARPENTIER, R.R. 2006. Geology and total petroleum systems of the Gulf of Guinea Province of west Africa. *US Geological Survey Bulletin*, **2207**-C.

BUROV, E. & CLOETINGH, S. 1997. Erosion and rift dynamics: new thermomechanical aspects of post-rift evolution of extensional basins. *Earth and Planetary Science Letters*, **150**, 7–26, https://doi.org/10.1016/S0012-821X(97)00069-1

BUROV, E. & POLIAKOV, A. 2003. Erosional forcing of basin dynamics: new aspects of syn-and post-rift evolution. *In*: NIEUWLAND, D.A. (ed.) *New Insights into Structural Interpretation and Modelling*. Geological Society, London, Special Publications, **212**, 209–223, https://doi.org/10.1144/GSL.SP.2003.212.01.14

CARTWRIGHT, J. & HUUSE, M. 2005. 3D seismic technology: the geological 'Hubble'. *Basin Research*, **17**, 1–20, https://doi.org/10.1111/j.1365-2117.2005.00252.x

CHIERICI, M.A. 1996. Stratigraphy, palaeoenvironments and geological evolution of the Ivory Coast–Ghana basin. *Géologie L'Afrique L'Atlantique Sud Actes Colloques Angers*, **1994**, 293–303.

CHOPRA, S. & MARFURT, K.J. 2007. *Seismic Attributes for Prospect Identification and Reservoir Characterization, Seg Geophysical Developments*. Society of Exploration Geophysicists, Tulsa, OK.

CHOPRA, S. & MARFURT, K.J. 2008. Introduction to this special section-Seismic Attributes. *Leading Edge*, **27**, 296–297, https://doi.org/10.1190/1.2896619

CHRISTIE-BLICK, N. & BIDDLE, K.T. 1985. Deformation and basin formation along strike-slip faults. *In*: BIDDLE, K.T. & CHRISTIE-BLICK, N. (eds) *Strike-Slip Deformation, Basin Formation, and Sedimentation, Deformation and Basin Formation along Strike-Slip Faults*. SEPM, Special Publications, Tulsa, Oklahoma, USA, **37**, 1–34.

CLIFT, P.D., LORENZO, J., CARTER, A., HURFORD, A.J. & ODP LEG 159 SCIENTIFIC PARTY 1997. Transform tectonics and thermal rejuvenation on the Côte d'Ivoire–Ghana margin, west Africa. *Journal of the Geological Society*, **154**, 483–489, https://doi.org/10.1144/gsjgs.154.3.0483

COOLE, P., TYRRELL, M. & ROCHE, C. 2015. Offshore Ivory Coast: reducing exploration risk. *GeoExpro*, **12**, 80–84.

CUNNINGHAM, W.D. & MANN, P. 2007. Tectonics of strike-slip restraining and releasing bends. *In*: CUNNINGHAM, W.D. & MANN, P. (eds) *Tectonics of Strike-Slip Restraining and Releasing Bends*. Geological Society, London, Special Publications, **290**, 1–12, https://doi.org/10.1144/SP290.1

DAILLY, P., HENDERSON, T., HUDGENS, E., KANSCHAT, K. & LOWRY, P. 2013. Exploration for Cretaceous stratigraphic traps in the Gulf of Guinea, West Africa and the discovery of the Jubilee Field: a play opening discovery in the Tano Basin, Offshore Ghana. In: MOHRIAK, W.U., DANFORTH, A., POST, P.J., BROWN, D.E., TARI, G.C., NEMČOK, M. & SINHA, S.T. (eds) Conjugate Divergent Margins. Geological Society, London, Special Publications, 369, 235–248, https://doi.org/10.1144/SP369.12

DAVISON, I., FAULL, T., GREENHALGH, J., BEIRNE, E.O. & STEEL, I. 2015. Transpressional structures and hydrocarbon potential along the Romanche Fracture Zone: a review. In: NEMČOK, M., RYBÁR, S., SINHA, S.T., HERMESTON, S.A. & LEDVÉNYIOVÁ, L. (eds) Transform Margins: Development, Controls and Petroleum Systems. Geological Society, London, Special Publications, 431, 235–248, https://doi.org/10.1144/SP431.2

DE GROOT, P. & BRIL, B. 2005. The open source model in geosciences and OpendTect in particular. SEG Technical Program Expanded Abstracts, 2005, 802–805, https://doi.org/10.1190/1.2148280

DORÉ, T. & LUNDIN, E. 2015. Hyperextended continental margins – knowns and unknowns. Geology, 43, 95–96, https://doi.org/10.1130/focus012015.1

DUMESTRE, M.A. 1985. Northern Gulf of Guinea shows promise. Oil and Gas Journal, 83, 154.

EAGLES, G. 2007. New angles on South Atlantic opening. Geophysical Journal International, 168, 353–361.

ELVSBORG, A. & DALODE, J. 1985. Benin hydrocarbon potential looks promising. Oil and Gas Journal, 6, 126–131.

FIRTH, J., HORSTAD, I. & SCHAKEL, M. 2014. Experiencing the full bandwidth of energy from exploration to production with the art of BroadSeis. First Break, 32, 89–97.

FRIEDRICH, O., NORRIS, R.D. & ERBACHER, J. 2012. Evolution of middle to Late Cretaceous oceans – a 55 m.y. record of Earth's temperature and carbon cycle. Geology, 40, 107–110, https://doi.org/10.1130/G32701.1

GRANOT, R. & DYMENT, J. 2015. The Cretaceous opening of the South Atlantic Ocean. Earth and Planetary Science Letters, 414, 156–163, https://doi.org/10.1016/j.epsl.2015.01.015

HAACK, R.C., SUNDARARAMAN, P., DIEDJOMAHOR, J.O., XIAO, H., GANT, N.J., MAY, E.D. & KELSCH, K. 2000. Niger Delta petroleum systems, Nigeria. In: MELLO, M.R. & KATZ, B.J. (eds) Petroleum Systems of South Atlantic Margins. AAPG, Memoirs, Tulsa, Oklahoma, USA, 73, 213–231.

HANSEN, D.MØ., CARTWRIGHT, J.A. & THOMAS, D. 2004. 3D seismic analysis of the geometry of igneous sills and sill junction relationships. In: DAVIES, R.J., CARTWRIGHT, J.A., STEWART, S.A., LAPPIN, M. & UNDERHILL, J.R. (eds) 3D Seismic Technology: Application to the Exploration of Sedimentary Basins. Geological Society, London, Memoirs, 29, 199–208, https://doi.org/10.1144/GSL.MEM.2004.029.01.19

HEINE, C., ZOETHOUT, J. & MÜLLER, R.D. 2013. Kinematics of the South Atlantic rift. Solid Earth, 4, 215–253, https://doi.org/10.5194/se-4-215-2013

HILL, D., COMBEE, C. & BACON, J. 2006. Over/under acquisition and data processing: the next quantum leap in seismic technology? First Break, 24, 81–96.

HOLMES, M.A. 1998. Thermal diagenesis of Cretaceous sediment recovered at the Côte d'Ivoire Ghana Transform Margin. In: MASCLE, J., LOHMANN, G.P. & MOULLADE, M. (eds) Proceedings of the Ocean Drilling Program, Scientific Results. College Station, TX, Ocean Drilling Program, TX, USA, 53–70.

IACOPINI, D. & BUTLER, R.W. 2011. Imaging deformation in submarine thrust belts using seismic attributes. Earth and Planetary Science Letters, 302, 414–422.

JONES, E.J.W., CANDE, S.C. & SPATHOPOULOS, F. 1995. Evolution of a major oceanographic pathway: the equatorial atlantic. In: SCRUTTON, R.A., STOKER, M.S., SHIMMIELD, G.B. & TUDHOPE, A.W. (eds) The Tectonics, Sedimentation and Palaeoceanography of the North Atlantic Region. Geological Society, London, Special Publications, 90, 199–213, https://doi.org/10.1144/GSL.SP.1995.090.01.12

KELLEY, D.S., BAROSS, J.A. & DELANEY, J.R. 2002. Volcanoes, fluids, and life at mid-ocean ridge spreading centers. Annual Reviews in Earth and Planetary Science, 30, 385–491, https://doi.org/10.1146/annurev.earth.30.091201.141331

KJEMPERUD, A., AGBESINYALE, W., AGDESTEIN, T., GUSTAFSSON, C. & YÜKLER, A. 1991. Tectono-stratigraphic history of the Keta basin, Ghana with emphasis on late erosional episodes. Géologie Africaine: Colloque de Géologie de Libreville, recueil des Communications, 6–8 May 1991, 55–69.

MACGREGOR, D.S., ROBINSON, J. & SPEAR, G. 2003. Play fairways of the Gulf of Guinea transform margin. In: ARTHUR, T.J., MACGREGOR, D.S. & CAMERON, N.R. (eds) Petroleum Geology of Africa: New Themes and Developing Technologies. Geological Society, London, Special Publications, 207, 131–150, https://doi.org/10.1144/GSL.SP.2003.207.7

MARFURT, K.J. & ALVES, T.M. 2015. Pitfalls and limitations in seismic attribute interpretation of tectonic features. Interpretation, 3, SB5–SB15, https://doi.org/10.1190/INT-2014-0122.1

MARFURT, K.J., KIRLIN, R.L., FARMER, S.L. & BAHORICH, M.S. 1998. 3-D seismic attributes using a semblance-based coherency algorithm. Geophysics, 63, 1150–1165.

MARTIN, J., DUVAL, G. & LAMOURETTE, L. 2015. What lies beneath the deepwater tano basin. GeoExpro, 12, 28–30.

MASCLE, J. & BLAREZ, E. 1987. Evidence for transform margin evolution from the Ivory Coast–Ghana continental margin. Nature, 326, 378–381, https://doi.org/10.1038/326378a0

MCCLAY, K.R. 1990. Extensional fault systems in sedimentary basins: a review of analogue model studies. Marine Petroleum Geology, 7, 206–233, https://doi.org/10.1016/0264-8172(90)90001-W

MCCLAY, K.R. & ELLIS, P.G. 1987. Geometries of extensional fault systems developed in model experiments. Geology, 15, 341–344, https://doi.org/10.1130/0091-7613(1987)15<341:GOEFSD>2.0.CO;2

MERCIER DE LÉPINAY, M., LONCKE, L., BASILE, C., ROEST, W.R., PATRIAT, M., MAILLARD, A. & DE CLARENS, P. 2016. Transform continental margins – Part 2: a worldwide review. Tectonophysics, 693(Part A), 96–115, https://doi.org/10.1016/j.tecto.2016.05.038

MORGAN, P. 1983. Constraints on rift thermal processes from heat flow and uplift. Tectonophysics, Processes *of Continental Rifting*, **94**, 277–298, https://doi.org/10.1016/0040-1951(83)90021-5

MORRISON, J., BURGESS, C., CORNFORD, C. & N'ZALASSE, B. 2000. Hydrocarbon systems of the Abidjan margin, Côte d'Ivoire. Offshore West Africa. *Fourth Annual Conference*, 21–23 March 2000, Abidjan. Pennwell Publishing, Tulsa, OK.

MOULIN, M., ASLANIAN, D. & UNTERNEHR, P. 2010. A new starting point for the South and Equatorial Atlantic Ocean. *Earth-Science Reviews*, **98**, 1–37, https://doi.org/10.1016/j.earscirev.2009.08.001

NEMČOK, M., HENK, A., ALLEN, R., SIKORA, P.J. & STUART, C. 2013. Continental break-up along strike-slip fault zones; observations from the Equatorial Atlantic. *In*: MOHRIAK, W.U., DANFORTH, W.U., POST, P.J., BROWN, D.E., TARI, G.C., NEMČOK, M. & SINHA, S.T. (eds) *Conjugate Divergent Margins*. Geological Society, London, Special Publications, **369**, 537–556, https://doi.org/10.1144/SP369.8

NEMČOK, M., RYBÁR, S., SINHA, S.T., HERMESTON, S.A. & LEDVÉNYIOVÁ, L. 2016. *Transform Margins: Development, Controls and Petroleum Systems*. Geological Society, London, Special Publications, **431**, https://doi.org/10.1144/SP431

NÜRNBERG, D. & MÜLLER, R.D. 1991. The tectonic evolution of the South Atlantic from Late Jurassic to present. *Tectonophysics*, **191**, 27–53.

PEIRCE, C., WHITMARSH, R.B., SCRUTTON, R.A., PONTOISE, B., SAGE, F. & MASCLE, J. 1996. Côte d'Ivoire–Ghana margin: seismic imaging of passive rifted crust adjacent to a transform continental margin. *Geophysical Journal International*, **125**, 781–795, https://doi.org/10.1111/j.1365-246X.1996.tb06023.x

PÉREZ-DÍAZ, L. & EAGLES, G. 2014. Constraining South Atlantic growth with seafloor spreading data. *Tectonics*, **33**, 1848–1873, https://doi.org/10.1002/2014TC003644

PLETSCH, T., ERBACHER, J. ET AL. 2001. Cretaceous separation of Africa and South America: the view from the West African margin (ODP Leg 159). *Journal of South American Earth Science, Mesozoic Palaeontology and Stratigraphy of South America and the South Atlantic*, **14**, 147–174, https://doi.org/10.1016/S0895-9811(01)00020-7

POSAMENTIER, H.W., DAVIES, R.J., CARTWRIGHT, J.A. & WOOD, L. 2007. Seismic geomorphology – an overview. *In*: DAVIES, R.J., POSAMENTIER, H.W. & WOOD, L.J. (eds) *Seismic Geomorphology: Applications to Hydrocarbon Exploration and Production*. Geological Society, London, Special Publications, **277**, 1–14, https://doi.org/10.1144/GSL.SP.2007.277.01.01

RABINOWITZ, P.D. & LABRECQUE, J. 1979. The Mesozoic South Atlantic Ocean and evolution of its continental margins. *Journal of Geophysical Research*, **84**, 5973–6002.

RANERO, C.R. & PÉREZ-GUSSINYÉ, M. 2010. Sequential faulting explains the asymmetry and extension discrepancy of conjugate margins. *Nature*, **468**, 294–299, https://doi.org/10.1038/nature09520

RESTON, T.J. & PÉREZ-GUSSINYÉ, M. 2007. Lithospheric extension from rifting to continental breakup at magma-poor margins: rheology, serpentinisation and symmetry. *International Journal of Earth Science (Geologische Rundschau)*, **96**, 1033–1046, https://doi.org/10.1007/s00531-006-0161-z

SABATO CERALDI, T., HODGKINSON, R. & BACKÉ, G. 2016. *The Petroleum Geology of the West Africa Margin: An Introduction*. Geological Society, London, Special Publications, **438**, https://doi.org/10.1144/SP438

SAGE, F., BASILE, C., MASCLE, J., PONTOISE, B. & WHITMARSH, R.B. 2000. Crustal structure of the continent–ocean transition off the Côte d'Ivoire–Ghana transform margin: implications for thermal exchanges across the palaeotransform boundary. *Geophysical Journal International*, **143**, 662–678, https://doi.org/10.1046/j.1365-246X.2000.00276.x

SANDWELL, D.T., MÜLLER, R.D., SMITH, W.H.F., GARCIA, E. & FRANCIS, R. 2014. New global marine gravity model from CryoSat-2 and Jason-1 reveals buried tectonic structure. *Science*, **346**, 65–67, https://doi.org/10.1126/science.1258213

SINTON, J.M. & DETRICK, R.S. 1992. Mid-ocean ridge magma chambers. *Journal of Geophysical Research: Solid Earth*, **97**, 197–216, https://doi.org/10.1029/91JB02508

TARI, G. 2006. Traditional and new play types of the offshore Tano Basin of Côte d'Ivoire and Ghana, West Africa. *Houston Geological Society Newsletter*, January, **48**, 27–34.

TORSVIK, T.H., ROUSSE, S., LABAILS, C. & SMETHURST, M.A. 2009. A new scheme for the opening of the South Atlantic Ocean and the dissection of an Aptian salt basin. *Geophysical Journal International*, **177**, 1315–1333.

WELLS, S., WARNER, M., GREENHALGH, J. & BORSATO, R. 2012. Offshore Côte d'Ivoire: a modern exploration frontier. *GeoExpro*, **9**, 36–40.

YE, J., CHARDON, D., ROUBY, D., GUILLOCHEAU, F., DALL'ASTA, M., FERRY, J.-N. & BROUCKE, O. 2017. Paleogeographic and structural evolution of northwestern Africa and its Atlantic margins since the early Mesozoic. *Geosphere*, **13**, 1–31.

Estimating palaeobathymetry with quantified uncertainties: a workflow illustrated with South Atlantic data

L. PÉREZ-DÍAZ[1]* & G. EAGLES[2]

[1]*Department of Earth Sciences, Royal Holloway, University of London, Egham, Surrey TW20 0EX, UK*

[2]*Alfred Wegener Institute, Helmholtz Centre for Polar and Marine Research, Am Alten Hafen 26, 27568, Bremerhaven, Germany*

L.P.D., 0000-0002-6922-3275
Correspondence: lucia.perezdiaz@rhul.ac.uk

Abstract: We present and illustrate a workflow to produce palaeobathymetric reconstructions, using examples from the South Atlantic Ocean. With a recent high-resolution plate kinematic model as the starting point, we calculate an idealized basement surface by applying plate-cooling theory to seafloor ages and integrating the results with depths along the extended continental margins. Then, we refine the depths of this basement surface to account for the effects of sedimentation, variations in crustal thickness and dynamic topography. Finally, the corrected idealized surface is cut along appropriate plate outlines for the desired time slice and reconstructed using appropriate Euler parameters.

In order to assess the applicability of modelled results, we critically examine the limitations and uncertainties resulting from the datasets used and assumptions made. Palaeobathymetry modelled with our approach is likely to be least reliable over parts of large igneous provinces close to the times of their eruption, and most reliable within the oceanic interiors for Neogene time slices. The uncertainty range is not smaller than 500 m for any significant region at any time, and its mean over 95% of locations in all time slices is close to 1800 m.

Palaeobathymetry is an essential boundary condition for studies and models of palaeocirculation, palaeoclimate and hydrocarbon prospectivity. By integrating published studies about plate kinematics and the thermal structure of oceanic lithosphere with subsidence models for continent–ocean transition zones, grids of sedimentary and crustal thickness, and dynamic topography estimates, we have produced a workflow that can be used for any oceanic basin for which tectonic motions are well constrained. Here, we describe this workflow using the South Atlantic (Fig. 1) as an example.

At first order, plate tectonics control palaeobathymetry both by determining the changes in the geographical location of the lithosphere and by the changes in its vertical level (through the mechanism of thermal subsidence). Using a kinematic model of the South Atlantic opening (Pérez-Díaz & Eagles 2014) as the starting point, we model palaeobathymetry following the steps below:

- We use a high-resolution seafloor-age grid (Pérez-Díaz & Eagles 2017*a*), derived from the plate kinematic model, to model the subsidence of oceanic lithosphere as a function of its age by applying plate-cooling theory (GDH1: Stein & Stein 1992).
- We implement a method for modelling the subsidence of continent–ocean transition zones (COTZs) through time, which allows us to extend the thermally subsiding surface as far as areas of unstretched continental crust at the ocean margins.
- We refine the resulting top-of-basement surface to account for other factors affecting bathymetry at smaller scales or amplitudes, both within the ocean and the COTZs (variations in sediment and crustal thickness; and topography of Large Igneous Provinces (LIPs), aseismic ridges and seamounts, and dynamic topography).

Some steps within this workflow account for processes that are relatively well understood and/or they use datasets whose uncertainties are well known, such as the age grid and its application in calculating oceanic thermal subsidence. Others are more susceptible to introducing errors due to large inherent uncertainties in the datasets (e.g. dynamic topography), or poor knowledge or understanding of the timing or identity of processes (e.g. in COTZ subsidence). We describe these uncertainties and a method for quantifying them that allows us to present deepest and shallowest palaeobathymetric error models for any given model age.

From: McCLAY, K. R. & HAMMERSTEIN, J. A. (eds) 2020. *Passive Margins: Tectonics, Sedimentation and Magmatism.* Geological Society, London, Special Publications, **476**, 135–162.
First published online March 1, 2018, https://doi.org/10.1144/SP476.1
© 2018 The Author(s). Published by The Geological Society of London. All rights reserved.
For permissions: http://www.geolsoc.org.uk/permissions. Publishing disclaimer: www.geolsoc.org.uk/pub_ethics

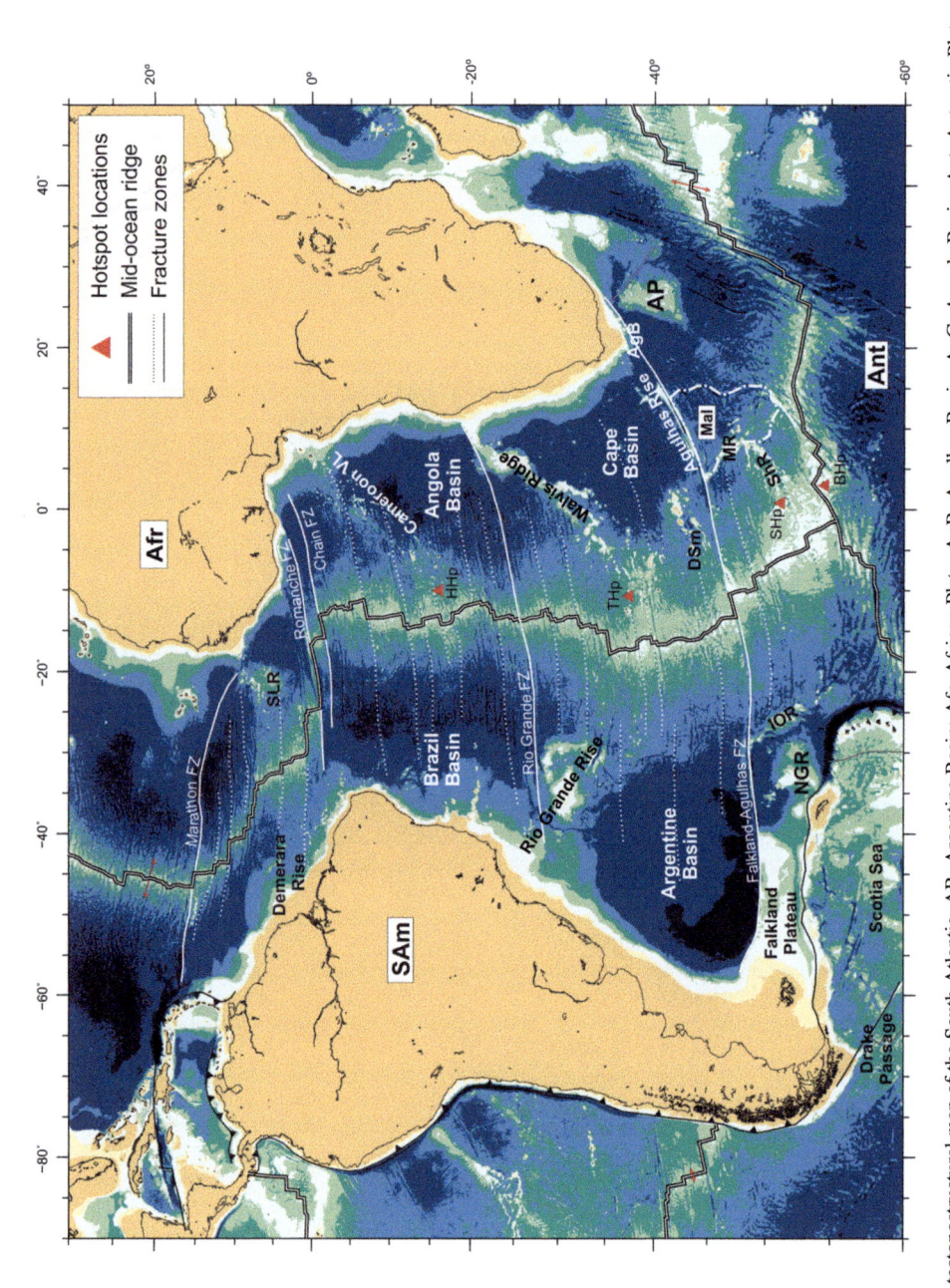

Fig. 1. General tectonostructural map of the South Atlantic. AB, Argentine Basin; Afr, African Plate; AgB, Agulhas Basin; AnG, Angola Basin; Ant, Antarctic Plate; AP, Agulhas Plateau; BHp, Bouvet Hotspot; Cameroon VL, Cameroon Volcanic Line; DSm, Discovery Seamounts; HHp, St Helena Hotspot; IOR, Islas Orcadas Rise; MR, Meteor Rise; NGR, North Georgian Rise; SAm, South American Plate; SHp, Shona Hotspot; ShR, Shona Ridge; SLR, Sierra Leona Rise; THp, Tristán da Cunha Hotspot.

Generating top-of-basement surfaces

Thermal subsidence

The thermal evolution of oceanic lithosphere through time is one of the most frequently revisited problems in geodynamics. Observations of the decrease in heat flow and increase in depth with seafloor age have prompted two main groups of models aimed at describing the way in which the oceanic lithosphere cools and subsides as it spreads away from mid-ocean ridges. In one, the lithosphere behaves as the cold upper boundary layer of a cooling half-space ('Half-space' models: e.g. Turcotte & Oxburgh 1967; Parker & Oldenburg 1973; Davis & Lister 1974). Comparisons of model predictions of heat flow and depth data show that half-space cooling models systematically overpredict depth and underpredict heat flow for older oceanic lithosphere, although small areas of seafloor following half-space subsidence trends can be found for almost all available ages of oceanic lithosphere (e.g. Adam & Vidal 2010). The second group, of so-called plate-cooling models, results from a desire to portray the more widespread observation of seafloor flattening with age. They are built by fitting curves to observations of the variability of seafloor depth or heat flow with age, assuming that they characterize the cooling and thermal contraction of a lithosphere whose isothermal lower boundary flattens with age (Langseth et al. 1966; McKenzie 1967). This flattening has been variously attributed to shear heating in the asthenosphere (Schubert et al. 1976), radioactive heating (Forsyth 1977; Jarvis & Peltier 1980), dynamic phenomena (Schubert & Turcotte 1972; Schubert et al. 1978; Morgan & Smith 1992) and thermal rejuvenation by hotspot reheating events (Crough 1978; Heestand & Crough 1981; Nagihara et al. 1996; Smith & Sandwell 1997) or smaller-scale convection in the uppermost mantle (Afonso et al. 2008). An in-depth review of these processes is provided by McNutt (1995), but here it is enough to note that attempts to improve thermal models by accounting explicitly for any of them have not produced significant improvements to predictive models for seafloor depth with age. With these considerations in mind, we have not attempted to generate a best-fitting depth–age curve for the South Atlantic, for which we find that plate-cooling models, in general, and GDH1, in particular (Stein & Stein 1992), adequately depict thermal subsidence where the seafloor age is known (Fig. 2). Other thermal models may be preferable for different ocean basins and should be given some consideration when modelling thermal subsidence.

For the present day, we use the seafloor age grid of Pérez-Díaz & Eagles (2017a) directly as input to calculate depths below sea level due to thermal subsidence as modelled by GDH1 (Stein & Stein 1992) after having adjusted the equations to account for a deeper average ridge depth in the South Atlantic than that in GDH1 (−2657 m).

For any given time before present day (t), we first generate a correction surface that, when subtracted from the present-day age grid, adjusts its ages to eliminate those younger than t (Fig. 3a). Then, we apply GDH1 (Stein & Stein 1992) equations to calculate a thermal subsidence surface for time t (Fig. 3b).

Continent–ocean transition zone (COTZ) depth through time

In order to achieve smooth palaeobathymetric reconstructions covering not only the oceanic parts of an ocean basin but also extending over the neighbouring extended continental crust, the shape through time of the COTZs needs to be modelled. To make this possible, we generated an idealized subsidence surface that crosses the COTZ, seamlessly covering the space between its oceanic and continental extremes according to the following scheme:

- The extent of the COTZ along the South American and African margins is defined by two lines: (1) a control line on land, located within undoubtedly continental and unstretched lithosphere ('outer line' or OL); and a control line beyond the distal edge of the extended continental margin (onwards 'inner line' or IL), within undoubtedly oceanic crust. These lines are conservative estimates that we have digitized by taking into account the locations of the inwards and outwards edges of the ensemble of the continent–ocean boundary (COB) identifications compiled by Eagles et al. (2015), as well as the locations of cratonic areas within South America and Africa.
- Depths along the IL for time t are determined using GDH1 and the age grid (Stein & Stein 1992; Pérez-Díaz & Eagles 2017a).
- Heights along the OL are fixed for times between the onset of seafloor spreading and the present day. These depths are sampled from a present-day topography map (Smith & Sandwell 1997) from which the isostatic contributions of the variable sediment and crustal thickness to topography, as well as those of the dynamic topography, have been removed.

With the IL and OL depths for time t set, depth profiles between these two control lines might be modelled in a number of ways. At long wavelengths, COTZs can be treated as thermally subsiding, or as flexural edge-of-plate or intraplate features. Figure 4 shows that simple flexural calculations produce, in many cases, theoretical bathymetric profiles across COTZs that, when the effects of sedimentation, stretching and dynamic topography are restored,

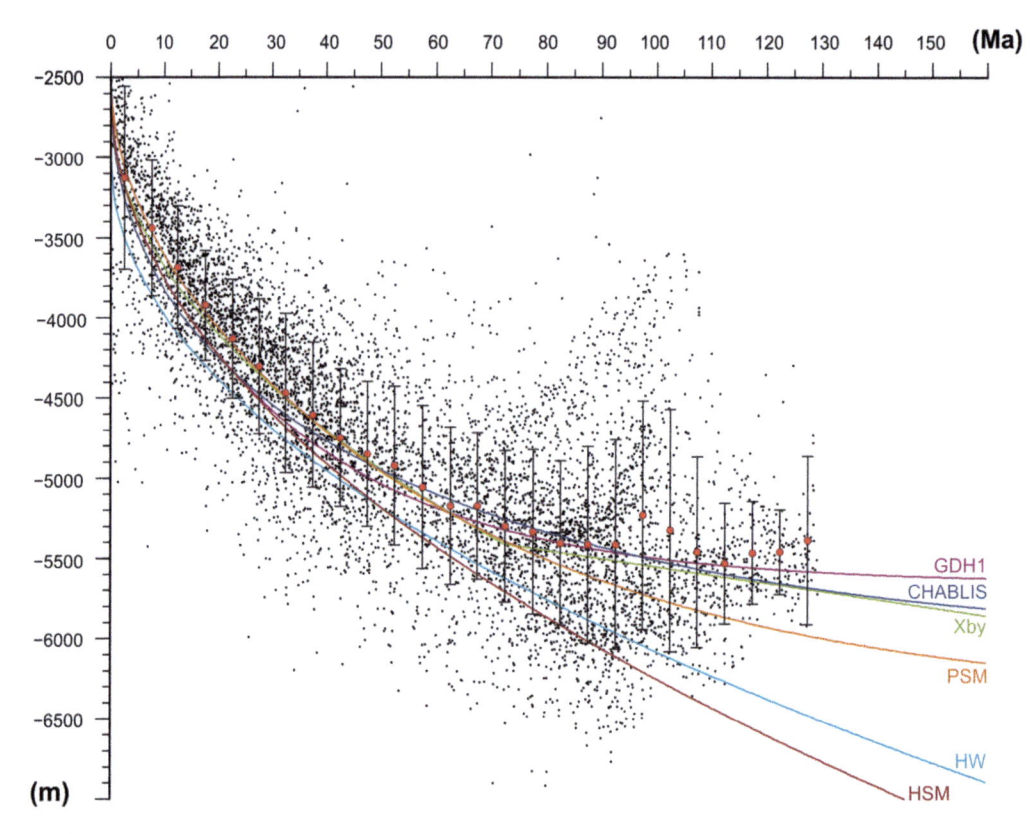

Fig. 2. Depth–age data for the South Atlantic plotted over various thermal model curves. Age data: seafloor age grid of Pérez-Díaz & Eagles (2017a). Depth data: extracted from a bathymetric map of the South Atlantic corrected for sedimentation, crustal thickness variations and dynamic topography. Red circles and black bars are averages and standard deviations calculated for each 5 myr bin, respectively. GDH1, Stein & Stein (1994); CHABLIS, Doin & Fleitout (1996); Xby, Crosby *et al.* (2006); PSM, Parsons & Sclater (1977); HW, Hillier & Watts (2005); HSM, Davis & Lister (1974).

resemble closely present-day observations. However, in some areas, a flexural curve does a poor job of replicating the shape of the margin. For this reason, we take an alternative approach that uses present-day bathymetry as a guide to the past shape of COTZs and is likely to be more applicable in ocean basins globally. We start by extracting depth information, at equally spaced points between IL and OL, from a map of present-day bathymetry corrected for the isostatic effects of sediment and crustal thickness variations and dynamic topography. By doing this, flow lines across COTZs become depth profiles independent of sedimentation and crustal stretching, whose effects vary with time. These depth profiles are then normalized and adjusted so that points along the IL always lie at depths predicted by GDH1 (Stein & Stein 1992).

This approach implies the assumption that long-wavelength depth profiles of COTZs only change in response to sedimentation, cooling of the oceanic lithosphere and dynamic support from the convecting mantle; but that at isostatically supported wavelengths, the shape of the underlying basement is largely a consequence of extensional tectonics in the upper crust and break-up volcanism, and therefore remains constant post-break-up. This assumption finds support in physical and numerical models of continental margin evolution (e.g. see Blaich *et al.* 2010; Huismans & Beaumont 2011; Brune *et al.* 2014). Because, currently, we make no attempt to palinspastically restore the extended continental margins, the assumption of stable post-break-up basement topography should not introduce large errors.

Goswami *et al.* (2015) presented a modelling method for reconstructing present-day global ocean bathymetry whose treatment of COTZs bears many similarities to the one we describe above. They calculated depth to basement at the seawards limit of COTZs by applying plate-cooling theory to oceanic

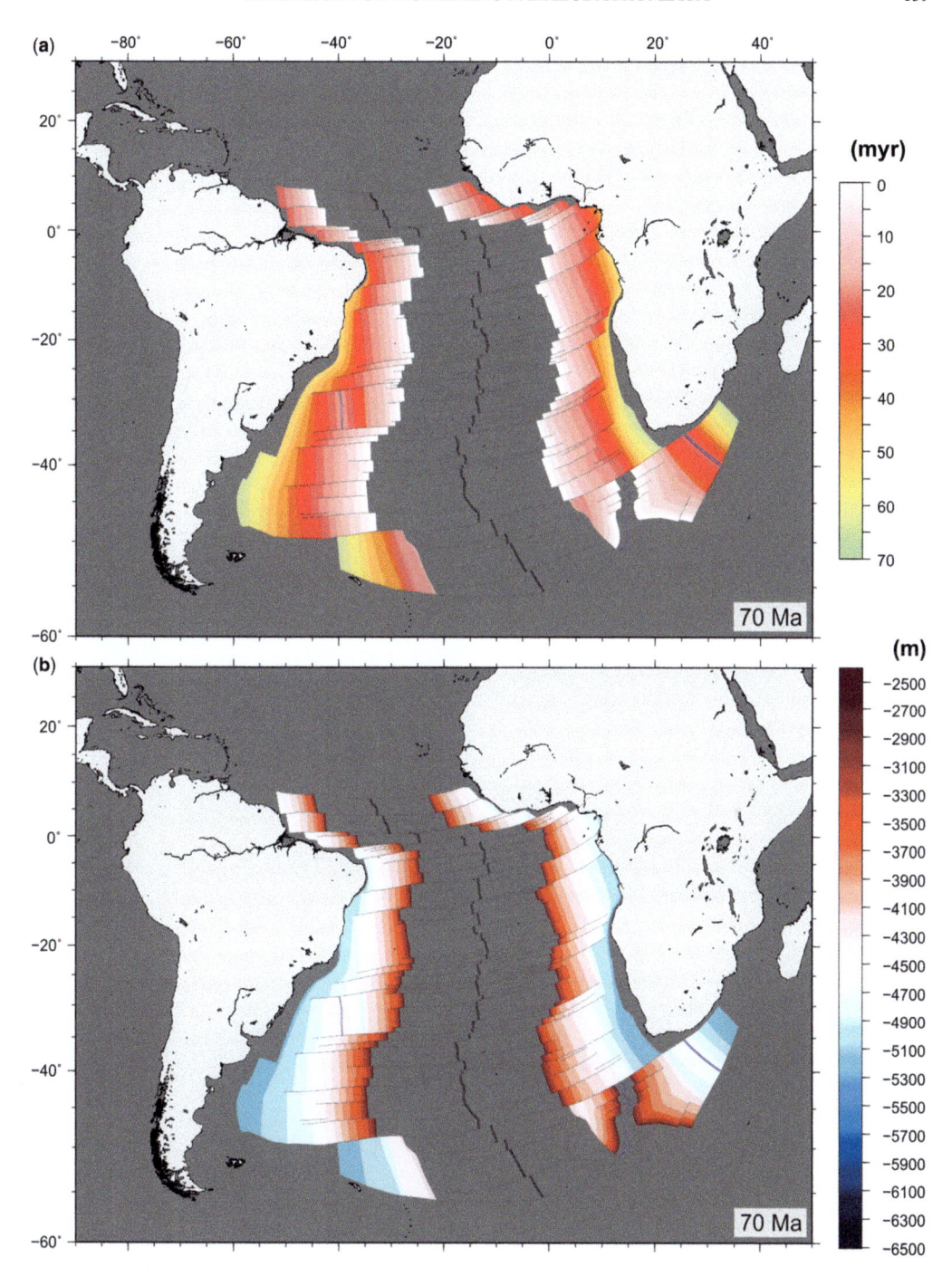

Fig. 3. For 70 Ma. (**a**) Palaeo-age grid and (**b**) seafloor depths as predicted by GDH1 (Stein & Stein 1992).

lithosphere, using the age grids of Müller *et al.* (2008*a*). Then, they adjusted these depths by accounting for an isostatically corrected sediment layer and generated margin profiles by identifying, across COTZs, three segments (shelf, slope and rise) to which they applied distinct gradients calibrated from stacked global bathymetry curves across several of the world's oceans. By following this

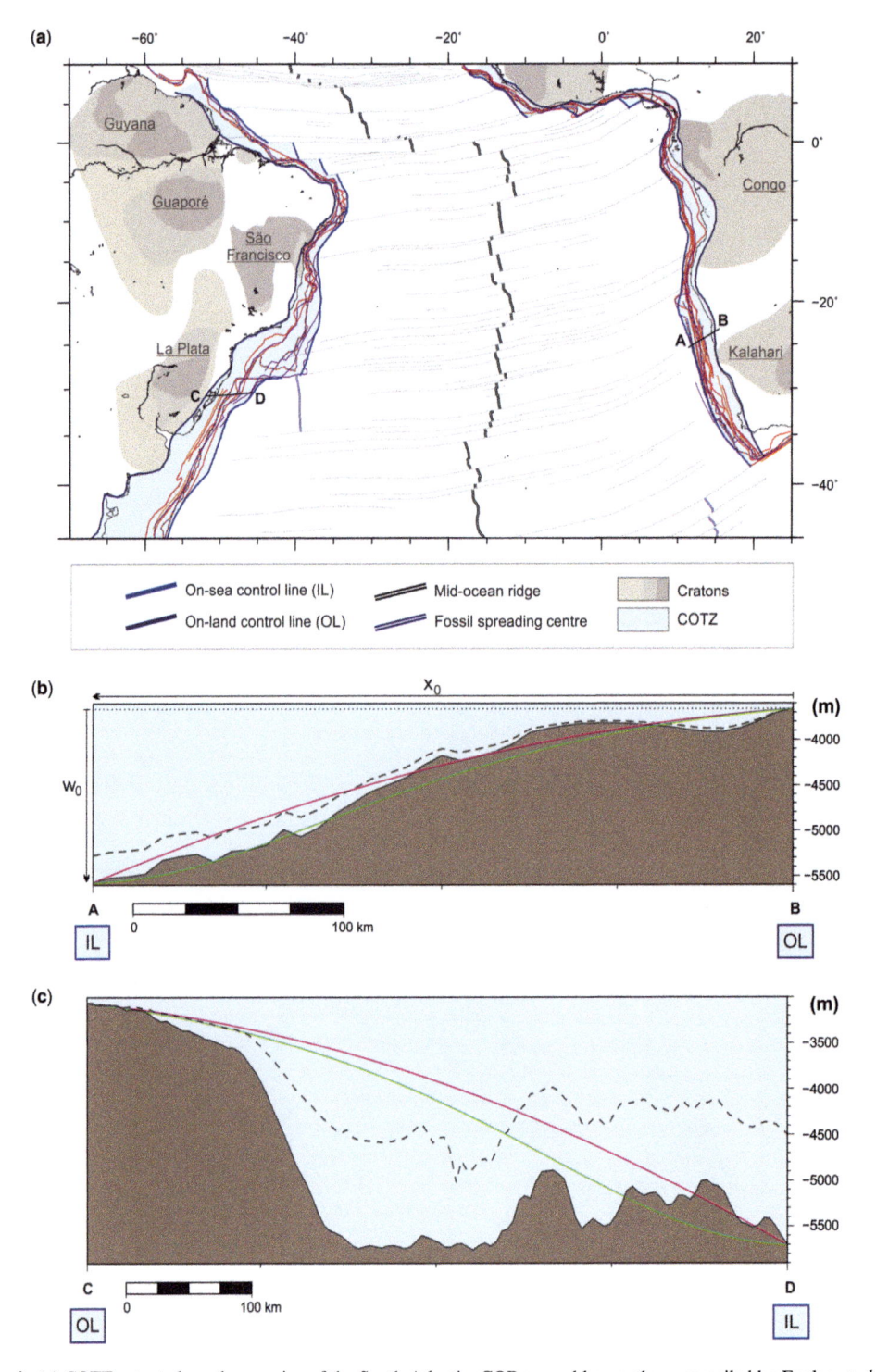

Fig. 4. (a) COTZ extent along the margins of the South Atlantic. COB ensembles are those compiled by Eagles *et al.* (2015). (b) & (c) COTZ cross-sections. Dashed lines are depths extracted from a corrected present-day bathymetry map. Solid lines are depths corrected to account for the depth of IL as predicted by GDH1 at time *t*. Magenta lines are edge-of-plate flexural curves. Green lines are intraplate flexural curves.

approach, they aimed to account for the heterogeneity of extended continental margins. Although they followed a process-based approach and their results are shown to closely replicate modern bathymetry as portrayed by the ETOPO1 dataset (Amante & Eakins 2009), they made no attempt at integrating the effects of dynamic topography and variable oceanic crustal thickness on depth, and used a different seafloor age dataset than the one we use here (Pérez-Díaz & Eagles 2017a). Further to this, by using global bathymetry curves instead of present-day COTZ observations, their approach is more likely to smooth over local features.

Other contributors to bathymetry

In order to reduce uncertainty in palaeobathymetric reconstructions, the contributions to depth of second-order processes need to be quantified and used to correct the basement surfaces described in the previous section (Fig. 5). An initial idea of the contribution of these processes to bathymetry can be obtained by subtracting the present-day modelled basement surface (Fig. 5a) from a map of present-day bathymetry derived from satellite altimetry data. When this is done, a series of residual bathymetry anomalies are revealed (Fig. 6). Positive anomalies (warm colours) arise when the seafloor is shallower than the modelled upper surface of the lithosphere. Sediment build-up and crustal thickening by volcanic/plutonic processes both give rise to positive anomalies. Negative anomalies (cold colours), such as those observed in the Argentine Basin, show a less localized character. They mark areas where the seafloor is deeper than predicted using the GDH1 model of a thermally subsiding lithospheric plate (Stein & Stein 1992). The use of a different thermal model for oceanic lithosphere would yield different residual anomalies.

In steps, we adjust the present-day modelled top-of-basement surface (Fig. 5a) to account for the depth effects of: (1) sedimentation; (2) variable crustal thickness; and (3) dynamic topography. The resulting further residual bathymetry anomalies, calculated by subtracting the top-of-basement surface adjusted for one or more of these processes from present-day satellite-derived bathymetry, are a useful context in which to interpret the uncertainties involved in the data used and the workflow itself. This is fundamental when applying the workflow to times before present day, in order to understand the limitations in palaeobathymetric reconstructions.

When referring to residual bathymetry anomalies (R), we will use a notation of the form $R_{x1...3}$, with the set $x_{1...3}$ consisting of one or more of the following: s (a correction for sediment thickness); c (a correction for crustal thickness); or d (a correction for dynamic topography). For example, R_{sd} are residual bathymetry anomalies after present-day dynamic topography and the isostatic effects of sediment thickness variations are corrected for. In other words, if we assume that we know sediment thickness and dynamic topography perfectly, then R_{sd} reveals the bathymetric signal of variations in crustal thickness alone.

Sedimentation

The density of sediment is greater than that of the water mass it replaces during sedimentation. Therefore, deposition of a sediment layer will cause the lithosphere to sink in response to the increased load. If the thickness of this sediment layer is known, the isostatic response of the lithosphere under it can be calculated. This isostatic correction (I) when applied to measured present-day bathymetry adjusts seafloor depth to account for a certain thickness of sediment (s) and its isostatic signal. We use the approximation from Sykes (1996) for the relationship between I and s, and the sediment thickness map of Laske et al. (2013) for the South Atlantic (Fig. 7a) to calculate the isostatic correction from sediment thickness, as follows:

$$I = 0.43422s - 0.010395s^2. \qquad (1)$$

Assuming that the sediment thickness map used is a reliable representation of reality, by applying this isostatic correction to the map of predicted basement depths (Fig. 5a) the effects of sedimentation are accounted for (Fig. 7b), and the residual bathymetry anomalies can be reduced accordingly (Fig. 7c).

For times in the past, it is necessary to undo the effects of sediment that had yet to be deposited. We calculate the sediment thickness from the present-day grids of Laske et al. (2013) by assuming a linear sedimentation rate.

Crustal thickness

In a crust of variable thickness, and assuming a value of average thickness which, for oceanic crust, will be something between 5 and 10 km (White et al. 1992), the lower density of the oceanic crust with respect to the underlying mantle that supports it means that areas thicker and thinner than average will give rise to positive and negative residual bathymetry anomalies. Accounting for the effects on bathymetry of variations in crustal thickness presents a greater challenge than doing so for variations in the thickness of the sediment cover. This is so because the best available crustal thickness grid (CRUST1.0: Laske et al. 2013) is of low resolution (1°: 111.2 × 111.2 km at

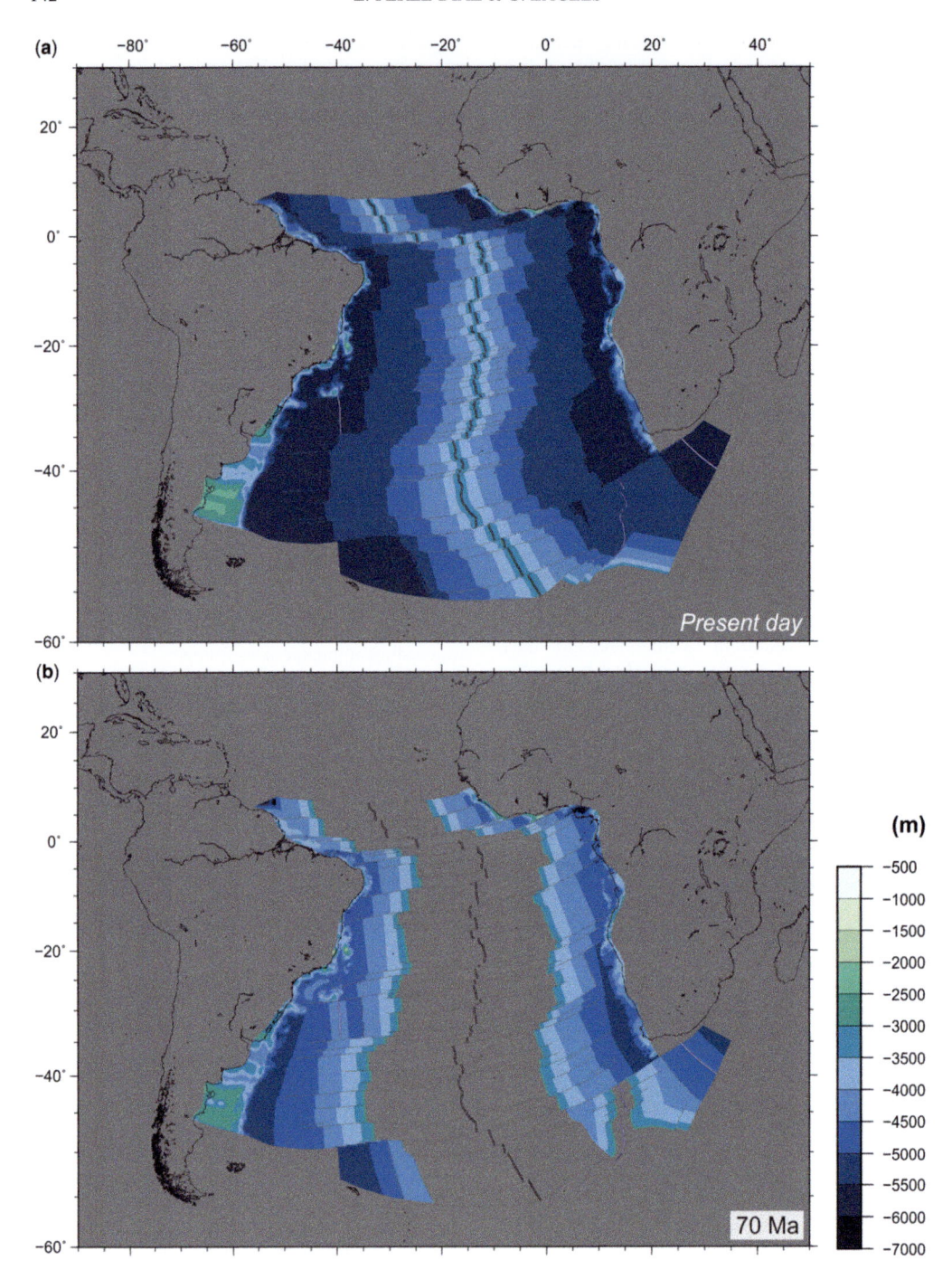

Fig. 5. Predicted basement depths at (**a**) the present day and (**b**) 70 Ma.

the equator), and in most cases does not image seamounts or other regional volcanic constructs where the oceanic crust is thicker than its surroundings. The relationship between crustal thickness and the residual anomaly it gives rise to (assuming Airy isostasy) can be written as:

$$Y = C + R + M. \qquad (2)$$

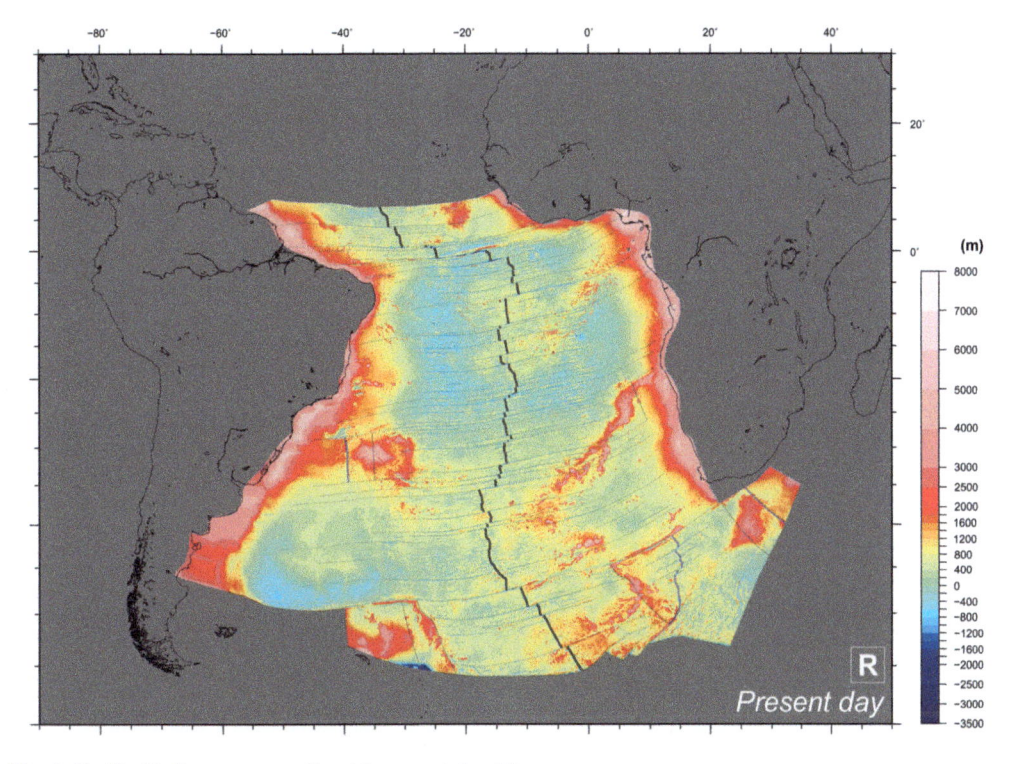

Fig. 6. Residual bathymetry anomalies at the present day (*R*).

Here, *Y*, the total crustal thickness, equals the sum of *C* (average oceanic crust thickness), *R* (a residual bathymetry signal which could be R_{sd}, R_s or R_d) and *M* (the depth of the crustal root below the base of the neighbouring average oceanic crust). Assuming Airy isostasy, a flat base of the crust, average oceanic crustal thickness of 7 km (which for South Atlantic spreading rates is a reasonable value: White *et al.* 1992), then ρ_c = 2950 kg m^{-3}, ρ_w = 1030 kg m^{-3} and ρ_m = 3300 kg m^{-3}, and rearranging:

$$R = \frac{(Y-7)}{6.4857}. \qquad (3)$$

We quantify the depth effects of crustal thickness variability by applying this equation to the crustal thickness estimates from CRUST1.0 (Fig. 8a) (Laske *et al.* 2013). Modelled depths, already accounting for the effect of sediment load (Fig. 7b), can subsequently be further adjusted to also account for the footprint of variations in crustal thickness (Fig. 8b). As a result, positive residual anomalies are reduced significantly along the extended continental margins (Fig. 8c). However, because CRUST1.0 fails to clearly image large areas of hotspot-related crustal thickening within the oceanic interiors, many strong local positive anomalies remain (e.g. the Rio Grande and Walvis ridges, Agulhas Rise, Shona Rise, Meteor Rise, Islas Orcadas Rise, NE Georgia Rise, and the Cameroon Volcanic Line).

For times in the past, we apply the crust correction to all time steps to compensate for instantaneous stretching that affected the COTZs prior to the onset of spreading modelled by Pérez-Díaz & Eagles (2014). A further step, necessary to account for post-break-up topography built as a result of hotspot activity whose effects on crustal thickness are not captured in CRUST1.0, is described in a later section.

Dynamic topography

Viscous stresses transmitted vertically to the lithosphere from zones of contrasting buoyancy in the Earth's mantle are known to be responsible for its long wavelength uplift or subsidence. The surface expressions of these mantle fluctuations are generally referred to as dynamic topography (Pekeris 1935; Morgan 1965; McKenzie 1977; Parsons & McKenzie 1978; Hager & O'Connell 1981; Parsons & Daly 1983; Richards & Hager 1984; Hager *et al.* 1985).

Models of dynamic topography, such as those developed by Bernhard Steinberger for Müller *et al.* (2008*b*) (Fig. 9a), can be used to further reduce the residual bathymetry anomalies in Figure 8c, as shown in Figure 9c. For times in the past, we account for the effects of dynamic topography by

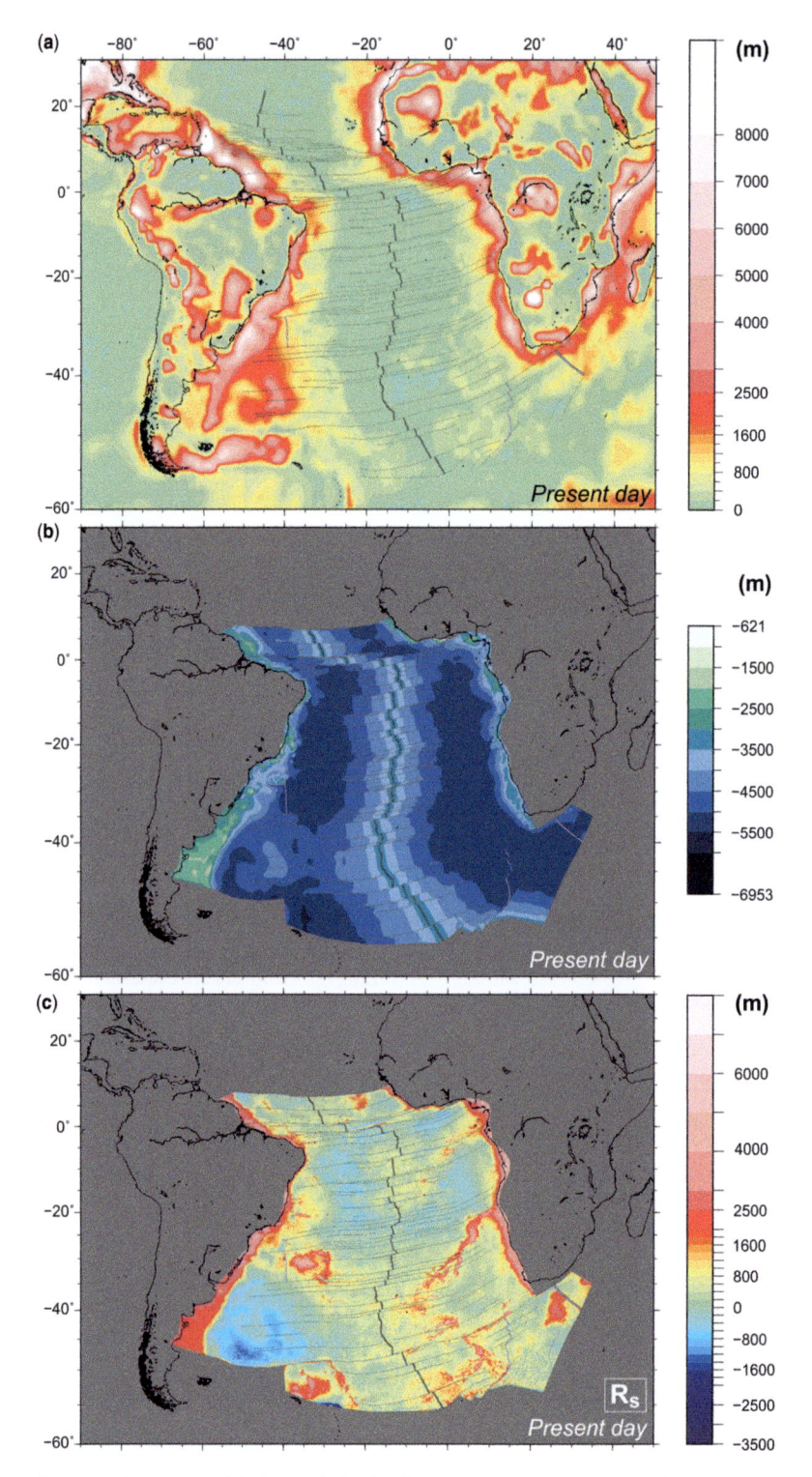

Fig. 7. (a) Sediment thickness map of Laske *et al.* (2013). (b) Modelled depths modified to include the effects of the variable sediment thickness. (c) R_s: residual bathymetry anomalies remaining after applying the sediment correction.

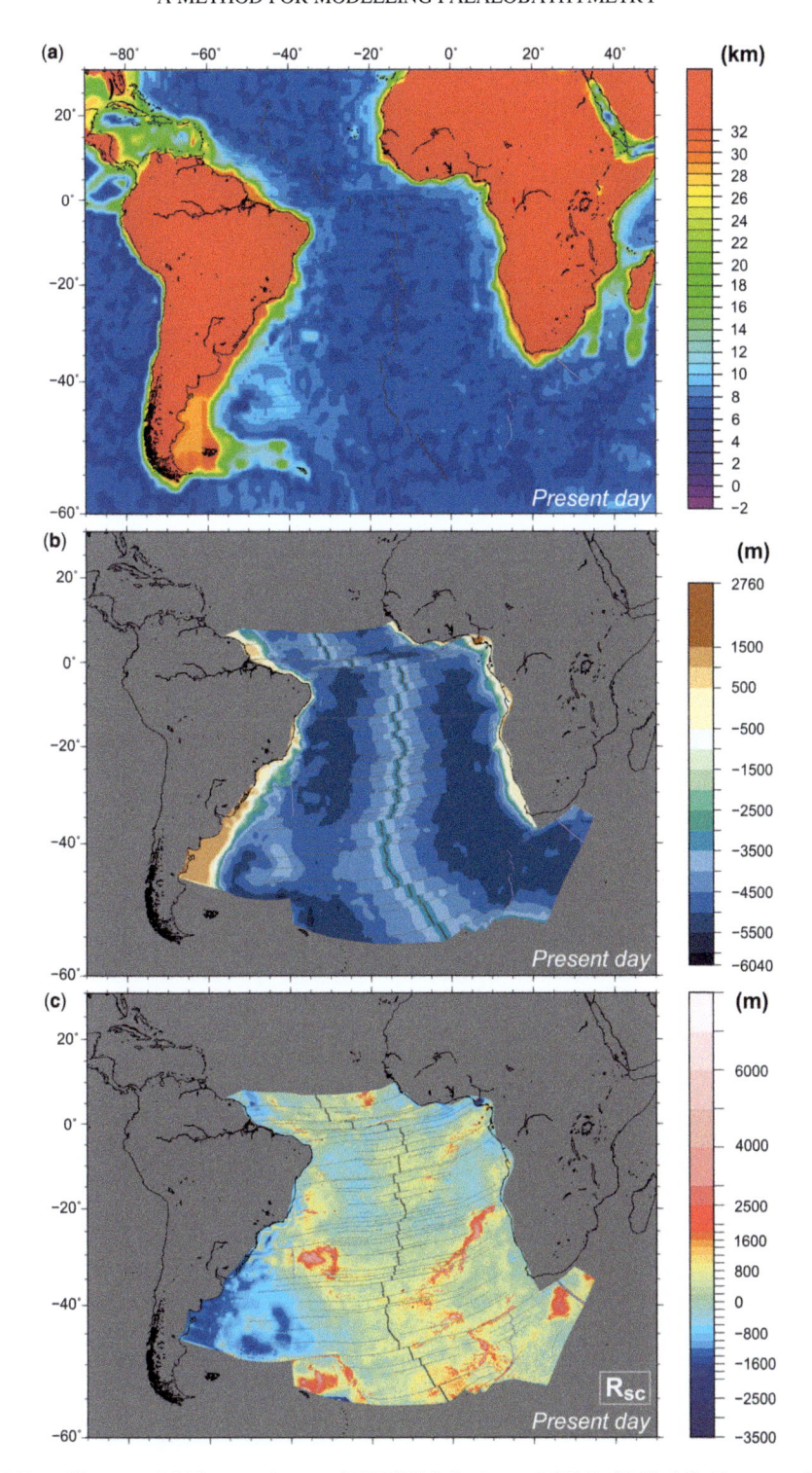

Fig. 8. (a) Crystalline crustal thickness estimates of CRUST1.0 (Laske *et al.* 2013). (b) Modelled basement depths modified to account for the variable sediment and crustal thicknesses. (c) R_{sc}: residual bathymetry anomalies remaining after applying the sediment and crustal thickness corrections.

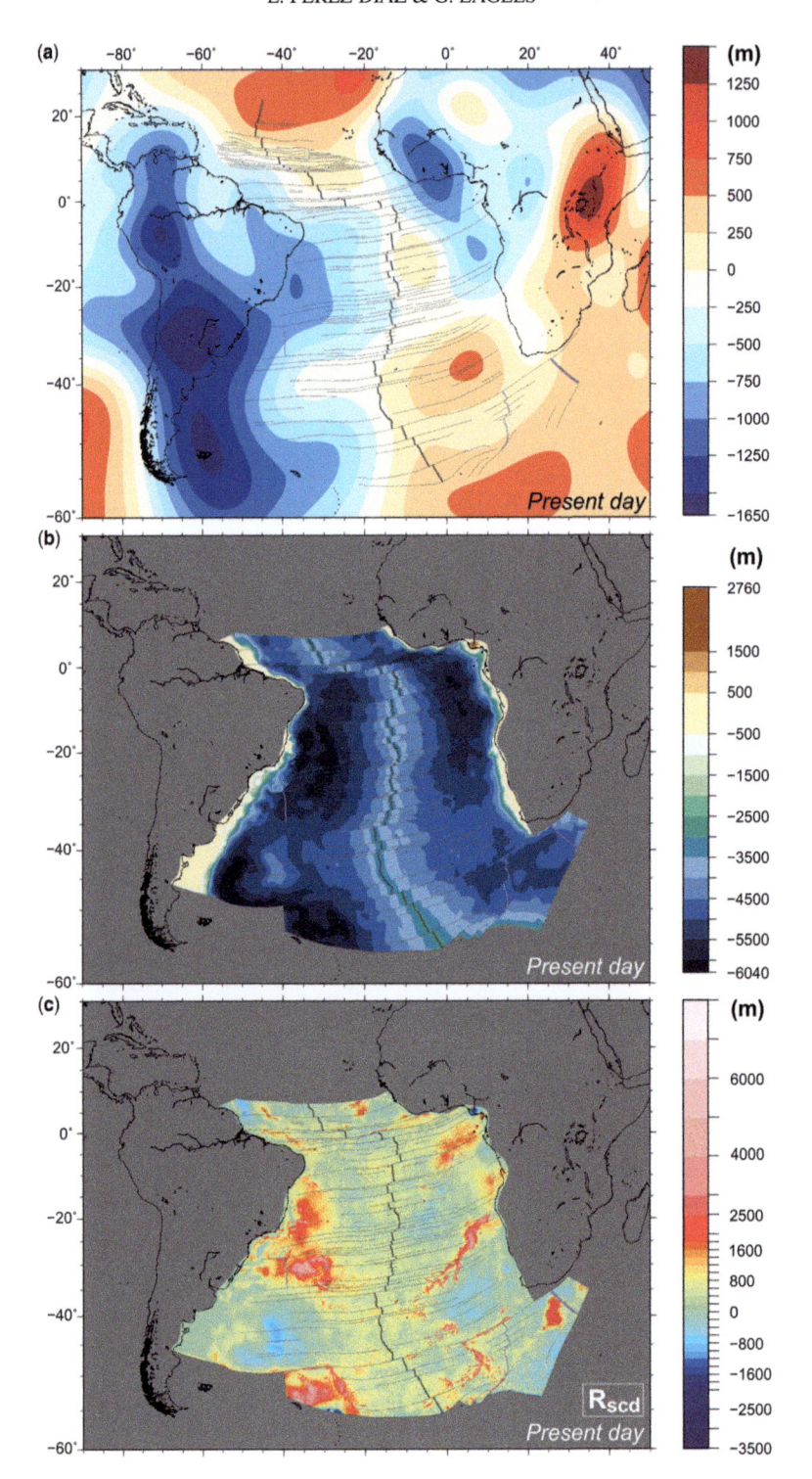

Fig. 9. (**a**) Dynamic topography at the present day (Müller *et al.* 2008*b*). (**b**) Modelled basement depths after incorporating the effects of loading, stretching and dynamic topography. (**c**) R_{scd}: residual bathymetry anomalies remaining after accounting for the sediment and crustal thickness variations and dynamic topography.

reconstructing plate positions into the mantle's absolute reference frame (Torsvik *et al.* 2008) and using Steinberger's dynamic topography reconstructions.

Using residual anomalies as a predictive tool

The differences between maps of modelled basement depths, such as those shown in Figures 7b or 8b, accounting for the effects of any two of the three processes outlined above and present-day bathymetry can be used as a predictive tool for the effects of the third of the processes. Figure 8b accounts for variable sediment and crustal thicknesses, so the residual anomalies resulting from subtracting these depths from present-day bathymetry will provide insights into dynamic topography. In this section, we model each of the three processes discussed by assuming that once two of them have been accounted for, the remaining residual bathymetry anomalies are solely a result of the third of the processes. How close predictions made in this way are to reality depends on how well the effect of the two processes from which the third is derived are known or modelled, as well as the uncertainties in GDH1 (Stein & Stein 1992) and the bathymetric data. This becomes evident in the following subsection, with sediment thickness predictions being strongly affected by the uncertainties in crustal thickness and dynamic topography grids.

Sediment thickness predictions

Sediment thickness (s) can be calculated substituting for I in Sykes' polynomial (Sykes 1996), so that:

$$s = \frac{0.43422 \pm \sqrt{0.1885 - 0.04159(R_{cd})}}{0.02079}. \quad (4)$$

R_{cd} here are the residual bathymetry anomalies resulting from subtracting a modelled top-of-basement surface, accounting for dynamic topography and variations in crustal thickness, from present-day satellite-derived bathymetry.

Figure 10 shows the differences between the sediment thickness grid of Laske *et al.* (2013) and that modelled following the steps outlined above. If we ignore the very thick false sediment signals resulting from the residual bathymetry anomalies attributable to the Rio Grande–Walvis pair and other LIPs, sediment distribution is similar in both grids. The largest differences in sediment thickness appear along the margins, with modelled thickness being much larger and covering a greater area. Again, this may partly be a result of a crustal thickness grid that poorly images changes in crustal structure near the continents. The Argentine Basin looks fairly different in both grids (compare the area labelled 'S2' in both panels of Fig. 10), which CRUST1.0 presents with nearly 5 km of sediment but the isostatic model suggest may be almost sediment free.

Crustal thickness predictions

Crustal thickness (Y) can be calculated from residual bathymetry anomalies (R_{sd}) by rearranging equation (2):

$$Y = 6.4857 R_{sd} + 7. \quad (5)$$

The residual bathymetry anomalies used here (R_{sd}) are the differences in depth between present-day bathymetry and a basement surface modified to account for sedimentation and dynamic topography. At first glance, the difference in resolution between CRUST1.0 (Laske *et al.* 2013) and the grid of predicted crustal thickness stands out (Fig. 11). Crustal thickness along the margins is similar in both grids, with the exception of the margin segment immediately north of Rio Grande Rise along the South American margin, where predicted crustal thicknesses are much larger than those shown by CRUST1.0 (C1 in Fig. 11). Frustratingly, few independent data exist to help assess the cause of this difference (Chulick *et al.* 2013). The Argentine Basin represents an example of the opposite, an area where predicted crustal thickness is smaller than the seismic-derived estimate shown by CRUST1.0. In this particular region, when one looks at the sediment thickness grid (Laske *et al.* 2013) (Fig. 10a), the similarity between the shape of the area of thicker crust and that of thicker sediment cover is noticeable (C2 and S2 in Figs 10a & 11a). It is possible that both sediment and crustal thickness really are greater in that part of the Argentine Basin, in which case the lack of an accompanying gravity anomaly (e.g. Sandwell *et al.* 2014) would need careful explanation. A more plausible possibility is that either sediment or crustal thickness (or both) have been used separately as interpretations for a particular seismic signature, resulting in overestimated values here.

Large Igneous Provinces (LIPs), aseismic ridges and seamounts

CRUST1.0 (Laske *et al.* 2013) does not portray the expected or more-recently proved variations in crustal thickness associated with many LIPs, aseismic ridges and seamounts. Because these features are not direct consequences of thermal subsidence of the ocean, they cannot be inferred from plate kinematic models. In order to include these features in reconstructions of palaeobathymetry, we therefore follow the steps below.

First, we compiled a dataset of longitude–latitude–age points along hotspot tracks in the

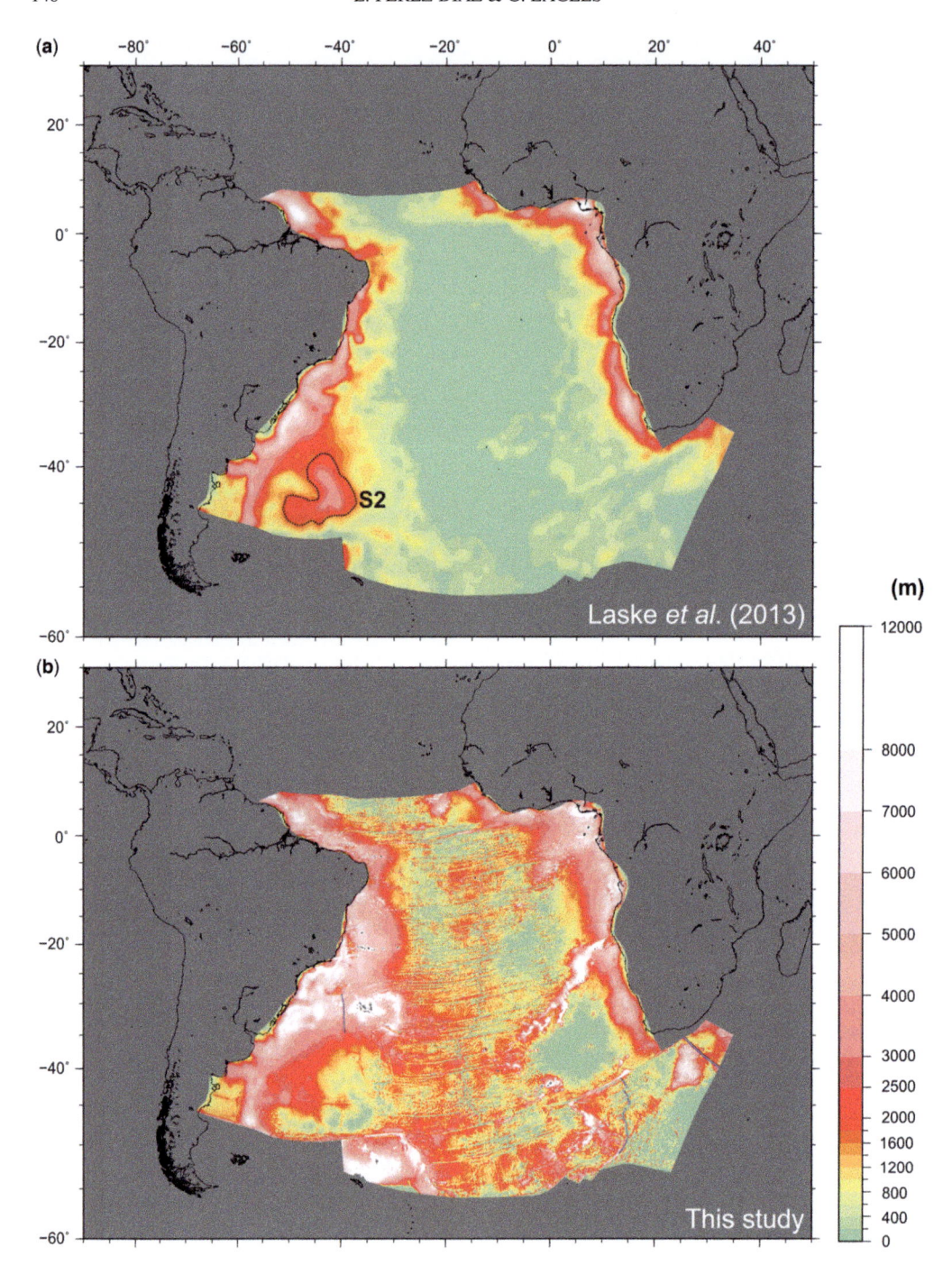

Fig. 10. (a) Sediment thickness map of Laske *et al.* (2013). (b) Sediment thickness as predicted from residual bathymetry anomalies (R_{cd}). S2 is an area of thick sediment within the Argentine Basin mentioned in text.

South Atlantic from published literature (O'Connor & Duncan 1990; O'Connor *et al.* 2012; O'Connor & Jokat 2015). This contains points along the Tristan, St Helena, Bouvet, Martin Cas, Ascension, Gough, Discovery and Shona hotspot trails. Some of the ages are based on radiometric dating of drilled

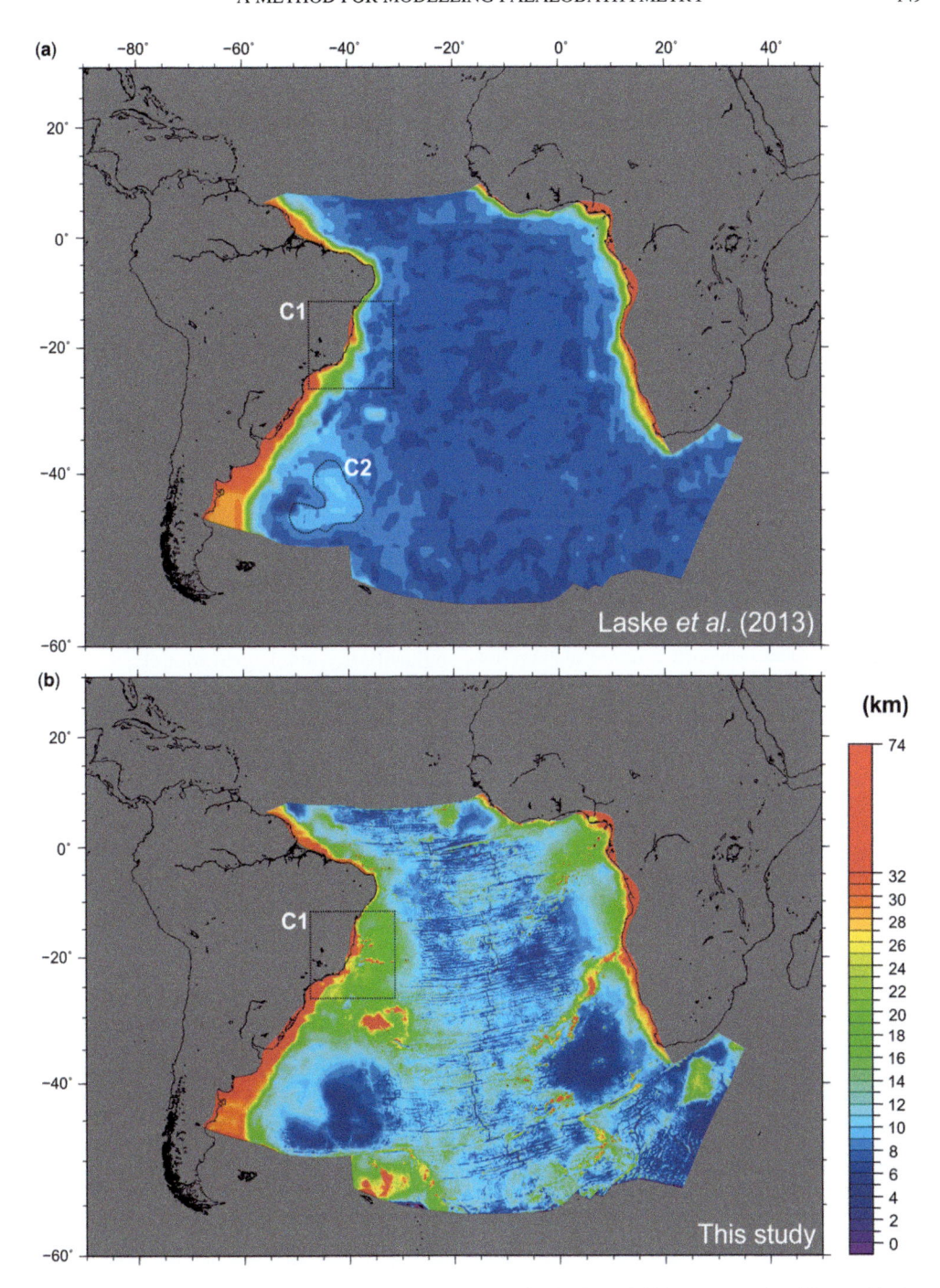

Fig. 11. (**a**) Crustal thickness map of CRUST1.0 (Laske *et al.* 2013). (**b**) Crustal thickness as predicted from residual bathymetry anomalies (R_{sd}). C1 and C2, see the text for details.

or dredged samples. Others are based on the modelling of plate motion by O'Connor & Duncan (1990) over a set of fixed hotspots in the mantle.

Second, for a reconstruction at time *t* Ma, points in the dataset dated as younger than *t* are filtered out. Areas within a 250 km radius of the remaining points (whose ages are older than or equal to t) are used to extract values of R_{scd} residual bathymetry (Fig. 9c) that we can reasonably expect to relate to crustal thicknesses that exceed those shown in CRUST1.0. This radius is intended to reflect the effects of a wide plume head or sublithospheric flow of melt away from the plume conduit.

Finally, a low-pass filter is applied to the extracted residual signals to ensure that wavelengths between 250 and 100 km are progressively weakened, and shorter wavelengths are cut out completely. The main aim in doing this is to smooth out any sharp edges at 250 km distance from the age-constrained points in the dataset being used. The result is a grid of excess topography with values that increase smoothly from 0 to the thickness shown by R_{scd} within the locus of grid cells that we might expect to have experienced crustal thickening as a result of hotspot activity by time t. These grids of excess topography related to hotspots are used as the fourth dataset to refine a thermally subsiding top-of basement surface (together with the sediment and crustal thickness, and dynamic topography datasets reviewed earlier) to produce palaeobathymetric models (Fig. 12).

Including LIPs is an important step in order to produce palaeobathymetric reconstructions for the purpose of palaeoceanographic interpretation, because LIPs have the potential to form barriers to water circulation at multiple depths (e.g. Wright & Miller 1996; Poore et al. 2006; Ehlers & Jokat 2013). The method we follow to include LIPs involves assuming that when one removes the effects of sedimentation, crustal thickness variations and dynamic topography from present-day bathymetry, the remaining residual bathymetry anomalies reflect the existence of volcanism-related excess topography. Therefore, predicted LIP topography may be either over- or underestimated, depending on the inaccuracies of the sediment, crustal thickness and dynamic topography datasets.

Dynamic topography predictions

After accounting for the isostatic effects of crustal thickness variations and sedimentation, and comparing R_{sc} to satellite-derived present-day bathymetry, we filtered the residual anomalies in order to extract signals whose wavelength is within the characteristic range for dynamic topography (Hoggard et al. 2016). In order to do this, we used a bandpass filter (second-order Butterworth polynomial filter) that passes wavelengths of between 2000 and 3000 km, and removes anomalies whose wavelength is shorter or longer than any of these cut-off values.

The result of doing this is shown together with the dynamic topography grid of Müller et al. (2008b) in Figure 13. In terms of the distribution of positive and negative anomalies, both grids are broadly comparable, with a strong negative anomaly in the Argentine Basin region and positive anomalies towards the African Plate. However, and similar to what happens with sediment thickness predictions from R_{cd}, the failure of CRUST1.0 to image many of the South Atlantic's aseismic ridges hinders the dynamic topography prediction. In this case, strong false-positive dynamic topography is predicted in areas where aseismic ridges are located (the Rio Grande–Walvis ridges, Agulhas Rise and the North East Georgia Rise are clear examples), as a result of a satellite-derived bathymetry which is much shallower than that depicted over an isostatically compensated cooling lithosphere (lacking any crustal thickening as a result of volcanism). At least some of the large apparent positive dynamic topography off the coast of southern Africa is therefore likely to result from a combination of underestimated sediment thickness in the Cape Basin and underestimated crustal thickness, with features such as the Walvis Ridge, Meteor Rise, and the Shona and Discovery seamounts unaccounted for by CRUST1.0 (Laske et al. 2013).

Quantification of total uncertainty in palaeobathymetric grid models

As described above, our palaeobathymetric estimates are generated by calculating the depth to the top surface of a lithosphere that forms by conductive cooling of the mantle, and then adjusting this surface for the isostatic effects of varying thicknesses of the crust and sediments overlying it, and for the effects of vertical stresses transmitted to its base during convection of the viscous mantle below. All of these considerations are affected by errors with various sources, whose effects are to produce an estimate of palaeobathymetry that is either deeper or shallower than the unknown true value. Table 1 gives a first idea of the expected order of magnitude, a few hundred metres, of uncertainty that might arise from the presence of these errors by comparing paleobathymetric estimates from DSDP drill cores to the models given. It also reveals that our attempts to account for a wider range of processes than by Sykes et al. (1998) are likely to have significantly reduced this uncertainty. Despite this, the number of comparison sites is too small for a meaningful predictive analysis of uncertainty. Instead, in the following, we therefore describe forward considerations of the effects of likely errors in our modelling procedure for uncertainties in the models.

Uncertainties in calculations of thermal subsidence

Uncertainty in the depth to the top surface of the thermally subsided lithosphere step might be

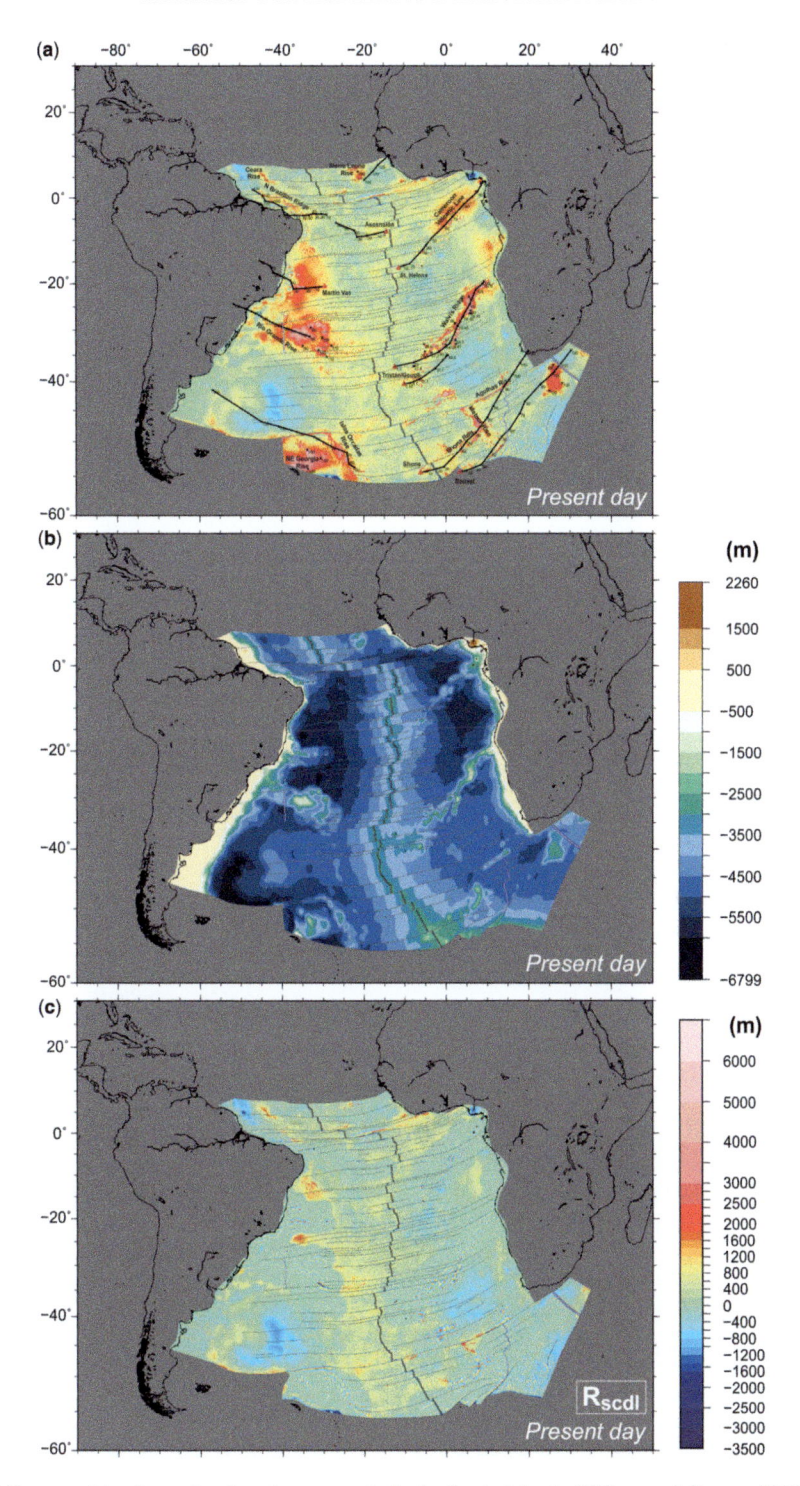

Fig. 12. (a) Dataset of dated samples along hotspot tracks in the South Atlantic (O'Connor & Duncan 1990; O'Connor et al. 2012; O'Connor & Jokat 2015). The background shows R_{scd}. (b) Modelled basement depths as in Figure 8b, modified to account for the topography of the aseismic ridges. (c) Residual bathymetry anomalies after subtracting (b) from present-day satellite-derived bathymetry.

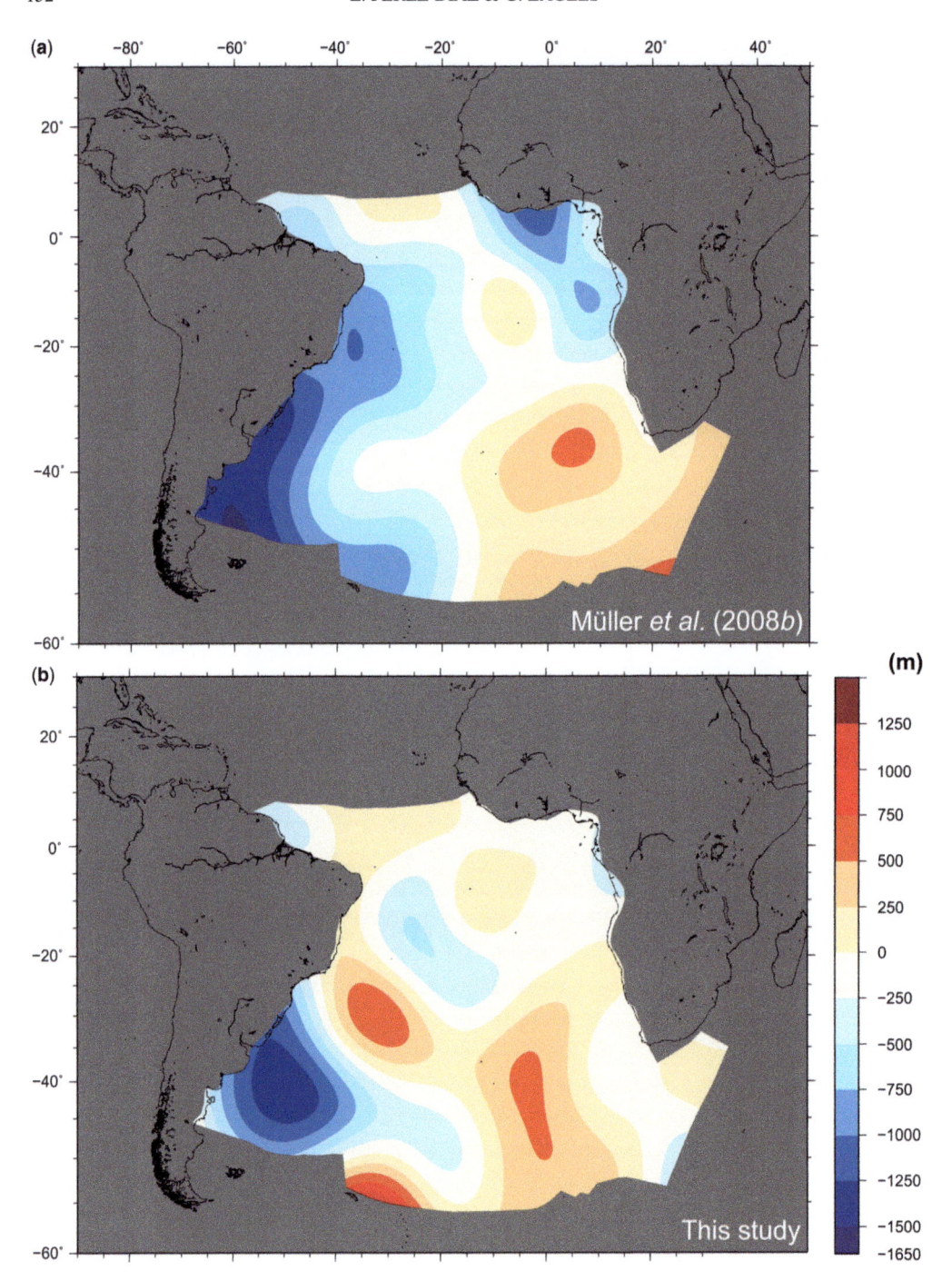

Fig. 13. (a) Dynamic topography in the South Atlantic as modelled by B. Steinberger (Müller *et al.* 2008b). (b) Dynamic topography as predicted from residual bathymetry anomalies (R_{sc}).

dominated by the choice of lithospheric thermal model or the uncertainties in the chosen model itself. The standard deviations of seafloor depths over same-aged areas in the South Atlantic show that so-called plate-cooling models are to be preferred over half-space models for predicting seafloor

Table 1. *Comparison of corrected water depth (CWD) values derived from DSDP drill core data with those obtained following the method described in this paper and those of Sykes et al. (1998)*

Site	Age (Ma)	Longitude (°)	Latitude (°)	CWD*[1]	CWD[2]	CWD[3]	Difference[†a]	Difference[b]
361	129	15.45	−35.07	−5101	−5597	−5150	496	49
513	36	−24.64	−47.58	−4536	−4845	−5158	309	622
516	108	−35.28	−30.28	−1839	−1411	−1944	428	105
698	118	−33.1	−51.46	−2228	−3875	−2755	1647	527
701	53	−23.21	−51.98	−4842	−4935	−4868	93	26
703	92	7.89	−47.05	−1952	−3189	−2202	1237	250

*[1]Corrected water depth in drill core data (DSDP); [2]CWD (Sykes *et al.* 1998); [3]CWD (this study).
†[a]CWD¹−CWD²; [b]CWD¹−CWD³.

depth, but are less prescriptive of any particular plate-cooling variant. We chose to use the GDH1 model of Stein & Stein (1994) because of its closest resemblance to the mean depths in the South Atlantic, which for most ages vary by less than 100 m from GDH1 predictions, and do not exceed 300 m for any age (Pérez-Díaz & Eagles 2017*a*).

A more significant and readily quantifiable estimate of the uncertainty in using GDH1 is that which propagates through it from uncertainty in the seafloor age. Pérez-Díaz & Eagles (2017*a*) provided their age grid with an accompanying set of quantified age uncertainties, which they showed to imply variable, but potentially large (600 m), long-wavelength errors in palaeobathymetry near mid-ocean ridges acting during the Cretaceous normal polarity superchron, but smaller errors in other settings. The age uncertainty is unsigned, meaning that these errors might have the effect of producing inappropriately shallow or inappropriately deep estimates of palaeobathymetry.

We extended the thermally subsided surface across the model COTZs simply by stretching the present-day basement surface between an undoubtedly oceanic inner line and an outer line on supposedly non-extended continental crust to fit the contemporary range between the thermally subsided depth of the inner line and the present-day height of the outer line in the absence of dynamic topography. To this, we applied an estimate of uncertainty appropriate for subsidence by thermal contraction using the relationships derived assuming one-dimensional (1D) heat conduction by McKenzie (1978). The potential depth error we calculated in this instance propagates from an assumed error in the age of the instantaneous rifting in those relationships. In our palaeobathymetric modelling process, this age is implicitly the same as the age of the COTZ's IL. In the uncertainty analysis, this serves as a minimum age estimate for the end of rifting in the COTZ because of the choice of an IL that is undoubtedly oceanic and, therefore, definitively post-rifting. The effect of this age being inappropriately young is to produce COTZ model depths at any time that are inappropriately shallow. For our analysis, we applied a potential error of 10 myr towards older ages for the end of instantaneous rifting at the outer line that varies smoothly to the value of the oceanic age grid error at the inner line. The depth uncertainty that this produces is largest close to the IL and for times close to the age of the IL.

We have not attempted to quantify other processes (e.g. flexure, gravity gliding) that are known to affect short-wavelength bathymetry in specific shelf and slope settings at the present day.

Uncertainty in crustal thickness estimates

The largest uncertainties related to variations of crustal thickness are the result of the dataset's low spatial resolution (Laske *et al.* 2013), which shows very little variation in oceanic crustal thickness. Compared to this, the natural variation of normal oceanic crustal thickness formed at plate divergence rates like those encountered in the South Atlantic is thought to occur within a tight, but nevertheless significant, range (4–8 km, mode near 7 km) as a consequence of the crust's formation by adiabatic compression of well-mixed upper-mantle rocks (White *et al.* 1992). A more recent study has shown that this variation is partially age dependent, and suggested a gradual cooling-related reduction in mantle fecundity as its cause (Van Avendonk *et al.* 2017). To account for possible effects of erroneous oceanic crustal thickness, we allowed the 7 km steady-state crustal thickness in equation (3) to vary with age according to Van Avendonk *et al.*'s (2017) regression, and permitted its subject to then vary by a further ±1.0 km, which captures nearly 100% of the remaining present-day off-axis variability in measured oceanic crustal thicknesses in Van Avendonk *et al.*'s (2017) compendium. Crustal thickness can vary from the gridded values in such a way that the palaeobathymetry produced using it is either inappropriately shallow (where the true thickness is at its maximum above the gridded thickness and the long-term average is at its minimum) or inappropriately deep (where the true thickness is at a minimum below the gridded thickness and the long-term average is at its maximum). Given these possibilities, we produced both deepening and shallowing error surfaces for crustal thickness uncertainties.

The crustal thickness grid of Laske et al. (2013) also fails to show the thicker igneous crust underlying many unstudied or less-studied oceanic LIPs, resulting in large areas of erroneously deep palaeobathymetry. Our modelling procedure accounts for this inadequacy by isolating the palaeo-residual topography along known hotspot tracks and restoring it to the reconstruction. The residual bathymetry used is derived from present-day bathymetry, the uncertainty of which might be in the range of 200 m (Smith & Sandwell 1997). A larger error is entailed in the assumption that the LIPs at the present-day are preserved products of magmatic–volcanic events dating from the instants of plume arrival beneath the lithospheric regions they are built on. This assumption is inadequate, as shown by the widespread determination of late-stage volcanism on submarine LIPs or the dated variability of lava ages exposed on Iceland, which suggests that the LIP there built up over the last 20 myr. Based on this, sets of our palaeobathymetric maps may be too shallow around active hotspots over 20 myr-long periods. To capture some of the uncertainty coming from this expectation, and in the absence of robust estimates of the rate of LIP growth, we calculated a linear proportion of the residual bathymetry that varies between zero (20 myr downstream of the hotspot) and 0.5 (at the hotspot location).

Uncertainty in sediment calculations

A further step in producing palaeobathymetry is to load the thermally subsided lithosphere with a pile of sediments whose thicknesses are estimated on the basis of a global compilation and as a linear proportion of the time elapsed between the time of the reconstruction and the age of the crust. The effect of this loading is calculated using an empirically derived isostatic correction (Sykes 1996). This correction uses densities derived mostly from seismic velocity analysis and which follow a depth-dependent trend, and represents an improvement over others that forwardly assign a uniform density to the entire sediment package, resulting in overestimated corrections (Sclater et al. 1977, 1985; Hayes 1988; Renkin & Sclater 1988; Kane & Hayes 1992). Nonetheless, for sediment loads like the majority of those shown in Figure 7a, the various isostatic corrections yield similar results, and so the uncertainties associated with the choice of isostatic correction scheme are not quantified here.

The global sediment thickness grid used (Laske et al. 2013) is based on large regional compilations of sediment thickness contours derived from reflection and refraction seismic velocity studies (Hayes 1991). Because, in many cases, seismic basement is not imaged and in those cases in which it is it may not represent the upper surface of the crystalline crust, the sediment thickness shown by Laske et al. (2013) is a minimum estimate. To illustrate this, a recent correlation of industry seismic datasets to global grids suggests a tendency for Laske et al. (2013) to systematically underestimate sediment thickness by as much as 20% (Hoggard et al. 2017), albeit within a broad scatter. In contrast, Whittaker et al. (2013) suggested that the effect of uncertainty in velocity solutions for sediment thickness estimates off southern Australia may be in the region of 25% of the minimum estimated thickness. With this in mind, for each time slice, we calculated the effect of a 25% increase in sediment load throughout the study area. This effect decreases with age because, as part of our modelling process, the uncertainty in isostatic correction to basement depth is calculated using ever-smaller proportions of the possible error in present-day sediment thickness.

A potentially large remaining uncertainty is related to the assumption, when reconstructing sediment thickness for times in the past, of a linear sedimentation rate. This is a simplification whose effect can be removed by a more appropriate approach for regions where chronostratigraphic-stage-scale isopach datasets exist. For now, in the absence of such data for most parts of the South Atlantic, we do not quantify the assumption's effects on palaeobathymetry for the uncertainty analysis.

Uncertainty in dynamic topography models

The effects of global mantle circulation on topography are modelled with inputs from mantle tomography and assumptions about the mantle viscosity profile (Müller et al. 2008b). Variations in S-wave velocity obtained with seismic tomography are used to make interpretations of the temperature and density heterogeneities within the mantle. Lateral variations of density cause convective flow and provide insights about the locations of dynamic topography highs and lows. The amplitudes of these depend heavily on the mantle viscosity profile, and so assumptions about this parameter have a strong effect on dynamic topography models. An overview of the errors that ought to be expected from viscosity profiles derived from geoid fits is given by Panasyuk & Hager (2000).

For the present day, when misfits with respect to residual bathymetry anomalies are calculated, dynamic topography predictions from geodynamic models are generally found to be too high

(Lithgow-Bertelloni & Silver 1998; Panasyuk & Hager 2000; Pari & Peltier 2000; Cadek & Fleitout 2003; Steinberger & Holme 2008).

The grids of dynamic topography that we used (Müller et al. 2008b) were tuned to portray dynamic topography within a ±1.5 km amplitude range. They used seismic tomography to infer density heterogeneities within a stratified mantle and account for the effects of latent heat release across the phase boundary at 660 km depth. Uncertainties in these models are largest for times in the past, with no dynamic topography estimates for times before 100 Ma and estimates for ages older than 70 Ma considered unlikely to be meaningful (Steinberger pers. comm.). A further source of error lies in the fact that the modelled dynamic topography depends on modelled mantle convection that responds to tractions calculated using a global plate kinematic model whose South Atlantic plate motions are different from those we use for our palaeobathymetries. The differences, in particular to the nature of the plate boundaries implied by those motions, however, are of small significance at the global scale, and the effects of the tractions are known to be of second-order significance even for the pattern of whole-mantle circulation (Steinberger et al. 2004).

The possible errors owing to dynamic topography in the modelling can be either positive (too much dynamic topography has been estimated and removed) or negative (too little estimated and removed). To quantify these, we compared the estimate of present-day dynamic topography by Müller et al. (2008a, b) to our own estimate of present-day South Atlantic residual topography, which ideally at long wavelengths should be equivalent surfaces. We applied two standard deviations (2 SDs) of the differences between these datasets (±288 m) as a plausible maximum error range at 0 Ma. By 70 Ma and later, we assumed that dynamic topography is essentially unknowable, and thus applied a larger maximum range equal to 2 SDs of the entire variation for that time slice. For times between 70 and 0 Ma, we applied a linear increase between the standard deviations used for those two ages.

Quantification of total uncertainty in palaeobathymetric grid models

Table 2 summarizes the error considerations described above and classifies them according to whether they imply the calculated palaeobathymetry to be too deep or too shallow. By summation of each of these two uncertainty classes, it is possible to produce: (i) a maximum likely deepening correction; and (ii) a maximum likely shallowing correction. Figure 14 shows examples of these corrections appropriate to the modelled palaeobathymetry at 60 Ma.

Table 2. *Summary of errors considered for the uncertainty analysis*

	Source and nature of error						Depiction of uncertainty
	Oceanic lithosphere age	Onset of post-rift in COTZ	Sediment thickness	Crustal thickness	Dynamic topography	Height of LIP or aseismic ridge	
Error that makes presented surface too deep	Too old	Negligible*	Too thin	Too thin	Too negative/not positive enough	Negligible[‡]	Add summed errors to generate shallowest palaeobathymetry for uncertainty
Error that makes presented surface too shallow	Too young	Too young	Negligible[†]	Too thick	Too positive/not negative enough	Too high for time slice	Subtract summed errors to generate deepest palaeobathymetry for uncertainty

*Ages assigned on basis of the oldest constrained oceanic age (Pérez-Díaz & Eagles 2017a), and therefore youngest estimate. Older post-rift onsets are therefore unreasonable.
[†]Laske et al. (2013) and Whittaker et al. (2013) state that their sediment thicknesses are minimum estimates, as the base of the sediment pile may not be interpretable in some seismic data, or the reflection interpreted as from crystalline basement may not be from basement rocks.
[‡]We consider it unlikely that LIPs lost considerable elevation over their lifetime. This is reasonable for submarine LIPs where erosion can be considered negligible.

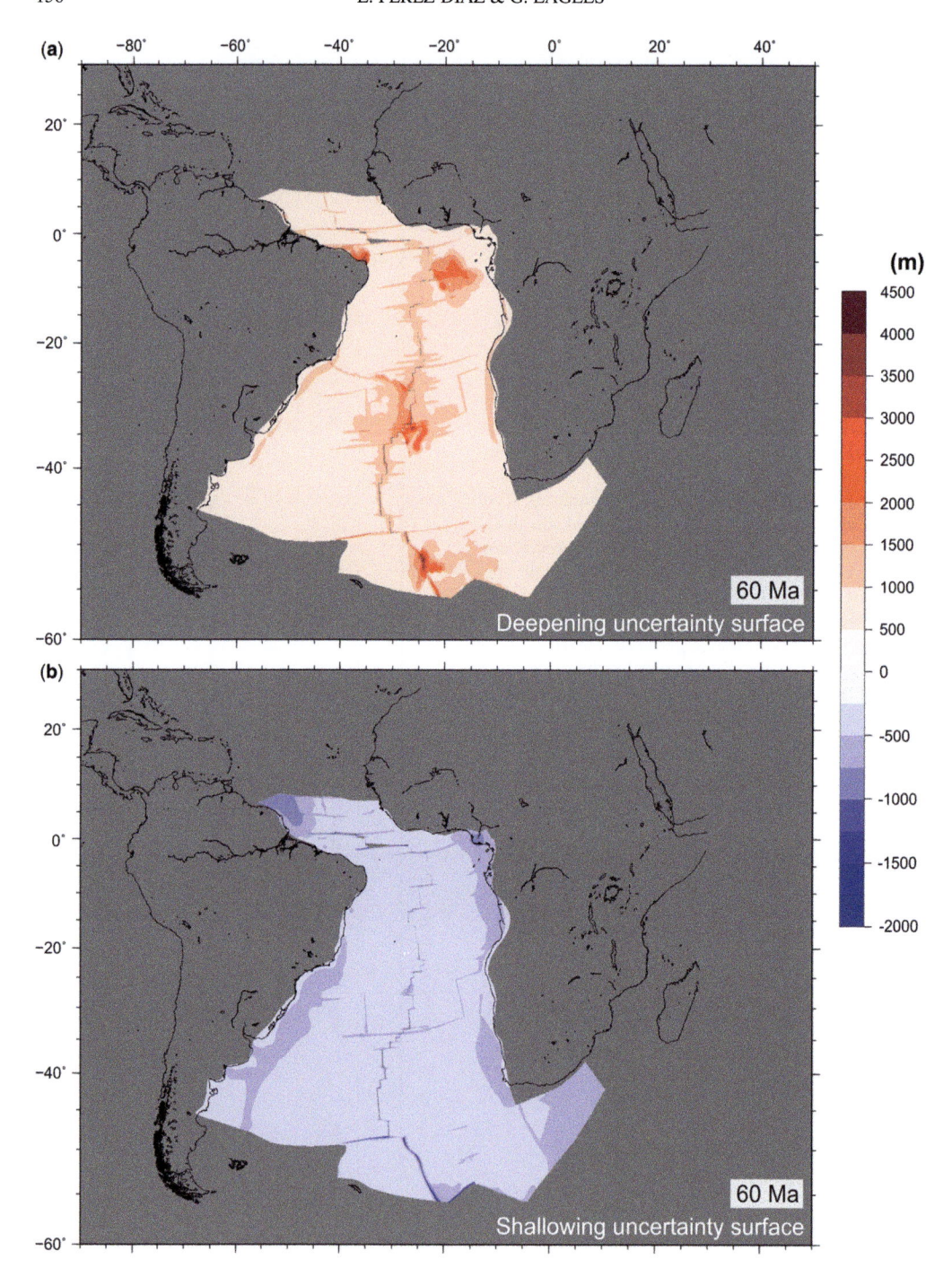

Fig. 14. Combined effects of all uncertainties that imply the modelled 60 Ma bathymetry might be (**a**) deeper or (**b**) shallower than a less uncertain model might show.

The maximum error range (the sum of the magnitudes of the shallowing and deepening components) implied in Figure 14 is 4908 m, which like all of the largest range values is encountered over parts of LIPs that are modelled to have been forming close to mid-ocean ridge crests at 60 Ma. This reflects our method's insensitivity to what we have assumed to be finite emplacement periods for those LIPs. This source of uncertainty dominates the upper end of the uncertainty ranges for all model ages, and should also be considered to dominate critical uncertainty in the precise timing of the production and removal of barriers and filters for palaeo-abyssal currents.

The mean and standard deviation of the range shown in Figure 14, however, are 1317 m and 231 m, respectively, reflecting the more modest uncertainty ranges (minimum 794 m) calculated over the large areas of abyssal plain with thin sediment cover and monotonous oceanic crustal thickness. For Neogene time slices, in which the proportion of such material is larger owing to widening of the ocean, the mean of the uncertainty range reduces to less than 1100 m and the standard deviation to 200 m. In older time slices, the opposite is the case, with uncertainty in the time of instantaneous rifting in the COTZs becoming more significant, leading the mean of the uncertainty range at 110 Ma, for example, to approach 2300 m and its standard deviation 600 m.

Overall, these considerations are consistent with the expectation that confidence in our older time slices should be considered to be less than in our younger ones. Analysis of the full set of uncertainty ranges for all modelled ages suggests a confidence range of 1800 m (mean and 2 SDs) may be appropriate and conservative for 95% of nodes. This range, however, is not symmetrical about our palaeobathymetric estimates because of the large range estimates over LIPs, which all imply the modelled palaeobathymetry to be too shallow, and because of the asymmetry of the GDH1 and McKenzie (1978) age–depth curves for oceanic lithosphere and instantaneously stretched COTZs. Given this, to best portray uncertainty, we sum our shallowing and deepening corrections with the modelled palaeobathymetry to produce shallowest and deepest plausible bathymetries within uncertainties.

Assessment of uncertainty appropriateness

Figure 15 compares a present-day bathymetry and its shallowest and deepest uncertainty surfaces that have been generated using the procedures described above to the present-day bathymetry in the GEBCO 2014 grid (version 20150318: http://www.gebco.net), which is based on a combination of sparse ship soundings and interpolations based on satellite gravimetry. In view of the fact that our procedure is not designed to model short-wavelength variations, the bottom part of the figure maps only those areas exceeding 50 km in diameter within which the GEBCO bathymetry completely lies outside the range implied by the shallowing and deepening uncertainties for their modelled counterparts. These areas amount to 5.6% of the modelled region, suggesting that, in terms of coverage at least, the volumes between our shallowest and deepest surfaces might well be considered as similar to 95% confidence estimates for the modelled palaeobathymetric surfaces.

The distribution of nodes that lie deeper in the GEBCO 2014 estimates than their modelled counterparts has a mean of 211 m and a standard deviation of 245 m. The majority of these areas coincide with estimates of thick crust and/or thick sediments in CRUST1.0; in particular in the outer Argentine Basin, where large disagreements with the crustal and sediment thickness predictions of residual bathymetry have already been noted (Figs 10 & 11). In the Cape Basin, a smaller area of deeper-than-modelled seafloor may hint at a local lithospheric cooling history that differs from GDH1. The distribution of GEBCO 2014 nodes lying shallower than our shallowest uncertainty estimates is more skewed to large values: a mean of 1106 m and standard deviation of 597 m. The locations of these mismatches are centred on R_{csd} highs that have been incompletely sampled by our procedure for isolating and restoring LIP topography. Given their size and their concentration around the central Atlantic gateway, to whose evolution the Albian and Cenomanian palaeoclimates are likely to have been sensitive, future work may be necessary to more fully represent these areas and/or their uncertainties in palaeobathymetry for those times.

Summary

To aid our summary, Figure 16 shows South Atlantic palaeobathymetry for a Paleocene time slice, modelled following the workflow described in this paper. This, and other time slices for the South Atlantic, are presented, interpreted and discussed in geological and palaeoceanographical terms by Pérez-Díaz & Eagles (2017b). At this time, the topography of the mid-ocean ridge lies at depths close to 2600 m, as is the case for its present-day counterpart. Away from the ridge crest, the seafloor gradually drops down to depths in the region of 5700 m in four distinct basins (the Brazil, Angola, Argentine and Cape basins). These variations reflect our use of plate-cooling theory to model thermal subsidence of the oceanic lithosphere from a high-resolution grid of seafloor ages derived from the kinematic model of Pérez-Díaz & Eagles (2014). Between and within these basins, rising up to several thousand metres above the modelled abyssal plains, a number of regional plateaus represent the forerunners of Large Igneous Provinces (LIPs) like today's Rio Grande Rise and Walvis Ridge, which we have modelled as the products of intraplate volcanism

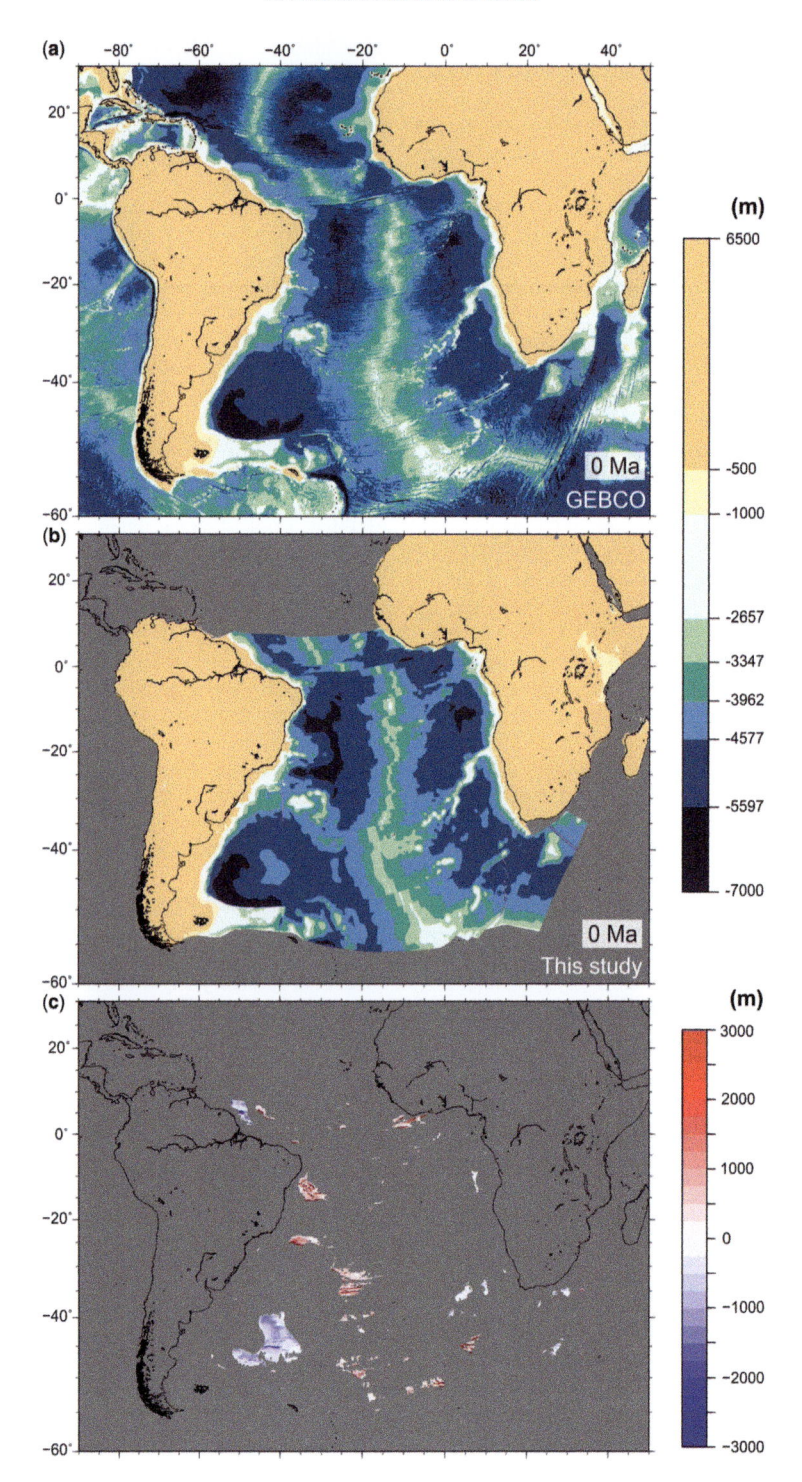

Fig. 15. (**a**) Satellite-altimetry-derived present-day bathymetry. (**b**) Present-day bathymetry modelled following the workflow presented in this paper. (**c**) Significant (>50 km diameter) areas in which measured present-day bathymetry lies deeper than its deepest modelled equivalent within uncertainty (blues) or shallower than its shallowest modelled equivalent within uncertainty.

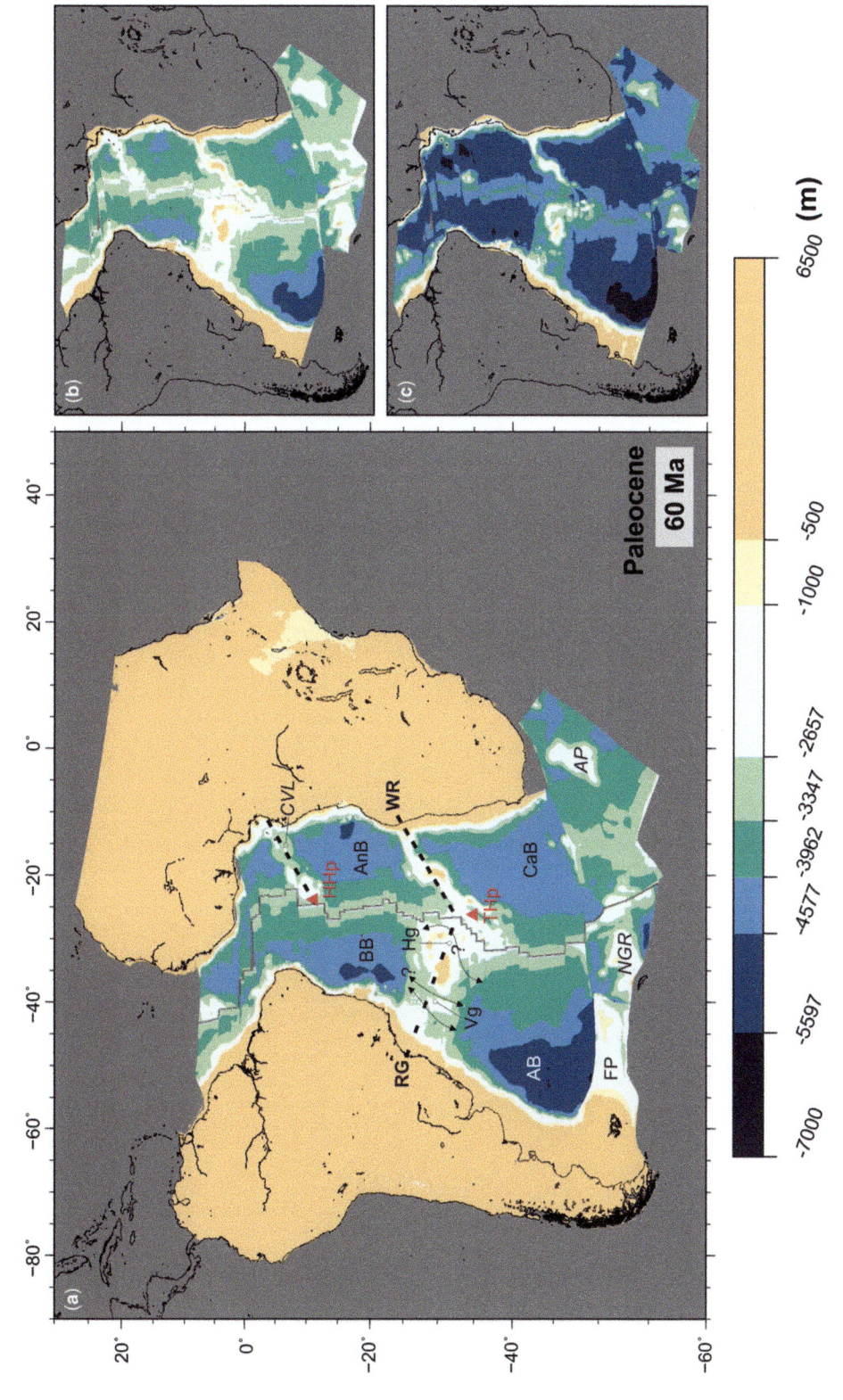

Fig. 16. (a) Palaeobathymetric reconstruction at 60 Ma, with (b) minimum and (c) maximum depth uncertainty estimates (Pérez-Díaz & Eagles 2017b).

related to hotspots over which the African and South American plates slowly moved. The deep ocean regions rise smoothly up towards continental shelves that rim the African and South American continents which lie much closer together than they do today. This variation reflects our application of isostatic corrections to model the bathymetric effects of large-scale sedimentation and changing crustal thickness at and across the extended margins of continents that have moved into their present relative positions as parts of two large lithospheric plates since early Cretaceous times. At very long wavelengths, modest and smooth deflections from the bathymetry predicted by these processes depict the effects of regional up- and downwarping of the lithosphere by slow convection of the viscous mantle rocks beneath the South Atlantic Ocean. By forward considerations and by comparison to published point estimates of palaeobathymetry at drill core sites, we show that the depths in this grid or grids like it for other time slices can conservatively be considered as accurate to within as little as 700 m over large oceanic parts of the map area, but much less so over short distances near LIPs and in early time slices. This accuracy approaches the vertical resolution of the model deep ocean in general circulation models, demonstrating that palaeobathymetric maps built using it are suitable for use in deep-time palaeoceanographic studies. Finally, our approach, being largely process- rather than data-based, can be expected to yield results of similar high confidence and quality for large areas of the world's palaeo-oceans.

Funding Both authors are grateful to Royal Holloway, University of London and the Alfred Wegener Institute, Helmholtz Centre for Polar and Marine Research for funding. LPD would like to thank the COMPASS Consortium for further funding support.

References

ADAM, C. & VIDAL, V. 2010. Mantle flow drives the subsidence of oceanic plates. *Science*, **328**, 83–85, https://doi.org/10.1126/science.1185906

AFONSO, J.C., ZLOTNIK, S. & FERNÀNDEZ, M. 2008. Effects of compositional and rheological stratifications on small-scale convection under the oceans: Implications for the thickness of oceanic lithosphere and seafloor flattening. *Geophysical Research Letters*, **35**, L20308, https://doi.org/10.1029/2008GL035419

AMANTE, C. & EAKINS, B. 2009. *ETOPO 1 Arc-minute Global Relief Model: Procedures, Data Sources and Analysis*. NOAA Technical Memorandum NESDIS NGDC-24. National Geophysical Data Center, Boulder, Colorado, USA.

BLAICH, O., FALEIDE, J.I., TSIKALAS, F., LILLETVEIT, R., CHIOSSI, D., BROCKBANK, P. & COBBOLD, P. 2010. Structural architecture and nature of the continent–ocean transitional domain at the Camamu and Almada Basins (NE Brazil) within a conjugate margin setting. *In*: VINING, B.A. & PICKERING, S.C. (eds) *Petroleum Geology: From Mature Basins to New Frontiers – Proceedings of the 7th Petroleum Geology Conference*. Geological Society, London, 867–883, https://doi.org/10.1144/0070867

BRUNE, S., HEINE, C., PÉREZ-GUSSINYÉ, M. & SOBOLEV, S.V. 2014. Rift migration explains continental margin asymmetry and crustal hyper-extension. *Nature Communications*, **5**, 1–9, https://doi.org/10.1038/ncomms5014

CADEK, O. & FLEITOUT, L. 2003. Effect of lateral viscosity variations in the top 300 km on the geoid and dynamic topography. *Geophysical Journal International*, **152**, 566–580.

CHULICK, G.S., DETWEILER, S. & MOONEY, W.D. 2013. Seismic structure of the crust and uppermost mantle of South America and surrounding oceanic basins. *Journal of South American Earth Sciences*, **42**, 260–276.

CROSBY, A.G., MCKENZIE, D. & SCLATER, J.G. 2006. The relationship between depth, age and gravity in the oceans. *Geophysical Journal International*, **166**, 553–573, https://doi.org/10.1111/j.1365-246X.2006.03015.x

CROUGH, S. 1978. Thermal origin of midplate hotspot swells. *Geophysical Journal International*, **55**, 451–469.

DAVIS, E. & LISTER, C. 1974. Fundamentals of ridge crest topography. *Earth and Planetary Science Letters*, **21**, 405–413.

DOIN, M. & FLEITOUT, L. 1996. Thermal evolution of the oceanic lithosphere: an alternative view. *Earth and Planetary Science Letters*, **142**, 121–136, https://doi.org/10.1016/0012-821X(96)00082-9

EAGLES, G., PÉREZ-DIAZ, L. & SCARSELLI, N. 2015. Getting over continent ocean boundaries. *Earth Science Reviews*, **151**, 244–265, https://doi.org/10.1016/j.earscirev.2015.10.009

EHLERS, B.M. & JOKAT, W. 2013. Paleobathymetry of the northern North Atlantic and consequences for the opening of the Fram Strait. *Marine Geophys. Res.*, **34**, 25, https://doi.org/10.1007/s11001-013-9165-9

FORSYTH, D.W. 1977. The evolution of the upper mantle beneath mid-ocean ridges. *Tectonophysics*, **38**, 89–118, https://doi.org/10.1016/0040-1951(77)90202-5

GOSWAMI, A., OLSON, P.L., HINNOV, L.A. & GNANADESIKAN, A. 2015. OESbathy version 1.0: a method for reconstructing ocean bathymetry with generalized continental shelf-slope-rise structures. *Geoscientific Model Development*, **8**, 2735–2748, https://doi.org/10.5194/gmd-8-2735-2015

HAGER, B.H. & O'CONNELL, R.J. 1981. A simple global model of plate dynamics and mantle convection. *Journal of Geophysical Research*, **86**, 4843–4867, https://doi.org/10.1029/JB086iB06p04843

HAGER, B.H., CLAYTON, R.W., RICHARDS, M.A., COMER, R.P. & DZIEWONSKI, A.M. 1985. Lower mantle heterogeneity, dynamic topography and the geoid. *Nature*, **313**, 541–545.

HAYES, D. 1988. Age-depth relationships and depth anomalies in the southeast Indian Ocean and South Atlantic Ocean. *Journal of Geophysical Research*, **93**, 2937–2954.

HAYES, D.E. 1991. *Marine Geological and Geophysical Atlas of the Circum-Antarctic to 30°S*. American Geophysical Union, Antarctic Research Series, **54**.

HEESTAND, R. & CROUGH, S. 1981. The effect of hot spots on the oceanic age–depth relation. *Journal of Geophysical Research*, **86**, 6107–6114.

HILLIER, J.K. & WATTS, A.B. 2005. Relationship between depth and age in the North Pacific Ocean. *Journal of Geophysical Research B: Solid Earth*, **110**, 1–22, https://doi.org/10.1029/2004JB003406

HOGGARD, M.J., WHITE, N. & AL-ATTAR, D. 2016. Global dynamic topography observations reveal limited influence of large-scale mantle flow. *Nature Geoscience*, **9**, 1–8, https://doi.org/10.1038/ngeo2709

HOGGARD, M., WINTERBOURNE, J., CZARNOTA, K. & WHITE, N. 2017. Oceanic residual depth measurements, the plate cooling model and global dynamic topography. *Journal of Geophysical Research: Solid Earth*, **122**, 2328–2372, https://doi.org/10.1002/2016JB013457

HUISMANS, R. & BEAUMONT, C. 2011. Depth-dependent extension, two-stage breakup and cratonic underplating at rifted margins. *Nature*, **473**, 74–78, https://doi.org/10.1038/nature09988

JARVIS, G. & PELTIER, W. 1980. Oceanic bathymetry profiles flattened by radiogenic heating in a convecting mantle. *Nature*, **285**, 649–651, https://doi.org/10.1073/pnas.0703993104

KANE, K.A. & HAYES, D.E. 1992. Long-lived mid-ocean ridge segmentation. *Journal of Geophysical Research*, **97**, 317–330.

LANGSETH, M., HOUTZ, R.E., DRAKE, C.L. & NAFE, J.E. 1966. Crustal structure of the mid-ocean ridges. *Journal of Geophysical Research*, **71**, 341–352.

LASKE, G., MASTERS, G., MA, Z. & PASYANOS, M. 2013. Update on CRUST1.0 – a 1-degree global model of Earth's crust. *Geophysical Research Abstracts*, **15**, 2658.

LITHGOW-BERTELLONI, C. & SILVER, P. 1998. Dynamic topography, plate driving forces and the African superswell. *Letters to Nature*, **395**, 345–348, https://doi.org/10.1038/26212

MCKENZIE, D.P. 1967. Some remarks on heat flow and gravity anomalies. *Journal of Geophysical Research*, **72**, 6261–6273, https://doi.org/10.1029/JZ072i024p06261

MCKENZIE, D. 1977. Surface deformation, gravity anomalies and convection. *Geophysical Journal of the Royal Astronomical Society*, **48**, 211–238.

MCKENZIE, D. 1978. Some remarks on the development of sedimentary basins. *Earth and Planetary Science Letters*, **40**, 25–32.

MCNUTT, M. 1995. Marine geodynamics: depth–age revisited. *Reviews of Geophysics*, **33**, (Suppl. 1), 413–418.

MORGAN, J. & SMITH, W. 1992. Flattening of the sea-floor depth–age curve as a response to asthenospheric flow. *Nature*, **359**, 524–527.

MORGAN, W.J. 1965. Gravity anomalies and convection currents: 1. A sphere and cylinder sinking beneath the surface of a viscous fluid. *Journal of Geophysical Research*, **70**, 6175–6187.

MÜLLER, R.D., SDROLIAS, M., GAINA, C. & ROEST, W.R. 2008a. Age, spreading rates, and spreading asymmetry of the world's ocean crust. *Geochemistry, Geophysics, Geosystems*, **9**, Q04006, https://doi.org/10.1029/2007GC001743

MÜLLER, R.D., SDROLIAS, M., GAINA, C., STEINBERGER, B. & HEINE, C. 2008b. Long-term sea-level fluctuations driven by ocean basin dynamics. *Science*, **319**, 1357–1362, https://doi.org/10.1126/science.1151540

NAGIHARA, S., LISTER, C.R. & SCLATER, J.G. 1996. Reheating of old oceanic lithosphere: Deductions from observations. *Earth and Planetary Science Letters*, **139**, 91–104, https://doi.org/10.1016/0012-821X(96)00010-6

O'CONNOR, J.M. & DUNCAN, R.A. 1990. Evolution of the Walvis Ridge–Rio Grande Rise Hot Spot System: Implications for African and South American Plate motions over plumes. *Journal of Geophysical Research*, **95**, 17 475–17 502, https://doi.org/10.1029/JB095iB11p17475

O'CONNOR, J.M. & JOKAT, W. 2015. Tracking the Tristan-Gough mantle plume using discrete chains of intraplate volcanic centers buried in theWalvis Ridge. *Geology*, **43**, 715–718, https://doi.org/10.1130/G36767.1

O'CONNOR, J.M., LE ROEX, A.P., CLASS, C., WIJBRANS, J.R., KEßLING, S., KUIPER, K.F. & NEBEL, O. 2012. Hotspot trails in the South Atlantic controlled by plume and plate tectonic processes. *Nature Geoscience*, **5**, 735–738, https://doi.org/10.1038/ngeo1583

PANASYUK, S.V. & HAGER, B.H. 2000. Inversion for mantle viscosity profiles constrained by dynamic topography and the geoid, and their estimated errors. *Geophysical Journal International*, **143**, 821–836.

PARI, G. & PELTIER, W.R. 2000. Subcontinental mantle dynamics: a further analysis based on the joint constraints of dynamic surface topography and free-air graviy. *Journal of Geophysical Research*, **105**, 5635–5662.

PARKER, R.L. & OLDENBURG, D.W. 1973. Thermal model of ocean ridges. *Nature Physical Science*, **242**, 137–139, https://doi.org/10.1038/physci242137a0

PARSONS, B. & DALY, S. 1983. The relationship between surface topography, gravity anomalies, and temperature structure of convection. *Journal of Geophysical Research*, **88**, 1129–1144, https://doi.org/10.1029/JB088iB02p01129

PARSONS, B. & MCKENZIE, D. 1978. Mantle convection and the thermal structure of the plates. *Journal of Geophysical Research*, **83**, 4485–4496, https://doi.org/10.1029/JB083iB09p04485

PARSONS, B. & SCLATER, J. 1977. An analysis of the variation of ocean floor bathymetry and heat flow with age. *Journal of Geophysical Research*, **82**, 803–827.

PEKERIS, C.L. 1935. Thermal convection in the interior of the Earth. *Geophysical Supplements to the Monthly Notices of the Royal Astronomical Society*, **3**, 343–367, https://doi.org/10.1111/j.1365-246X.1935.tb01742.x

PÉREZ-DÍAZ, L. & EAGLES, G. 2014. Constraining South Atlantic growth with seafloor spreading data. *Tectonics*, **33**, 1848–1873, https://doi.org/10.1002/2014TC003644

PÉREZ-DÍAZ, L. & EAGLES, G. 2017a. A new high-resolution seafloor age grid for the South Atlantic Geochemistry. *Geophysics, Geosystems*, **18**, 1–14, https://doi.org/10.1002/2016GC006750

PÉREZ-DÍAZ, L. & EAGLES, G. 2017b. South Atlantic paleobathymetry since early Cretaceous. *Scientific reports*, **7**, 11819. https://doi.org/10.1038/s41598-017-11959-7

POORE, H.R., SAMWORTH, R., WHITE, N.J., JONES, S.M. & MCCAVE, I.N. 2006. Neogene overflow of Northern Component Water at the Greenland-Scotland Ridge. *Geochemistry, Geophysics, Geosystems*, **7**, Q06010, https://doi.org/10.1029/2005GC001085

RENKIN, M.L. & SCLATER, J.G. 1988. Depth and age in the North Pacific. *Journal of Geophysical Research*, **93**,

2919–2935, https://doi.org/10.1029/JB093iB04p0 2919

Richards, M.A. & Hager, B.H. 1984. Geoid anomalies in a dynamic Earth. *Journal of Geophysical Research: Solid Earth*, **89**, 5987–6002, https://doi.org/10.1029/JB089iB07p05987

Sandwell, D.T., Müller, R.D., Smith, W.H., Garcia, E. & Francis, R. 2014. New global marine gravity model from CryoSat-2 and Jason-1 reveals buried tectonic structure. *Science*, **346**, 65–67.

Schubert, G. & Turcotte, D. 1972. One-dimensional model of shallow mantle convection. *In*: *Plate Tectonics*. Collected Reprint Series. American Geophysical Union, Washington DC, 945–951, https://doi.org/10.1002/9781118782149.ch30

Schubert, G., Froidevaux, C. & Yuen, D.A. 1976. Oceanic lithosphere and astenosphere: thermal and mechanical structure. *Journal of Geophysical Research*, **81**, 3525–3540.

Schubert, G., Yuen, D.A., Froidevaux, C., Fleitout, L. & Souriau, M. 1978. Mantle circulation with partial shallow return flow: Effects on stresses in oceanic plates and topography of the sea floor. *Journal of Geophysical Research*, **83**, 745, https://doi.org/10.1029/JB083iB02p00745

Sclater, J.G., Hellinger, S. & Tapscott, C. 1977. Paleobathymetry of Atlantic Ocean from the Jurassic to present. *Journal of Geology*, **85**, 509–552.

Sclater, J.G., Meinke, L. & Murphy, C. 1985. The depth of the ocean through the Neogene. *In*: Kennett, J.P. (ed.) *The Miocene Ocean: Paleoceanography and Biogeography*. Geological Society of America, Memoirs, **163**, 1–19.

Smith, W.H. & Sandwell, D.T. 1997. Global sea floor topography from satellite altimetry and ship depth soundings. *Science*, **277**, 1956–1962, https://doi.org/10.1126/science.277.5334.1956

Stein, C. & Stein, S. 1992. A model for the global variation in oceanic depth and heat flow with lithospheric age. *Nature*, **359**, 123–129.

Stein, C.A. & Stein, S. 1994. Thermal evolution of oceanic lithosphere. *Geophysical Research Letters*, **21**, 709–712.

Steinberger, B. & Holme, R. 2008. Mantle flow models with core-mantle boundary constraints and chemical heterogeneities in the lowermost mantle. *Journal of Geophysical Research: Solid Earth*, **113**, 1–16, https://doi.org/10.1029/2007JB005080

Steinberger, B., Sutherland, R. & O'Connell, R.J. 2004. Prediction of Emperor–Hawaii seamount locations from a revised model of global plate motion and mantle flow. *Nature*, **430**, 167–173.

Sykes, T.J. 1996. A correction for sediment load upon the ocean floor: uniform v. varying sediment density estimations – implications for isostatic correction. *Marine Geology*, **133**, 35–49, https://doi.org/10.1016/0025-3227(96)00016-3

Sykes, T.J.S. *et al.* 1998. Southern hemisphere palaeobathymetry. In: Geological Society, London, Special Publications **131**, 1–42, https://doi.org/10.1144/GSL.SP.1998.131.01.02

Torsvik, T.H., Müller, R.D., Van der Voo, R., Steinberger, B. & Gaina, C. 2008. Global plate motion frames: Toward a unified model. *Reviews of Geophysics*, **46**, 1–44, https://doi.org/10.1029/2007RG000227

Turcotte, D. & Oxburgh, E. 1967. Finite amplitude convection cells and continental drift. *Journal of Fluid Mechanics*, **28**, 29–42.

Van Avendonk, H.J., Davis, J.K., Harding, J.L. & Lawver, L.A. 2017. Decrease in oceanic crustal thickness since the breakup of Pangaea. *Nature Geoscience*, **10**, 58–61.

White, R.S., McKenzie, D. & O'Nions, R.K. 1992. Oceanic crustal thickness from seismic measurements and rare earth element inversions. *Journal of Geophysical Research*, **97**, 19 683–19 715, https://doi.org/10.1029/92JB01749

Whittaker, J., Goncharov, A., Williams, S., Müller, R.D. & Leitchenkov, G. 2013. Global sediment thickness data set updated for the Australian–Antarctic Southern Ocean. *Geochemistry, Geophysics, Geosystems*, **14**, 3297–3305, https://doi.org/10.1002/ggge.20181

Wright, J.D. & Miller, K.G. 1996. Control of North Atlantic Deep Water circulation by the Greenland–Scotland Ridge. *Paleoceanography*, **11**, 157–170.

Crustal structure of the central sector of the NE Brazilian equatorial margin

JAVIER TAMARA[1]*, KEN R. McCLAY[1] & NEIL HODGSON[2]

[1]*Fault Dynamics Research Group, Department of Earth Sciences, Royal Holloway University of London, Egham Hill, Egham, Surrey TW20 0EX, UK*

[2]*Spectrum Geo Ltd, Dukes Court, Duke Street, Woking GU21 5BH, UK*

JT, 0000-0001-9015-9928

*Correspondence: jatamarag@gmail.com

Abstract: The central equatorial Brazilian margin is divided into the Amazon and Barreirinhas divergent segments separated by the Pará-Maranhão transform segment. Analysis of regional 2D seismic lines allowed the definition of the crustal architecture of the margin. In the study area, the Barreirinhas segment has a proximal domain with a 30–35 km-thick continental crust, a 20–40 km-wide necked domain where the crust thins to 10 km, and an outboard domain with hyperextended continental crust. The Pará-Maranhão and Amazon segments consist of exhumation domains and their transition to ocean crust. Their structural styles indicate that this is a magma-poor passive margin with oceanic crust formed in a slow spreading centre. The Pará-Maranhão segment is bounded by two branches of the Saint Paul Fracture Zone that displace crustal domains with structures that document the transition from the distal part of a transform margin to an oceanic fracture zone. Two groups of post-rift volcanic complexes have been identified in the exhumation and oceanic domains, and whose distribution is controlled by the fracture zones. Late Cretaceous–Recent gravitationally-driven slide systems and mass-transport deposits indicate long-lived margin collapse and sediment redistribution fundamentally controlled by the underlying crustal structure of this part of the northeastern Brasilian passive margin.

The equatorial Atlantic margin of northeastern Brazil is a magma-poor passive margin characterized by NW–SE divergent and intervening ENE–WSW transform segments formed during the Early Cretaceous (Fig. 1) (Mascle & Blarez 1987; Mascle et al. 1988; Basile et al. 1993; Darros De Matos 2000; Antobreh et al. 2009). This margin displays the combined effect of oblique rifting and extension, together with the reactivation of inherited Proterozoic fabrics (Mascle & Blarez 1987; Mascle et al. 1988; Darros De Matos 2000; Basile et al. 2005; Antobreh et al. 2009; Nemčok et al. 2013; Heine et al. 2013; Mercier de Lépinay et al. 2016). From north to south, it consists of the Amazon divergent segment, the Pará-Maranhão transform segment, that connects oceanwards with the Saint Paul Fracture Zone, the Barreirinhas divergent segment and the Ceará transform segment that links to the Romanche Fracture Zone (Figs 1 & 2) (Darros De Matos 2000; Mercier de Lépinay et al. 2016).

The crustal architecture of the NE Brazilian margin shares many similarities with other divergent and transform margins (Watts et al. 2009; Zalán 2015; Mercier de Lépinay et al. 2016). Domains of thinned continental crust and presumed exhumed continental mantle have been described, and significant lateral variations in the width of the zone of thinned continental crust and the ocean–continental transition have been postulated (Houtz et al. 1977; Greenroyd et al. 2008b; Watts et al. 2009; Zalán 2015; Sapin et al. 2016; Kusznir et al. 2018). There are few published studies, and the crustal structure and distribution of crustal domains are yet to be defined (Darros De Matos 2000; Bizzi et al. 2003; Henry et al. 2011; Krueger et al. 2014; Davison et al. 2015; Zalán 2015).

The existence of the transform margin has been proposed to be a continuation of the Saint Paul Fracture Zone in order to fit the conjugate margins into plate kinematic models (Moulin et al. 2010; Heine et al. 2013; Pérez-Díaz & Eagles 2014; Mercier de Lépinay et al. 2016). This has recently been inferred from gravity-inversion models that suggest a similarity to the French Guyana transform margin to the north (Kusznir et al. 2018). The morphology of the Pará-Maranhão transform margin is different to other transform margins in the region (e.g. Côte d'Ivoire-Ghana margin–Romache Fracture Zone–Ceará margin (Fig. 1): Mascle et al. 1988; Attoh et al. 2004; Antobreh et al. 2009; Davison et al. 2015; Mercier de Lépinay et al. 2016), and little detail is known about its crustal architecture.

From: McClay, K. R. & Hammerstein, J. A. (eds) 2020. *Passive Margins: Tectonics, Sedimentation and Magmatism*. Geological Society, London, Special Publications, **476**, 163–191.
First published online March 4, 2020, https://doi.org/10.1144/SP476-2019-54
© 2020 The Author(s). Published by The Geological Society of London. All rights reserved.
For permissions: http://www.geolsoc.org.uk/permissions. Publishing disclaimer: www.geolsoc.org.uk/pub_ethics

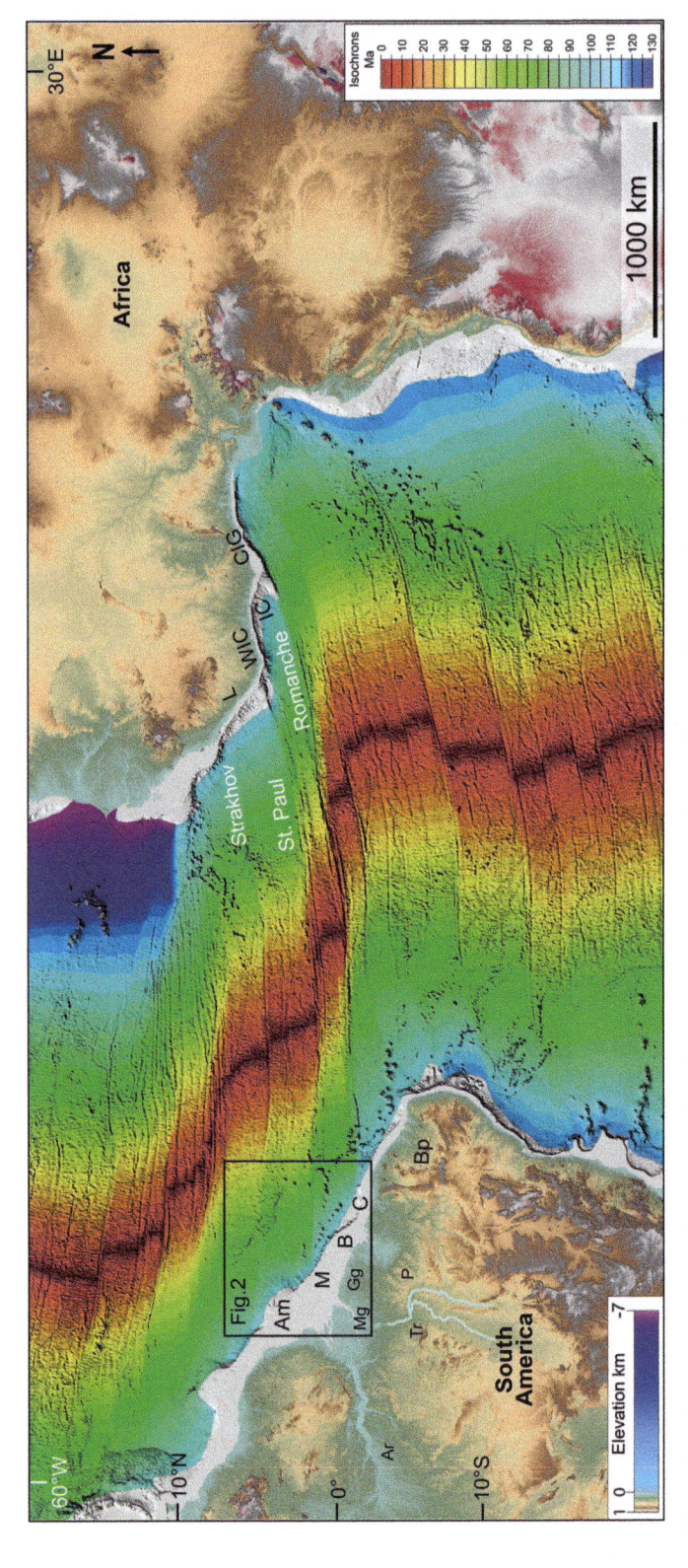

Fig. 1. Digital elevation model of the equatorial Atlantic showing oceanic crust age from Müller *et al.* (2008), and the location of divergent and transform margin segments and main fracture zones. Divergent margin segments are Amazon (Am), Barreirinhas (B), Liberia (L) and Ivory Coast Basin (IC). Transform margin segments are Ceará (C), Ivory coast-Ghana (CIG), Pará-Maranhão (M) and western Ivory Coast (WIC). Drainage elements are Amazon River (Ar) and Tocantins River (Tr). Borborema Province (Bp), Gurupi–São Luis graben systems (Gg), Marajó graben (Mg) and Parnaíba Basin (P).

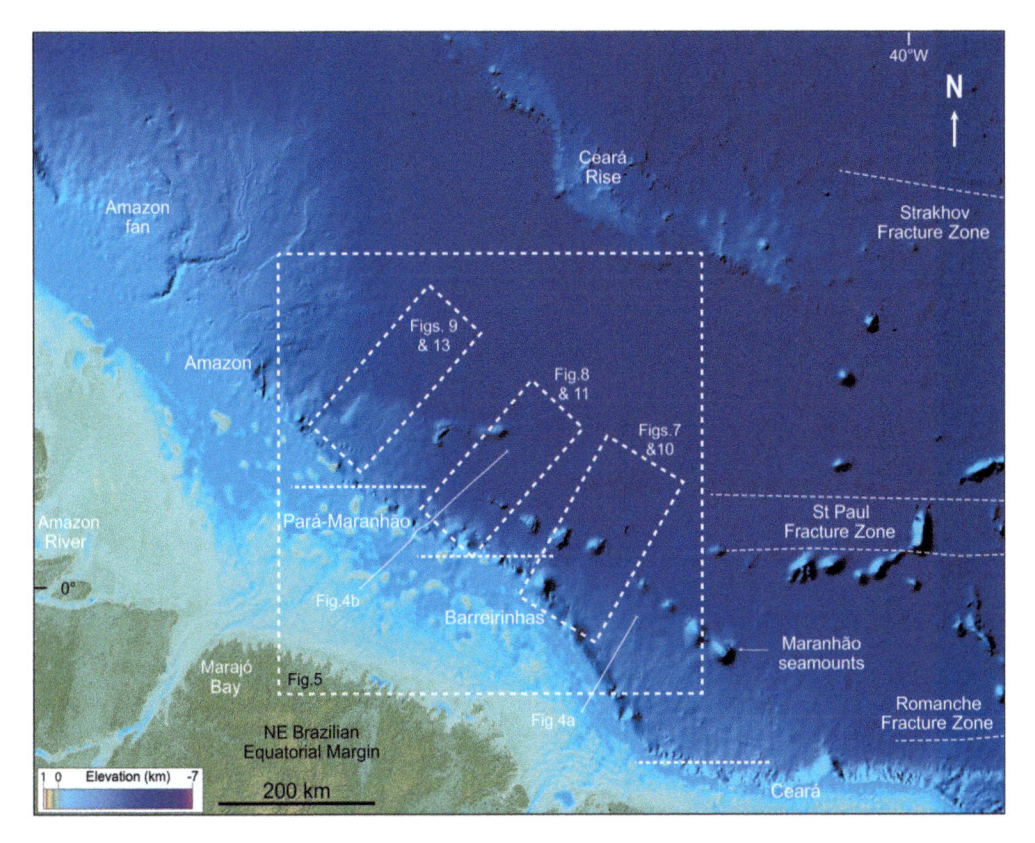

Fig. 2. Location map showing the digital elevation model of the central sector of the equatorial margin of Brazil, and locations of Amazon and Barreirinhas divergent segments, and Pará-Maranhão and Ceará transform margins segments. The traces of fracture zones and main volcanic complexes are highlighted. The boxes represent the approximate locations of 2D lines. Precise locations have been omitted due to confidentiality reasons.

This research used 27 high-quality pre-stack time migration (PSTM) 2D seismic sections that were recorded to 14 s two-way-travel time (TWT), and the survey extends for c. 600 km along the margin from the northern part of Barreirinhas segment to the southern part of Amazon segment (Fig. 2). The dataset permits analysis of the crustal structure underlying the continental slope to the abyssal plain, and allowed definition of the crustal architecture of this part of Brazilian equatorial margin. Seismic horizons were correlated with published sections on the continental platform and upper slope in order to define a regional stratigraphic context.

The aims of this paper are to describe the along-strike variations in the architecture of the central sector of the NE Brazilian equatorial margin, and to present, to the best of our knowledge, the first documentation of the crustal structure of the distal part of the Pará-Maranhão transform margin and its oceanward transition to the Saint Paul Fracture Zone. Multiple magmatic events have also been recognized and a general evolutionary model of the Pará-Maranhão margin is proposed and compared to models for other passive-margin systems.

Geological setting

The central sector of the equatorial margin of Brazil extends for c. 800 km along a N45–20W direction, from the intersection of the Romanche Fracture Zone with the continental margin in the south to the Amazon fan system in the north (Fig. 2). The bathymetric profiles show a c. 30–60 km-wide upper slope dipping between 4° and 8°, and a c. 60–100 km-wide lower slope dipping 1.3°–1.6° that connects with a 0.2°–0.3° dipping abyssal plain. The change between the upper and lower slopes is marked by an inflexion at approximately 3.5–4 km depth. Several seamounts of the Maranhão chain occur on the abyssal plain and close to the continental slope.

The NE Brazilian margin has a stratigraphic record ranging from Early Cretaceous to Recent

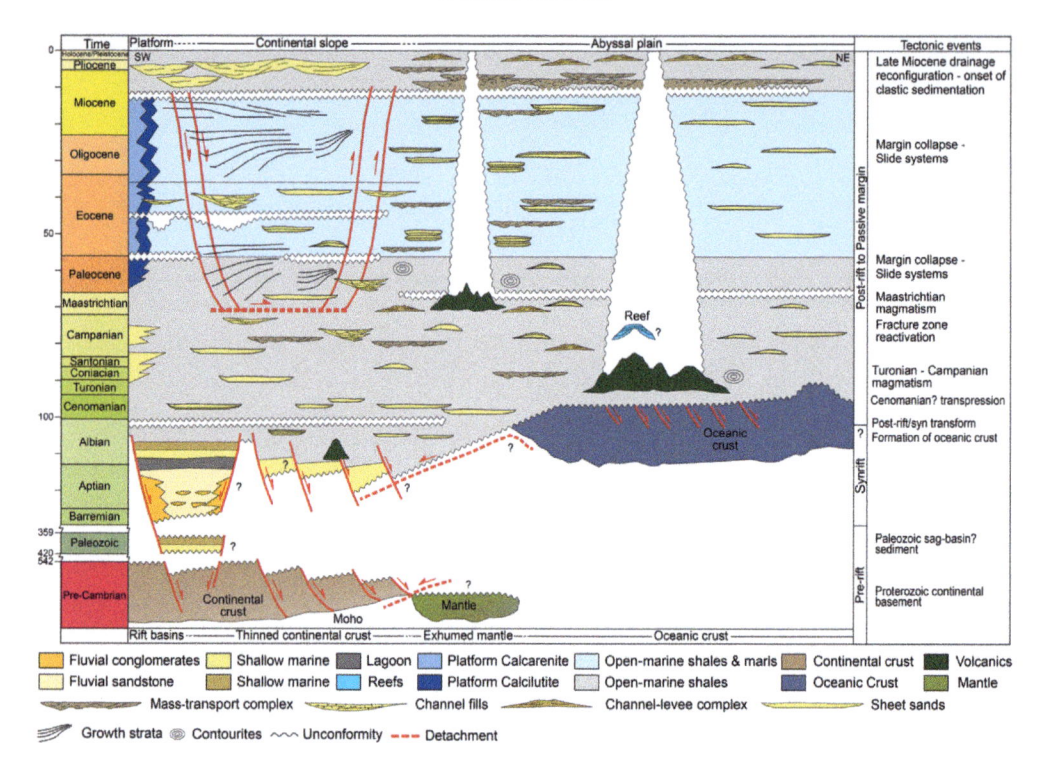

Fig. 3. Regional tectonostratigraphic framework of the central sector of the equatorial margin of Brazil from the continental slope to the abyssal plain.

(Figs 3 & 4) (Soares et al. 2007). Figure 3 shows a simplified tectonostratigraphic framework of the three segments of the central sector of the Brazilian equatorial margin identified in this study.

Amazon divergent segment

The onshore basement of the Amazon segment consists of Archean and Proterozoic granitoids and gneisses, as well as Proterozoic meta-sedimentary and meta-volcanic sequences of the Amazon Craton (Fig. 3) (Mohriak 2003; Schobbenhaus & Brito Neves 2003). The continental crust is 34–37 km thick and thins oceanwards (eastwards) to less than 10 km over a zone about 200 km wide (Greenroyd et al. 2007, 2008a; Watts et al. 2009; Mercier de Lépinay et al. 2016). The proximal part includes half-graben with Triassic and Early Cretaceous synrift strata (Bizzi et al. 2003; Mohriak 2003; Schobbenhaus & Brito Neves 2003; Schobbenhaus et al. 2004). The ocean–continental boundary is narrow and connects to a 4–5 km-thick oceanic crust interpreted to have formed by a slow spreading centre (Houtz et al. 1977; Rodger et al. 2006; Watts et al. 2009). Eruptive volcanic complexes with possible associated carbonate build-ups were also described by Houtz et al. (1977).

Pará-Maranhão transform segment

The central transform segment is identified as a bend in the continental slope where the trace of the Saint Paul Fracture Zone projects into the continental margin (Figs 2 & 5). It is bounded by two parallel fracture zones, 80 km apart, which make an angle of c. 50° at the intersection with the margin (Fig. 2). Their bathymetric expressions are longitudinal ridges, canyons and aligned seamounts from the mid-ocean ridge up to c. 42° W (Figs 1 & 2). Near the continental margin, the fracture zones can be mapped in the bathymetric and gravity anomaly images of the abyssal plain (Kumar & Embley 1977; Matthews et al. 2011). The transition from fracture zones in the oceanic crust into the transform margin has not been clearly documented (Darros De Matos 2000; Moulin et al. 2010; Mercier de Lépinay et al. 2016).

Barreirinhas divergent segment

The onshore basement of the Barreirinhas segment is formed by Archean granitoids, schists and meta-volcanic rocks of the Sao Luis Craton, and Proterozoic meta-sedimentary and meta-volcanic sequences, granitoids and gneisses of the Gurupi

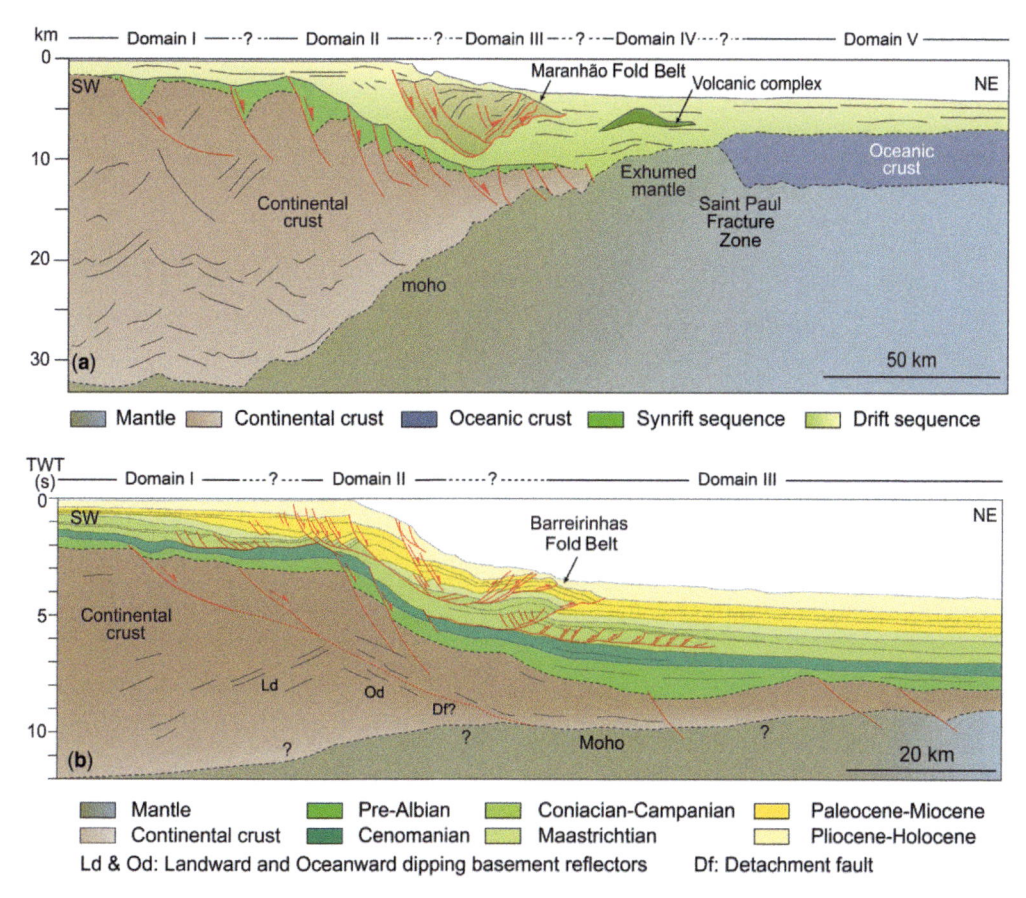

Fig. 4. Regional crustal sections of the Pará-Maranhão and Barreirinhas segments. (**a**) Interpretation of line GB1-4500 from Zalán (2015) showing the crustal architecture of the Pará-Maranhão Basin and the crustal domains (I–V) defined in this study. (**b**) Line drawing of the crust structure of the Barreirinhas Basin modified from the seismic section in figure 7 of Krueger *et al.* (2012) and the crustal domains (I and II) defined in this study.

Belt (Fig. 3) (Klein & Moura 2008; Klein *et al.* 2013). The continental crust is 35–40 km thick and has NW–SE-trending graben containing Early Cretaceous fluvial to shallow-marine strata (Soares Júnior *et al.* 2008; Watts *et al.* 2009; Henry *et al.* 2011; Daly *et al.* 2014; Zalán 2015). The crust thins eastwards to less than 10 km over a 30 km-wide zone characterized by tilted, transparent basement fault blocks that deepen and decrease in thickness towards the ocean (Fig. 4) (Henry *et al.* 2011; Zalán 2015; Kusznir *et al.* 2018). The ocean–continental boundary has been interpreted as being exhumed continental mantle (Zalán 2015). Here the oceanic crust is *c.* 5 km thick but is locally thickened by magmatic complexes (Houtz *et al.* 1977; Davison *et al.* 2015; Kusznir *et al.* 2018). This segment has multiple volcanic complexes that occur within the post-rift sequence and are characterized by discontinuous high-amplitude reflections embedded within low-amplitude to chaotic reflectors (Zalán 2015). Some of the volcanic complexes produce a strong bathymetric expression along NE–SW and east–west trends (Fig. 2).

Evolution of the Cretaceous equatorial margin of Brazil

Rifting of the equatorial Atlantic probably started in the Barremian(?)–Early Aptian(?) with distributed deformation in intracratonic basins that later became focused on the Amazon, Pará-Maranhão and Barreirinhas basins (Fig. 3) (Darros De Matos 2000; Costa *et al.* 2002; Soares *et al.* 2007; Trosdtorf Junior *et al.* 2007). In the Ivory Coast Basin, the conjugate margin of the Barreirinhas segment, rifting is interpreted to have started in the Middle Aptian with thinning of the continental lithosphere and mantle exhumation during the Late Albian(?) (Gillard

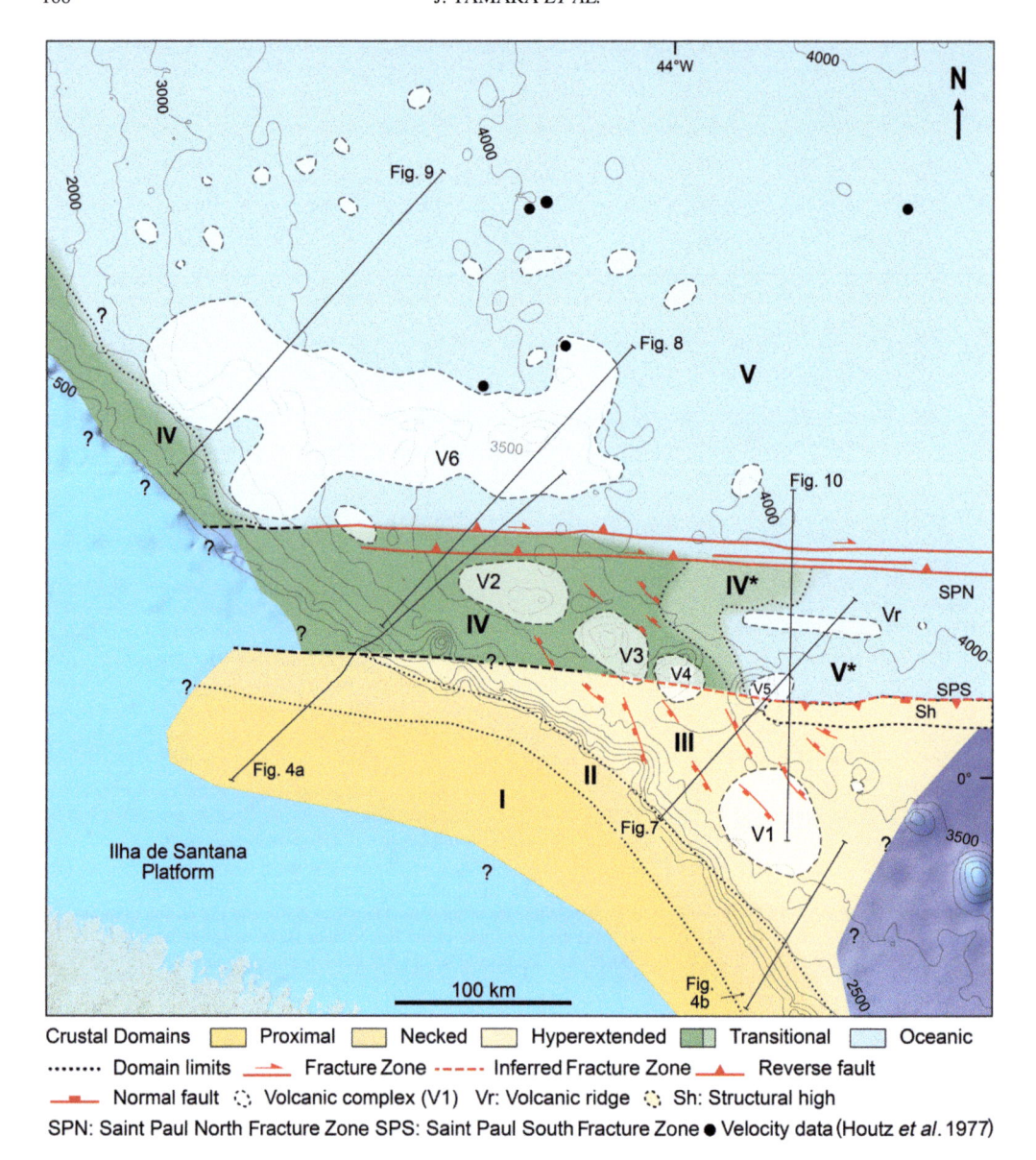

Fig. 5. Map showing the distribution of crustal domains and fracture zones in the central sector of the equatorial margin of Brazil. The location of volcanic complexes and the structural high are included. The boxes represent the approximate locations of 2D lines. Precise locations have been omitted due to confidentiality reasons.

et al. 2017; Ye et al. 2019). Plate kinematic models suggest that accretion of oceanic crust in isolated basins started in the Albian (120–105 Ma: Moulin et al. 2010; Heine et al. 2013) and was coeval with transtension along the transform margins (e.g. Ceará margin: Darros De Matos 2000; Davison et al. 2015). Inversion of transtensional basins and the formation of transpressional belts along the Ceará and Demerara margins of South America (Darros De Matos 2000; Basile et al. 2013; Davison et al. 2015), as well as along the Ghana–Ivory Coast and western Ivory Coast margins in western Africa (Mascle & Blarez 1987; Basile et al. 1993; Mascle et al. 1997; Attoh et al. 2004; Antobreh et al. 2009; Scarselli et al. 2018), has been attributed to Late Albian–Cenomanian contraction driven by changes in the extension direction, and relative movement between South America and west Africa

(Fig. 3) (Darros De Matos 2000; Heine et al. 2013; Nemčok et al. 2013; Davison et al. 2015). Final separation of the continental lithosphere of South America and Africa occurred at 104 Ma along the western Ivory Coast margin and at 94 Ma along the Ghana–Ivory Coast margin, together with migration of the oceanic spreading centres along the respective transform margins at 96 and 74 Ma, respectively (Heine et al. 2013; Nemčok et al. 2013). Changes in extension direction from NE–SW to east–west at 108–92, 89–65 and 65–52 Ma have also been proposed by Nemčok et al. (2013) along the west African margin.

Seismic stratigraphy

Figures 3 and 6 show the general stratigraphic framework for the Pará-Maranhão margin. Four regional Barremian(?)–present-day mega-sequences can be recognized. The basal mega-sequence in the Barreirinhas and Pará-Maranhão segments consists of Barremian(?)–Late Cretaceous(?) sediments from above the basement to top Cenomanian(?) synrift–early post-rift strata. In the Amazon segment, this mega-sequence of late synrift to post-rift strata, including several volcanic complexes, occurs on top of transparent and oceanic basement to the Upper Cretaceous unconformity.

The Late Cretaceous(?), Paleocene(?)–Miocene (?) and Miocene(?)–present-day mega-sequences occur throughout the study area, and in the Pará-Maranhão segment also contain Campanian(?) and Maastrichtian(?) volcanic complexes.

The thicknesses of the sedimentary sequences and the depth of the basement were estimated using a time–depth function for depth-conversion in Decision Space Desktop. The model is based on seismic velocity data from Houtz et al. (1977) and Watts et al. (2009) in the continental slope and the abyssal plain that range from 1.7 km s^{-1} close to the seafloor to 4.5 km s^{-1} near to the top of the basement in the continental slope.

Basal mega-sequence: Barremian(?)–Late Cretaceous(?) synrift to post-rift

In the Barreirinhas segment, the basal mega-sequence contains two units defined by the top basement and the top Cenomanian? (Fig. 6a). The lower unit consists of synkinematic growth wedges above the basement in the half-grabens. It includes discontinuous high-amplitude reflections within a package of low-amplitude reflections, and intercalations of continuous strata of low- and high-amplitude reflectivity. The top of the sequence is marked by a broad high-amplitude reflection below the top Albian? (Horizon B in Fig. 6a). This sequence has been interpreted to include the transition from fluvial to deep-marine deposits (Fig. 3) (Soares et al. 2007; Trosdtorf Junior et al. 2007).

The Cenomanian(?) strata onlap the underlying synkinematic unit and basement highs in the footwalls of normal faults. It displays thick continuous and discontinuous low- amplitude reflections, alternating with continuous high-amplitude reflections and thin discontinuous high-amplitude reflectors with sub-tabular to channelized geometries that are either isolated or stacked vertically (Fig. 6). Towards the top, there are strata with low-amplitude, chaotic reflectivity and erosional bases. In the Barreirinhas segment, the basal mega-sequence is 0.8–1 s thick TWT (c. 1.6–c. 2 km) but thickens to up to 1.8 s TWT (c. 3.6 km) in the hanging walls of normal faults.

In the Amazon segment, the basal sequence occurs on top of transitional and oceanic crust, and shows high-amplitude continuous reflections interbedded within thick low-amplitude to semi-transparent reflections, and isolated or vertically stacked high-amplitude sub-tabular to channelized reflections with erosional bases (Fig. 6b). This unit is covered by volcanic complexes and laterally equivalent deposits (described in the section 'Volcanic complexes' later in this paper). Some of these have subhorizontal high-amplitude reflections at the top and clinoforms on their flanks, such that they are interpreted as carbonate build-ups (Burgess et al. 2013). The sections are 1–1.5 s TWT (c. 2–3 km) thick, and thin to between 0.2 and 0.5 s TWT (c. 0.5–1 km) towards the abyssal plain.

These basal units are interpreted to include turbidity deposits with sheet sands, channel systems and mass-transport complexes (cf. Posamentier & Kolla 2003; Weimer et al. 2006).

Late Cretaceous(?) post-rift mega-sequence

The base of this mega-sequence is defined by the top Cenomanian in the Barreirinhas and Pará-Maranhão segments, and by the Upper Cretaceous unconformity in the Amazon segment (Fig. 3). It consists of a basal unit of low-amplitude continuous and parallel layers, a middle section with chaotic deposits of channelized base or mounded geometries, and a series of very thin, low-amplitude and continuous reflections in the top (Fig. 6b). The overlying Maastrichtian(?) strata consist of intercalations of low- and high-amplitude continuous and parallel reflections with local occurrences of chaotic units with channelized bases or mounded geometries (Fig. 6b). The seismic character of these units is interpreted to indicate turbidity sequences with sheet sands, channel systems and mass-transport complexes (e.g. Posamentier & Kolla 2003; Weimer et al. 2006). Close to volcanic edifices, asymmetrical

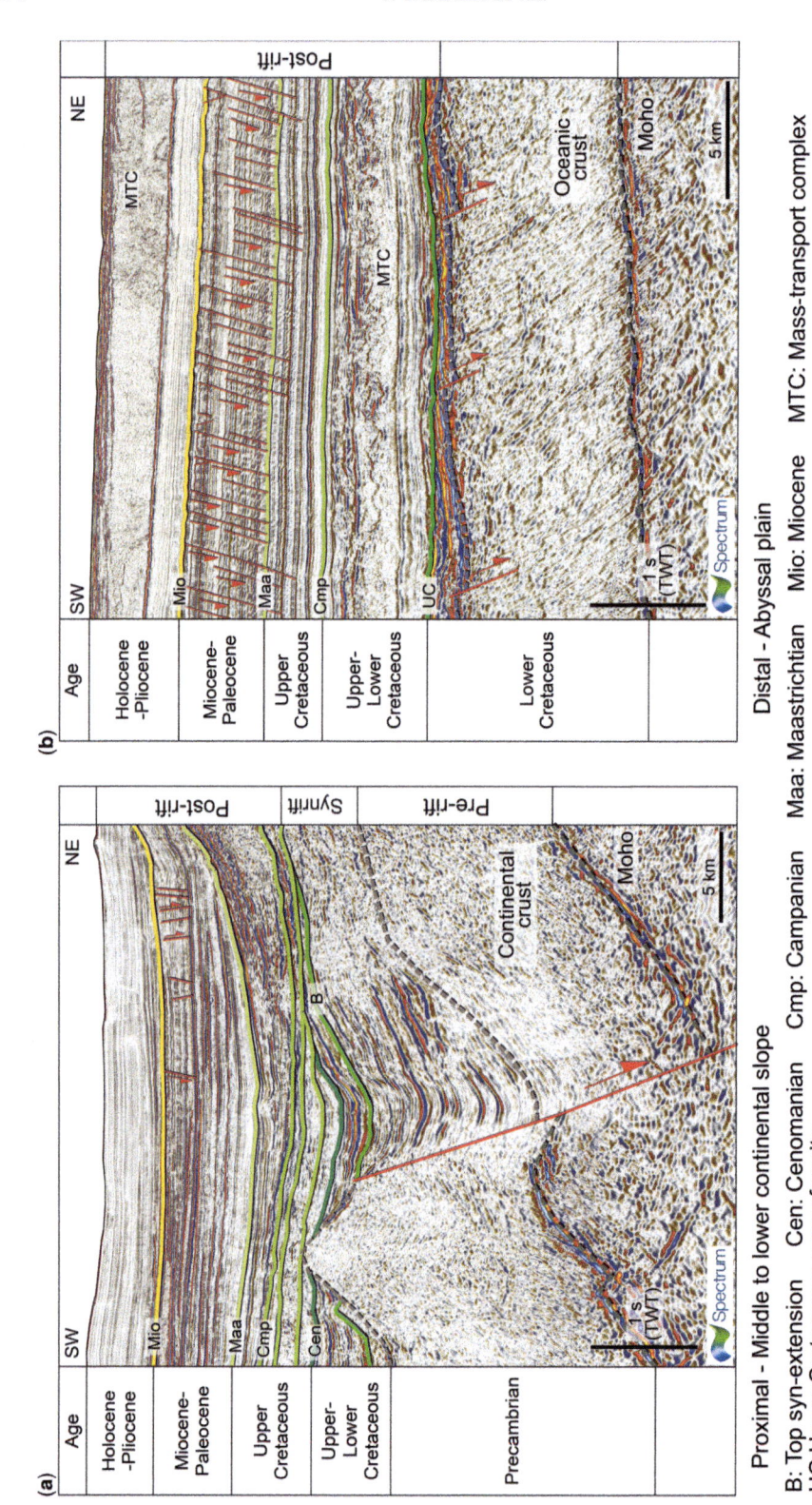

Fig. 6. Seismic stratigraphic character of the sedimentary filling in the middle and lower continental slope (**a**) and in the abyssal plain (**b**). Ages of key horizons are based on correlation with the seismic sections of Krueger *et al.* (2012). Spectrum is acknowledged for publication permission.

low-amplitude mounded deposits associated with channels systems are interpreted as contourites deposits (e.g. Rebesco *et al.* 2014). The sequence is 2–2.2 s TWT (*c.* 3.5 km) thick in the proximal areas of the Amazon segment, and thins to 0.5–0.7 s TWT (*c.* 0.8–1 km) on the abyssal plain and in the Barreirinhas segment.

Paleocene(?)–Miocene(?) passive-margin mega-sequence

In the proximal areas, the succession is 0.6–1 s TWT (*c.* 0.9–1.5 km) thick, and includes high-amplitude reflectors with irregular bases and sub-tabular geometries that laterally change to low-amplitude reflections and chaotic deposits (Fig. 6b). In the distal sectors, this is a 0.5–0.6 s TWT (*c.* 0.7–1 km) thick homogeneous package of low- to medium-amplitude parallel reflections with aggradational patterns (Figs 4 & 6).

Clusters of NE- and SW-dipping normal faults with displacements of less than 50 ms TWT, and spacings of between 0.5 and 1 km are mainly confined to the Paleocene–Miocene mega-sequence. The presence of upper and basal tips, their spacing and displacements, and their stratabound vertical extent are characteristic of polygonal faults (Cartwright *et al.* 2003).

Pliocene(?)–Holocene passive-margin mega-sequence

This mega-sequence is 0.6–0.7 s TWT (*c.* 0.6–1.0 km) thick, and is characterized by low- and high-amplitude reflectors (Fig. 6). It includes a massive, up to 0.5 s TWT (*c.* 0.5 km), series of layers with chaotic low-amplitude and discontinuous low- to medium-amplitude reflectors with varying dips, and erosional bases. These are interpreted as Early Pliocene–Late Pleistocene (Silva *et al.* 2010; Reis *et al.* 2016) amalgamated mass-transport complexes. To the north, these are overlain by channel–levee complexes and turbidity systems of the deep Amazon fan system that increase the thickness up to 4 km (Watts *et al.* 2009). In the south, this mega-sequence consists of low- and medium-amplitude parallel reflectors interbedded with high-amplitude reflectors with tabular, channelized and mounded geometries, such that they are interpreted as channel–levee complexes and turbidity systems.

Crustal architecture of NE Brazilian margin

Figure 5 shows the distribution of crustal domains in the central sector of the NE Brazilian margin, and Figures 7–10 show regional cross-sections that illustrate the main structural elements and crustal architectures that characterize each domain.

Five crustal domains have been differentiated based on the geometry, thickness and seismic character of the basement, the structural style, geometries and depths to detachment of faults, and the geometries of synkinematic sequences, as well as their relationships with adjacent domains and position within the margin. They have been classified as proximal, necked, hyperextended, transitional (exhumed) and oceanic domains following the classification scheme and models of Peron-Pinvidic *et al.* (2013), Péron-Pinvidic *et al.* (2015) and Osmundsen & Péron-Pinvidic (2018). The proximal and necked domains occur in the continental platform and upper slope, and were defined based on published seismic sections by Krueger *et al.* (2012) and Zalán (2015).

Proximal domain I

Figure 4 shows two geo-seismic sections across the continental platform and slope of the Pará-Maranhão segments with interpretation of Zalán (2015), and from the Barreirinhas segment with reinterpretation of a seismic section from Krueger *et al.* (2012). Gravity and seismic data suggest that the continental crust in the inboard area is 30–35 km thick (Zalán 2015; Kusznir *et al.* 2018). The top of the continental basement is subhorizontal to slightly tilted oceanwards below the outboard continental platform (Fig. 4). The crust has low–medium reflectivity and multiple reflectors with variable amplitudes and dips (Zalán 2015). A series of mid-crustal high-amplitude discontinuous reflections form a band of landward-dipping basement reflectors that deepen westwards from 6 to 8 s TWT (*c.* 6.5–9 km) (Fig. 4b).

Visible extensional faults appear to be restricted to the upper crust and form multiple tilted fault blocks that have pre-Albian(?) growth sequences. The Moho reflection is subhorizontal, and is estimated from seismic and gravimetric data to be at a depth of 30–35 km (de Castro *et al.* 2014; Zalán 2015).

Necked domain II

In the necked domain, the top of the continental basement dips eastwards and basinwards from 3 to 7 s TWT (*c.* 4–8 km) depth (Figs 4 & 7). The thickness of the crust is reduced from 30–35 km to less than 10 km over a distance of 30–40 km (Fig. 4). Although the structure is obscured by the lack of definition in the seismic image, reinterpretation of profiles from Krueger *et al.* (2012) indicates that thinning of the continental crust occurs over a narrow zone close to the upper continental slope and is

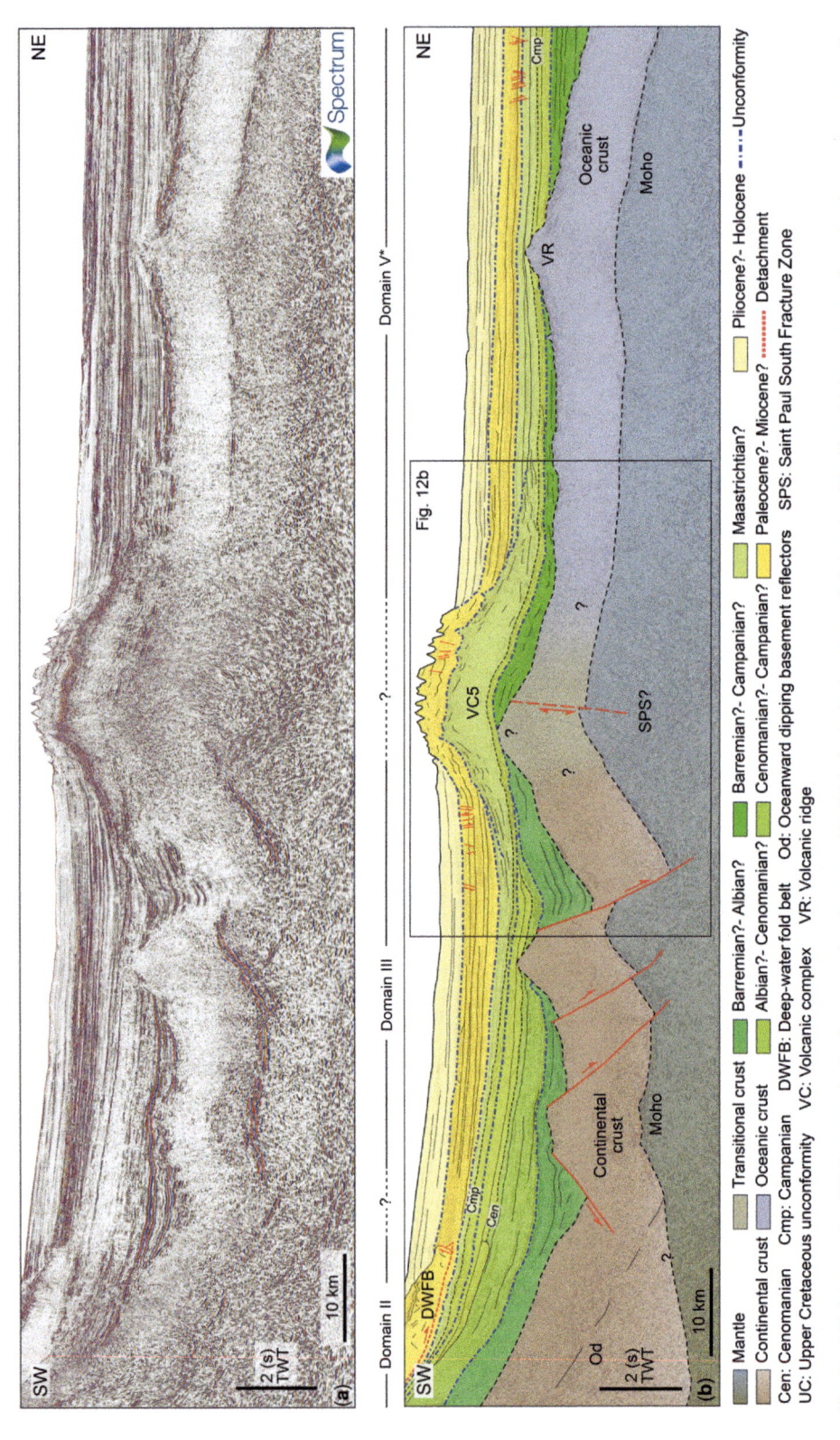

Fig. 7. a) Uninterpreted and b) interpreted seismic section showing rotated fault blocks in the hyperextended domain of the Barreirinhas segment, the volcanic complex on top of the southern branch of the Saint Paul Fracture Zone and the volcanic ridge in the flexure of the oceanic crust in the Pará-Maranhão segment. Spectrum is acknowledged for publication permission.

STRUCTURE OF THE NE BRAZILIAN EQUATORIAL MARGIN 173

Fig. 8. a) Uninterpreted and b) interpreted seismic section showing transparent basement without Moho reflection in the transitional domain of the Pará–Maranhão segment and the oceanic domain in the Amazon segment. The limit between domains is a high-angle reverse-fault system of the northern branch of the Saint Paul Fracture Zone. In the Amazon segment, volcanic complexes occur on top of the basal sequence and are covered by Campanian(?)–Maastrichtian(?) sediments with mounded contourite deposits in the NE. Spectrum is acknowledged for publication permission.

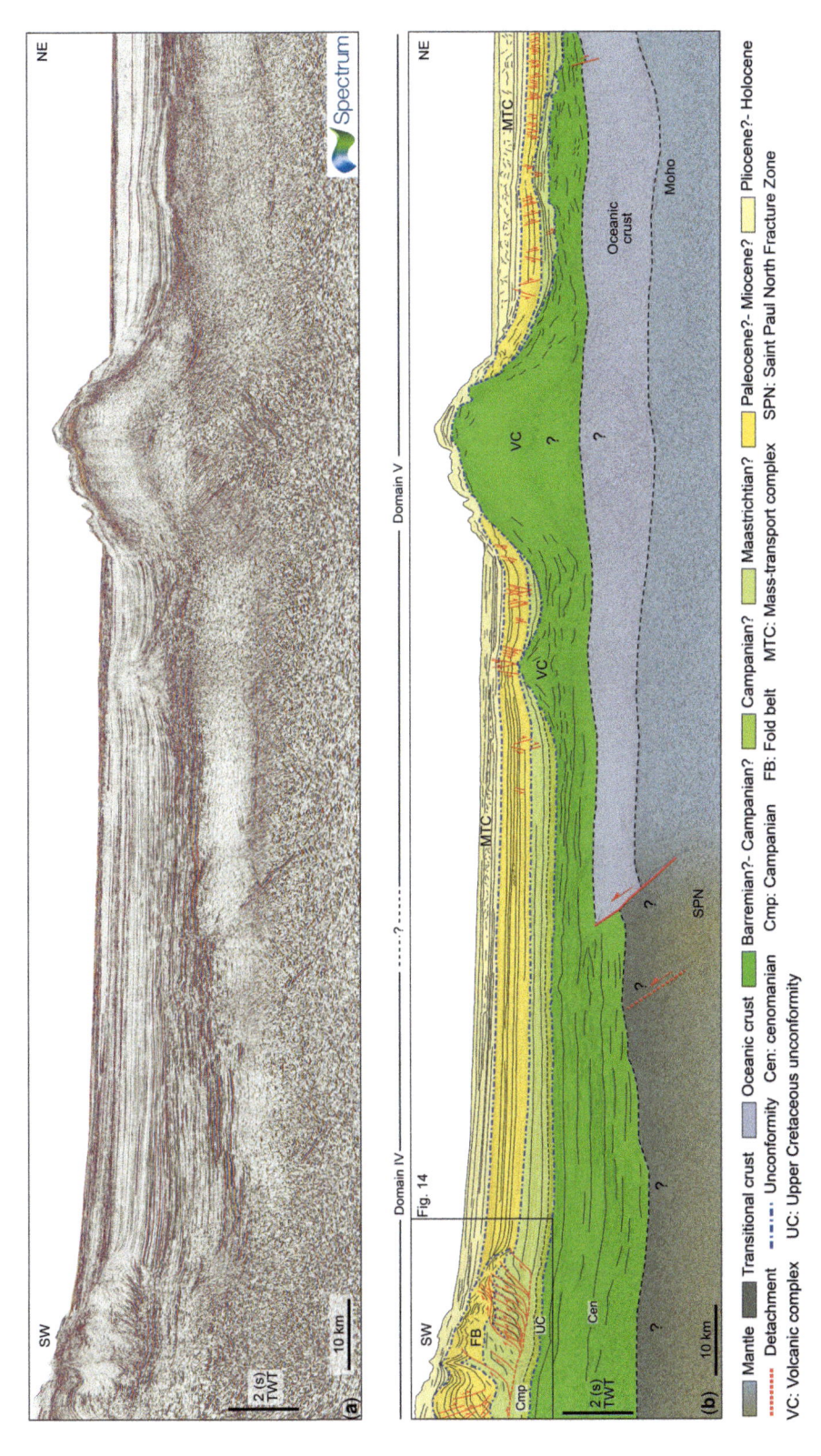

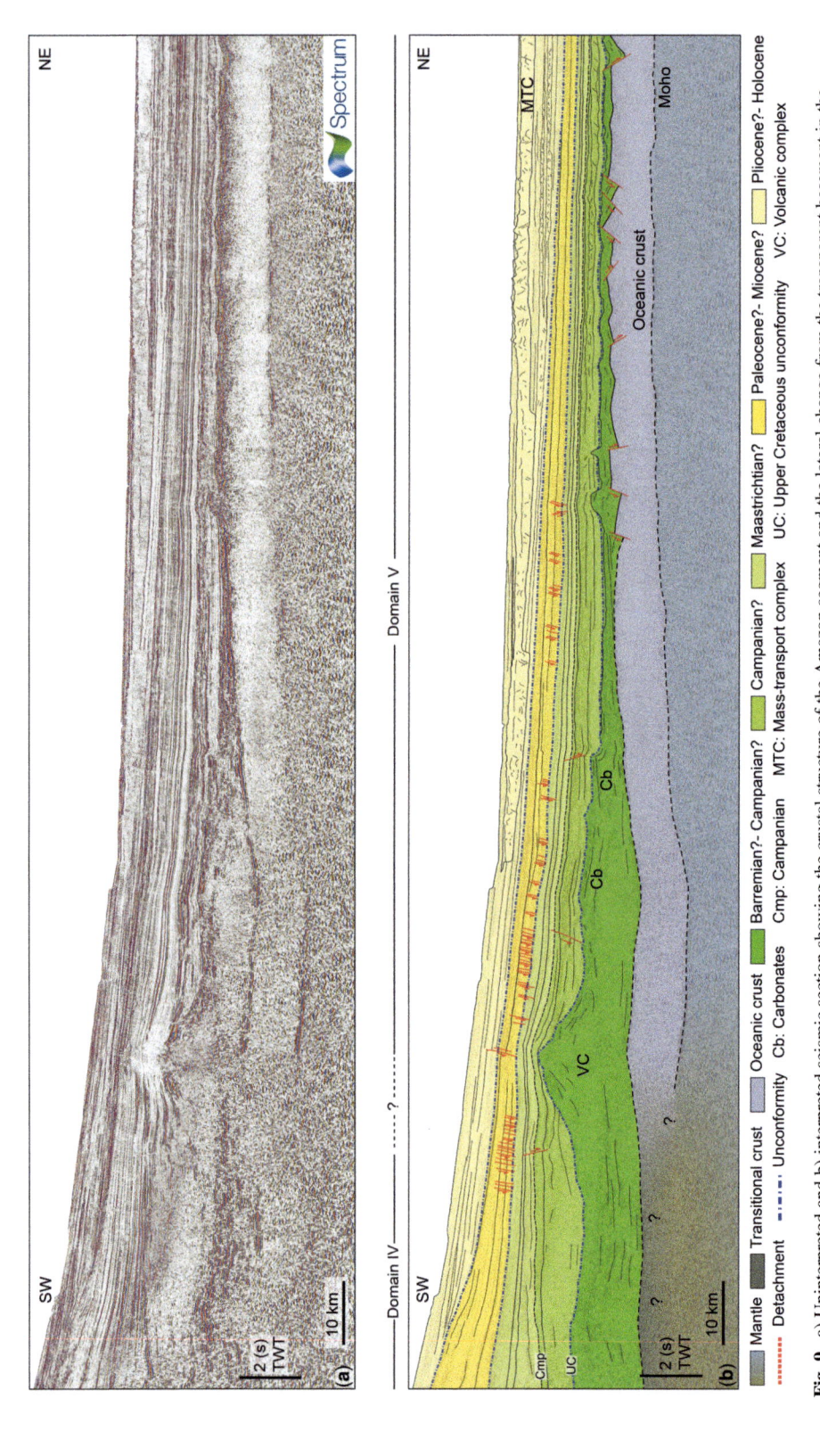

Fig. 9. a) Uninterpreted and b) interpreted seismic section showing the crustal structure of the Amazon segment and the lateral change from the transparent basement in the transitional domain to oceanic crust. The volcanic complex post-dates the sediment on top of the basement in the transitional domain. The oceanic crust bends below the volcanic complexes. Spectrum is acknowledged for publication permission.

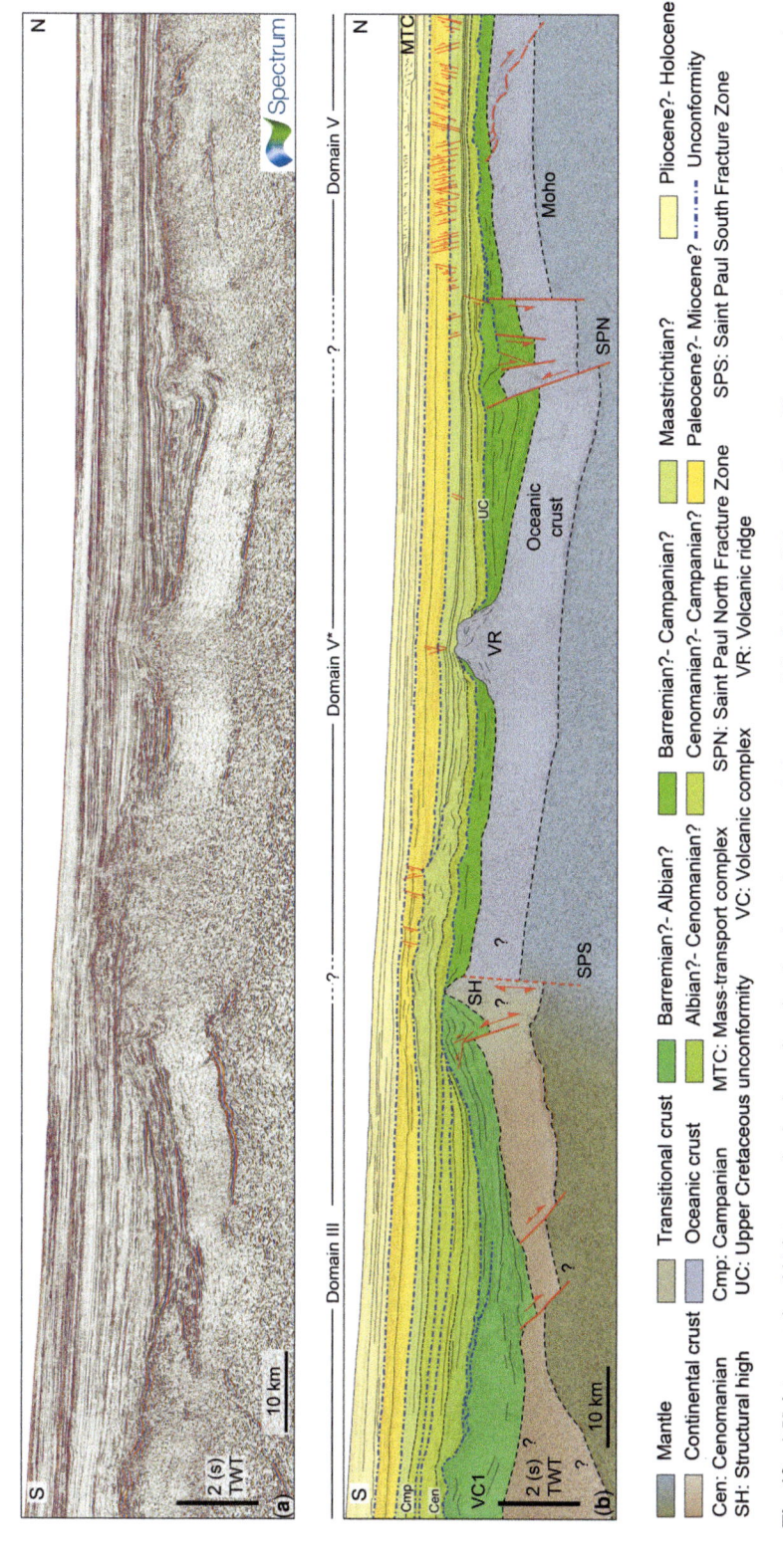

Fig. 10. a) Uninterpreted and b) interpreted seismic section showing the main structural elements of the Saint Paul Fracture Zone. The southern fracture zone shows a structural high where the basement rises towards the fracture zone and is covered with growth strata with multiple unconformities. The reverse fault in the northern fracture zone limits a flexural basin with a volcanic ridge in the monocline hinge. Spectrum is acknowledged for publication permission.

controlled by an array of normal faults. The domain shows a set of oceanward, high-amplitude basement reflectors that increase in depth from 6 to 9 s TWT (c. 7–13 km) over a 20 km distance (Figs 4b & 6). Normal faults in the outboard continental platform and in the continental necked domain appear to detach onto this group of high-amplitude reflectors, and these in turn appear to connect to the Moho reflection (Fig. 4b).

Hyperextended domain III

Outboard and oceanwards from domain II, the basement has a transparent and homogeneous reflectivity continuous with the seismic character of the basement of the necked domain, and is thus interpreted as very thinned continental crust (Figs 5, 7 & 10). Crustal blocks are 10–20 km wide and 0.5–3 s TWT (c. 3–9 km) thick. They show progressive eastward and southeastward thinning with an increased number of faults, and a decreased fault spacing (Fig. 10). The top of the continental basement is marked by a high-amplitude reflection and its base by a well-defined Moho reflection. This is locally amplified by additional high-amplitude reflectors (Fig. 7). Both basinward- and landward-dipping normal faults form tilted fault blocks accompanied by growth stratal architectures. The faults are slightly listric and show apparent dips of 30°–40°. In these sections most faults cut the entire crust and terminate within the mantle (Fig. 10). The larger faults in this domain have displacements of 0.5–1.0 s TWT (c. 1.5–3 km) and displace the Moho reflection, producing a local footwall rise of the mantle (Fig. 7). Small displacement faults also occur within the crustal blocks. Where the crustal thickness is reduced, the top of the basement changes from 7 s TWT (c. 8 km) close to the continental slope to 8.5–9 s TWT (c. 9–10 km) in the southeastern part of the Maranhao segment (Figs 7 & 10).

The northern limit of the hyperextended domain in the Barreirinhas segment is marked by a structural high where the continental basement is thickened and its top rises above its regional level (Fig. 10). The basement is covered by the southward-dipping strata of the basal sequence bounded by multiple unconformities and forms a wedge that broadens to the north. This wedge is in turn covered by a northward-thinning sequence of Cenomanian(?) strata. The contact of the continental basement with the oceanic crust to the north is abrupt.

Transitional domain IV

Figure 8 shows the overall geometry of the transitional domain, and Figure 5 shows a detailed section of the transitional domain in a section near Figure 8. The basement of the transitional domain has transparent to chaotic low-amplitude reflectivity (Figs 8, 9 & 11). The top is defined by a medium- to high-amplitude reflection or by the change from the overlying layered section to low-amplitude reflectivities but there are no reflections that can be interpreted as the base of the basement or as a detachment level.

The normal faults have apparent dips of 30°–50°. These form 10–15 km-wide and <1 km-high tilted faults blocks, together with growth stratal wedges and horst blocks with irregular subhorizontal tops (Fig. 11). The basement morphology is also defined by angular ridges 5 km wide and c. 0.5 km high, and these are interpreted to be tilted fault blocks. They occur in c. 15 km-wide fault-bounded sub-basins and are covered by sediments that infill the topography.

In the Pará-Maranhão central segment, the top of the basement is 9–9.5 s TWT (c. 11–12 km) deep and has an overall subhorizontal geometry forming an east–west basin bounded by the fracture zones (Fig. 8). Figure 8 shows the main reverse fault of the northern fracture zone that limits the basin in the transitional domain of the Pará-Maranhão segment from the oceanic domain of the Amazon segment. The basement morphology has been infilled by subhorizontal strata that buried the synkinematic sediments and basement highs (Fig. 8).

The change from the basement in the transitional domain to oceanic crust is gradual. In the upper and middle sections, the basement shows a progressive eastward shift from transparent reflectivities to discontinuous reflections that increase in amplitude and continuity towards the ocean basin. To the west, the top of the basement is locally displaced by normal faults with growth wedges but to the east shows a subhorizontal geometry. The most oceanward occurrence of faulted blocks coincide with a zone where a high-amplitude reflection at the base of the crust becomes prominent, and the top of the basement is irregular and formed by discontinuous and locally continuous high-amplitude reflections. The basement reflectivity changes until its seismic character appears to be similar to that of the oceanic crust. Within the study area, the overall change in reflectivity occurs over a distance of 40 km where the top of the basement rises to 8–8.5 s TWT (c. 7–8 km) before connecting with the oceanic domain and without the development of basement highs or large volcanic complexes.

In the Amazon segment, the transitional domain has an NW–SE orientation and is 9–9.5 s TWT (c. 11–12 km) deep (Fig. 9). The change from transparent transitional crust to oceanic crust occurs over a narrow, 10–20 km-wide, zone where the top of the basement rises to 8.5 s TWT (c. 10 km) west of the volcanic complex (Fig. 9). The sequence of high-amplitude sub-parallel reflections at the top of

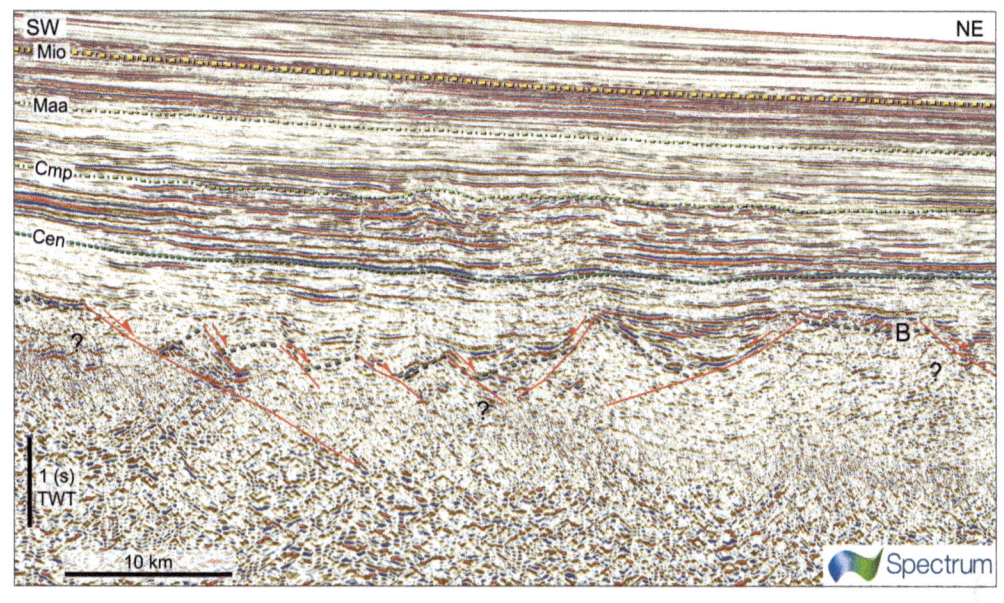

Fig. 11. Seismic section showing the seismic character and structural style of the basement in the transitional domain of the Pará-Maranhão segment. Faulted blocks with growth strata are covered by subhorizontal sediments. No Moho reflection is observed. Spectrum is acknowledged for publication permission.

the transitional domain thins towards the east. The top basement reflector increases in amplitude and continuity in the transition zone, and further to the east a high-amplitude reflector defines the base of the oceanic crust (Fig. 9). Some volcanic complexes occur in the transition zone but they cover and post-date the sediments above the oceanic crust, and appear not to be directly connected to the oceanic crust or the transparent basement. As in the Pará-Maranhão segment, the change in basement architecture in the transition zone is gradual and without the development of a significant basement high.

Oceanic domain V

The oceanic crust is sub-tabular in shape, is c. 1.3–1.7 s TWT (c. 4.2–5.8 km) thick and has an irregular morphology of the upper surface (Figs 6b, 9 & 10). The base of the oceanic basement displays a semi-transparent to low-amplitude chaotic reflectivity above the flat, high-amplitude Moho reflector. The middle of the oceanic crust shows an upward increase in the amplitude, and a number of discontinuous and undulatory reflections. The top of the oceanic crust shows high-amplitude continuous and discontinuous reflectors that can be a thin layer, or shows thickening in the hanging walls of normal faults (Figs 6b & 9).

The oceanic crust thins to c. 1 s TWT (c. 2.5 km) close to the fracture zones and thickens to 2 s TWT (c. 5 km) near some volcanic complexes. The oceanic basement has domino-style tilted fault blocks that are 5–7 km wide with fault offsets up to 0.5 s TWT (c. 1 km) (Figs 6b & 9). Growth strata wedges in the upper layer of the crust show a fanning of dips and thickening towards the faults (Fig. 9). These extensional faults detach within the oceanic crust and do not offset the Moho as in the hyperextended domain described above.

In the Amazon segment, the depth of the oceanic crust is approximately 7.5–8 s TWT (c. 7–8 km) in the eastern part, whereas, to the west, it bends and deepens to 9 s TWT (c. 10 km) over a distance of 50 km, and becomes subhorizontal to slightly tilted towards the west (Fig. 9).

In the central Pará-Maranhão segment, the oceanic crust is 7.5–8 s TWT (c. 7–8 km) deep and forms a 30–35 km-wide north-dipping monocline where the top of the basement deepens to 8.5 s TWT (c. 9 km) close to the northern fracture zone (Fig. 10). The monoclonal hinge zone is parallel to the fracture zone, and coincides with a volcanic ridge that is 15 km wide, 0.5 s TWT (c. 1 km) high and extends for more than 100 km parallel to the monoclinal flexure.

Volcanic complexes

Six main volcanic complexes (V1–V6) with conical or trapezoidal geometries in 2D cross-sections occur in the study area (Fig. 5). The cores of these volcanic build-ups have a transparent seismic facies that vary laterally to chaotic low-amplitude reflectors with isolated discontinuous, medium- and high-amplitude reflectors (Fig. 12). Away from the core of the volcanic complexes, the seismic facies gradually change to layered continuous and discontinuous high-amplitude reflections, interbedded with medium- to low-amplitude continuous reflections. The tops of volcanic edifices are characterized by a continuous high-amplitude reflector onlapped by younger sediments. The bases of the volcanic complexes exhibit downlap patterns over the underlying strata. The flanks of the complexes are steep at the top, with lower dips down the flanks (Fig. 12). Some volcanic complexes have multiple edifices that form 40–60 km-wide and 1.5–3 s TWT (c. 4–8 km)-high amalgamated build-ups from several volcanic vents (Figs 8 & 12).

Complex V1 occurs above the hyperextended domain of the Pará-Maranhão segment (V1 in Figs 5 & 10). This overlies thinned continental crustal blocks and synkinematic sediments in the half-graben. The top surface of this complex is onlapped by the top Albian(?) horizon and appears to be fully overlapped by the top Cenomanian(?) strata. Volcanic complexes V2, V3, V4 and V5 occur above transparent and oceanic basement in the region between fracture zones and form seamounts on the present-day bathymetry (Figs 2, 5 & 12). V2 post-dates the top Cenomanian(?) and is onlapped by the top Campanian(?). V3, V4 and V5 post-date the top Campanian(?), are onlapped by the top Cretaceous(?) horizon, and form an east–west chain that is sub-parallel to the southern branch of the Saint Paul Fracture Zone (Figs 5 & 12b).

Volcanic complex V6 in the Amazon segment forms a WNW–ESE chain of more than 10 edifices 1.5–3 s TWT (c. 4–8 km) high (Figs 8 & 9) that extends for more than 250 km and is 70 km wide (Fig. 5). Some complexes are covered by sub-horizontal and parallel low- to medium-amplitude reflectors or by high-amplitude reflectors with clinoform geometries on the tops of the volcanic constructs (Fig. 13), and are interpreted as hemipelagic sequences and carbonate build-ups (cf. Burgess et al. 2013). They contrast with the inclined discontinuous high-amplitude reflections on the flanks of volcanic complexes (Fig. 12b) that are interpreted as volcanic flows and volcanoclastics. The carbonate build-ups mainly occur on relatively small (2 s TWT high) volcanic edifices close to the continental slope. The V6 complex occurs on top of the basal mega-sequence, and is onlapped by the top Campanian(?) and younger horizons (Fig. 12a).

Additional volcanic complexes occur as isolated edifices in the distal part of the oceanic domain (Fig. 5). The seismic character of the core of the volcanic complexes is similar to that of the base of the oceanic crust, and their associated volcanoclastic sediments downlap and merge with the upper layer of the oceanic crust.

Slides and mass-transport complexes

Slide complexes

A number of shallow, gravity-driven slides systems occur along the NE Brazilian margin including the Pará-Maranhão and Barreirinhas slide systems on the Pará-Maranhão margin (Zalán 2005, 2011; Krueger et al. 2012; Oliveira et al. 2012). These are linked updip extensional and downdip contractional fault systems located on the upper and lower continental slopes in water depths ranging from 50 to 3330 m (Figs 7, 8 & 14). They occur for 500 km along strike and are typically 50–80 km wide (Fig. 14), detached on Late Cretaceous 'overpressured' shales, and deform 3–4 km of Late Cretaceous–Recent slope to deep-marine strata. The Pará-Maranhão slide complex includes updip listric growth faults and downdip thrust systems that form complex 3D imbricate fans and duplex structures that impinge onto a downdip volcanic complex (Zalán 2011; Tamara & McClay 2015). The Barreirinhas system consists of updip listric faults that link downdip to imbricate systems and shear fault-bend folds. The system has been interpreted to include layer-parallel and concave detachments (Zalán 2011; Krueger et al. 2012).

Mass-transport complexes

Mass-transport complexes have been widely recognized along the NE Brasilian passive margin and are abundant in the Campanian(?)–Recent strata of the Pará-Maranhão margin (Figs 3 & 6). These form c. 50–200 m-thick units of low-amplitude chaotic reflectivity with high-amplitude discontinuous reflections with variable dips. They commonly have erosional bases and irregular to mounded top surfaces (Figs 6, 8 & 9). The mass-transport complexes can thin and pinch out laterally or terminate at erosional boundaries. Giant mass-transport systems (mega-slides) can be more than 200–800 m thick and extend over thousands of square kilometres (cf. fig. 12 of Reis et al. 2016). They have basal erosional surfaces and chaotic internal reflectivity, as well as internal erosional surfaces that indicate amalgamated systems. These complexes terminate downslope in thrust-related folds. Silva et al. (2010) and Reis et al. (2016) documented at least four regional mega-slide/mass-transport systems on the NE Brazilian margin.

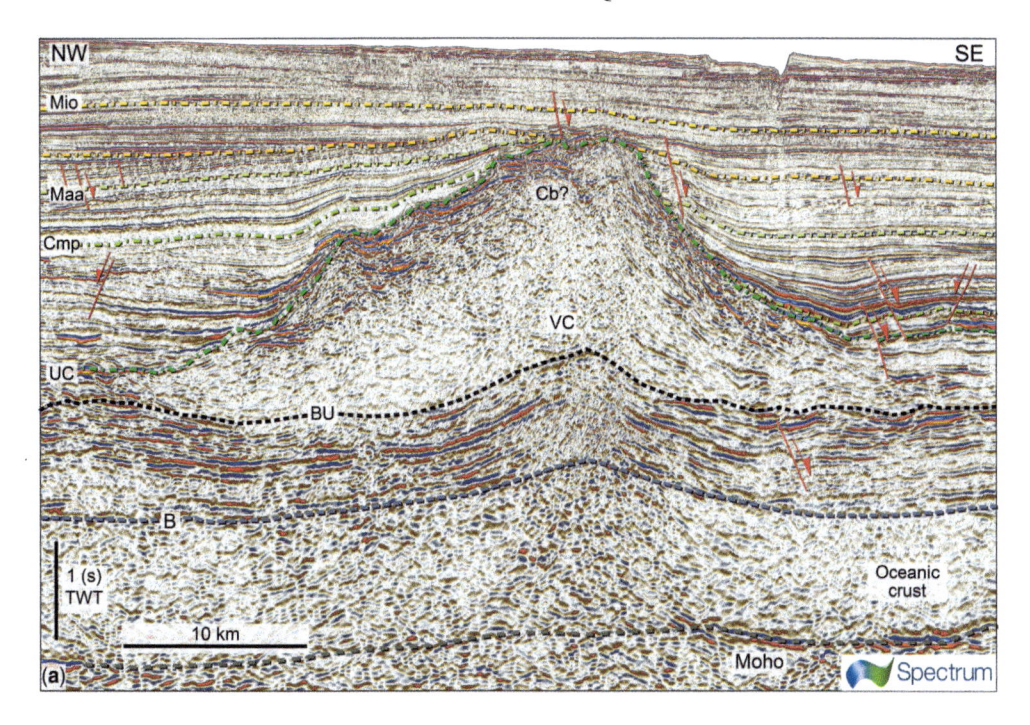

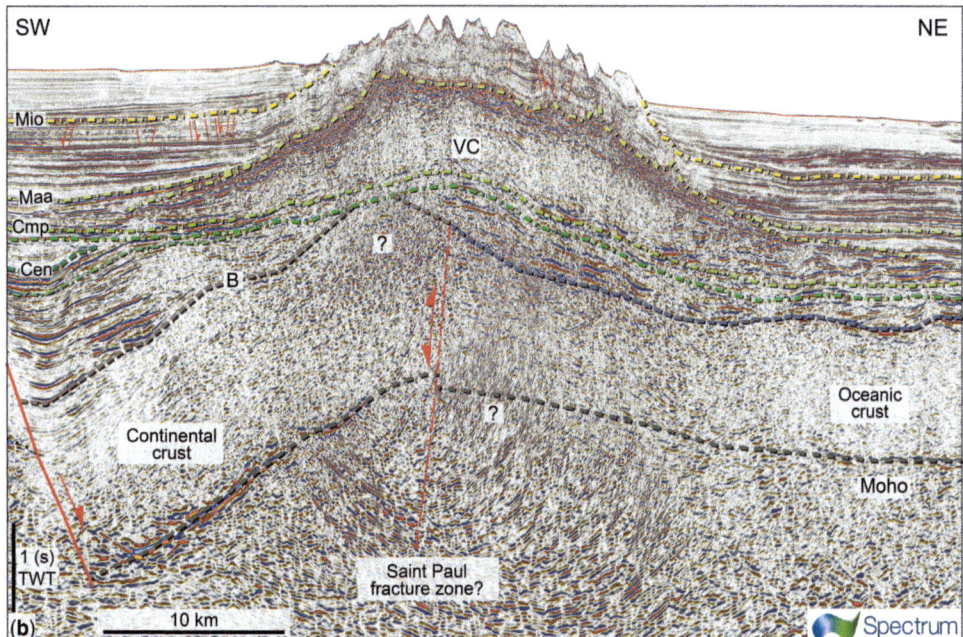

B: Top basement   BU: Top basal unit   Cb: Carbonates   Cen: Cenomanian   Cmp: Campanian
Maa: Maastrichtian   Mio: Miocene   UC: Upper Cretaceous unconformity   VC: Volcanic complex

Fig. 12. Detailed seismic sections showing the seismic character of volcanic complexes. (**a**) Volcanic complex on top of the basal unit and onlapped by the top Campanian(?) horizon in the Amazon segment. (**b**) Volcanic complex on top of the southern branch of the Saint Paul Fracture Zone. Volcanoclastic sequences on the flanks downlap over underlying Campanian? strata. The top is onlapped by Maastrichtian? and younger sequences. Spectrum is acknowledged for publication permission.

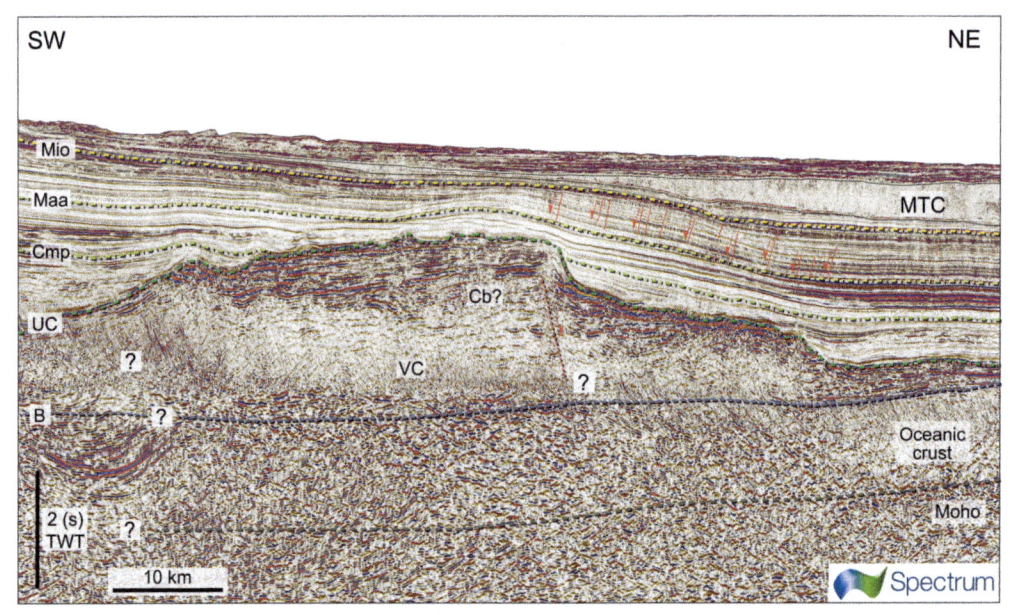

B: Top basement   BU: Top basal unit   Cb: Carbonates   Cmp: Campanian   Maa: Maastrichtian
Mio: Miocene   MTC: Mass-transport complex   UC: Upper Cretaceous unconformity   VC: Volcanic complex

Fig. 13. Detailed section showing carbonate build-up on top of volcanic complex (V6) in the Amazon segment. The build-up is covered by Campanian(?) strata. Pliocene(?)–Recent mass-transport deposits filled the pre-existing bathymetry to the NE of the section. Spectrum is acknowledged for publication permission.

Discussion

Crustal architectures

The central equatorial Brazilian margin is divided into the Amazon and Barreirinhas divergent segments separated by the Pará-Maranhão transform segment. Within these segments, five crustal domains (I–V) have been recognized. From proximal to the continent to the offshore oceanic crust these are:

(I) Proximal domain: continental platform and inboard rift basins; crustal thickness of 30–35 km.

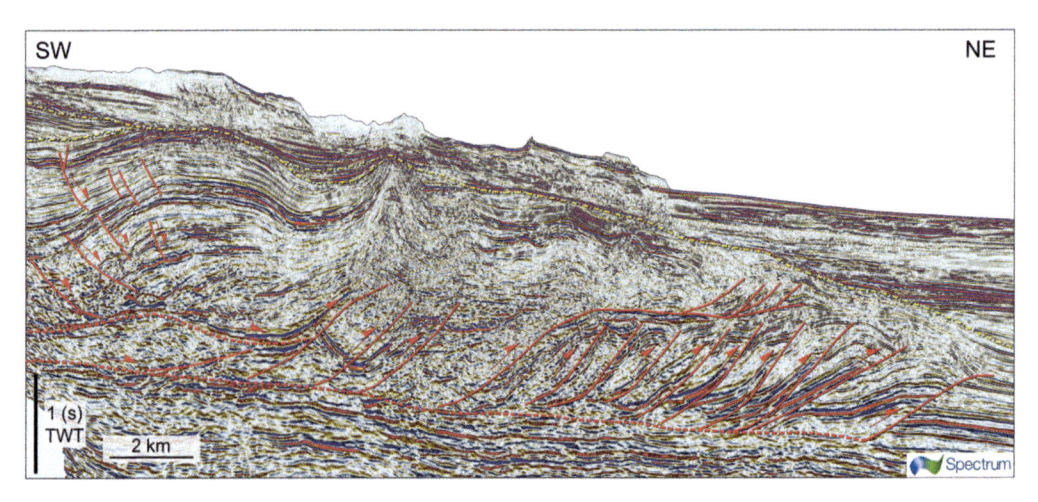

Fig. 14. Detailed section of the Maranhao slide complex showing a fold and thrust belt with fault-related folds in the contractional domain and listric normal faults with rollover anticlines in the extensional domain. Spectrum is acknowledged for publication permission.

(II) Necking domain: thinning of continental crust, faulted blocks; crustal thickness of <10–30 km.
(III) Hyperextended domain: highly thinned continental crust, highly rotated domino faults; crustal thickness of 5–10 km.
(IV) Exhumation domain: exhumed continental mantle, faulted blocks below the sag basin; 'transitional basement' bordering oceanic crust.
(V) Oceanic crust: faulted upper crust, overlain by deep ocean sediments; thickness of 5–6 km.

Proximal to hyperextended domains – continental crust domains. Seismic sections across the external platform and continental slope show that the continental basement thins from 30 to 35 km in the proximal domains to *c.* 5–10 km in the hyperextended domain (Figs 4 & 7). Tapering of the crust occurs in a narrow, *c.* 20–40 km-wide, zone where the top of the basements and the top of the Moho form a wedge and the accommodation space increases (Sutra & Manatschal 2012; Peron-Pinvidic *et al.* 2013). Normal faults in the necked domain sole into the zone of oceanward-dipping basement reflectors, suggesting a shear zone, whereas landward-dipping basement reflectors occur at mid-crustal levels (Fig. 4b & 15). Similar arrays of landward and oceanward basement reflectors have been described in the Campos Basin as detachment faults, and as reflections at the boundary between the upper and lower crust (Fig. 16) (Unternehr *et al.* 2010).

In the hyperextended domain, tilted fault blocks of thinned continental, growth strata with a fanning of dips, together with a shallowing of the Moho reflections in the footwalls (Figs 7 & 10), indicate the rotation of crustal blocks during extension. It is likely that most faults were connected along a detachment within the upper mantle but no lower-crust reflections, such as those described in the Iberian or Norwegian margins (Reston *et al.* 2007; Osmundsen & Ebbing 2008), have been identified in the study area. To the east of the intersection of the shear zone with the Moho reflector, most faults in the hyperextended domain cut the entire crust (Figs 4b, 7 & 10), suggesting that crust deformed as a single rheological unit (Pérez-Gussinyé *et al.* 2003). This is consistent with the crustal models that suggest a change in the structural style from decoupled in the necked domain to coupled in the hyperextended domains (Pérez-Gussinyé *et al.* 2003; Sutra *et al.* 2013; Peron-Pinvidic *et al.* 2013). The effects of transtensional deformation on the structure of the necked and hyperextended domains remain to be studied.

Transitional domain – ocean–continent boundary. In the transitional domain, the transparent reflectivity of the basement and the lack of a reflection to mark the base of the crust contrasts strongly with the seismic features seen in other crustal domains (Figs 8, 9 & 11). The rotated fault blocks form a regionally subhorizontal top of seismic basement that indicates a detachment within the basement,

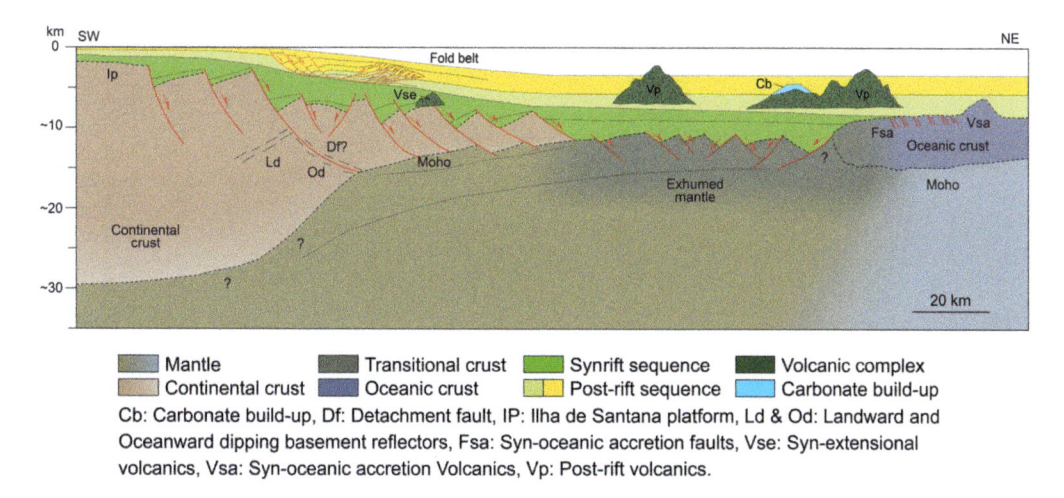

Fig. 15. Schematic model showing the main structural elements of the northeastern Brazilian margin. The crustal domains show the transition from a proximal to a hyperextended domain, with a necked domain with oceanward-dipping reflections interpreted as a detachment fault. The ocean–continental boundary is characterized by exhumed mantle covered by subhorizontal sediments, and showing a transition to oceanic crust. An Albian(?) magmatic complex occurs over thinned continental crust, whereas Turonian(?)–Campanian(?) and Maastrichtian(?) complexes favoured exhumed mantle and oceanic domains. Slide complexes are located on the continental slope. The crustal domains follow the classifications of Peron-Pinvidic *et al.* (2013) and Osmundsen & Péron-Pinvidic (2018).

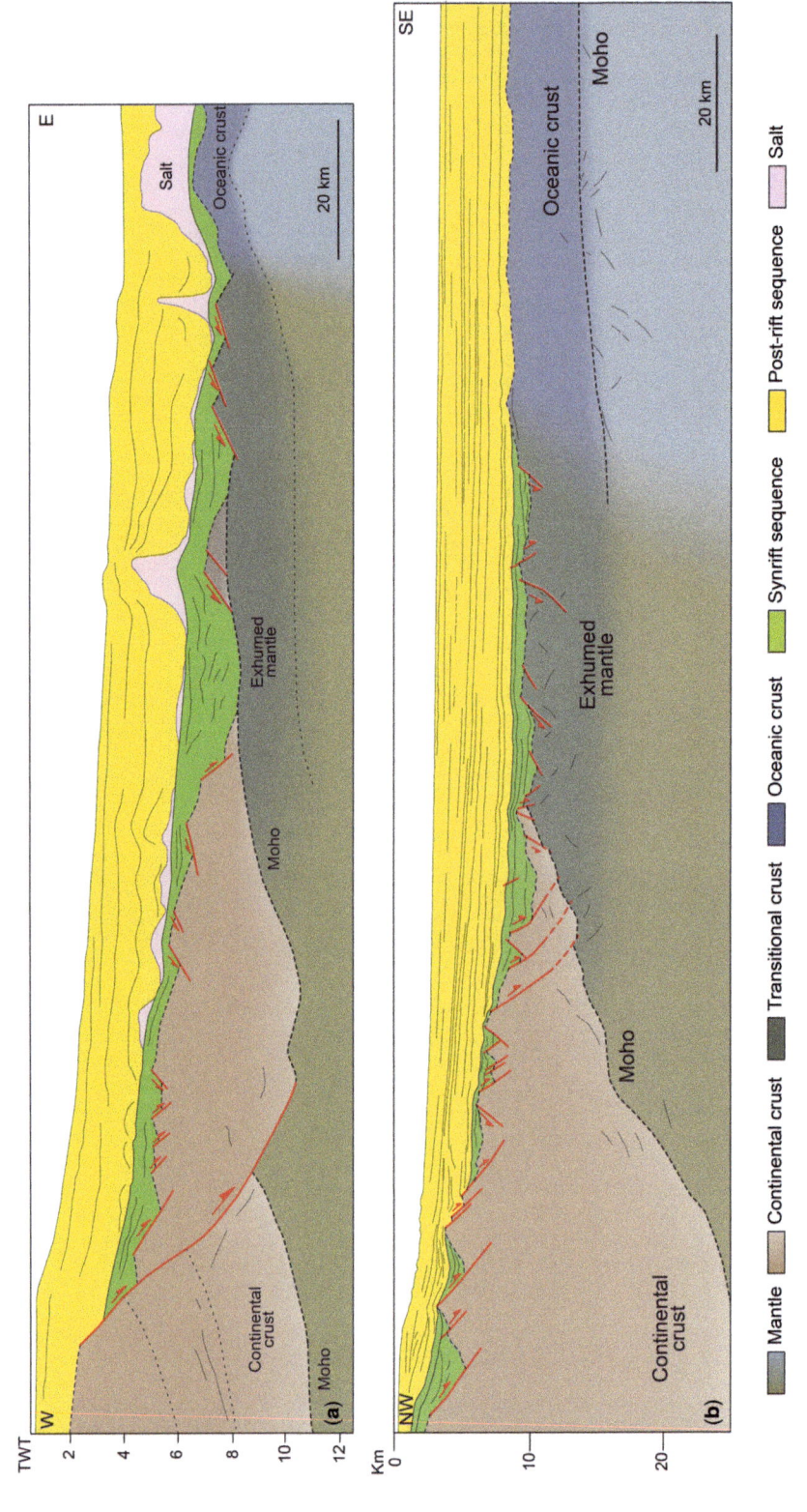

Fig. 16. Comparative examples of passive margins with similar crustal architectures to the central sector of the equatorial margin of Brazil. (**a**) Line drawing of a seismic section from the Campos Basin (Unternehr *et al.* 2010). (**b**) Line drawing of the East Indian passive margin (Haupert *et al.* 2016).

with the subhorizontal overlying sediments resembling a sag basin (Figs 8 & 11). These characteristics, together with the observation that the top of the basement in the transitional domain is deeper than that in the oceanic and the hyperextended domains, indicates that this may be exhumed continental mantle (cf. Peron-Pinvidic et al. 2013; Zalán 2015).

In the distal part of the transitional domain, the lateral change in seismic character from the transparent basement to the oceanic crust is marked by a gradual increase in amplitude and continuity of reflections which is interpreted to indicate a proto-oceanic crust. This is consistent with the lateral increase in continuity of the reflections at the base of the crust that further to the east define the Moho at the base of the oceanic crust (Fig. 9). The top of the seismic basement rises gradually oceanwards without any evidence of major fault steps. The lack of large magmatic complexes coeval with the formation of the oceanic crust or seaward-dipping reflectors may indicate a magma-poor mantle (cf. Peron-Pinvidic et al. 2013; Zalán 2015). The presence of normal faults at the edges of the exhumation domain, the increase in magmatic material at the base of the crust and the gradual change in basement reflectivity suggest that the transition from exhumed mantle to oceanic crust may be similar to the hybrid transitional crust described Gillard et al. (2017) in the conjugate Ivory Coast margin.

The structural style, morphology and progressive change to oceanic crust is similar to transitional domains in the ocean–continental boundary of the East Indian margin (Radhakrishna et al. 2012; Pindell et al. 2014; Haupert et al. 2016), the Angolan margin (Unternehr et al. 2010) and the Labrador margin (Keen et al. 2018).

Oceanic domain. In this research the ocean domain is characterized by a rather thin, 4–6 km-thick, subtabular oceanic crustal slab with a layered seismic character, and a well-defined top basement and Moho reflections (Figs 9 & 10). This is similar to the 4–5 km-thick crust proposed in previous studies of the Amazon segment further to the north (Houtz et al. 1977; Rodger et al. 2006; Watts et al. 2009). The layered seismic character of the crust may correspond to the velocity structure found by Watts et al. (2009) and is interpreted as a thinned version of the typical 6–8 km-thick normal oceanic crust (White et al. 1992; Mutter & Mutter 1993). The irregular and rugose morphology of the top of the crust, the upper volcaniclastic layer and the domino-style rotated blocks found in this domain are similar to those described by Watts et al. (2009) and found in thin oceanic crust in the Newfoundland and Labrador margins (Funck 2003; Hopper et al. 2004, 2007; Delescluse et al. 2015; Keen et al. 2018). Thinned oceanic crust is interpreted as resulting from accretion at a ultraslow–slow spreading centre such as the Gakkel Ridge (Coakley & Cochran 1998; Schmidt-Aursch & Jokat 2016; Nikishin et al. 2018), the Southeast Indian Ridge (Dick et al. 2003) and the Mohns Ridge (Klingelhöfer et al. 2000).

The oceanic basement in the Amazon segment shows bending as it deepens to the west below the volcanic complex V6 (Figs 5, 8 & 9). Due to the area covered by the volcanic complex (250 km long and 70 km wide) and the height of the volcanic edifices (1.5–3 s TWT (c. 4–8 km) high), it is interpreted that flexure of the oceanic crust resulted from loading after emplacement of the volcanic complex. In contrast, the north-dipping monocline to the south of the northern fracture zone (Fig. 10) and the sediment on top of it suggest that the monocline resulted from flexure of the crust due to reverse movement on the northern fracture zone during the Albian(?). Detailed analysis of the structures associated with the fracture zones will be discussed in a separate publication.

The Pará-Maranhão transform margin segment

The central Pará-Maranhão transform segment is bounded by two branches of the Saint Paul Fracture Zone (Fig. 5). The northern branch can be followed for more than 400 km in the study area. To the west, the fracture zone limits a zone of exhumed continental mantle in the Pará-Maranhão segment from oceanic crust in the Amazon segment (Fig. 8). To the east, the main structure of the fracture zone is a reverse fault that limits the north-dipping oceanic crust of the Pará-Maranhão segment on the south from the oceanic crust of the Amazon segment (Fig. 10). The landward limit of oceanic crust on both sides of the fracture zone has more than 300 km of horizontal separation (Fig. 5).

The southern fracture zone separates exhumed mantle and oceanic crust in the Pará-Maranhão segment from thinned continental crust in the Barreirinhas segment (Fig. 5). The structural high of thickened continental crust is parallel to the margin (Figs 5 & 10). This increase in crustal thickness is also found in the gravity-inversion models (cf. A in profile 10 in fig. 7 of Kusznir et al. 2018), where it may probably be enhanced by the magmatic addition of the pre-Albian(?) volcanic complexes (V1). Growth strata within the Barremian(?) to Albian(?) sequence that thicken towards the fracture zone are uplifted and deformed, indicating the possible inversion of an older extensional basin (Fig. 10). Erosion of the pre-Cenomanian(?) sequences at the top of this structural high and tilting of regional uniformities suggest significant uplift and deformation.

The fracture zones highlight the structural transition from a transform margin to an oceanic fracture zone. In the southern fracture zone, the contact between thinned continental crust and oceanic crust is sub-perpendicular to the spreading centre for more than 100 km, indicating that this part of the fracture zone forms a distal transform margin (Figs 5, 7 & 10). Structural highs or basement ridges with similar characteristics are also found along other transform margins: for example, along the Ghana and Falkland transform margins (Lorenzo & Wessel 1997; Antobreh et al. 2009; Basile et al. 2013; Mercier de Lépinay et al. 2016). In a similar way, the northern fracture zone shows an interaction of distal crustal domains of exhumed mantle with oceanic crust that forms the transition from the more distal part of a transform margin to an oceanic fracture zone (Fig. 5 & 8).

The Pará-Maranhão transform segment lacks the bathymetric expression found in other equatorial transform margins, despite having a 300 km horizontal separation of crustal domains (Fig. 5) that is of similar magnitude as the 500 km offset of the mid-ocean ridge and the 600 km-long western Ivory Coast conjugate margin (Antobreh et al. 2009; Davison et al. 2015) (Figs 1 & 2). In northeastern Brazil, it is suggested that the full geometry of the transform margin is masked by high sedimentation close to the intersection of the Amazon and Pará-Maranhão segments. This hypothesis is consistent with the sediment thickness maps of Watts et al. (2009) which show a thick Early Cretaceous–Miocene section close to this intersection. As the Amazon River was formed during the Miocene (Figueiredo et al. 2009; Gorini et al. 2014), it is suggested, as an initial hypothesis, that a possible source for the thick sedimentary section may be related to drainage systems following Early Cretaceous graben like the Marajo, Sao Luis and Gurupi in the proximal domains of the Brazilian margin (Costa et al. 2002; Soares Júnior et al. 2008), connected to a palaeo-Tocantins river system between the Amazon Craton and the Parnaiba Basin, together with high rates of erosion of the continental margin (Figs 1 & 2).

Timing and distributions of magmatism

Three magmatic events were identified in the margin: a Late Albian(?) magmatic event in the Barreirinhas segment (Figs 5 & 10), a Turonian(?)–Campanian(?) event that formed the volcanic chain in the Amazon segment and partially extended to the Pará-Maranhão segment (volcanic complexes V6 and V2: Figs 5, 8, 9 & 12a), and a Maastrichtian(?) event with localized magmatism along the southern fracture zone (V3–V5: Figs 5 & 12b).

The pre-Albian event occurred within thinned continental crust and overlies the synkinematic strata in the half-graben, indicating formation after the initial phase of hyperextension or during its late stages (Fig. 10). Similar volcanic complexes on top of hyperextended crust have been described in the conjugate Ivory Coast margin (Gillard et al. 2017; Scarselli et al. 2018); however, the origin and importance of this magmatic event are yet to be defined.

The complexes of the Turonian(?)–Campanian (?) event form an east–west chain, and cut through and cover the basal mega-sequence on top of the oceanic basement, indicating a post-rift emplacement (Figs 5, 8, 9 & 12a). In the oceanic crust (domain V), the magmatic complexes are postulated to have been controlled by pre-existing fracture zones that are parallel to the Saint Paul Fracture Zone. However, in the transitional crust (exhumation domain IV), the Turonian(?)–Campanian(?) volcanics seem to be aligned along the ocean–continental boundary (Fig. 5). In a similar way, the Maastrichtian(?) complexes occur on transitional crust close to the southern fracture zone. This distribution indicates that the emplacement of volcanic complexes may have been favoured by the southern fracture zone and weaker serpentinized mantle in the transitional domain (Fig. 5).

The source for the volcanic complexes may be related to excess magmatism at the mid-ocean ridge, as interpreted for the Late Cretaceous Ceará Rise further to the NE (Houtz et al. 1977; Kumar & Embley 1977) (Fig. 2), where the fracture zones most likely acted as conduits for magma remobilization towards the continental margin (Watts et al. 2009). Alternatively, the age ranges and distributions of volcanic complexes close to the margin may have resulted from localized magmatism produced by small convection cells formed at abrupt changes in crustal thicknesses (Vogt 1991; King & Anderson 1998; King 2007), such as in narrow necked domains, as well as along the transform margin. This mechanism has been proposed to explain magmatism in the Borborema province further to the east (Knesel et al. 2011), as well as a similar distribution of volcanics close to the edge of continental lithosphere in the western Antarctic margin (Kipf et al. 2014).

Regional uplift and margin collapse

The distribution and timing of deformation of the Pará-Maranhão and Barreirinhas slide complexes (Zalán 2005, 2011; Krueger et al. 2012; Oliveira et al. 2012) indicate significant long-lived and regionally-controlled margin-collapse processes, which contrast with localized gravitational collapse associated with the formation of delta systems like the deep Amazon fan (Cobbold et al. 2004). In a similar way, the abundance of Late Cretaceous–Recent mass-transport deposits indicates that instability of the continental margin with slope failure and sediment remobilization to the lower slope and basin occurred at multiple times during the post-rift evolution of the margin.

Clinoforms truncated by subhorizontal unconformities and subhorizontal strata on the top of some volcanic complexes indicate that the crests of the volcanoes were in relatively shallow water (Figs 12 & 13). These volcanic complexes had at least *c.* 1 km of relief above the basin seafloor and are now covered by *c.* 1–2 km of Maastrichtian(?)–Recent strata, and occur at water depths of a 3 km; thus indicating at least 4 km of subsidence since the Campanian(?).

Inboard, Late Cretaceous–Paleogene shallow-marine and coastal-plain sequences are bounded by Middle–Late Campanian, Maastrichtian and Middle–Late Eocene unconformities that mark episodes of uplift, exposure and erosion on the continental shelf, together with the development of incised valleys (Rossetti 2001, 2004; Rossetti & Santos 2004; Trosdtorf Junior *et al.* 2007; Pakulski 2011). Periods of regional palaeosol and laterite development during the Eocene and Plio-Pleistocene alternated with periods of shallow-marine sedimentation in the Oligo-Miocene (Rossetti *et al.* 2013; da Costa *et al.* 2014, 2016). These deposits are currently found at *c.* 25–65 m above sea level, showing post-Miocene uplift (Rossetti *et al.* 2013; da Costa *et al.* 2014, 2016).

Exposures of the Proterozoic basement with remnants of the synrift and post-rift sequences also indicate periods of uplift, exhumation, erosion and subsidence. The low-relief geomorphology of this sector of the Brazilian margin indicates significant erosion, and contrasts with the relatively high topographical relief of the Borborema province to the east and with the onshore areas of southeastern Brazilian margin, particularly the Campos and Santos basins where coastal mountains exceed 2000 m (de Almeida *et al.* 2000; Peulvast & De Claudino Sales 2004; Peulvast *et al.* 2008; Morais Neto *et al.* 2009; Japsen *et al.* 2012*a*; Jelinek *et al.* 2014). These features indicate that proximal areas of the equatorial margin differ from other sectors of the Brazilian margin, and experienced multiple episodes of subsidence, deposition, uplift and erosion during its post-rift evolution, instead of continuous denudation of the pre-break-up morphology.

The structural and stratigraphic elements described above indicate that margin uplift, exhumation and basinward tilting, together with the underlying crustal architectures, have played a key role in the post-rift evolution of the margin. These produced steep continental slopes (5°–8°) that, together with high sedimentation on the platform, generated gravitational collapse with the formation of slides and mass-transport complexes. The causes of the margin uplift in the equatorial margin are still to be defined, including the reasons for its present-day topography. In elevated passive margins, multiple mechanisms such as crustal structure, flexural uplift, mantle dynamics and crustal-scale folding have been invoked to explain passive-margin uplift (Osmundsen & Redfield 2011; Japsen *et al.* 2012*b*; Green *et al.* 2018). Osmundsen & Redfield (2011) proposed that elevated areas of the Norwegian passive margin correspond with zones of abrupt changes in the crustal thickness and that these areas are susceptible to vertical movement (flexural uplift). A similar model may possibly be used to explain the uplift history of the northeastern Brazilian margin, where a flexural uplift associated with a narrow necked domain was enhanced by loading by volcanic complexes emplaced during the Late Cretaceous.

Evolution of the Brazilian margin

In the Barreirinhas segment, the top Albian(?) horizon has covered the tilted fault blocks in the hyperextended and exhumed domains, indicating that crustal thinning and oceanization had already occurred, and that accretion of oceanic crust had probably started by the Late Albian(?). This is consistent with the interpretation of Late Albian deep-marine sedimentation in the Barreirinhas Basin (Pakulski 2011) related to an increase in accommodation space by crustal thinning (Sutra *et al.* 2013). The quality of the seismic data does not permit a delineation of pre-Albian sequences; however, the synkinematic fluvial and shallow-marine sediments in onshore basins (Darros De Matos 2000; Costa *et al.* 2002; Soares *et al.* 2007) indicate that the interval from lithospheric stretching to oceanisation occurred in the Barremian and continued to the Late Albian when oceanic crust formed.

On the western Ivory Coast margin, Scarselli *et al.* (2018) also suggested that the top Albian units post-date the formation of tilted fault blocks in the hyperextended crustal region, and that seafloor spreading and fully marine conditions were established by the Cenomanian. In the Pará-Maranhão and Amazon segments of the Brazilian margin, the top of the Cenomanian(?) strata covers the oceanic crust, which indicates that migration of oceanic accretion axes along the transform margin occurred during the Late Albian(?) or Early Cenomanian(?).

Turonian(?)–Campanian(?) and Maastrichtian(?) magmatism along leaky fracture zones and weak exhumed serpentinized mantle resulted in volcanic edifices more than 2 km high that became barriers for sediment distribution from the continental slope and potentially changed deep-water circulation patterns, generating contourites systems. After the emplacement of the volcanic complexes, the Late Cretaceous–present-day evolution of the margin shows a progressive deepening and tilting towards the basin. This, together with high sedimentation rates on the continental slope, induced multiple slide complexes and recurrent development of mass-transport complexes. Margin instability and sediment remobilization appears to have been produced by episodic periods of uplift, exposure, erosion and subsidence in the continental platform (Rossetti 2001, 2004; Rossetti & Santos 2004; Trosdtorf Junior et al. 2007; Pakulski 2011; Rossetti et al. 2013).

Comparative examples of crustal structure of the margin

This research shows that the northeastern equatorial margin of Brazil has many similarities to other magma-poor passive continental margins. Figure 15 shows a conceptual composite model, based on this study, of the crustal structure of the margin following the models of Peron-Pinvidic et al. (2013) and Osmundsen & Péron-Pinvidic (2018). The five distinct domains show the characteristics of a rift to necking to hyperextensional transition and of exhumed continental mantle similar to those proposed for many margins (e.g. Peron-Pinvidic et al. 2013; Péron-Pinvidic et al. 2015). It must be emphasized, however, that considerable along-strike variations in margin architectures may be expected, as illustrated in the domain map in Figure 5.

Similar rift to hyperextended crustal architectures have been proposed for the Campos Basin (Unternehr et al. 2010) and the East Indian margin (Fig. 16) (Radhakrishna et al. 2012; Pindell et al. 2014; Haupert et al. 2016). In the Campos Basin, the narrow necked domain, and the architecture of oceanward- and landward-dipping reflectors interpreted as mid-crustal reflections and detachment faults, respectively (Unternehr et al. 2010), are used as analogues for the proximal and necked domains of the Barreirinhas divergent segment (Fig. 4).

The hyperextended and exhumation domains are similar to those extensively studied in the Iberian margin (Boillot et al. 1980, 1987; Peron-Pinvidic et al. 2013; Sutra et al. 2013; Lymer et al. 2019). In the East Indian margin, tilted fault blocks of thinned continental crust show a progressive basinward thinning (Fig. 16b), whereas the exhumed domain is faulted, covered by subhorizontal strata, and its top is deeper than in the hyperextended and oceanic domains (Radhakrishna et al. 2012; Pindell et al. 2014; Haupert et al. 2016). The transition to the oceanic crust occurs without major contemporaneous development of basement highs or large magmatic complexes, and the seismic character of the basement progressively changes to that of normal oceanic crust (Radhakrishna et al. 2012; Pindell et al. 2014; Haupert et al. 2016). It may be inferred that similar processes of differential stretching of continental crust leading to oceanization that explain the crustal structure of other magma-poor continental margin may have also occurred on the equatorial margin of northeastern brazil.

Conclusions

- This study documents the along-strike variation and segmentation of the central sector of the northeastern Brazilian equatorial margin. It is a magma-poor passive margin formed by Early Cretaceous oblique rifting.
- Five tectonic domains have been recognized, from west to east, across and along this part of the NE Brazilian passive margin:
  o Domain I: an inboard continental platform and rift margin where the continental crust is c. 30–35 km thick;
  o Domain II: a c. 20–40 km-wide necked domain formed by extensionally faulted continental crust with a major detachment fault leading to;
  o Domain III: a hyperextended domain with faulted, rotated and thinned continental crust c. 5–10 km thick;
  o Domain IV: an outer zone of faulted, seismically transparent 'basement' with no Moho reflection, interpreted to be exhumed and serpentinized continental mantle with a transition to;
  o Domain V: tabular oceanic crust 5–6 km thick with a clear Moho reflection and domino-style extensional fault blocks in the oceanic crust.
- The three distinct segments of this part of the passive margin are delineated by two branches of the Saint Paul Fracture Zone that obliquely intersect the continental margin. These bound the Pará-Maranhão segment and show the transition from a transform margin to an oceanic fracture zone.
- At least six volcanic complexes associated with three main episodes of post-rift magmatism have been identified in this sector of the Brazilian margin. An Albian(?) volcanic complex occurs above the hyperextended continental crust of Domain III, indicating localized synextensional magmatism. Two other Turonian–Campanian(?) and Maastrichtian(?) volcanic complexes form a 200 km-long east–west volcanic chain. These volcanic complexes occur above presumed exhumed mantle and oceanic crust, and were controlled by fracture zones parallel to the transform sector of the passive margin.
- The transition from the platform to the slope and deep water is characterized by very steep surface slopes ranging from 5° to 8°. As the margin

evolved with significant uplift and exhumation, the resultant steep slopes on the continental margin gave rise to multiple surface and near-surface failures that produced numerous mass-transport complexes at different stratigraphic levels, as well as large slide complexes with downslope fold and thrust belts. Multiple unconformities in the coastal plain and on the continental platform attest to significant uplift, exhumation and erosion during the post-rift evolution of the margin. It is suggested that the narrow necked domain and the emplacement of the volcanic complex favoured continental margin uplift and instability.
- The northeastern equatorial margin of Brazil shows many similarities to the crustal architectures of the passive margins of the Campos Basin (Brazil), as well as with the East Indian margin. However, given the wide spacing of the 2D seismic lines, further research using regional 3D seismic data is needed to delineate and understand the 3D architecture of these passive margin to oceanic basin systems.

Acknowledgements Spectrum is thanked for providing the seismic data used in this research and for kindly allowing us to publish the sections included in this paper. The Fault Dynamics Research Group is acknowledged for funding Javier Tamara's PhD research and for providing the workstations and research facilities. Halliburton is thanked for providing the Decision Space Desktop software for seismic interpretation. We thank Webster Mohriak and an anonymous reviewer for their careful review, as well for their useful and constructive comments and advice.

Funding The Fault Dynamics Research Group, Royal Holloway University of London, is acknowledged for providing financial support to Javier Tamara.

Author contributions JT: conceptualization (lead), formal analysis (lead), investigation (lead), methodology (lead), visualization (lead), writing – original draft (lead), writing – review & editing (lead); **KRM**: conceptualization (equal), formal analysis (equal), funding acquisition (lead), investigation (equal), methodology (supporting), resources (lead), supervision (equal), writing – original draft (supporting), writing – review & editing (equal); **NH**: conceptualization (supporting), investigation (supporting), methodology (supporting), resources (equal), visualization (supporting), writing – review & editing (supporting).

References

ANTOBREH, A.A., FALEIDE, J.I., TSIKALAS, F. & PLANKE, S. 2009. Rift-shear architecture and tectonic development of the Ghana margin deduced from multichannel seismic reflection and potential field data. *Marine and Petroleum Geology*, **26**, 345–368, https://doi.org/10.1016/j.marpetgeo.2008.04.005

ATTOH, K., BROWN, L., GUO, J. & HEANLEIN, J. 2004. Seismic stratigraphic record of transpression and uplift on the Romanche transform margin, offshore Ghana. *Tectonophysics*, **378**, 1–16, https://doi.org/10.1016/j.tecto.2003.09.026

BASILE, C., MASCLE, J., POPOFF, M., BOUILLIN, J.P. & MASCLE, G. 1993. The Ivory Coast–Ghana transform margin: A marginal ridge structure deduced from seismic data. *Tectonophysics*, **222**, 1–19, https://doi.org/10.1016/0040-1951(93)90186-N

BASILE, C., MASCLE, J. & GUIRAUD, R. 2005. Phanerozoic geological evolution of the Equatorial Atlantic domain. *Journal of African Earth Sciences*, **43**, 275–282, https://doi.org/10.1016/j.jafrearsci.2005.07.011

BASILE, C., MAILLARD, A. ET AL. 2013. Structure and evolution of the Demerara plateau, offshore French Guiana: Rifting, tectonic, inversion and post-rift tilting at transform-divergent margins intersection. *Tectonophysics*, **591**, 16–29, https://doi.org/10.1016/j.tecto.2012.01.010

BIZZI, L.A., SCHOBBENHAUS, C., VIDOTTI, R.M. & GONÇALVES, J.H. (eds) 2003. *Geologia, Tectônica e Recursos Minerais Do Brasil: Texto, Mapas & SIG*. CPRM – Serviço Geológico do Brasil, Belo Horizonte, Brazil.

BOILLOT, G., GRIMAUD, S., MAUFFRET, A., MOUGENOT, D., KORNPROBST, J., MERGOIL-DANIEL, J. & TORRENT, G. 1980. Ocean-continent boundary off the Iberian margin: A serpentinite diapir west of the Galicia Bank. *Earth and Planetary Science Letters*, **48**, 23–34, https://doi.org/10.1016/0012-821X(80)90166-1

BOILLOT, G., RECQ, M. ET AL. 1987. Tectonic denudation of the upper mantle along passive margins: a model based on drilling results (ODP leg 103, western Galicia margin, Spain). *Tectonophysics*, **132**, 335–342, https://doi.org/10.1016/0040-1951(87)90352-0

BURGESS, P.M., WINEFIELD, P., MINZONI, M. & ELDERS, C. 2013. Methods for identification of isolated carbonate buildups from seismic reflection data. *AAPG Bulletin*, **97**, 1071–1098, https://doi.org/10.1306/12051212011

CARTWRIGHT, J., JAMES, D. & BOLON, A. 2003. The genesis of polygonal fault systems: a review. *In*: VAN RENSBERGEN, P., HILLIS, R., MALTMAN, A.J. & MORLEY, C.K. (eds) 2003. *Subsurface Sediment Mobilization*. Geological Society, London, Special Publications, **216**, 223–243, https://doi.org/10.1144/GSL.SP.2003.216.01.15

COAKLEY, B.J. & COCHRAN, J.R. 1998. Gravity evidence of very thin crust at the Gakkel Ridge (Arctic Ocean). *Earth and Planetary Science Letters*, **162**, 81–95, https://doi.org/10.1016/S0012-821X(98)00158-7

COBBOLD, P.R., MOURGUES, R. & BOYD, K. 2004. Mechanism of thin-skinned detachment in the Amazon Fan: Assessing the importance of fluid overpressure and hydrocarbon generation. *Marine and Petroleum Geology*, **21**, 1013–1025, https://doi.org/10.1016/j.marpetgeo.2004.05.003

COSTA, J.B.S., HASUI, Y., BEMERGUY, R., SOARES JÚNIOR, A. & VILLEGAS, J. 2002. Tectonics and paleogeography of the Marajo Basin, northern Brazil. *Anais da Academia Brasileira de Ciencias*, **74**, 519–531, https://doi.org/10.1590/S0001-37652002000300013

DA COSTA, M.L., DA CRUZ, G.S., DE ALMEIDA, H.D.F. & POELLMANN, H. 2014. On the geology, mineralogy and geochemistry of the bauxite-bearing regolith in the lower Amazon basin: Evidence of genetic relationships. *Journal of Geochemical Exploration*, **146**, 58–74, https://doi.org/10.1016/j.gexplo.2014.07.021

DA COSTA, M.L., LEITE, A.S. & PÖLLMANN, H. 2016. A laterite-hosted APS deposit in the Amazon region, Brazil: The physical–chemical regime and environment of formation. *Journal of Geochemical Exploration*, **170**, 107–124, https://doi.org/10.1016/j.gexplo.2016.08.015

DALY, M.C., ANDRADE, V., BAROUSSE, C.A., COSTA, R., MCDOWELL, K., PIGGOTT, N. & POOLE, A.J. 2014. Brasiliano crustal structure and the tectonic setting of the Parnaíba basin of NE Brazil: Results of a deep seismic reflection profile. *Tectonics*, **33**, 2102–2120, https://doi.org/10.1002/2014TC003632

DARROS DE MATOS, R.M. 2000. Tectonic evolution of the equatorial South Atlantic. *In*: MOHRIAK, W.U. & TALWANI, M. (eds) *Atlantic Rifts and Continental Margins*. American Geophysical Union Geophysical Monograph Series, **115**, 331–354, https://doi.org/10.1029/GM115p0331

DAVISON, I., FAULL, T., GREENHALGH, J., BEIRNE, E.O. & STEEL, I. 2015. Transpressional structures and hydrocarbon potential along the Romanche Fracture Zone: a review. *In*: NEMČOK, M., RYBÁR, S., SINHA, S.T., HERMESTON, S.A. & LEDVÉNYIOVÁ, L. (eds) 2016. *Transform Margins: Development, Controls and Petroleum Systems*. Geological Society, London, Special Publications, **431**, 235–248, https://doi.org/10.1144/SP431.2

DE ALMEIDA, F.F.M., DE BRITO NEVES, B.B. & CARNEIRO, C.D.R. 2000. The origin and evolution of the South American Platform. *Earth-Science Reviews*, **50**, 77–111, https://doi.org/10.1016/S0012-8252(99)00072-0

DE CASTRO, D.L., FUCK, R.A., PHILLIPS, J.D., VIDOTTI, R.M., BEZERRA, F.H.R. & DANTAS, E.L. 2014. Crustal structure beneath the Paleozoic Parnaíba Basin revealed by airborne gravity and magnetic data, Brazil. *Tectonophysics*, **614**, 128–145, https://doi.org/10.1016/j.tecto.2013.12.009

DELESCLUSE, M., FUNCK, T., DEHLER, S.A., LOUDEN, K.E. & WATREMEZ, L. 2015. The oceanic crustal structure at the extinct, slow to ultraslow Labrador Sea spreading center. *Journal of Geophysical Research: Solid Earth*, **120**, 5249–5272, https://doi.org/10.1002/2014JB011739

DICK, H.J.B., LIN, J. & SCHOUTEN, H. 2003. An ultraslow-spreading class of ocean ridge. *Nature*, **426**, 405–412, https://doi.org/10.1038/nature02128

FIGUEIREDO, J.J.P., HOORN, C., VAN DER VEN, P. & SOARES, E.F. 2009. Late Miocene onset of the Amazon River and the Amazon deep-sea fan: Evidence from the Foz do Amazonas Basin. *Geology*, **37**, 619–622, https://doi.org/10.1130/G25567A.1

FUNCK, T. 2003. Crustal structure of the ocean–continent transition at Flemish Cap: Seismic refraction results. *Journal of Geophysical Research: Solid Earth*, **108**, 2531, https://doi.org/10.1029/2003JB002434

GILLARD, M., SAUTER, D., TUGEND, J., TOMASI, S. & EPIN, M. 2017. Birth of an oceanic spreading centre at a magma-poor rift system. *Scientific Reports*, **7**, 15072, https://doi.org/10.1038/s41598-017-15522-2

GORINI, C., HAQ, B.U., REIS, A.T., SILVA, C.G., CRUZ, A., SOARES, E.F. & GRANGEON, D. 2014. Late Neogene sequence stratigraphic evolution of the Foz do Amazonas Basin, Brazil. *Terra Nova*, **26**, 179–185, https://doi.org/10.1111/ter.12083

GREEN, P.F., JAPSEN, P., CHALMERS, J.A., BONOW, J.M. & DUDDY, I.R. 2018. Post-breakup burial and exhumation of passive continental margins: Seven propositions to inform geodynamic models. *Gondwana Research*, **53**, 58–81, https://doi.org/10.1016/j.gr.2017.03.007

GREENROYD, C.J., PEIRCE, C., RODGER, M., WATTS, A.B. & HOBBS, R.W. 2007. Crustal structure of the French Guiana margin, West Equatorial Atlantic. *Geophysical Journal International*, **169**, 964–987, https://doi.org/10.1111/j.1365-246X.2007.03372.x

GREENROYD, C.J., PEIRCE, C., RODGER, M., WATTS, A.B. & HOBBS, R.W. 2008a. Demerara Plateau – The structure and evolution of a transform passive margin. *Geophysical Journal International*, **172**, 549–564, https://doi.org/10.1111/j.1365-246X.2007.03662.x

GREENROYD, C.J., PEIRCE, C., RODGER, M., WATTS, A.B. & HOBBS, R.W. 2008b. Do fracture zones define continental margin segmentation? – Evidence from the French Guiana margin. *Earth and Planetary Science Letters*, **272**, 553–566, https://doi.org/10.1016/j.epsl.2008.05.022

HAUPERT, I., MANATSCHAL, G., DECARLIS, A. & UNTERNEHR, P. 2016. Upper-plate magma-poor rifted margins: Stratigraphic architecture and structural evolution. *Marine and Petroleum Geology*, **69**, 241–261, https://doi.org/10.1016/j.marpetgeo.2015.10.020

HEINE, C., ZOETHOUT, J. & MÜLLER, R.D. 2013. Kinematics of the South Atlantic rift. *Solid Earth*, **4**, 215–253, https://doi.org/10.5194/se-4-215-2013

HENRY, S., KUMAR, N., DANFORTH, A., NUTTALL, P. & VENKATRAMAN, S. 2011. Ghana/Sierra Leone lookalike plays in northern Brazil. *GeoExPro*, **8**, 36–41.

HOPPER, J.R., FUNCK, T. *ET AL.* 2004. Continental breakup and the onset of ultraslow seafloor spreading off Flemish Cap on the Newfoundland rifted margin. *Geology*, **32**, 93, https://doi.org/10.1130/G19694.1

HOPPER, J.R., FUNCK, T. & TUCHOLKE, B.E. 2007. Structure of the Flemish Cap margin, Newfoundland: insights into mantle and crustal processes during continental breakup. *In*: KARNER, G.D., MANATSCHAL, G. & PINHEIRO, L.M. (eds) *Imaging, Mapping and Modelling Continental Lithosphere Extension and Breakup*. Geological Society, London, Special Publications, **282**, 47–61, https://doi.org/10.1144/SP282.3

HOUTZ, R.E., LUDWIG, W.J., MILLIMAN, J.D. & GROW, J.A. 1977. Structure of the northern Brazilian continental margin. *Bulletin of the Geological Society of America*, **88**, 711–719, https://doi.org/10.1130/0016-7606(1977)88<711:SOTNBC>2.0.CO;2

JAPSEN, P., BONOW, J.M. *ET AL.* 2012a. Episodic burial and exhumation in NE Brazil after opening of the South Atlantic. *Bulletin of the Geological Society of America*, **124**, 800–816, https://doi.org/10.1130/B30515.1

JAPSEN, P., CHALMERS, J.A., GREEN, P.F. & BONOW, J.M. 2012b. Elevated, passive continental margins: Not rift shoulders, but expressions of episodic, post-rift burial and exhumation. *Global and Planetary Change*, **90–91**, 73–86, https://doi.org/10.1016/j.gloplacha.2011.05.004

JELINEK, A.R., CHEMALE, F., VAN DER BEEK, P.A., GUADAGNIN, F., CUPERTINO, J.A. & VIANA, A. 2014. Denudation history and landscape evolution of the northern East-Brazilian continental margin from apatite fission-track thermochronology. *Journal of South American Earth Sciences*, **54**, 158–181, https://doi.org/10.1016/j.jsames.2014.06.001

KEEN, C.E., DICKIE, K. & DAFOE, L.T. 2018. Structural characteristics of the ocean–continent transition along the rifted continental margin, offshore central Labrador. *Marine and Petroleum Geology*, **89**, 443–463, https://doi.org/10.1016/j.marpetgeo.2017.10.012

KING, S.D. 2007. Hotspots and edge-driven convection. *Geology*, **35**, 223, https://doi.org/10.1130/G23291A.1

KING, S.D. & ANDERSON, D.L. 1998. Edge-driven convection. *Earth and Planetary Science Letters*, **160**, 289–296, https://doi.org/10.1016/S0012-821X(98)00089-2

KIPF, A., HAUFF, F. *ET AL.* 2014. Seamounts off the West Antarctic margin: A case for non-hotspot driven intraplate volcanism. *Gondwana Research*, **25**, 1660–1679, https://doi.org/10.1016/j.gr.2013.06.013

KLEIN, E.L. & MOURA, C.A.V. 2008. São Luís Craton and Gurupi Belt (Brazil): possible links with the West African Craton and surrounding Pan-African belts. *In*: PANKHURST, R.J., TROUW, R.A.J., DE BRITO NEVES, B.B. & DE WIT, M.J. (eds) 2008. *West Gondwana: Pre-Cenozoic Correlations Across the South Atlantic Region*. Geological Society, London, Special Publications, **294**, 137–151, https://doi.org/10.1144/SP294.8

KLEIN, E.L., ANGÉLICA, R.S., HARRIS, C., JOURDAN, F. & BABINSKI, M. 2013. Mafic dykes intrusive into Pre-Cambrian rocks of the São Luís cratonic fragment and Gurupi Belt (Parnaíba Province), north-northeastern Brazil: Geochemistry, Sr–Nd–Pb–O isotopes, $^{40}Ar/^{39}Ar$ geochronology, and relationships to CAMP magmatis. *Lithos*, **172–173**, 222–242, https://doi.org/10.1016/j.lithos.2013.04.015

KLINGELHÖFER, F., GÉLI, L., MATIAS, L., STEINSLAND, N. & MOHR, J. 2000. Crustal structure of a super-slow spreading centre: A seismic refraction study of Mohns Ridge, 72°N. *Geophysical Journal International*, **141**, 509–526, https://doi.org/10.1046/j.1365-246X.2000.00098.x

KNESEL, K.M., SOUZA, Z.S., VASCONCELOS, P.M., COHEN, B.E. & SILVEIRA, F.V. 2011. Young volcanism in the Borborema Province, NE Brazil, shows no evidence for a trace of the Fernando de Noronha plume on the continent. *Earth and Planetary Science Letters*, **302**, 38–50, https://doi.org/10.1016/j.epsl.2010.11.036

KRUEGER, A., MURPHY, M., GILBERT, E. & BURKE, K. 2012. Deposition and deformation in the deepwater sediment of the offshore Barreirinhas Basin, Brazil. *Geosphere*, **8**, 1606–1631, https://doi.org/10.1130/GES00805.1

KRUEGER, A., MURPHY, M., BURKE, K. & GILBERT, E. 2014. The Brazilian equatorial margin: A snapshot in time of an oblique rifted margin. Search and Discovery Article #30325 presented at the AAPG 2014 Annual Convention and Exhibition, April 6–9, 2014, Houston, Texas, USA.

KUMAR, N. & EMBLEY, R.W. 1977. Evolution and origin of Ceara Rise: An aseismic rise in the western equatorial Atlantic. *Bulletin of the Geological Society of America*, **88**, 683.

KUSZNIR, N.J., ROBERTS, A.M. & ALVEY, A.D. 2018. Crustal structure of the conjugate Equatorial Atlantic Margins, derived by gravity anomaly inversion. *In*: MCCLAY, K. & HAMMERSTEIN, J. (eds) *Passive Margins: Tectonic, Sedimentation and Magmatism*. Geological Society, London, Special Publications, **476**, https://doi.org/10.1144/SP476.5

LORENZO, J.M. & WESSEL, P. 1997. Flexure across a continent–ocean fracture zone: the northern Falkland/Malvinas Plateau, South Atlantic. *Geo-Marine Letters*, **17**, 110–118, https://doi.org/10.1007/s003670050015

LYMER, G., CRESSWELL, D.J.F. *ET AL.* 2019. 3D development of detachment faulting during continental breakup. *Earth and Planetary Science Letters*, **515**, 90–99, https://doi.org/10.1016/j.epsl.2019.03.018

MASCLE, J. & BLAREZ, E. 1987. Evidence for transform margin evolution from the Ivory Coast–Ghana continental margin. *Nature*, **326**, 378–381, https://doi.org/10.1038/326378a0

MASCLE, J., BLAREZ, E. & MARINHO, M. 1988. The shallow structures of the Guinea and Ivory Coast–Ghana transform margins: Their bearing on the Equatorial Atlantic Mesozoic evolution. *Tectonophysics*, **155**, 193–209, https://doi.org/10.1016/0040-1951(88)90266-1

MASCLE, J., LOHMANN, P. & CLIFT, P. 1997. Development of a passive transform margin: Côte d'Ivoire–Ghana transform margin – ODP Leg 159 preliminary results. *Geo-Marine Letters*, **17**, 4–11, https://doi.org/10.1007/PL00007205

MATTHEWS, K.J., MÜLLER, R.D., WESSEL, P. & WHITTAKER, J.M. 2011. The tectonic fabric of the ocean basins. *Journal of Geophysical Research: Solid Earth*, **116**, B12109, https://doi.org/10.1029/2011JB008413

MERCIER DE LÉPINAY, M., LONCKE, L., BASILE, C., ROEST, W.R., PATRIAT, M., MAILLARD, A. & DE CLARENS, P. 2016. Transform continental margins – Part 2: A worldwide review. *Tectonophysics*, **693**, 96–115, https://doi.org/10.1016/j.tecto.2016.05.038

MOHRIAK, W.U. 2003. Bacias Sedimentares da Margem Continental Brasileira. *In*: BIZZI, L.A., SCHOBBENHAUS, C., VIDOTTI, R.M. & GONÇALVES, J.H. (eds) *Geologia, Tectônica e Recursos Minerais Do Brasil: Texto, Mapas & SIG*. CPRM – Serviço Geológico do Brasil, Belo Horizonte, Brazil, 87–168.

MORAIS NETO, J.M., HEGARTY, K.A., KARNER, G.D. & DE ALKMIM, F.F. 2009. Timing and mechanisms for the generation and modification of the anomalous topography of the Borborema Province, northeastern Brazil. *Marine and Petroleum Geology*, **26**, 1070–1086, https://doi.org/10.1016/j.marpetgeo.2008.07.002

MOULIN, M., ASLANIAN, D. & UNTERNEHR, P. 2010. A new starting point for the South and Equatorial Atlantic Ocean. *Earth-Science Reviews*, **98**, 1–37, https://doi.org/10.1016/j.earscirev.2009.08.001

MÜLLER, R.D., SDROLIAS, M., GAINA, C. & ROEST, W.R. 2008. Age, spreading rates, and spreading asymmetry of the world's ocean crust. *Geochemistry, Geophysics, Geosystems*, **9**, 1–19, https://doi.org/10.1029/2007GC001743

MUTTER, C.Z. & MUTTER, J.C. 1993. Variations in thickness of layer 3 dominate oceanic crustal structure. *Earth and*

Planetary Science Letters, **117**, 295–317, https://doi.org/10.1016/0012-821X(93)90134-U

NEMČOK, M., HENK, A., ALLEN, R., SIKORA, P.J. & STUART, C. 2013. Continental break-up along strike-slip fault zones; observations from the Equatorial Atlantic. *In*: MOHRIAK, W.U., DANFORTH, A., POST, P.J., BROWN, D.E., TARI, G.C., NEMČOK, M. & SINHA, S.T. (eds) 2013. *Conjugate Divergent Margins*. Geological Society, London, Special Publications, **369**, 537–556, https://doi.org/10.1144/SP369.8

NIKISHIN, A.M., GAINA, C., PETROV, E.I., MALYSHEV, N.A. & FREIMAN, S.I. 2018. Eurasia Basin and Gakkel Ridge, Arctic Ocean: Crustal asymmetry, ultra-slow spreading and continental rifting revealed by new seismic data. *Tectonophysics*, **746**, 64–82, https://doi.org/10.1016/j.tecto.2017.09.006

OLIVEIRA, M.J., SANTAREM, P., MORAES, A., ZALÁN, P.V., CALDEIRA, J.L., TANAKA, A. & TROSDTORF JUNIOR, I. 2012. Linked extensional–compressional tectonics in gravitational systems in the equatorial margin of Brazil. *In*: GAO, D. (ed.) *Tectonics and Sedimentation: Implications for Petroleum Systems*. AAPG Memoirs, **100**, 159–178, https://doi.org/10.1306/13351552M1003532

OSMUNDSEN, P.T. & EBBING, J. 2008. Styles of extension offshore mid-Norway and implications for mechanisms of crustal thinning at passive margins. *Tectonics*, **27**, TC6016, https://doi.org/10.1029/2007TC002242

OSMUNDSEN, P.T. & PÉRON-PINVIDIC, G. 2018. Crustal-scale fault interaction at rifted margins and the formation of domain-bounding breakaway complexes: insights from offshore Norway. *Tectonics*, **37**, 935–964, https://doi.org/10.1002/2017TC004792

OSMUNDSEN, P.T. & REDFIELD, T.F. 2011. Crustal taper and topography at passive continental margins. *Terra Nova*, **23**, 349–361, https://doi.org/10.1111/j.1365-3121.2011.01014.x

PAKULSKI, C. 2011. *Bioestratigrafia E Paleoecologia De Foraminíferos Da Bacia De Barreirinhas, Cretáceo, Margem*. PhD thesis, Universidade Federal do Rio Grande do Sul, Brazil.

PÉREZ-DÍAZ, L. & EAGLES, G. 2014. Constraining South Atlantic growth with seafloor spreading data. *Tectonics*, **33**, 1848–1873, https://doi.org/10.1002/2014TC003644

PÉREZ-GUSSINYÉ, M., RANERO, C.R., RESTON, T.J. & SAWYER, D. 2003. Mechanisms of extension at nonvolcanic margins: Evidence from the Galicia interior basin, west of Iberia. *Journal of Geophysical Research: Solid Earth*, **108**, 1–19, https://doi.org/10.1029/2001JB000901

PERON-PINVIDIC, G., MANATSCHAL, G. & OSMUNDSEN, P.T. 2013. Structural comparison of archetypal Atlantic rifted margins: A review of observations and concepts. *Marine and Petroleum Geology*, **43**, 21–47, https://doi.org/10.1016/j.marpetgeo.2013.02.002

PÉRON-PINVIDIC, G., MANATSCHAL, G., MASINI, E., SUTRA, E., FLAMENT, J.M., HAUPERT, I. & UNTERNEHR, P. 2015. Unravelling the along-strike variability of the Angola–Gabon rifted margin: a mapping approach. *In*: SABATO CERALDI, T., HODGKINSON, R.A. & BACKE, G. (eds) 2017. *Petroleum Geoscience of the West Africa Margin*. Geological Society, London, Special Publications, **438**, 49–76, https://doi.org/10.1144/SP438.1

PEULVAST, J.P. & DE CLAUDINO SALES, V. 2004. Stepped surfaces and palaeolandforms in the northern Brazilian <<Nordeste>>: Constraints on models of morphotectonic evolution. *Geomorphology*, **62**, 89–122, https://doi.org/10.1016/j.geomorph.2004.02.006

PEULVAST, J.P., CLAUDINO SALES, V., BÉTARD, F. & GUNNELL, Y. 2008. Low post-Cenomanian denudation depths across the Brazilian Northeast: Implications for long-term landscape evolution at a transform continental margin. *Global and Planetary Change*, **62**, 39–60, https://doi.org/10.1016/j.gloplacha.2007.11.005

PINDELL, J., GRAHAM, R. & HORN, B. 2014. Rapid outer marginal collapse at the rift to drift transition of passive margin evolution, with a Gulf of Mexico case study. *Basin Research*, **26**, 701–725, https://doi.org/10.1111/bre.12059

POSAMENTIER, H.W. & KOLLA, V. 2003. Seismic geomorphology and stratigraphy of depositional elements in deep-water settings. *Journal of Sedimentary Research*, **73**, 367–388, https://doi.org/10.1306/111302730367

RADHAKRISHNA, M., TWINKLE, D., NAYAK, S., BASTIA, R. & RAO, G.S. 2012. Crustal structure and rift architecture across the Krishna-Godavari basin in the central Eastern Continental Margin of India based on analysis of gravity and seismic data. *Marine and Petroleum Geology*, **37**, 129–146, https://doi.org/10.1016/j.marpetgeo.2012.05.005

REBESCO, M., HERNÁNDEZ-MOLINA, F.J., VAN ROOIJ, D. & WÅHLIN, A. 2014. Contourites and associated sediments controlled by deep-water circulation processes: State-of-the-art and future considerations. *Marine Geology*, **352**, 111–154, https://doi.org/10.1016/j.margeo.2014.03.011

REIS, A.T., DA ARAÚJO, É.F.S. ET AL. 2016. Effects of a regional décollement level for gravity tectonics on late Neogene to recent large-scale slope instabilities in the Foz do Amazonas Basin, Brazil. *Marine and Petroleum Geology*, **75**, 29–52, https://doi.org/10.1016/j.marpetgeo.2016.04.011

RESTON, T.J., LEYTHAEUSER, T., BOOTH-REA, G., SAWYER, D.S., KLAESCHEN, D. & LONG, C. 2007. Movement along a low-angle normal fault: The S reflector west of Spain. *Geochemistry, Geophysics, Geosystems*, **8**, Q06002, https://doi.org/10.1029/2006GC001437

RODGER, M., WATTS, A.B., GREENROYD, C.J., PEIRCE, C. & HOBBS, R.W. 2006. Evidence for unusually thin oceanic crust and strong mantle beneath the Amazon Fan. *Geology*, **34**, 1081–1084, https://doi.org/10.1130/G22966A.1

ROSSETTI, D.F. 2001. Late Cenozoic sedimentary evolution in northeastern Pará, Brazil, within the context of sea level changes. *Journal of South American Earth Sciences*, **14**, 77–89, https://doi.org/10.1016/S0895-9811(01)00008-6

ROSSETTI, D.F. 2004. Paleosurfaces from northeastern Amazonia as a key for reconstructing paleolandscapes and understanding weathering products. *Sedimentary Geology*, **169**, 151–174, https://doi.org/10.1016/j.sedgeo.2004.05.003

ROSSETTI, D.F. & SANTOS, A.E. 2004. Facies architecture in a tectonically influenced estuarine incised valley fill of Miocene age, northern Brazil. *Journal of South American Earth Sciences*, **17**, 267–284, https://doi.org/10.1016/j.jsames.2004.08.003

Rossetti, D.F., Bezerra, F.H.R. & Dominguez, J.M.L. 2013. Late oligocene-miocene transgressions along the equatorial and eastern margins of Brazil. *Earth-Science Reviews*, **123**, 87–112, https://doi.org/10.1016/j.earscirev.2013.04.005

Sapin, F., Davaux, M., Dall'Asta, M., Lahmi, M., Baudot, G. & Ringenbach, J.-C. 2016. Post-rift subsidence of the French Guiana hyper-oblique margin: from rift-inherited subsidence to Amazon deposition effect. *In*: Nemčok, M., Rybár, S., Sinha, S.T., Hermeston, S.A. & Ledvényiová, L. (eds) 2016. *Transform Margins: Development, Controls and Petroleum Systems*. Geological Society, London, Special Publications, **431**, 125–144, https://doi.org/10.1144/SP431.11

Scarselli, N., Duval, G., Martin, J., McClay, K. & Toothill, S. 2018. Insights into the early evolution of the Côte d'Ivoire margin (west Africa). *In*: McClay, K. & Hammerstein, J. (eds) *Passive Margins: Tectonic, Sedimentation and Magmatism*. Geological Society, London, Special Publications, **476**, https://doi.org/10.1144/SP476.8

Schmidt-Aursch, M.C. & Jokat, W. 2016. 3D gravity modelling reveals off-axis crustal thickness variations along the western Gakkel Ridge (Arctic Ocean). *Tectonophysics*, **691**, 85–97, https://doi.org/10.1016/j.tecto.2016.03.021

Schobbenhaus, C. & Brito Neves, B.B. 2003. A Geologia do Brasil no Contexto da Plataforms Sul-Amreicana. *In*: Bizzi, L.A., Schobbenhaus, C., Vidotti, R.M. & Gonçalves, J.H. (eds) *Geologia, Tectônica e Recursos Minerais Do Brasil: Texto, Mapas & SIG*. CPRM – Serviço Geológico do Brasil, Belo Horizonte, Brazil, 5–54.

Schobbenhaus, C., Gonçalves, J.H. ET AL. 2004. *Carta Geológica Do Brasil Ao Milionésimo, Sistema de Informações Geográficas-SIG*. Programa Geologia Do Brasil, Brasilia, Brazil.

Silva, C.G., Araújo, É., Reis, A.T., Perovano, R., Gorini, C., Vendeville, B.C. & Albuquerque, N.C. 2010. Megaslides in the Foz do Amazonas Basin, Brazilian Equatorial Margin. *In*: Mosher, D.C. (eds) *Submarine Mass Movements and Their Consequences*. Advances in Natural and Technological Hazards Research, **28**. Springer, Dordrecht, The Netherlands, 581–591, https://doi.org/10.1007/978-90-481-3071-9_47

Soares, E.F., Zalán, P.V., de Jesus, J., Figueiredo, J.J.P. & Trosdtorf Junior, I. 2007. Bacia do Pará-Maranhão. *Boletim de Geociências da Petrobrás*, **15**, 321–329.

Soares Júnior, A.V., Costa, J.B.S. & Hasui, Y. 2008. Evolução da margem atlântica equatorial do Brasil: Três fases distensivas. *Geociências*, **27**, 427–437.

Sutra, E. & Manatschal, G. 2012. How does the continental crust thin in a hyperextended rifted margin? Insights from the Iberia margin. *Geology*, **40**, 139–142, https://doi.org/10.1130/G32786.1

Sutra, E., Manatschal, G., Mohn, G. & Unternehr, P. 2013. Quantification and restoration of extensional deformation along the Western Iberia and Newfoundland rifted margins. *Geochemistry, Geophysics, Geosystems*, **14**, 2575–2597, https://doi.org/10.1002/ggge.20135

Tamara, J. & McClay, K.R. 2015. 4-D Evolution of the Maranhao deepwater fold belt, Offshore NE Brasil. Search and Discovery Article #90217 presented at the AAPG International Conference & Exhibition, September 13–16, 2015, Melbourne, Australia.

Trosdtorf Junior, I., Zalán, P.V., Figueiredo, J.J.P. & Soares, E.F. 2007. Bacia de Barreirinhas. *Boletim de Geociências da Petrobrás*, **15**, 331–339.

Unternehr, P., Peron-Pinvidic, G., Manatschal, G. & Sutra, E. 2010. Hyper-extended crust in the South Atlantic: in search of a model. *Petroleum Geoscience*, **16**, 207–215, https://doi.org/10.1144/1354-079309-904

Vogt, P.R. 1991. Bermuda and Appalachian-Labrador rises: Common non-hotspot processes? *Geology*, **19**, 41, https://doi.org/10.1130/0091-7613(1991)019<0041:BAALRC>2.3.CO;2

Watts, A.B., Rodger, M., Peirce, C., Greenroyd, C.J. & Hobbs, R.W. 2009. Seismic structure, gravity anomalies, and flexure of the amazon continental margin, NE Brazil. *Journal of Geophysical Research: Solid Earth*, **114**, 1–23, https://doi.org/10.1029/2008JB006259

Weimer, P., Slatt, R.M., Bouroullec, R., Fillon, R., Pettingill, H., Pranter, M. & Tari, G. 2006. *Introduction to the Petroleum Geology of Deepwater Setting*. AAPG Studies in Geology, **57**, https://doi.org/10.1306/St571314

White, R.S., McKenzie, D. & O'Nions, R.K. 1992. Oceanic crustal thickness from seismic measurements and rare earth element inversions. *Journal of Geophysical Research: Solid Earth*, **97**, 19 683–19 715, https://doi.org/10.1029/92JB01749

Ye, J., Rouby, D., Chardon, D., Dall'Asta, M., Guillocheau, F., Robin, C. & Ferry, J.N. 2019. Post-rift stratigraphic architectures along the African margin of the Equatorial Atlantic: Part I the influence of extension obliquity. *Tectonophysics*, **753**, 49–62, https://doi.org/10.1016/j.tecto.2019.01.003

Zalán, P.V. 2005. End members of gravitational fold and thrust belts (GFTBs) in the deep waters of Brazil. *In*: Shaw, J., Connors, C. & Suppe, J. (eds) *Seismic Interpretation of Contractional Fault-Related Folds*. AAPG Studies in Geology, **53**, 147–156.

Zalán, P.V. 2011. Fault-related folding in the deep waters of the Equatorial margin of Brazil. *In*: McClay, K., Shaw, J. & Suppe, J. (eds) *Thrust Fault-Related Folding*. AAPG Memoirs, **94**, 335–355, https://doi.org/10.1306/13251344M943089

Zalán, P.V. 2015. Re-interpretation of an Ultra-Deep Seismic Section in the Pará-Maranhão Basin – Implications for the petroleum potential of the ultra-deep waters. Paper OTC-26134 presented at the Offshore Technology Conference, 27–29 October 2015, Rio de Janeiro, Brazil.

'Ductile v. Brittle' – Alternative structural interpretations for the Niger Delta

PEDRO A RESTREPO-PACE

Oil Search Limited, 1 Bligh Street, level 23, Sydney, NSW 2000
pedro.restrepo@oilsearch.com

Abstract: A wealth of subsurface information gathered from over 60 years of hydrocarbon exploration offshore Nigeria provides a reference study area on the interaction between sedimentation, structure and overpressure in a large delta system. The current geological paradigm is that structuration and synkinematic sedimentation is governed by shale mobility from the deeper parts of the delta. This concept is largely the result of interpretations derived from vintage seismic data and insufficient calibration of the deeper parts of the delta. Long-cable seismic data are providing new insights into this interpretation conundrum. A first-order problem, which is of particular interest here, relates to the linkage between the extensional structures updip with the compressional structures downdip. The translational zone between extension and compression is key to unravelling the nature of this link and any associated structural material balance discrepancies. The primary focus of the current paper is to interrogate seismic data and to provide alternative interpretations to the accepted paradigm. Two end-member interpretations of the Niger Delta regional seismic dip lines – referred to here as the 'ductile model' and the 'brittle model' – are presented. Aside from their internal geometrical dissimilarities, these interpretations suggest fundamentally different kinematic and geomechanical models. The latter may offer a wider scope for deep – largely neglected – hydrocarbon exploration targets. Ultimately, these ideas could provide the conceptual framework that, in conjunction with improved seismic efforts, could lead to rejuvenated exploration portfolios.

Deformation of large passive margin clastic wedges is driven primarily by near-field stresses in the form of gravity-glide and/or gravity-spread mechanisms (Morley *et al.* 2011). Of particular interest to hydrocarbon exploration/exploitation are deltas, many of which constitute extremely prolific petroleum provinces. The Niger Delta is an impressive 12 km-thick sedimentary wedge that tapers outboard from the continental margin over a distance of 350 km. Commercial oil was first discovered in the Niger Delta in 1956. Estimated reserves thus far encountered are estimated to be above 35 BBOE (billion barrels of oil equivalent). An extensive volume of literature has been published on the regional aspects of the Niger Delta (e.g. refer to Doust & Omatsola 1989; Morley & Guerin 1996; Hooper *et al.* 2002; Morley 2003). The delta has developed over the Benue Trough since the Cretaceous and was structurally constrained by features associated with the opening of the Equatorial Atlantic (Doust & Omatsola 1989; Hooper *et al.* 2002). The stage of major delta build-up followed the early Cretaceous post-rift sedimentation and took place as a prograding wedge from Late Eocene to present. In general terms, the Niger Delta is composed of two geomechanical units: the Paleogene basal and overpressured Akata Formation slope to abyssal shales; and the Oligocene–Recent Agbada Formation, consisting of nearshore to prodelta sediments. Overpressure at depth developed primarily as a result of its high sedimentation rates (c. 35–600 m/1 kyr: Pastouret *et al.* 1978). The Akata Formation is the loci of basal slip for the deltaic wedge in the form of discrete detachments that are seismically well imaged in the outer toe-thrust region (refer to seismic examples later in the text). Overpressure acted as a 'catalyst' assisting gravity-driven deformation of this clastic wedge (e.g. Morley 2003; Cobbold *et al.* 2009). Kinematically, the deformation progressed in tandem with sedimentation (Mourgues *et al.* 2009; Rouby *et al.* 2011) and mostly in break-forward sequence, although local variations in timing occurred possibly as a result of taper adjustments and localized fluid-pressure transfers (Krueger & Grant 2013).

The Niger Delta is unique in having extensive seismic and well data. Despite ample calibration, the overall structural framework and kinematics of the delta remain uncertain. It comprises an extensional domain updip and a compressional domain downdip (toe-thrust system). The toe-thrust system is well imaged seismically, and the geometries and kinematics are well understood (e.g. Bilotti & Shaw 2005; Corredor *et al.* 2005). The extensional domain updip consists of regional and counter-regional growth systems; these are poorly imaged at depth. Linking the extensional and compressional systems is a zone of considerable structural complexity consisting of buried thrust systems, detachment

From: MCCLAY, K. R. & HAMMERSTEIN, J. A. (eds) 2020. *Passive Margins: Tectonics, Sedimentation and Magmatism.*
Geological Society, London, Special Publications, **476**, 193–204.
First published online March 7, 2018, https://doi.org/10.1144/SP476.2
© 2018 The Author(s). Published by The Geological Society of London. All rights reserved.
For permissions: http://www.geolsoc.org.uk/permissions. Publishing disclaimer: www.geolsoc.org.uk/pub_ethics

folds and mud diapirs, and is known as the 'translational domain' (Doust & Omatsola 1989; Hooper et al. 2002) or 'inner deformation front' (Fig. 1).

The deep overpressured zones have typically been regarded as a region dominated by mobile shale deformation (argillokinesis), with the shallow stratigraphy involved in detached folding/faulting combined with growth-'withdrawal' sedimentation (e.g. Hooper et al. 2002; Morley 2003; Morley et al. 2011). The latter invokes protracted shale mobility, a condition valid as long as the overpressure is sustained. This structural framework paradigm has prevailed in Niger Delta interpretations, and has persisted because of poor seismic imaging at depth and from the evidence of overpressure detected while drilling, from sea-bottom pock marks or inferred from seismic velocity inversions. Additionally, geophysicists with salt-basin experience (e.g. Gulf of Mexico) introduced interpretation biases that resulted in a preferential mobile shale model. An alternative interpretation to shale-driven deformation has emerged from the careful examination of long-cable seismic data (e.g. Restrepo-Pace et al. 2006; Bellingham et al. 2014). The deep seismically 'bland' areas, distinctive in vintage datasets, now depict coherent reflections in regional long-cable data. This led to increased efforts in seismic acquisition (broadband and/or 3D) that resolved complex areas, high dip panels and deeper low-frequency reflectors. As a result, a picture of more extensive 'brittle' and restricted 'ductile' deformation is now developing. Analogue models (Mourgues et al. 2009; Lacoste et al. 2012) have simulated the newly interpreted structural geometries derived from better seismic datasets. Such a paradigm shift may have profound implications for oil and gas exploration, opening up the possibility of new, deeper plays and the rejuvenation of maturing shallow-play portfolios. It is the objective of this paper to illustrate possible interpretations – primary input to mapping, prospecting, modelling and hydrocarbon exploration de-risking – using key regional seismic line examples that may provide new insights into the kinematics and dynamics of the Niger Delta deformation.

The Niger Delta seismic interpretation conundrum

The basis of the ductile interpretational model

It is not surprising that the first regional cross-sectional interpretations for oil and gas exploration of the Niger Delta incorporated a large core of ductile material at depth. These depicted a shallow seismically coherent blanket of strata deformed over a seismically chaotic/wipe-out zone considered to be mobile shale. The top of the mobile shale has been picked by interpreters at the boundary where a non-interpretable wipe-out zone lies (vague boundary at best, data quality dependent) and/or selected at the top velocity reversals considered a proxy for overpressure. Gravity data are used to derive the base of the delta. Similarities between Gulf of Mexico structures and minibasin geometries with those observed in the Niger Delta led geophysical interpreters to transpose their interpretation bias. This conceptual transposition extended to the kinematic framework: that is, in which the accommodation space updip and the compressional toe developed as mud is 'squeezed' upwards and outwards (Wu & Bally 2000; Ajakaiye & Bally 2002; McClay et al. 2003; Rowan et al. 2004). Thus, the conventional interpretation model for the Niger Delta implies that the extension from the present-day shelf area is taken as volume changes of the mobile shale and, to a much lesser extent, as structural shortening in the deep-water toe-thrust area. The mobile shale interpretation has been successfully applied in the delineation of shallow hydrocarbon drilling traps. At the same time, this somewhat simplistic fixed approach employed by exploration companies for years has limited the possibility of conceptually formulating deeper traps or chasing deeper reservoirs. The mobile shale or 'ductile' interpretation in principle requires protracted and continuous shale mobility for it to be viable; a rather unlikely scenario as shale bleeds-off its overpressure into more porous layers/faults or as mud fluids make their way upsection. Moreover, chemical transformations experienced by shale above 80°C inhibit shale mobility even under overpressure conditions (Day-Stirrat et al. 2010). Poor-quality seismic data are partly responsible for this interpretation bias. Overpressure, which is prevalent in many of these large clastic systems, tends to degrade the seismic signal. Shale masses lack the impedance contrast that permits apparent lateral boundaries with the 'host rock' in seismic sections to be determined (one distinctive feature from salt provinces – although newer seismic does define these mobile shale boundaries better) (Fig. 2). Steep dips at minibasin flanks and complex structures that have not been adequately resolved in processing seem to lie within many of these transparent zones. Observations regarding similarities and differences between shale and salt-dominated provinces were initially described by Morley & Guerin (1996) and subsequently in greater detail in Morley (2003).

Despite recent advances in our understanding of argillokinesis (i.e. sand-box and numerical models built to replicate the mobile shale interpretations), they have fallen short of adequately addressing the intrinsic anisotropies of shale and its pore-fluid-pressure variations. Many early simulations continue

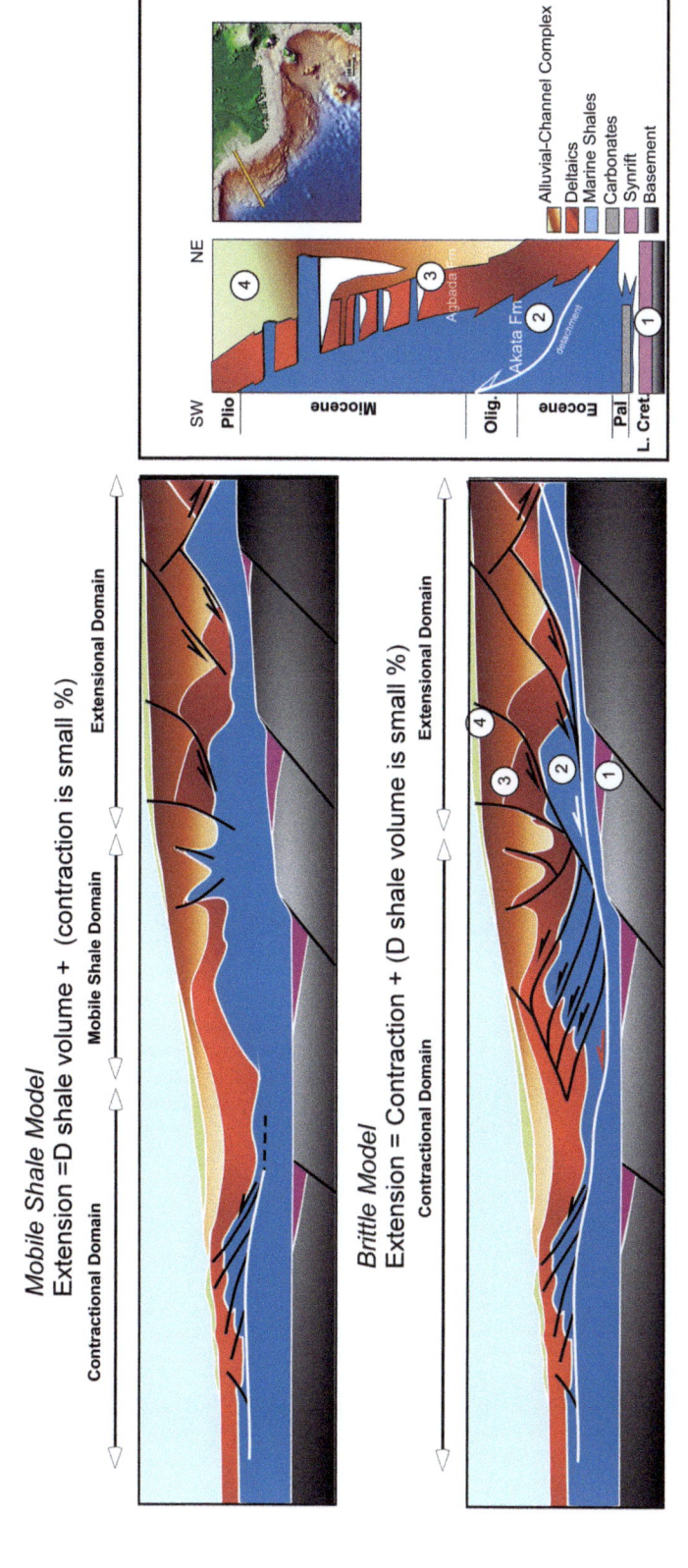

Fig. 1. Stratigraphy (modified from Corredor *et al.* 2005) and schematic conceptual alternative models for dip sections of a typical modern delta. In the upper section, the large extension updip is accommodated by a small contraction at the toe thrust, plus a shale volume change. In the lower section, the large amounts of extension updip are accommodated by shortening downdip in the translational and toe-thrust zones, via a basal detachment.

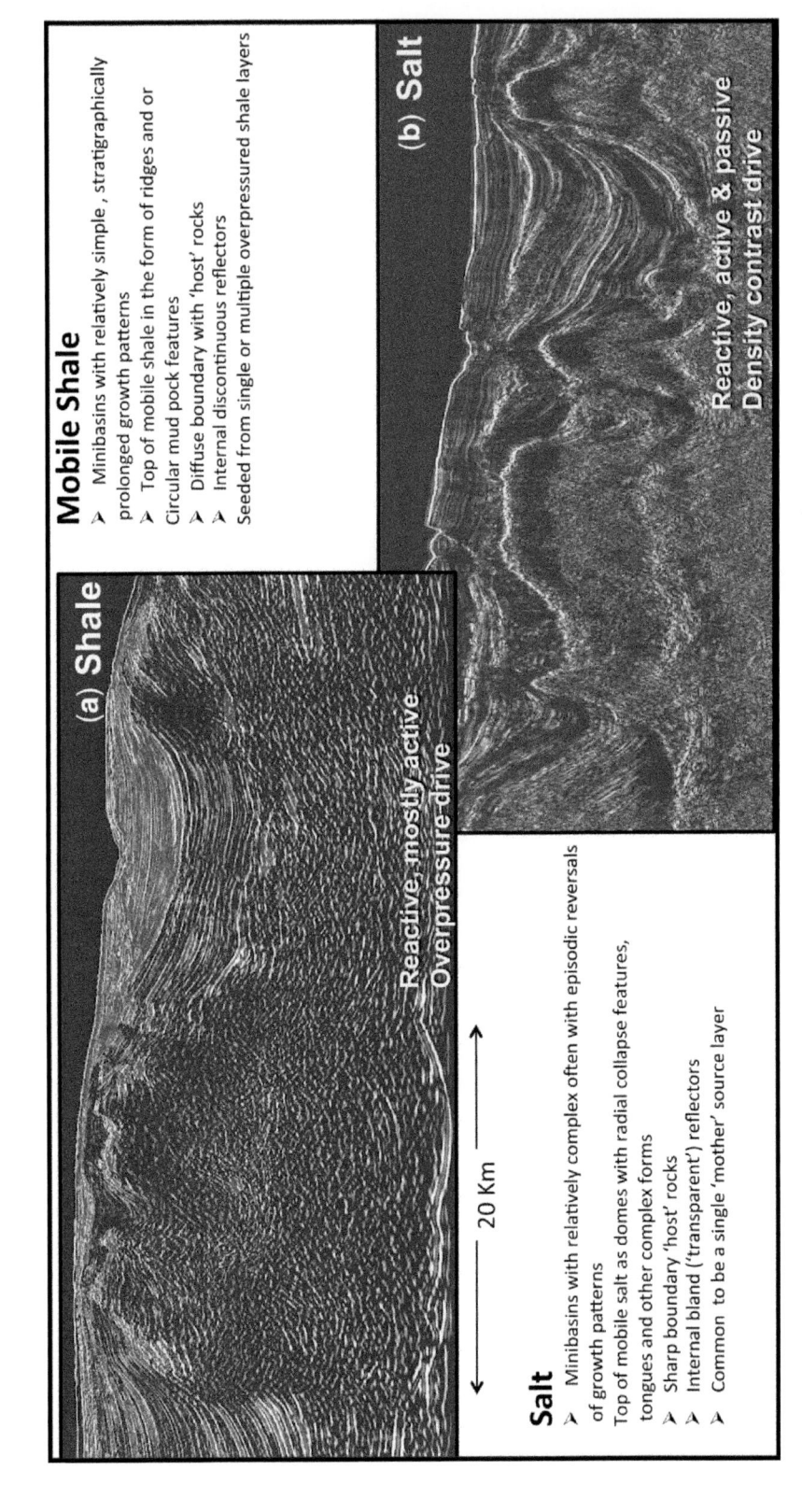

Fig. 2. Example of seismic lines from the Niger Delta and Campos Basin to describe similarities and contrasting data/features of (**a**) mobile shale and (**b**) salt following observations by Morley & Guerin (1996). The apparent geometrical similarities with salt provinces have greatly influenced Niger Delta seismic interpretations, yet the rheology of shale and salt are fundamentally different from each other (refer to the text for further details).

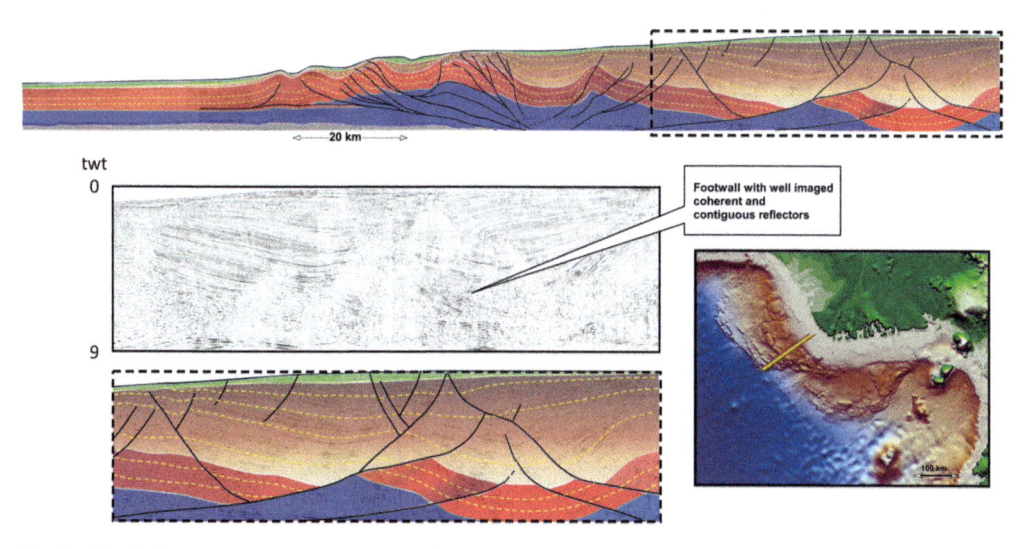

Fig. 3. Niger Delta examples from the extensional domain. Regional and counter-regional or back-to-back structures generate substantial accommodation space. The footwall area and deeper sections are poorly imaged; in vintage data, these are largely transparent and have been considered to consist mostly of mobile shale. Coherent, contiguous reflectors can be observed in the footwall area. The resolved dip panels provide the basis for the interpretation shown here. TWT, two-way travel time.

to draw upon analogies from salt rheology (Suppe 2011). Surface mud volcanoes and sea-bottom pockmarks provide us with direct evidence of movement of fluid/mud, yet these are generally of limited dimensions – of the order of a few hundred metres. The larger features rarely get drilled as they pose potential drilling hazards. There are documented examples of wells drilled into the transparent chaotic 'mobile shale' zones in the Barbados accretionary prism that register orderly stratification rather than the chaotic mud piles inferred from seismic data (Deville *et al.* 2006).

The alternative case for a brittle interpretational model

Transparent and discontinuous seismic zones from vintage data can be discerned as organized contiguous, coherent reflectors in more modern long-offset data (Fig. 3). High-quality 3D datasets have improved the definition of footwall structures within the extensional domain in the Niger Delta. Normal fault footwalls in the extensional domain are particularly poorly imaged in vintage datasets. As a result, these areas have been prone to be interpreted as mud escape features, and thus assumed to be acting as the principal mechanism for the creation of large regional and counter-regional depocentres (Doust & Omatsola 1989). Pockmarks and mud vents are commonly arranged spatially along faults or at fold crests. The latter may indicate that mobile shale zones are largely controlled by pre-existing structures and/or are the result of pressure transfer (Krueger & Grant 2013) and pressure-induced breaching, rather than being the mechanism responsible for creating the structures themselves (Cobbold *et al.* 2009).

The translational domain *sensu latto* (also referred to as the inner deformation front) may hold the key to resolving the kinematic link between the extensional and compressional domains. It is this zone that has been fraught with the greatest difficulty in unravelling its internal structural architecture, mostly because of poor seismic imaging and insufficient calibration. Here, again, this domain has been habitually interpreted as being dominated by mobile-shale-related structures (e.g. thick mobile shale and mobile shale thrust belts in Weiner *et al.* 2010; Morley *et al.* 2011, amongst others). Long-cable data depict zones of imbrication, duplexing, antiformal stacking and complex zones (Fig. 4a–c). Mud escape structures are commonly associated with faults and folds in the inner compressional deformation front, and are likely to be associated with critically stressed structures resulting in trap breaching. Geological interpretations of older seismic data depict large volumes of mobile shale under the inner compressional domain. Modern and improved-quality seismic data better define the mobile shale geobodies. The reduced mud volumes portrayed in these datasets, and the structural geometries depicted, suggest that diapirism is more of a by-product than a causative for the overall structuration. The latter is not only relevant regarding

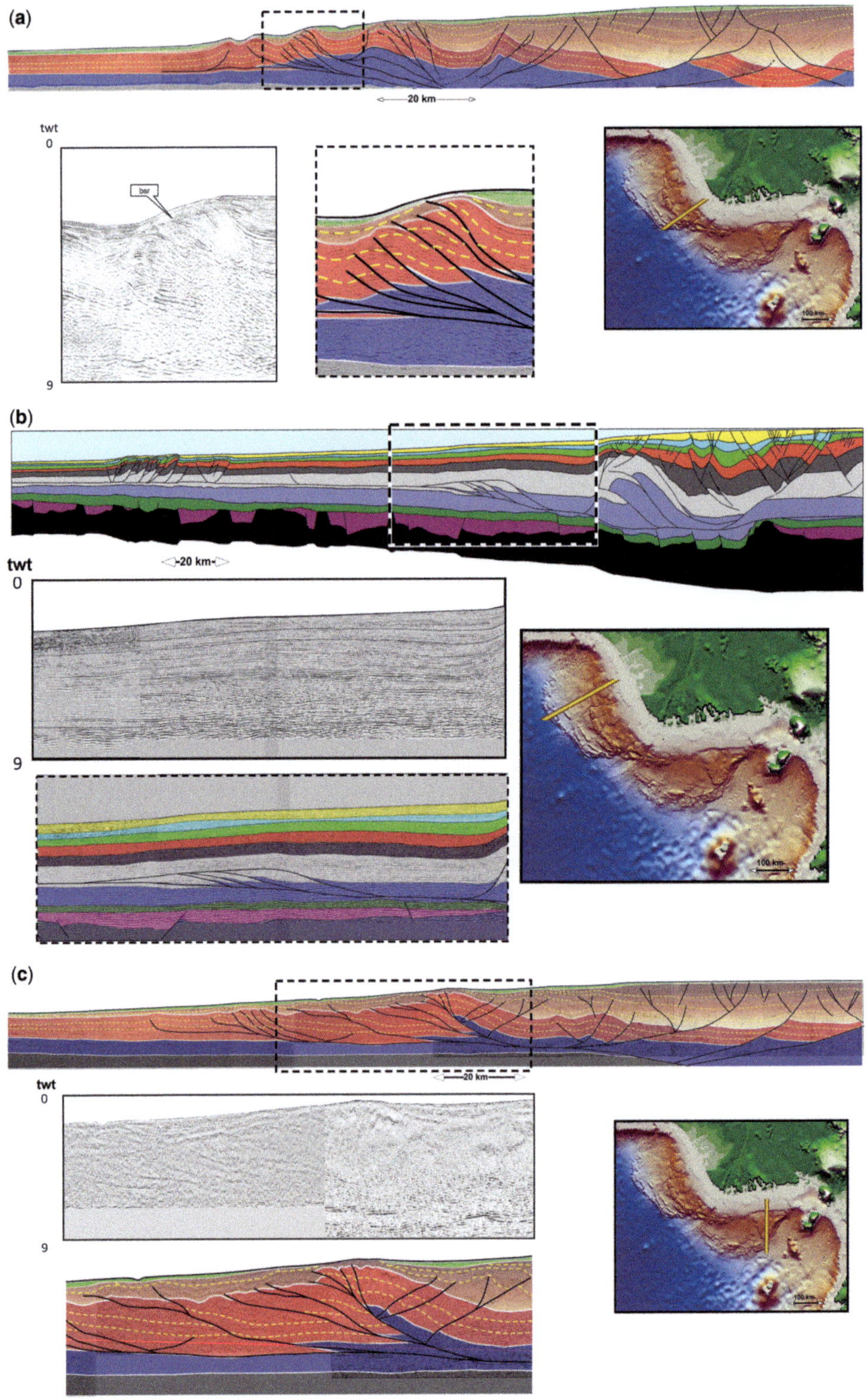
(a)
20 km
twt
0
bsr
9
100 km
(b)
20 km
twt
0
9
100 km
(c)
20 km
twt
0
9
100 km

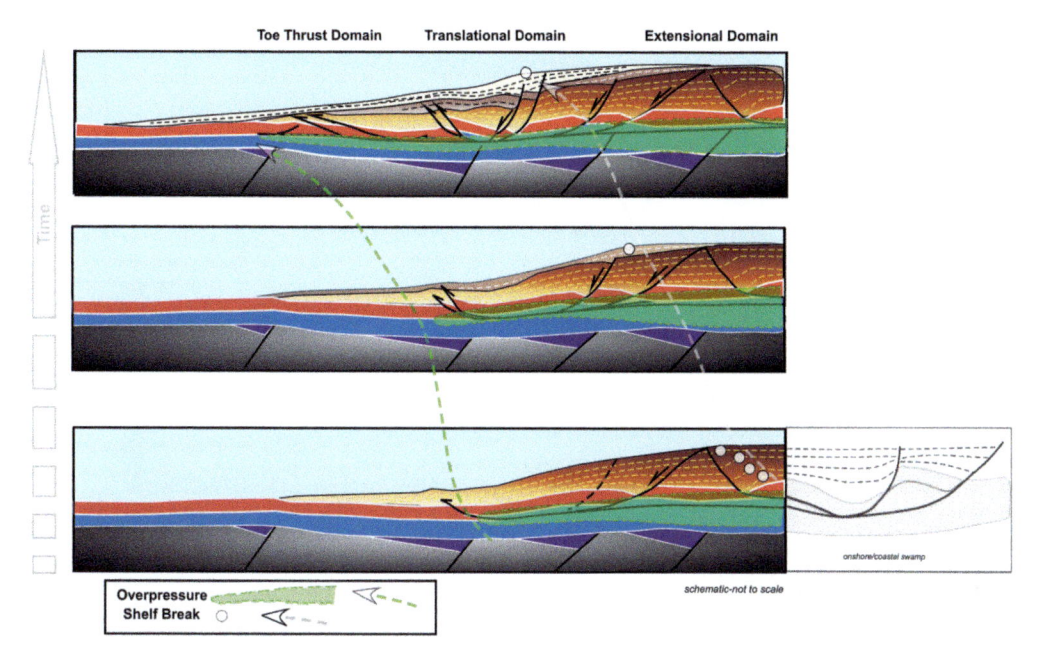

Fig. 5. Schematic representation of the interplay between the prograding delta wedge, overpressure and structuration (modified from Mourgues *et al.* 2009). Rapid sedimentation and outboard build-up of the delta wedge drives structuration. The extension nucleated near the shelf break advances outboard (Rouby *et al.* 2011; Sapin *et al.* 2012). Rapid sedimentation causes overpressure, and the thickening of the advancing wedge pushes the Akata shales into the hydrocarbon maturity window. An early extension–compression couple (inner deformation front) develops, followed by a more outboard extension–compression (toe-thrust) pair.

structural style and kinematics but is also significant for charge modelling. Mobile shale systems can deliver hydrocarbons via efficient vertical/lateral fracture permeability conduits (e.g. as documented in the Barbados accretionary prism: Deville *et al.* 2003).

There are relatively few contentious issues regarding the well-imaged toe-thrust system, its internal geometries or kinematics; even less debatable when analysed in isolation from the overall delta system. But unravelling the connection of the compressional toe with the extensional domain is essential to the understanding of the Niger Delta's deformation. With regards to the latter, the central issue is material balance: that is, how extension updip gets compensated downdip (Suppe 2011) as the toe-thrust system often depicts small amounts of shortening relative to the extension updip.

Analogue models indicate that extension is triggered by loading and overpressure as the delta advances outboard (Mourgues *et al.* 2009; Lacoste *et al.* 2012). Calibrated mapping of synkinematic strata, in the context of temporal and spatial distributions of structures as the delta progrades, supports the dynamic interplay of delta progradation, overpressure and structural timing (Fig. 5). This framework is instrumental in understanding the development of the translational zone. Most of the shortening of the Niger Delta appears to have taken place here – not in the present toe-thrust area – in the form of tightly spaced fault imbricates, horses, antiformal thrust stacks and associated mud vents.

Connecting updip extension with downdip contraction

Sediments that lie beyond the deformation edge of the Niger Delta have limited calibration. The improved imaging of long-offset regional seismic data allows this pre-kinematic stratigraphy to be

Fig. 4. Seismic lines depicting the structural complexities within the compressional (inner) deformation front, mud diapirs and the translational provinces of Krueger & Grant (2013). (**a**) Coherent dip panels and discontinuities suggest a complex structural stacking potential for stacked pay. Synkinematic and growth stratigraphy portray highly rotated onlapping reflectors at the edges of the piggy-back basins. (**b**) Low-relief, large-wavelength structure of the translational domain generated by a low-angle tightly spaced imbricate thrust system. (**c**) Tightly spaced imbricate thrust stack of the inner deformation front. bsr, bottom simulating reflector.

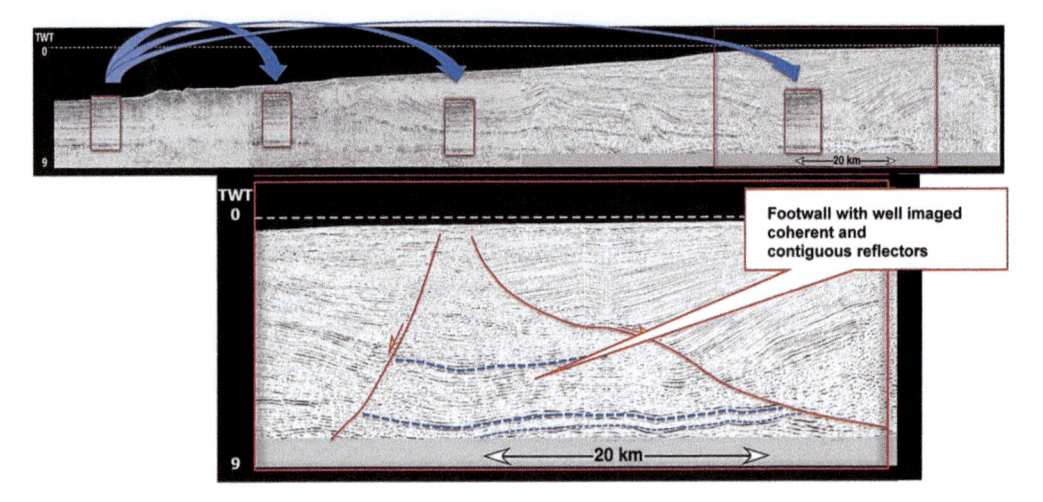

Fig. 6. Jump tie of deep-water undeformed prekinematic stratigraphy into various structural zones of the delta. This correlation provides the basic framework to construct viable 'brittle'-style structural interpretations in the absence of the calibration of large areas formerly interpreted as mobile shale (e.g. extensional domain footwall structures).

traced from the undeformed deep-water sediments towards the 'hinterland' regions of the delta wedge (Fig. 6). Notwithstanding the limitations of jump-tie correlations, a framework can be constructed that allows for viable interpretations. Figure 6 illustrates two end-member solutions to a single regional line. The 'brittle'-type interpretation links updip with downdip structures via a basal detachment. A simple three-stage retro-deformation of the 'brittle' interpretation provides context for the structural connection between structural domains and the structural timing constrained by synkinematic sedimentation (Fig. 7). Large extensional growth systems are linked to contractional structures that developed by the late Mid-Miocene, with most of the shortening taking place in the inner deformation front. The link between the extensional and contractional domains occurs via a basal detachment, and

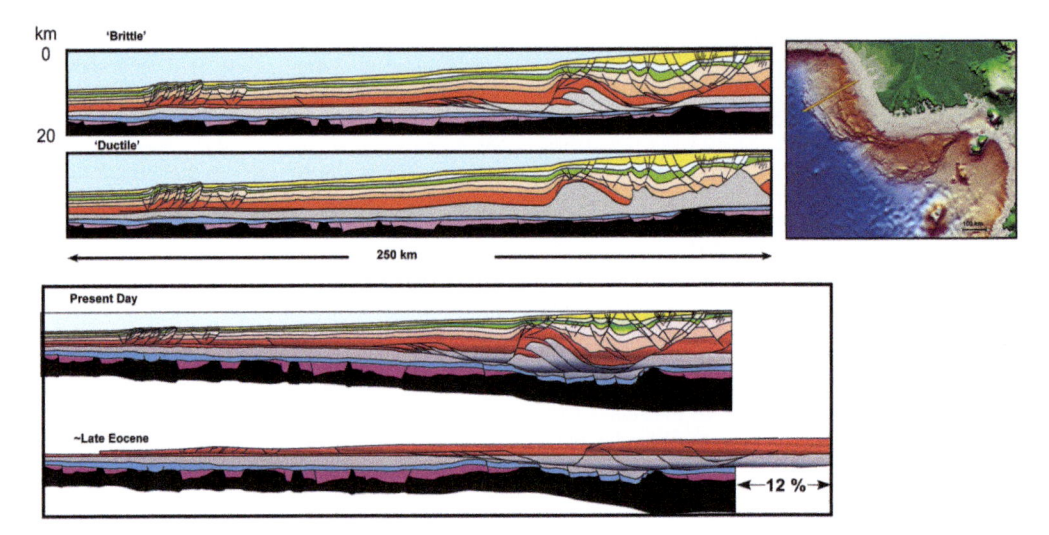

Fig. 7. Alternative interpretations of a single regional seismic line in the Niger Delta referred to here as the 'ductile' and 'brittle' models. These are fundamentally different kinematically as the brittle model implies that the updip extension is connected to the downdip compressive domain via a discrete basal detachment system. Below, a simple three-stage restoration (no de-compaction) indicates that an area change of 12% had taken place by the late Mid-Miocene and most of the updip extension is taken up by the inner deformation front.

the timing relationships can be determined from mapping in detail the Late Miocene–present synkinematic sediments. What is most relevant here is that the 'brittle'-type interpretation provides the kinematic framework that may adequately constrain sediment dispersal, structural timing, charge and containment: that is, it can provide a template for modelling and geological underpinning to play concepts verifiable with the drill bit. Sequential palaeostructure maps constructed within this framework can provide new insights regarding the current distribution of hydrocarbon discoveries. It must be noted that seismic data quality is variable even in a single vintage set, the result of complex structuring, high dips, tight imbrications, overpressure, reduced acoustic impedance and more. But it is also the result of the accepted paradigm of a massive mud substratum, which ultimately has hindered efforts to improve seismic data (both in acquisition and processing).

Discussion points

- There is still a lingering question regarding the amount of extension updip and its balance with contraction downdip. Toe-thrust shortening is of the order of 25% (Corredor *et al.* 2005). Extension updip is on the order of 10–40% (Restrepo-Pace *et al.* 2006). The discrepancies between the amount of extension and contraction at any given section lie within the translational zone. As stated, this area is the location of complex structures/palaeo-fold belts.
- Characterization of updip extensional structures and their linkage with the contractional toe are fundamental to understanding plumbing (i.e. migration modelling/de-risking). It also provides a template through time to investigate sediment dispersal patterns.
- Cobbold *et al.* (2004, 2009) described the role of elevated pore pressure in controlling deformation and, in particular, the relevance of source-rock maturity in developing detachments and controlling structuration in the outer toe-thrust system. In addition to providing insights into fluid migration, the latter can, in principle, be used to conceptualize and map the extent of effective source rock in the delta or to model the migration of fluid into the toe-thrust area. Ultimately, this concept can be applied in evaluating sensitivities around hydrocarbon maturation, migration and charge in the absence of direct rock data. Source rocks rarely get drilled.

Along-strike variability

Depobelts have been defined to help describe structural and depositional subprovinces of the Niger Delta (e.g. Hooper *et al.* 2002; Krueger & Grant 2013). Additional along-strike variability can be determined from sequential seismic interpretations. The sections investigated here extend further into the extensional domain than those described by Bilotti & Shaw (2005) and therefore can characterize more adequately the Delta taper as a whole. In addition to the depobelts and dip-orientated structural domains, along-strike structural groups can be determined from the Delta taper, fault spacing and variable structural style. Notwithstanding the limitations associated with depth conversion and the apparent dip or orientation of the sections relative to the delta sediment input, the measured taper angles α (bathymetric slope) and β (detachment dip) are relatively low, such as those calculated for the toe-thrust region alone (Bilotti & Shaw 2005). Domain I in Figure 8 is characterized by low taper and domains IIa, IIb–III with relatively higher taper values. Both domains I and IIa, and sections 12 and 13 (IIb) have the extensional domain separated a considerable distance from the compressional toe. This area of separation between the outer and inner deformation fronts comprises low-frequency and low relief structures, often associated with a ramp on the basal detachment. Domain III, with the highest taper, has characteristic extensional and compressional zones closely spaced, and the latter with amalgamated inner and outer deformation fronts. The spread in taper values suggests that the lateral variability is due to variations in the mechanical behaviour of the clastic wedge, most likely to be related to gross lithology changes and variable geo-pressure conditions. The taper and structural style of Domain III are likely to be linked to basal rugosities at the basement level associated with the Charcot Fracture Zone. The deformation zones with a greater complexity lie in the shale diapir/inner thrust belt and, in some instances, the translational domain zones of Krueger & Grant (2013). The complex structures consist of antiformal stacks and tightly spaced thrust imbricate systems. Within this highly structured domain may lie the greatest potential to develop new structural plays. First-order de-risking requires high-quality seismic data to determine more accurately the extent and role of argillokinesis here. In present-day shallow waters and the coastal-swamp area, three-way extensional tilt blocks and deeper, poorly imaged, overpressured zones are being realized as the remaining exploration targets (sections 8–15), but operational considerations related to overpressure have hindered a more aggressive exploration effort. As better seismic data and calibration becomes available, along-strike structural variations can provide insight into uncalibrated geology, which in conjunction with alternative interpretations can lead to the testing of untapped deeper reservoirs. Furthermore, it may

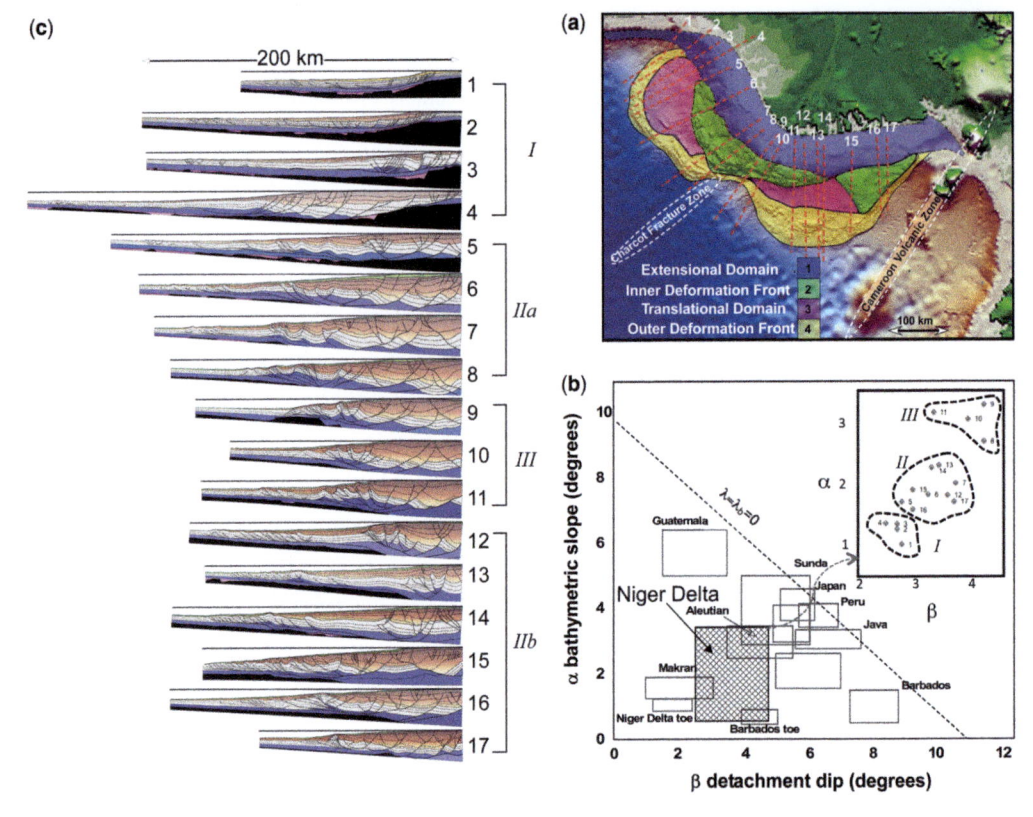

Fig. 8. (a) Structural provinces as along-strike variations derived from wedge-taper measurements (modified from Krueger & Grant 2013). (b) Plot of measured taper angles for 17 sections of the Niger Delta (modified from Bilotti & Shaw 2005) with along-strike structural domains. (c) Along-strike structural variations of the Niger Delta based on wedge taper, structural spacing and style (refer to the text for a further description).

provide 3D predictive insight into fluid plumbing, hydrocarbon charge and phase.

Potential underexplored plays

The translational zone, inner thrust system and the 'shale diapir' zone *sensu* Krueger & Grant (2013) may hold the potential to generate new play concepts in the compressional domain. The majority of interpretations of this area have mostly written off their exploration potential by suggesting these are made mostly of thick mobile shale and/or a possible reservoir bypass zone. It is possible that older sand systems may have been caught up in the palaeo-toe-thrust systems. Stacked thrust systems may allow for stacked pay. Conceptual structural models in such areas of challenged seismic data are needed to entice better seismic acquisition efforts that can lead to the generation of new mappable plays.

In the extensional domain, footwall structures previously regarded as mud escape features can now be interpreted as blocks with coherent layered events when using improved seismic data. New traps are also beginning to emerge in revised interpretations (e.g. 'three-way' fault-dependent closures). This domain has drilling challenges as a result of high overpressures. Again, with improved seismic datasets, geomechanical studies and with the aid of modern pressure-while-drilling tools, new commercial plays in this domain can be brought to fruition.

Conclusions

Inspecting regional datasets and high-effort 3D images provide the interpreter with alternative structural interpretations to the mobile shale scenario. This paper includes all but a very limited sample of data supporting the 'brittle model'. In the long run, the aim is to validate these concepts by acquiring higher-quality data, which can result in upgraded hydrocarbon play and prospect portfolios. Evidence for the 'ductile' or mobile shale model has been supported by the poorly imaged geology of the Niger

Delta and its conceptual connection to overpressure at depth. The 'ductile model' interpretations have been influenced by interpretation bias based on geometrical analogies from salt-dominated provinces. Direct evidence, for the most part, comes for shale diapirism in the form of sea-bottom pockmarks, overpressure and some examples of well-imaged shallow mud pipes.

Evidence for the brittle model as presented here is mostly geometrically based on the reinterpretation of better-resolved dip panels within the otherwise 'mud diapir' chaotic/wipe-out seismic domain. For the Niger Delta play portfolio rejuvenation, there is an iterative process required to break the widely accepted paradigm: that of providing new interpretations on improved-quality seismic data that yield new exploration concepts. The dominantly 'brittle' model presented here provides an alternative template for mapping, modelling and de-risking. A revision of sand dispersal, maturation and migration in the context of the dynamics of an outboard advancing structural system – aided by overpressure – should provide the basis for a new exploration portfolio in a mature province. As interpretation is data-driven, emphasis should be placed on improved imaging of complex zones that entice concept validation via drilling.

Acknowledgement The author would like to thank Neil Grant and Jose de Vera for the thorough revision of the manuscript. Their comments and observations substantially improved this paper.

References

AJAKAIYE, D.E. & BALLY, A.W. 2002. *Manual and Atlas of Structural Styles. Niger Delta*. AAPG Continuing Education Course Notes Series, **41**.

BELLINGHAM, P., CONNORS, C., HAWORTH, R., RADOVICH, B. & DANFORTH, A. 2014. The deepwater Niger Delta: an unexplored world-class petroleum province. *GeoExpro*, **11**, 54–56.

BILOTTI, F. & SHAW, J.H. 2005. Deep-water Niger Delta fold and thrust belt modeled as a critical-taper wedge; the influence of elevated basal fluid pressure on structural styles. *AAPG Bulletin*, **89**, 1475–1491.

COBBOLD, P.R., MOURGUES, R. & BOYD, K. 2004. Mechanism of thin-skinned detachment in the Amazon fan: assessing the importance of fluid overpressure and hydrocarbon generation. *Marine and Petroleum Geology*, **21**, 1013–1025.

COBBOLD, P.R., CLARKE, J.B. & LOSETH, H. 2009. Structural consequences of fluid overpressure and seepage forces in the outer thrust belt of the Niger Delta. *Petroleum Geoscience*, **15**, 3–15, https://doi.org/10.1144/1354-079309-784

CORREDOR, F., SHAW, J.F. & BILOTTI, F. 2005. Structural styles in the deep-water fold and thrust belts of the Niger Delta. *AAPG Bulletin*, **89**, 753–780.

DAY-STIRRAT, R.J., MCDONNELL, A. & WOOD, L.J. 2010. Diagenetic and seismic concerns associated with interpretation of deeply 'mobile shales'. *In*: WOOD, L. (ed.) *Shale Tectonics*. AAPG Memoirs, **93**, 5–27.

DEVILLE, E., BATTANI, A. *ET AL.* 2003. Processes of mud volcanism in the Barbados–Trinidad compressional system: New structural, thermal and geochemical data. Search and discovery article #30017, AAPG Annual Meeting, May 11–14, 2003, Salt Lake City, Utah, USA.

DEVILLE, E., GUERLAIS, S.H. *ET AL.* 2006. Liquefied vs stratified sediment mobilization processes: insight from the south of the Barbados accretionary prism. *Teconophysics*, **428**, 33–47.

DOUST, H. & OMATSOLA, E. 1989. Niger Delta. *In*: EDWARDS, J.D. & SANTOGROSSI, P.A. (eds) *Divergent/Passive Margins Basins*. AAPG Memoirs, **48**, 201–238.

HOOPER, J.R., FITZSIMMONDS, R.J., GRANT, N. & VENDEVILLE, B.C. 2002. The role of deformation in controlling depositional patterns in the south-central Niger Delta, West Africa. *Journal of Structural Geology*, **24**, 847–859.

KRUEGER, W.S. & GRANT, N.T. 2013. The growth history of toe thrusts of the Niger Delta and the role of pore pressure. *In*: MCCLAY, K., SHAW, J. & SUPPE, J. (eds) *Thrust Fault-Related Folding*. AAPG Memoirs, **94**, 357–390.

LACOSTE, A., VENDEVILLE, B.C., MOURGUES, R., LONKTE, L. & LEBACQ, M. 2012. Gravitational instabilities by fluid overpressure and downslope incision – insights from analytical and analog modeling. *Journal of Structural Geology*, **42**, 151–162.

MCCLAY, K., DOOLEY, T. & ZAMORA, G. 2003. Analogue models of delta systems above ductile substrates. *In*: VAN RENSBERGEN, P., HILLIS, R.R., MALTMAN, A.J. & MORLEY, C.K. (eds) *Subsurface Sediment Mobilization*. Geological Society, London, Special Publications, **216**, 411–428, https://doi.org/10.1144/GSL.SP.2003.216.01.27

MORLEY, C.K. 2003. Mobile shale-related deformation in large deltas developed on passive and active margins. *In*: VAN RENSBERGEN, P., HILLIS, R.R., MALTMAN, A.J. & MORLEY, C.K. (eds) *Subsurface Sediment Mobilization*. Geological Society, London, Special Publications, **216**, 335–357, https://doi.org/10.1144/GSL.SP.2003.216.01.22

MORLEY, C.K. & GUERIN, G. 1996. Comparison of gravity driven deformation styles and behaviour associated with mobile shales and salt. *Tectonics*, **15**, 1154–1170.

MORLEY, C.K., KING, R., HILLIS, R., TINGAY, M. & BACKE, G. 2011. Deepwater fold and thrust belt classification, tectonics, structure and hydrocarbon prospectivity: a review. *Earth-Science Reviews*, **104**, 41–91.

MOURGUES, R., LECOMTE, E., VENDEVILLE, B. & RAILLARD, S. 2009. An experimental investigation of gravity-driven shale tectonics in a progradational delta. *Tectonophysics*, **474**, 643–656.

PASTOURET, L., CHAMLEY, H., DELIBRIAS, G., DUPLESSY, J.-C. & THIEDE, J. 1978. Late Quaternary deep-sea sedimentation off the Niger Delta. *PANGAEA*, https://doi.org/10.1594/PANGAEA.707458, supplement to Pastouret, L., Chamley, H., Delibrias, G., Duplessy, J.-C. & Thiede, J. 1978. Late Quaternary climatic changes in western tropical Africa deduced from deep-sea sedimentation off Niger Delta. *Oceanologica Acta*, **1**, 217–232.

RESTREPO-PACE, P.A., FILBRANDT, J., LAPIDO, K., ONICHABOR, G.S. & RICHARD, P. 2006. Structural styles II; deltas. Global exploration newsletter. *Shell Exploration and Production*, **2**, 3–8.

ROUBY, D., NALPAS, T., JERMANNAUD, P., ROBIN, C., GUILLOCHEAU, F. & RAILLARD, S. 2011. Gravity driven deformation controlled by the migration of the delta front: the Plio-Pleistocene of the eastern Niger Delta. *Tectonophysics*, **513**, 54–67.

ROWAN, M.G., PEEL, F.G. & VANDERVILLE, B.C. 2004. Gravity driven fold belts on passive margins. Thrust tectonics and hydrocarbon systems. *In*: MCCLAY, K.R. (ed.) *Thrust Tectonics and Hydrocarbon Systems*. AAPG Memoirs, **82**, 157–182.

SAPIN, F., RINGENBACH, J.C., RIVES, T. & PUBELLIER, M. 2012. Counter-regional faults in shale-dominated deltas: Origin, mechanism, and evolution. *Marine and Petroleum Geology*, **37**, 121–128.

SUPPE, J. 2011. Mass balance and thrusting in detachment folds. *In*: MCCLAY, K., SHAW, J. & SUPPE, J. (eds) *Thrust-Related Folding*. AAPG Memoirs, **94**, 21–37.

WEINER, R.W., MANN, M.G., ANGELICH, M.T. & MOLYNEUX, J.B. 2010. Mobile shale in the Niger Delta: Characteristics, structure, and evolution. *In*: WOOD, L. (ed.) *Shale Tectonics*. AAPG Memoirs, **95**, 145–161.

WU, S. & BALLY, A.W. 2000. Slope tectonics – Comparison and contrasts in structural styles of salt and shale tectonics in the Gulf of Mexico with Shale Tectonics in the Niger Delta in the Gulf of Guinea. *In*: MOHRIAK, W. & TALWANI, M. (eds) *Atlantic Rifts and Continental Margins*. American Geophysical Union, Geophysical Monographs, **115**, 151–172.

Orthorhombic faulting in the Beagle Sub-basin, North West Shelf, Australia

K. D. McCORMACK[1]* & K. R. McCLAY[2]

[1]*Woodside Energy Ltd, 240 St Georges Terrace, Perth, WA 6000, Australia*

[2]*Fault Dynamics Research Group, Geology Department, Royal Holloway, University of London, Egham, Surrey TW2 0EX, UK*

K.D.M., 0000-0002-2034-907X

Correspondence: kenneth.mccormack@woodside.com.au

Abstract: The Beagle Platform forms a structurally complex outboard boundary to the Mesozoic Beagle Sub-basin in the Northern Carnarvon Basin in 50–1000 m of water *c.* 250 km offshore on Australia's North West Shelf. North- and NE-trending conjugate fault systems link to form *c.* 5–10 km-wide rhomboidal horsts bound by *c.* 5 km-wide graben in complex orthorhombic symmetry.

Interpretation of the Canning TQ3D three-dimensional (3D) seismic survey identified four populations of faults comprising: (1) latest Triassic–Early Cretaceous north–south-striking normal faults (Fault Population I); (2) latest Triassic–Early Cretaceous NE–SW-striking normal faults (Fault Population II); (3) Cretaceous polygonal faults (Fault Population III); and (4) Neogene–present-day north–south- and NE–SW-trending en echelon conjugate fault arrays (Fault Population IV). Structural interpretation, comprising fault seismic interpretation, displacement analyses and fault orientation analysis, illustrates that orthorhombic extensional faulting occurred penecontemporaneously.

Three-dimensional non-plane strain with non-zero intermediate (σ_2) extension magnitude controls the near-synchronous displacement of fault population I and II conjugate fault sets in orthorhombic symmetry to create the characteristic rhomboidal fault geometry. Neogene–present-day north–south- and NE–SW-striking en echelon conjugate fault arrays (Fault Population IV) form in response to oblique reactivation of these subjacent latest Triassic–Early Cretaceous polymodal faults (Fault Population I and Fault Population II). Fault Population I, II and IV together form a vertically decoupled (soft-linked) pseudo-conjugate fault system partitioned by a Cretaceous interval characterized by polygonal fault systems (Fault Population III) and monoclinic draping.

The structural interpretation of the Beagle Platform illustrates a seismic-scale orthorhombic fault symmetry accommodating 3D strain and the insufficiency of plane-strain 'Andersonian' conjugate fault theory to resolve complex polymodal faulting evolved penecontemporaneously under a single stress regime. Subsequent oblique extension reactivation of these polymodal fault systems also demonstrates that complex soft-linked en echelon extensional conjugate fault arrays form in response to a single stress regime. This research supports the evolution of complex polymodal faulting under simple states of stress, and has implications for understanding the distribution, linkage and age of extensional faults within the Beagle Sub-basin and other extensional basins.

Faults in brittle rock form within complex active process zones by coalescence of randomly orientated and distributed tensile and shear microfractures that propagate to form macrofractures. The Griffith criterion (Griffith 1924), Coulomb–Mohr failure criterion (Anderson 1951) and slipline theory of plasticity (Odé 1960) models of fault development assume simplified axisymmetrical strain boundary conditions comprising principle stress/strain axes aligned vertically and faults that align with principal stress/strain axes. These models predict the development of two fault sets in conjugate (bimodal) symmetry parallel to the intermediate stress axis (σ_2); preclude strain due to slip in the direction of the intermediate stress axis (σ_2); and result in plane strain accommodated by deformation parallel to the maximum (σ_1) and minimum (σ_3) stress directions (Fig. 1a). However, these experiments apply axial stress (σ_1) greater than the balanced radial stress ($\sigma_2 = \sigma_3$) applied through a confining jacket to a cylindrical geometry that precludes explanation of three-dimensional (3D) polymodal fault patterns widely observed in rocks subjected to 3D stress states ($\sigma_1 > \sigma_2 > \sigma_3$). Triaxial experiments of Aydin & Reches (1982) and Reches & Dietrich (1983) with $\sigma_1 > \sigma_2 > \sigma_3$ generated polymodal shear fractures with varying degrees of obliquity with respect to the principal stress axes and magnitudes of principal strain rates in both sedimentary and igneous rock. Modelling by Oertel (1965), Aydin (1977), Reches (1978), Reches & Dietrich (1983), Mitra (1979), Aydin & Reches (1982), Krantz (1986) and Healy *et al.* (2006) demonstrated that four penecontemporaneous fault sets in orthorhombic symmetry

From: McCLAY, K. R. & HAMMERSTEIN, J. A. (eds) 2020. *Passive Margins: Tectonics, Sedimentation and Magmatism.* Geological Society, London, Special Publications, **476**, 205–230.
First published online March 23, 2018, https://doi.org/10.1144/SP476.3
© 2018 The Author(s). Published by The Geological Society of London. All rights reserved.
For permissions: http://www.geolsoc.org.uk/permissions. Publishing disclaimer: www.geolsoc.org.uk/pub_ethics

(a) Conjugate Fault Plane Strain

(b) Orthorhombic Fault 3D Strain

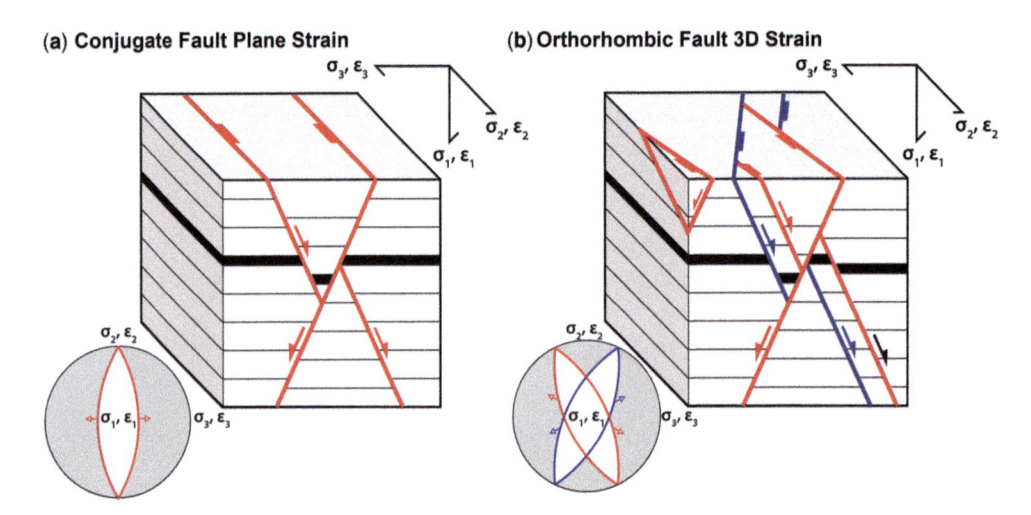

Fig. 1. Model for (**a**) plane-strain conjugate fault symmetry about σ_1 that bisects the acute dihedral angle to intersect parallel to σ_2 following Anderson (1951); compared to (**b**) non-plane-strain (3D) orthorhombic fault symmetry about both σ_1 and σ_2 bisecting the acute dihedral angle of two conjugate failure planes attributed to the subjection of rock to complex stress that manifests as $\sigma_1 > \sigma_2 > \sigma_3$ where significant (non-zero) intermediate or minimum horizontal stress (σ_2) exists contemporaneously with maximum horizontal stress (σ_3). Arrows indicate slip directions in both diagrams.

relative to the principal stress/strain axes develop under general non-plane 3D strain. Furthermore, the odd-axis model (Krantz 1986) quantifies this relationship by linking fault set orientation and the differential strain ratio (k) between the maximum (σ_1) and intermediate (σ_2) strain directions subject to 3D strain that predicts two fault sets in conjugate symmetry as two end members within a continuum of orthorhombic fault symmetry. Therefore, plane-strain conditions producing two fault sets in conjugate symmetry represent exceptional end-member approximations and 3D strain conditions producing four coeval fault sets in orthorhombic fault symmetry characterizes the general deformation of rocks (Fig. 1b).

Field and laboratory investigations confirm that polymodal fault sets forming orthorhombic arrays exist at the microfracture scale (<1 mm: e.g. Oertel 1965; Aydin 1977; Reches 1978; Reches & Dietrich 1983; Mitra 1979; Aydin & Reches 1982) and mesofracture scale (1–10 mm: e.g. Krantz 1986, 1988, 1989; Healy et al. 2006). However, observational paucity of macrofractures (100–10 000 m) comprising seismically resolvable faults (Krantz 1989; Oesterlen & Blenkinsop 1994; Sagy et al. 2003; Miller et al. 2007) limits the universal scale-independent application of the non-plane-strain theory of structural evolution.

This study constrains the geometrical and temporal evolution of an orthorhombic fault array on the Beagle Platform situated between the inboard Beagle Sub-basin and outboard northern Exmouth Plateau on the North West Shelf (NWS) of Australia (Fig. 2). This offshore platform forms a NE-orientated regional structural high on the NW margin of a Late Triassic–Early Cretaceous failed intra-cratonic rift that developed into a passive margin following the break-up of Australia and Greater India. Analysis of 3D seismic reflection data (the Canning TQ3D) characterizes the geometry, displacement and linkages of Triassic–Cretaceous and superjacent Neogene faults to demonstrate the penecontemporaneous evolution of four fault sets in orthorhombic symmetry. These analyses illustrate the development of complex polymodal strain phenomena under simple states of stress ($\sigma_1 > \sigma_2 > \sigma_1$) that precludes polyphase deformation and may be applied to the evaluation of fault-architecture evolution in other extensional basins.

Location and geological setting

The Mesozoic Beagle Sub-basin covers c. 30 000 km^2 offshore, and comprises a series of north-trending horsts and graben bifurcated by NE-striking faults found SW of NW-striking transform faults defining a diffuse continent–ocean boundary partitioning the Middle Jurassic Argo Abyssal Plain from the latest Jurassic–Lower Cretaceous Gascoyne Abyssal Plain (Blevin et al. 1994). The Argo Abyssal plain to the north, Roebuck Basin to the NE, Pilbara Craton to the south, Dampier Sub-basin to the SW and Exmouth Plateau to the west bound the Beagle Sub-basin (Fig. 2). The Beagle Platform comprises the peripheral and structurally transitional

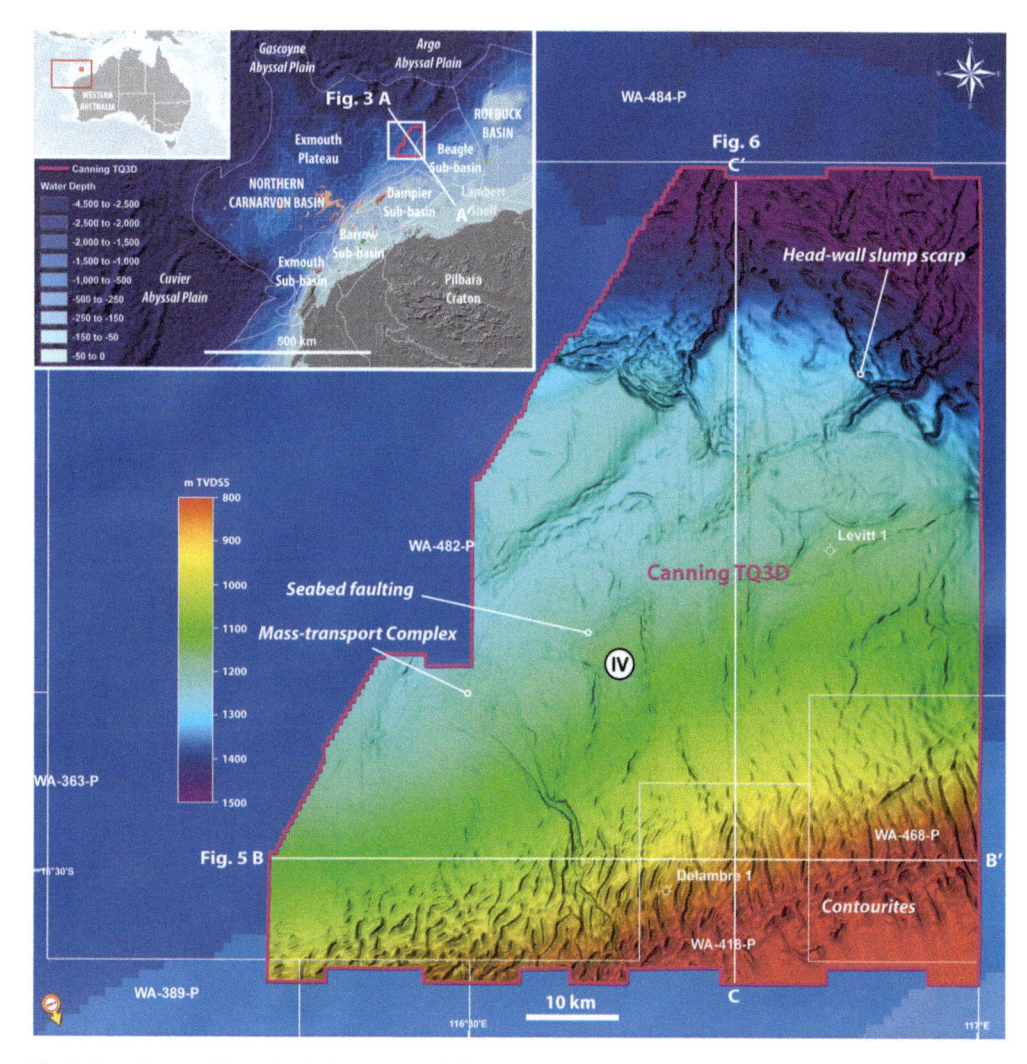

Fig. 2. Location map illustrating the bathymetry within the area-of-interest imaged by the Canning TQ3D seismic survey; regional present-day bathymetry and location (shown in the inset); and the locations of geoseismic sections A–A′ (Fig. 3), B–B′ (Fig. 5) and C–C′ (Fig. 6). m TVDSS, metres true vertical depth subsea.

NW boundary of the Beagle Sub-basin characterized by the cessation of north–south-trending faults (Blevin et al. 1994; Geoscience Australia 2010). Figure 3 illustrates a regional composite seismic section (A–A′) imaging the Beagle Platform bound by the inboard Lambert Shelf and outboard Exmouth Plateau that displays normal faults accommodating the dominantly extensional margin.

A small oil accumulation at Nebo 1 (Osborne 1994), and minor shows in Picard 1 and Depuch 1, support the presence of a petroleum system within the Beagle Sub-basin. Principal reservoir facies occur within complex structural traps formed by Jurassic–Cretaceous fault systems subsequently reactivated in the Neogene (Blevin et al. 1994).

Neogene reactivation of the pre-existing structural fabric reduces fault-seal and trap integrity elsewhere on the NWS (Gartrell et al. 2005; Bailey et al. 2006). Fault reactivation may compromise fault-seal integrity and therefore presents a major exploration risk within the Beagle Sub-Basin. Development of a detailed evolutionary model that examines Mesozoic–Cenozoic fault inception, distribution, segmentation and linkage remains critical in reducing exploration uncertainty.

Basin evolution

Australia's NWS comprises the offshore Bonaparte, Browse, offshore Canning and Northern Carnarvon

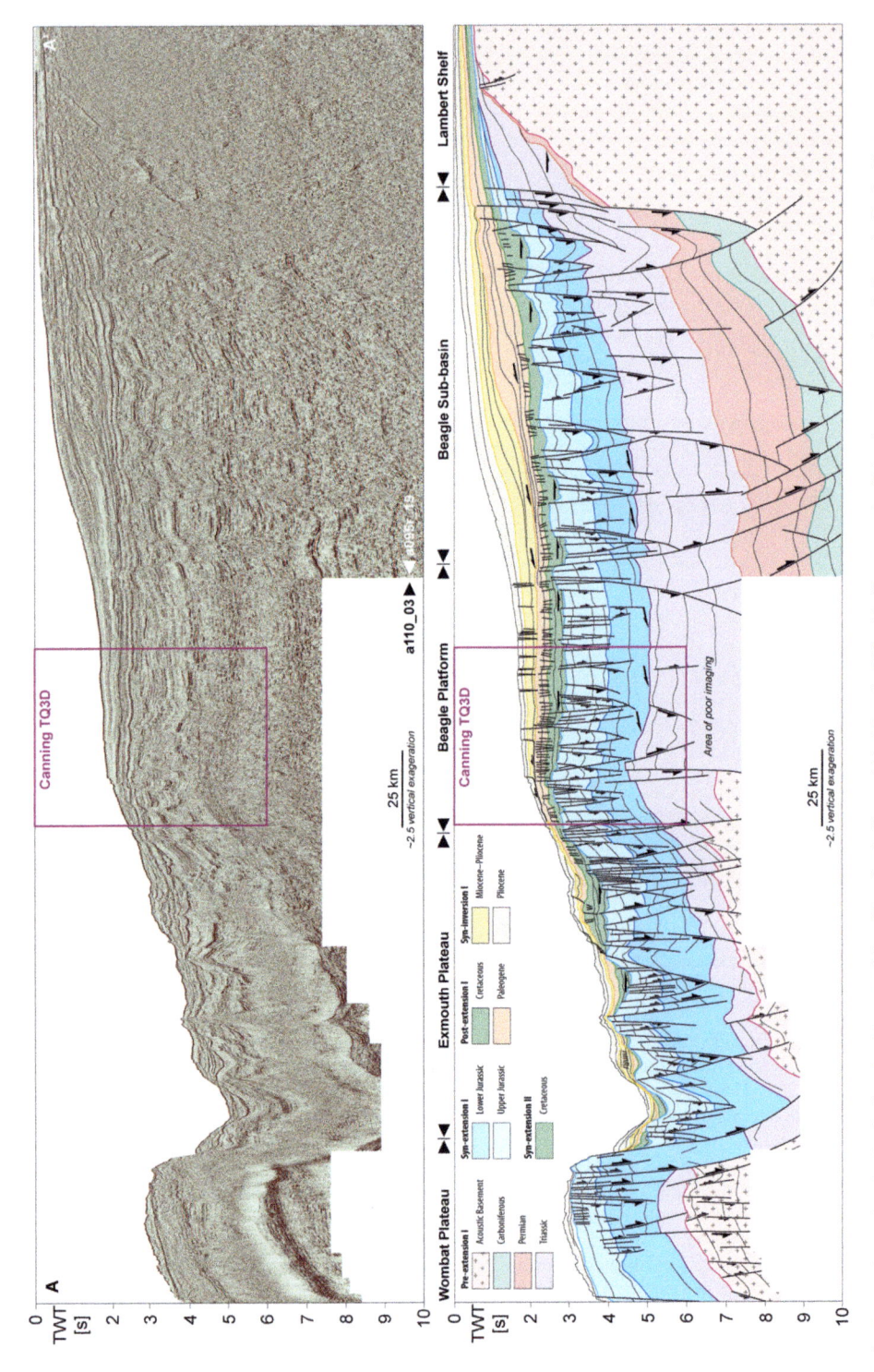

Fig. 3. Composite regional seismic line A–A', comprising 2D seismic lines a110_03 and a095r_19, illustrates the Triassic pre-extensional, Jurassic–Early Cretaceous syn-extensional and Cretaceous–present-day post-extensional architecture across a Lambert Shelf–Beagle Sub-basin–Exmouth Plateau–Wombat Plateau transect. Figure 2 illustrates the location of this seismic transect.

basins, which were collectively termed the Westralian Super-basin by Yeates *et al.* (1987). The Northern Carnarvon Basin (NCB) was initiated during the Permian–Triassic break-up of Pangea and the contemporaneous opening of the Neotethys Ocean (Driscoll & Karner 1998; Golonka *et al.* 2006). Late Triassic–Early Cretaceous extension created the Beagle Sub-basin that was subsequently reactivated obliquely during Neogene deformation (Mitchelmore & Smith 1994; Driscoll & Karner 1998; Longley *et al.* 2002). The chronostratigraphy of the Beagle Sub-basin follows Marshall & Lang (2013), with sequence intervals defined by transgressive surfaces (TS) or sequence boundaries (SB) (Fig. 4).

Triassic pre-extension

Middle–Late Triassic (TR10–TR20) subsidence accommodated >3000 m-thick Anisian–Norian (TR10–TR20) post-Permian extension intervals characterized by a regressive transition from marine mudstone (Locker Shale) to fluvial-deltaic (Mungaroo Formation) strata hosting subordinate fining-upwards sequences (Kopsen & McGann 1985; Blevin *et al.* 1994; Driscoll & Karner 1998; Longley *et al.* 2002). An Anisian–Ladinian (TR10) marine transgression comprising regional carbonate deposition (Cossigny Member) punctuates this otherwise regressive infilling of the purported extant Paleozoic intra-cratonic rift (Fig. 3). These sequences effectively comprise the pre-extension megasequence within the Jurassic Beagle Sub-basin.

Latest Triassic–Late Jurassic extension

Rhaetian (TR30.1TS)–Oxfordian (J40.0SB) extension initiated the en echelon intra-cratonic Exmouth, Dampier, Barrow and Beagle sub-basins (Driscoll & Karner 1998; Longley *et al.* 2002). The regional Oxfordian Unconformity (J40.0SB) demarks the rift cessation attributed to the inception and spreading of the Argo seafloor at *c.* 155 Ma, separating a continental fragment (Argoland) from Australia (Longley *et al.* 2002; Heine & Müller 2005). The Delambre 1 (Woodside 1981) well penetrates Rhaetian (TR30)–Oxfordian (J40) intervals that records an abrupt marine transgression from fluvial-deltaic coarse clastics (TR20) to ooltic limestone (TR30); followed by a deltaic progradation that manifests as coarsening-upwards marine siliciclastic (J10–J40) environments of deposition (Woodside 1981; Stephenson *et al.* 1998).

Latest Jurassic–Early Cretaceous extension

Renewed Late Jurassic (J50.0SB)–Early Cretaceous (K20.0SB) extension, attributed to obliquely reactivated pre-existing faults, accommodated nominal Kimmeridigian–Valanginian (J40–K20) hanging-wall growth wedges adjacent to peneplaned structural highs (Blevin *et al.* 1994; McCormack & McClay 2013; Black *et al.* 2017). The Delambre 1 well penetrates a condensed Middle Jurassic–Lower Cretaceous interval comprising composite unconformities removing *c.* 15 myr of Callovian (J30) and Berriasian (K10) intervals preserved within adjacent graben, and characterized by transgressive prograding fan deltas on a broad shallow-marine shelf (Stephenson *et al.* 1998). The Late Jurassic–Early Cretaceous cessation of extension and correlative transgression resulted in the drowning of emergent and/or shoaling structural relief (i.e. Delambre 1 horst) and progressive coastal retreat to establish middle-shelf environments of deposition (Woodside 1981; Stephenson *et al.* 1998). The Valanginian (K20.0SB) Unconformity demarks the cessation of late syn-extension sedimentation within the Beagle Sub-basin attributed to the break-up of Greater India from Australia.

Early Cretaceous–Late Cretaceous post-extension

Early–Late Cretaceous (K30.2MFS–K60.0SB) marine transgression, attributed to post-extension subsidence, resulted in non-calcareous claystone (Muderong Shale) deposition preceding hemipelagic marls and carbonates upon establishment of the incipient Indian Ocean circulation. The structurally high Beagle Platform precluded deposition of uniformly thick post-rift succession, with <300 m of Cretaceous section penetrated at the Delambre 1 well location (Woodside 1981).

Cenozoic–present-day passive margin

The Late Miocene–Pliocene (8–3 Ma: Keep *et al.* 2002) collision of the Australian and SE Asian plates reactivated pre-existing fault systems and interrupted progradational carbonate wedge deposition (Longley *et al.* 2002). This deformation significantly impacted the integrity and hydrocarbon retention in extant traps (e.g. Castillo *et al.* 2000; Keep *et al.* 2002; Gartrell *et al.* 2005; Frankowicz & McClay 2010). Hemipelagic carbonate environments of deposition dominate the Pliocene–present-day passive margin of the Beagle Sub-basin within the NCB.

Data and research methodology

The Canning TQ3D survey acquired between 1999 and 2000 images 4385 km^2, which comprises 26 676 line kilometres of subsurface data in 500–1500 m of water depth spanning portions of the WA-482-P, WA-418-P and WA-468-P permits of the NCB (WesternGeco 2000) (Fig. 2). The

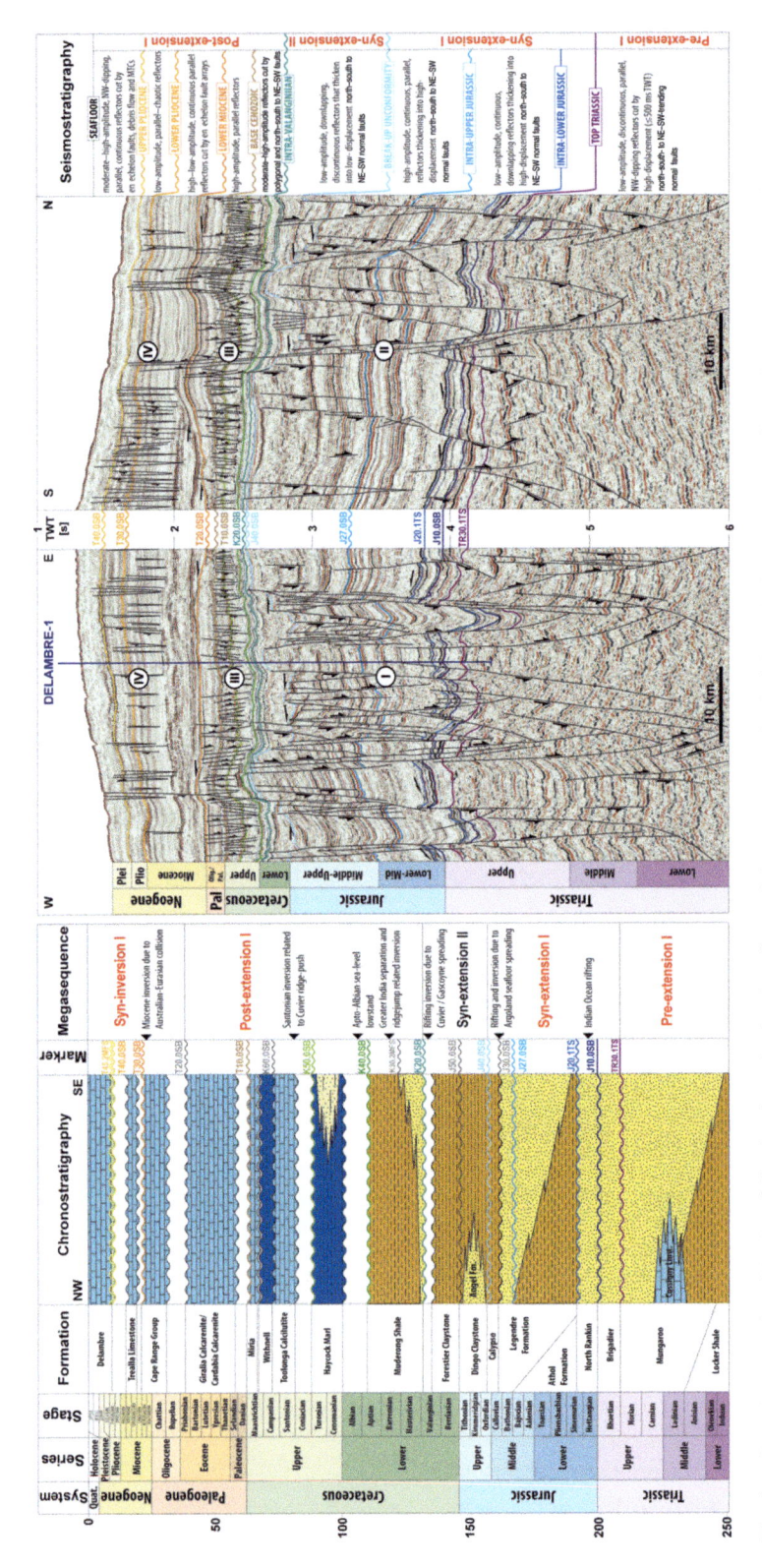

Fig. 4. Seismochronostratigraphic chart following the Delambre 1 well completion report convolved with Marshall & Lang (2013) illustrating the lithostratigraphic nomenclature; sequence stratigraphic nomenclature; tectonic megasequence nomenclature; and the east–west and north–south seismic manifestation of these sequences within the Canning TQ3D seismic survey.

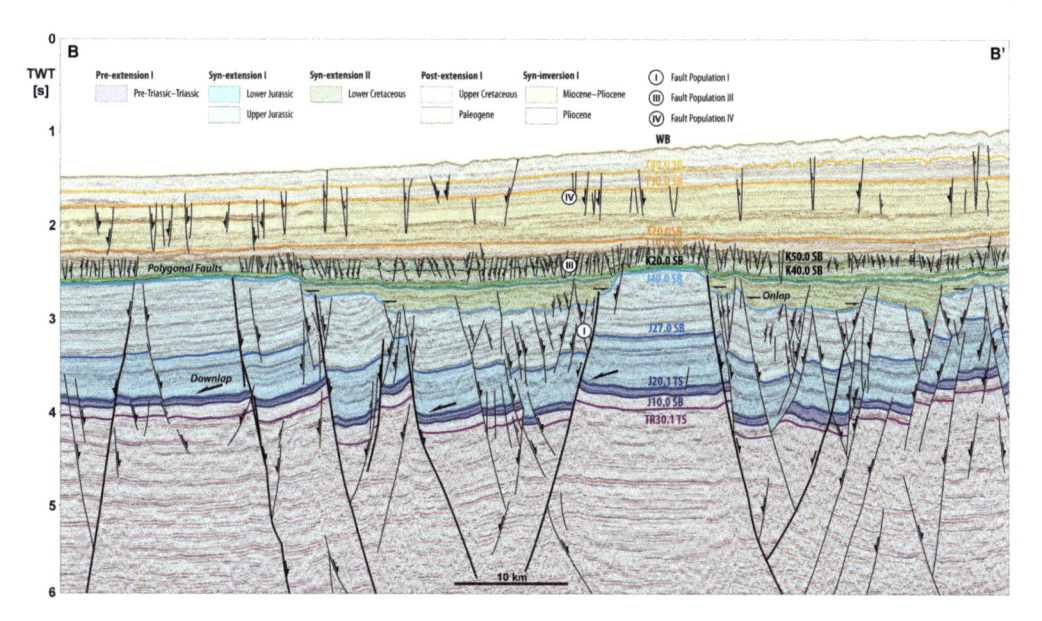

Fig. 5. Geoseismic section B–B' of cross-line 1900 illustrating the structural and stratigraphic architecture in an east–west orientation within the Canning TQ3D seismic survey. TWT, two-way time.

post-stack time-migrated data comprise 4628 in-lines and 6789 cross-lines at 16 m line spacing, extending to 6000 ms two-way travel-time (TWT). Seven 4600 m streamers with a 12.5 m group interval and dual 2368 cubic inch (cu. in.) airgun source arrays recorded flip/flop for 7000 ms at a 2 ms sampling rate with 180 Hz/72 dB Hi-cut and 3 Hz/18 dB low-cut analogue filters at 25 m (50 m per array) in-line and cross-line common depth point spacing. Frequencies of 40 and 32 Hz dominate the Cenozoic and Cretaceous sequences, and provide a maximum vertical seismic resolution ($\lambda/4$) of 19 m at 3100 m s^{-1} and 20 m at 2500 m s^{-1}, respectively. Frequencies of 28 and 20 Hz dominate the Jurassic and Triassic sequences, and provide a maximum vertical seismic resolution ($\lambda/4$) of 36 m at 4000 m s^{-1} and 53 m at 4200 m s^{-1}, respectively. The Delambre 1 well provides petrophysical, lithological, age and velocity data that quantitatively constrain the interpretation of the Canning survey. The presence of water-bearing sequences in Delambre-1 corresponds with an absence of proximal seismically observed direct hydrocarbon indications. Depth conversion of the seismic data utilized an interval velocity model derived from the Delambre 1 velocities.

Interpretation of the nine horizons comprising one conformity, four regional unconformities and four paraconformities using Schlumberger's Geoframe 4.4 and PETREL 2016 software constrains the Pre-extension, Syn-extension I, Syn-extension II and Post-extension/Syn-inversion megasequences (Figs 4–6). Megasequences host unique reflector amplitudes, reflector continuity, depositional architecture and fault structure. Amplitude, coherency, reflection dip and reflection azimuth attributes calculated from depth-converted seismic facilitated construction of structure and isopach maps used to define the geological evolution within the Canning survey. Orientation analyses were produced using Badleys T7 software and plotted using Stereonet 9.5 (Allmendinger et al. 2012; Cardozo & Allmendinger 2013).

Two-dimensional calculations of fault displacement are expressed as an expansion index (EI) following Thorsen (1963), defined simply by:

$$\text{expansion index (EI)} = \frac{\text{hanging-wall thickness}}{\text{footwall thickness}}.$$

Indices indicate the magnitude of thickness change across faults, with >1.0 demonstrating hanging-wall expansion. Depth measurements were plotted against the centre of the footwall for both growth intervals and displacement measurements.

These detailed analyses of the Canning survey illustrate two dominant strike orientations of high-displacement (>100 m) Triassic–Lower Cretaceous faults (Fault Population I and Fault Population II); a layer-bound low-displacement (<15 m) Cretaceous–Paleogene polygonal fault system; and en echelon conjugate Neogene–present-day fault arrays with bathymetric expression (Fault Population IV) (Figs 5 & 6).

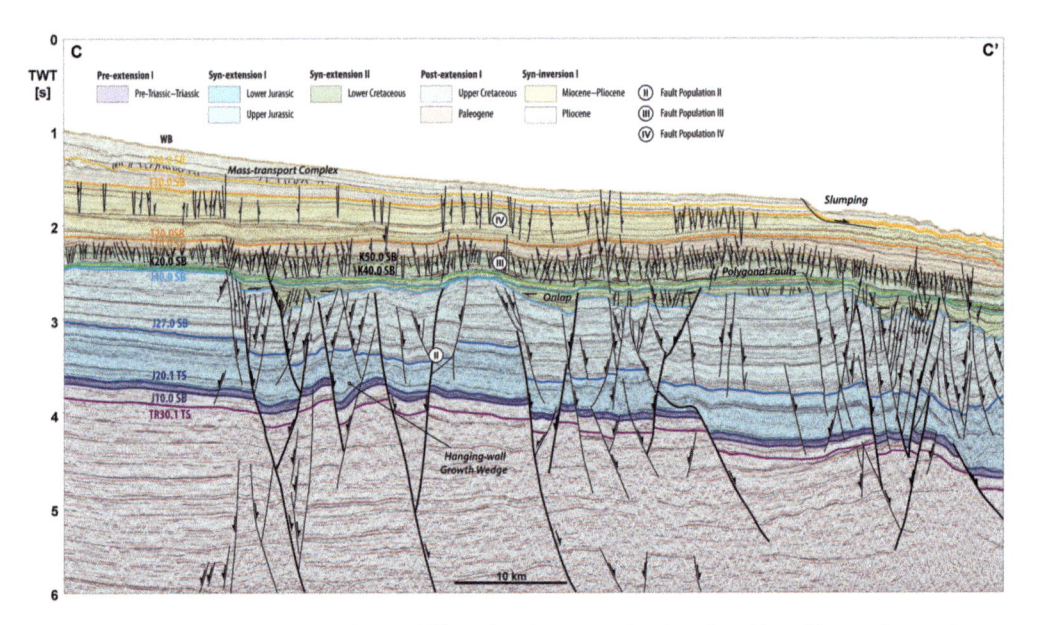

Fig. 6. Geoseismic section C–C' of in-line 4900 illustrating the structural and stratigraphic architecture in a north–south orientation within the Canning TQ3D seismic survey. TWT, two-way time.

Seismic stratigraphy

Seismostratigraphic evaluation of the Canning survey identified five tectonic megasequences (Fig. 4) with stratal geometries that record the Triassic–present-day tectonostratigraphic evolution and comprise:

- The Pre-extension Megasequence: Palaeozoic–Late Triassic (TR30.1TS) isopachous fluvial-deltaic sequences attributed to regional post-Permian-extension subsidence.
- The Syn-extension I Megasequence: Latest Triassic (TR30.1TS)–Upper Jurassic (J40.0SB) transgressive marine and progradational deltaic sequences hosting syn-extension geometries.
- The Syn-extension II Megasequence: Upper Jurassic (J40.0SB)–Lower Cretaceous (K20.0SB) transgressive marine sequences hosting syn-extension geometries attributed to renewed extension.
- The Post-extension Megasequence: Lower Cretaceous (K20.0SB)–Eocene (T20.0SB) marine transgressive mudstone and marl attributed to post-extension subsidence.
- The Syn-inversion Megasequence: Eocene (T20.0SB)–present-day (WB) progradational carbonate sequences modified by normal to oblique reactivation attributed to regional contraction.

The Triassic Pre-extension Megasequence

Thick (≤40 m), semi-continuous, parallel reflectors characterize the Pre-extension Megasequence within the Canning survey. Reflectors dip gently (<5°) west in the western part of the survey; remain subhorizontal in the central part of the survey; dip east in the eastern part of the survey; and dip progressively steeper from south to north (0°–15°: Fig. 5). North–south-trending, east–west-dipping and NE-striking, NW–SE-dipping keystone faulting (Fig. 7) rotates and variably offsets (>300 m) reflectors within this megasequence (Figs 5 & 6). Hanging-wall reflector thickening isolated to NE-striking, NW-dipping faults in the northern periphery of the survey records restricted Upper Triassic growth. Delambre 1 penetrates 1208 m of Upper Triassic strata characterized by claystone, siltstone and minor sandstone intervals below potentially sealing Upper Triassic shale. The Triassic Pre-extension Megasequence represents deposition of a basinwards-dipping distal delta on the periphery of an intra-cratonic sag basin attributed to Paleozoic extension.

The Late Triassic–Late Jurassic Syn-extension I Megasequence

Thin (≤20 m), continuous, parallel reflectors, downlapping onto the subjacent Rhaetian conformity (TR30.1TS) and truncated by a low-angle intersection with the superjacent Oxfordian Unconformity (J40.0SB) characterize the Syn-extension I Megasequence. Lateral amplitude variation and hanging-wall reflector thickening within keystone graben bound by north–south-trending, east–west-dipping and NE–SW-trending, NW–SE-dipping faults (Fig. 8) record syn-extension growth (Figs 5 & 6). To the north,

Fig. 7. Rhaetian (TR30.1TS) top Pre-extension I Megasequence depth structure map illustrating the plan-view rhombohedral horsts bound by north- and NE-striking conjugate Fault Population I and II fault sets. The Delambre 1 and Levitt 1 wells penetrated horsts. Fault linkages accommodated displacement by hard-linkage at fault intersections and by soft-linkage observed as relay ramps. The structural interpretation illustrates synclinal folding within narrow north- and NE-trending graben, and low-displacement intra-horst faults sympathetic to high-displacement inter-horst faults.

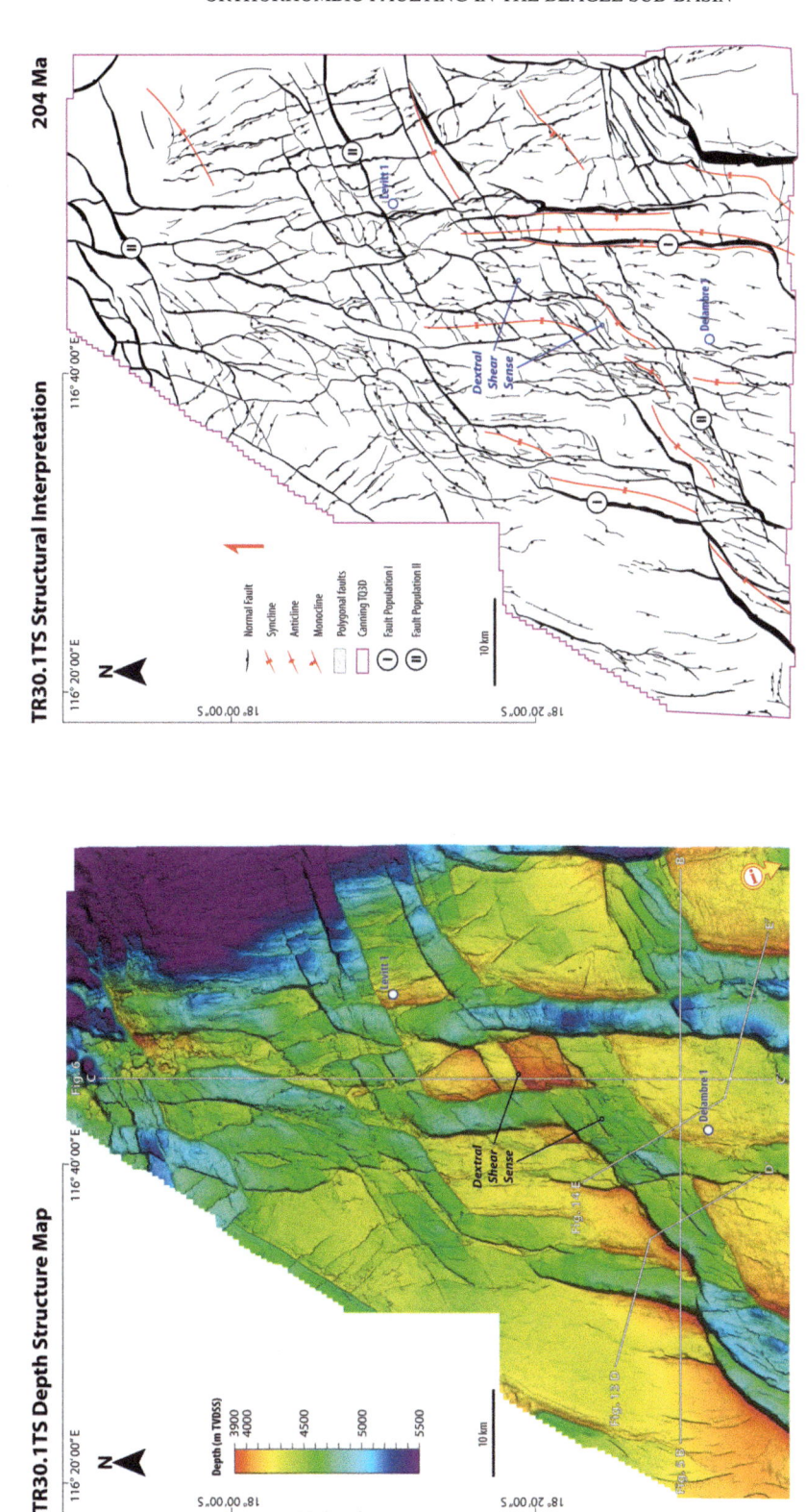

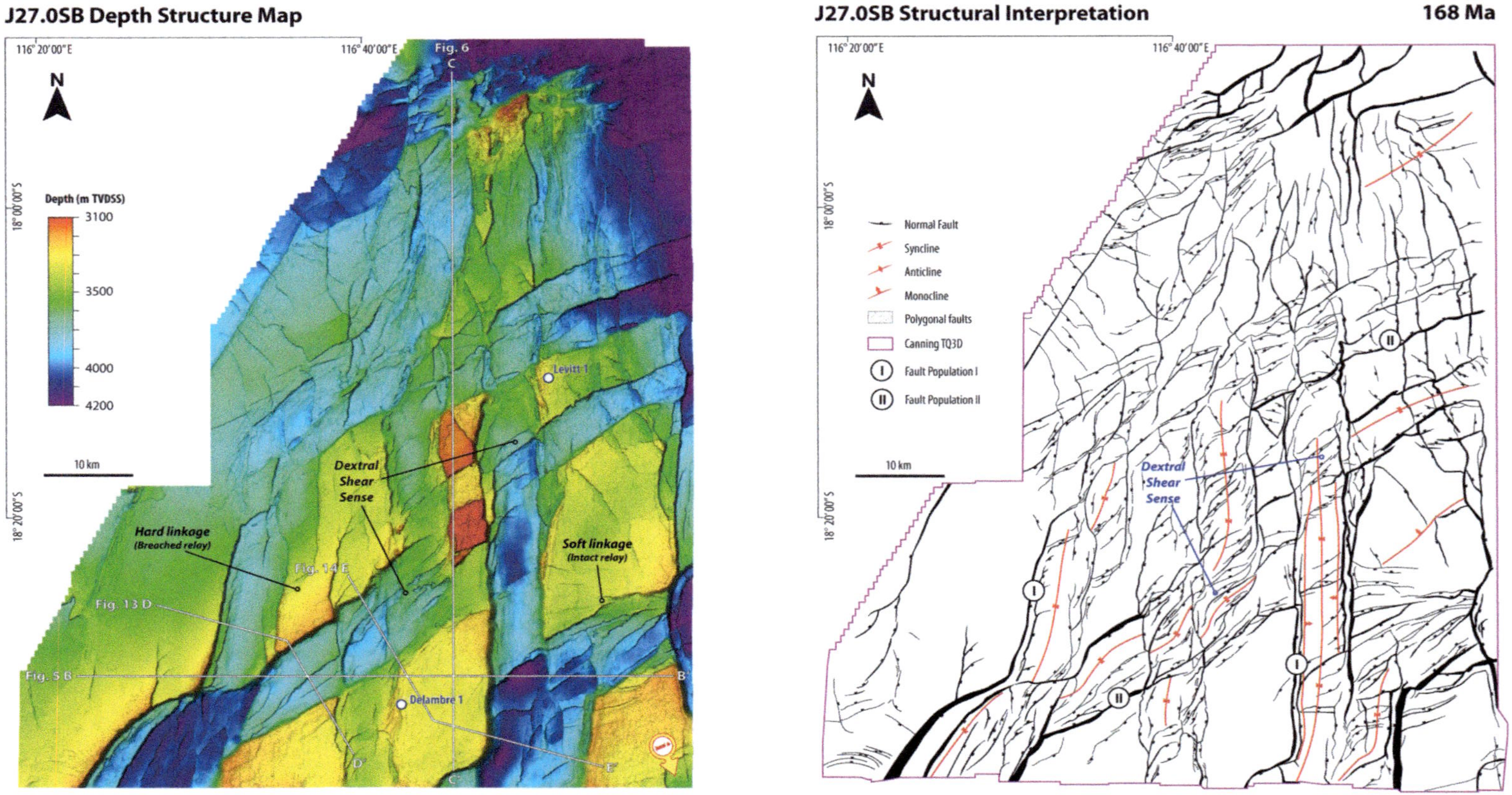

Fig. 8. Bajocian (J27.0SB) intra Syn-extension I Megasequence depth structure map illustrating plan-view rhombohedral horsts bound by north- and NE-striking conjugate fault sets. Graben appear absolutely wider and horsts diminished due to the geometrical effect of the downwards-converging Fault Population I and II conjugate fault sets. The limited areal extent of the Canning TQ3D seismic survey precludes comments on the spatial variation in fault dominance and apparent diminution of the north-striking faults northwards which is likely to represent increasing displacement and the presence of a complex overburden degrading the seismic image quality.

the intra-Middle Jurassic paraconformity (J27.0 SB) demarks a transition from subjacent Lower Jurassic low-impedance contrast, discontinuous and parallel reflector facies to superjacent strong and continuous Middle Jurassic reflector facies attributed to primary lithological differences consistent with deltaic progradation (Figs 5 & 6). Erosion of relatively uplifted horst footwalls results in reflector truncation by aggradational fault scarps that are subsequently onlapped. Reflectors progressively dip steeper from south to north (5–20°), and locally dip (5°–25°) east to SE and west to SW towards faults. Reflectors define hanging-wall synclines within 5–7 km-wide and 20–60 km-long north-trending graben (Fig. 8). Depositional architecture, accommodation and peneplanation by the Oxfordian Unconformity (J40.0SB) reduce the thickness of the Syn-extension I Megasequence from c. 2000 to 1050 m from north to south (Fig. 6). Delambre 1 penetrated the Bathonian–Hettangian interval, with the Upper Jurassic erosional vacuity attributed to the Oxfordian (J40.0SB) and Valanginian (K20.0SB) composite unconformity.

The Late Jurassic–Early Cretaceous Syn-extension Megasequence II

The Syn-extension II Megasequence comprises thin (≤20 m), discontinuous, non-parallel reflectors contained within discrete north–south- and NE–SW-trending graben bound by the Oxfordian Unconformity (J40.0SB) and upper Valanginian Unconformity (K20.0SB). Reflectors disconformably downlap onto the Oxfordian Unconformity (J40.0SB) and onlap denuded structurally juxtaposed Jurassic footwalls (Figs 5 & 6). Lower Cretaceous growth wedges thicken into both north–south- and NE–SW-striking graben bordering faults, comprise <500 m-thick synrift sequences, record structural extension, and infill pre-existing erosional topography attributed to the Oxfordian Unconformity (J40.0SB).

The Early Cretaceous–present-day Post-extension Megasequence

The Early Cretaceous (K20.0SB)–present-day (WB) Post-extension Megasequence, attributed to thermal subsidence and transition to a passive margin, comprises two seismically distinct super-sequences subdivided by the Eocene paraconformity (T20.0SB: Fig. 4). The Aptian Unconformity (K40.0SB: Fig. 9), Early Cretaceous Unconformity (K50.0SB), Base Cenozoic paraconformity (T10.0SB) and Eocene paraconformity (T20.0SB) subdivide the Post-extension super-sequence. The Miocene (T30.0SB), Pliocene (T40.0SB) and seabed (WB) subdivide the Syn-inversion super-sequence (Fig. 4).

Discontinuous high-amplitude reflectors onlapping and draping the Valanginian Unconformity (K20.0SB) characterize the 750–1050 m-thick super-sequence beneath the Eocene Unconformity (T20.0SB). Claystone identified within Delambre 1 corresponds with the Muderong Shale regional seal deposited in response to the Valanginian–Aptian marine transgression that drowned pre-Hauterivian sequences confined to north- and NE-trending graben (Fig. 9). Closely spaced (<200 m), small-scale (<5–25 m), layer-bound, normal faults that form polygons in plan-view dissect strata between the Base Cretaceous (K10.1 MFS) and Base Cenozoic (T10.0SB) unconformities, and comprise a regional polygonal fault array (Figs 5 & 6). Occasionally, polygonal faults penetrate the Top Cretaceous Unconformity (T10.0SB) to remain confined below the Eocene Unconformity (T20.0SB: Fig. 6).

The Eocene–present-day Syn-inversion Megasequence

The superjacent Eocene (T20)–present-day Syn-inversion super-sequence comprises NW-dipping (≤5°), regionally continuous, sub-parallel, high-amplitude reflector sequences that taper 2200–900 m south to north (Fig. 6). Seismically translucent facies comprising laterally continuous chaotic reflector packages (10–50 m thick) above discrete high-amplitude reflectors illustrate mass-transport complexes (MTCs) attributed to slope failure of partially consolidated sediments (Fig. 6). Laterally discontinuous high-amplitude anomalies comprising chaotic reflectors illustrate low-sinuosity debris flows during the Cenozoic. Predominately north- and NE-trending conjugate fault pairs nominally offset Upper Cretaceous–present-day reflectors and spatially correlate with the subjacent structural fabric (Fig. 10).

North–south-trending carbonate mound contourite drift deposits and NW-trending incised low-sinuosity channelling with multiple debris fans and overbank deposits comprise the present-day slope topography towards the south of the survey (Fig. 2). East–west- to north–south-trending, north-dipping, 2000–5000 m-long cuspate slump scarp segments link and form updip from MTCs towards the north of the survey (Figs 2 & 10). Cuspate slump depressions adhering to subjacent Triassic–Cretaceous structural orientations comprise bathymetry within the survey centre.

Fault architecture

Evaluation of the Canning survey identified four populations (I, II, III and IV: Fig. 4) of extensional

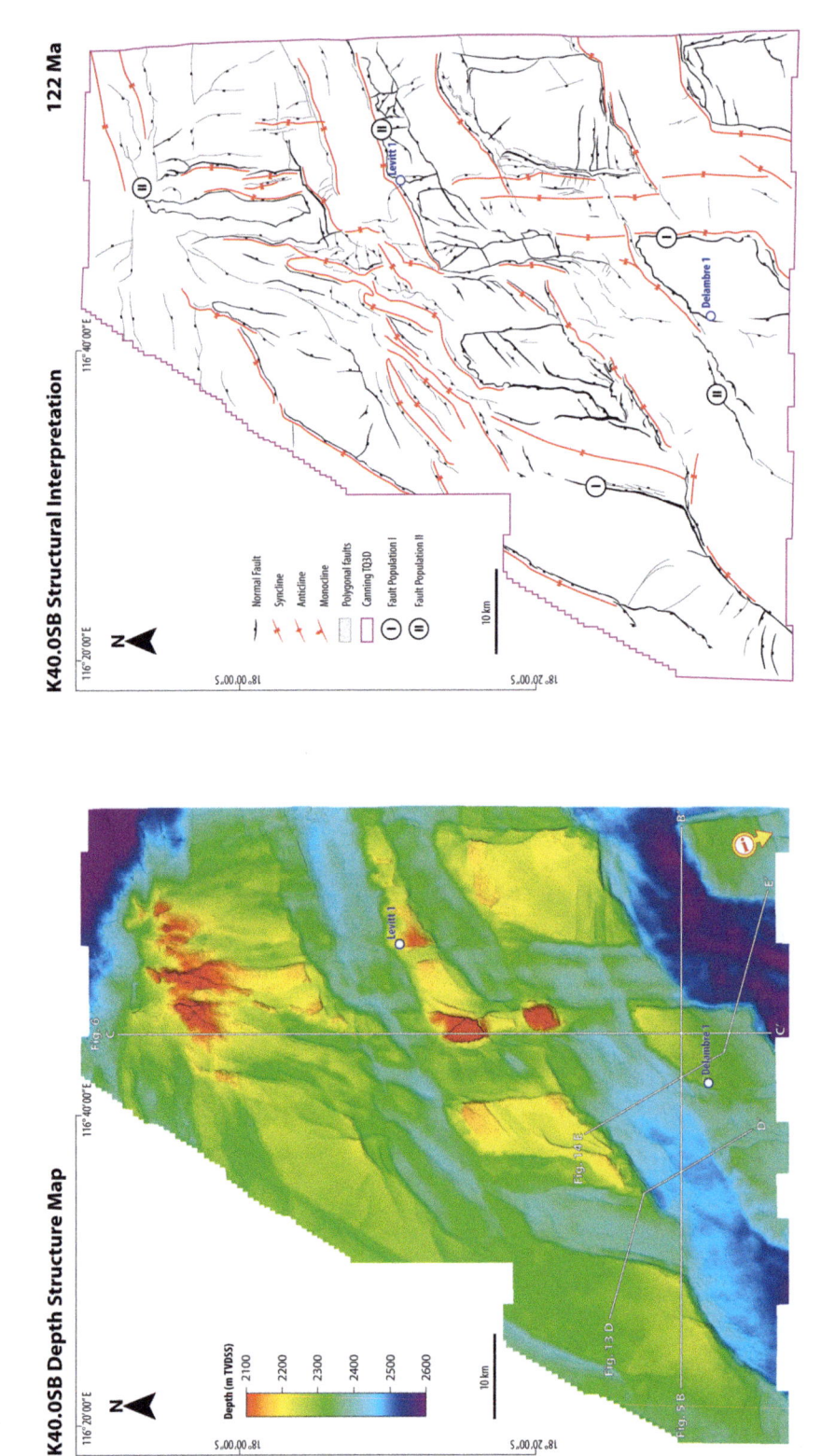

Fig. 9. Aptian (K40.0SB) base Post-extension I Megasequence depth structure map illustrating the persistent expression of the extant and subjacent Fault Population I and II fault sets. Differential compaction of Jurassic–Early Cretaceous Syn-extension I and II megasequences results in monoclinic folds orientated sympathetically to underlying faults and occasional fault penetration of the Aptian (K40.0SB) sequence boundary. Fault planes that appear denuded and regressive, wide graben with reduced horsts, and topographical infilling supports the cessation of extension.

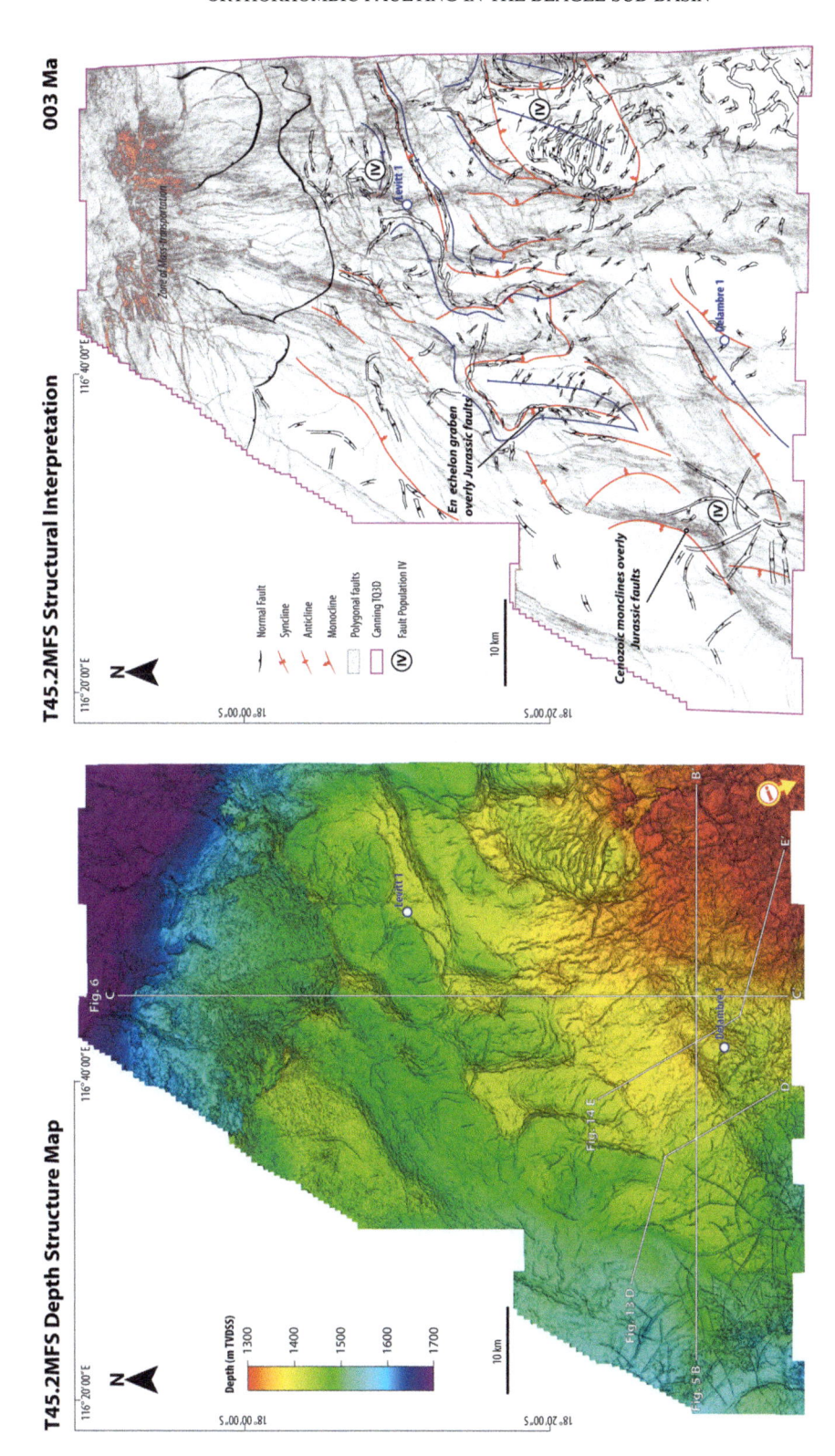

Fig. 10. Mio-Pliocene (T45.2MFS) intra-Syn-inversion I Megasequence depth structure map illustrating the continued expression of subjacent Fault Population I and II fault sets due to differential compaction, monoclinic folding and probable reactivation attributed to continent–arc collision between Australia and SE Asia.

faults that control the Triassic–present-day structural evolution and comprise:

- Fault Population I: Late Triassic–Early Cretaceous normal faults striking north–south and dipping east–west with syn-extension growth packages.
- Fault Population II: Late Triassic–Early Cretaceous normal faults striking NE–SW and dipping NW–SE with syn-extension growth packages.
- Fault Population III: Cretaceous areally extensive and vertically restricted polygonal fault arrays.
- Fault Population IV: Neogene north–south- and NE–SW-trending conjugate arrays of steeply-dipping en echelon normal faults.

Latest Triassic–Late Jurassic Fault Population I

North–south-trending, east–west-dipping extensional faults cut Triassic–Lower Cretaceous strata beneath the Valanginian Unconformity (K20.0SB), and form five north–south-orientated graben within the Canning survey (Figs 7 & 8). These faults characteristically display rotated hanging-wall reflectors, subordinate antithetic/synthetic faults, antiformal proximal hanging-wall deformation, central synformal deformation, and define 5–7 km-wide, 20–60 km-long linear graben adjacent to 5–15 km-wide undeformed horsts (Figs 7 & 8). Fault segments range from $c.$ 2 to 15 km in strike length, and displacement exceeds $c.$ 1025 m that diminishes northwards. These faults remain seismically well imaged between 2.5 and 6 s TWT (Figs 5 & 6). NE–SW-trending (Fault Population II) faults occasionally cross-cut north–south-trending (Fault Population I) faults and commonly link parallel to form sigmoidal fault blocks with a dextral shear sense (Figs 7 & 8). These Jurassic–Cretaceous faults cut the intra-Jurassic paraconformity (J27.0SB), and form segmented en echelon fault arrays and relay ramps (Fig. 8). At the intra-Jurassic paraconformity (J27.0SB) level, these faults range from 1 to 5 km in length in conjugate arrays adjacent to east–west-dipping, north–south-trending faults, and transfer displacement by hard and soft linkage (Fig. 8).

Latest Triassic–Late Jurassic Fault Population II

NE–SW-trending, NW–SE-dipping planar extensional faults cut Triassic–Cretaceous strata beneath the Valanginian Unconformity (K20.0SB), and form a discrete graben towards the south and diffuse faulting towards the north of the survey (Figs 7 & 8). These faults characteristically display rotated ($\leq 5°$) hanging-wall reflectors, hanging-wall synclines, antithetic faults and occasionally penetrate the Base Upper Cretaceous Unconformity (K50.0SB) to terminate within Upper Cretaceous strata (Figs 5 & 6). NE–SW-trending (Fault Population II) faults both cross-cut and link to north–south-trending (Fault Population I) faults. Fault linkage defines a regional rhomboid structural fabric comprising $c.$ 60° acute and $c.$ 120° obtuse angles, and subordinate sigmoidal fault-bound blocks describing clockwise rotation (i.e. dextral: Figs 7 & 8). Individual fault displacement exceeds $c.$ 615 m, which diminishes northwards.

Early–Middle Cretaceous Fault Population III

Two tiers of stacked and coupled polygonal faults (Fault Population III) comprising 100–200 m of spaced, planar, concave-up, layer-bound normal faults with <10–25 m displacements form within the Lower Cretaceous Syn-extension II and Upper Cretaceous–Cenozoic Post-rift megasequences (Figs 5 & 6). Linked fault segments (200–500 m in length) define characteristic polygonal patterns in plan-view confined and truncated by the Base Cretaceous (K10.1 MFS)–Base Eocene (T20.0SB) unconformities. Polygonal faults occasionally dip-link with refracted north–south- and NE–SW-trending tectonic Fault Population I and Fault Population II that penetrate the Cenozoic–present-day Post-extension Megasequence (Figs 5 & 6).

Neogene–present-day Fault Population IV

Numerous, planar, 1–7 km-long fault segments characterize the 200–400 m-wide en echelon conjugate graben arrays hosting 16–23 m displacements that commonly terminate at the seafloor (Figs 2, 5, 6 & 10). These form complex north- to NE-, north- to NW-, NW-, and east- to NE-trending linked fault arrays directly above the north–south- and NE–SW-trending Upper Cretaceous Post-extension deformation fabric attributed to the underlying Jurassic–Cretaceous fault systems (Fig. 10). Individual faults dip steeply ($c.$ 80°) and intersect within the Neogene Syn-inversion Megasequence to form conjugate fault arrays in cross-section (Figs 5 & 6). Occasionally, single faults extend downwards and offset the Base Cenozoic Unconformity (T10.0SB), and link directly or indirectly with Jurassic–Cretaceous faults that penetrate the Valanginian Unconformity (K20.0SB: Fig. 6).

Mass-transport complexes and debris flows imaged by the Canning survey occur between the Base Cenozoic Unconformity (T10.0SB) and the seafloor (WB) to disrupt high-amplitude parallel bedding within broadly prograding carbonate wedges of the Paleogene–present day Giralia Calcarenite, Cape Range Group and Delambre Formation (Figs 2 & 6).

ORTHORHOMBIC FAULTING IN THE BEAGLE SUB-BASIN 219

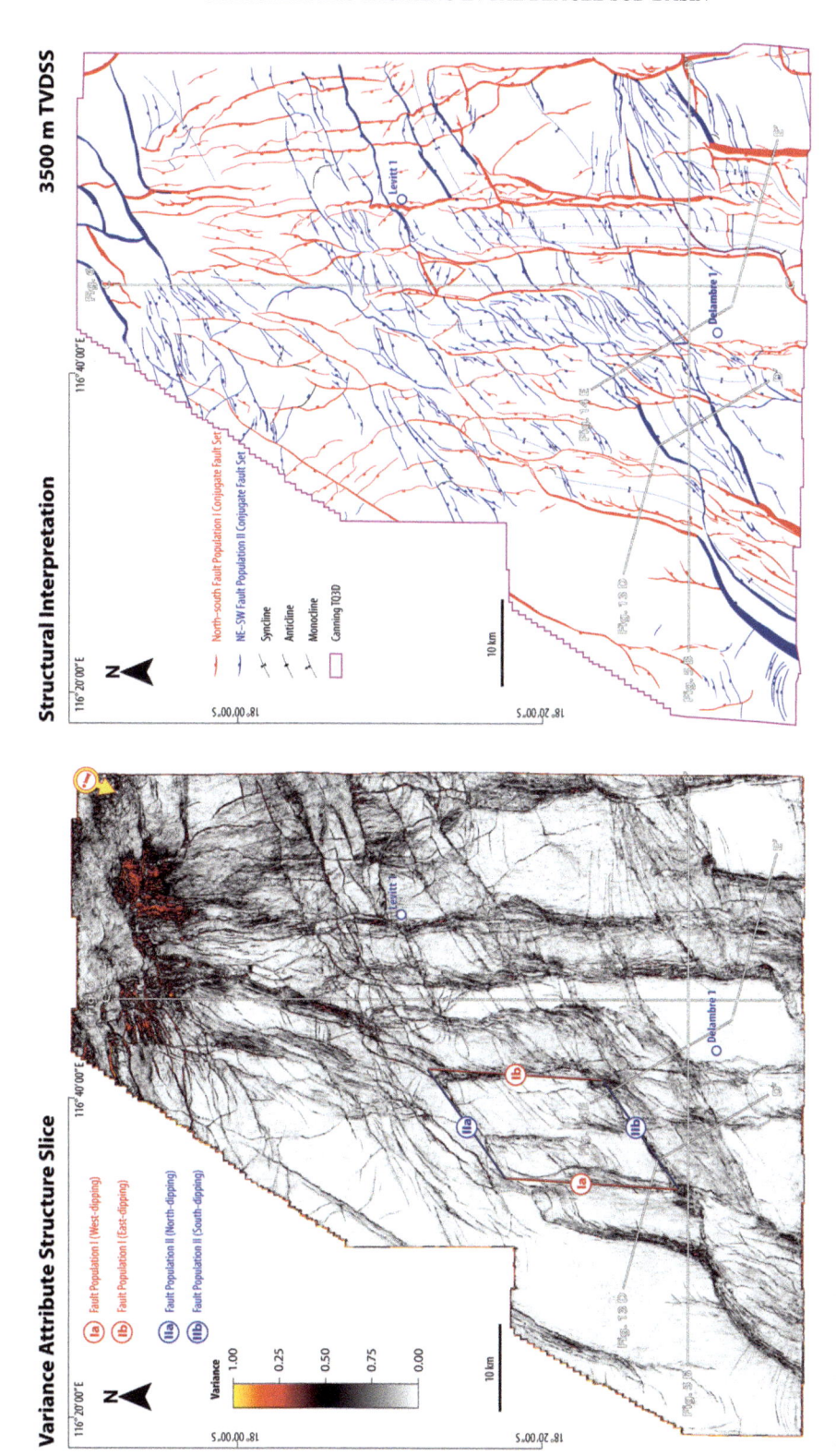

Fig. 11. Summary fault architecture illustrating characteristic orthorhombic fault patterns in a section view of the variance attribute (depth slice at 3000 m TVDSS (true vertical depth subsea)).

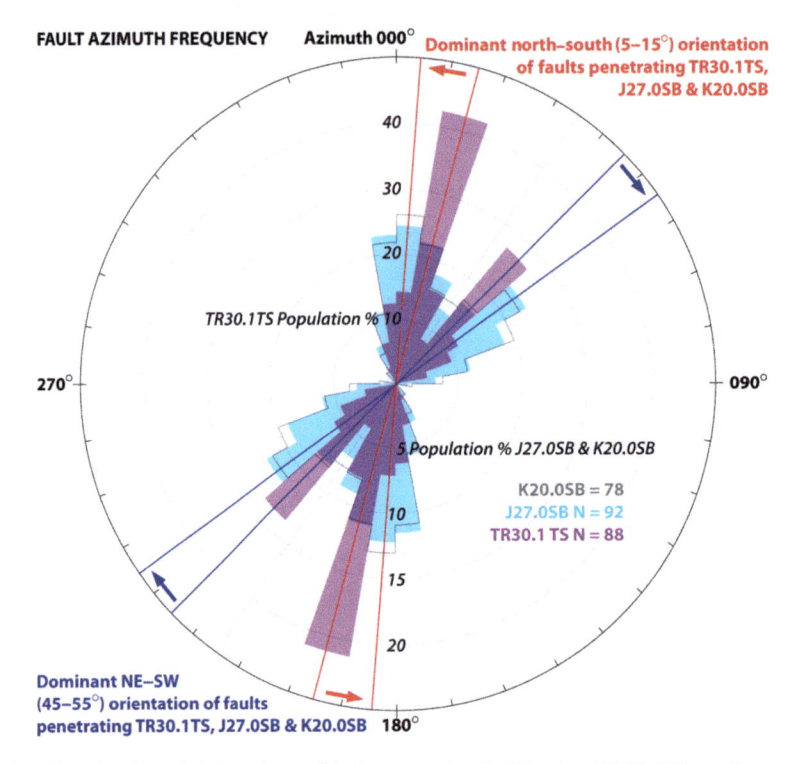

Fig. 12. Fault-strike azimuth statistical analyses of faults penetrating the Rhaetian (TR30.1TS) top Pre-extension Megasequence, Bajocian (J27.0SB) intra-Syn-extension I Megasequence and Valanginian (K20.0SB) top Syn-extension II Megasequence surfaces.

Discussion

Fault geometry and kinematic evolution

Extensional tectonics control the structural evolution and plethora of normal faults within the Beagle Sub-basin. North–south- (Fault Population I) and NE–SW-striking (Fault Population II) conjugate growth faults dissecting the Syn-extension I and II megasequences comprise the oldest faults in the Canning survey (Fig. 4). Late Triassic–Early Cretaceous rifting attributed to the break-up of the NW Australian margin generated these faults (Baillie et al. 1994; O'Brien et al. 1996; Longley et al. 2002; McCormack & McClay 2013; Black et al. 2017). These normal faults develop as two oppositely dipping sets at dihedral angles approximating 60° that form conjugate arrays (Fig. 11).

Fault-displacement history demonstrates the penecontemporaneous activation of both Fault Population I and Fault Population II arrays that demands either crossing conjugate normal faulting or non-plane-strain faulting to accommodate the observed polymodal architecture. The absence of sequentially activated conjugate normal faults and the temporal variance in the dominant displacement accruing fault set precludes a simple cross-cutting relationship. Characteristic rhombohedral fault architecture in section-view indicates a non-plane-strain genesis of Late Triassic–Late Jurassic Fault Population I and Fault Population II that is subsequently reactivated during renewed latest Jurassic–Early Cretaceous extension.

Kinematic analysis of the Canning survey supports a 3D strain field comprising two horizontal principal extension directions (ε_2 and ε_3) and a third vertical contraction direction (ε_1). Four sets of symmetrically arranged faults demonstrate a true orthorhombic geometry that formed penecontemporaneously with a near-ideal strain distribution that manifests as equal-displacement faults (Fig. 11). North–south- (Fault Population I) and NE–SW-striking (Fault Population II) conjugate growth faults display a conjugate symmetry about the 275° and 125° axes, respectively (Fig. 12). Fault Population I and Fault Population II link obliquely at 50° acute and 130° obtuse dihedral angles to form a rhombohedral lattice system in section-view that equates to a 3D rhombus-based oblique pyramidal horst characteristic of non-plane-strain deformation (Fig. 1b). The self-similar replication of this fault architecture at scales

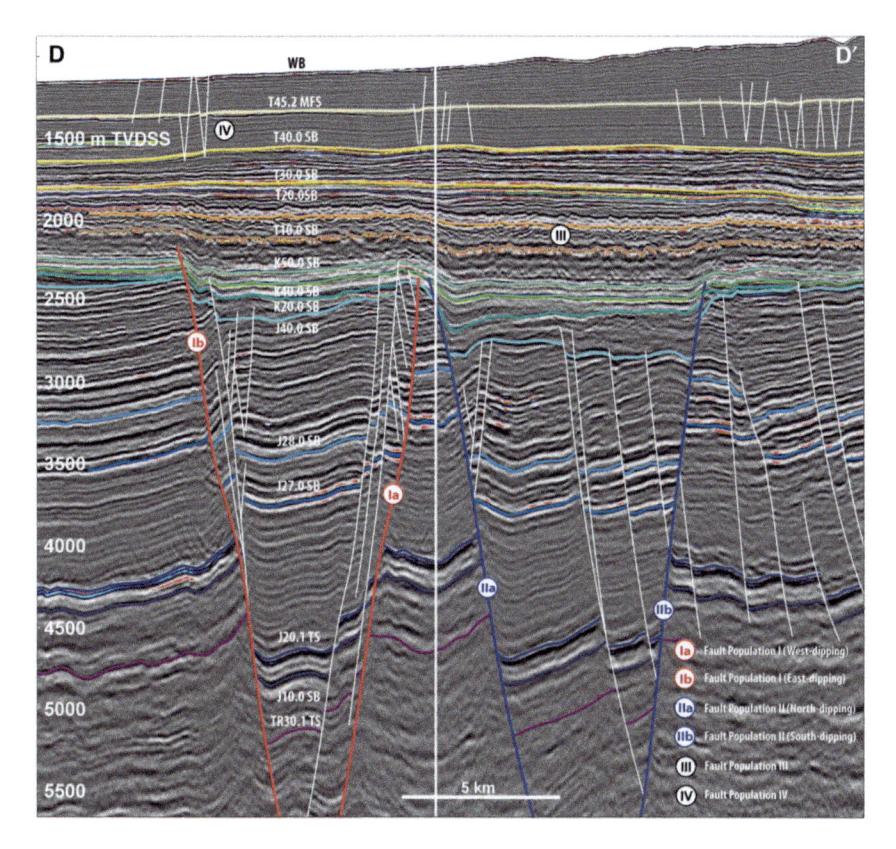

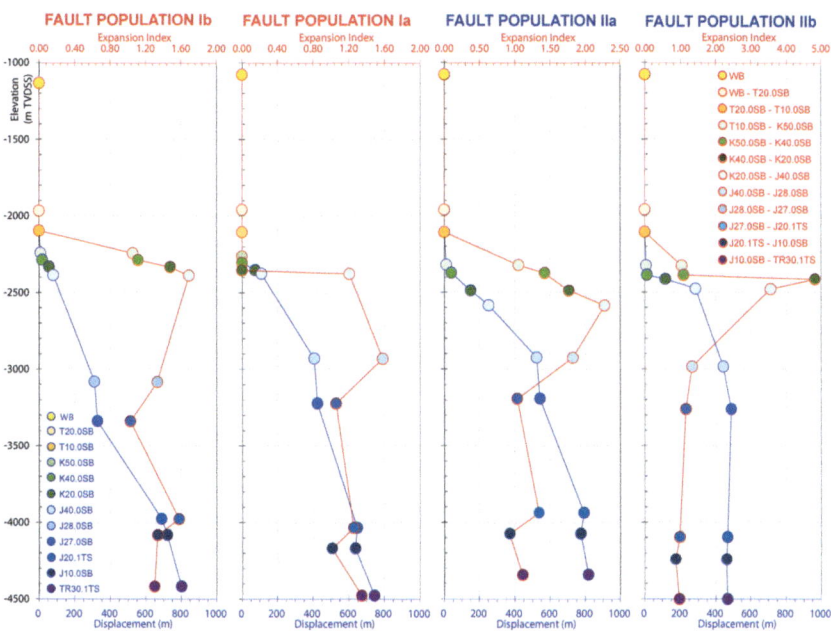

Fig. 13. Fault displacement and growth index analyses transect I illustrating the penecontemporaneous of Fault Population I and II conjugate fault sets comprising two sets of oppositely dipping, 60° dihedral angle, downwards-converging, and north- and NE-striking faults. Figures 7–11 illustrate the location of the seismic transect D–D'.

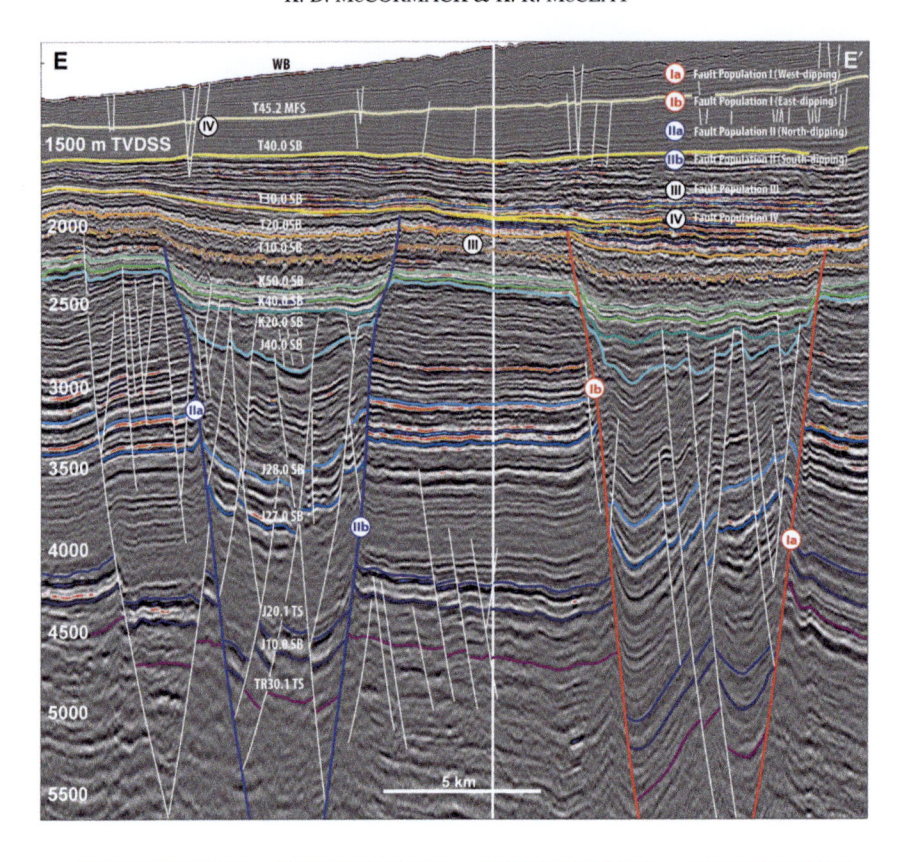

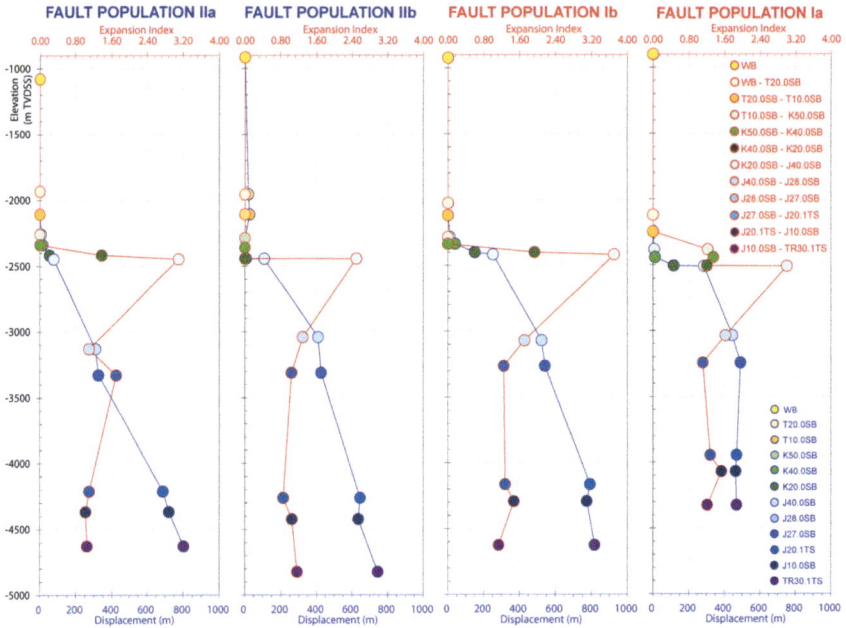

Fig. 14. Fault displacement and growth index analyses transect II illustrating the penecontemporaneous of Fault Population I and II conjugate fault sets comprising two sets of oppositely dipping, 60° dihedral angle, downwards-converging, and north- and NE-striking faults. Figures 7–11 illustrate the location of the seismic transect E–E′.

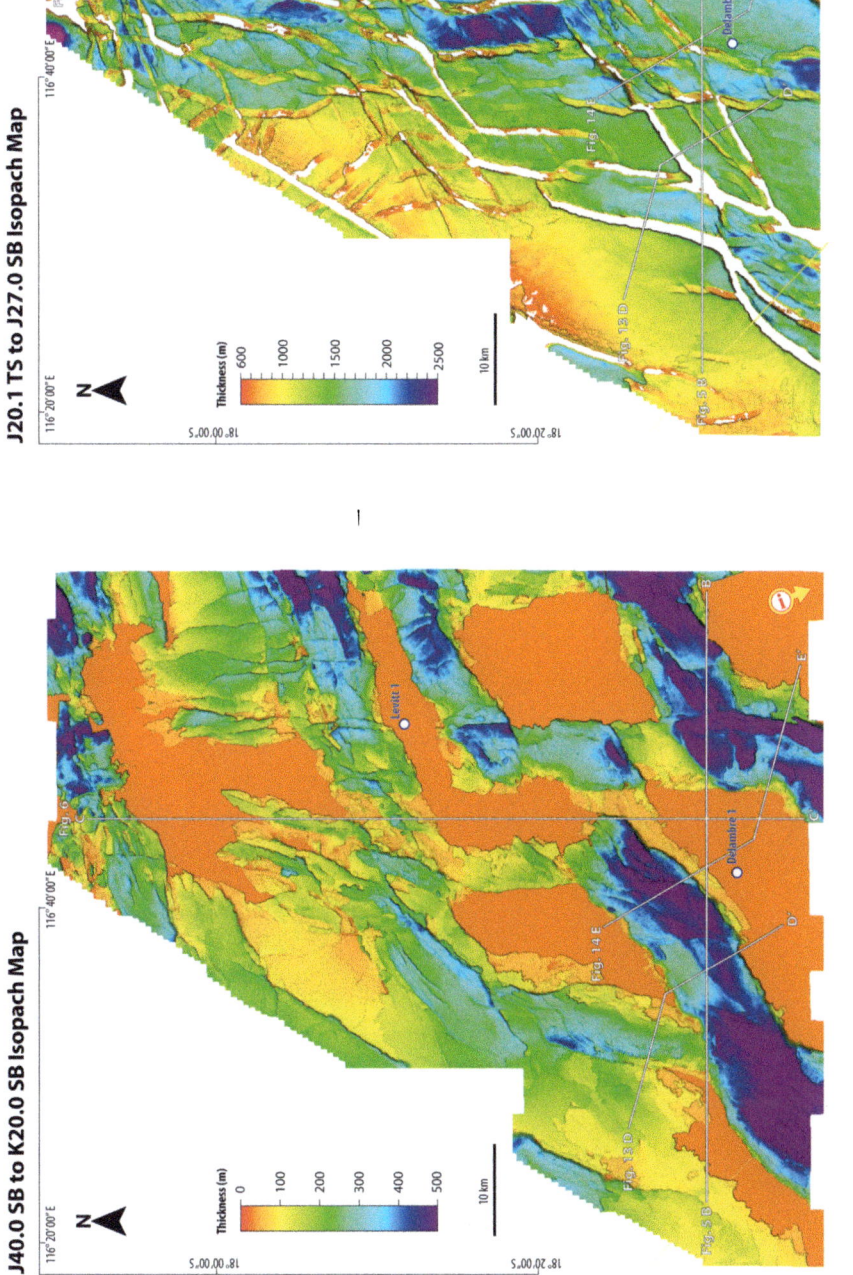

Fig. 15. Oxfordian (J40.0 SB)–Valanginian (K20.0SB) Syn-extension II Megasequence isopach demonstrating the non-ideal orthorhombic strain field that manifests as asymmetrical differential stratal thickening accommodated by growth within NE-trending and north-dipping faults; and Sinemurian (J20.1TS)–Bajocian (J27.0SB) Syn-extension I Megasequence isopach demonstrating the non-ideal orthorhombic strain field that manifests as the differential preservation of keystone fault-accommodated growth within north- and NE-trending graben.

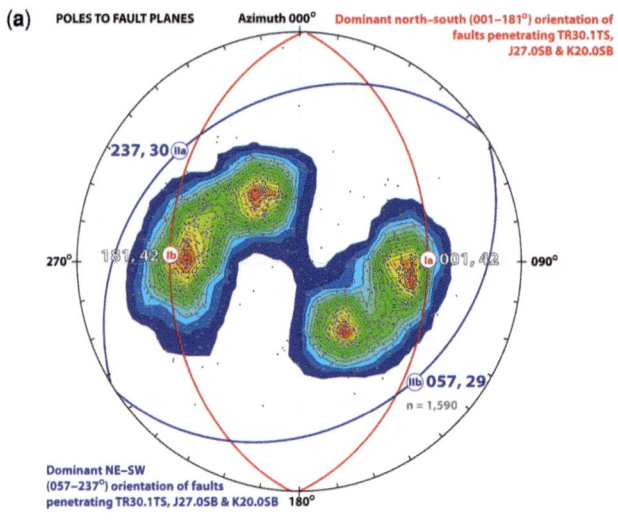

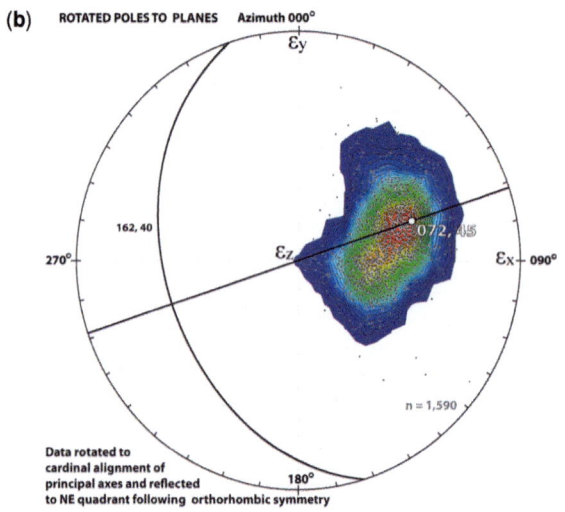

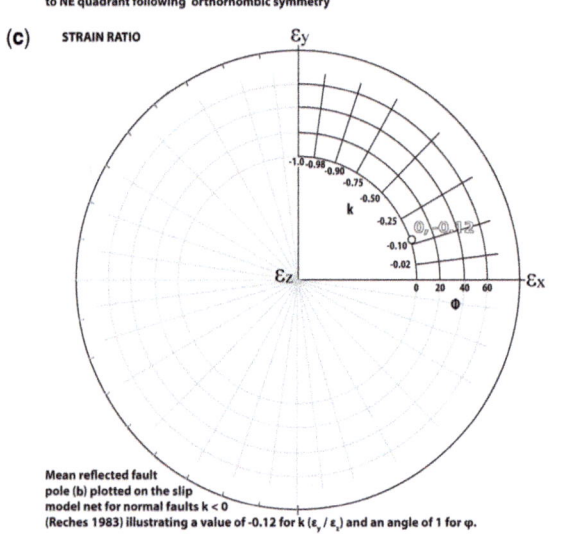

(a) POLES TO FAULT PLANES Azimuth 000° — Dominant north–south (001–181°) orientation of faults penetrating TR30.1TS, J27.0SB & K20.0SB

237, 30 (IIa)
181, 42 (Ib)
001, 42 (Ia)
(IIb) 057, 29
n = 1,590

Dominant NE–SW (057–237°) orientation of faults penetrating TR30.1TS, J27.0SB & K20.0SB

(b) ROTATED POLES TO PLANES Azimuth 000°

ε_y
162, 40
072, 45
ε_z
ε_x — 090°
n = 1,590

Data rotated to cardinal alignment of principal axes and reflected to NE quadrant following orthorhombic symmetry

(c) STRAIN RATIO

ε_y
-1.0 -0.98 -0.90 -0.75 -0.50 -0.25 -0.10 -0.02
k
0, -0.12
ε_z 0 20 40 60 ε_x
φ

Mean reflected fault pole (b) plotted on the slip model net for normal faults k < 0 (Reches 1983) illustrating a value of -0.12 for k ($\varepsilon_y/\varepsilon_x$) and an angle of 1 for φ.

Displacement analyses

Analyses of Fault Population I displacement at J20.1TS and J27.0SB illustrates Syn-extension I linkage of north- and NE-trending conjugate fault set segments to form fault-bound pyramidal horsts. Lateral fault linkages comprise relay (soft-linked) and fault-intersection (hard-linked) relationships comprising predominantly single-mode displacement profiles with maxima located in the centre of fault surfaces. Growth indices calculated across seven key intervals spanning Pre-extension I, Syn-extension I and Syn-extension II illustrate fault-controlled sediment growth (EI > 1.20) between the Rhaetian (TR30.1TS) and Valanginian (K20.0SB) along both north- and NE-striking fault conjugate fault populations (Figs 13 & 14).

Figures 13 and 14 illustrate fault displacement and growth indices across transect D–D' and E–E' (Fig. 11) that display penecontemporaneous activation of Fault Population I and Fault Population II as conjugate fault sets comprising two sets of oppositely dipping, 60° dihedral angle, downwards-converging and north- and NE-striking faults. Growth indices for both Fault Population I and Fault Population II display a consistent positive inflection at J40.0SB attributed to extension climax, break-up and subsequent cessation of rifting. Gross differences between the western (D–D') and eastern (E–E') transect growth indices suggest non-ideal strain partitioning that imparts spatially heterogeneous fault displacement. Syn-extension I (J20.1TS–J27.0SB) and II (J40.0SB–K20.0SB) isopachs (Fig. 15) illustrate that this non-ideal orthorhombic structural accommodation manifests as spatially variant stratal thicknesses.

Orthorhombic fault symmetry

Polymodal fault architecture observed within the Canning seismic data conform to kinematic and mechanical models of non-plane-strain deformation. Evaluation of fault orientation, displacement and linkage demonstrates penecontemporaneous evolution of the north- (Fault Population I) and NE-orientated (Fault Population II) conjugate fault sets in orthorhombic symmetry oblique to remote principal stress orientations comprising $\sigma_1 > \sigma_2 > \sigma_3$. Orthorhombic symmetry exists about the NW (c. 315°), NNE (c. 028°) and vertical axes of symmetry that correspond with maximum extensional strain (ε_x), minimum extensional strain (ε_y) and vertical contraction (ε_z). Calculation of the maximum (ε_x) and minimum (ε_y) extensional strain differential follows the Reches & Dietrich (1983) slip-model stereonet construction (Fig. 16a). Cardinal rotation of the principal fault-strain orientation interpreted from fault poles-to-planes followed by reflection into a common quadrant define a point cluster with a mean trend of 072° plunging 45° (Fig. 16b). Figure 16c illustrates this mean reflected fault pole on the slip-model net for normal faults ($k < 0$: Reches & Dietrich 1983), and the corresponding value of strain ($k = -0.12$) and frictional slip angle (1°).

This orthorhombic symmetrical construction confirms 3D strain as a possible mechanism of genesis for the rhombohedral fault pattern in section and rhombus-based oblique pyramidal horsts in three dimensions. However, the implausible angle of internal friction may be attributed to the analytical derivation of dip from seismic data; successive plane-strain phases of fault inception (i.e. not orthorhombic); or a deviation from the empirically derived model of Reches & Dietrich (1983). Reches & Dietrich (1983) noted a wide dispersion in physical tests for states of stress where $k < 0$ (i.e. extension) that may obviate the application of this conceptual proof to natural datasets hosting extensional faults.

The prominent manifestation of orthorhombic symmetry observed within the Canning survey remains unique within the Beagle Sub-basin and the broader NWS. This complex strain may be attributed to:

- the unique location within a structurally transitional terrane on the northern cusp of the en echelon failed intra-cratonic rifted basin;
- diachronous extensional events during intervallic fragmentation of the Gondwanan margin during global NeoTethyan dispersal;
- competing coeval stress regimes attributed to the contemporaneous extension that culminated in Argo and Gascoyne seafloor spreading may generate complex antecedent stress with balanced or, at least, non-zero intermediate stress (σ_2) relative to the maximum stress (σ_1).

Fig. 16. Calculation of the maximum (ε_x) and minimum (ε_y) extensional strain differential following Reches & Dietrich (1983) slip-model stereonet construction. The cardinal rotation of the principal fault-strain orientation interpreted from fault poles-to-planes followed by reflection into a common quadrant define a point cluster with a mean trend of 072° plunging 45°. This figure illustrates the mean reflected fault pole on the slip-model net for normal faults ($k < 0$: Reches & Dietrich 1983) and the corresponding value of strain ($k = -0.12$).

Polygonal fault systems

Polygonal Fault Population III developed within Lower Cretaceous–Paleocene strata and resembles polygonal fault systems identified in fine-grained sediments on Australia's NWS (Frankowicz & McClay 2010; McCormack & McClay 2013) and within other basins (Henriet et al. 1991; Cartwright & Dewhurst 1998). These non-tectonic faults form under horizontal isostatic stress states ($\sigma_2 = \sigma_3$) attributed to syneresis, the volumetric contraction due to molecular de-watering (Cartwright & Lonergan 1996; Cartwright et al. 2003). These polygonal faults develop within low-amplitude packages that correlate with marine mudstone penetrated by Delambre 1 which accommodate ductile monoclinal deformation, trending sympathetically to subjacent Mesozoic north- (Fault Population I) and NE-striking (Fault Population II) extensional faults.

Vertical fault linkage

The Neogene–present-day Fault Population IV form NNE-, NNW-, NW- and ENE-trending fault sets in conjugate symmetry orientated sympathetically to, but partitioned from, subjacent Fault Population I and Fault Population II displaying orthorhombic fault symmetry (Fig. 10). Partitioning of these fault sets by intervening Late Cretaceous–Paleocene intervals hosting disorientated polygonal fault arrays supports the soft-linkage of strain which manifests as monoclinal deformation between brittle lithologies. Neogene–recent reactivation of Fault Population I and Fault Population II fault sets results in the present-day bathymetric expression of Fault Population IV, with trends mimicking the subjacent Mesozoic extension architecture.

These conjugate fault sets form complex en echelon arrays with different geometry compared with those observed within the Sahul Platform (Frankowicz & McClay 2010) and Laminaria High (De Ruig et al. 2000) areas of the Bonaparte Basin (Fig. 2). Additional complexity, attributed to the reactivation of a pre-existing orthorhombic fault architecture, manifests as a deviation from the pseudo-conjugate arrays orientated in a single principle direction observed elsewhere on the margin. Ultimately, the linear, radial and disorientated fault patterns observed to penetrate Pliocene (Fig. 10) up to present-day (Fig. 2) strata suggest multiple causal agents comprising:

- reactivation by far-field tectonics (i.e. Australia–SE Asia collision);
- gravity collapse (e.g. outboard slump scarps observed at the seafloor);
- polygonal faulting attributed to syneresis (Cartwright & Dewhurst 1998).

Structural evolution of the Beagle Sub-basin

Figure 17 graphically summarizes the four principal tectonostratigraphic stages of evolution evident within the area of the Beagle Sub-basin imaged by the Canning survey.

Latest Triassic–Late Jurassic extension. North- (Fault Population I) and NE-striking (Fault Population II) conjugate normal fault arrays with nominally 60° dihedral dips nucleate and displace penecontemporaneously (Fig. 17a, b). Symmetrical displacement facilitates keystone faulting that accommodates expanded Syn-extension I Megasequence thickness within narrow linear graben that bound correlatively condensed intervals across broad rhombus-based oblique pyramidal horsts. Oxfordian (J40.0SB) uplift and denudation demarks extension cessation attributed to continental break-up and Gascoyne seafloor spreading outboard of the Beagle Sub-basin.

Late Jurassic–Early Cretaceous extension. Reactivation of antecedent Fault Population I and Fault Population II accommodate expanded Syn-extension II Megasequence thickness adjacent to optimally orientated and high-displacement graben-bounding faults (Fig. 17c). Renewed and subsequent cessation of this second phase of extension represents the regional termination of tectonism attributed to the break-up of Greater India from Australia demarked by the Valanginian (K20.0SB) Unconformity.

Early Cretaceous–Neogene post-extension. Post-extension I Megasequence marine transgression deposition comprised regionally thick and uniform marine mudstone hosting post-depositional polygonal faulting (Fault Population III: Fig. 17d). The megasequence remains <500 m thick within the Outer Beagle Platform. Condensation of this regional seal supports deposition influenced by the prevailing structural relief or basin underfilling.

Neogene–present-day passive margin. Paleogene–present-day carbonate platform progradation buried extant sequences (Fig. 17e). Segmented extensional conjugate fault arrays (Fault Population IV) nucleated during the Eocene–present-day as a result of far-field outer-arc compression attributed to the deformation associated with the subduction and collision of the Australian Plate with SE Asia in the region of Timor (Fig. 17f) (Keep et al. 2002; Longley et al. 2002). These faults developed above monoclinal folds sympathetic to subjacent north- (Fault Population I) and NE-striking (Fault Population II) faults. Consistent en echelon segmentation of these fault arrays indicates oblique NNE-orientated reactivation of extant orthorhombic fault orientations.

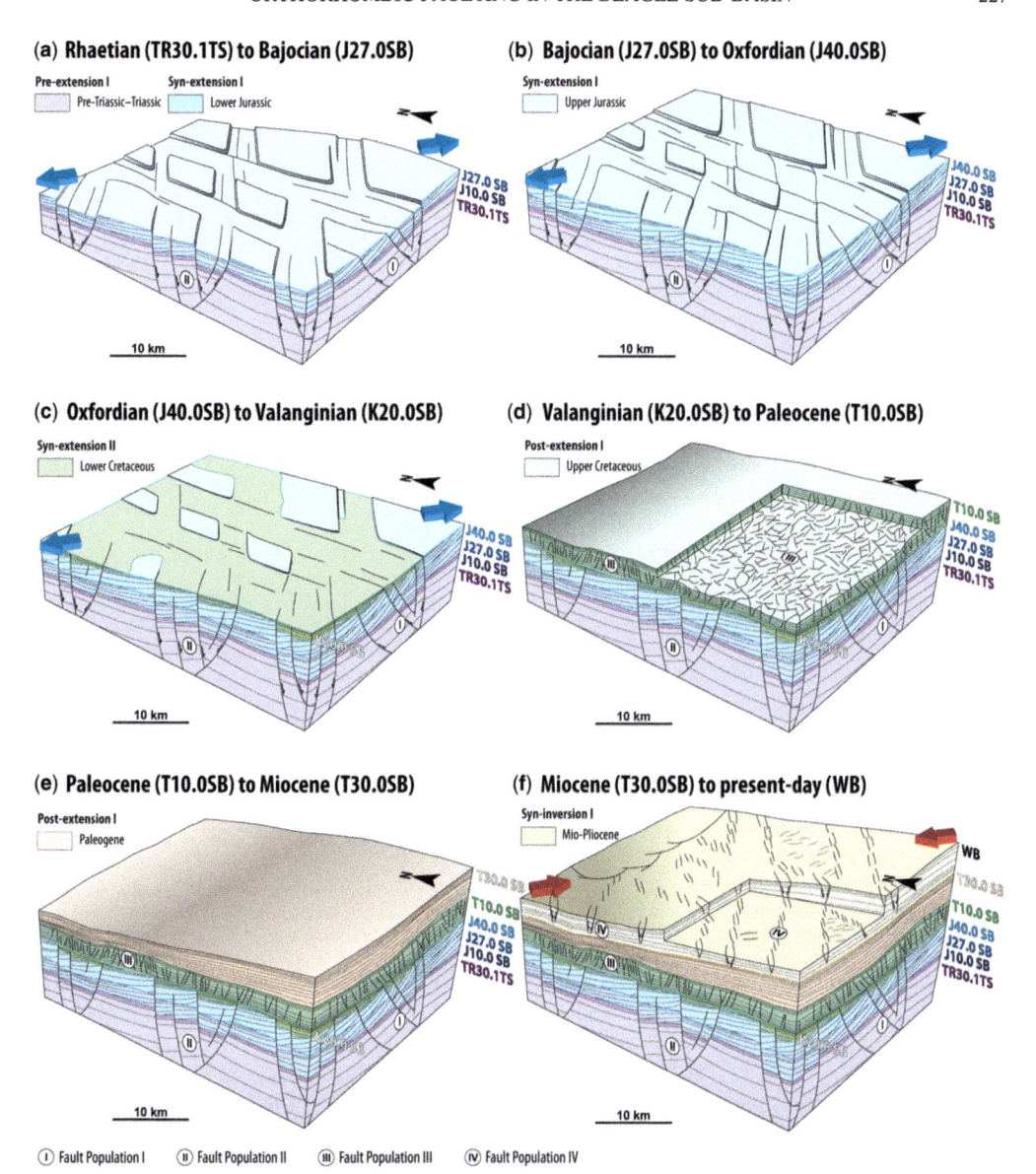

Fig. 17. Evolutionary models of the six key stages of structural evolution interpreted from the Canning seismic survey comprising: (**a**) Rhaetian (TR30.1TS)–Bajocian (J27.0SB) early Syn-extensional Megasequence I growth accommodated within north- and NE-trending keystone faulted graben (Fault Population I and Fault Population II) recording a non-ideal orthorhombic symmetry; (**b**) Bajocian (J27.0SB)–Oxfordian (J40.0SB) Syn-extensional Megasequence I continued accommodation of growth within upwards-broadening keystone graben; (**c**) Oxfordian (J40.0SB)–Valanginian (K20.0SB) Syn-extension II renewed extension and reactivation or extant orthorhombic fault fabric (Fault Population I and Fault Population II), and denudation of horsts which manifests as regressive fault scarps bounding graben hosting asymmetrical growth wedges comprising proximal fan deltas that interrupt the hitherto NW-prograding marine-influenced deltaic environments of deposition; (**d**) Valanginian (K20.0SB)–Paleocene (T10.0) Post-extension I Megasequence deposition of deep-water clay and mudstone hosting polygonal faults (Fault Population III) attributed to syneresis; (**e**) Paleocene (T10.0SB)–Miocene (T30.0SB) carbonate progradation; and (**f**) Miocene (T30.0SB)–present-day (WB) Syn-inversion Megasequence hosting low-displacement Fault Population IV faults nucleated above, and orientated sympathetically to, the extant extensional fault fabric of Fault Population I and Fault Population II.

Implications for the evolution of fault systems in extensional basins

The results of this study demonstrate that four fault sets in orthorhombic symmetry may develop penecontemporaneously within a single stress field attributed to tectonic extension. Identification of non-plane-strain deformation obviates polyphase tectonism attributed to the enigmatic intervallic rotation of stress fields. The revaluation of seismically imaged polymodal faulting within other extensional basins may indicate a similar fault-architecture genesis.

Strain partitioning controlled by mechanically incompetent ductile Cretaceous shale produces vertical fault segmentation. Impartation of enduring orthorhombic fault architecture controls the monocline hinge and en echelon oblique extension fault orientation. Strain partitioning by décollement, separating reactivated deeper faults from faults nucleated within cover sections, may facilitate the maintenance of trap integrity in reactivated extensional basins.

Conclusions

Structural architecture of the Canning survey area on the Outer Beagle Platform, in the Beagle Sub-basin, comprises north- to NE-trending, complex, segmented and orthorhombic extensional fault systems. From oldest to youngest, these include:

- Late Triassic–Early Cretaceous north-striking syn-extension faults (Fault Population I);
- Late Triassic–Early Cretaceous NE-striking syn-extension faults (Fault Population II);
- Cretaceous polygonal fault arrays (Fault Population III);
- Neogene–present-day north- and NE-trending conjugate fault arrays (Fault Population IV).

Fault Population I and II conjugate fault sets displace penecontemporaneously and display an orthorhombic fault architecture characteristic of 3D non-plane strain with non-zero intermediate (σ_2) extension magnitude (i.e. $\sigma_1 > \sigma_2 > \sigma_3$). Neogene–present-day north–south- and NE–SW-striking en echelon conjugate fault arrays (Fault Population IV) form in response to the oblique reactivation of Late-Triassic–Early Cretaceous polymodal faults (Fault Population I and Fault Population II). Fault Population I, II and IV form a vertically decoupled (soft-linked) pseudo-conjugate fault system partitioned by Fault Population III. This model of evolution supports the development of complex polymodal faulting under simple states of stress that hosts implications for understanding the distribution, linkage and age of extensional faults within the Beagle Sub-basin and other extensional basins.

Acknowledgements The authors thank Geoscience Australia for providing the seismic data and Royal Holloway, University of London, in partnership with Schlumberger and Badleys, for logistical support, and for the use of Petrel and TrapTester software, respectively. We thank Hannah Rogers, Malcolm Black, Nicola Scarselli and the Fault Dynamics Research Group for helpful comments during the preparation of this manuscript. Thanks also to Woodside Energy Ltd for their support in presenting an early conceptual thesis at The Roberts Conference (2016), and the anonymous reviewers for their efforts in reviewing this paper.

References

ALLMENDINGER, R.W., CARDOZO, N. & FISHER, D. 2012. *Structural Geology Algorithms: Vectors and Tensors in Structural Geology*. Cambridge University Press, Cambridge, UK.

ANDERSON, E.M. 1951. *The Dynamics of Faulting*, 2nd edn. Oliver and Boyd, Edinburgh.

AYDIN, A. 1977. *Faulting in sandstone*. PhD thesis, Stanford University, Stanford, California, USA.

AYDIN, A. & RECHES, Z. 1982. Number and orientation of fault sets in the field and in experiments. *Geology*, **10**, 107–112.

BAILEY, W.R., UNDERSCHULTZ, J., DEWHURST, D.N., KOVACK, G., MILDREN, S. & RAVEN, M. 2006. Multi-disciplinary approach to fault and top seal appraisal; Pyrenees–Macedon oil and gas fields, Exmouth Sub-basin, Australian NW Shelf. *Marine and Petroleum Geology*, **23**, 241–259.

BAILLIE, P.W., POWELL, C.McA., LI, Z.X. & RYALL, A.M. 1994. The tectonic framework of Western Australia's Neoproterozoic to recent sedimentary basins. *In*: PURCELL, P.G. & PURCELL, R.R. (eds) *The Sedimentary Basins of Western Australia: Proceedings of the Petroleum Exploration Society of Australia Symposium*. Petroleum Exploration Society of Australia, Perth, Australia, 45–62.

BLACK, M., MCCORMACK, K.D., ELDERS, C. & ROBERTSON, D. 2017. Extensional fault evolution within the southern Exmouth Sub-basin, North West Shelf, Australia. *Marine and Petroleum Geology*, **85**, 301–315.

BLEVIN, J.E., STEPHENSON, A.E. & WEST, B.G. 1994. Mesozoic structural development of the Beagle Sub-basin: Implications for the petroleum potential of the Northern Carnarvon Basin. *In*: PURCELL, P.G. & PURCELL, R.R. (eds) *The Sedimentary Basins of Western Australia: Proceedings of the Petroleum Exploration Society of Australia Symposium*. Petroleum Exploration Society of Australia, Perth, Australia, 479–496.

CARDOZO, N. & ALLMENDINGER, R.W. 2013. Spherical projections with OSXStereonet. *Computers & Geosciences*, **51**, 193–205.

CARTWRIGHT, J.A. & DEWHURST, D. 1998. Layer-bound compaction faults in fine-grained sediments. *Geological Society of America Bulletin*, **110**, 1242–1257.

CARTWRIGHT, J.A. & LONERGAN, L. 1996. Volumetric contraction during the compaction of mudrocks: a mechanism for the development of regional scale polygonal fault systems. *Basin Research*, **8**, 183–193.

CARTWRIGHT, J.A., JAMES, D. & BOLTON, A. 2003. The genesis of polygonal fault systems: a review. In: VAN RENSBERGEN, P., HILLIS, R.R., MALTMAN, A.J. & MORLEY, C.K. (eds) Subsurface Sediment Mobilization. Geological Society, London, Special Publications, 216, 223–243, https://doi.org/10.1144/GSL.SP.2003.216.01.15

CASTILLO, D.A., BISHOP, D.J., DONALDSON, I., KUEK, D., DE RUIG, M., TRUPP, M. & SHUSTER, M.W. 2000. Trap integrity in the Laminaria High-Nancar Trough region, Timor Sea: prediction of fault seal failure using well-constrained stress tensors and fault surfaces interpreted from 3-D seismic. Australian Petroleum Production and Exploration Association Journal, 40, 151–173.

DE RUIG, M.J., TRUPP, M., BISHOP, D.J., KUEK, D. & CASTILLO, D.A. 2000. Fault architecture and the mechanics of fault reactivation in the Nancar Trough/Laminaria area of the Timor Sea, northern Australia. Australian Petroleum Production and Exploration Association Journal, 40, 174–193.

DRISCOLL, N.W. & KARNER, G.D. 1998. Lower crustal extension across the Northern Carnarvon Basin, Australia: Evidence for an eastward dipping detachment. Journal of Geophysical Research: Solid Earth, 103, 4975–4991.

FRANKOWICZ, E. & MCCLAY, K.R. 2010. Extensional fault segmentation and linkages, Bonaparte Basin, outer North West Shelf, Australia. American Association of Petroleum Geologists Bulletin, 94, 977–1010.

GARTRELL, A., BAILEY, W.R. & BRINCAT, M. 2005. Strain localization and trap geometry as key controls on hydrocarbon preservation in the Laminaria High area. Australian Petroleum Production and Exploration Association Journal, 45, 477–492.

GEOSCIENCE AUSTRALIA 2010. Beagle Sub-basin, Northern Carnarvon Basin Release Areas W10-7, W10-8 and W10-9: Offshore Petroleum Exploration Acreage Release. Geoscience Australia, Melbourne, Australia.

GOLONKA, J., KROBICKI, M., PAJAK, J., GLANG, N.V. & ZUXHIEWICZ, W. 2006. Global Plate Tectonics and the Paleogeography of SE Asia. AGH University of Science and Technology, Kraków, Poland.

GRIFFITH, A.A. 1924. Theory and rupture. In: BIEZENO, C.B. & BURGERS, J.M. (eds) First International Congress of Applied Mechanics. Waltman, Delft, The Netherlands, 55–63.

HEALY, D., JONES, R.R. & HOLDSWORTH, R.E. 2006. Three-dimensional brittle shear fracturing by tensile crack interaction. Nature Letters, 439, 64–67.

HEINE, C. & MÜLLER, R. 2005. Late Jurassic rifting along the Australian North West Shelf: margin geometry and spreading ridge configuration. Australian Journal of Earth Sciences, 52, 27–39.

HENRIET, J.P., DE BATIST, M. & VERSCHUREN, M. 1991. Early fracturing of Paleogene clays, southernmost North Sea: Relevance to mechanism of primary hydrocarbon migration. In: SPENCER, A.M. (ed.) Generation, Accumulation and Production of Europe's Hydrocarbons. European Association of Petroleum Geologists (EAGE), Special Publications, 1, 217–227.

KEEP, M., CLOUGH, M. & LANGHI, L. 2002. Neogene tectonic and structural evolution of the Timor Sea region, NW Australia. In: KEEP, M. & MOSS, S. (eds) The Sedimentary Basins of Western Australia 3: Proceedings of the Petroleum Exploration Society of Australia Symposium, Perth, Australia. Petroleum Exploration Society of Australia, Perth, Australia, 341–353.

KOPSEN, E. & MCGANN, G. 1985. A review of the hydrocarbon habitat of the eastern and central Barrow-Dampier sub-basin, Western Australia. Australian Petroleum Exploration Association Journal, 25, 154–176.

KRANTZ, R.W. 1986. The odd-axis model: Orthorhombic fault patterns and three-dimensional strain fields. PhD thesis, The University of Arizona, Tucson, Arizona, USA.

KRANTZ, R.W. 1988. Multiple fault sets and three-dimensional strain: theory and application. Journal of Structural Geology, 10, 225–237.

KRANTZ, R.W. 1989. Orthorhombic fault patterns: the odd axis model and slip vector orientations. Tectonics, 8, 483–495.

LONGLEY, I., BUESSENSCHUETT, C. ET AL. 2002. The North West Shelf of Australia – a Woodside perspective. In: KEEP, M. & MOSS, S. (eds) The Sedimentary Basins of Western Australia 3: Proceedings of the Petroleum Exploration Society of Australia Symposium, Perth, Australia. Petroleum Exploration Society of Australia, Perth, Australia, 27–88.

MARSHALL, N.G. & LANG, S.C. 2013. A new sequence straigraphic framework for the North West Shelf, Australia. In: KEEP, M. & MOSS, S.J. (eds) The Sedimentary Basins of Western Australia 4: Proceedings of the Petroleum Exploration Society of Australia Symposium, Perth, Australia. Petroleum Exploration Society of Australia, Perth, Australia, 1–32.

MCCORMACK, K. & MCCLAY, K. 2013. Structural architecture of the Gorgon Platform, North West Shelf, Australia. In: KEEP, M. & MOSS, S.J. (eds) The Sedimentary Basins of Western Australia 4: Proceedings of the Petroleum Exploration Society of Australia Symposium, Perth, Australia. Petroleum Exploration Society of Australia, Perth, Australia, 1–24.

MILLER, J.MCL., NELSON, E.P., HITZMAN, M., MUCCILLI, P. & HALL, W.D.M. 2007. Orthorhombic fault-fracture patterns and non-plane strain in a synthetic transfer zone during rifting: Lennard shelf, Canning basin, Western Australia. Journal of Structural Geology, 29, 1002–1021.

MITCHELMORE, L. & SMITH, N. 1994. West Muiron discovery, WA-155-P: New life for an old prospect. In: PURCELL, P.G. & PURCELL, R.R. (eds) The Sedimentary Basins of Western Australia: Proceedings of the Petroleum Exploration Society of Australia Symposium. Petroleum Exploration Society of Australia, Perth, Australia, 583–596.

MITRA, G. 1979. Ductile deformation zones in Blue Ridge basement rocks and estimation of finite strains. Geological Society of America Bulletin, 90, 935–951.

O'BRIEN, G.W., LISK, M., DUDDY, I., EADINGTON, P.J., CADMAN, S. & FELLOWS, M. 1996. Late Tertiary fluid migration in the Timor Sea: a key control on thermal and diagenetic histories? Australian Petroleum Production and Exploration Association Journal, 36, 399–426.

ODÉ, H. 1960. Faulting as a velocity discontinuity in plastic deformation. In: GRIGGS, D. & HANDIN, J. (eds) Rock Deformation. Geological Society of America, Memoirs, **79**, 293–321.

OERTEL, P.P. 1965. The mechanics of faulting in clay experiments. Tectonophysics, **2**, 343–393.

OESTERLEN, P.M. & BLENKINSOP, T.G. 1994. Extension directions and strain near the failed triple junction of the Zambezi and Luangwa Rift Zones, southern Africa. Journal of African Sciences, **18**, 175–180.

OSBORNE, D.G. 1994. Nebo oil discovery, Beagle Sub-basin. In: PURCELL, P.G. & PURCELL, R.R. (eds) The Sedimentary Basins of Western Australia: Proceedings of the Petroleum Exploration Society of Australia Symposium. Petroleum Exploration Society of Australia, Perth, Australia, 653–654.

RECHES, Z. 1978. Analysis of faulting in a three-dimensional strain field. Tectonophysics, **47**, 109–129.

RECHES, Z. & DIETRICH, J.H. 1983. Faulting of rocks in three-dimensional strain fields. I. Failure of rocks in polyaxial, servo-controlled experiments. Tectonophysics, **95**, 111–132.

SAGY, A., RECHES, Z. & AGNON, A. 2003. Hierarchic three-dimensional structure and slip partitioning in the western Dead Sea pull-apart. Tectonics, **22**, 4–17.

STEPHENSON, A.E., BLEVIN, J.E. & WEST, B.G. 1998. The paleogeography of the Beagle Sub-basin, Northern Carnarvon Basin, Australia. Journal of Sedimentary Research, **68**, 1131–1145.

THORSEN 1963. Age of growth faulting in southeast Louisiana. Gulf Coast Association of Geological Societies Transactions, **13**, 103–110.

WESTERNGECO. 2000. Block WA-294-P 3D Seismic Program Final Field Operations Report, Perth, Australia.

WOODSIDE. 1981. Delambre No. 1 Completion Report. WA-90-P Beagle Sub-basin, Western Australia, Australia. Unpublished Report Woodside Offshore Petroleum Pty Ltd, Perth, Australia.

YEATES, A.N., BRADSHAW, M.T. ET AL. 1987. The Westralian Superbasin, an Australian link with Tethys. In: McKENZIE, K.G. (ed.) Shallow Tethys 2. Proceedings of the International Symposium on Shallow Tethys 2, Wagga Wagga. A.A. Balkema, Rotterdam, The Netherlands, 199–213.

Fault-scarp degradation in the central Exmouth Plateau, North West Shelf, Australia

AWAD BILAL*, KEN McCLAY & NICOLA SCARSELLI

Fault Dynamics Research Group, Department of Earth Sciences, Royal Holloway, University of London, Egham, Surrey TW20 0EX, UK

A.B., 0000-0002-4903-2682
Correspondence: awad.bilal@gmail.com

Abstract: Latest Triassic–earliest Late Jurassic domino-style extensional faulting in the central Exmouth Plateau, North West Shelf of Australia, exhibits footwall degradation scarps with up to 1.8 km of scarp retreat of the Upper Triassic Mungaroo Formation. Extensional fault-propagation folding, rotation and uplift produced gravitationally driven scarp collapse of the incompetent and mudstone-dominated uppermost Mungaroo Formation. Scarp degradation occurred along the entire extent of the footwalls of three major faults within the research area. Individual segments display listric fault surfaces in cross-section and scoop-shaped scars in three dimensions. The listric collapse faults dip towards the erosional scarp and sole out at different levels within the upper Triassic Mungaroo Formation.

Footwall crestal collapse formed coalesced, scoop-shaped degradation scarps with Mungaroo Formation debris deposited as wedges within the adjacent hanging-wall synclines. Maximum scarp degradation occurred at the fault centres and decreased towards the fault tips. This study proposes new three-dimensional evolutionary structural models for the fault-scarp degradation in the central Exmouth Plateau.

Footwall fault-scarp degradation has been widely documented at the crests of tilted extensional fault blocks in rifts and passive margins (e.g. Hesthammer & Fossen 1999; Sharp *et al.* 2000; Morley *et al.* 2007; Welbon *et al.* 2007; Elliott *et al.* 2012, 2017; Henstra *et al.* 2016). Syn-extensional scarp collapse commonly occurs due to gravitational instabilities with scoop-shaped listric faults and rotated blocks (Hesthammer & Fossen 1999; Stewart & Reeds 2003; Morley *et al.* 2007). They have been described from: the Gulf of Suez (e.g. Sharp *et al.* 2000; Afifi *et al.* 2016); the Gulf of Corinth, Greece (Ferentinos *et al.* 1988; Kokkalas & Koukouvelas 2005); the East Greenland passive continental margin (Surlyk 1978; Henstra *et al.* 2016); the Sirikit Field in the Phitsanulok Basin, Thailand (Morley *et al.* 2007); in the Bremstein Fault Complex and Vingleia Fault Complex, offshore mid-Norway (Elliott *et al.* 2012, 2017); and from the Viking Graben, northern North Sea (Struijk & Green 1991; Coutts *et al.* 1996; Underhill *et al.* 1997; Hesthammer & Fossen 1999; McLeod & Underhill 1999; McLeod *et al.* 2002; Gibbons *et al.* 2003; Welbon *et al.* 2007).

Footwall collapse structures occur in both subaerial and submarine environments, and at a variety of scales, ranging from hundreds of metres to kilometres (e.g. Dahl & Solli 1993; Hesthammer & Fossen 1999; McLeod & Underhill 1999; Morley *et al.* 2007; Elliott *et al.* 2012). Partially submerged scarps with subaerially exposed sections have also been documented in the footwalls of active normal faults (Dahl & Solli 1993; Patton *et al.* 1994; McLeod *et al.* 2002; McArthur *et al.* 2016b). Fully submerged fault blocks may undergo submarine erosion as well as submarine landslide collapse, with the products of footwall-scarp degradation incorporated into the synrift megasequences (e.g. Hesthammer & Fossen 1999; McLeod & Underhill 1999; Stewart & Reeds 2003). The hanging-wall debris preserves a record of both the fault activity and the degradation of the fault scarp (Stewart & Reeds 2003).

The development of extensional fault-propagation folds within rift basins during early fault growth stages has a significant effect on the extent of footwall uplift, incision and the characteristics of hanging-wall infill (Schlische 1995; Gawthorpe & Leeder 2000; Sharp *et al.* 2000; Khalil & McClay 2002, 2016). Fold-related strain is typically accommodated by folding and bedding-plane slip in mechanically incompetent units. These folds are characterized by footwall anticlines and hanging-wall synclines (Schlische 1995; Hardy & McClay 1999; Gawthorpe & Hardy 2002; Khalil & McClay 2002, 2016; Jackson *et al.* 2006; Zhao *et al.* 2017). Distributed zones of secondary normal faults may also develop in mechanically competent layers (e.g. Paul & Mitra 2015). Such extensional fault-propagation folding commonly leads to footwall instabilities and scarp degradation; particularly in weakly anisotropic strata such as claystones and siltstones (Khalil & McClay 2006; Leppard & Gawthorpe 2006).

On the Exmouth Plateau, the pre-extension Upper Triassic strata in the footwalls of major faults are significantly affected by gravity-driven degradation processes. This research uses 3D seismic data calibrated by well data from the central Exmouth Plateau, North West Shelf of Australia, to interpret footwall scarps and their associated hanging-wall debris developed during the latest Triassic–earliest Late Jurassic extension of the Exmouth Plateau. The main objectives of this study are to describe, for the first time, the 3D characteristics of the extensional fault-scarp degradation in the central Exmouth Plateau and to propose 4D evolutionary models for these systems.

Regional geology of the Exmouth Plateau and the Northern Carnarvon Basin

The Northern Carnarvon Basin (NCB) covers an area of 535 000 km^2 (Hocking 1988) and forms the southern part of Australia's NW continental margin – the North West Shelf (Fig. 1). The NCB is the southernmost in a series of NE-trending Paleozoic–Cenozoic basins that include the offshore Canning, Browse and Bonaparte basins; collectively termed the Westralian Super-basin (Yeates *et al.* 1987; Pryer *et al.* 2002; Geoscience Australia 2017). The NCB consists of the inboard NE-trending Exmouth, Dampier, Barrow and Beagle sub-basins, and the outboard

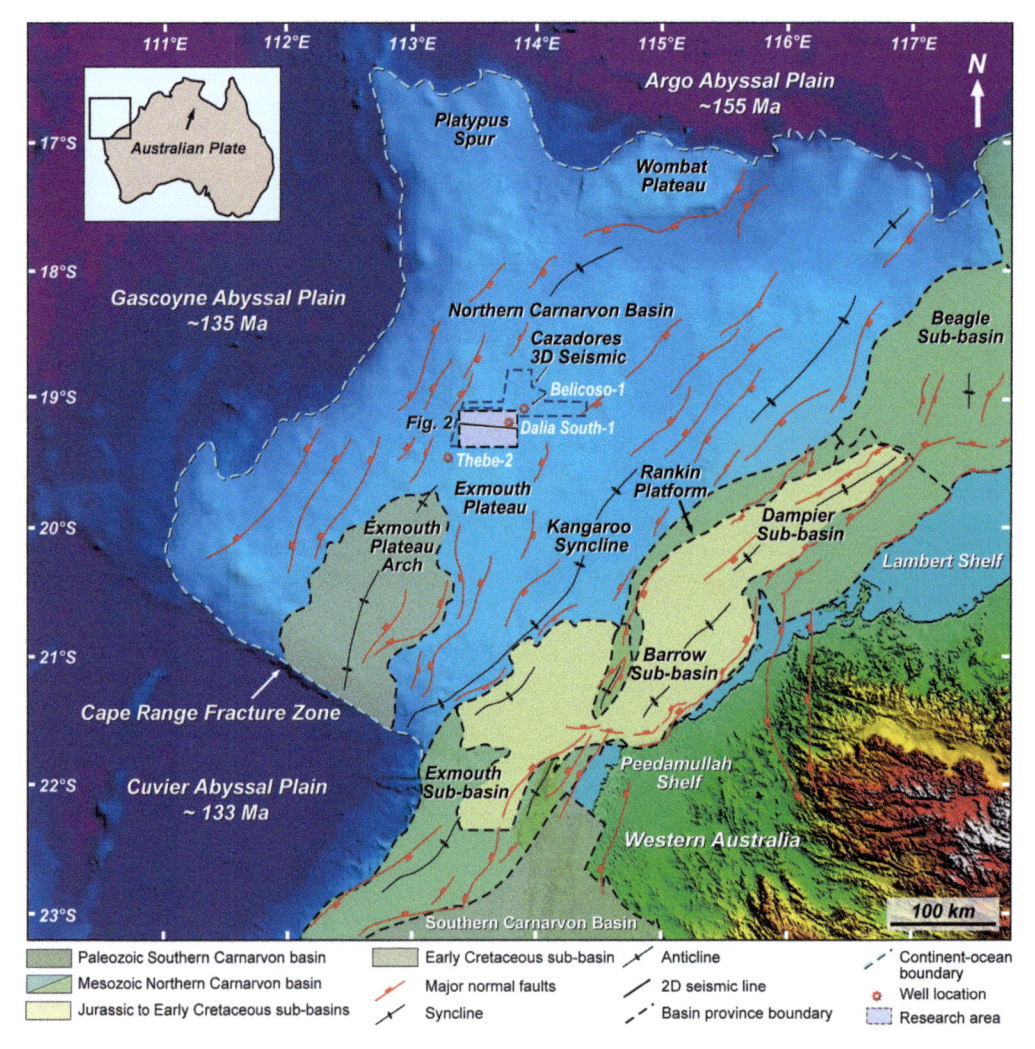

Fig. 1. Location map of the research area showing the main structural elements of the Northern Carnarvon Basin. The locations of the Cazadores 3D seismic survey and the area of research (purple square) are also shown. Interpretations and outlines of the sub-basins were compiled from Stagg & Colwell (1994) and Goncharov *et al.* (2006). The elevation and bathymetric data are from Geoscience Australia.

zone of the Exmouth Plateau (Stagg & Colwell 1994; Longley *et al.* 2002). The main structural elements of the NCB are illustrated in Figure 1. The latest Triassic–earliest Late Jurassic extension in the research area in the central Exmouth Plateau (Fig. 1) is characterized by domino-style rotational extensional faults with prominent footwall degradation of the uppermost Triassic strata (Fig. 2).

Late Carboniferous–Early Permian

The North West Shelf passive margin was initiated during the Late Carboniferous–Early Permian break-up of Pangea and the opening of the Neo-Tethys ocean (Driscoll & Karner 1998; Longley *et al.* 2002; Gibbons *et al.* 2012). This initiated the NW-trending inboard rift basins along the margin (Fig. 1).

Triassic subsidence

The Permo-Carboniferous rifting was followed by post-rift Triassic subsidence with the deposition of thick Triassic sections of predominantly marine claystones and siltstones with minor sandstones of the Locker Shale and their shelfal limestone equivalents. These grade upwards into the >7 km-thick Carnian–Norian succession of sandstones and mudstones of the marginal marine to fluvio-deltaic Mungaroo Formation (Driscoll & Karner 1998; Longley *et al.* 2002; Adamson *et al.* 2013; Marshall & Lang 2013) (Figs 3 & 4).

Latest Triassic–earliest Late Jurassic extension

Late Triassic (Rhaetian)–earliest Late Jurassic (Oxfordian) WNW-directed extension occurred as the Argo block separated from Gondwana (Heine & Müller 2005; Metcalfe 2013). Strain focused within the older rift fabrics to create the inboard Exmouth, Dampier, Barrow and Beagle sub-basins, and the outboard Exmouth Plateau to the NW of the NCB (Fig. 1) (Gartrell 2000; Pryer *et al.* 2002; Jitmahantakul & McClay 2013; McClay *et al.* 2013; McCormack & McClay 2013; Black *et al.* 2017). On the Exmouth Plateau, this extension was minor (cf. Fig. 2), and its onset is marked by the Rhaetian (TR30.1 TS) regional transgression (Marshall & Lang 2013; McCormack & McClay 2013; Gartrell *et al.* 2016). The main extension occurred in the Early Jurassic (Marshall & Lang 2013) in the inboard rift basins that accumulated thick (>6 km) Jurassic synrift sediments, whereas the outboard Exmouth Plateau was largely sediment starved and accumulated only a thin (<400 m) highly condensed Early–Mid-Jurassic section (Gartrell 2000; Longley *et al.* 2002; Gartrell *et al.* 2016).

The syn-extension sequence consists of the shelfal–offshore marine carbonate and associated reef build-ups of the Brigadier Formation (TR30) (Adamson *et al.* 2013; Grain *et al.* 2013), as well as the siltstones of the Murat and Athol formations that indicate deposition within a restricted marine setting (Hocking & Preston 1998; Longley *et al.* 2002) (Figs 3 & 4). The cessation of fault activity is marked by the key Oxfordian (J40.0 SB) unconformity (Heine & Müller 2005; McCormack & McClay 2013; Yang & Elders 2016).

Early Cretaceous extension

In the Exmouth Plateau, tectonic quiescence occurred from Oxfordian to Tithonian time. In the Berriasian (*c.* 144.2 Ma), Greater India began to separate from the western sector of the NCB, and this culminated in seafloor spreading and the formation of the Gascoyne (*c.* 135 Ma) and Cuvier (*c.* 133 Ma) abyssal plains to the west and south of the Exmouth Plateau, respectively (Fig. 1) (Müller *et al.* 1998; Heine *et al.* 2004; Direen *et al.* 2008; Gibbons *et al.* 2012). The Early Cretaceous break-up, coupled with a possible impingement of a Late Jurassic mantle plume in the southern Exmouth Plateau region (Rohrman 2015), produced uplift in the southern part of the NCB and generated the northwards progradation of the Barrow Delta across the Exmouth Plateau (Longley *et al.* 2002; Rohrman 2015; Reeve *et al.* 2016). Break-up-related volcanics and intrusives are widespread in the southern part of the NCB, particularly along the outer margins of the Exmouth Plateau (Figs 1 & 2) (Stagg *et al.* 2004; McClay *et al.* 2013; Rohrman 2013, 2015; Black *et al.* 2017; Magee *et al.* 2017).

Early Cretaceous–present-day passive margin

The passive margin in the NCB is marked by the regional Valanginian (K20.0 SB) erosional unconformity (Longley *et al.* 2002; Stagg *et al.* 2004). Passive-margin subsidence accommodated the Valanginian–Aptian deep-marine Muderong Shale that forms a regional seal (Tindale *et al.* 1998). The Late Oligocene (*c.* 25 Ma: Keep *et al.* 1998; Hall 2012) to present-day collision between the Australian plate and SE Asia microplates produced local inversion and reactivation of some of the larger faults in the NCB (e.g. Hengesh & Whitney 2016).

The Exmouth Plateau

The Exmouth Plateau is separated from the inboard rift sub-basins by the Kangaroo Syncline, and bounded to the north, NW and south by the Jurassic and Early Cretaceous oceanic crust of the Argo,

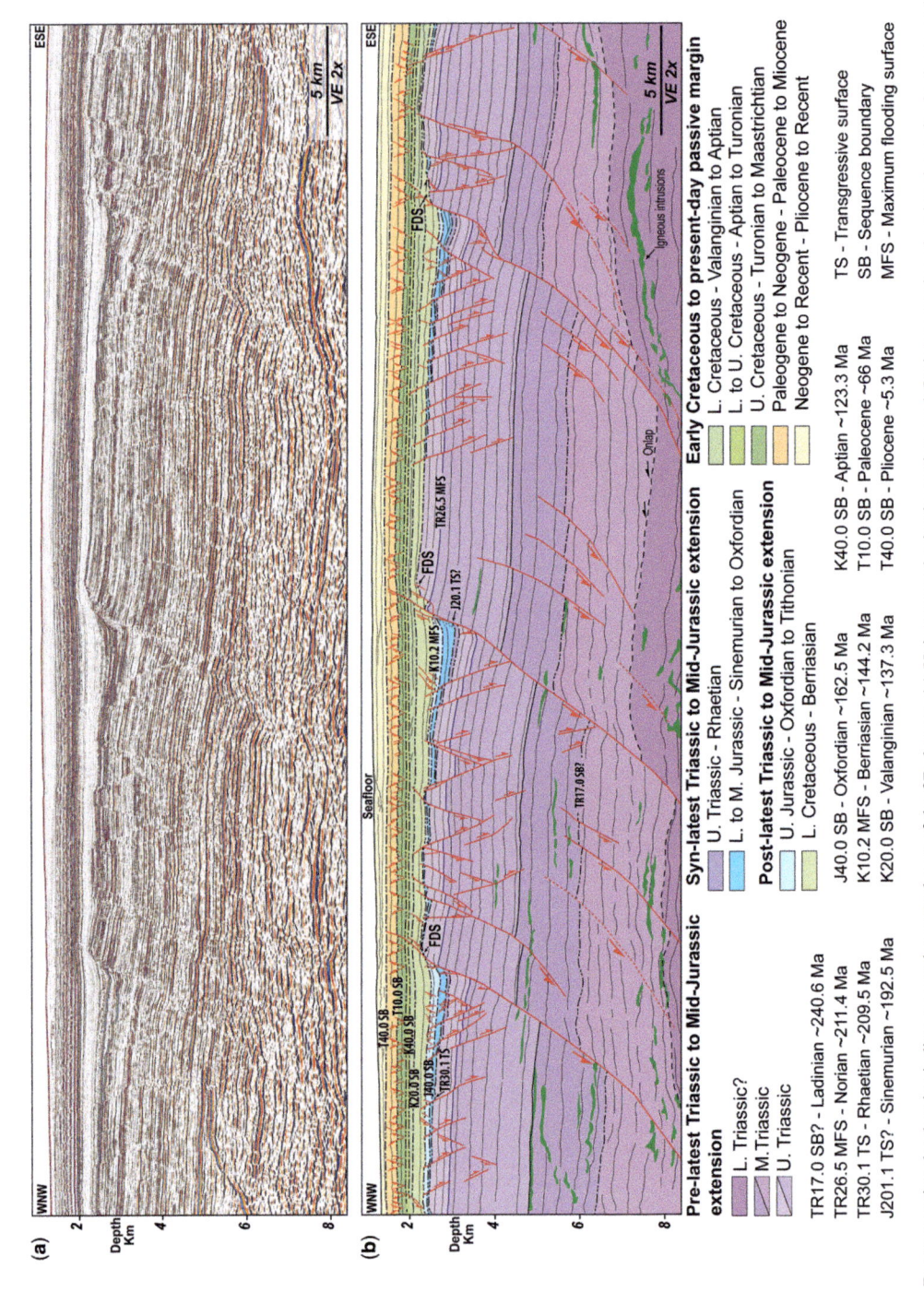

Fig. 2. Representative depth seismic line across the central part of the 3D dataset. (**a**) Uninterpreted seismic section; and (**b**) interpretation showing the major horizons and three asymmetrical half-graben bounded by three west-dipping planar extensional faults. The major faults are affected by footwall degradation scarps (FDS). The line location is shown in Figures 1, 5 and 6. VE, vertical exaggeration.

Fig. 3. Seismic stratigraphy of the Cazadores 3D survey comprising a simplified stratigraphy (modified from Lewis & Sircombe 2013) of the central Exmouth Plateau, key interpreted horizons and their ages. The main phases of the scarp degradation are also shown. Horizon names are based on Marshall & Lang (2013) and the interpretation was constrained by available wells. Major tectonic elements were compiled from Longley *et al.* (2002) and Gibbons *et al.* (2012). VE, vertical exaggeration.

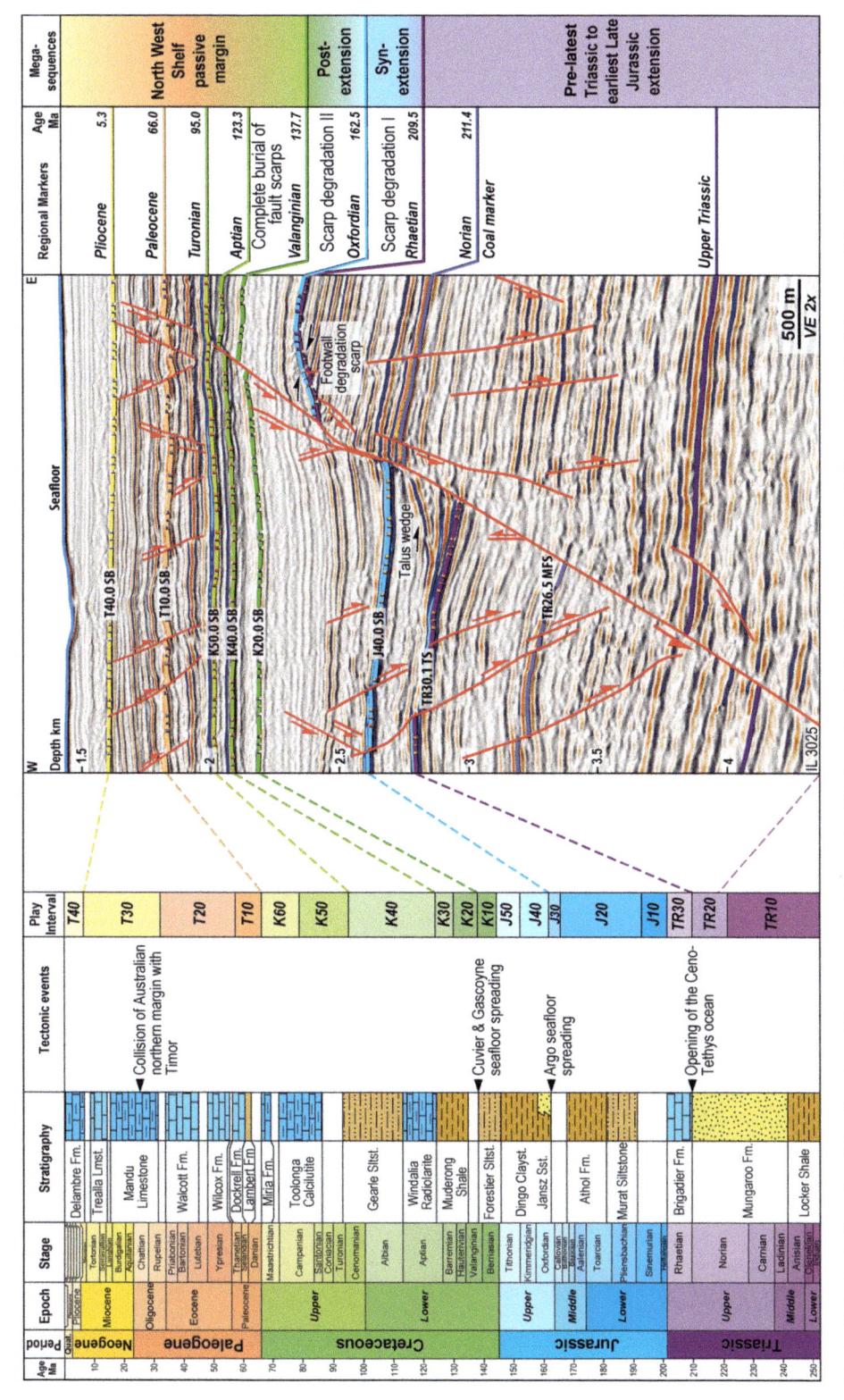

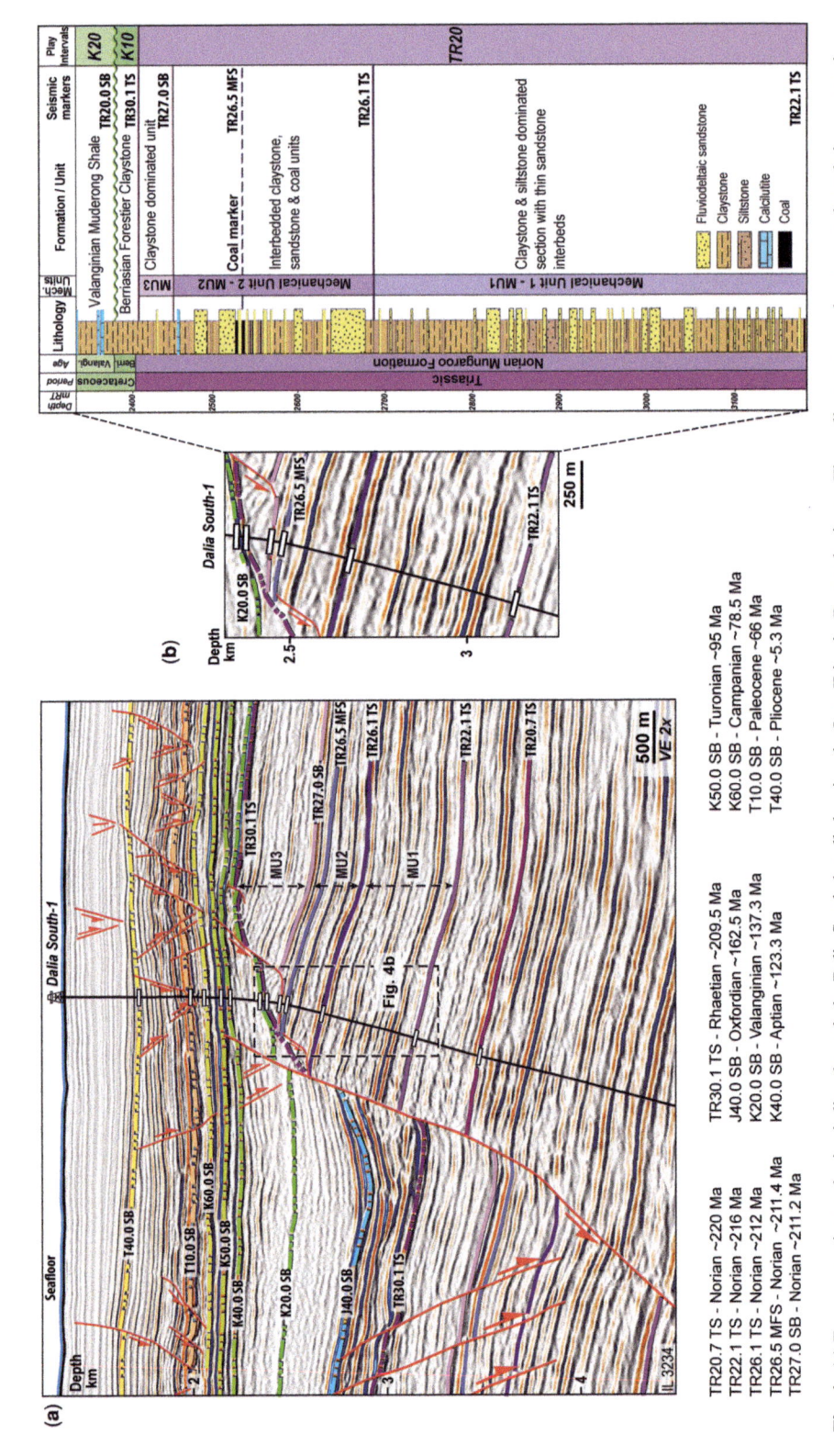

Fig. 4. (**a**) East–west-orientated seismic line through the Dalia South-1 well showing the Late Triassic–Recent horizons. The well penetrates a large scarp-degradation system in the footwall of a west-dipping major fault. No Jurassic section was encountered in the well. (**b**) Three mechanical-stratigraphic units (MU1–MU3) of the uppermost Norian section of the Mungaroo Formation. The well location is shown in Figure 1.

Gascoyne and Cuvier abyssal plains, respectively (Hopper et al. 1992; Stagg et al. 2004; Gibbons et al. 2012) (Fig. 1). It is approximately 400 km wide and 600 km long with an average water depth of c. 1.6 km (Fig. 1). The Exmouth Plateau underwent only relatively minor extension with a stretching factor of $\beta < 1.1$ during the latest Triassic–earliest Late Jurassic rifting. At this stage 10–18 km-spaced north- to NNE-trending domino-style rotational extensional faults were formed (Fig. 2). The Exmouth Plateau has at least 10–15 km of Mesozoic–present-day sediments consisting of >7 km of weakly deformed Mungaroo Formation and Locker Shale strata above inferred Permo-Carboniferous carbonates(?) that overlie either hyper-extended continental crust (Pryer et al. 2014) or even possibly Permian oceanic crust (Belgarde et al. 2015).

Research methodologies

Seismic data

The Cazadores 3D (2009) pre-stack time-migrated (PreSTM) survey covering an area of 4551 km² in the central Exmouth Plateau (Fig. 1) was used for this study. Wireline logs, check-shot data and formation tops from the Belicoso-1, Thebe-2 and Dalia South-1 wells were used to constrain the seismic interpretation (Fig. 1). Seismic interpretation and depth conversion were carried out using Landmark's DecisionSpace Desktop software. The stratigraphic scheme (Figs 2–4) used in this research follows that defined by Marshall & Lang (2013) for the North West Shelf of Australia.

Frequency analyses were carried out at a depth interval of c. 2200–3500 m, which corresponds to the Upper Triassic Norian (TR27.0 SB c. 211.2 Ma) to the Lower Cretaceous Aptian (K40.0 SB c. 123.3 Ma) succession where the fault-scarp degradation occurs. A 35 Hz dominant frequency and 2450 m s^{-1} average velocity gave a vertical and horizontal seismic resolution of 17.5 and 23 m, respectively.

Depth conversion of the seismic data was performed using time–depth data from the key wells in the research area. In the Exmouth Plateau, wells have only penetrated the upper part of the thick Mungaroo Formation. In order to provide velocity control at depths for strata below well penetrations, below the time–depth curves a simple analytical velocity model was derived using DecisionSpace Desktop with an assumption of a linear increase in velocity with depth.

Seismic interpretation

This study focused on a subarea of 1380 km² located in the westernmost part of the Cazadores survey (Fig. 1). Five key horizons spanning the pre-, syn- and post-extension megasequences were mapped in detail to define the faults and their degradation systems. Stratigraphic picks using 2D and 3D data were tied to the Belicoso-1 and Dalia South-1 wells as well as to the Thebe-2 well, which was outside the Cazadores 3D survey (Fig. 1). All wells were drilled in the footwall of major fault blocks. The Dalia South-1 well penetrated a large fault scarp in a west-dipping major fault (Fig. 4). The footwall erosion surfaces are mainly defined by the Late Triassic Rhaetian (TR30.1 TS) transgressive surface that was manually interpreted every two inlines (37.5 m) across the footwall blocks.

Interpreted faults and horizons were imported into Badleys TrapTester software for computing fault displacements. Displacement–distance profiles for the faults were constructed using the base syn-extension Rhaetian (TR30.1 TS) transgressive surface. In the fault footwalls, the interpreted Rhaetian surface was projected across and above the degradation complexes, as well as, where required, fault surfaces being projected updip in order to calculate the fault throws. The pre-extension Norian (TR26.5 MFS) coal marker was used as a guide for the projected Rhaetian (TR30.1 TS) surface. Uncertainties in defining the positions of the pre-erosion strata and palaeo-footwall crests are estimated to be ± 20 m.

Stratigraphic framework

The strata within the Cazadores survey varies from the Mid- to Upper Triassic pre-extension megasequence, the Upper Triassic–Upper Jurassic syn-extension megasequence and the Early Cretaceous post-extension to present-day passive-margin megasequences (Figs 3 & 4). Nine regional seismic horizons define the key tectonostratigraphic units in the research area (Fig. 3). For this paper, five key horizons from the upper part of the Triassic Mungaroo Formation to the Aptian (Figs 2–4) were mapped in great detail to define the fault systems and degradation complexes. The surfaces are:

- c. 211.4 Ma Late Triassic Norian (TR26.5 MFS) maximum flooding suface; known here as semi-regional coal marker;
- c. 209.5 Ma latest Triassic Rhaetian (TR30.1 TS) transgressive surface;
- c. 162.5 Ma earliest Late Jurassic Oxfordian (J40.0 SB) unconformity;
- c. 137.7 Ma Early Cretaceous Valanginian (K20.0 SB) unconformity;
- c. 123.3 Ma latest Early Cretaceous Aptian (K40.0 SB) sequence boundary (Figs 3 & 4).

Upper Triassic pre-extension megasequence c. 228.3–209.5 Ma

The pre-extension megasequence is the Carnian–Norian (c. 228.3 to c. 209.5 Ma) fluvio-deltaic Mungaroo (TR10–TR20) Formation (Adamson et al. 2013; Marshall & Lang 2013; Ford et al. 2015; Heldreich et al. 2017) (Figs 2 & 3). In the research area, the Mungaroo Formation is >4 km thick, and is bounded at the top by the strong, high-amplitude, Rhaetian (TR30.1 TS) transgressive surface that displays erosional truncations of the footwall crests of tilted fault blocks (Figs 2–4). The Mungaroo Formation consists of non-marine interbedded sandstones, siltstones, claystones and thin (c. 4 m) coal beds (Fig. 4).

Three stratigraphic units with different mechanical properties, MU1, MU2 and MU3, may be inferred for the Norian upper section of the Mungaroo Formation as intersected in the Dalia South-1 well (Fig. 4). Mechanical unit 1 (MU1: Fig. 4) is the claystone- and siltstone-dominated section between surfaces TR22.1 TS and TR26.1 TS that would be expected to be incompetent (i.e. weak) during deformation. The sand-rich, and therefore inferred competent, mechanical unit 2 (MU2: Fig. 4) occurs between the Norian TR26.1 TS transgressive surface and TR27.0 SB sequence boundary, and incorporates the Norian semi-regional coal marker (TR26.5 MFS) near the top (Fig. 4). The coal marker unit is a medium- to high-amplitude, continuous reflector that occurs c. 400–450 m below the Rhaetian (TR30.1 TS) transgressive surface. The overlying c. 100–300 m mechanical unit 3 (MU3) of the Norian Mungaroo Formation is dominated by weak, poorly lithified claystones and siltstones (Fig. 4). The majority of degraded fault scarps in the research area occur in MU3, and some of the listric slump scarp faults detach at the coal marker unit (TR26.5 MFS) near the top of MU2. Analyses of well data and seismic attributes indicate that the mechanical units (MU1–MU3) show variations across the central Exmouth Plateau, with a general landwards (eastwards) dominance of sandstone packages and basinwards (westwards) thickening of the mud-dominated units. These lithological variations in the Norian Mungaroo Formation are inferred to have exerted strong controls on the formation of fault-scarp degradation complexes in this part of the central Exmouth Plateau.

Upper Triassic–Mid-Jurassic syn-extension megasequence c. 209.5–162.5 Ma

The syn-extension megasequence is defined by the Rhaetian (TR30.1 TS) transgressive surface at the base, as shown in map view in Figure 5, and by the Oxfordian (J40.0 SB) unconformity at the top, as shown in map view in Figure 6 (see also Figs 2–4). In the study area, the syn-extension megasequence can be up to 400 m thick and consists of:

- offshore marine marls and calcilutites of the Rhaetian (c. 209.5 to c. 201.3 Ma) Brigadier (TR30) Formation (Fig. 3);
- shelfal–offshore marine siltstones and claystones of the Sinemurian–Callovian (c. 192.5–165.6 Ma) Athol and Murat (J20) formations (Fig. 3).

Syn-extension strata are thinnest (c. 20 m) in the degraded footwall scarps and these onlap the major Triassic footwall blocks as shown in Figure 6 (Dalia South-1 well: Fig. 4). The Dalia South-1 well penetrated a major fault scarp where the Jurassic section was missing and the Early Cretaceous post-extension Forestier (K10) Claystone immediately overlies the Rhaetian (TR30.1 TS) (Fig. 4).

The syn-extension megasequence is faulted by north- to NNE-trending domino-style planar growth faults that are truncated by the post-extension Valanginian (K20.0 SB) unconformity (Figs 2–4). Syn-extension strata display an overall wedge-shape geometry within hanging walls (Figs 3 & 4). The top syn-extension Oxfordian (J40.0 SB) unconformity is a regionally continuous, high-amplitude reflector that locally converges with the Rhaetian (TR30.1 TS) over the tilted fault blocks (Fig. 2).

Latest Jurassic–Early Cretaceous post-extension megasequence c. 162.5–123.3 Ma

The post-extension megasequence is dominated by the Early Cretaceous Barrow (K10–K20) Group (Figs 3 & 4). It consists of:

- thin (c. 30 m) Upper Jurassic marine marl and claystones interbedded with thin (c. 7 m) sandstones;
- the pro-delta Berriasian Forestier Claystone;
- the unconformable Valanginian offshore marine Muderong Shale that forms a regional seal.

The base of the post-extension megasequence is defined by the Oxfordian (J40.0 SB) unconformity. The Forestier Claystone (K10) is separated from the overlying Muderong Shale (K20–K30) by the Valanginian (K20.0 SB) unconformity (Figs 3 & 4). The top of the Muderong Shale is marked by the Aptian (K40.0 SB) sequence boundary (Fig. 3). Approximately 200 m of this post-extension megasequence was encountered in the Dalia South-1 well (Fig. 4). A number of the major domino faults were locally reactivated in the Early Cretaceous, displacing the Valanginian and Aptian surfaces (Figs 2–4).

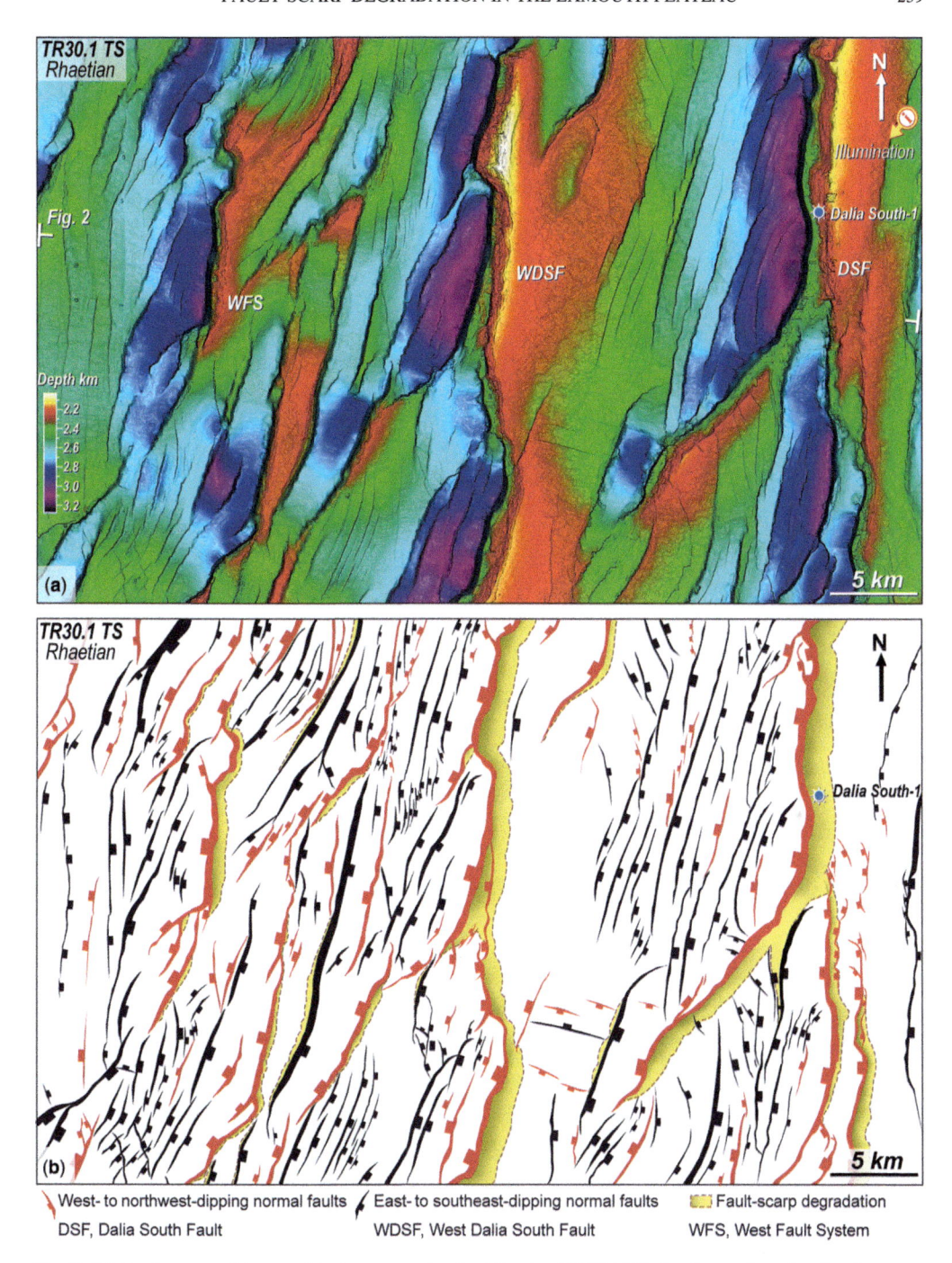

Fig. 5. (a) Depth structural image of the base syn-extension Rhaetian (TR30.1 TS) flooding surface showing the dominant north- to NNE-trending latest Triassic–earliest Late Jurassic extensional faults; and (b) fault interpretation map. Footwall degradation scarps are coloured yellow. Major fault systems are strongly affected by footwall degradation along their entire length.

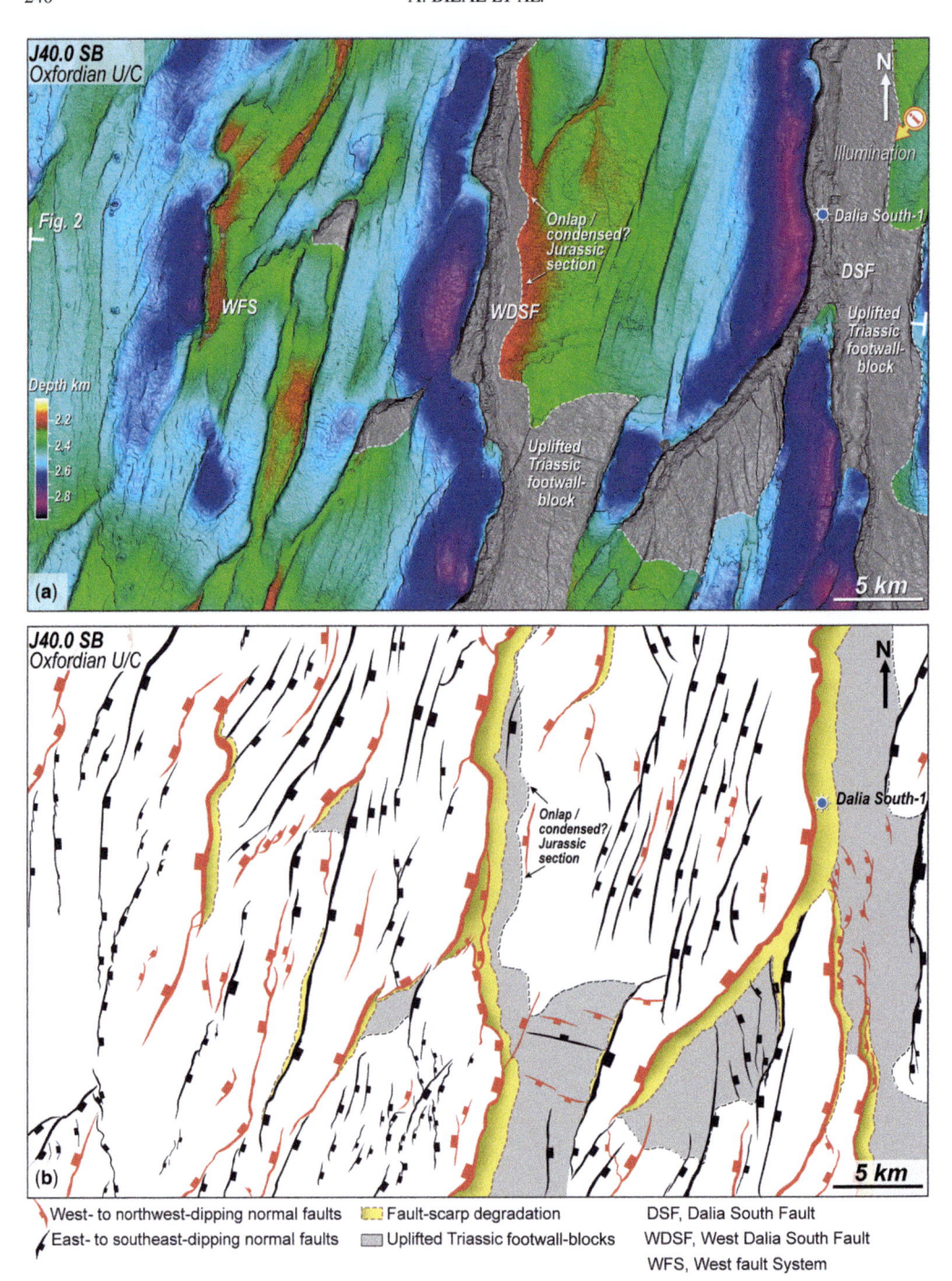

Fig. 6. (**a**) Depth structural image of the top syn-extension Oxfordian (J40.0 SB) unconformity showing the dominant north- to NNE-trending latest Triassic–earliest Late Jurassic extensional faults. The onlap of the Middle Jurassic syn-extension sequence on the uplifted and tilted Triassic fault blocks (grey surface) is shown by the dashed white line; and (**b**) fault interpretation map. Footwall degradation scarps are coloured yellow.

Structural framework

The research area is characterized by three major west-dipping planar extensional faults forming two central east-dipping, tilted footwall blocks in the thick Mungaroo Formation together with asymmetrical, hanging-wall half-graben infilled with thin asymmetrical wedges of uppermost Rhaetian–upper Jurassic strata (Fig. 2). The large faults have accumulated up to 1.2 km of displacement and are c. 35 km in length along strike. They are 16–18 km apart and extend below the base of the seismic data (Figs 2 & 5). In plan view, at the top of the pre-extension Mungaroo Formation, the Rhaetian (TR30.1 TS) surface shows a complex pattern of en echelon and offset depocentres bounded by kinked and clearly linked major faults such as the West Dalia South Fault (WDSF) and the Dalia South Fault (DSF) (Fig. 5a, b). The uplifted footwall blocks display metre to kilometre scale, concave erosional scarps and footwall degradation, as highlighted in Figure 5b. Smaller displacement (<200 m), largely east-dipping, antithetic faults occur in the hanging walls of these major faults (Figs 2 & 5). The antithetic faults are closely spaced (<1 km) and mainly truncated by the Oxfordian (J40.0 SB) unconformity.

This extensional fault pattern at the top of the pre-extension sequence (TR30.1 TS: Fig. 5) is complex, with the long major faults displaying relay ramps, breached relays and distinct kinks at along-strike fault linkage positions of originally separate, commonly en-echelon offset, fault segments (Fig. 5a, b). The lower displacement, generally antithetic, hanging-wall accommodation faults range from 1 to 16 km in strike length, and display complex patterns of individual faults and early-stage linkage geometries (the black faults in Fig. 5b). Similar

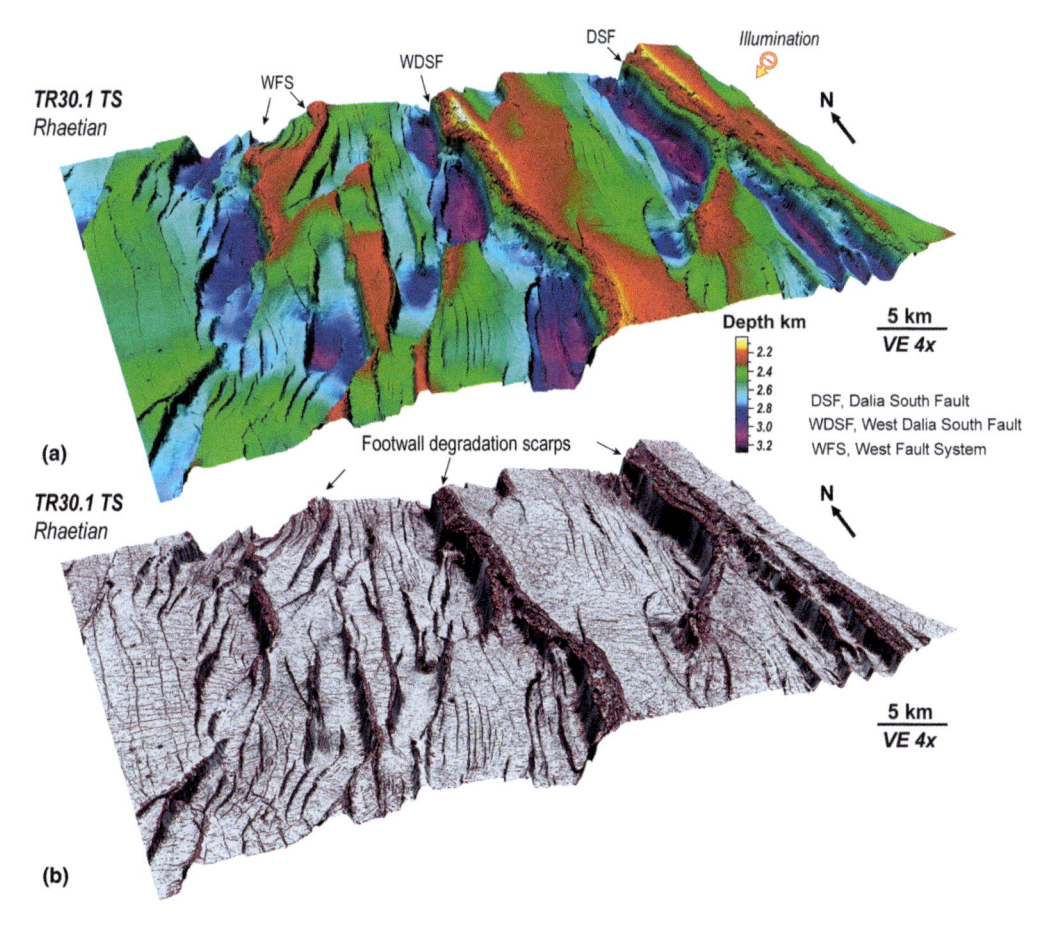

Fig. 7. 3D oblique visualization of the extensional faults and degraded fault scarps at the Rhaetian (TR30.1 TS). (**a**) Depth surface image; and (**b**) maximum curvature extraction on the depth surface highlighting the fault-scarp degradation zones. VE, vertical exaggeration.

fault architectures are shown at the top of the syn-extension sequence at the Oxfordian (J40.0 SB) unconformity but with less topography on both the main faults and the minor antithetic faults (Fig. 6a, b). The dominant fault strikes on both surfaces are N–S to NNE–SSW and the regional extension direction is inferred to be WNW; perpendicular to the dominant strikes of the smaller individual hanging-wall faults.

Fault-scarp degradation complexes

Prominent footwall degradation scarps are found along the three major west-dipping fault systems (Figs 5–8). They are characterized by low-angle truncations (c. 8°–15°) of the Mungaroo Formation, with up to 1.8 km of scarp retreat at the corners of the uplifted footwalls of the tilted fault blocks. The degradation scarps are best imaged in plan view at the level of the Rhaetian (TR30.1 TS) surface (Figs 5, 7 & 8). Significant scarp degradation has occurred where the footwall uplift was greatest, particularly on the WDSF and the DSF (Figs 5a, 6a, 7 & 8). Minor scarp degradation is only found on three of the larger antithetic faults at the Rhaetian (TR 30.1 TS) level in Figures 5 and 8.

The uplifted tilted Triassic footwall blocks are onlapped by the Jurassic offshore marine syn-extension strata, leaving the corners of these high-standing blocks with no syn-extension strata (Figs 4 & 6a, b). The fault-degradation scarps themselves are onlapped by the post-extension strata of the pro-deltaic Forestier Claystone and overlapped by the Valanginian (K20.0 SB) unconformity (Fig. 9a–d).

The degradation scarps dominantly occur in the uppermost unit (MU3) of the Mungaroo Formation (Figs 4 & 9a–d). Figure 9a and b shows that the northern part of the DSF (location shown in Fig. 8) has accumulated up to 900 m of throw. Here scarp degradation has removed nearly 500 m of the Norian (MU2-3) Mungaroo Formation from the corner of the tilted fault block (Fig. 9a, b). The scarp is onlapped by Valanginian strata and overlapped by the Lower Cretaceous Muderong Shale. The chaotic seismic facies in the syn-extension hanging-wall strata near the fault are interpreted to be footwall-derived slumped units (Fig. 9a, b).

The southern horst block of the DSF (Fig. 8) shows footwall-degradational scarps on both the eastern and western bounding faults (Fig. 9c, d). These are characterized by listric slump faults truncated by the Rhaetian (TR30.1 TS) surface (Fig. 9d). Here the southern part of the fault system has undergone less extensional offset (c. 600 m compared to c. 900 m on the northern part of the fault system) with less footwall uplift, such that the scarp

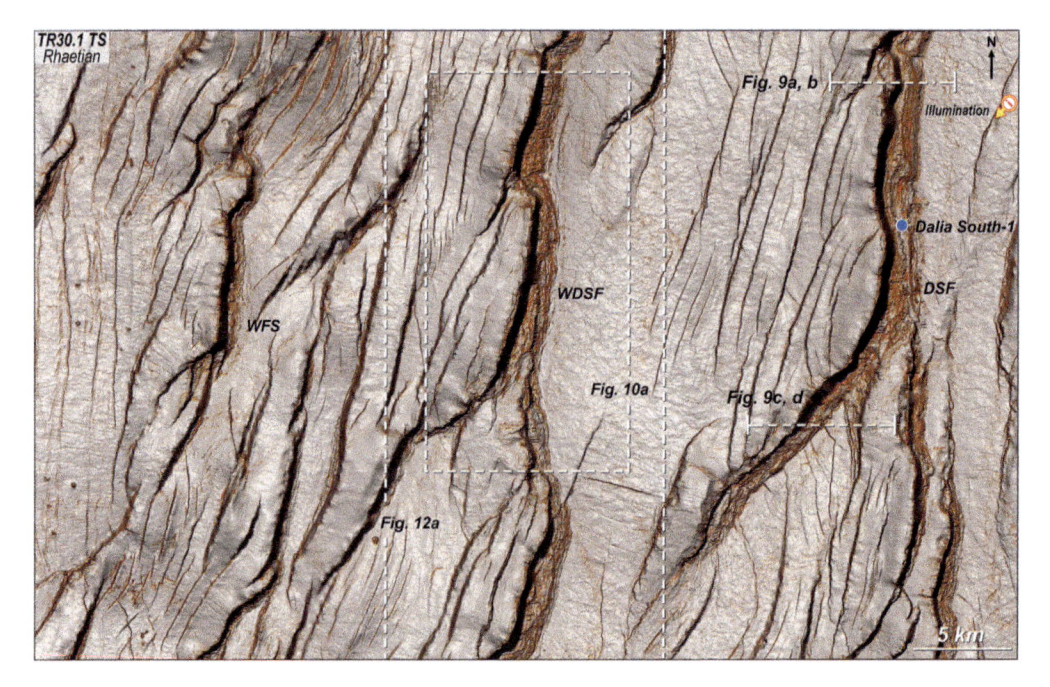

Fig. 8. Variance extraction image of the Rhaetian (TR30.1 TS) surface highlighting the degradation zones. DSF, Dalia South Fault; WDSF, West Dalia South Fault; WFS, West Fault System.

degradation only affected the upper c. 300 m of the top Mungaroo Formation with less than c. 200 m of the mechanical unit MU3 being removed from the footwall crest (Fig. 9d).

Similar footwall degradation scarps occur on the WDSF (Fig. 8) where listric slump faults are well developed (Fig. 10). Here they have formed on the footwall corners of en-echelon offset, and overlapping extension faults that became linked by breaching of the intervening relay ramps. The degradation scarps display well-developed, scoop-shaped fault surfaces that have coalesced along strike (Fig. 10a). Most of the listric scarp-collapse faults shown in Figures 9 and 10 detach within the shale-dominated Norian strata of mechanical unit MU3 in the upper Mungaroo Formation. Some also penetrate deeper to detach within the coal-rich units of the coal marker sequence (TR26.5 MFS; Fig. 4), giving rise to steps in the degradation surfaces (Figs 9 & 10). Footwall-scarp degradation features shown in Figures 9a, b, 10a, b and 11 exhibit distinctly benched topography that occurs within a narrow zone (<400 m) at the front of the scarp within the coal-rich unit (TR26.5 MFS) of mechanical unit MU2.

Hanging-wall talus wedges

As no wells penetrate the hanging walls of the large fault systems in the research area, interpretation of the Late Triassic–earliest Late Jurassic syn-extension (c. 209.5–162.5 Ma) megasequence has relied on the seismic facies and depositional geometries. Divergent reflectors and wedge-shaped bodies within the syn-extensional strata have been interpreted to be footwall-derived fans, and talus formed in the hanging walls adjacent to an active fault scarp (Figs 9 & 11). These talus wedges exhibit along-strike variations in thickness and seismic facies. A number of potential talus wedges have been identified along both the DSF and the WDSF systems, intercalated within the Lower Jurassic syn-extension strata (Figs 4, 10b & 11). The wedges have a low angle of repose (6°–8°) with onlap terminations onto the top boundary. The wedge widths can be up to 1.3 km in an east–west section and reach c. 240 m in thickness.

In several localities along strike on the major faults, the Oxfordian (J40.0 SB) unconformity is overlain by small post-extension talus wedges with a maximum thickness of c. 100 m (e.g. talus wedge II: Fig. 11). These are characterized by low- to medium-amplitude chaotic reflectors and the tops of the wedges are locally overlain by the Early Cretaceous Berriasian (K10.2 MFS c. 144.2 Ma) flooding surface. Their positions and nature suggest that they are local products of post-extension footwall erosion during the latest Jurassic before the remnant fault scarps were buried by the overlapping Muderong Shale (Fig. 11).

Fault segments, displacements and footwall-scarp degradation

The sector of the WDSF that extends across the study area is c. 35 km long and is formed of six, west-dipping normal fault segments (F1–F6: Fig. 12a) that are linked at the overlap zones via footwall-breached relay ramps (Fig. 12a, b). Three synclinal, en-echelon hanging-wall depocentres occur along the fault system (Fig. 12a). Fault displacements at the base of the syn-extension sequence (Late Triassic Rhaetian TR30.1 TS) were analysed using Badleys TrapTester software in order to determine the throw distributions and the amount of footwall-scarp degradation on the fault system.

The post-degradation and projected throw distributions for the six segments of the WDSF system are shown in Figure 12c. The majority of the faults exhibit asymmetrical, broadly flat-topped displacement profiles with along-strike steps, locally of large magnitude (c. 500 m) near branch lines. Maximum throw occurs approximately near the centres of the faults and decreases towards the fault tips. F3b, however, exhibits a peak-shaped profile. The more evolved faults, F1, F2 and F3a, are strongly linked and exhibit inactive fault-tip splays with up to c. 1200 m of projected maximum throw. Faults F4 and F5 are less linked, with a maximum projected throw of c. 550 m. F6 appears to be composed of four subsegments with a general increase in throw towards the southern tip of the fault. The projected throw graphs (Fig. 12c) partially mimic the post-degradation curves but show less scarp degradation towards the fault tips, indicating maximum degradation in the region of maximum displacement and hence maximum footwall uplift. There is also a strong stratigraphic control exerted by the incompetent >300 m-thick upper Norian mudstones (MU3) that are preferentially degraded and eroded on the footwalls of the major fault systems.

Discussion

Fault-scarp degradation in the central Exmouth Plateau

This study of footwall-scarp degradation complexes in the central Exmouth Plateau has shown that they are focused along the major, linked extensional faults where the displacements are greater than 250–300 m and as much as 1200 m or more (Figs 5, 6 & 12). They are interpreted to have formed in response to the latest Triassic–Mid-Jurassic faulting (Figs 3 & 5) that produced widespread extension across the central Exmouth Plateau. Minor, low-displacement extensional faults do not show footwall degradation features (Figs 5 & 6). The corners of the

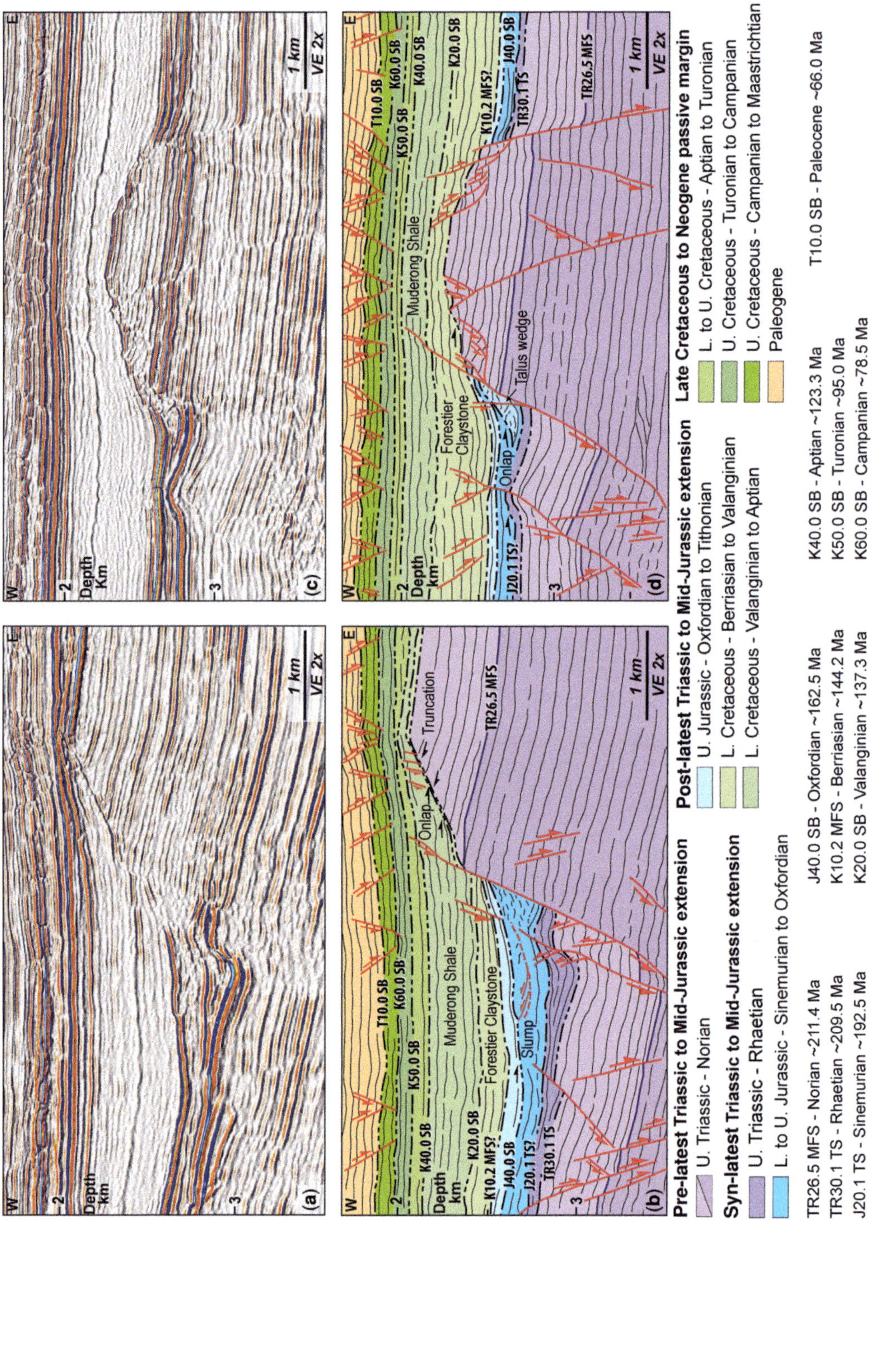
W
E
Depth km
2
3
(a)
1 km
VE 2x
(c)
(b)
(d)
T10.0 SB
K60.0 SB
K50.0 SB
K40.0 SB
K20.0 SB
K10.2 MFS?
J40.0 SB
J20.1 TS?
TR30.1 TS
TR26.5 MFS
Muderong Shale
Forestier Claystone
Slump
Onlap
Truncation
Talus wedge
Pre-latest Triassic to Mid-Jurassic extension
U. Triassic - Norian
Syn-latest Triassic to Mid-Jurassic extension
U. Triassic - Rhaetian
L. to U. Jurassic - Sinemurian to Oxfordian
Post-latest Triassic to Mid-Jurassic extension
U. Jurassic - Oxfordian to Tithonian
L. Cretaceous - Berriasian to Valanginian
L. Cretaceous - Valanginian to Aptian
Late Cretaceous to Neogene passive margin
L. to U. Cretaceous - Aptian to Turonian
U. Cretaceous - Turonian to Campanian
U. Cretaceous - Campanian to Maastrichtian
Paleogene
TR26.5 MFS - Norian ~211.4 Ma
TR30.1 TS - Rhaetian ~209.5 Ma
J20.1 TS - Sinemurian ~192.5 Ma
J40.0 SB - Oxfordian ~162.5 Ma
K10.2 MFS - Berriasian ~144.2 Ma
K20.0 SB - Valanginian ~137.3 Ma
K40.0 SB - Aptian ~123.3 Ma
K50.0 SB - Turonian ~95.0 Ma
K60.0 SB - Campanian ~78.5 Ma
T10.0 SB - Paleocene ~66.0 Ma

major tilted and uplifted footwall fault blocks underwent erosion and significant scarp collapse on scoop-shaped listric faults that detached within the incompetent, mud-rock dominated, upper Norian strata of the Triassic Mungaroo delta (MU3: Figs 4, 9, 10 & 11). In places, the listric collapse faults sole out on the Norian TR26.5 MFS coal units in the upper part of the mechanical unit MU2 (Figs 4, 9 & 10b). Scarp retreat as great as 1.8 km has been measured.

The main large-displacement extensional faults systems are characterized by extensional fault-propagation folds (Figs 2, 9, 10 & 11). These produced a narrow footwall anticline where the frontal limb would have been tilted to dip towards the hanging-wall syncline, such that bed-parallel slip and failure would have been enhanced during gravitational collapse and scarp degradation.

Syn-extension to early post-extension scarp collapse and erosion produced local talus wedges up to 240 m thick in the adjacent hanging-wall basins (Figs 4, 9 & 11). Hanging-wall fans and talus wedges have been widely documented in rift basins such as the Gulf of Suez (e.g. Sharp et al. 2000) and in the East African rift systems (Morley 1999), as well as in the southern Viking Graben, North Sea (Fraser et al. 2002).

The degraded fault scarps of the large west-dipping extensional fault systems are interpreted to have not been completely buried until the Aptian (c. 123.3 Ma; Figs 4, 9a, b & 11). The onlap of the open-marine Lower Cretaceous Berriasian and Valanginian reflectors against most of these scarps, in particular across the centre of the faults, indicate that they were exposed at these times. However, at the fault tips, the degraded scarps were completely buried by the Valanginian (K20.0 SB) unconformity (e.g. Fig. 9c, d). The smaller displacement faults experienced relatively minor footwall uplift focused on the central sector of the faults and most of them were completely buried by the latest Berriasian (c. 137 Ma).

Previous studies of footwall fault-scarp degradation in rift basins indicate that footwall erosion dominantly occurred during the rift climax where the fault slip and footwall uplift and rotation are typically high (Dahl & Solli 1993; Berger & Roberts 1999; Hesthammer & Fossen 1999; McLeod et al. 2002; Densmore 2004; McArthur et al. 2016a). Subaerially exposed islands may have formed in the footwalls of active normal faults particularly during the rift climax (Morley et al. 2007; McArthur et al. 2016b).

This study suggests that scarp degradation in the central Exmouth Plateau occurred in a fully-marine setting. This interpretation is supported by the occurrence of a basin-wide transition from the pre-extension fluvio-deltaic Mungaroo Formation to the syn-extension sequences of the shelf–offshore marine Brigadier Formation carbonates and the Jurassic offshore marine Murat and Athol formations. Well data also suggest that there is no evidence of subaerial exposure (e.g. rootlets) during the development of these scarp-degradation complexes. Moreover, in subaerial environments, scarp canyons are commonly widespread and modify the whole scarp; whereas in submarine settings, canyons are relatively localized (Stewart & Reeds 2003; Elliott et al. 2012). In this study there is a marked lack of canyons both on the scarp faces and on the back-limbs of the uplifted fault blocks, indicating a submarine environment during degradation similar to the interpretations of Hesthammer & Fossen (1999) and McLeod & Underhill (1999) for scarp degradation in the Viking Graben, northern North Sea.

Controls on extensional fault-scarp degradation

The relationships between variations in mechanical stratigraphy and the development of fault-related footwall topography during extension have been documented in many rifts and passive margins (Densmore 2004; Baran et al. 2010; Elliott et al. 2012). Lithological variations are thought to play a crucial role in the development of fault-scarp degradation, including the amount of scarp retreat and the overall scarp morphologies (Eliet & Gawthorpe 1995; Elliott et al. 2012).

In the research area, the degradation features mostly affect the c. 300 m uppermost mud-rock-dominated section (mechanical unit MU3) of the Norian Mungaroo Formation (Figs 4 & 9a, b), with a stepped and corrugated scarp morphology (e.g. Figs 10 & 11). Fault displacements vary along strike from no displacement at fault tips to maximum displacement and footwall uplift in the centre of the fault system, such that the central sectors are likely to undergo more degradation and scarp retreat as observed along the WDSF (Figs 8, 10 & 12). Positive relationships between fault throw and footwall-scarp erosional patterns have also been recognized from both outcrops and subsurface examples of fault-scarp erosion (e.g. Elliott et al. 2012).

Fig. 9. Detailed east–west depth seismic sections and their interpretations showing the structural styles associated with fault-scarp degradation. Uninterpreted section (**a**) and interpretation (**b**) showing a characteristic smooth topography of footwall-scarp degradation with no evidence of gravitational structures. Note the erosional truncation of the pre-extensional units. Uninterpreted seismic section (**c**) and interpretation (**d**) illustrating a degraded horst block bounded by two oppositely dipping extensional faults. Note the preserved slide blocks on the footwall crest and the talus wedge in the adjacent hanging-wall depocentre. VE, vertical exaggeration.

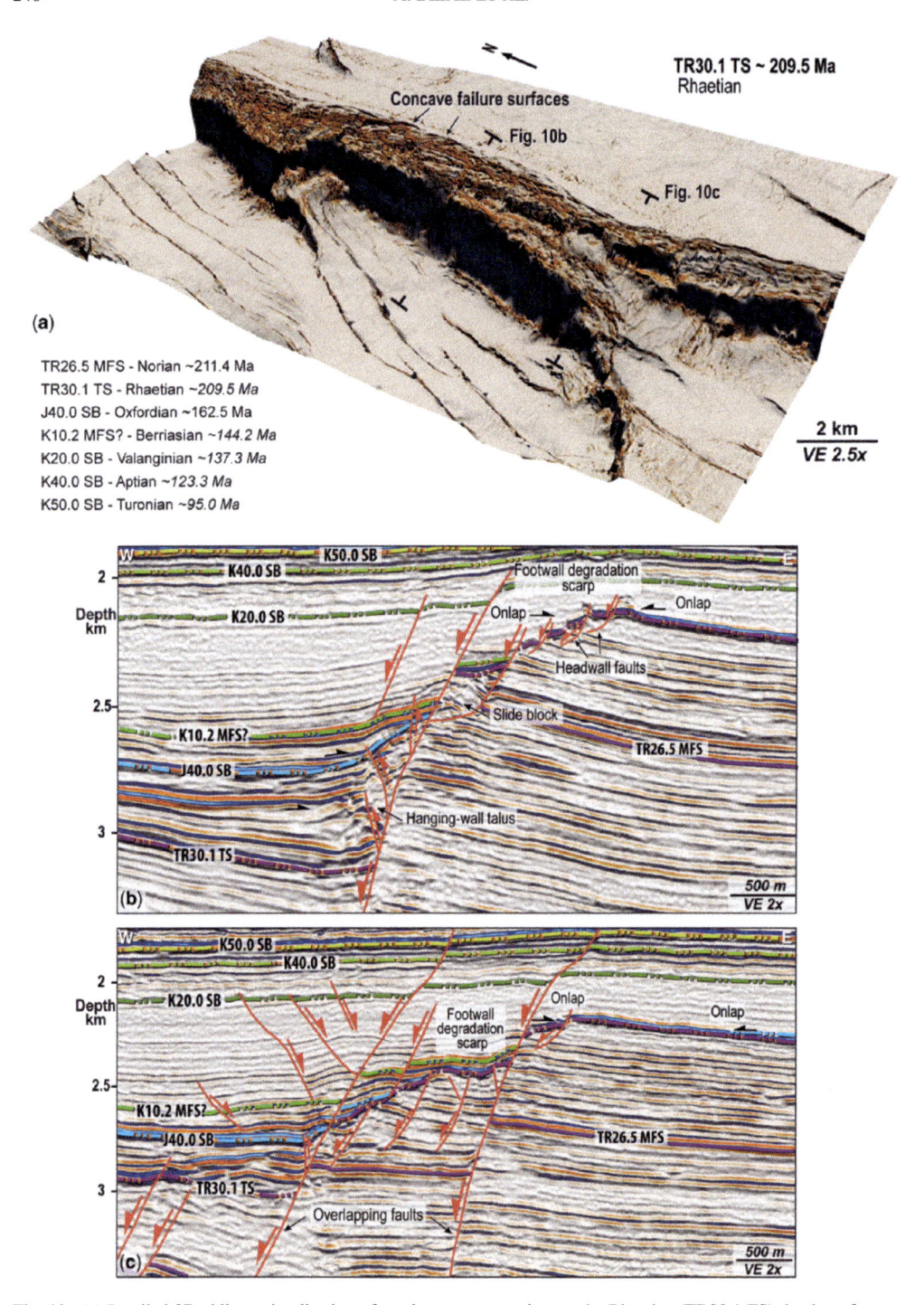

Fig. 10. (**a**) Detailed 3D oblique visualization of a coherency extraction on the Rhaetian (TR30.1 TS) depth surface showing coalesced scoop-shaped failure surfaces and along-strike variation of fault-scarp degradation. (**b**) East–west seismic section across the main fault showing successive deepening of listric faults on the footwall scarp. The deepest failure surface detaches at the coal marker (TR26.5 MFS) level and is linked to the main fault. (**c**) East–west seismic section across a footwall-breached relay ramp illustrating a west-dipping extensional fault system below the failure surface within the overlapping fault zones. VE, vertical exaggeration.

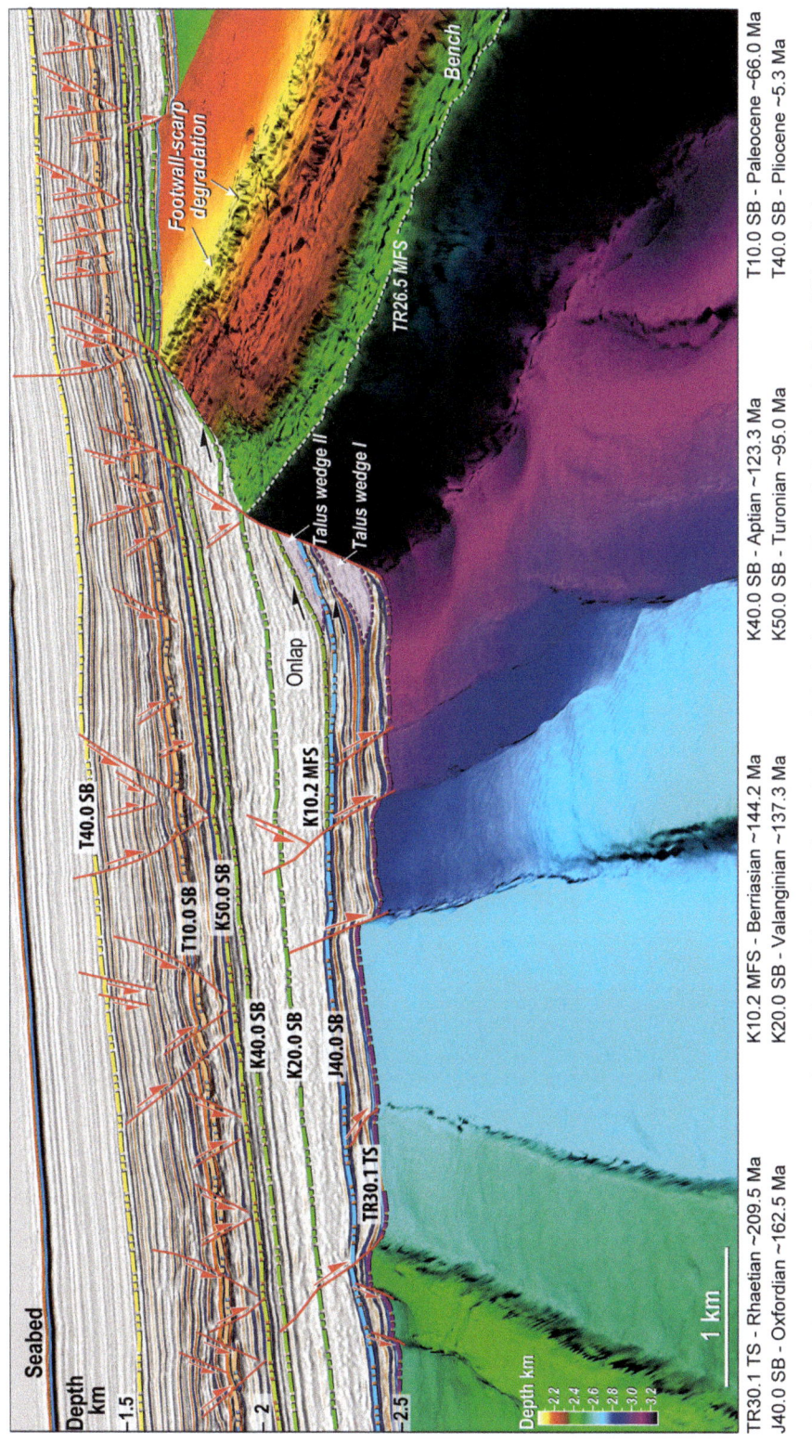

FAULT-SCARP DEGRADATION IN THE EXMOUTH PLATEAU 247

Fig. 11. 3D chair display in depth showing the Rhaetian (TR30.1 TS) surface and the vertical seismic section. The Dalia South Fault (DSF) exhibits a distinct fault-scarp-and-bench topography. The bench occurs within the incompetent coal-rich units (MU2: TR26.5 MF) of the Mungaroo Formation. Two hanging-wall talus wedges are also shown – syn-extension (talus wedge I) and earliest post-extension (talus wedge II).

TR30.1 TS - Rhaetian ~209.5 Ma   K10.2 MFS - Berriasian ~144.2 Ma   K40.0 SB - Aptian ~123.3 Ma   T10.0 SB - Paleocene ~66.0 Ma
J40.0 SB - Oxfordian ~162.5 Ma   K20.0 SB - Valanginian ~137.3 Ma   K50.0 SB - Turonian ~95.0 Ma   T40.0 SB - Pliocene ~5.3 Ma

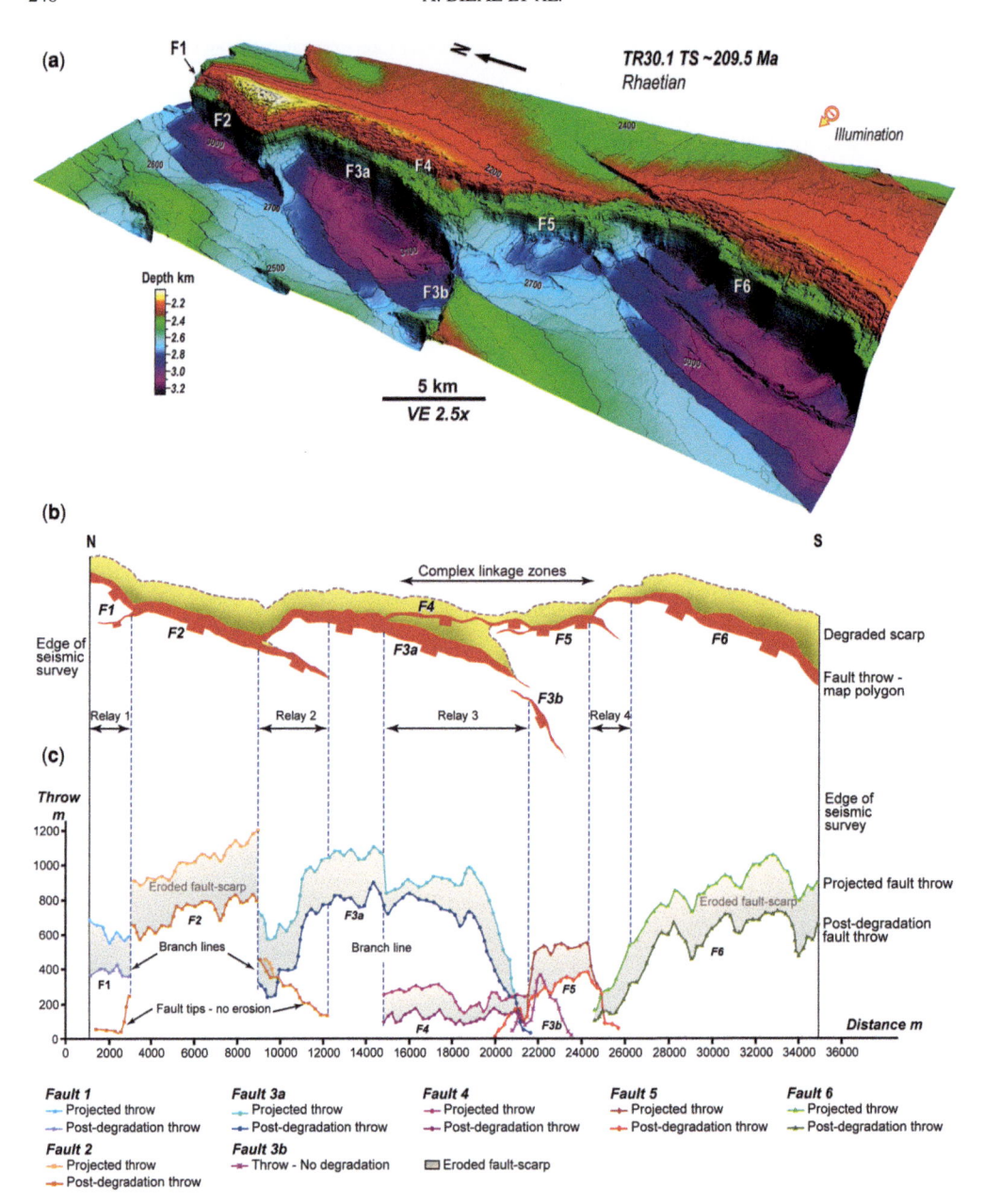

Fig. 12. (a) 3D oblique view of the Rhaetian (TR30.1 TS) surface showing six distinct fault segments (F1–F6) of the West Dalia South Fault (WDSF) system. The segments are linked via footwall-breached relay ramps and are strongly affected by footwall degradation. (b) Fault map of the fault segments (F1–F6) and major relay zones (Relay 1–Relay 4); footwall degradation is shown in yellow. (c) Fault throw–distance plot along the entire *c.* 35 km of the fault system showing both post-degradation and projected (pre-degradation) throw distributions. The majority of the faults exhibit broadly flat-topped displacement profiles with maximum throws near the fault centres and decreasing throws towards the fault tips. Projected fault throws partially mimic the post-degradation throws, with maximum degradation commonly associated with the regions of maximum fault displacements and footwall uplift. VE, vertical exaggeration.

Models for fault-scarp degradation: Exmouth Plateau

Figures 13 (2D evolution) and 14 (3D evolution) show proposed models for the progressive evolution of footwall fault-scarp degradation on the Exmouth Plateau.

Earliest Rhaetian (c. 209.5 Ma): initiation of extensional faulting. WNW-directed weak extension began in the Rhaetian with probable reactivation of pre-existing faults in the Mungaroo Formation to produce fault-propagation flexures (footwall anticlines and hanging-wall synclines) above the tips of blind faults at depth (Figs 13a & 14a).

Rhaetian: extension, fault-propagation folding, rotation and footwall uplift. With increased extension the faults propagated to the surface, as well as along strike, to develop extensional fault-propagation folds with prominent hanging-wall synclines (Figs 13b & 14b). Increased displacement, footwall rotation and uplift, coupled with low sedimentation rates in this part of the Exmouth Plateau, produced significant scarp relief (Figs 13b & 14b).

Sinemurian–Oxfordian (c. 192.5–162.5 Ma): syn-extension footwall degradation I. Figures 13c, d and 14c illustrate the progressive development of footwall collapse of the uppermost Mungaroo Formation (incompetent mechanical unit MU3) as fault displacement increased. The synclinal hanging-wall basins became partly infilled with predominantly Lower–Middle Jurassic syn-extension sediments. These incorporate the reworked uppermost Triassic marine carbonates of the Brigadier Formation, as well as the Jurassic marine claystones and siltstones of the Athol and Murat formations, with minor thickening into the fault hanging walls. Scarp degradation produced listric faults, submarine landslides and mass flows that formed wedge-shaped fans up to 240 m thick in the fault hanging walls (talus wedge I: Figs 13c, d & 14c). Continued starved regional sedimentation produced a condensed Mid-Jurassic section on the degraded fault scarps and on top of the footwall blocks (Figs 13d & 14c).

Oxfordian–Berriasian (c. 162.5–144.2 Ma) post-extension: footwall degradation II. The Oxfordian (J40.0 SB c. 162.5 Ma) unconformity marks the cessation of fault activity and post-extension deposition from the Oxfordian to the Berriasian. At this stage a second phase of scarp erosion occurred with hanging-wall talus wedges deposited in narrow zones adjacent to the master fault zone and intercalated with the uppermost Jurassic marine marls and claystones (talus wedge II: Figs 13e & 14d). These post-extension wedges are overlain by the Early Cretaceous Berriasian (K10.2 MFS) flooding surface.

Berriasian–earliest Valanginian (c. 144.2–139 Ma) post-extension: fault sealing. Deposition of the Berriasian Forestier Claystone (K10) occurred with more focused sedimentation in starved hanging-wall locations and onlap terminations onto the base of the degraded scarps. Burial of some faults by the early Valanginian sealed the faults and left eroded footwall scarps exposed (Figs 13f & 14e).

Valanginian (c. 137.7 Ma) post-extension: fault reactivation. Minor fault reactivation and breakthrough of the degraded scarps occurred in the Valanginian and continued into the Aptian, offsetting the Valanginian (K20.0 SB) unconformity. Deposition of the Muderong Shale (K20–K30) infilled the remnant topography (Figs 13g & 14f).

Valanginian–Aptian (c. 137.7–123.3 Ma) post-extension: complete burial of the main fault scarps. Figures 13h and 14g show continued deposition of the Muderong Shale over the degraded footwall scarps during Valanginian–Barremian times. The Aptian (K40.0 SB c. 123.3 Ma) sequence boundary marks the top of the Muderong Shale and seals many major fault scarps across the Exmouth Plateau. Continued minor reactivation of the main faults occurred together with new extensional faults that nucleated at the upper break in slope of the degraded scarps and propagated upwards into the passive-margin megasequence (Figs 13h & 14g).

Comparative examples

Fault-scarp degradation has been well documented in many rift systems and in particular in numerous examples from the Viking Graben, northern North Sea (Hesthammer & Fossen 1999; McLeod & Underhill 1999; Fraser et al. 2002; McLeod et al. 2002; Gibbons et al. 2003; Stewart & Reeds 2003; Welbon et al. 2007; Elliott et al. 2012) (e.g. Fig. 15). Fault-scarp degradation slump fault systems that are cuspate both in map view and in cross-section have been mapped in detail in both the corner of the Brent fault block (Fig. 15a) as well as in the Statfjord Field (Fig. 15b) where productive reservoirs have been found in the listric fault blocks (Welbon et al. 2007). The Alwyn Field example shows an erosional truncation across the corner of the tilted fault block (Fig. 15c). The Brae fields of the southern Viking Graben display well-documented fault-scarp retreat and degradation with significant hanging-wall fans formed in the hanging walls adjacent to the faults (Fraser et al. 2002). The above features are very similar in architecture and depositional systems to the degradation features documented in this paper

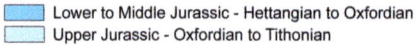

Fig. 13. 2D evolutionary models of fault-scarp degradation in the central Exmouth Plateau. (**a**) Earliest Rhaetian extension and development of monoclinal flexure above a blind fault. (**b**) Rhaetian extension, uplift and fault-propagation folding enhances footwall gravitational instabilities. (**c**) Continued displacement on the fault resulted in gravitational collapse of the uppermost Mungaroo claystone unit (MU3) and early deposition of hanging-wall debris (talus wedge I). (**d**) With increased fault offset, deeper units (MU2) are exposed to gravitational instability together with footwall erosion. Mid-Jurassic strata onlap and bury syn-extension talus wedge I. (**e**) Post-extension scarp degradation with eroded sediments re-deposited in narrow zones in the hanging-wall depocentres (talus wedge II). (**f**) Sealing of the main fault and onlap of the Berriasian Forestier Claystone onto the lower part of the scarp. (**g**) Valanginian minor reactivation and fault breakthrough of the degraded scarp. A new fault nucleates at the breakaway zone. (**h**) Continued minor reactivation and complete burial of the scarp occurred by the Aptian (K40.0 SB) sequence boundary.

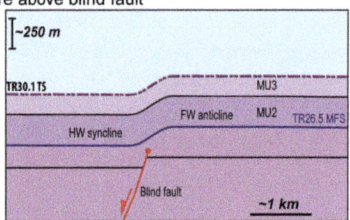

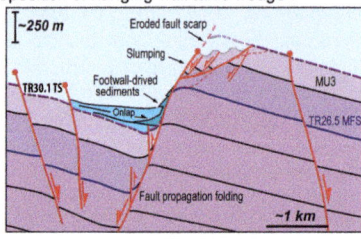

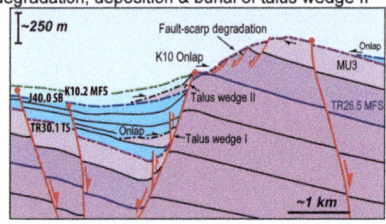

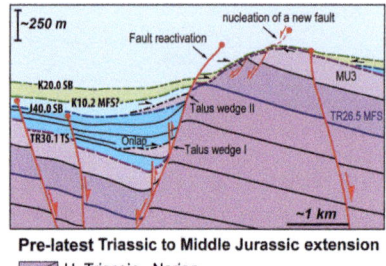

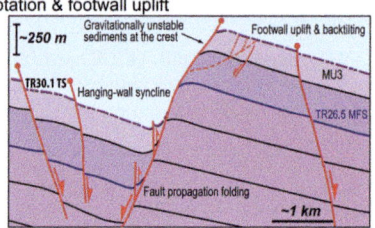

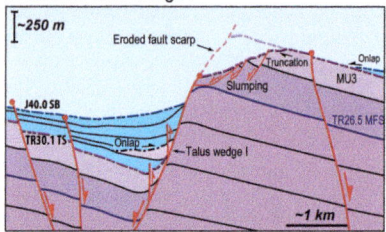

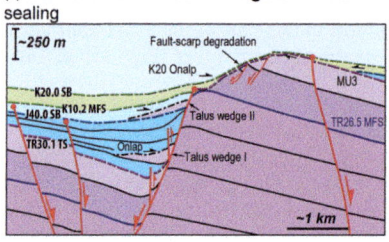

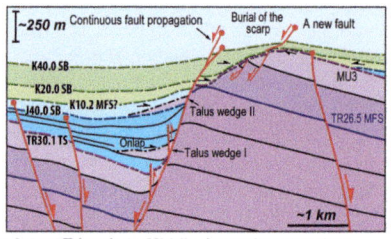

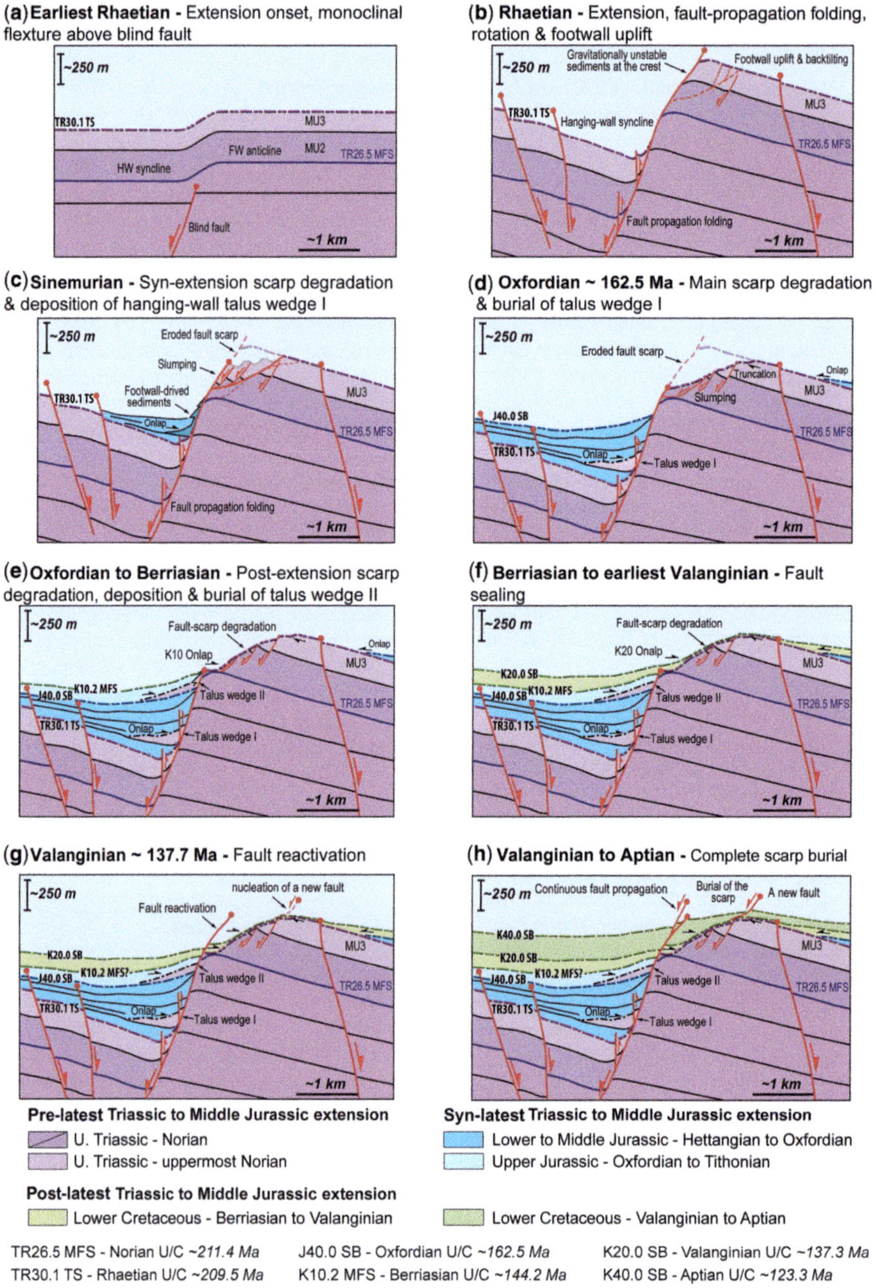

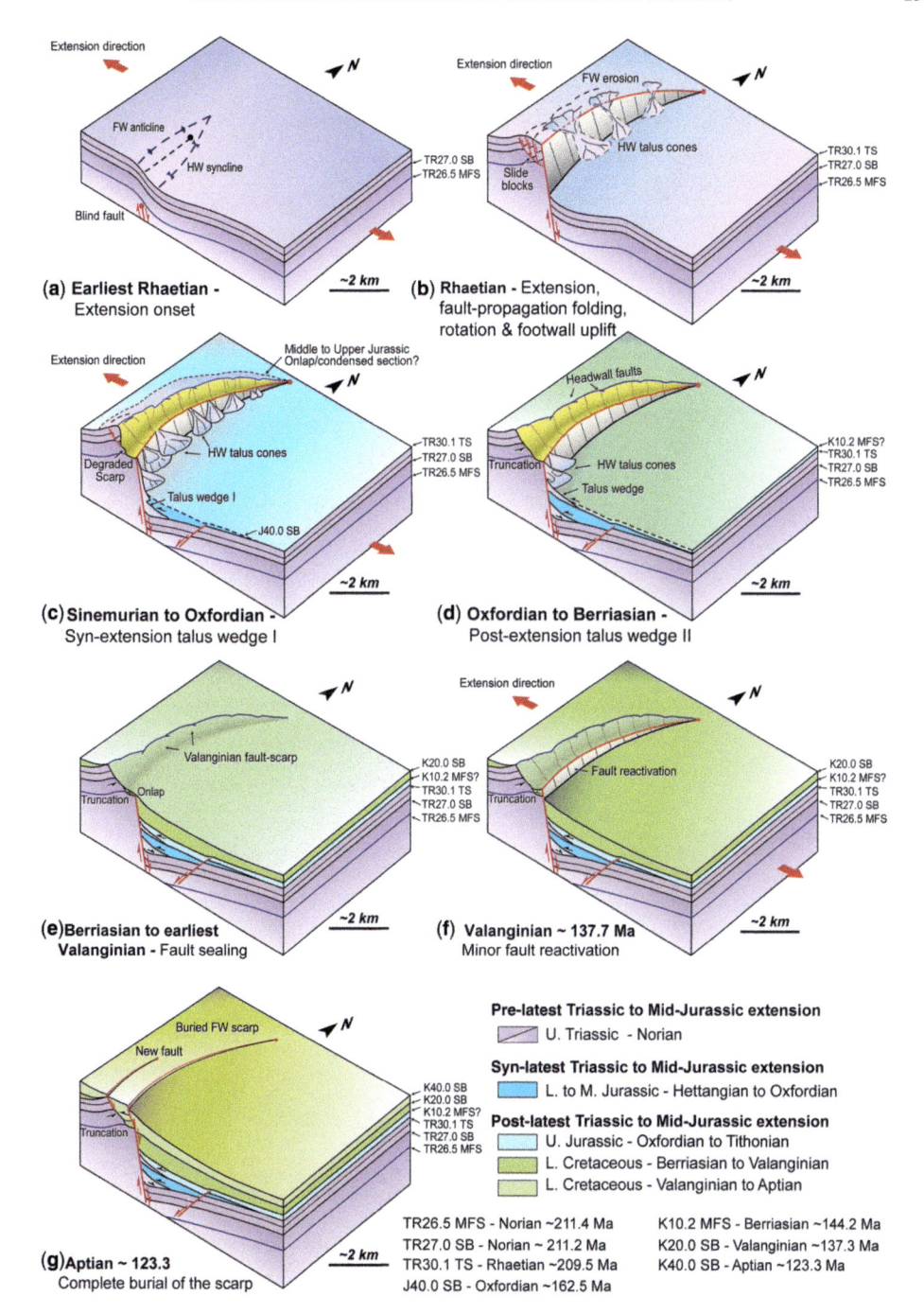

Fig. 14. 3D evolutionary block models of extensional fault-scarp degradation in the central Exmouth Plateau. (**a**) Monoclinal flexure above a blind fault; (**b**) fault breakthrough during earliest syn-extension with an uplifted footwall crest subjected to both gravitational collapse and footwall erosion. (**c**) Syn-extension footwall degradation with eroded sediments re-deposited as hanging-wall talus wedges I. A few slide blocks may be preserved in the footwall. (**d**) Minor footwall erosion during earliest post-extension in the regions of high displacement. (**e**) Sealing of the main fault and early burial of the lower part of the degraded slope. (**f**) Early Cretaceous fault reactivation associated with sediment drape over the degraded footwall slope. (**g**) Complete burial of the degraded scarp with nucleation of a new fault at the upper breaks in slopes of the degraded scarps.

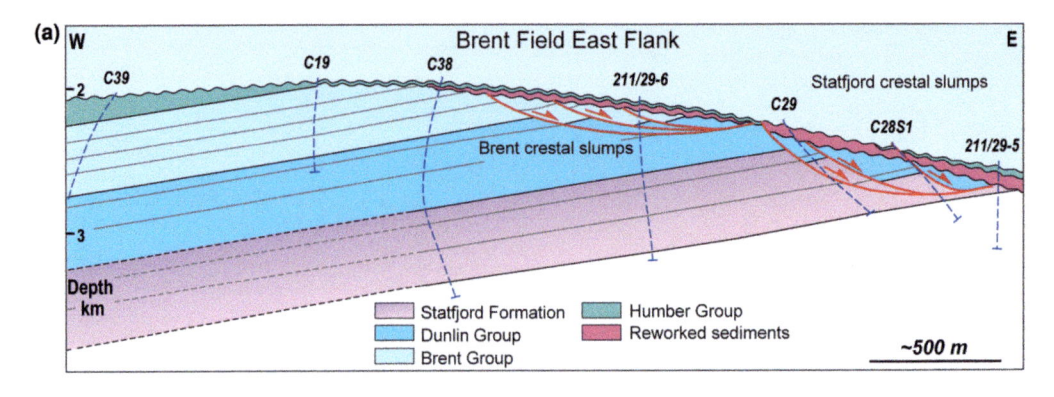

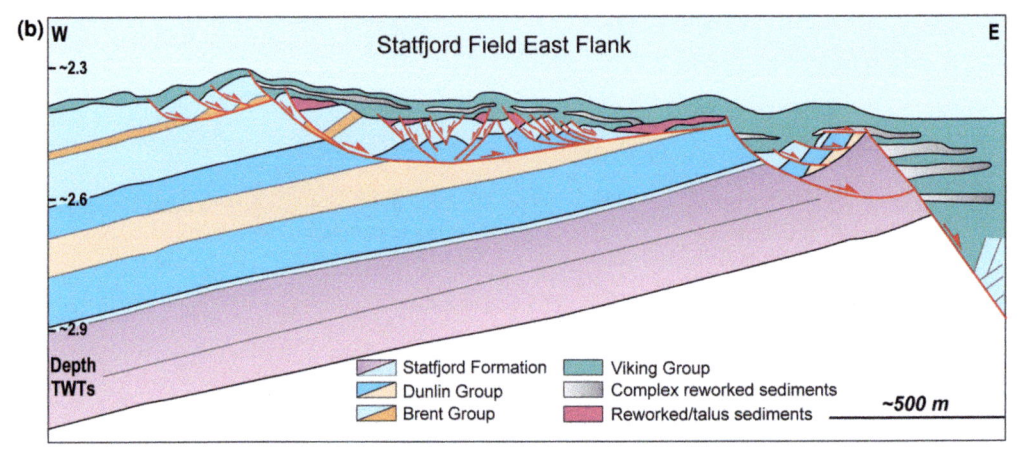

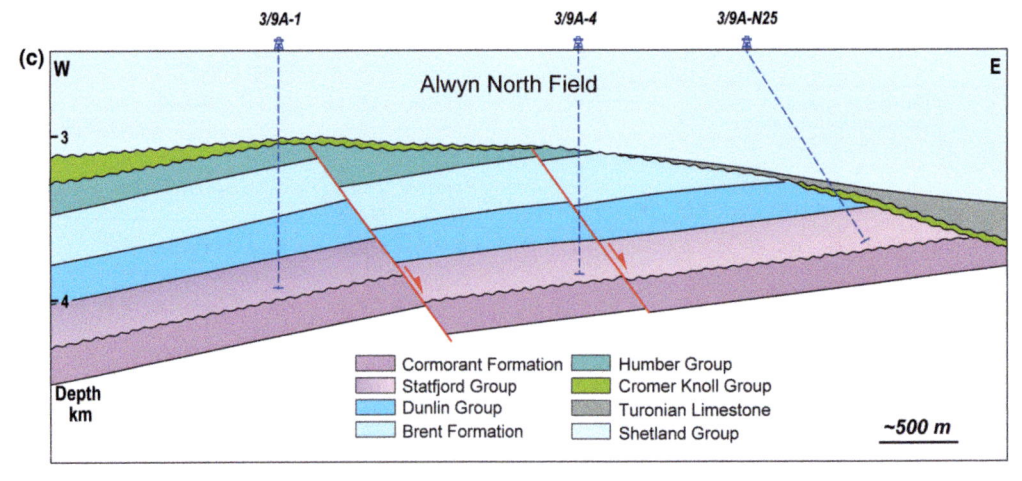

Fig. 15. Detailed east–west gravitational collapse profiles from the North Sea region: (**a**) the east flank of the Brent Field (modified from Struijk & Green 1991); (**b**) the Statfjord Field east flank (modified from Welbon *et al.* 2007); and (**c**) a profile from the Alwyn Field (modified from Inglis & Gerard 1991).

for the central Exmouth Plateau. Footwall-scarp degradation in rifts and passive margins can produce significant scarp retreat, in places in excess of 1 km, and strongly affect the geometries of footwall traps, the effectiveness of seals and the distribution reservoirs (e.g. Welbon *et al.* 2007).

Conclusions

The conclusions from this research are:

- Latest Triassic (Rhaetian)–earliest Late Jurassic (Oxfordian) WNW-directed extension in this part of the Exmouth Plateau produced north–south- to NNE–SSW-striking, long and linked, mainly west-dipping, domino-style fault systems in the Triassic Mungaroo Formation. The extensional faults have displacements on the large linked fault systems in excess of 1200 m at the level of the Late Triassic Rhaetian (TR30.1 TS) transgressive surface. The hanging-wall basins are strongly segmented, underfilled and synclinal in nature, with prominent extensional fault-propagation folds that formed in the upper Norian units and in the thin, Lower–Mid-Jurassic syn-extension strata.
- The major extensional fault systems display strong, probably submarine, fault-scarp degradation characterized by listric slump faulting combined with erosion. Fault-scarp retreat in places exceeded 1800 m. The greatest degradation occurred where the fault displacement and footwall uplift was the largest.
- Fault-scarp degradation occurred mainly during the latest Triassic–earliest Late Jurassic extension, producing talus wedges, up to 240 m thick, that partly infilled the hanging-wall sub-basins in front of the degraded fault scarps. A second minor period of scarp degradation produced small hanging-wall talus wedges <100 m thick during post-extension subsidence in the Late Jurassic (post the Oxfordian J40.0 SB unconformity).
- Upper Jurassic–Lower Cretaceous strata infilled the remnant post-extension accommodation space and overlapped onto the degraded fault scarps. Minor subsidence-related compaction and fault reactivation caused some of the larger faults to penetrate upwards, and to displace the Valanginian (K20.0 SB) and Aptian (K40.0 SB) surfaces.
- The fault-scarp degradation features in this study are interpreted to have been strongly controlled by the mechanical stratigraphy and anisotropy of the incompetent, shale-dominated strata of the Norian Mungaroo Formation – mechanical unit MU3. In places, listric slump faults in the degradation scarps sole out onto the incompetent coal marker unit (TR26.5 MFS) in the upper Mungaroo Formation – mechanical unit MU2.
- The extensional fault-scarp degradation structures analysed in this paper are similar to those found in the Viking Graben, northern North Sea and in many other rift systems, as well as in buried passive margins. Scarp-degradation features may have a significant impact on seismic images, as well as on both reservoir and seal distributions and characteristics in these terranes.

Acknowledgements This research forms part of the PhD thesis of A. Bilal. Seismic data were kindly provided by Geoscience Australia as part of a collaboration agreement with the Fault Dynamics Research Group (FDRG), Department of Earth Sciences, Royal Holloway University of London. Halliburton Ltd is thanked for providing the seismic interpretation software, and Badley Geoscience Ltd are thanked for the use of TrapTester software. Research resources and equipment were kindly provided by the FDRG. Personnel of Fault Dynamics Research Group are acknowledged for their assistance and insightful comments. The authors also thank N. Marshall and K. McCormack for their constructive comments and suggestions, and editor J. Hammerstein for his helpful comments.

Funding The Libyan Ministry of Higher Education and Scientific Research is thanked for the financial support to A. Bilal.

References

ADAMSON, K.R., LANG, S.C., MARSHALL, N.G., SEGGIE, R.J., ADAMSON, N.J. & BANN, K.L. 2013. Understanding the Late Triassic Mungaroo and Brigadier Deltas of the Northern Carnarvon Basin, North West Shelf, Australia. In: KEEP, M. & MOSS, S.J. (eds) *The Sedimentary Basins of Western Australia IV. Proceedings of Petroleum Exploration Society of Australia Symposium*. Petroleum Exploration Society of Australia, Perth, Australia, 18–21.

AFIFI, A.S., MOUSTAFA, A.R. & HELMY, H.M. 2016. Fault block rotation and footwall erosion in the southern Suez rift: implications for hydrocarbon exploration. *Marine and Petroleum Geology*, **76**, 377–396, https://doi.org/10.1016/j.marpetgeo.2016.05.029

BARAN, R., GUEST, B. & FRIEDRICH, A.M. 2010. High-resolution spatial rupture pattern of a multiphase flower structure, Rex Hills, Nevada: new insights on scarp evolution in complex topography based on 3-D laser scanning. *Geological Society of America Bulletin*, **122**, 897–914, https://doi.org/10.1130/B26536.1

BELGARDE, C., MANATSHHAL, G., KUZNIR, N., SCARSELLI, S. & RUDER, M. 2015. Rift processes in the Westralian Superbasin, North West Shelf, Australia: insights from 2D deep reflection seismic interpretation and potential fields modelling. *APPEA Journal*, **55**, 400–400, https://doi.org/10.1071/AJ14035

BERGER, M. & ROBERTS, A.M. 1999. The Zeta Structure: a footwall degradation complex formed by gravity sliding on the western margin of the Tampen Spur, northern North Sea. In: FLEET, A.J. & BOLDY, S.A.R. (eds) *Petroleum Geology of Northwest Europe: Proceedings of the 5th Conference, Volume 1*. Geological Society, London, 107–116, https://doi.org/10.1144/0050107

BLACK, M., MCCORMACK, K.D., ELDERS, C. & ROBERTSON, D. 2017. Extensional fault evolution within the Exmouth Sub-basin, North West Shelf, Australia. *Marine and Petroleum Geology*, **85**, 301–315, https://doi.org/10.1016/j.marpetgeo.2017.05.022

COUTTS, S.D., LARSSON, S.Y. & ROSMAN, R. 1996. Development of the slumped crestal area of the Brent reservoir,

Brent Field; an integrated approach. *Petroleum Geoscience*, **2**, 219–229, https://doi.org/10.1144/petgeo.2.3.219

DAHL, N. & SOLLI, T. 1993. The structural evolution of the Snorre Field and surrounding areas. *In*: PARKER, J.R. (ed.) *Petroleum Geology of Northwest Europe: Proceedings of the 4th Conference on Petroleum Geology of NW, Europe*. Geological Society, London, 1159–1166, https://doi.org/10.1144/0041159

DENSMORE, A.L. 2004. Footwall topographic development during continental extension. *Journal of Geophysical Research*, **109**, F03001, https://doi.org/10.1029/2003JF000115

DIREEN, N.G., STAGG, H.M.J., SYMONDS, P.A. & COLWELL, J.B. 2008. Architecture of volcanic rifted margins: new insights from the Exmouth–Gascoyne margin, Western Australia. *Australian Journal of Earth Sciences*, **55**, 341–363, https://doi.org/10.1080/08120090701769472

DRISCOLL, N. & KARNER, G. 1998. Lower crustal extension across the Northern Carnarvon basin, evidence for an eastward dipping detachment. *Journal of Geophysical Research: Solid Earth*, **103**, 4975–4991.

ELIET, P.P. & GAWTHORPE, R.L. 1995. Drainage development and sediment supply within rifts, examples from the Sperchios basin, central Greece. *Journal of the Geological Society, London*, **152**, 883–893, https://doi.org/10.1144/gsjgs.152.5.0883

ELLIOTT, G.M., WILSON, P., JACKSON, C.A.L., GAWTHORPE, R.L., MICHELSEN, L. & SHARP, I.R. 2012. The linkage between fault throw and footwall scarp erosion patterns: an example from the Bremstein Fault Complex, offshore Mid-Norway. *Basin Research*, **24**, 180–197, https://doi.org/10.1111/j.1365-2117.2011.00524.x

ELLIOTT, G.M., JACKSON, C.A.L., GAWTHORPE, R.L., WILSON, P., SHARP, I.R. & MICHELSEN, L. 2017. Late syn-rift evolution of the Vingleia Fault Complex, Halten Terrace, offshore Mid-Norway; a test of rift basin tectono-stratigraphic models. *Basin Research*, **29**, 465–487, https://doi.org/10.1111/bre.12158

FERENTINOS, G., PAPATHEODOROU, G. & COLLINS, M.B. 1988. Sediment Transport processes on an active submarine fault escarpment: Gulf of Corinth, Greece. *Marine Geology*, **83**, 43–61, https://doi.org/10.1016/0025-3227(88)90051-5

FORD, C.C., DIRSTEIN, J.K. & STANLEY, A.J. 2015. Prospectivity insights from automated pre- interpretation processing of open-file 3D seismic data: characterising the Late Triassic Mungaroo Formation of the Carnarvon Basin, North West Shelf of Australia. *AAPEA Journal*, **55**, 15–34.

FRASER, S.I., ROBINSON, A.M. ET AL. 2002. Upper Jurassic. *In*: ARMOUR, A., EVANS, D. & HICKEY, C. (eds) *The Millennium Atlas: Petroleum Geology of the Central and Northern North Sea*. The Geological Society, London, 157–189.

GARTRELL, A., TORRES, J., DIXON, M. & KEEP, M. 2016. Mesozoic rift onset and its impact on the sequence stratigraphic architecture of the Northern Carnarvon Basin. *AAPEA Journal*, **56**, 143–158, https://doi.org/10.1071/AJ15012

GARTRELL, A.P. 2000. Rheological controls on extensional styles and the structural evolution of the Northern Carnarvon Basin, North West Shelf, Australia. *Australian Journal of Earth Sciences*, **47**, 231–244, https://doi.org/10.1046/j.1440-0952.2000.00776.x

GAWTHORPE, R. & HARDY, S. 2002. Extensional fault-propagation folding and base-level change as controls on growth-strata geometries. *Sedimentary Geology*, **146**, 47–56, https://doi.org/10.1016/S0037-0738(01)00165-8

GAWTHORPE, R.L. & LEEDER, M.R. 2000. Tectono-sedimentary evolution of active extensional basins. *Basin Research*, **12**, 195–218, https://doi.org/10.1111/j.1365-2117.2000.00121.x

GEOSCIENCE AUSTRALIA. 2017. *Offshore Petroleum Acreage Release. Northern Carnarvon Regional Geology. Western Australia*. Geoscience Australia, Canberra, Australia.

GIBBONS, A.D., BARCKHAUSEN, U., VAN DEN BOGAARD, P., HOERNLE, K., WERNER, R., WHITTAKER, J.M. & MÜLLER, R.D. 2012. Constraining the Jurassic extent of Greater India: tectonic evolution of the West Australian margin. *Geochemistry, Geophysics, Geosystems*, **13**, 1–25, https://doi.org/10.1029/2011GC003919

GIBBONS, K.A., JOURDAN, C.A. & HESTHAMMER, J. 2003. The Statfjord Field, Blocks 33/9, 33/12 Norwegian sector, Blocks 211/24, 211/25 UK sector, Northern North Sea. *In*: GLUYAS, J.G. & HICHENS, H.M. (eds) *United Kingdom Oil and Gas Fields: Commemorative Millennium Volume*. Geological Society, London, Memoirs, **20**, 335–353, https://doi.org/10.1144/GSL.MEM.2003.020.01.29

GONCHAROV, A., DEIGHTON, I., DUFFY, L., MCLAREN, S., TISCHER, M. & HEINE, C. 2006. Basement and crustal controls on hydrocarbons maturation on the Exmouth Plateau, North West Australian Margin. Search and discovery article 10119 presented at the AAPG 2006 International Conference and Exhibition, 5–8 November 2006, Perth, Australia.

GRAIN, S.L., PEACE, W.M., HOOPER, E.C.D., MCCARTAIN, E., MASSARA, P.J., MARSHALL, N.G. & LANG, S.C. 2013. Beyond the deltas: Late Triassic isolated carbonate build-ups on the Exmouth Plateau, Carnarvon Basin, Western Australia. *In*: KEEP, M. & MOSS, S.J. (eds) *The Sedimentary Basins of Western Australia IV. Proceedings of Petroleum Exploration Society of Australia Symposium*. Petroleum Exploration Society of Australia, Perth, Australia, 18–21.

HALL, R. 2012. Late Jurassic–Cenozoic reconstructions of the Indonesian region and the Indian Ocean. *Tectonophysics*, **570**, 1–41, https://doi.org/10.1016/j.tecto.2012.04.021

HARDY, S. & MCCLAY, K.R. 1999. Kinematic modelling of extensional forced folding. *Journal of Structural Geology*, **21**, 695–702.

HEINE, C. & MÜLLER, R. 2005. Late Jurassic rifting along the Australian North West Shelf: margin geometry and spreading ridge configuration. *Australian Journal of Earth Sciences*, **52**, 27–39, https://doi.org/10.1080/08120090500100077

HEINE, C., MÜLLER, R.D. & GAINA, C. 2004. Reconstructing the lost eastern Tethys Ocean Basin: convergence history of the SE Asian margin and marine gateways. *In*: CLIFT, P., HAYES, D., KUHNT, W. & WANG, P. (eds) *Continent–Ocean Interactions within East Asian Marginal Seas*. American Geophysical Union, Geophysical Monographs, **149**, 37–54, https://doi.org/10.1029/149GM03

HELDREICH, G., REDFERN, J., LEGLER, B., GERDES, K. & WILLIAMS, B.P.J. 2017. Challenges in characterizing subsurface paralic reservoir geometries: a detailed case study of the Mungaroo Formation, North West Shelf, Australia. In: HAMPSON, G.J., REYNOLDS, A.D., KOSTIC, B. & WELLS, M.R. (eds) *Sedimentology of Paralic Reservoirs: Recent Advances*. Geological Society, London, Special Publications, **444**, 59–108, https://doi.org/10.1144/SP444.13

HENGESH, J.V. & WHITNEY, B.B. 2016. Transcurrent reactivation of Australia's western passive margin: an example of intraplate deformation from the central Indo-Australian plate. *Tectonics*, **35**, 1066–1089, https://doi.org/10.1002/2015TC004103

HENSTRA, G.A., GRUNDVÅG, S.-A. ET AL. 2016. Depositional processes and stratigraphic architecture within a coarse grained rift-margin turbidite system: the Wollaston Forland Group, east Greenland. *Marine and Petroleum Geology*, **76**, 187–209, https://doi.org/10.1016/j.marpetgeo.2016.05.018

HESTHAMMER, J. & FOSSEN, H. 1999. Evolution and geometries of gravitational collapse structures with examples from the Statfjord Field, northern North Sea. *Marine and Petroleum Geology*, **16**, 259–281, https://doi.org/10.1016/S0264-8172(98)00071-3

HOCKING, R.M. 1988. Regional geology of the Northern Carnarvon Basin. In: PURCELL, P.G. & PURCELL, R.R. (eds) *The North West Shelf, Australia. Proceedings of Petroleum Exploration Society of Australia Symposium*. Petroleum Exploration Society of Australia, Perth, Australia, 97–114.

HOCKING, R.M. & PRESTON, W.A. 1998. Western Australia: Phanerozoic geology and mineral resources. *AGSO Journal of Australian Geology and Geophysics*, **17**, 245–260.

HOPPER, J.R., MUTTER, J.C., LARSON, R.L. & MUTTER, C.Z. 1992. Magmatism and rift margin evolution: evidence from northwest Australia. *Geology*, **20**, 853–857, https://doi.org/10.1130/0091-7613(1992)020<0853:MARMEE>2.3.CO;2

INGLIS, I. & GERARD, J. 1991. The Alwyn North Field, Blocks 3/9a, 3/4a, UK North Sea. In: ABBOTTS, I.L. (ed.) *United Kingdom Oil and Gas Fields, 25 Years Commemorative Volume*. Geological Society, London, Memoirs, **14**, 21–32, https://doi.org/10.1144/GSL.MEM.1991.014.01.03

JACKSON, C.A.L., GAWTHORPE, R.L. & SHARP, I.R. 2006. Style and sequence of deformation during extensional fault-propagation folding: examples from the Hammam Faraun and El-Qaa fault blocks, Suez Rift, Egypt. *Journal of Structural Geology*, **28**, 519–535, https://doi.org/10.1016/j.jsg.2005.11.009

JITMAHANTAKUL, S. & MCCLAY, K. 2013. Late Triassic–Mid-Jurassic to Neogene extensional fault systems in the Exmouth Sub-Basin, Northern Carnarvon Basin, North West Shelf, Western Australia. In: KEEP, M. & MOSS, S.J. (eds) *The Sedimentary Basins of Western Australia IV. Proceedings of Petroleum Exploration Society of Australia Symposium*. Petroleum Exploration Society of Australia, Perth, Australia, 18–21.

KEEP, M., POWELL, C.M. & BAILLIE, P.W. 1998. Neogene deformation of the North West Shelf, Australia. In: PURCELL, P.G. & PURCELL, R.R. (eds) *The Sedimentary Basins of Western Australia II. Proceedings of Petroleum Exploration Society Australia Symposium*. Petroleum Exploration Society of Australia, Perth, Australia, 81–91.

KHALIL, S.M. & MCCLAY, K.R. 2002. Extensional fault-related folding, northwestern Red Sea, Egypt. *Journal of Structural Geology*, **24**, 743–762, https://doi.org/10.1016/S0191-8141(01)00118-3

KHALIL, S.M. & MCCLAY, K.R. 2006. Extensional fault-related folding, Gulf of Suez, Egypt. *Middle East Research Center, Ain Shams University, Earth Science Series*, **20**, 1–12.

KHALIL, S.M. & MCCLAY, K.R. 2016. 3D geometry and kinematic evolution of extensional fault-related folds, NW Red Sea, Egypt. In: CHILDS, C., HOLDSWORTH, R.E., JACKSON, C.A.-L., MANZOCCHI, T., WALSH, J.J. & YIELDING, G. (eds) *The Geometry and Growth of Normal Faults*. Geological Society, London, Special Publications, **439**, 109–130, https://doi.org/10.1144/SP439.11

KOKKALAS, S. & KOUKOUVELAS, I.K. 2005. Fault-scarp degradation modeling in central Greece: the Kaparelli and Eliki faults (Gulf of Corinth) as a case study. *Journal of Geodynamics*, **40**, 200–215, https://doi.org/10.1016/j.jog.2005.07.006

LEPPARD, C.W. & GAWTHORPE, R.L. 2006. Sedimentology of rift climax deep water systems; Lower Rudeis Formation, Hammam Faraun Fault Block, Suez Rift, Egypt. *Sedimentary Geology*, **191**, 67–87, https://doi.org/10.1016/j.sedgeo.2006.01.006

LEWIS, C.J. & SIRCOMBE, K.N. 2013. Use of U–Pb geochronology to delineate provenance of North West Shelf sediments, Australia. In: KEEP, M. & MOSS, S.J. (eds) *The Sedimentary Basins of Western Australia IV. Proceedings of Petroleum Exploration Society of Australia Symposium*. Petroleum Exploration Society of Australia, Perth, Australia, 18–21.

LONGLEY, I.M., BUESSENSCHUETT, C. ET AL. 2002. The North West Shelf of Australia–a Woodside perspective. In: KEEP, M. & MOSS, S.J. (eds) *The Sedimentary Basins of Western Australia 3. Proceedings of Petroleum Exploration Society of Australia Symposium*. Petroleum Exploration Society of Australia, Perth, Australia, 27–88.

MAGEE, C., JACKSON, C.A.-L., HARDMAN, J.P. & REEVE, M.T. 2017. Decoding sill emplacement and forced fold growth in the Exmouth Sub-basin, offshore northwest Australia: implications for hydrocarbon exploration. *Interpretation*, **5**, SK11–SK22, https://doi.org/10.1190/INT-2016-0133.1

MARSHALL, N.G. & LANG, S.C. 2013. A New Sequence Stratigraphic Framework for the North West Shelf, Australia. In: KEEP, M. & MOSS, S.J. (eds) *The Sedimentary Basins of Western Australia IV. Proceedings of Petroleum Exploration Society of Australia Symposium*. Petroleum Exploration Society of Australia, Perth, Australia, 18–21.

MCARTHUR, A.D., HARTLEY, A.J., ARCHER, S.G., JOLLEY, D.W. & LAWRENCE, H.M. 2016a. Spatiotemporal relationships of deep-marine, axial, and transverse depositional systems from the synrift Upper Jurassic of the central North Sea. *AAPG Bulletin*, **100**, 1469–1500, https://doi.org/10.1306/04041615125

MCARTHUR, A.D., JOLLEY, D.W., HARTLEY, A.J., ARCHER, S.G. & LAWRENCE, H.M. 2016b. Palaeoecology of syn-rift topography: a Late Jurassic footwall island on the Josephine Ridge, Central Graben, North Sea. *Palaeogeography, Palaeoclimatology, Palaeoecology*, **459**, 63–75, https://doi.org/10.1016/j.palaeo.2016.06.033

McCLAY, K., SCARSELLI, N. & JITMAHANTAKUL, S. 2013. Igneous Intrusions in the Carnarvon Basin, NW Shelf, Australia. *In*: KEEP, M. & MOSS, S.J. (eds) *The Sedimentary Basins of Western Australia IV. Proceedings of Petroleum Exploration Society of Australia Symposium*. Petroleum Exploration Society of Australia, Perth, Australia, 18–21.

McCORMACK, K.D. & McCLAY, K. 2013. Structural architecture of the Gorgon Platform, North West Shelf, Australia. *In*: KEEP, M. & MOSS, S.J. (eds) *The Sedimentary Basins of Western Australia IV. Proceedings of Petroleum Exploration Society of Australia Symposium*. Petroleum Exploration Society of Australia, Perth, Australia, 18–21.

McLEOD, A.E. & UNDERHILL, J.R. 1999. Processes and products of footwall degradation, northern Brent Field, Northern North Sea. *In*: FLEET, A.J. & BOLDY, S.A.R. (eds) *Petroleum Geology of Northwest Europe: Proceedings of the 5th Conference, Volume 1*. Geological Society, London, 91–106, https://doi.org/10.1144/0050091

McLEOD, A.E., UNDERHILL, J.R., DAVIES, S.J. & DAWERS, N.H. 2002. The influence of fault array evolution on synrift sedimentation patterns: controls on deposition in the Strathspey-Brent-Statfjord half graben, northern North Sea. *AAPG Bulletin*, **86**, 1061–1093, https://doi.org/10.1306/61eedc24-173e-11d7-8645000102c1865d

METCALFE, I. 2013. Gondwana dispersion and Asian accretion: tectonic and palaeogeographic evolution of eastern Tethys. *Journal of Asian Earth Sciences*, **66**, 1–33, https://doi.org/10.1016/j.jseaes.2012.12.020

MORLEY, C.K. 1999. Basin evolution trends in East Africa. *In*: MORLEY, C.K. (ed.) *Geoscience of Rift Systems – Evolution of East Africa*. AAPG, Studies in Geology, **44**, 131–150.

MORLEY, C.K., IONNIKOFF, Y., PINYOCHON, N. & SEUSUTTHIYA, K. 2007. Degradation of a footwall fault block with hanging-wall fault propagation in a continental-lacustrine setting: how a new structural model impacted field development plans, the Sirikit field, Thailand. *AAPG Bulletin*, **91**, 1637–1661, https://doi.org/10.1306/06280707014

MÜLLER, R.D., MIHUT, D. & BALDWIN, S. 1998. A new kinematic model for the formation and evolution of the west and northwest Australian margin. *In*: KEEP, M. & MOSS, S.J. (eds) *The Sedimentary Basins of Western Australia II. Proceedings of Petroleum Exploration Society Australia Symposium*. Petroleum Exploration Society of Australia, Perth, Australia, 55–72.

PATTON, T.L., MOUSTAFA, A.R., NELSON, R.A. & ABDINE, S.A. 1994. Tectonic evolution and structural setting of the Suez Rift. *In*: LANDON, S.M. (ed.) *Interior Rift Basins*. AAPG Memoirs, **59**, 9–55.

PAUL, D. & MITRA, S. 2015. Fault patterns associated with extensional fault-propagation folding. *Marine and Petroleum Geology*, **67**, 120–143, https://doi.org/10.1016/j.marpetgeo.2015.04.020

PRYER, L., BLEVIN, J. *ET AL*. 2014. Structural architecture and basin evaluation of the North West Shelf (Abstract). *APPEA Journal*, **54**, 474.

PRYER, L.L., ROMINE, K.K., LOUTIT, T.S. & BARNES, R.G. 2002. Carnarvon Basin architecture and structure defined by the integration of mineral and petroleum exploration tools and techniques. *APPEA Journal*, **42**, 287–309.

REEVE, M.T., JACKSON, C.A.L., BELL, R.E., MAGEE, C. & BASTOW, I.D. 2016. The stratigraphic record of pre-breakup geodynamics: evidence from the Barrow Delta, offshore Northwest Australia. *Tectonics*, **35**, 1–34, https://doi.org/10.1002/2016TC004172

ROHRMAN, M. 2013. Intrusive large igneous provinces below sedimentary basins: an example from the Exmouth Plateau (NW Australia). *Journal of Geophysical Research: Solid Earth*, **118**, 4477–4487, https://doi.org/10.1002/jgrb.50298

ROHRMAN, M. 2015. Delineating the Exmouth mantle plume (NW Australia) from denudation and magmatic addition estimates. *Lithosphere*, **7**, 589–600, https://doi.org/10.1130/L445.1

SCHLISCHE, R.W. 1995. Geometry and origin of fault-related folds in extensional settings. *AAPG Bulletin*, **79**, 1661–1678, https://doi.org/10.1306/7834DE4A-1721-11D7-8645000102C1865D

SHARP, I.R., GAWTHORPE, R.L., UNDERHILL, J.R. & GUPTA, S. 2000. Fault-propagation folding in extensional settings: examples of structural style and synrift sedimentary response from the Suez rift, Sinai, Egypt. *Geological Society of America Bulletin*, **112**, 1877–1899, https://doi.org/10.1130/0016-7606(2000)112<1877:FPFIES>2.0.CO;2

STAGG, H.M.J. & COLWELL, J.B. 1994. The structural foundations of the Northern Carnarvon Basin. *In*: PURCELL, G.P. & PURCELL, R.R. (eds) *The Sedimentary Basins of Western Australia. Proceedings of Petroleum Exploration Society of Australia Symposium*. Petroleum Exploration Society of Australia, Perth, Australia, 349–372.

STAGG, H.M.J., ALCOCK, M.B., BERNARDEL, G., MOORE, A.M.G., SYMONDS, P.A. & EXON, N.F. 2004. *Geological Framework of the Outer Exmouth Plateau and Adjacent Ocean Basins*. Geoscience Australia Record, **2004/13**.

STEWART, S.A. & REEDS, A. 2003. Geomorphology of kilometre-scale extensional fault scarps: factors that impact seismic interpretation. *AAPG Bulletin*, **87**, 251–272, https://doi.org/10.1306/08190201041

STRUIJK, A.P. & GREEN, R.T. 1991. The Brent Field, Block 211/29, UK North Sea. *In*: ABBOTTS, I.L. (ed.) *United Kingdom Oil and Gas Fields, 25 Years Commemorative Volume*. Geological Society, London, Memoirs, **14**, 63–72, https://doi.org/10.1144/GSL.MEM.1991.014.01.08

SURLYK, F. 1978. *Submarine Fan Sedimentation along Fault Scarps on Tilted Fault Blocks (Jurassic–Cretaceous boundary, East Greenland)*. Grønlands Geologiske Undersøgelse, **128**.

TINDALE, K., NEWELL, N., KEALL, J. & SMITH, N. 1998. Structural evolution and charge history of the Exmouth Sub-basin, Northern Carnarvon Basin, Western Australia. *In*: PURCELL, P.G. & PURCELL, R.R. (eds) *The Sedimentary Basins of Western Australia II. Proceedings* of *Petroleum Exploration Society Australia Symposium*. Petroleum Exploration Society of Australia, Perth, Australia, 447–472.

UNDERHILL, J.R., SAWYER, M.J., HODGSON, P., SHALLCROSS, M.D. & GAWTHORPE, R.L. 1997. Implications of fault scarp degradation for Brent Group prospectivity, Ninian Field, northern North Sea. *AAPG Bulletin*, **81**, 999–1022.

WELBON, A.I.F., BROCKBANK, P.J., BRUNSDEN, D. & OLSEN, T.S. 2007. Characterizing and producing from reservoirs in landslides: challenges and opportunities. *In*: JOLLEY, S.J., BARR, D., WALSH, J.J. & KNIPE, R.J. (eds) *Structurally Complex Reservoirs*. Geological Society, London, Special Publications, **292**, 49–74, https://doi.org/10.1144/SP292.3

YANG, X.-M. & ELDERS, C. 2016. The Mesozoic structural evolution of the Gorgon Platform, North Carnarvon Basin, Australia. *Australian Journal of Earth Sciences*, **63**, 755–770, https://doi.org/10.1080/08120099.2016.1243579

YEATES, A., BRADSHAW, M. ET AL. 1987. The Westralian superbasin: an Australian link with Tethys. *In*: MCKENZIE, K.G. (ed.) *Shallow Tethys 2, International Symposium Proceedings*. A.A. Balkema, Rotterdam, The Netherlands, 199–213.

ZHAO, H., GUO, Z. & YU, X. 2017. Strain modelling of extensional fault-propagation folds based on an improved non-linear trishear model: a numerical simulation analysis. *Journal of Structural Geology*, **95**, 60–76, https://doi.org/10.1016/j.jsg.2016.12.009

Tectono-stratigraphy of the Dampier Sub-basin, North West Shelf of Australia

HONGDAN DENG* & KEN McCLAY

Fault Dynamics Research Group, Department of Earth Sciences, Royal Holloway University of London, Egham, Surrey TW20 0EX, UK

HD, 0000-0002-5546-3884
Correspondence: denghongdan@gmail.com

Abstract: The Dampier Sub-basin, an inboard rift system of the Northern Carnarvon Basin in the North West Shelf, Australia, underwent two major phases of continental rifting in the Late Paleozoic and in the Latest Triassic to Late Jurassic. Six tectono-stratigraphic megasequences separated by regional unconformities have been identified: (1) Pre-Late Carboniferous Pre-rift 1; (2) Late Carboniferous to Late Permian Syn-rift 1; (3) Early to Latest Triassic Post-rift 1; (4) Latest Triassic to Early Late Jurassic Syn-rift 2; (5) Late Jurassic to Early Cretaceous Post-rift 2; and (6) Early Cretaceous to Present-day passive margin megasequences. The Late Paleozoic rifting produced a series of planar extensional faults on the eastern flank, some of which were later rotated by the Latest Triassic to Late Jurassic WNW–ESE extension to low-angles (c. 30°). This Mesozoic extension was localized above a NE–SW-trending basement structure, resulting in en échelon inboard rift basins and overlapping boundary fault systems. The study outlines an updated basin tectono-stratigraphic model for the Dampier Sub-basin and provides new insights for structural evolution associated with the development of the North West Shelf passive continental margin of Australia.

Supplementary material: Additional information on seismic and well location, depth–time conversion, structural maps, and thickness maps used in this paper are available at https://doi.org/10.6084/m9.figshare.c.4529987

The Dampier Sub-basin is a Mesozoic inboard rift system of the Northern Carnarvon Basin, North West Shelf of Australia (Fig. 1). The basin is c. 150 km in length and c. 60 km wide, flanked by basement-high margins (Fig. 2). It is located between 19 and 21° S and 115–117° E (Fig. 1), covering an area of c. 36 000 km^2 and contains thick (>10 km) Late Paleozoic to Present-day sediment (Stagg & Colwell 1994; Longley et al. 2002; Jablonski & Saitta 2004).

The Dampier Sub-basin is an important hydrocarbon basin with significant oil (>633 MMbls) and gas (>34 Tcf) discoveries (Barber 2013). Many papers on the structure and stratigraphy of the Dampier Sub-basin have been published (e.g. Hill 1994; Jablonski 1997; Longley et al. 2002; Pryer et al. 2002; Langhi & Borel 2005; Marshall & Lang 2013); however, most of these studies focused either on regional stratigraphy definition (e.g. Marshall & Lang 2013) or on regional structural evolution (e.g. Pryer et al. 2002; Jablonski & Saitta 2004), and there is a lack of detailed and integrated basin-scale analysis.

This paper has integrated 2-D and 3-D seismic datasets with well data (Fig. 4) to develop a coherent tectono-stratigraphic and structural analysis of the Dampier Sub-basin from the Late Paleozoic to the Present day. The aim is to constrain the timing and distribution of the tectono-stratigraphic sequences combined with a structural analysis of the Dampier Sub-basin to provide new insights into the development of the Northern Carnarvon Basin and the Australian North West Shelf passive margin in a larger picture.

Geological background

Regional tectonics

The North West Shelf of Australia has undergone Late Paleozoic and Mesozoic rifting (Stagg & Colwell 1994; Longley et al. 2002). The Late Paleozoic NW–SE rifting created the so-called 'Westralian Superbasin' that encompasses the Northern Carnarvon, Roebuck, Browse and Bonaparte basins and that forms the wide continental margin of the North West Shelf (Fig. 1a) (Yeates et al. 1987; Gartrell 2000; Longley et al. 2002).

The Latest Triassic to Late Jurassic WNW–ESE-directed extension created a series of inboard en échelon sub-basins in the Northern Carnarvon Basin (Tindale et al. 1998; Longley et al. 2002; Jablonski & Saitta 2004; Fig. 1). These deep sub-basins trapped, in places, greater than 6 km of Mesozoic siliciclastic sediments derived mainly from the Australian continental landmass (Longley et al. 2002). To the west of the inboard sub-basins, the Exmouth Plateau

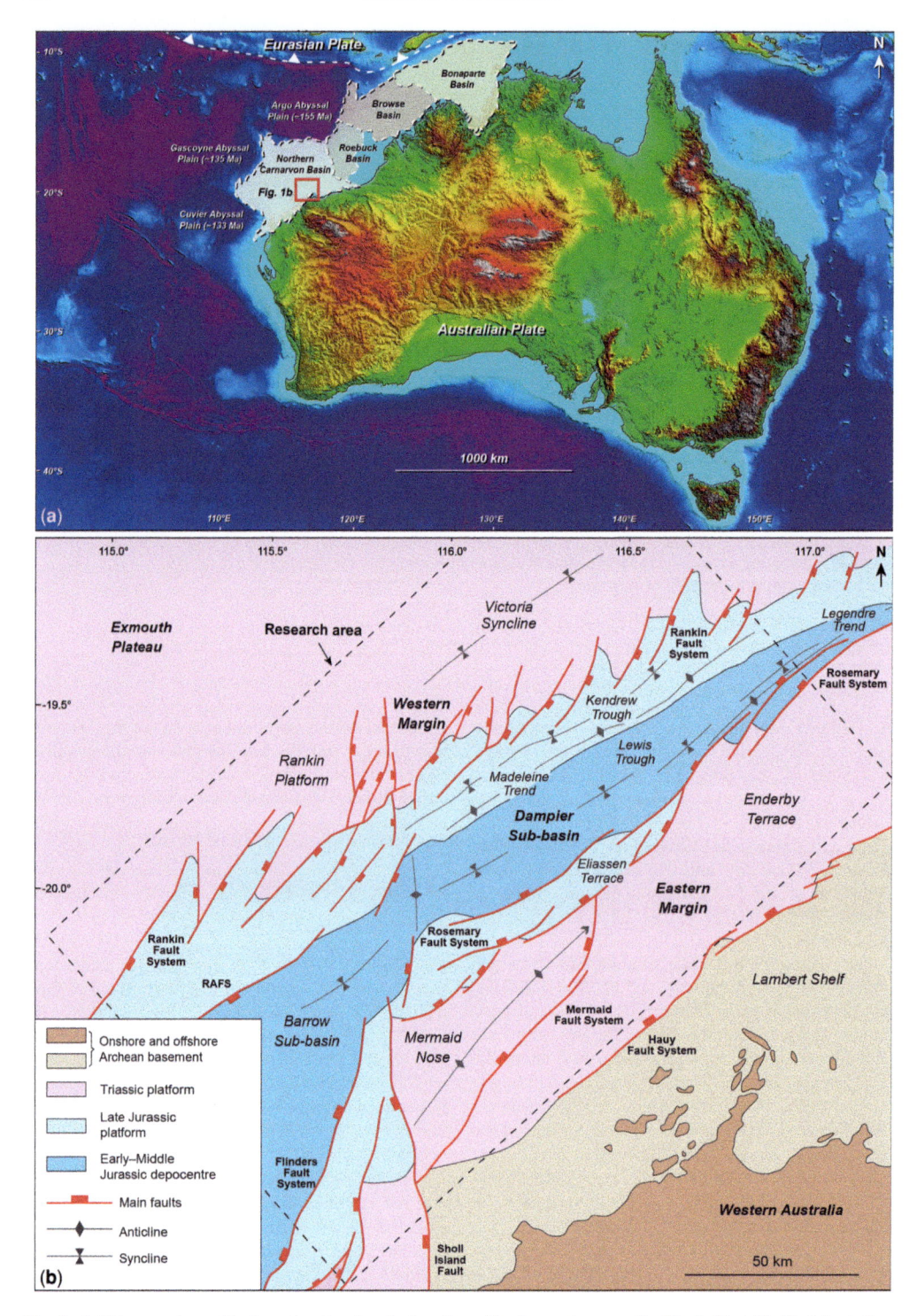

Fig. 1. (**a**) Topography and bathymetry showing the location of basin provinces on the North West Shelf of Australia with respect to major plate boundaries. (**b**) Major tectonic elements of the Dampier Sub-basin and basin margins. Compiled and modified from Phillips Australian Oil Company (1992), Australian Geological Survey Organisation *et al.* (1993) and Saitta *et al.* (2003). Map location is shown in Figure 1a.

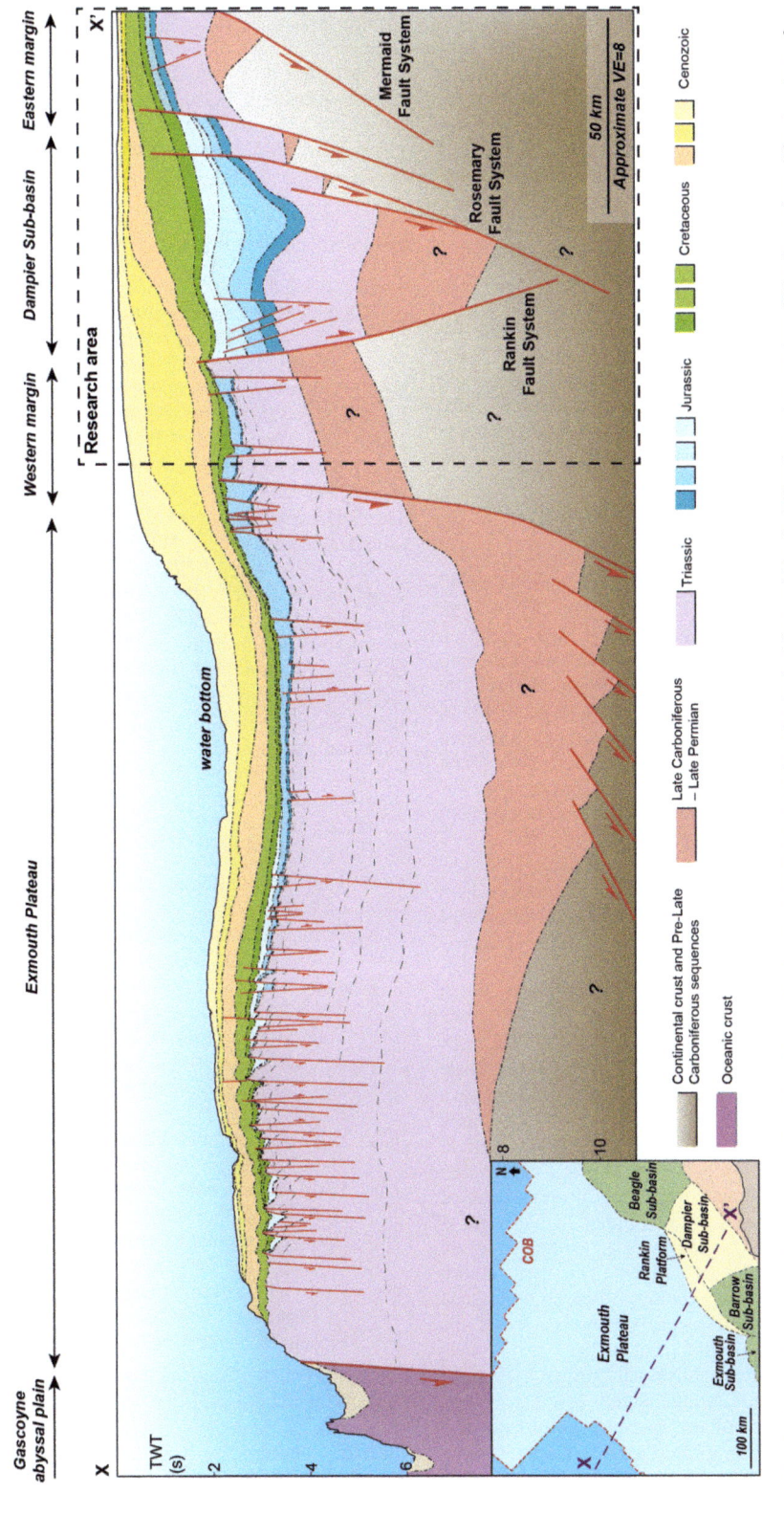

Fig. 2. Schematic cross-section showing the rift system of the Northern Carnarvon Basin. The dashed lines within the Triassic section are intra-Triassic markers. The structure of the Exmouth Plateau is modified from Marshall & Lang (2013).

(Fig. 1b) only underwent limited extension with relatively small displacement 'domino-style' faults (Karner & Driscoll 1999; Yang & Elders 2016; Black et al. 2017; Bilal et al. 2018). In the Late Jurassic to Early Cretaceous, cumulative extension resulted in continental breakup at the extremities of the Exmouth Plateau, with Argo abyssal plain developed at c. 155 Ma; the Gascoyne abyssal plain developed at c. 135 Ma; and the Cuvier abyssal plain developed at c. 133 Ma (Fig. 1a; Mihut & Müller 1998; Stagg et al. 2004; Heine & Müller 2005; Gibbons et al. 2012).

In the Mid-Miocene (c. 16 Ma), collision of the outer northern margin of the North West Shelf with the Eurasian plate (Fig. 1a; Keep et al. 2007; Gibbons et al. 2012; Hall 2012; Saqab et al. 2016) generated far-field stress that triggered local reactivation of some fault systems (Hull & Griffiths 2002; Hengesh & Whitney 2016).

Stratigraphy

The Late Paleozoic strata have a maximum thicknesses of c. 5 km in the onshore area (Eyles et al. 2003). In the offshore area, they have not been confidently identified in seismic data across the Dampier Sub-basin and the Rankin Platform owing to deep burial and lack of well penetrations (Thomas et al. 2004). On the eastern margin, seismic reflection together with wells (Kybra-1, Arabella-1, and Roebuck-1) reveal the Paleozoic sediments in half-grabens (e.g. Jablonski 1997; Langhi & Borel 2005; Fig. 3).

The Triassic Locker Shale and Mungaroo Formation (Adamson et al. 2013) thicken from <2 km on the eastern margin (Mermaid Nose) to at least 10 km in places on the Exmouth Plateau, characterized by largely flat-lying, shallow marine to deltaic–fluvial succession, broadly distributed across the North West Shelf (Karner & Driscoll 1999; Longley et al. 2002; Stagg et al. 2004; Heldreich et al. 2017; McGee et al. 2017). The Locker Shale is inferred to be mechanically weak and to have acted as a detachment unit for some of the young rift faults (Gartrell et al. 2016; McHarg et al. 2018, 2019; Deng & McClay 2019). Recent exploration successes have gained momentum in more detailed investigation of the Lower to Upper Triassic stratigraphy and palaeogeography (e.g. Molyneux et al. 2016; McGee et al. 2017; Marsh et al. 2018; Woodward et al. 2018), which unravelled large deltaic–fluvial or canyoning systems mainly sourced the Australia continent and directed to the Northern Carnarvon Basin. Our research here primarily focuses on the first-order sedimentation associated with tectonic activities of the Dampier Sub-basin. The detailed stratigraphy of the Lower to Upper Triassic is beyond the scope of this paper.

The Latest Triassic to Late Jurassic syn-rift strata (Fig. 3) are up to 4 km thick in the sub-basin axis and thin onto the rift flanks (Jablonski 1997; Karner & Driscoll 1999). The Post-rift Jurassic and Cretaceous shallow-marine passive margin sediments were deposited above the inboard rift basins (Longley et al. 2002).

Research methodologies

This study used c. 8600 km of 2-D seismic lines and c. 16 000 km^2 3-D seismic volumes comprising seven surveys (Fig. 4) to build an unbroken and basin-wide interpretation and analysis. The 2-D seismic lines were acquired between 1992 and 1995. Seismic record length is 4.5–8.0 s, and line spacing varies between 2 and 30 km. The merged 3-D seismic surveys were acquired between 1998 and 2003 with record length varying between 3.5 and 6.0 s TWT. Integrated 2-D and 3-D seismic interpretation was undertaken using Halliburton's Landmark software.

Well data were collected from the Western Australian Petroleum & Geothermal Information Management System (https://wapims.dmp.wa.gov.au/wapims) and from the National Offshore Petroleum Information Management System (https://nopims.dmp.wa.gov.au/nopims). Biostratigraphic ages, lithologies, well-tops, and velocity information were used to determine key seismic horizons. Seismic data were depth-converted using check-shots from 16 wells (shown in Supplementary Materials)

Ten key horizons (Fig. 5) were interpreted across the Dampier Sub-basin:

(1) Late Carboniferous, Gzhelian, P10.0 SB (c. 302 Ma?);
(2) Base Triassic, Induan, TR10.0 SB (c. 252.2 Ma);
(3) Early Jurassic, Sinemurian, J20.1 TS (c. 192.5 Ma);
(4) Late Middle Jurassic, Callovian, J30.1 TS (c. 165.6 Ma);
(5) Late Jurassic, Oxfordian, J40.0 SB (c. 162.5 Ma);
(6) Early Cretaceous, Valanginian, K20.0 SB (c. 137.9 Ma);
(7) Early Cretaceous, Aptian, K40.0 SB (c. 123.3 Ma);
(8) Late Cretaceous, Campanian, K60.0 SB (c. 78.5 Ma);
(9) Base Cenozoic, Danian, T10.0 SB (c. 66.0 Ma);
(10) Present-day seafloor (c. 0.0 Ma).

The Mesozoic and Cenozoic horizons were chronologically calibrated with the regional stratigraphic framework of Marshall & Lang (2013), and the Paleozoic stratigraphy was calibrated with that described by Crostella et al. (2000).

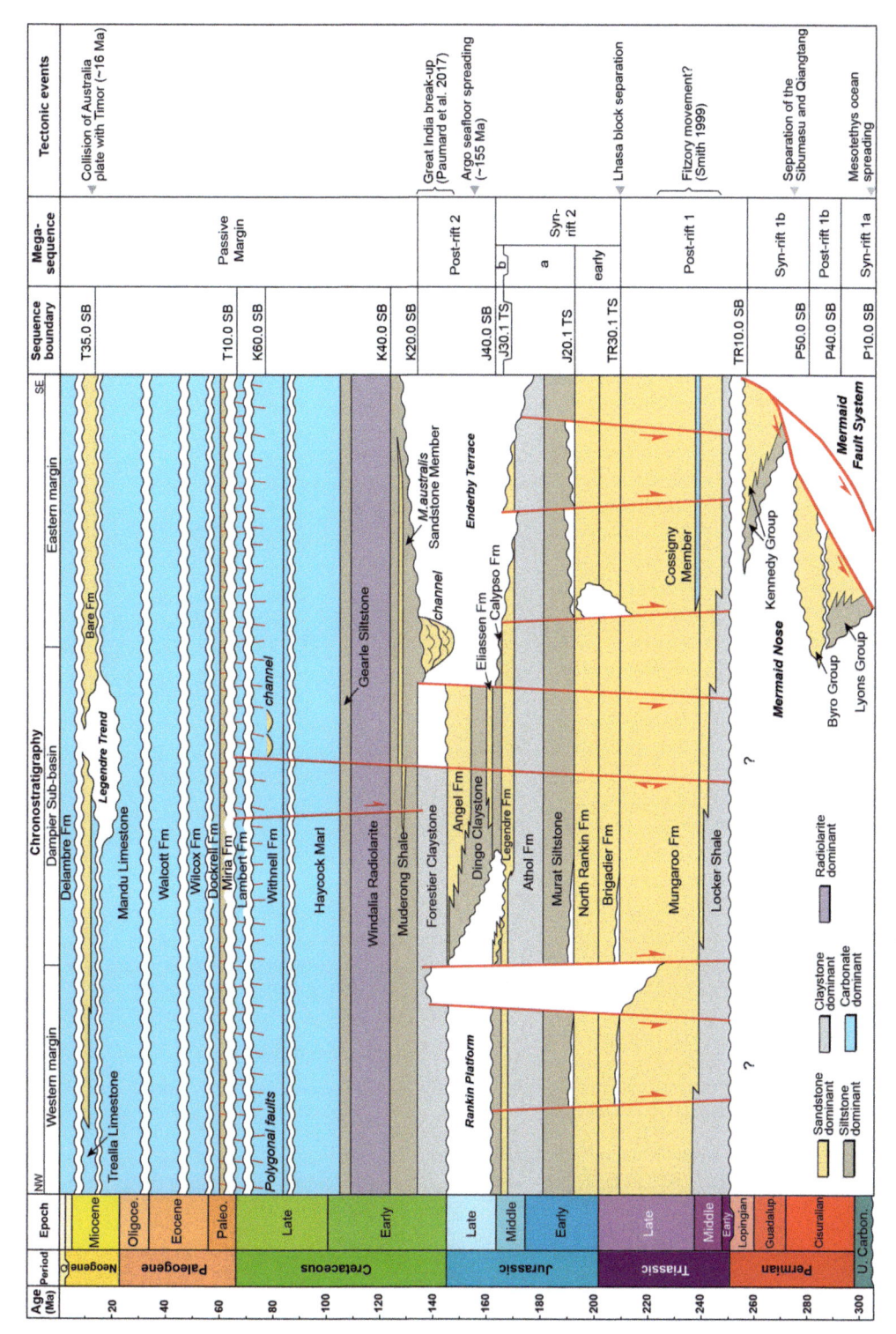

Fig. 3. Simplified tectono-stratigraphic chart of the Dampier Sub-basin. Geological timescale and sequence boundaries are adapted from Marshall & Lang (2013). Tectonic events are modified from Longley *et al.* (2002), Jablonski & Saitta (2004) and Smith (1999).

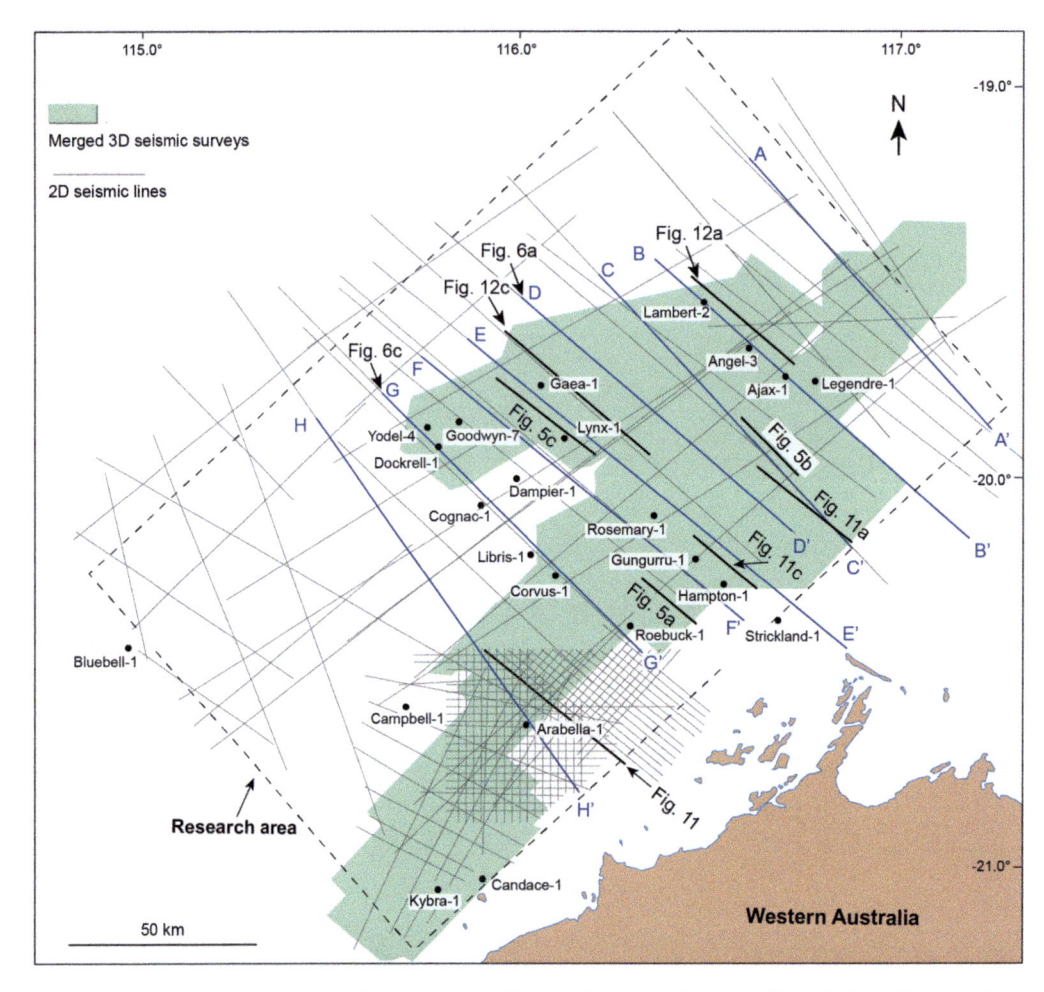

Fig. 4. 2-D & 3-D seismic datasets used in the research. The 2-D lines (grey) have a total length of c. 8600 km and the merged 3-D seismic data (light-green) covers an area of 16 000 km^2. Wells were tied to seismic sections of this research. Labelled sections (A–H) are shown in Figure 13.

Structural domains of the Dampier Sub-basin

The research area consists of the Dampier Sub-basin and flanking margins separated by basin boundary fault systems – the Rosemary Fault System in the east and Rankin Fault System in the west (Figs 1b, 2 & 6).

The Dampier Sub-basin includes the Lewis Trough, Madeleine Trend, Kendrew Trough, and Eliassen Terrace (Figs 1b & 6) with up to 4 km of syn-rift strata (Fig. 5b). The eastern margin consists of the Enderby Terrace in the NE and Mermaid Nose in the SW and preserves the Late Paleozoic structures and strata. The western margin (Rankin Platform) contains thick (>4 km) Triassic sediments (Jablonski *et al.* 2013) and a thin (<1 km) veneer of Jurassic units in the NNE-trending graben systems (Newman 1994).

Basin tectono-stratigraphy

Six major megasequences with regionally calibrated boundaries have been identified in the Dampier Sub-basin (Fig. 5).

Pre-Late Carboniferous Pre-rift 1 megasequence

The Pre-rift 1 megasequence (Fig. 5a) includes the Precambrian basement and the Early Carboniferous units. The basement units are characterized by discontinuous reflectors below the marked reflection of the Precambrian basement–cover boundary and

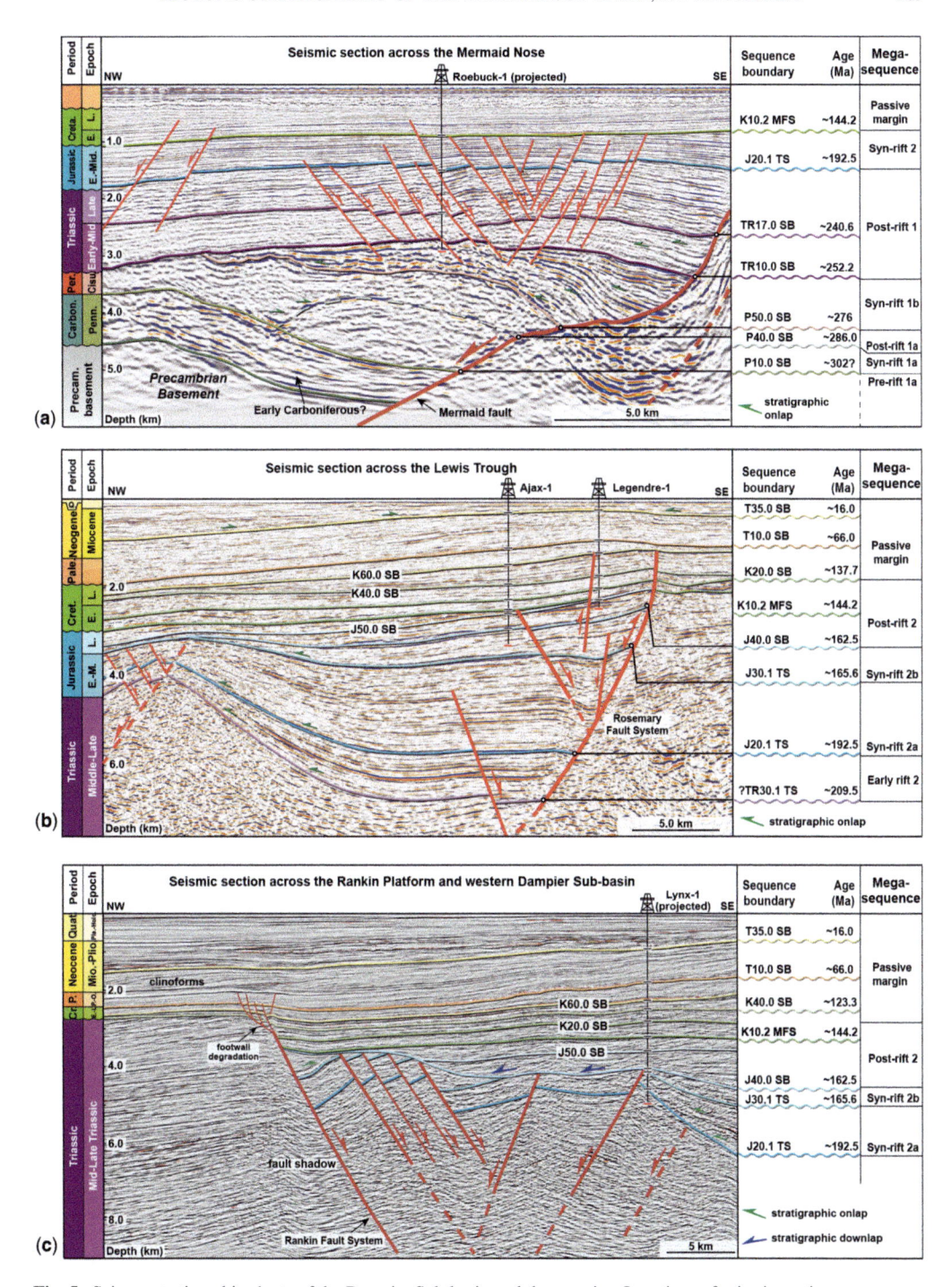

Fig. 5. Seismo-stratigraphic charts of the Dampier Sub-basin and the margins. Locations of seismic sections are shown in Figure 4. (**a**) Seismic section showing the Late Paleozoic Syn-rift 1 megasequence; (**b**) seismic section showing the Latest Triassic to Late Jurassic Syn-rift 2 megasequence; and (**c**) seismic section showing the Late Middle Jurassic Syn-rift 2b and Early Cretaceous to Present-day Post-rift 2 megasequences. Sequence boundary ages are based on Crostella *et al.* (2000), Marshall & Lang (2013).

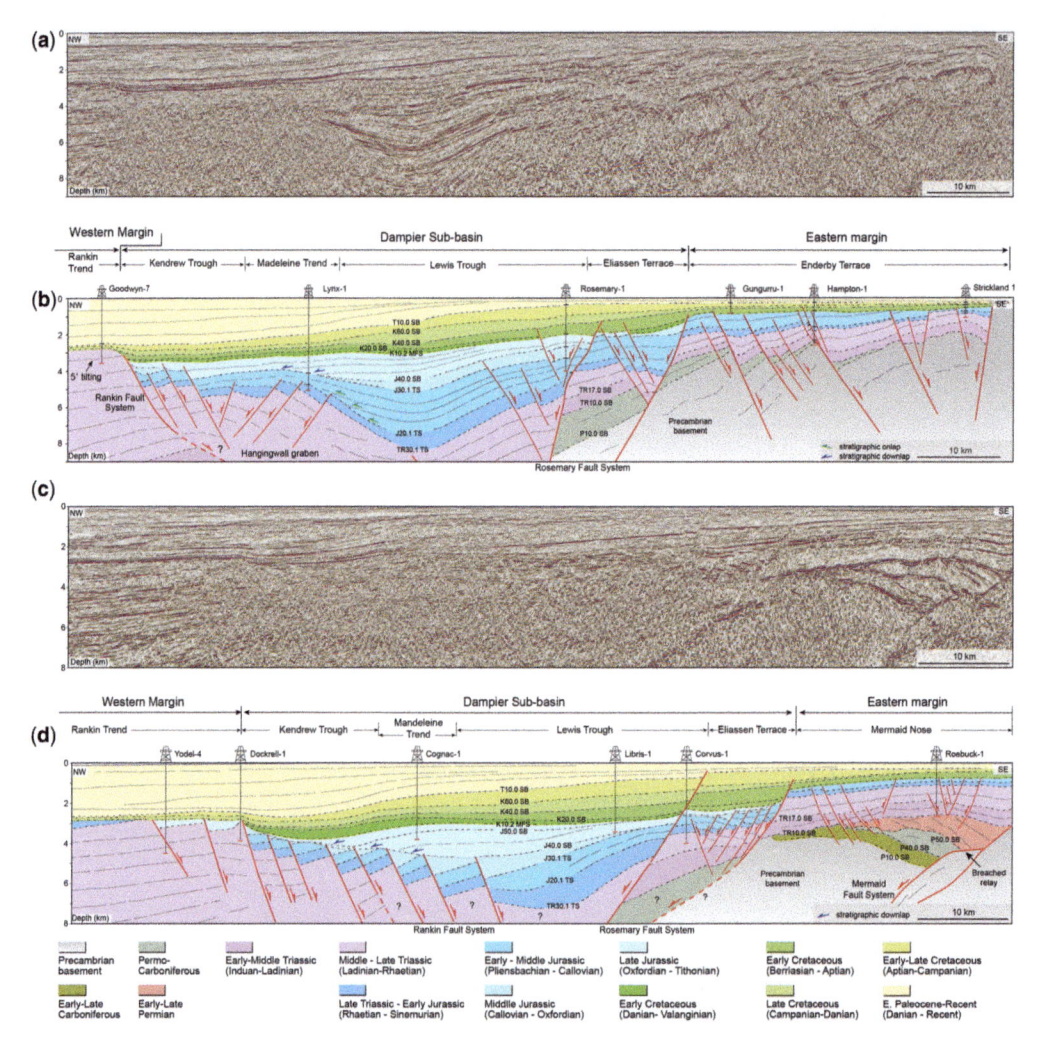

Fig. 6. Uninterpreted and interpreted cross-section of the central (**a–b**) and southern (**c–d**) Dampier Sub-basin. Seismic lines and wells location are shown in Figure 4. Note that in the eastern margin of the Dampier Sub-basin, the Enderby Terrace is characterized by SE-dipping, basement-involved normal fault systems and the Mermaid Nose is characterized by NW-dipping normal fault systems, which consist of planar normal faults that are linked at depth. The Lewis Trough is a Latest Triassic to Late Jurassic depocentre.

can be observed only on the eastern margin (Fig. 6) where the cover sequence is significantly thinner than that of the basin area to the west. The Early Carboniferous units are identified on the Mermaid Nose (Fig. 5a), where it shows moderate- to high-amplitude reflectors at depths of 4.5–5.5 km.

Late Carboniferous to Latest Permian Syn-rift 1 megasequence

The Late Carboniferous to Latest Permian Syn-rift 1 megasequence is present in the hanging-wall half-graben of the NW-dipping Mermaid Fault System (Fig. 7a), constrained by the P10.0 SB at the base and TR10.0 SB at the top (Fig. 5a). This half-graben is about 20 km wide and 2–3 km deep, and the sediments display two growth wedges (Syn-rift 1a and 1b) separated by parallel reflectors (Post-rift 1a) (Fig. 5a).

The lower growth wedge (Syn-rift 1a) (Fig. 5a) shows an elongated lens-shaped geometry and is characterized by low-amplitude discontinuous reflections. This sediment package is probably the poorly sorted Permo-Carboniferous glacial units, as has been drilled by the Kybra-1 well (Bond Corporation Petroleum Division 1988) and recorded by

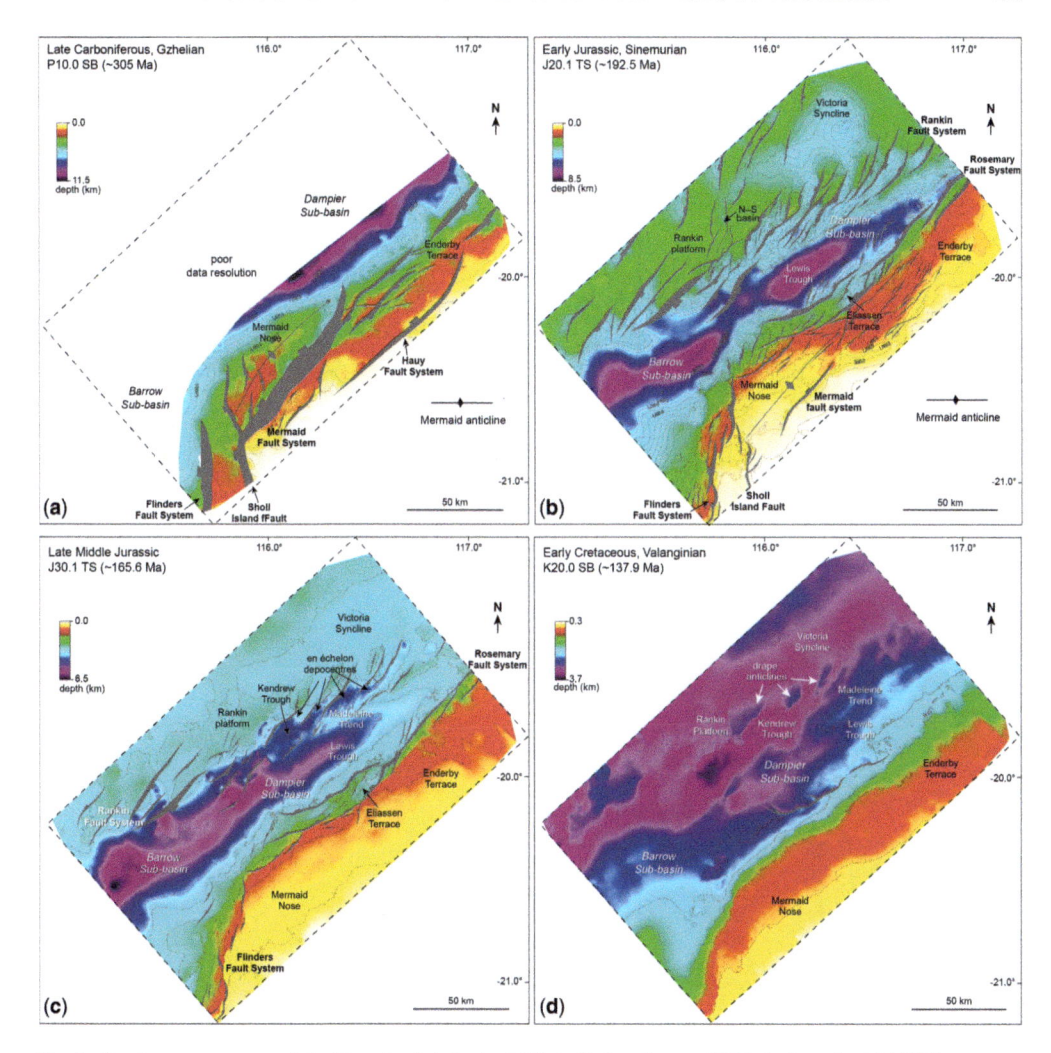

Fig. 7. Depth structure maps of the Dampier Sub-basin. (**a**) Late Carboniferous; (**b**) Early Jurassic; (**c**) Middle Jurassic; and (**d**) Early Cretaceous. Contour interval = 500 m.

Langhi & Borel (2005). Biostratigraphic information from the Candace-1 well (Australian Occidental Petroleum Pty. Ltd. 1983) indicates a Late Carboniferous to Early Permian age for Syn-rift 1a, constrained by the presence of *Microbaculispora tentula*. Therefore, the top of this growth package is P40.0 SB (286 Ma) and the base is P10.0 SB (302 Ma?) (Crostella *et al.* 2000; Marshall & Lang 2013).

The upper growth wedge (Syn-rift 1b) has high-amplitude reflections (Fig. 5a). This growth package consists of the Kennedy Group and was intersected and identified by the Roebuck-1 well that, with the presence of the *P. microcorpus* and *D. parvithola* zones, indicates a minimum age of 252–264 Ma (Kelman *et al.* 2013; Marshall & Lang 2013)

although the maximum age of the Kennedy Group is 274 Ma (Crostella *et al.* 2000). Thus, the top and base boundaries of syn-rift 1b are TR10.0 SB and P50.0 SB, respectively, based on the ages correlation of the stratigraphy of Marshall & Lang (2013).

Sediment distributions. The distribution of the Syn-rift 1 megasequence is confined within the half-grabens of the Mermaid Nose and Enderby Terrace (Figs 8, 10 & 11). Maximum thickness of this growth package exceeds 3 km. On the Mermaid Nose, the grabens are bounded by NE-striking and NW-dipping Mermaid Fault System (Fig. 7a). On the Enderby Terrace, the grabens are mainly bounded by NE-striking and SE-dipping fault systems. In the Dampier Sub-basin and on the Rankin

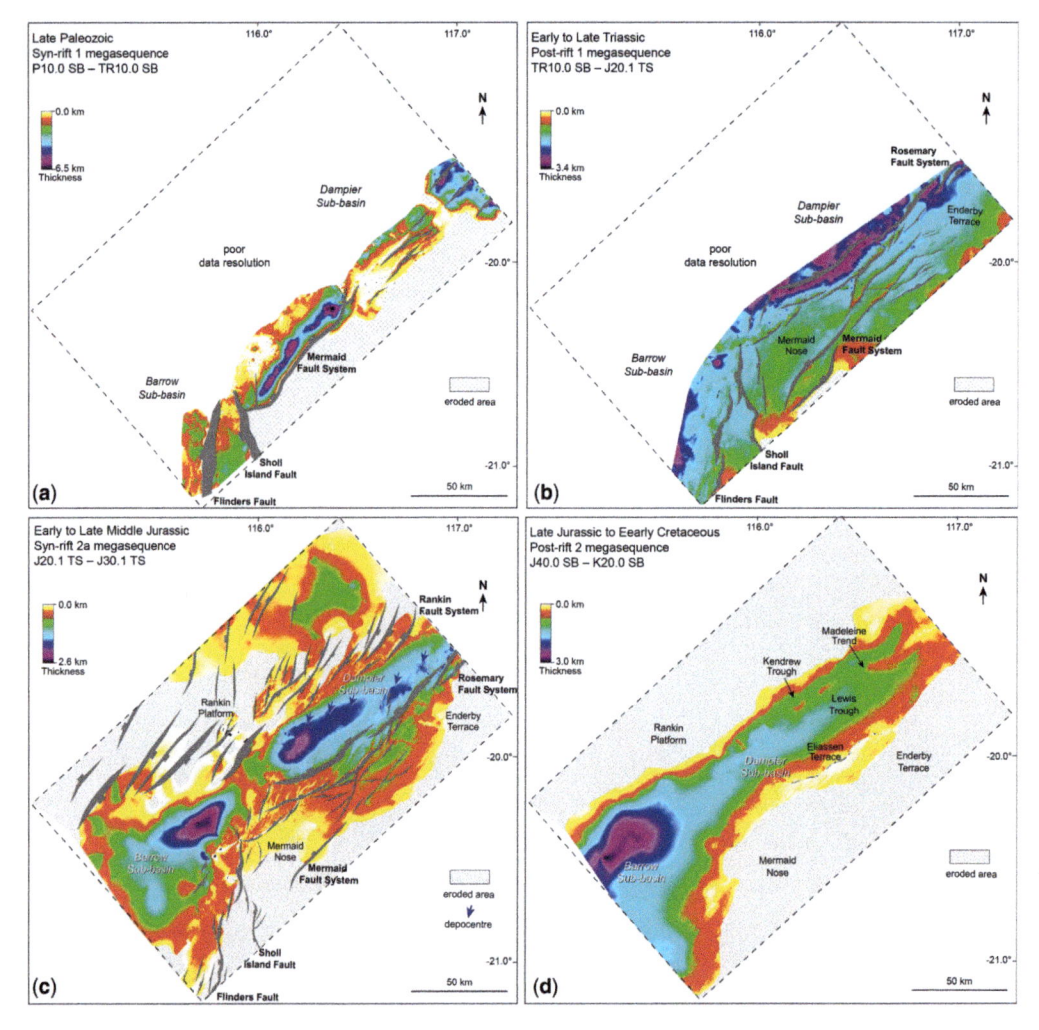

Fig. 8. Thickness maps of the Dampier Sub-basin. (**a**) Late Paleozoic; (**b**) Early to Latest Triassic; (**c**) Early to Late Middle Jurassic; and (**d**) Late Jurassic to Early Cretaceous. Hatched regions are eroded areas.

Platform, the Syn-rift 1 megasequences are deeply buried and poorly imaged in seismic sections.

Early to Latest Triassic Post-rift 1 megasequence

The Early to Latest Triassic Post-rift 1 megasequence includes the Locker Shale and the Mungaroo Formation (Fig. 3), defined by the Lower Triassic TR10.0 SB (base) and Latest Triassic TR30.1 TS (top) (Fig. 5a). The Locker Shale (TR10.0 SB–TR17.0 SB?) is characterized by low-amplitude reflections, whereas the Mungaroo Formation (TR17.0 SB?–TR30.1 TS) is dominated by moderate- to high-amplitude reflections. The Locker Shale is mainly a marine succession with lithologies varying from shale, through siltstone, to fine-grained sandstone in some wells (e.g. Hampton-1 well; Fig. 4).

Figure 8b shows Post-rift 1 megasequence (1–2 km) on the eastern margin where there are moderate changes of sediment thickness across the main faults. Although not all of the stratigraphic section of this megasequence has been penetrated by wells, it can be reasonably constrained based on the marked basement-cover reflection at the base (e.g. Figs 6, 10 & 11) and a well-constrained Mungaroo Formation at the top (Fig. 5a).

The Provenance of the Triassic sediment has been studied through the minerals dating of the drilled Triassic sediments (e.g. Lewis & Sircombe 2013; Zimmermann & Hall 2016) and through the attribute analysis of the seismic data (Marsh et al. 2018). All of these analyses indicate that most of the sediments were derived from the land mass of the Australia Plate, with some from the microplates that were attached to the Australia Plate. Sediments were transported via transcontinental channel systems.

Latest Triassic to Late Jurassic Syn-rift 2 megasequence

The Latest Triassic to Late Jurassic Syn-rift 2 megasequence is observed in the Dampier Sub-basin, bounded by the Rosemary and Rankin fault systems (Figs 1, 2 & 6). Within this succession, the moderate- to high-amplitude reflections show syn-kinematic growth (Fig. 5b) that thickens towards the basin-border fault. Based on the stratigraphic architecture, the Syn-rift 2 megasequence has been subdivided into three units: (1) Early rift 2; (2) Syn-rift 2a; and (3) Syn-rift 2b. The Early rift 2 sedimentary units are defined by the TR30.1 TS (base) and J20.1 TS (top). The growth wedge is obvious, and although the stratigraphic interval has not been directly encountered in wells, it has been reasonably extrapolated based on Lynx-1 and Gnu-1 wells (Fig. 12).

The Syn-rift 2a (J20.1 TS–J30.1 TS) hosted c. 3 km of syn-kinematic deposition (Fig. 5), and the basin bounding faults are highly segmented, consisting of a series of left-stepping fault segments (Fig. 7b). Notably, the depocentres along the Lewis Trough were not directly bounded by basin boundary faults except for the northernmost one (sections A–A' and B–B' in Fig. 13). Depocentres in the centre and SE of the Dampier Sub-basin show synclinal forms, aligning in en échelon pattern along the Lewis Trough (Fig. 8c). Apart from the Dampier Sub-basin, Syn-rift 2a growth faulting is also seen inside the Perseus and Eagle grabens on the Rankin Platform and hanging wall of reactivated basement fault on the Enderby Terrace (Deng 2017).

The Syn-rift 2b (J30.1 TS–J40.0 SB) is the second extension episode during which the Rosemary and the Rankin fault systems accrued c. 1 and c. 3 km displacements, respectively (Fig. 7c). In the east of the basin, the associated sedimentary growth wedge is constrained by the Ajax-1 and Legrendre-1 wells (Fig. 5b); in the western part of the basin a small growth wedge is imaged in seismic data (Fig. 5c), with sequence boundaries constrained by the Fisher-1 and Haycock-1 wells (Veenstra 1985). The activation of the Rosemary and Rankin fault systems at this stage significantly expanded the overall basin width from 30 to 60 km (Fig. 7b, c), accompanied by the Kendrew Trough development and characterized by a series of right-stepping depocentres (Fig. 7c).

Figure 8c shows that the Syn-rift 2a reaches up to 2.6 km in thickness in the centre of the Dampier Sub-basin, whereas on the Mermaid Nose, the Syn-rift 2a is thin (<300 m) and partly eroded (Fig. 6b). On the Enderby Terrace, the Syn-rift 2a units are <1 km thick, preserved in the hanging walls of half-grabens of the reactivated basement faults (Fig. 6a). On the western margin, the NNE-trending graben structures were active and accumulated <500 m Early Jurassic strata (Fig. 6b).

The Syn-rift 2b sequence has a maximum thickness of 1 km in the hanging wall of the Rosemary Fault System and is thin (<200 m) in the hanging wall of the Rankin Fault System. In the central Lewis Trough, the Syn-rift 2b is c. 600 m thick, not bounded by main faults (Fig. 6a).

Late Jurassic to Early Cretaceous Post-rift 2 megasequence

The Late Jurassic to Early Cretaceous Post-rift 2 megasequence is defined by the Oxfordian J40.0 SB sequence boundary at the base and the Valanginian K20.0 SB sequence boundary at the top (Fig. 5b, c). The J40.0 SB is largely an erosive unconformity on the basin shoulders and partly erosive within the basin (Fig. 6), marking the end of the syn-rift 2 package. Most large fault systems in the Dampier Sub-basin show some, but limited, displacement in the sequences above the J40.0 SB (Fig. 11). This is further supported by the parallel reflectors of the post-rift 2 units. The thickness map (Fig. 8d) shows thin (<300 m) or absent Post-rift 2 strata at the basin margins and thick (2 km) strata within the Dampier Sub-basin, indicative of differential Post-rift 2 thermal subsidence associated with burial and compaction.

Early Cretaceous to Holocene passive margin megasequence

The passive margin megasequence ranging from Valanginian K20.0 SB to the Holocene T10.0 SB was strongly affected by the thermal subsidence. In this sequence, westward propagation of clinoforms indicates that sediments were sourced mainly from the Australia continent and an open-marine environment might have formed to host large accumulation of carbonate rocks that reached up to 2 km along the axis of the Lewis Trough (Fig. 2).

The Cenozoic sequence increases in thickness westward from <500 m on the eastern margin to c. 2.7 km on the Rankin Platform (Fig. 6), showing little influence of the thermal subsidence inherited from the Mesozoic rifting. In the Miocene, minor to moderate inversion along the Rosemary Fault System created a tip-line anticline (Fig. 5b).

Basin structure

Basin bounding faults of the Dampier Sub-basin

The Dampier Sub-basin is bounded by the Rosemary Fault System to the east and the Rankin Fault System to the west (Fig. 2). The Rosemary Fault System is >150 km long and *c.* 10 km wide, comprising a number of linked fault segments that are mostly NE and NNE trending (Fig. 9). These right-stepping segments are linked through breached relay ramps at the Early to Middle Jurassic level (Fig. 7b). To the north of the Rosemary Fault System, fault growth initiated in the Latest Triassic (Rheatian) to Late Jurassic (Fig. 6a), whereas in the south the fault system started mainly from the Early Jurassic (Fig. 8c) and branched into two fault arrays that are 10 km apart.

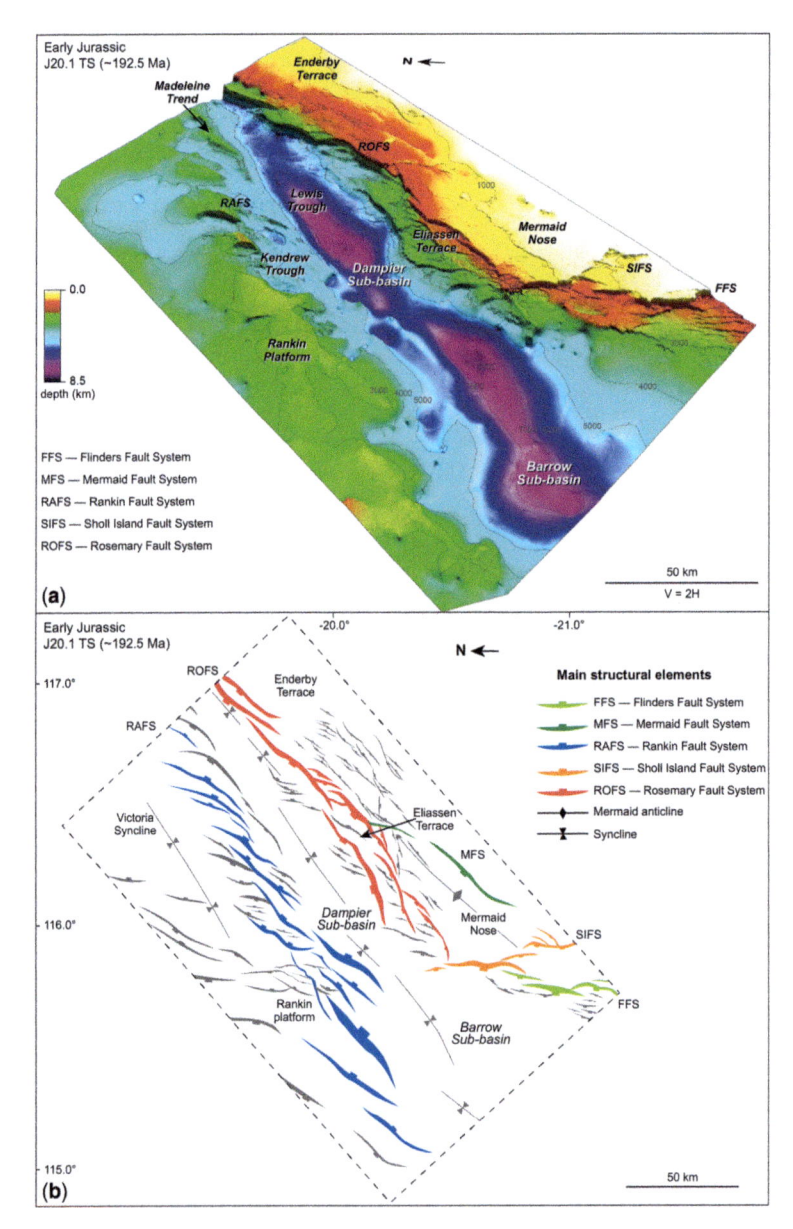

Fig. 9. Major structural elements of the Dampier Sub-basin, basin margins, and surrounding areas. (**a**) 3-D visualization of the basin structure; and (**b**) structural map of major fault systems at the J20.1 TS.

The SE-dipping Rankin Fault System is c. 200 km in extent, comprising six right-stepping en échelon fault segments at the Middle Jurassic level (Fig. 7b). The main fault has listric shape at depth, with a hanging wall anticline and crestal-collapse graben that forms the Madeleine Trend and Kendrew Trough (Fig. 6a). Growth stratal patterns indicate Callovian deformation on these fault systems (Fig. 5c) that is accompanied by accommodation space generation in the immediate hanging wall (Kendrew Trough; Fig. 6).

Eastern margin of the Dampier Sub-basin

The eastern margin comprises the Enderby Terrace in the NE and the Mermaid Nose in the SW (Figs 9 & 10).

Enderby Terrace. The Enderby Terrace is characterized by basement-involved and 'domino-style' extensional faults (Fig. 10) with the largest displacement (up to c. 2 km) at the top-basement level, the P10.0 SB in this case. The Syn-rift 1 and Syn-rift 2 growth stratal patterns (Fig. 10) indicate Late Paleozoic and Early and Middle Jurassic fault activities. The basement-involved faults are 30–70 km long (Fig. 7a) with linked fault traces at the P10.0 SB and TR10.0 SB sequence boundaries and systematic fault segmentation at the J20.1 TS stratigraphic horizon (Fig. 7b), showing synthetic footwall splays that branch upward at (or near) the top basement. In addition, hanging-wall extensional fault-related folds are observed over the apparent concave-upward fault geometries (Deng & McClay 2019).

Mermaid Nose and Mermaid Fault System. The Mermaid Nose is a NE-plunging and broadly folded anticline with an aspect of c. 80 × c. 30 km (Fig. 7a, b) in the hanging wall of the curved Mermaid fault system (Figs 5a & 6b). Analysis of the Mermaid Fault System suggests that the 'ramp–flat–ramp' fault is a composite geometry constituted by a series of planar normal faults (Fig. 11). Within this system, the planar fault in the NW is bounded by the Late Carboniferous to Early Permian growth strata, and the planar fault in the SE is bounded by Early to Late Permian growth strata (Figs 6b & 11). In addition, the Early Jurassic growth wedge bounding the upper part of the Mermaid fault suggests the third episode of fault reactivation. Therefore, the formation of this 'ramp–flat–ramp' fault geometry may have undergone at least three stages of extension, which contributed to the formation of subparallel faults and relay terrace in Late Paleozoic extensions and subsequent fault reactivation in the Early Jurassic extension that linked the previous faults (Fig. 11).

Fault displacement of the Mermaid Fault System is about 10 km (Fig. 6b). The fault dip is c. 30° with hanging wall rotated by 30° (Deng 2017). This large displacement, low-angle geometry and rotation of hanging wall fault blocks are records of Paleozoic rifting that took place at the edge of the North West Shelf continental margin.

The high-amplitude reflections at c. 8 km depth (Figs 6b & 11) may represent igneous intrusions hosted within the Precambrian basement. The anomalously high reflection has a thickness of 1–2 km and is subparallel to the basement top (P10.0 SB). It is possible that the high-amplitude reflection represents intrusive rocks or sills, which migrated along the stratified basement anisotropies of the Precambrian basement.

Western margin: Rankin Platform

The Rankin Platform is the eastern part of the Exmouth Plateau, characterized by a series of NNE-trending horsts and grabens. Growth strata in the hanging wall of these NNE-trending faults indicate that these faults accrued displacement in the Latest Triassic (Rheatian) to Early Jurassic extension (Fig. 12), which is different from the Late Middle Jurassic (syn-rift 2b) Rankin Fault System (Fig. 5c). A cross-cutting relationship (Fig. 12b) between fault systems on the Rankin Platform and the western bounding fault system (Rankin fault) further suggests that they have distinct deformation histories – the former developed earlier and was truncated by the latter. In addition, fault systems on the Rankin Platform consist of a north–south-trending basin (Fig. 7b), characterized by en échelon fault segments at the basin boundaries and NNE-trending sigmoidal faults within the basin.

In the footwall of the Rankin Fault System (Fig. 6a), Triassic-dominated successions were tilted 5° westward, capped by the Oxfordian Unconformity (J40.0 SB). The tilting of rift shoulder indicates footwall flexural uplift (e.g. Braun & Beaumont 1989). Fault segment bounding the Goodwyn horst has >3 km fault displacement, and the amount of uplift and erosion reaches c. 1.6 km based on the Goodwyn-6 and Goodwyn-9 wells (Jablonski *et al.* 2013). However, except for the Goodwyn horst, apparent fault block rotation is not recorded on the entire Rankin Platform (Fig. 7c), probably owing to the segmented nature of the Rankin Fault System and the lack of uniform uplift.

Discussion

Late Paleozoic Rifting

Previous work postulated that the Late Paleozoic rifting occurred in the Late Carboniferous to Early Permian and that the syn-rift basin infill passes upward into a thick Permian and Triassic basin as a

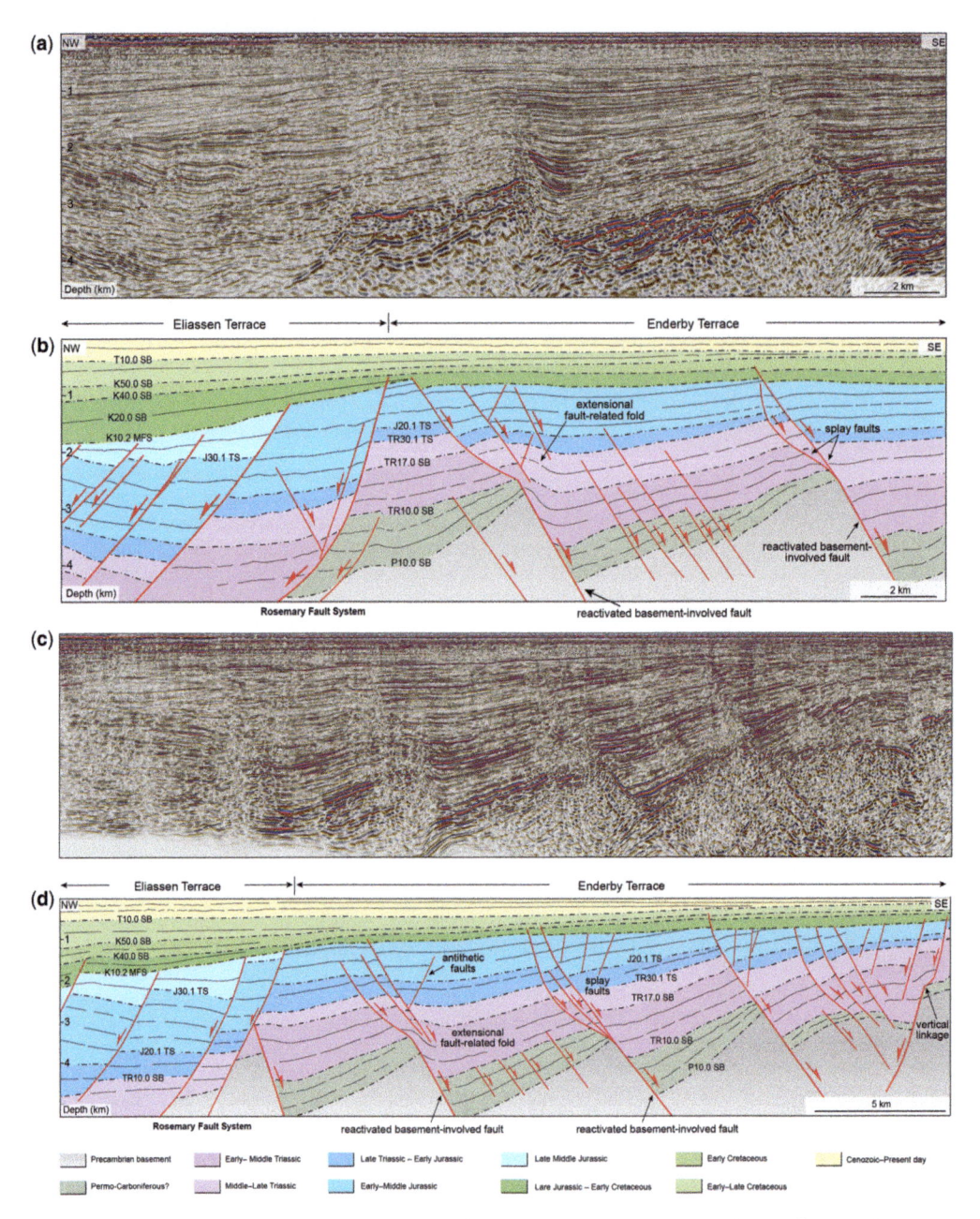

Fig. 10. Uninterpreted and interpreted cross-sections (**a–b**) and (**c–d**) showing the structural style of the reactivated basement-involved fault systems on the Enderby Terrace of the eastern margin. Seismic lines location is shown in Figure 4.

form of thermal subsidence (Yeates *et al.* 1987; Williamson *et al.* 1990; Etheridge & O'Brien 1994; Gartrell 2000; Longley *et al.* 2002; Eyles *et al.* 2003; Langhi & Borel 2005). Our interpretation of the Mermaid Nose structure shows a relatively continuous (except for P40.0 SB–P50.0 SB) growth wedge from the Late Carboniferous to Late Permian (Fig. 5a) and subsequent thermal subsidence from the Early to Latest Triassic. This is in agreement with the start of widespread deposition of the Locker Shale and Mungaroo Sandstone across the North West Shelf. In addition, this time frame is consistent

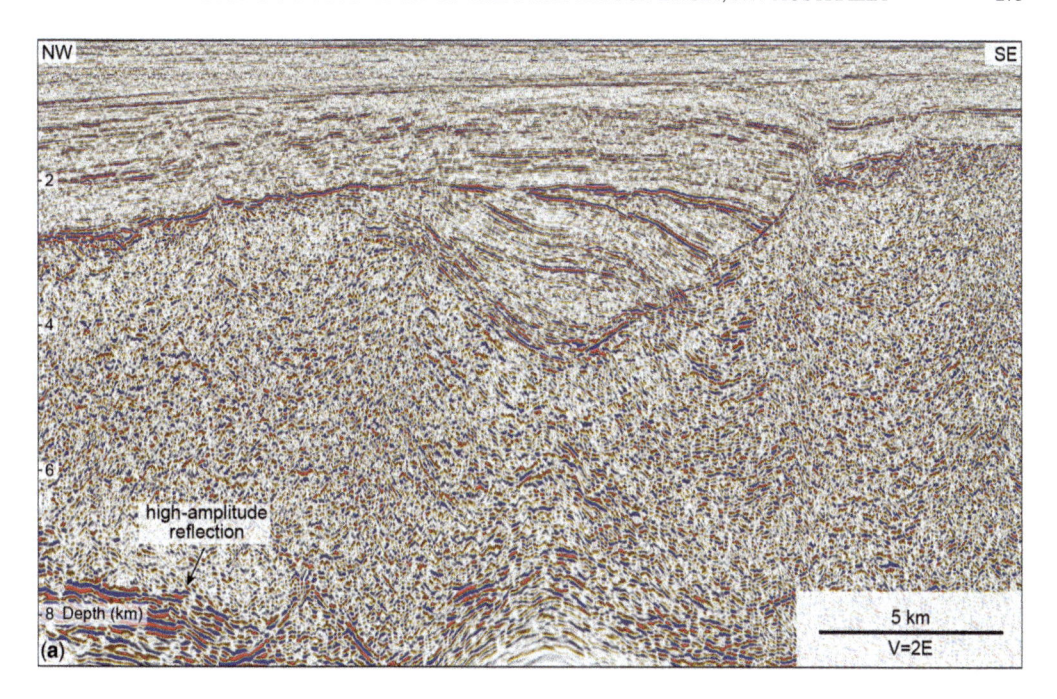

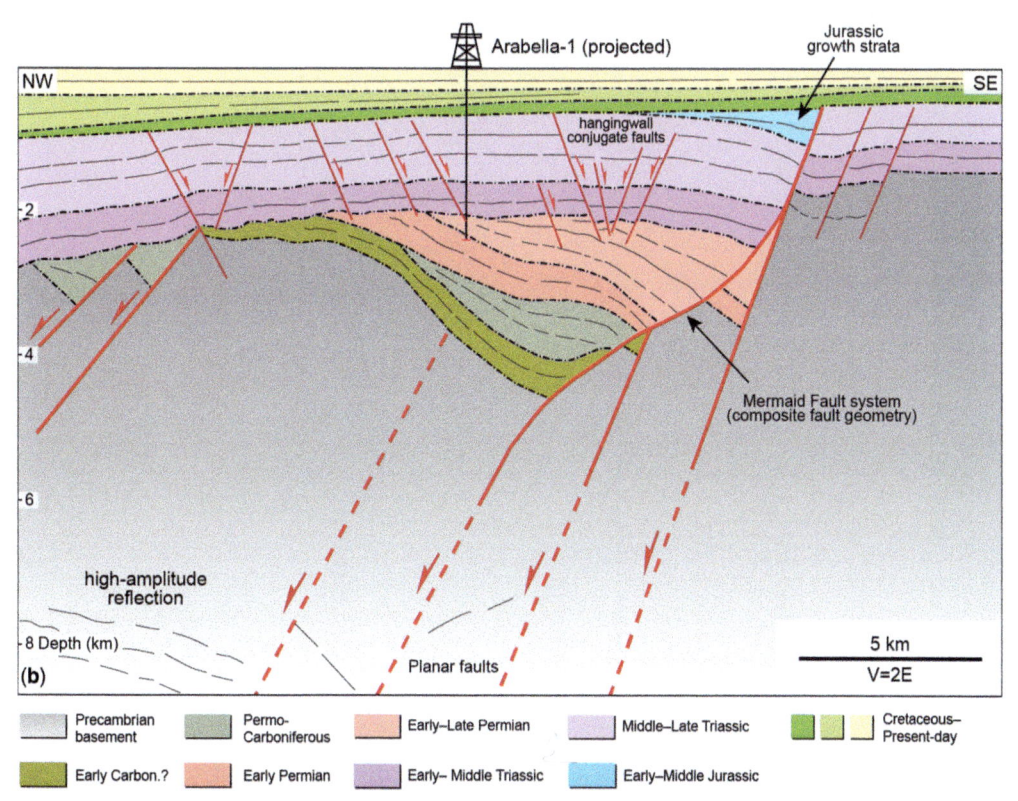

Fig. 11. Uninterpreted (**a**) and interpreted (**b**) seismic section showing the structure of the Mermaid Fault System. Note that this fault system consists of a series of linked planar faults and the reactivation of the fault system resulted in a curved composite geometry in the Late Paleozoic and basement units.

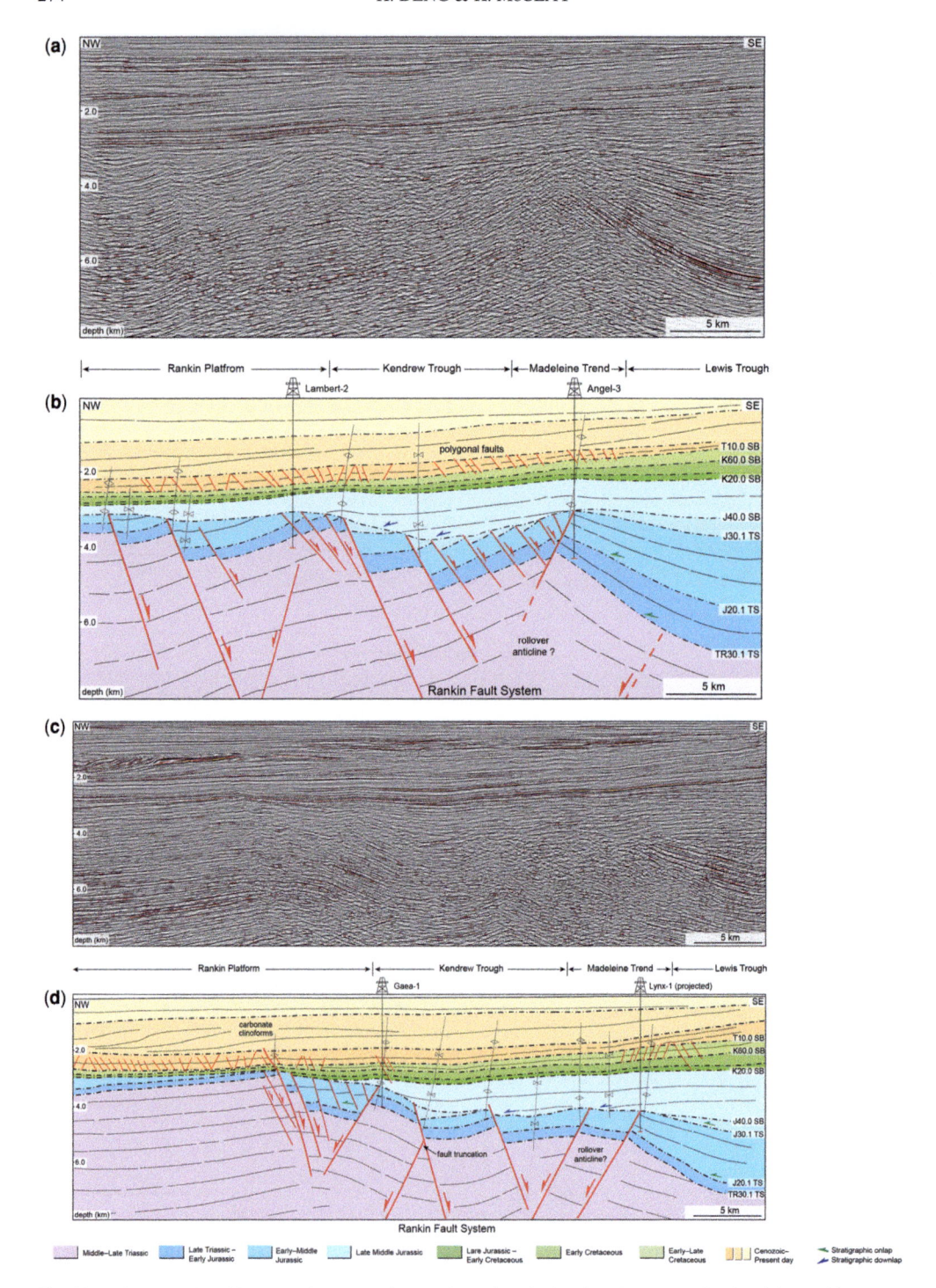

Fig. 12. Uninterpreted and interpreted cross-sections (**a–b**) and (**c–d**) showing the structural style of the Rankin Platform. Section locations are shown in the Figure 4. Note that J40.0 SB is an erosional unconformity that separates the underlying fault blocks from the overlying units. The Late Jurassic to Cenozoic sequences are deformed over the rotated fault blocks, suggesting that they are drape structures. In section (c)–(d), the truncation of Jurassic graben by the Rankin Fault System indicates that the latter postdate the former.

with regional extension tectonics of the west Australia offshore basin (Hall *et al.* 2013; Rollet *et al.* 2013) and igneous activities on the North West Shelf and adjacent areas (Reeckmann & Mebberson 1984; Charlton 2001; Gorter & Deighton 2002).

The Late Paleozoic extension developed a series of planar normal faults (Figs 6, 10 & 11) at a time when the North West Shelf was significantly thinned in the Late Paleozoic (Williamson *et al.* 1990; Etheridge & O'Brien 1994; Stagg & Colwell 1994; Gartrell 2000). Although recent studies (e.g. Pryer *et al.* 2014) of deep reflection data suggest the possible existence of a detachment fault beneath the thick package of Mesozoic sequence, the fault systems on the eastern margin of the Dampier Sub-basin show no evidence of basin-scale low-angle faults. In our case, the structural style of the Mermaid Nose and the Enderby Terrace is more comparable with the proximal domain of the passive margin defined by Péron-Pinvidic *et al.* (2013). Beneath the Dampier Sub-basin, the lower crust has been significantly thinned to *c.* 10 km (Hill 1994), which suggests that the proximal domain to the necking domain transition occurred in only 50 km.

Latest Triassic to Late Jurassic rifting

The Triassic to Late Jurassic rifting consists of three episodes of extension: Early Syn-rift 2, Syn-rift 2a and Syn-rift 2b. The Early Syn-rift 2 is the rift initiation stage that has been observed in the north part of the Dampier Sub-basin. This stage of fault activity is weaker and displacement is smaller than those of the following episodic extensions in the Early and Middle Jurassic, as is evidenced by the thickness of the syn-kinematic growth strata (Fig. 5a) and literally no fault reactivation on the eastern margin (Deng & McClay 2019). In the Exmouth Sub-basin (Jitmahantakul & McClay 2013; Black *et al.* 2017) and Exmouth Plateau (Gartrell *et al.* 2016; Yang & Elders 2016; Bilal *et al.* 2018), a stratigraphic growth wedge resting on the hanging wall of faulted blocks is obvious, suggesting widespread fault activities across the Northern Carnarvon Basin.

The Syn-rift 2a is the rift climax stage that prevails across the entire Dampier Sub-basin and flanking margins, which in the Dampier Sub-basin is the development of a series of en échelon depocentres (<2.6 km thick) (Fig. 8c) and in the eastern margin is characterized by basement fault reactivation (Deng & McClay 2019). The Rosemary Fault System was the dominant basin bounding structure that was active and achieved *c.* 2 km fault displacement. This episode of extension is also recorded in other inboard rift basins of the Northern Carnarvon Basin, such as the Exmouth, Barrow and Beagle sub-basins (Longley *et al.* 2002).

The Syn-rift 2b is a rift climax stage mainly localized along in the Dampier Sub-basin and no large fault activation or reactivation occurred on the flanking margins. This phase of extension is probably revoked by the reactivation of basement structure underlying the Dampier Sub-basin (Jablonski & Saitta 2004). The Syn-rift 2b extension in Callovian has uplifted the Rankin Platform, resulting in the titling of the footwall fault block and crest erosion marked as angular unconformities on the eastern edge of the Rankin Platform (Figs 6 & 13). Sediments eroded from tilted footwall crests were transported transversely across structural highs and redeposited in the proximal hanging wall basins along the Rankin Fault System. The footwall uplift occurring in response to isostatic imbalance is an important basin geomorphology feature (Braun & Beaumont 1989; Gawthorpe & Leeder 2000) commonly recorded in rift systems, such as the Red Sea (Bosworth 2015), East African rift system (Chorowicz 2005) and North West Shelf (Reeve *et al.* 2016; Paumard *et al.* 2018).

Rift basin localization

The Early to Late Jurassic rifting was mainly localized in the Dampier Sub-basin, where it attained thick (*c.* 4 km) syn-kinematic sediments in the synclinal basin (Fig. 13). One possible explanation for the rift basin localization and symmetrical geometry is a 'ramp syncline' in the hanging wall of a ramp–flat fault system (e.g. McClay & Scott 1991). This type of 'ramp syncline' has been reported in places where salt mobility facilitates large amount of hanging-wall rotation (e.g. Ferrer *et al.* 2016; Roma *et al.* 2018). However, the only candidate for ductile deformation in upper crustal level, the Locker Shale, is not as ductile as the salt that allows material flow to accommodate gravity loading. In addition, the segmented nature of the basin boundary fault systems (Rosemary and Rankin fault systems) is unlikely to be the large continuous detachment faults upon which the ramp basin develops.

An alternative explanation of the Lewis Trough formation is associated with the presence of an Archean–Proterozoic weakness zone (Pryer *et al.* 2002) that significantly controlled the basin localization. Numerical modelling by Chenin & Beaumont (2013) suggests that extension strain preferentially localizes at the crustal weak zones, which leads to rapid necking instabilities and offset basin development (Fig. 14). This modelling shows a direct comparison of tectonic elements with that of the Northern Carnarvon Basin – the narrow inboard rift system, little deformed rigid plateau, and rifted margin are comparable with Dampier Sub-basin, Exmouth Plateau and Late Jurassic and Cretaceous breakup margins, respectively. Although Chenin

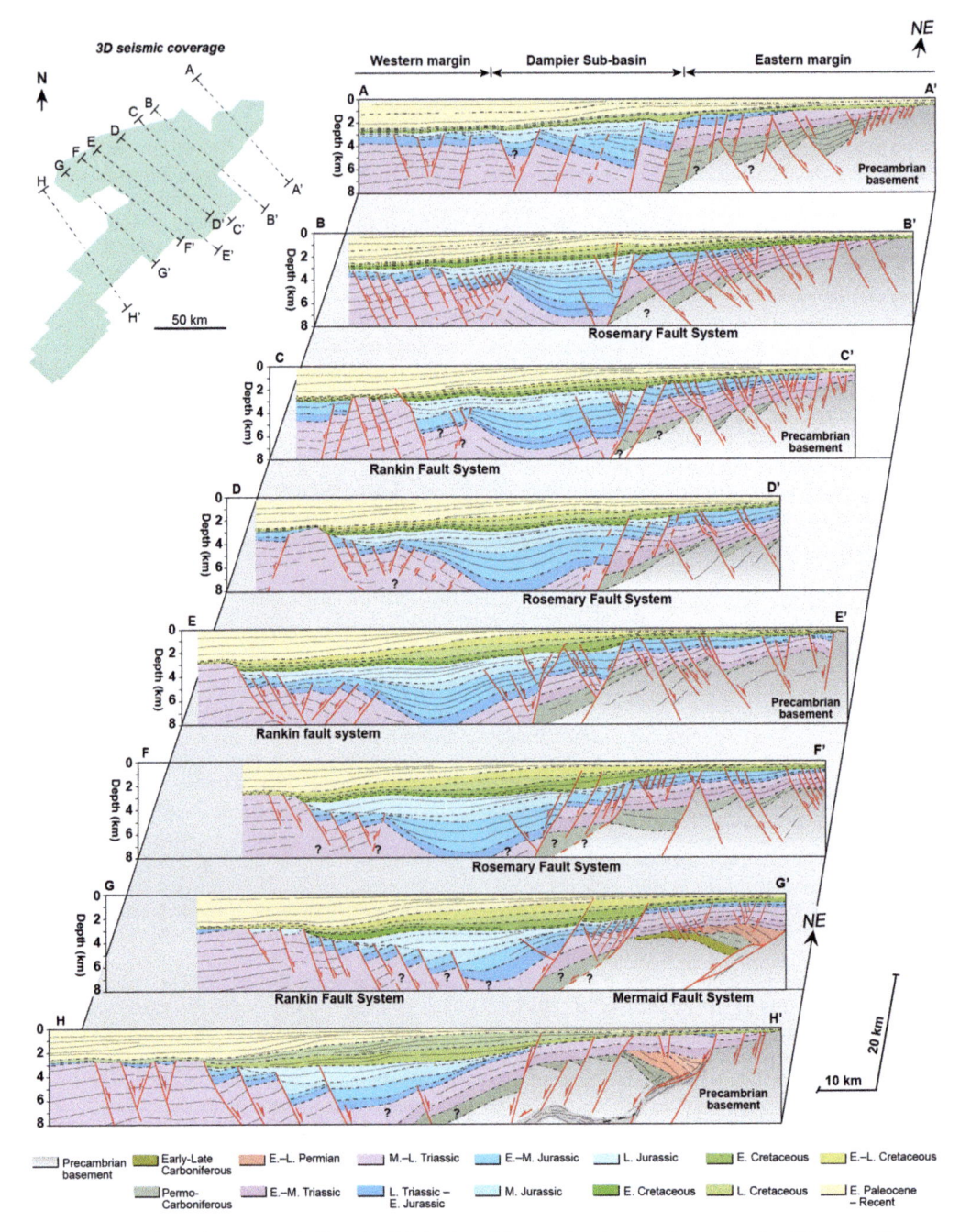

Fig. 13. Isometric sections across the Dampier Sub-basin and the margins showing the structural architecture of the studied area. The inset map shows locations of the 2-D sections (also see Fig. 4).

and Beaumont's numerical modelling (Fig. 14) did not show detailed basin architecture of the abandoned rift structure, it demonstrates the evolution of the first-order basin in the presence of inherited weakness. The modelling process is very similar to the evolution of Dampier Sub-basin, which became the focus of continental rifting in Latest Triassic to Late Jurassic and was then aborted after basin localization shift to the west (Mihut & Müller 1998; Karner & Driscoll 1999). Therefore, we suggest

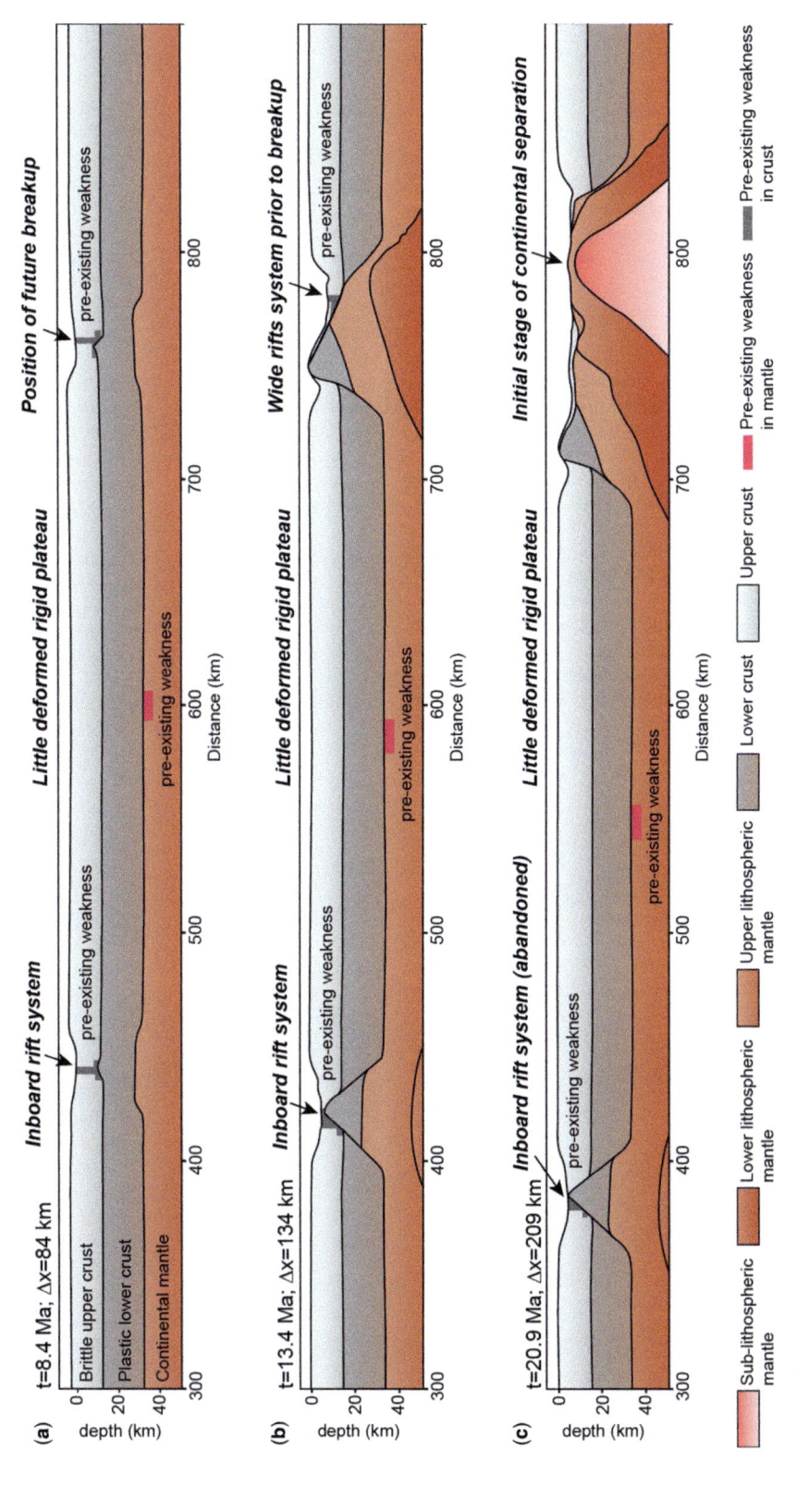

Fig. 14. Numerical modelling example of the presence of inherited weaknesses in continental lithosphere including coupled strong crust and weak mantle lithosphere, after Chenin & Beaumont (2013). Note that (a) after 8.4 km of extension, two rift basins developed along the pre-existing weakness zone in the crust. (b) After 13.4 Ma, the rift basin became localised along the crustal weakness on the right. (c) After 20.9 Ma of evolution the rift basin on the left was abandoned and the rift basin on the right continues, resulting in continental lithosphere breakup.

that the development of Dampier Sub-basin was controlled by the presence of underlying weakness. We also believe that more accurate modelling results could be achieved for the simulation of Dampier Sub-basin by taking account of the thinned lower crust (Stagg & Colwell 1994), the wider weak zone size (similar to the width of the Lewis Trough), extension duration (47 Ma), and the heterogenous mechanical composition of sedimentary succession (thick Locker Shale and Mungaroo Sandstone).

Oblique rifting

The structural element map of the Dampier sub-basin demonstrates right-stepping segmentation of the Rosemary and Rankin fault systems (Fig. 9) and en échelon distribution of the depocentres (Fig. 8c). This deformation pattern very much resembles the analogue modelling of McClay & White (1995) and Corti et al. (2013), which showed that depocentre distribution and basin border faults are subperpendicular to the regional stress field and oblique to the heterogeneous basin boundaries. The basin subsidence is largely confined by a pre-existing weak zone, where strain was compartmentalized from the lower basement and flanking margins.

The natural analogue of oblique rift is the Late Jurassic basin of the Central North Sea (Erratt et al. 1999) that shows widespread occurrences of basin boundary fault segmentation and en échelon distribution of the basin depocentres (Fig. 15), developed in response to oblique extension and reactivation of an underlying Paleozoic structure (Bartholomew et al. 1993). The Late Jurassic structural elements of the Central North Sea are generally similar in character to those of the Jurassic Dampier Sub-basin (Fig. 7b, c). The difference is the obliquity of the extension. The oblique angle of the Late Jurassic Central North Sea is c. 45° (NW–SE basement fault with east–west extension), but the oblique angle of the Dampier Sub-basin is c. 15° (NE–SW basement fault with NE–SW extension). As a result, the lateral offset of basin boundary faults of the Central North Sea (up to 70 km) and the Dampier Sub-basin (c. 10 km) becomes different. Apart from the Central North Sea, similar oblique rift systems can also be observed in the Main Ethiopian Rift (Corti 2012) and the Exmouth Sub-basin in the North West Shelf Australia (Jitmahantakul & McClay 2013).

Rift basin expansion

The Dampier Sub-basin expanded from the Early Jurassic (30 km wide) to Late Middle Jurassic (60 km wide) to include the Kendrew Trough, Madeleine Trend, and Eliassen Terrace (Fig. 8c). The development of these subplatforms may have been controlled by thick (>400 m) mechanically weak

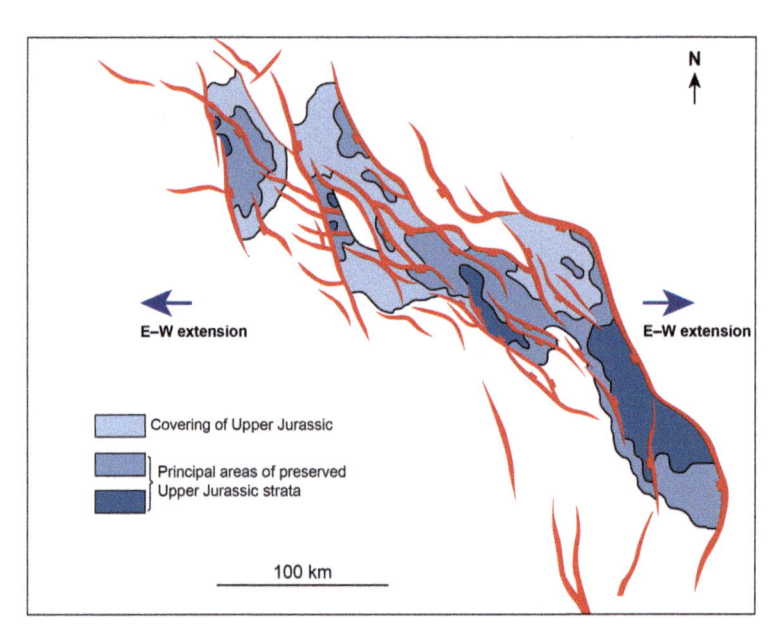

Fig. 15. Late Jurassic basin of the Central North Sea (after Erratt et al. 1999). The Jurassic basin formed in east–west regional extension with the presence of pre-existing NW–SE tectonic fabrics developed in Caledonian and Variscan, which resulted in high segmentation of the basin boundary fault systems and the discrete arrangement of the depocentres.

layer of the Locker Shale during basement fault reactivation (McHarg et al. 2019). It is shown by analogue modelling (e.g. Dooley et al. 2003; Lewis et al. 2013; Vasquez et al. 2018) that weak stratigraphy hampers upward propagation of the basin boundary faults and promotes lateral expansion (e.g. Deng & McClay 2019). In the western Dampier Sub-basin, the segmented Rankin Fault System may locally detach along the weak interface of the base Locker Shale and vertically connect to the underlying weakness zone. This process initially resulted in the formation of monocline, which explains the Early to Middle Jurassic stratigraphic onlaps over the western part of the Lewis Trough (Figs 6b, 12 & 13). Continued extension in the Callovian was accompanied by fault plane flattening along the Locker Shale, which finally linked with pre-existing and young faults to form a 'ramp–flat–ramp' geometry. In the hanging wall of the curved fault plane, crestal collapse graben and intrabasinal highs correspond to the Kendrew Trough and Madeleine Trend in the west of the Dampier Sub-basin (Fig. 6a).

Post-rift and passive margin

The Dampier Sub-basin reached thermal equilibrium by the Early Cenozoic (Fig. 13). This is contrary to Hull & Griffiths (2002) and Romine et al. (1997) who suggested that the thermal subsidence continued until Oligocene or even later. Our results show that the basin thickening to the west has not focused on the previous rifting centre since Cenozoic (Figs 6 & 13). In the Miocene, inversion of the Rosemary Fault System (Fig. 5b) reflects the onset of transpressional structural growth of basin boundary fault system. This pulse of inversion with up to 300 m uplift (Fig. 5b) in the Middle Miocene to Pliocene partly contributed to the deposition of the siliciclastic sediments (Bare Formation; Fig. 3) that represents a long-lived (11 myr) break of otherwise carbonate-rich units of the Northern Carnarvon Basin (Sanchez et al. 2012).

Structural evolution of the Dampier Sub-basin

The structural evolution of the Dampier Sub-basin can be summarized in eight stages (Fig. 16):

(1) Permo-Carboniferous rifting (Syn-rift 1a) – the NE-trending basement-involved fault systems were developed from the eastern margin, throughout the Lewis Trough, to the Rankin Platform (Fig. 16a). On the Enderby Terrace these planar normal fault systems dip to SE, and on the Mermaid Nose the planar faults dip to NW. The Dampier Sub-basin was influenced by the basement weak zone and developed segmented faults along the basin margins. The north–south-trending Sholl Island Fault (Fig. 1b) extended to the Rankin Platform.

(2) Early to Late Permian rifting (Syn-rift 1b) – the newly developed NW-dipping normal faults on the Mermaid Nose were linked with previous faults as a single large fault system through the breach of relay terrace at depth (Fig. 16b). The Mermaid Fault System become linked with the Rosemary Fault System and the Rankin Fault System is also linked as a large fault. Basin subsidence locally accommodates the sediments derived from the margins.

(3) The Early to Latest Triassic deposition (Post-rift 1) – following the Syn-rift 1, thermal subsidence created large (1–2 km depth) accommodation space in the eastern margin (Fig. 16c). Larger sedimentation occurs along the eastern basin boundary fault, across which basin thickness increases to the west. Our research could not figure out the exact thickness of the Early to Latest Triassic deposition, partly limited by the deep burial of the sediment and the deteriorated resolution of seismic at depth. Fault activities were weaker compared with the syn-rift stage.

(4) Latest Triassic rifting (Early Syn-rift 2) – the basement faults on the Enderby Terrace were reactivated and the upward propagation accompanied by negative refraction along the Locker Shale created a non-planar fault step that demanded hanging-wall deformation (Fig. 16d). The Rosemary Fault was reactivated and mainly localized in the NE of the Dampier Sub-basin. NNE-trending fault systems started to develop, not controlled by basement structures.

(5) Early to Middle Jurassic rifting (Syn-rift 2a) – extension mainly localized in the Dampier Sub-basin owing to reactivation of the basement weak zone (Fig. 16e). It also reactivated most basement faults on the Enderby Terrace. These faults truncated the previous monoclines and started to link as a segmented fault system. The Mermaid Fault was also reactivated with limited displacement. The Rosemary Fault System was reactivated along the pre-existing weak zone, characterized by right-stepping fault segmentation. NNE-trending fault systems developed in the Dampier Sub-basin and on the Rankin Platform.

(6) Late Middle Jurassic rifting (Syn-rift 2b) – localized extension in the Dampier Sub-basin reactivated the Rankin Fault System and expanded the basin boundary to the west and east (Fig. 16f). This reactivation developed right-stepping fault segments of the Rankin Fault System and Kendrew Trough and

Fig. 16. Schematic block diagrams showing the evolution of the Dampier Sub-basin and the flanking areas; no scale implied. (**a**) Late Carboniferous to Early Permian Syn-rift 1a, 'domino-style' normal faults development. (**b**) Early to Late Permian Syn-rift 1b, strong fault rotation and development of the low-angle normal fault system. (**c**) Early to Latest Triassic Post-rift 1, during the thermal subsidence, deposition of widespread Locker Shale and Mungaroo Formation in weak extension environment. (**d**) Latest Triassic Early Rift 2, limited fault initiation in the north part of the Dampier Sub-basin. (**e**) Early to Middle Jurassic Syn-rift 2a, 'sag-like' subsidence along the Lewis Trough. Fault activity mainly occurred along the Rosemary fault. NNE-trending fault system developed on the Rankin Platform and reactivated fault system developed on the Enderby Terrace. Fault reactivation on the Mermaid Nose is characterized by breaching the relay terrace between planar faults. (**f**) Late Middle Jurassic Syn-rift 2b, Rankin fault system becomes active and probably detach along the Locker Shale Formation and characterized by syncline and hangingwall structure and en échelon fault segments. The Eliassen Terrace developed in the eastern margin. (**g**) Late Jurassic to Early Cretaceous Post-Rift 2, thermal subsidence created accommodation space that mainly focused on the Dampier Sub-basin. (**h**) Early Cretaceous to Present-day passive margin stage, inversion of the fault segments of the pre-existing ROFS produced fault-propagation folds at the tip of these faults. ET, Enderby Terrace; ELT, Eliassen Terrace; LT, Lewis Trough; MN, Mermaid Nose; RP, Rankin Platform; RAFS, Rankin Fault System; ROFS, Rosemary Fault System; SIFS, Sholl Island Fault System.

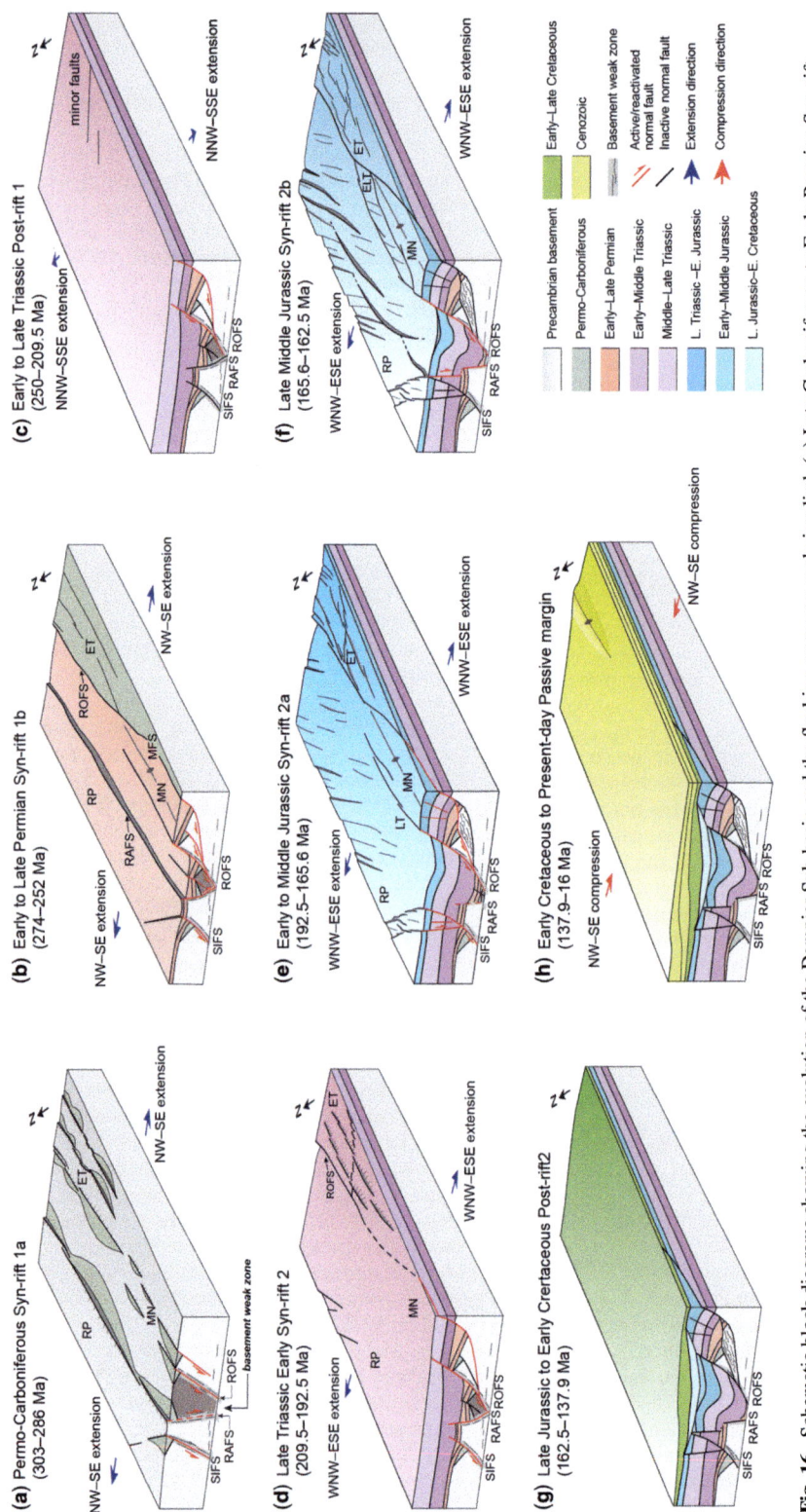

Madeleine Trend in the hanging wall of the 'ramp-flat' fault. The Rankin fault system also truncated the NNE-trending faults, resulting in a series of rhomboidal horst and graben structures on the Rankin Platform.

(7) Late Jurassic to Early Cretaceous deposition (Post-rift 2) – thermal subsidence following the Latest Triassic to Late Jurassic rifting created large accommodation space that accumulated thick (3–4 km) sediments along the Dampier Sub-basin (Fig. 16g). Sediments on the eastern and western margins were relatively thin (c. 1 km).

(8) Early Cretaceous to Present-day deposition (Passive margin) – thick sediments were transported from the eastern landmass and developed the westward propagation delta (Fig. 16h). Most strata were carbonate rich. In the Miocene, local inversion along the NE of the Rosemary Fault System produced moderate hanging wall (300 m fold amplitude) inversion anticlines.

Conclusions

The main conclusions are:

(1) Late Paleozoic basin developed by 'domino-style' normal faults and the Mesozoic rift system focused on the pre-existing NE-trending basement weak zone.

(2) The six tectonically controlled megasequences are (a) Pre-Late Carboniferous Pre-rift 1, (b) Late Carboniferous to Latest Permian Syn-rift 1, (c) Early–Latest Triassic Post-rift 1, (d) Latest Triassic to early Late Jurassic Syn-rift 2, (e) Late Jurassic to Early Cretaceous Post-rift 2 and (f) Early Cretaceous to Present-day Passive margin megasequences.

(3) Fault architecture of the Late Paleozoic rifting is dominated by a series of planar extensional faults on the eastern flank, some of which were reactivated in the Latest Triassic (Rhaetian) to Late Jurassic, resulting in a large-scale (80 km long, 10 km displacement, 30° rotation) low-angle normal fault. The structural style of the Mesozoic rifting is defined by localized rift basin, bordered by segmented basin boundary fault systems. At this stage, the Dampier Sub-basin underwent rift initiation (few faults), rift localization (30 km wide) and rift expansion (60 km wide).

The timing and structural style of the Dampier Sub-basin are crucial for future basin modelling. The tectono-stratigraphic evolution of the Dampier Sub-basin also provides new insights for the evolution of the North West Shelf passive margin and continental margin rift basin worldwide.

Acknowledgements The research presented in this paper is part of Hongdan's PhD research. Seismic data were kindly provided by Geoscience Australia through our collaborative research agreement. The staff and students of the Department of Earth Sciences at Royal Holloway University of London who provided the necessary support that allowed this project to be completed are gratefully thanked. The authors also thank Chris Elders and Martin Insley for the thoughtful discussions during the early stage preparation of this paper. Neil Marshall critically read an earlier version of this manuscript, and his comments are appreciated. The two reviewers Simon Lang and Ken McCormack are gratefully acknowledged for helping us improve the manuscript. Halliburton is thanked for kindly providing the academic licence for Landmark seismic interpretation software packages.

Funding Funded by the Fault Dynamics Research Group Royal Holloway University of London.

Author contributions HD: Conceptualization (Lead), Data curation (Lead), Formal analysis (Lead), Funding acquisition (Supporting), Investigation (Lead), Methodology (Equal), Project administration (Lead), Software (Lead), Validation (Lead), Visualization (Lead), Writing - Original Draft (Lead), Writing - Review & Editing (Supporting); KM: Conceptualization (Supporting), Data curation (Supporting), Funding acquisition (Lead), Investigation (Supporting), Methodology (Supporting), Project administration (Supporting), Resources (Equal), Software (Supporting), Supervision (Lead), Validation (Equal), Writing - Review & Editing (Lead).

References

ADAMSON, K.R., LANG, S.C., MARSHALL, N.G., SEGGIE, R.J., ADAMSON, N.J. & BANN, K.L. 2013. Understanding the Late Triassic Mungaroo and Brigadier Deltas of the Northern Carnarvon Basin, North West Shelf, Australia. In: KEEP, M. & MOSS, S.J. (eds) *The Sedimentary Basins of Western Australia IV: Proceedings of the Petroleum Exploration Society of Australia Symposium, Perth, Australia*. Petroleum Society of Australia, Perth, Australia, 1–29.

AUSTRALIAN GEOLOGICAL SURVEY ORGANISATION, STAGG, H.M.J., FITZGERALD, C.J., STAGG, H.M.J. & NOPEC AUSTRALIA. 1993. *Tectonic Elements of the North West Shelf*. Australia.

AUSTRALIAN OCCIDENTAL PETROLEUM PTY. LTD. 1983. Candace 1 Well Completion Report, Unpublished. Perth, WA.

BARBER, P.M. 2013. Oil exploration potential in the greater northern Australian – New Guinea Super Gas Province. In: KEEP, M. & MOSS, S. (eds) *The Sedimentary Basins of Western Australia IV: Proceedings of the Petroleum Exploration Society of Australia Symposium, Perth, WA, 2013*, Petroleum Society of Australia, Perth, Australia, 1–32.

BARTHOLOMEW, I.D., PETERS, J.M. & POWELL, C.M. 1993. Regional structural evolution of the North Sea: oblique slip and the reactivation of basement lineaments. *Petroleum Geology of Northwest Europe: Proceedings of the 4th Conference*. Geological Society, London, 1109–1122, https://doi.org/10.1144/0041109

BILAL, A., MCCLAY, K. & SCARSELLI, N. 2018. Fault-scarp degradation in the central Exmouth Plateau, North West Shelf, Australia. *In*: MCCLAY, K.R. & HAMMERSTEIN, J.A. (eds) *Passive Margins: Tectonics, Sedimentation and Magmatism*. Geological Society, London, Special Publications, **476**, First published online July 5, 2018, https://doi.org/10.1144/SP476.11

BLACK, M., MCCORMACK, K.D., ELDERS, C. & ROBERTSON, D. 2017. Extensional fault evolution within the Exmouth Sub-basin, North West Shelf, Australia. *Marine and Petroleum Geology*, **85**, 301–315, https://doi.org/10.1016/j.marpetgeo.2017.05.022

BOND CORPORATION PETROLEUM DIVISION 1988. Kybra 1 Well Completion Report, Unpublished. Perth, WA.

BOSWORTH, W. 2015. The Red Sea. *In*: RASUL, N.M.A. & STEWART, I.C.F. (eds) *Springer Earth System Sciences*. Springer, Berlin, 1–128.

BRAUN, J. & BEAUMONT, C. 1989. A physical explanation of the relation between flank uplifts and the breakup unconformity at rifted continental margins. *Geology*, **17**, 760–764, https://doi.org/10.1130/0091-7613(1989)017<0760:APEOTR>2.3.CO;2

CHARLTON, T.R. 2001. Permo-Triassic evolution of Gondwanan eastern Indonesia, and the final Mesozoic separation of SE Asia from Australia. *Journal of Asian Earth Sciences*, **19**, 595–617, https://doi.org/10.1016/S1367-9120(00)00054-7

CHENIN, P. & BEAUMONT, C. 2013. Influence of offset weak zones on the development of rift basins: activation and abandonment during continental extension and breakup. *Journal of Geophysical Research: Solid Earth*, **118**, 1698–1720, https://doi.org/10.1002/jgrb.50138

CHOROWICZ, J. 2005. The East African rift system. *Journal of African Earth Sciences*, **43**, 379–410, https://doi.org/10.1016/j.jafrearsci.2005.07.019

CORTI, G. 2012. Evolution and characteristics of continental rifting: analog modeling-inspired view and comparison with examples from the East African Rift System. *Tectonophysics*, **522**, 1–33, https://doi.org/10.1016/j.tecto.2011.06.010

CORTI, G., RANALLI, G., AGOSTINI, A. & SOKOUTIS, D. 2013. Inward migration of faulting during continental rifting: effects of pre-existing lithospheric structure and extension rate. *Tectonophysics*, **594**, 137–148, https://doi.org/10.1016/j.tecto.2013.03.028

CROSTELLA, A., LASKY, R.P., BLUNDELL, K.A., YASIN, A.R. & GHORI, K.A.R. 2000. *Petroleum Geology of the Peedamullah Shelf and Onslow Terrace, Northern Carnarvon Basin, Western Australia: Australia Geological Survey*.

DENG, H. 2017. *Tectono-Stratigraphic Evolution of the Dampier Sub-Basin and the Rankin Platform, NW Shelf of Australia*, Unpublished PhD thesis. Royal Holloway University of London.

DENG, H. & MCCLAY, K. 2019. Development of extensional fault and fold system: insights from 3D seismic interpretation of the Enderby Terrace, NW Shelf of Australia. *Marine and Petroleum Geology*, **104**, 11–28, https://doi.org/10.1016/j.marpetgeo.2019.03.003

DOOLEY, T., MCCLAY, K.R. & PASCOE, R. 2003. 3D analogue models of variable displacement extensional faults: applications to the Revfallet Fault system, offshore mid-Norway. *New Insights into Structural Interpretation and Modelling*, **212**, 151–167.

ERRATT, D., THOMAS, G.M. & WALL, G.R.T. 1999. The evolution of the Central North Sea Rift. *Petroleum Geology of Northwest Europe: Proceedings of the 5th Conference*. Geological Society, London, 63–82, https://doi.org/10.1144/0050063

ETHERIDGE, M.A. & O'BRIEN, G.W. 1994. Structural and tectonic evolution of the Western Australian margin basin system. *Petrolum Exploration Society of Australia (PESA) Journal*, **22**, 45–63.

EYLES, C.H., MORY, A.J. & EYLES, N. 2003. Carboniferous–Permian facies and tectono-stratigraphic successions of the glacially influenced and rifted Carnarvon Basin, western Australia. *Sedimentary Geology*, **155**, 63–86, https://doi.org/10.1016/S0037-0738(02)00160-4

FERRER, O., MCCLAY, K. & SELLIER, N.C. 2016. Influence of fault geometries and mechanical anisotropies on the growth and inversion of hanging-wall synclinal basins: insights from sandbox models and natural examples. *In*: CHILDS, C, HOLDSWORTH, R.E., JACKSON, C.A.-L., MANZOCCHI, T., WALSH, J.J. & YIELDING, G. (eds) *The Geometry and Growth of Normal Faults*. Geological Society, London, Special Publications, **439**, 487–509, https://doi.org/10.1144/SP439.8

GARTRELL, A.P. 2000. Rheological controls on extensional styles and the structural evolution of the Northern Carnarvon Basin, North West Shelf, Australia. *Australian Journal of Earth Sciences*, **47**, 231–244, https://doi.org/10.1046/j.1440-0952.2000.00776.x

GARTRELL, A., TORRES, J., DIXON, M. & KEEP, M. 2016. Mesozoic rift onset and its impact on the sequence stratigraphic architecture of the Northern Carnarvon Basin. *The APPEA Journal*, **56**, 143–158, https://doi.org/10.1071/AJ15012

GAWTHORPE, R.L. & LEEDER, M.R. 2000. Tectono-sedimentary evolution of active extensional basins. *Basin Research*, **12**, 195–218, https://doi.wiley.com/10.1111/j.1365-2117.2000.00121.x

GIBBONS, A.D., BARCKHAUSEN, U., VAN DEN BOGAARD, P., HOERNLE, K., WERNER, R., WHITTAKER, J.M. & MÜLLER, R.D. 2012. Constraining the Jurassic extent of Greater India: Tectonic evolution of the West Australian margin. *Geochemistry Geophysics Geosystems*, **13**, 1–25, https://doi.org/10.1029/2011GC003955

GORTER, J.D. & DEIGHTON, I. 2002. Effects of igneous activity in the offshore northern Perth Basin – evidence from petroleum exploration wells, 2D seismic and magnetic surveys. *In*: KEEP, M. & MOSS, S. (eds) *The Sedimentary Basins of Western Australia 3: Proceedings of the Petroleum Exploration Society of Australia Symposium, Perth, Australia*. Petroleum Exploration Society of Australia, Perth, Australia, 875–899.

HALL, R. 2012. Late Jurassic–Cenozoic reconstructions of the Indonesian region and the Indian Ocean. *Tectonophysics*, **570–571**, 1–41, https://doi.org/10.1016/j.tecto.2012.04.021

HALL, L.S., GIBBONS, A.D., BERNARDEL, G., WHITTAKER, J.M., NICHOLSON, C., ROLLET, N. & MÜLLER, R.D. 2013. Structural Architecture of Australia's Southwest Continental Margin and Implications for Early Cretaceous Basin Evolution. *The Sedimentary Basins of Western Australia IV: Proceedings of the Petroleum Exploration Society of Australia Symposium, Perth, Australia*. Petroleum Society of Australia, Perth, Australia, 1–22.

HEINE, C. & MÜLLER, R. 2005. Late Jurassic rifting along the Australian North West Shelf: margin geometry and spreading ridge configuration. *Australian Journal of Earth Sciences*, **52**, 27–39, https://doi.org/10.1080/08120090500100077

HELDREICH, G., REDFERN, J., LEGLER, B., GERDES, K. & WILLIAMS, B.P.J. 2017. Challenges in characterizing subsurface paralic reservoir geometries: a detailed case study of the Mungaroo Formation, North West Shelf, Australia. *In*: NORRIS, S., NEEFT, E.A.C. & VAN GEET, M. (eds) *Multiple Roles of Clays in Radioactive Waste Confinement*. Geological Society, London, Special Publications, **444**, 59–108, https://doi.org/10.1144/SP444.13

HENGESH, J.V. & WHITNEY, B.B. 2016. Transcurrent reactivation of Australia's western passive margin: An example of intraplate deformation from the central Indo-Australian plate. *Tectonics*, **35**, 1066–1089, https://doi.org/10.1002/2015TC004103

HILL, G. 1994. The role of the pre-rift structure in the architecture of the Dampier Basin area, North West Shelf, Australia. *The APPEA Journal*, **34**, 602–613, https://doi.org/10.1071/AJ93046

HULL, J.N.F. & GRIFFITHS, C.M. 2002. Sequence stratigraphic evolution of the Albian to Recent section of the Dampier Sub-basin, North West Shelf, Australia. *In*: KEEP, M. & MOSS, S. (eds) *The Sedimentary Basins of Western Australia 3: Proceedings of the Petroleum Exploration Society of Australia Symposium, Perth, Australia*. Petroleum Exploration Society of Australia, Perth, Australia, 617–639.

JABLONSKI, D. 1997. Recent advances in the sequence stratigraphy of the Triassic to Lower. Cretaceous succession in the Northern Carnarvon Basin. *The APPEA Journal*, **37**, 429–454, https://doi.org/10.1071/AJ96026

JABLONSKI, D. & SAITTA, A.J. 2004. Permian to Lower Cretaceous plate tectonics and its impact on the tectonostratigraphic development of the Western Australian margin. *The APPEA Journal*, **44**, 287–328, https://doi.org/10.1071/AJ03011

JABLONSKI, D., PRESTON, J., WESTLAKE, S. & GUMLEY, C.M. 2013. Unlocking the origin of hydrocarbons in the central part of the Rankin Trend, Northern Carnarvon Basin, Australia. *In*: KEEP, M. & MOSS, S. (eds) *The Sedimentary Basins of Western Australia IV: Proceedings of the Petroleum Exploration Society of Australia Symposium, Perth, Australia*. Petroleum Society of Australia, Perth, Australia, 1–31.

JITMAHANTAKUL, S. & MCCLAY, K. 2013. Late Triassic–Mid-Jurassic to Neogene extensional fault systems in the Exmouth Sub-Basin, Northern Carnarvon Basin, North West Shelf, Western Australia. *In*: KEEP, M. & MOSS, S.J. (eds) *The Sedimentary Basins of Western Australia IV: Proceedings of the Petroleum Exploration Society of Australia Symposium, Perth, Australia*. Petroleum Society of Australia, Perth, Australia, 1–22.

KARNER, G.D. & DRISCOLL, N.W. 1999. Style, timing and distribution of tectonic deformation across the Exmouth Plateau, northwest Australia, determined from stratal architecture and quantitative basin modelling. *In*: MAC NIOCAILL, C. & RYAN, P.D. (eds) *Continental Tectonics*. Geological Society, London, Special Publications, 271–311, https://doi.org/10.1144/GSL.SP.1999.164.01.14

KEEP, M., HARROWFIELD, M. & CROWE, W. 2007. The Neogene tectonic history of the North West Shelf, Australia. *Exploration Geophysics*, **38**, 151–174, https://doi.org/10.1071/EG07022

KELMAN, A.P., NICOLL, R.S. ET AL. 2013. *Northern Carnarvon Basin Biozonation and Stratigraphy, 2013, Chart 36*.

LANGHI, L. & BOREL, G.D. 2005. Influence of the Neotethys rifting on the development of the Dampier Sub-basin (North West Shelf of Australia), highlighted by subsidence modelling. *Tectonophysics*, **397**, 93–111, https://doi.org/10.1016/j.tecto.2004.10.005

LEWIS, C.J. & SIRCOMBE, K.N. 2013. Use of U-Pb geochronology to delineate provenance of North West Shelf sediments, Australia. *In*: KEEP, M. & MOSS, S.J. (eds) *The Sedimentary Basins of Western Australia IV: Proceedings of the Petroleum Exploration Society of Australia Symposium, Perth, Australia*. Petroleum Society of Australia, Perth, Australia, 1–27.

LEWIS, M.M., JACKSON, C.A.-L. & GAWTHORPE, R.L. 2013. Salt-influenced normal fault growth and forced folding: the Stavanger Fault System, North Sea. *Journal of Structural Geology*, **54**, 156–173, https://doi.org/10.1016/j.jsg.2013.07.015

LONGLEY, I.M., BUESSENSCHUETT, C. ET AL. 2002. The North West Shelf of Australia – a Woodside perspective. *In*: KEEP, M. & MOSS, S. (eds) *The Sedimentary Basins of Western Australia 3: Proceedings of the Petroleum Exploration Society of Australia Symposium, Perth, Australia*. Petroleum Exploration Society of Australia, Perth, Australia, 27–88.

MARSH, T., KOWALIK, B., WELCH, R., POWELL, A., HOWE, H. & HALLAGER, B. 2018. Unravelling the Triassic Mungaroo Formation within North Carnarvon Basin using regional stratal slice volumes. *The APPEA Journal*, **58**, 833, https://doi.org/10.1071/AJ17232

MARSHALL, N.G. & LANG, S.C. 2013. A new sequence stratigraphic framework for the North West Shelf, Australia. *In*: KEEP, M. & MOSS, S. (eds) *The Sedimentary Basins of Western Australia IV: Proceedings of the Petroleum Exploration Society of Australia Symposium, Perth, Australia*. Petroleum Society of Australia, Perth, Australia, 1–32.

MCCLAY, K.R. & SCOTT, A.D. 1991. Experimental models of hangingwall deformation in ramp–flat listric extensional fault systems. *Tectonophysics*, **188**, 85–96, https://doi.org/10.1016/0040-1951(91)90316-K

MCCLAY, K.R. & WHITE, M.J. 1995. Analogue modelling of orthogonal and oblique rifting. *Marine and Petroleum Geology*, **12**, 137–151, https://doi.org/10.1016/0264-8172(95)92835-K

MCGEE, R., GOODALL, J. & MOLYNEUX, S. 2017. A re-evaluation of the Lower to Middle Triassic on the

Candace Terrace, Northern Carnarvon Basin. *The APPEA Journal*, **57**, 263–276, https://doi.org/10.1071/AJ16003

McHarg, S., I'Anson, A. & Elders, C. 2018. The Permian and Carboniferous extensional history of the Northern Carnarvon Basin and its influence on Mesozoic extension. *ASEG Extended Abstracts*, **2018**, 1, https://doi.org/10.1071/ASEG2018abM3_1B

McHarg, S., Elders, C. & Cunneen, J. 2019. Origin of basin-scale syn-extensional synclines on the southern margin of the Northern Carnarvon Basin, Western Australia. *Journal of the Geological Society*, **176**, 115–128, https://doi.org/10.1144/jgs2018-043

Mihut, D. & Müller, R.D. 1998. Volcanic margin formation and Mesozoic rift propagators in the Cuvier Abyssal Plain off Western Australia. *Journal of Geophysical Research: Solid Earth*, **103**, 27135–27149, https://doi.org/10.1029/97JB02672

Molyneux, S., Goodall, J., McGee, R., Mills, G. & Hartung-Kagi, B. 2016. Observations on the Lower Triassic petroleum prospectivity of the offshore Carnarvon and Roebuck basins. *The APPEA Journal*, **56**, 173, https://doi.org/10.1071/AJ15014

Newman, S. 1994. Clues to the structural history of the Rankin Trend, from 3D seismic data. *In*: Purcell, P.G. & Purcell, R.R. (eds) *The Sedimentary Basins of Western Australia: Proceedings of the Petroleum Exploration Society of Australia Symposium, Perth, Australia*. Petroleum Exploration Society of Australia, Perth, WA, Perth, Australia, 497–507.

Paumard, V., Bourget, J. et al. 2018. Controls on shelf-margin architecture and sediment partitioning during a syn-rift to post-rift transition: Insights from the Barrow Group (Northern Carnarvon Basin, North West Shelf, Australia). *Earth-Science Reviews*, **177**, 643–677, https://doi.org/10.1016/j.earscirev.2017.11.026

Péron-Pinvidic, G., Manatschal, G. & Osmundsen, P.T. 2013. Structural comparison of archetypal Atlantic rifted margins: a review of observations and concepts. *Marine and Petroleum Geology*, **43**, 21–47, https://doi.org/10.1016/j.marpetgeo.2013.02.002

Phillips Australian Oil Company 1992. Forrest 1 Well Completion Report, Unpublished. Perth, WA.

Pryer, L.L., Romine, K.K., Loutit, T.S. & Barnes, R.G. 2002. Carnarvon basin architecture and structure defined by the integration of mineral and petroleum exploration tools and techniques. *The APPEA Journal*, **42**, 287–309, https://doi.org/10.1071/AJ01016

Pryer, L., Blevin, J. et al. 2014. Structural architecture and basin evaluation of the North West Shelf. *The APPEA Journal*, **54**, 474–474, https://doi.org/10.1071/AJ13047

Reeckmann, S.A. & Mebberson, A.J. 1984. Igneous intrusions in the northwest Canning Basin and their impact on oil exploration. *In*: Purcell, P.G. (ed.) *The Canning Basin WA, Proceedings: GSA/PESA Canning Basin Symposium*. Petroleum Exploration Society of Australia, Perth, WA, 389–400.

Reeve, M.T., Jackson, C.A.-L., Bell, R.E., Magee, C. & Bastow, I.D. 2016. The stratigraphic record of pre-breakup geodynamics: evidence from the Barrow Delta, offshore Northwest Australia. *Tectonics*, **35**, 1935–1968, https://doi.org/10.1002/2016TC004172

Rollet, N., Pfahl, M. et al. 2013. Northern extension of active petroleum systems in the Offshore Perth Basin – an integrated stratigraphic, geochemical, geomechanical and seepage study. *In*: Keep, M. & Moss, S.J. (eds) *The Sedimentary Basins of Western Australia IV: Proceedings of the Petroleum Exploration Society of Australia Symposium, Perth, Australia*. Petroleum Society of Australia, Perth, Australia, 1–39.

Roma, M., Vidal-Royo, O., McClay, K., Ferrer, O. & Muñoz, J.A. 2018. Tectonic inversion of salt-detached ramp–syncline basins as illustrated by analog modeling and kinematic restoration. *Interpretation*, **6**, T127–T144, https://doi.org/10.1190/INT-2017-0073.1

Romine, K.K., Durrant, J.M., Cathro, D.L. & Bernardel, G. 1997. Petroleum play element prediction for the Cretaceous-Tertiary basin phase, Northern Carnarvon Basin. *The APPEA Journal*, **37**, 315–339, https://doi.org/10.1071/AJ96020

Saitta, A.J., Singleton, V.C. & Jablonski, D. 2003. WA-312-P Technical Review Report for Victoria Petroleum N.L. Perth, WA.

Sanchez, C.M., Fulthorpe, C.S. & Steel, R.J. 2012. Middle Miocene–Pliocene siliciclastic influx across a carbonate shelf and influence of deltaic sedimentation on shelf construction, Northern Carnarvon Basin, Northwest Shelf of Australia. *Basin Research*, **24**, 664–682, https://doi.org/10.1111/j.1365-2117.2012.00546.x

Saqab, M.M., Bourget, J., Trotter, J. & Keep, M. 2016. New constraints on the timing of flexural deformation along the northern Australian margin: implications for arc-continent collision and the development of the Timor Trough. *Tectonophysics*, **696–697**, 14–36, https://doi.org/10.1016/j.tecto.2016.12.020

Smith, S.A. 1999. *The Phanerozoic Basin-fill History of the Roebuck Basin*. University of Adelaide.

Stagg, H.M.J. & Colwell, J.B. 1994. The structural foundations of the Northern Carnarvon Basin. *In*: Purcell, P.G. & Purcell, R.R. (eds) *The Sedimentary Basins of Western Australia: Proceedings of the Petroleum Exploration Society of Australia Symposium, Perth, Australia*. Petroleum Exploration Society of Australia, Perth, Australia, 349–364.

Stagg, H.M.J., Alcock, M.B., Bernardel, G., Moore, A.M.G., Symonds, P.A. & Exon, N.F. 2004. *Geological Framework of the Outer Exmouth Plateau and Adjacent Ocean Basins, Geoscience Australia Record 2004/13*. Petroleum & Marine Division, Geoscience Australia, Canberra.

Thomas, G.P., Lennane, M.R., Glass, F., Walker, T., Partington, M., Leischner, K.R. & Davis, R.C. 2004. Breathing new life into the eastern Dampier Sub-basin: an integrated review based on geophysical, stratigraphic and basin modelling evaluation. *The APPEA Journal*, **44**, 123–150, https://doi.org/10.1071/AJ03004

Tindale, K., Newell, N., Keall, J. & Smith, N. 1998. Structural evolution and charge history of the Exmouth Sub-basin, northern Carnarvon Basin, Western Australia. *In*: Purcell, P.G. & Purcell, R.R. (eds) *The Sedimentary Basins of Western Australia 2: Proceedings of the Petroleum Exploration Society of Australia Symposium, Perth, Australia*. Petroleum Exploration Society of Australia, Perth, Australia, 447–472.

Vasquez, L., Nalpas, T., Ballard, J.-F., Le Carlier De Veslud, C., Simon, B., Dauteuil, O. & Bernard,

X.D.U. 2018. 3D geometries of normal faults in a brittle–ductile sedimentary cover: analogue modelling. *Journal of Structural Geology*, **112**, 29–38, https://doi.org/10.1016/j.jsg.2018.04.009

VEENSTRA, E. 1985. Rift and drift in the Dampier Sub-basin, a seismic and structural interpretation. *The APPEA Journal*, **25**, 177–189, https://doi.org/10.1071/AJ84016

WILLIAMSON, P.E., SWIFT, M.G., KRAVIS, S.P., FALVEY, D.A. & BRASSIL, F. 1990. Permo-Carboniferous rifting of the Exmouth Plateau region (Australia): an intermediate plate model. *In*: PINET, B. & BOIS, C. (eds) *The Potential of Deep Seismic Profiling for Hydrocarbon Exploration*. Technip Editions, Paris, 237–248.

WOODWARD, J., MINKEN, J., THOMPSON, M., KONGAWOIN, M., HANSEN, L. & FABRICI, R. 2018. The Lower Triassic Caley Member: depositional facies, reservoir quality and seismic expression. *The APPEA Journal*, **58**, 878, https://doi.org/10.1071/AJ17172

YANG, X.-M. & ELDERS, C. 2016. The Mesozoic structural evolution of the Gorgon Platform, North Carnarvon Basin, Australia. *Australian Journal of Earth Sciences*, **63**, 755–770, https://doi.org/10.1080/08120099.2016.1243579

YEATES, A.N., BRADSHAW, M.T. *ET AL*. 1987. The Westralian superbasin: an Australian link with Tethys. *In*: MCKENZIE, K.G. (ed.) *International Symposium on Shallow Tethys 2*. A.A. Balkema, Rotterdam, 199–213.

ZIMMERMANN, S. & HALL, R. 2016. Provenance of Triassic and Jurassic sandstones in the Banda Arc: Petrography, heavy minerals and zircon geochronology. *Gondwana Research*, **37**, 1–19, https://doi.org/10.1016/j.gr.2016.06.001

The impact of base-salt relief on salt flow and suprasalt deformation patterns at the autochthonous, paraautochthonous and allochthonous level: insights from physical models

TIM P. DOOLEY[1]*, MICHAEL R. HUDEC[1], LEONARDO M. PICHEL[2] & MARTIN P. A. JACKSON[1][†]

[1]Applied Geodynamics Laboratory, Bureau of Economic Geology,
The University of Texas at Austin, Austin, TX, USA

[2]School of Earth and Environmental Sciences, University of Manchester,
Oxford Road, Manchester M13 9PL, UK

*Correspondence: tim.dooley@beg.utexas.edu

Abstract: Salt is sensitive to the geometry of the substrate it flows across. We use physical models to investigate the impact of base-salt relief on deformation patterns. First, we investigate early-stage gravity gliding across base-salt relief. Salt flowing onto structural high blocks forms a zone of thickened salt and associated shortening owing to a flux mismatch. On the downdip edge of the basement high another flux mismatch generates a topographic monocline (ramp-syncline basin) with associated extensional and contractional hinges. With multiple base-salt high blocks, this structural pattern was repeated down the entire slope. Laterally discontinuous base-salt relief generated additional complexities such as major rotations of raft blocks and intervening diapirs as salt is channelled around and between base-salt relief. At the allochthonous level, regional dip, salt budget and base-salt relief influence flow patterns. Individual salt sheets spread sub-radially with streamlines skewed down the regional slope. As the canopy coalesced along allosutures, inward flow from the canopy peripheries dominated, driven by more vigorous flow from outer feeders owing to less competition for source-layer salt. Subsequent shortening returned flow patterns to grossly dip-parallel. However, salt flows fastest where it is thickest and thus chains of feeders channel rapid intracanopy flow.

Broadly speaking, passive margins developed above a salt detachment are typically large (100–500 km wide) gravity-driven systems comprising kinematically linked domains of proximal extension and distal shortening connected by a zone of passive translation (Fig. 1a; e.g. Fort et al. 2004; Hudec & Jackson 2004; Brun & Fort 2011; Quirk et al. 2012; Peel 2014). A simple, gravity-driven, physical model with a smoothly dipping base of salt graphically illustrates this three-domain structural division (Fig. 1b, c). Previously published physical-modelling studies questioned this simple and rather static three-domain structural zonation. For example, Brun & Fort (2011) illustrate the migration and overprinting of extensional and contractional provinces during the evolution of margin-scale physical models. McClay et al. (1998) also documented the overprinting of contractional provinces by a seaward-migrating upper-slope and shelf-extensional zone. Jackson et al. (2011) and Dooley et al. (2013) showed how coastal uplift in the Gulf of Mexico led to the development of a shortening zone, a pillow fold belt, far inboard from the toe-thrust province. All of these processes may complicate the structural zonation in these systems.

A unique characteristic of rock salt is the ability to flow under its own weight over geological time scales (Jackson & Hudec 2017), and thus it is sensitive to the geometry of the surface it flows across. Early-stage gravity gliding induced by basin tilt on salt-bearing passive margins may be complicated by base-salt topography owing to folding and faulting, or to erosional rugosity. For example, Hudec et al. (2013) proposed that, in the Gulf of Mexico, the Jurassic Louann Salt was deposited in depressions above hyperextended continental and transitional crust. After deposition of the salt, the rifting is interpreted to have continued for 7–12 myr before

---
[†]Deceased

From: McClay, K. R. & Hammerstein, J. A. (eds) 2020. *Passive Margins: Tectonics, Sedimentation and Magmatism.*
Geological Society, London, Special Publications, **476**, 287–315.
First published online January 1, 2018, https://doi.org/10.1144/SP476.13
© 2018 The Author(s). Published by The Geological Society of London. All rights reserved.
For permissions: http://www.geolsoc.org.uk/permissions. Publishing disclaimer: www.geolsoc.org.uk/pub_ethics

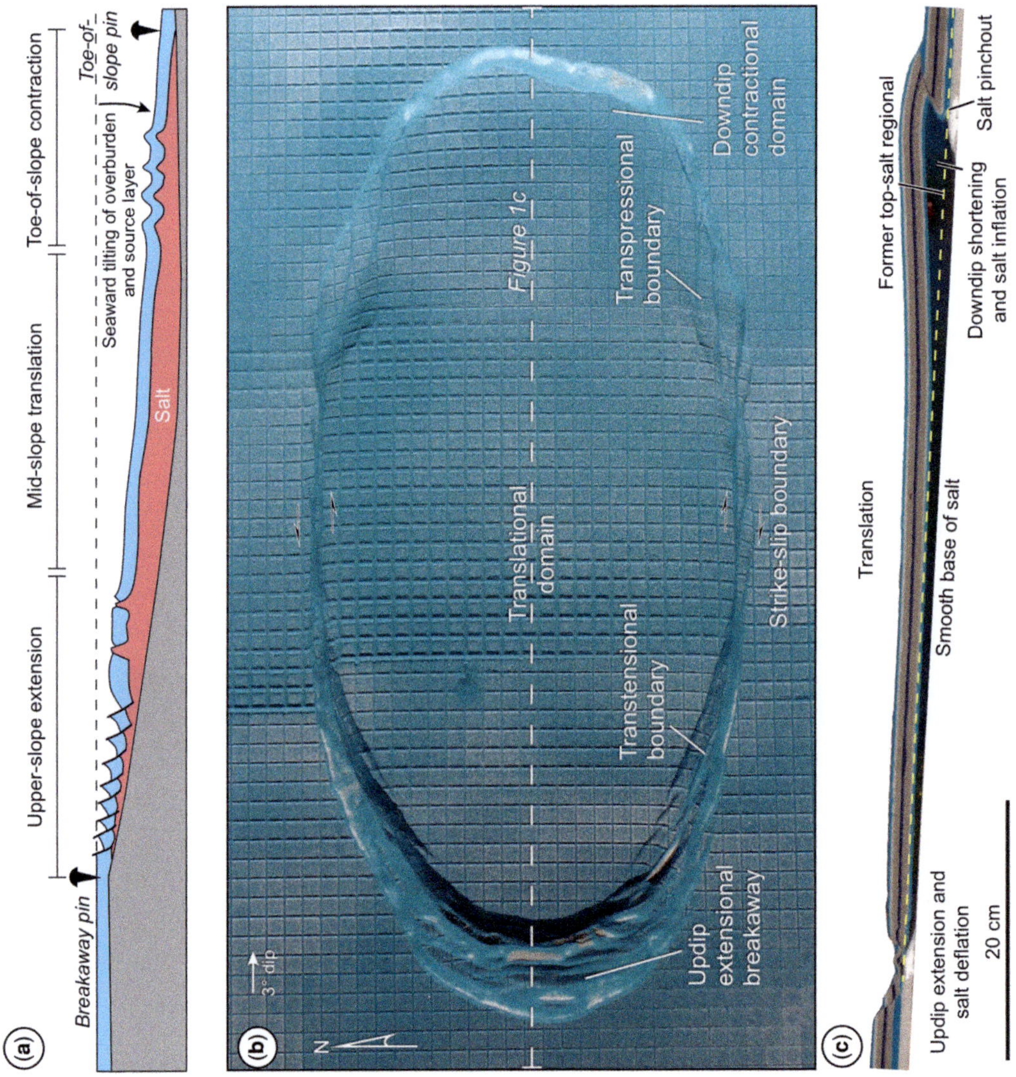
(a)
Breakaway pin
Upper-slope extension
Mid-slope translation
Toe-of-slope contraction
Seaward tilting of overburden and source layer
Toe-of-slope pin
Salt
(b)
N
3° dip
Updip extensional breakaway
Transtensional boundary
Translational domain
Strike-slip boundary
Transpressional boundary
Figure 1c
Downdip contractional domain
(c)
Updip extension and salt deflation
20 cm
Translation
Smooth base of salt
Former top-salt regional
Downdip shortening and salt inflation
Salt pinchout

initiation of seafloor spreading and basinward flow of salt across a rugose substrate. Even greater base-salt relief is possible in basins where salt was deposited during crustal extension, syn-rift salt, where salt may thin dramatically or be absent across intrarift highs such as horsts or transfer zones (e.g. the Scotian Margin (Deptuck & Kendell 2017) and the Atlantic margin of Morocco (Tari & Jabour 2013)). In these cases, deformation observed in sediments overlying salt is directly related to salt flow over base-salt relief, rather than being the result of an outside force (e.g. the interior uplift of Dooley et al. 2013) or a gradual migration of structural domains during margin evolution (e.g. Fort et al. 2004). Local perturbations are also likely to be transitory as the salt and its thin overburden move downdip across, and beyond, base-salt rugosity.

Examples of previous modelling studies incorporating base-salt relief include the convergent and divergent radial-gliding models of Cobbold & Szatmari (1991), the studies of Gaullier et al. (1993) and Maillard et al. (2003) on the effects of residual topography below a salt décollement on fault orientation in the overburden, and the basin-scale models of Adam & Krézsek (2012). More recently, Dooley et al. (2017) and Dooley & Hudec (2017) used a series of physical models to examine the role of base-salt relief at low and high angles to the margin tilt direction, as well as examining the roles of more complex salt isopachs in the formation, translation and rotation of structures formed during early-stage gravity gliding. Ferrer et al. (2017) also documented disruption of the extension–translation–contraction pattern by base-salt relief in their study of the complexities associated with volcanic seamounts that partition a salt basin. All of these studies illustrated surprising complexities in salt flow and suprasalt deformation patterns that segmented the translation zones, despite the relatively simple model configurations.

In this paper we first revisit the effects of a single sub-salt high block at a high angle to the mean salt-flow direction at the autochthonous or paraautochthonous level from Dooley et al. (2017), before ramping up the complexity and moving on to describe physical models with multiple high blocks at a high angle to the mean salt-flow direction, and the impact of spatially limited base-salt high blocks. These models were run with and without syn-kinematic sedimentation. Our goal is to primarily document how flow over base-salt relief can pre-structure the suprasalt sequence owing to multiple instances of localized extension and contraction, eventually translating these weak zones outboard into deep water where they may be later exploited as the margin evolves. The effects of base-salt relief are not confined to flow at the deep autochthonous or paraautochthonous level, but occur wherever salt flows or spreads out over an uneven surface. To investigate this we also present some new results on the impact of regional dip, salt budget and base-salt rugosity on salt flow at the allochthonous level during the coalescence and subsequent squeezing of a salt canopy.

Model methodology

Two sets of models are presented in this study: Set 1 (Models 1–5), comprising experiments investigating flow and roof deformation at the early stage of margin evolution where salt and its thin roof flow downslope across a variety of irregularities at the base of salt (Fig. 2; Table 1); and Set 2 (Model 6), comprising experiments investigating the impact of regional dip, salt budget and sub-salt relief during a more mature stage of the passive margin where vertical salt rise and subsequent spreading of allochthonous sheets form a canopy (Fig. 3). In Set 1 experiments, sub-salt topography was built using silica sands or wooden blocks to create the basement steps, across which the salt thickness changed. Our model salt was deposited across basement topography and allowed to settle for several days. In all Set 1 experiments, deformation was initiated by simply tilting the rig to 3° and removing the confining downdip endwall to create an open-toe system, similar to the models of Gaullier et al. (1993). Models 1–5 in Set 1 were run with and without syn-kinematic sedimentation. Where present, syn-kinematic sediments were added as a series of prograding wedges, inducing a degree of spreading within the system (Fig. 2).

Set 2 experiments were far more complex in their construction and evolution, and only portions of these models are shown in this study. Model salt

Fig. 1. (a) Simplified schematic cross-section of a linked kinematic system above a gravitationally unstable, seaward-dipping salt detachment (modified after Brun & Fort 2011). (b) Overhead view, and (c) dip line of a simple model illustrating gravity-driven deformation above an elliptical salt sheet (from Dooley et al. 2017). Model was tilted to 3° to initiate deformation, and the base of salt was smooth. Extension dominates at the updip breakaway, whereas the downdip domain consists of a thrust fault that formed at the salt pinchout. These domains are separated by a translational domain in the model centre. Lateral margins of the salt sheet consist of strike-slip, transtensional and transpressional boundaries. Note the salt inflation at the downdip edge of the salt basin owing to downdip flow within the salt layer.

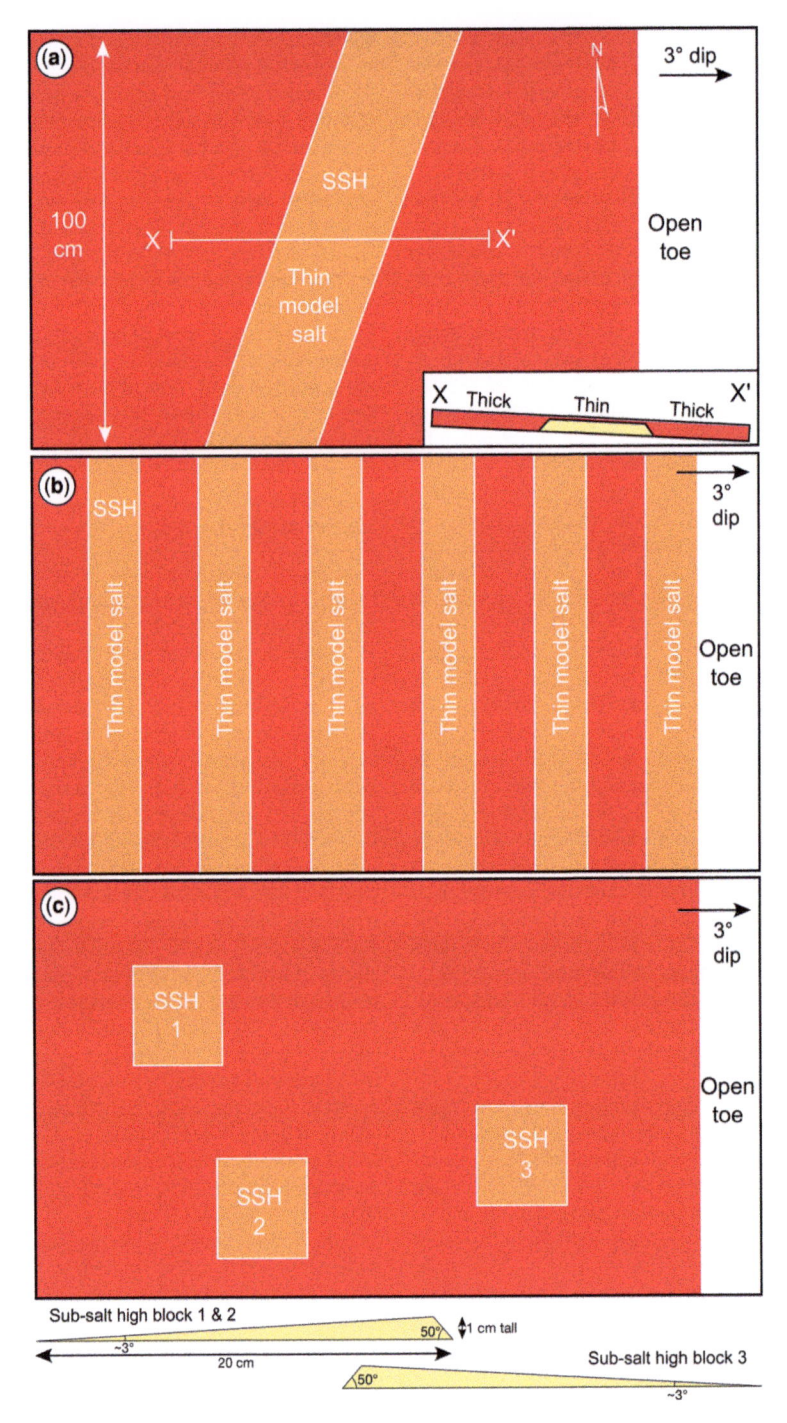

Fig. 2. Model configurations for Set 1 models. (**a**) Setup for Model 1, containing a sub-salt high block oriented at a high angle to regional dip. Our salt analogue thins across the high block. (**b**) Setup for Models 2–4 containing a series of six sub-salt high blocks orthogonal to regional dip direction. See Table 1 for details of stratigraphy of the models. (**c**) Setup for Model 5. Sub-salt topography in this model consisted of a series of three isolated high blocks with triangular profiles in the dip direction. High blocks 1 and 2 were oriented with the steep step at their downdip edges, whereas the steep step of high block 3 was on its updip edge. SSH, Sub-salt high block.

Table 1. *Experimental parameters for Set 1 experiments*

Model no.	Model summary	Angle of sub-salt to regional dip	Number of sub-salt steps	Model salt thickness Minimum (mm)	Model salt thickness Maximum (mm)	Prekinematic overburden thickness (mm)	Syn-kinematic sedimentation	Model dip
Model 1	High-angle sub-salt step – model salt thins across step	70°	1	5	15	10	No	3°
Model 2	High-angle sub-salt step – model salt thins across step	90°	6	3	8	6	No	3°
Model 3	High-angle sub-salt step – model salt thins across step	90°	6	3	8	24	Yes – 1° wedges	3°
Model 4	High-angle sub-salt step – model salt thickens thins step	90°	6	3	8	6	Yes – 1° wedges	3°
Model 5	Isolated sub-salt 'allochthons' with triangular profiles	90°	3	3	8	6	No	3°

was deposited as a uniform-thickness layer that pinched out seawards. A thin layer of sand covered the model salt and in predetermined locations this sand layer was vacuumed off to initiate diapiric rise owing to differential loading (Fig. 3a). The diapirs gradually grew upwards with continued sedimentation – downbuilding – until salt rise rate exceeded the sedimentation rate and sheets began to spread radially (Fig. 3b–d). At this stage the rig was also tilted basinwards by 1° in order to produce asymmetric spreading of the surface salt sheets (Fig. 3b–d). Individual salt sheets spread laterally along base-salt flats and climbed along base-salt ramps (Fig. 3d) until they eventually sutured to form a canopy in Model 6. Once the canopy had formed it was subjected to a phase of shortening. In natural systems such as the Gulf of Mexico this shortening was gravity driven at the toe of slope, or related to interior uplift that increased the elevation-head gradient and destabilized the slope (e.g. Dooley *et al.* 2013). To simplify matters in our experiments, shortening was driven by a moving endwall. Total model shortening was 12 cm.

Models were recorded by computer-controlled time-lapse digital cameras. A stereo CCD system and associated DIC (digital image correlation) software also captured the upper surface of the models, allowing us to monitor changes in height, displacements and surface strains as the model evolved (see Adam *et al.* 2005, for further details of DIC analysis). Archival imagery from experiments that were not actively monitored by this stereo DIC system was processed in 2D by software (DaVis by LA Vision) in order to produce maps of surface displacements (*x* and *y*, but not *z*), and surface strains.

As is common in physical-model studies of salt tectonics, our salt analogue was a ductile silicone and its brittle overburden was modelled by dry, granular materials. The silicone used in these experiments was a near-Newtonian viscous polydimethylsiloxane (PDMS, equivalent to DowCorning SGM-36). This long-chain polymer has a density of 950–980 kg m^{-3} and a dynamic shear viscosity of 2.5×10^4 Pa s at a strain rate of 3×10^{-1} s^{-1} (Weijermars 1986; Weijermars *et al.* 1993). The thin overburdens were composed of a mixture of silica sand (bulk density of 1700 kg m^{-3}; grain size of 300 μm; internal coefficient of friction, μ = 0.55–0.65; see material properties in, for example, McClay (1990), Krantz (1991) and Schellart (2000)) and hollow ceramic microspheres (bulk density of 600 kg m^{-3}, average grain size of 90 μm, and typical μ = 0.45; see Rossi & Storti 2003, for details). Silica sands and ceramic beads are mixed together to produce overburdens of varying densities, with an additional advantage of reducing the bulk grainsize of the granular overburden. The goal of Set 1 models is to document the structural styles of overburden

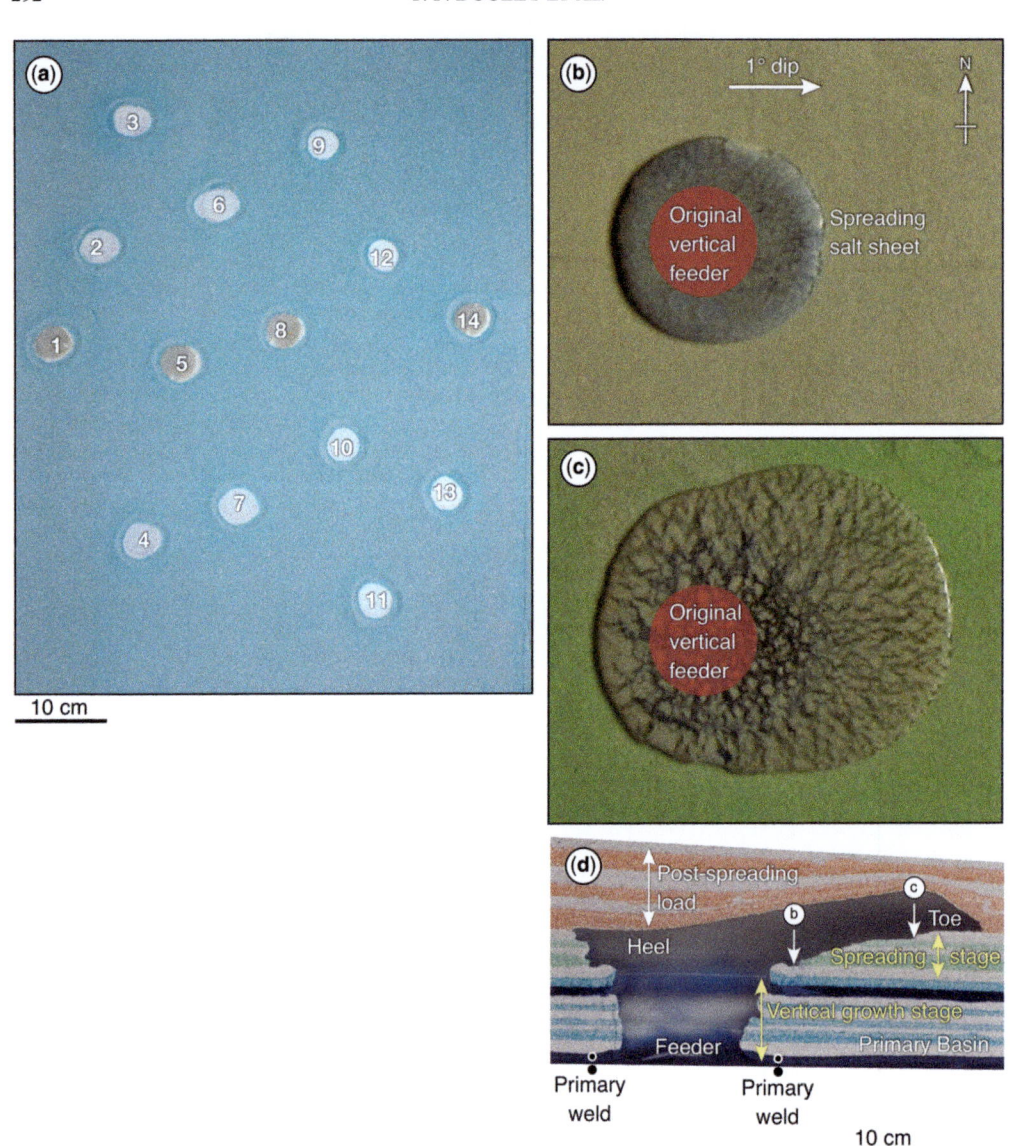

Fig. 3. Configuration for Set 2 models showing initial diapir array and growth stages prior to salt-sheet suturing and canopy formation. (**a**) Our salt analogue was covered by a uniform-thickness granular layer with a density 1.4 times that of the model salt. Vacuuming off this granular layer at predetermined locations seeded a series of 14 diapirs. Diapirs are numbered 1–14, identical to the numbering system in Figure 19. (**b, c**) Initially diapirs grew vertically upwards by down-building before being allowed to flare and spread laterally as diapiric growth outpaced model sedimentation rate. Asymmetric spreading was further encouraged by imposing a 1° regional dip. (**d**) Cross-section through an isolated salt sheet. The initial vertical diapir is seen in the lower section before flaring out to form an asymmetric salt sheet that climbed seaward along a series of ramps and flats. Layers seen in (b) and (c) are indicated by arrows. In this model subsequent loading by a prograding sedimentary wedge expelled model salt toward the toe of the sheet. Primary welds ring the diapir, inhibiting further growth.

deformation as it flows across the rugose basement, and not to ascertain the specific components that gliding and spreading impart to the system (e.g. Peel 2014) owing to the variable densities. In Set 2 experiments only the lowermost layer was denser than our model salt. This was done to encourage initial diapiric rise, and continued rise during subsequent downbuilding.

Set 1 models: base-salt relief at a high angle to salt-flow direction – complex salt-flow dynamics and superposed deformations

Effect of a single base-salt high block oriented at a high angle to mean salt-flow direction

Overhead views and selected DIC strain analyses illustrate the evolution of Model 1 (Model 6 of Dooley *et al.* 2017; Figs 4 & 5). Model design is shown in Figure 2a. After 10 h of model runtime, five distinct deformation zones are seen from the updip breakaway to the downdip open toe (Figs 4a & 5a). These are: (i) an updip breakaway zone with extensional graben orthogonal to model dip; (ii) a zone of shortening and uplift, with associated outer-arc extension on the crest of this uplift that paralleled the trend of the sub-salt high block; (iii) a distinctive topographic low or monocline above the downdip edge of the sub-salt high block; (iv) a zone of extension characterized by graben that were parallel to the strike of the sub-salt high block; and (v) a zone of open-toe extension with well-developed graben orthogonal to the regional slope.

With continued evolution, a reversal in strain patterns was observed above the updip margin of the sub-salt high block (zone ii in Figs 4a, b & 5a, b). The area of initial uplift and associated shortening

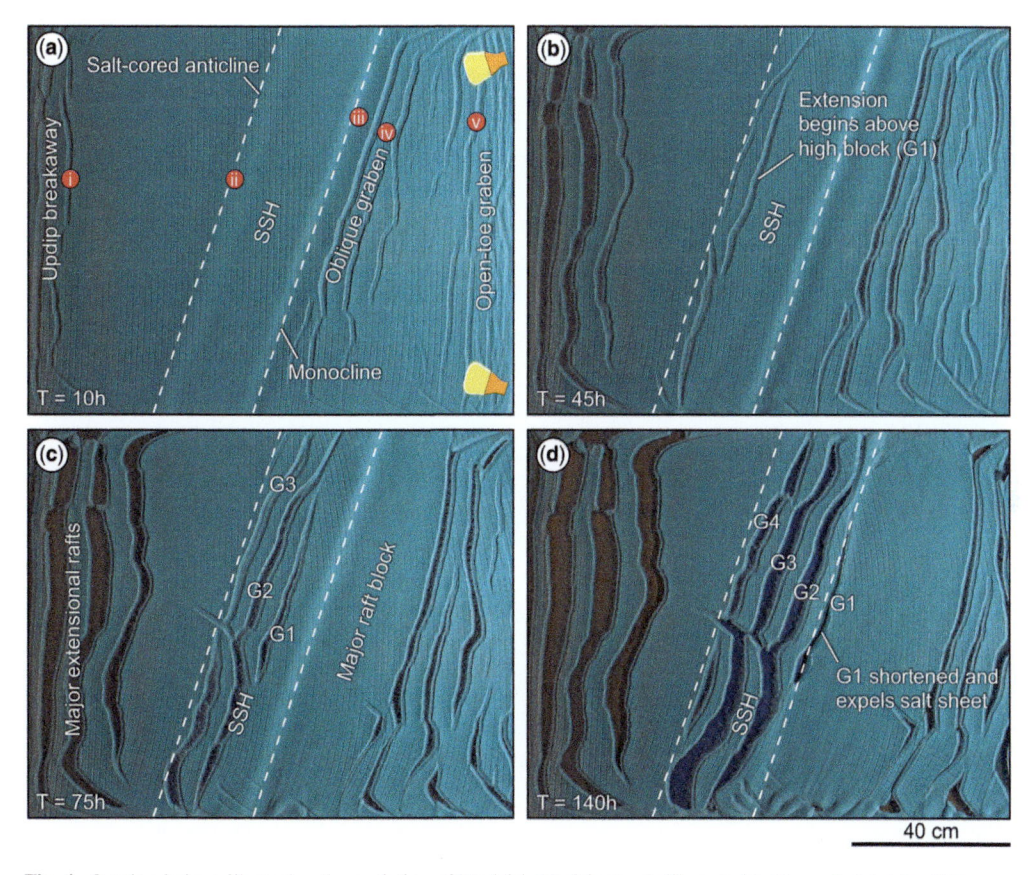

Fig. 4. Overhead views illustrating the evolution of Model 1. Model setup is illustrated in Figure 2. (**a**) After 10 h, five distinct structural zones are seen in the overhead views and strain maps: (i) an updip breakaway zone with extensional structures almost orthogonal to model dip; (ii) a poorly defined salt-cored anticline parallel to the updip edge of the basement high (see Fig. 16 for strain map); (iii) a topographic hinge or monocline that marks the downdip edge of the base-salt high block; (iv) an oblique extensional system that parallels the downdip edge of the base-salt high block; and (v) open-toe extension structures orthogonal to model dip. (**b–d**) Overhead views at 45, 75 and 140 h into the model runtime. The updip extension within the breakaway zone broadens with time, forming major raft blocks. The site of early contraction above the base-salt high block was superseded by extension; these oblique graben were gradually translated downdip, were shortened and expelled salt sheets as they moved into thicker salt at the downdip edge of the base-salt high block.

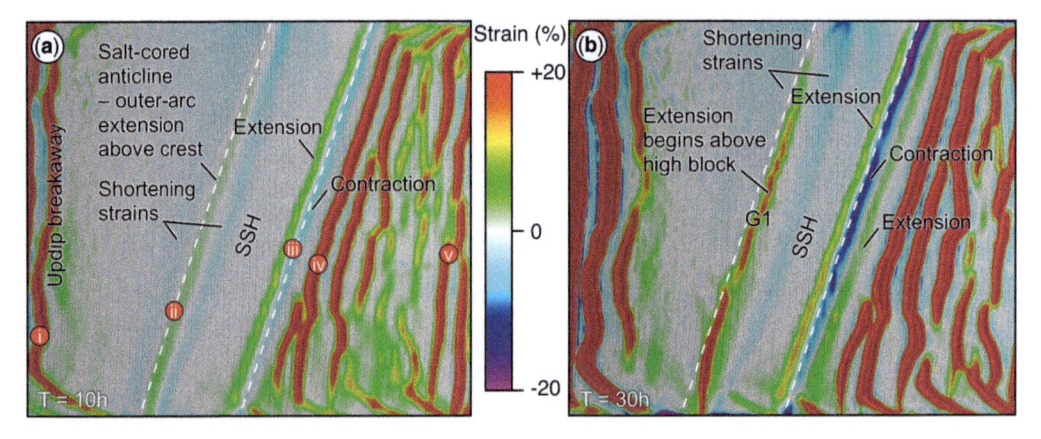

Fig. 5. 2D DIC analysis of archival imagery from Model 1 after (**a**) 10 and (**b**) 30 h model runtime. The early-stage shortening zone – consisting of outer-arc extension above the uplift (green) flanked by zones of contraction (blue) – is clearly seen in (a). This zone is superseded by a zone of extension after 30 h of model runtime as in (b). Extensional and contractional hinges characterize the updip and downdip edges of the topographic hinge that marks the transition from thin to thick salt. Note that extensional strains have been clipped in order to highlight contractional strains and hinge zones.

strains was subsequently overprinted by extension (Figs 4b & 5b). Over time, a series of extensional graben, cored by reactive diapirs, with strikes that paralleled the trend of the sub-salt high block formed and were translated downslope (Fig. 4b, c). The first formed of these graben (G1 in Fig. 4b–d) was subsequently shortened and ejected a salt sheet onto the model surface as it moved downslope off the sub-salt high block. At the updip breakaway, major extension resulted in significant separation of rafts. The early-formed graben of zones iv and v (Fig. 4a) were translated downslope and separated from the mid-slope extensional zone by a major raft block (Fig. 4b–d).

Strain patterns highlight these intriguing processes at the updip and downdip edges of the sub-salt high block (Fig. 5). During the early stages, an anticline developed above the updip edge of the sub-salt high block, with surface strains indicating outer-arc extension on its crest and flanked by zones of shortening (Fig. 5a). With increasing runtime, this salt-cored anticline began to collapse under extension, forming a series of downdip-younging graben that were translated downslope (Figs 4b–d, 5b & 6). The zone of shortening strains on the downdip side of the original uplift was translated downslope ahead of the first graben (G1 in Fig. 5b). Above the downdip edge of the sub-salt high block, the strain pattern remained consistent throughout the model runtime (Fig. 5). This margin was defined by a topographic low that paralleled the sub-salt high block (Fig. 6). Linear zones of extension and contraction that flanked this structure defined the updip and downdip topographic hinges, respectively, of this monocline (green and blue zones in Fig. 5a). Graben that formed above the updip edge of the sub-salt high block were translated downslope, extended further across the extensional hinge, and then shortened and partially inverted as they passed through the contractional hinge of this system (Figs 4d & 6).

The complex overburden strain history associated with this relatively simple model prompts the question: what factors contribute to the structural variation in time and space of overburden translated over a sub-salt high block oriented at a high angle to the flow direction? The following section attempts to answer this.

Streamlines, basal drag and salt-flux mismatches across a stepped base of salt

When dealing with a viscous fluid such as salt, it is useful to consider streamlines, flux and the effects of basal drag, itself dependent on the salt-thickness ratio T/t (Dooley et al. 2017; Fig. 7), on flow velocities within the system. First we consider the case where salt flowing downdip encounters a sub-salt high block (Fig. 7a, b). In general, a decrease in salt thickness means that streamlines converge above the updip edge of the base-salt high block and the salt flow accelerates (Bernoulli's principle). This is the case where basal drag is not a factor, i.e. where the ratio between the normal salt thickness and thickness above a base-salt high block (T/t) is small. Acceleration compensates for the lesser salt thickness above the base-salt high block, and there is no flux mismatch (Fig. 7a). However, if the thickness ratio (T/t) is large, then drag is important and

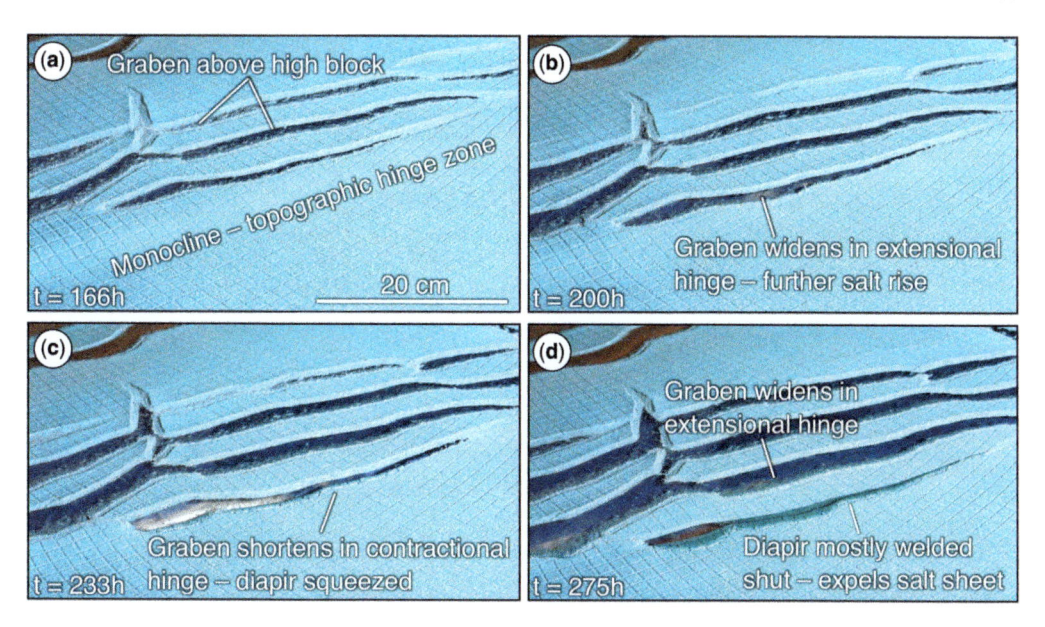

Fig. 6. (**a–d**) Oblique views of Model 1 focused on the sub-salt high block after 166, 200, 233 and 275 h runtime. Oblique views clearly illustrate the monocline above the seaward edge of the sub-salt high block. Scale is accurate for the midground in these images.

there is a resultant flux mismatch (Fig. 7b). In this scenario, more salt is being fed onto the base-salt high block than can be accommodated by flow within and through the thinner salt body, resulting in early-stage salt thickening and contraction above the high block, as we saw in Model 1 (Figs 4 & 5).

Now consider the case where the salt flows off this high block and into thicker salt across a basinward-dipping ramp (a base-salt low block; Fig. 7c, d). Once more, if the salt-thickness ratio (T/t) is low, then the effects of basal drag are minimal. In this scenario, streamlines diverge into the low block, and any flux mismatch is accommodated by deceleration and salt thickening (Fig. 7c). However, if the thickness ratio (T/t; Fig. 7d) is high, then basal drag comes into play, resulting in a major flux mismatch between the high and low block. The zone of thicker salt downdip of the base-salt high block initially pulls away from the high block under extension. The flux mismatch results in the creation of (1) an extensional monocline with an updip extensional hinge as material is pulled rapidly across the sub-salt step, and (2) a contractional hinge as material then encounters the thicker and slower-moving salt downdip of the sub-salt high block (Fig. 7d). This result explains the enhanced extension and subsequent inversion and salt expulsion seen in Model 1 (Figs 4 & 5).

The evolution of overburden deformation patterns in a step-up/step-down system is illustrated in Figure 8a. In the early stages, a thickened body of salt builds up above the updip margin of the high block over time because of the flux mismatch. At the downdip edge of the high block, the thicker salt body pulls away from the step to form a graben parallel to the high block (Figs 4a, b & 8a). Contractional thickening of the salt above the updip edge of the high block allowed its velocity to gradually increase over time as the effects of basal drag became proportionally less important. Eventually this velocity increase results in extension of this thickened body of salt and the formation of a new breakaway zone and associated graben in the brittle overburden (Figs 4b–d & 8a). As these graben are translated across and off the sub-salt high, they undergo rapid extension across the extensional hinge and shortening/inversion through the contractional hinge of the monocline (Figs 4c, d, 6 & 8a).

The opposite case is considered in Figure 8b, in which flowing salt encounters a base-salt low block (step down) and then a downdip high block (step up). At the updip edge of the low block, a monocline begins to develop as the greater flux from the thick-salt region pulls away from the thin-salt region (Fig. 8b). As the salt steps up further downdip, the greater flux from the thick-salt region results in contractionally thickened salt because of the aforementioned flux mismatch (Figs 7 & 8) and an associated fold belt. Over time, graben that formed

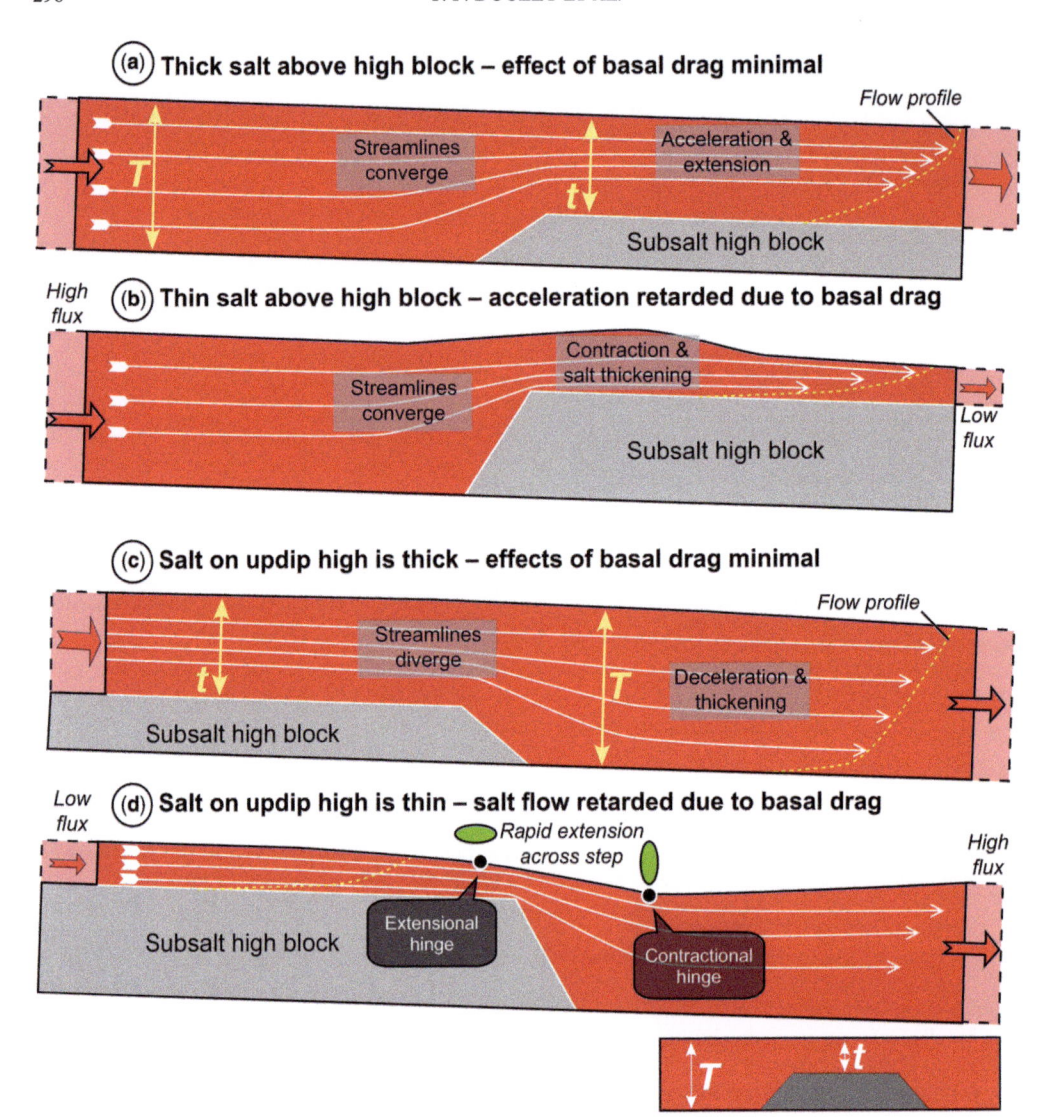

Fig. 7. Synoptic figure explaining the great structural variation in time and space seen in Model 1 (modified from Dooley *et al.* 2017). (**a**) If salt above a downdip high block is relatively thick, then the effects of basal drag are minimal. Streamlines converge, and flow accelerates and promotes extension. (**b**) However, if salt above the high block is thin, then this acceleration is retarded owing to basal drag, resulting in an initial stage of flux mismatch across the high. The streamlines still converge but now promote compression and salt thickening. (**c**) When flowing off a high block into a low block, the effects of basal drag are again minimal if salt on the updip high block is thick. In this case, the streamlines diverge, the salt flow decelerates and the salt mass thickens. (**d**) If the salt above the updip is thin, then, once more, salt flow is retarded owing to basal drag. The flux mismatch results in a monocline across the sub-salt step consisting of an updip extensional hinge and a downdip compressional hinge. Initial tilting produces extension downdip of this hinge owing to the flux mismatch. Later, materials crossing the high block experience rapid extension followed by compression.

updip of the low block were translated through the contractional hinge and inverted within the low block (Fig. 8b). As the salt velocity gradually increased above the downdip step because of thickening, the plateau of thickened salt collapsed under extension (Fig. 8b). A model with a single sub-salt low block was presented in Dooley *et al.* (2017; their Model 7). With these analyses in mind, we can set a framework for interpreting our other models with more complex sub-salt topographies.

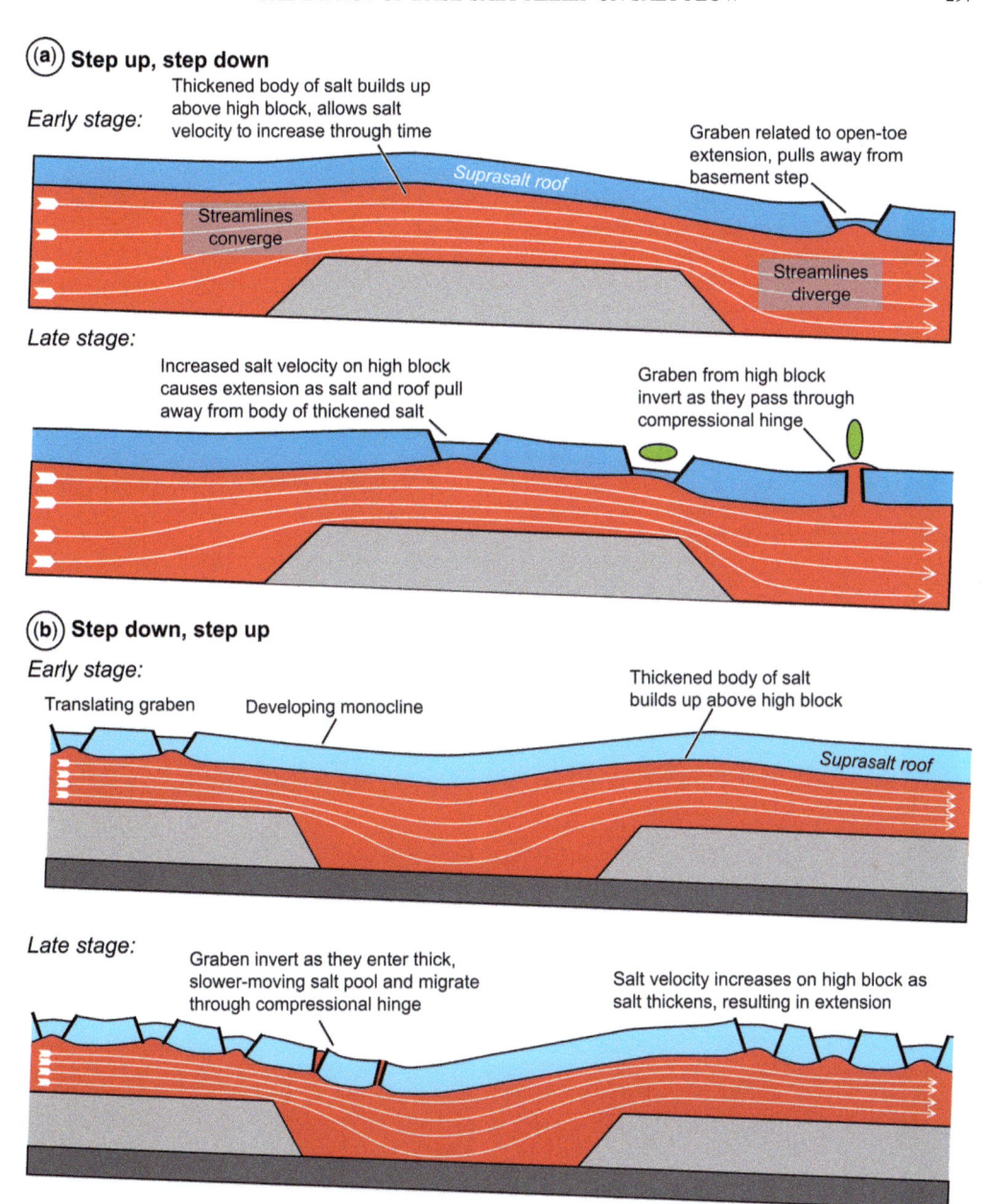

Fig. 8. Synoptic figure explaining the temporal evolution of salt flow on overburden deformation (modified from Dooley *et al.* 2017). (**a**) In a step-up/step-down scenario, initial compressional thickening of salt occurs on the updip side of the high block because of the flux mismatch. This salt thickening allows salt velocity to increase through time as the effects of basal drag are minimized, eventually resulting in extensional collapse. At the downdip edge of the high block, rapid extension initially occurs because of the flux mismatch, the proximity of the open toe and the lack of a buttress in this location. Graben formed during collapse of the thickened body of salt are extended further as they pass through the extensional hinge and then are inverted as they pass through the compressional hinge. (**b**) In a step-down/step-up scenario, a monocline develops as material translates into the thickened salt. Diapirs formed updip are gradually translated toward the low block. At a later stage, the translated grabens are inverted as they pass through the contractional hinge once they enter thick salt. On the downdip side of the low, a thickened body of salt builds up above the high block. This thickened body of salt at the downdip edge of the low eventually collapses under extension as the salt velocity increases enough to overcome the basal friction.

Effects of multiple, laterally continuous, base-salt high blocks orthogonal to mean salt-flow direction

Figures 9–11 illustrate overhead views, height change and strain data and detailed overhead views of Model 2 during the various stages of its evolution. Base-salt relief in this model consisted of six sub-salt high blocks orthogonal to regional dip (Fig. 2b). Our model salt thinned dramatically across these high blocks and a thin prekinematic roof was added before tilting the rig to 3° (Fig. 2b and Table 1). Overhead views show that during the early stages of the model runtime a series of graben formed above the updip edges of the high blocks that were slowly translated downslope (Fig. 9a–c). Height change after 35 h of runtime illustrates uplift above the edges of the sub-salt high blocks and the development of topographic hinges at the downdip edges of each high block (Fig. 10a, b). Strain data after 35 h of runtime show a zone of outer-arc extension and a zone of shortening at the updip edge of each high block as well as extensional and contractional hinges at the downdip edge of the high blocks (Fig. 10c). This pattern is repeated consistently across the model surface, and is identical to the patterns seen in Model 1, where contractional uplift occurs initially at the updip edges of the high block, and thicker salt pulls rapidly away from the downdip edges of the high block, creating a topographic hinge zone or monocline (Figs 5 & 10). Both of these features are a result of the flux mismatches and flow-velocity perturbations generated by the presence of a stepped base of salt (Figs 7 & 8).

As Model 2 evolved, graben formed above all of the sub-salt high blocks and the earliest formed of these extensional structures were translated through the monocline at the downdip edges of the sub-salt high blocks where they passed through the extensional and contractional hinges of each monocline (Figs 9c–h, 10b & 11a, b). New graben were generated above the high blocks as model salt

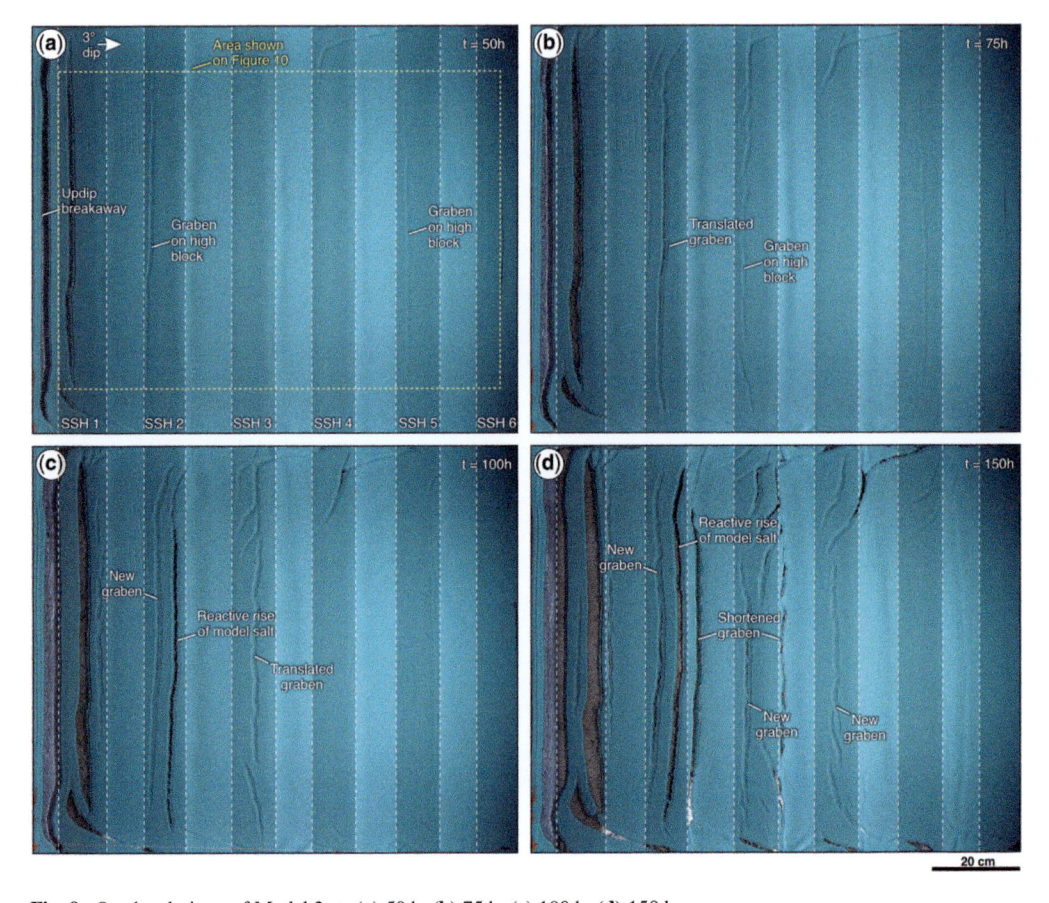

Fig. 9. Overhead views of Model 2 at: (**a**) 50 h; (**b**) 75 h; (**c**) 100 h; (**d**) 150 h.

continually flowed up onto the downdip high blocks, eventually dissecting the entire slope into a series of predominantly high-block-parallel graben and associated reactive diapirs, and shortened graben located within the intervening low blocks (Fig. 9h). The multi-stage deformation history of the suprasalt roof is illustrated in the detailed surface views of Figure 11. In the upper view, graben G1 and G5, originally formed near the updip edges of sub-salt high blocks 2 and 3 respectively, were shortened and welded shut by expelling small salt sheets as they passed through the contractional hinge at the downdip edges of these high blocks (Fig. 11a). At this time graben G2 was just entering this contractional hinge zone and was beginning to invert and expel a salt sheet (Fig. 11a). Graben G3, G4, G6 and G7 were all extending at this time (Fig. 11a). After a further 75 h, graben G1 and G5 were extended once more as they entered zones of faster-moving salt downdip (Fig. 11b). Graben G5 was mildly shortened once more as it moved up onto the downdip sub-salt high block, where it began to extend once more. Graben G2 and G6 were welded shut as they passed through the contractional hinge, and graben G3 was beginning to shorten as it crossed off the high block into this contractional hinge zone (Fig. 11b). Graben G4 and G7 widened considerably as they moved downslope and began to enter the extensional hinge close to the downdip edges of the sub-salt high blocks (Fig. 11b).

Model 3 was run with a pre-kinematic roof three times as thick as that in Model 2, but with the same array of sub-salt blocks and salt isopach (Figs 2b & 12 and Table 1). With a thicker and stronger roof, the most obvious features from an overhead view after 24 h model runtime are a series of graben above high blocks near the updip breakaway and downdip open toe of this system (Fig. 12a). Height change data shed more light on the deformation patterns, revealing a consistent pattern of uplift above the base-salt high blocks (contractional uplift) and topographic lows at the downdip edges of these high

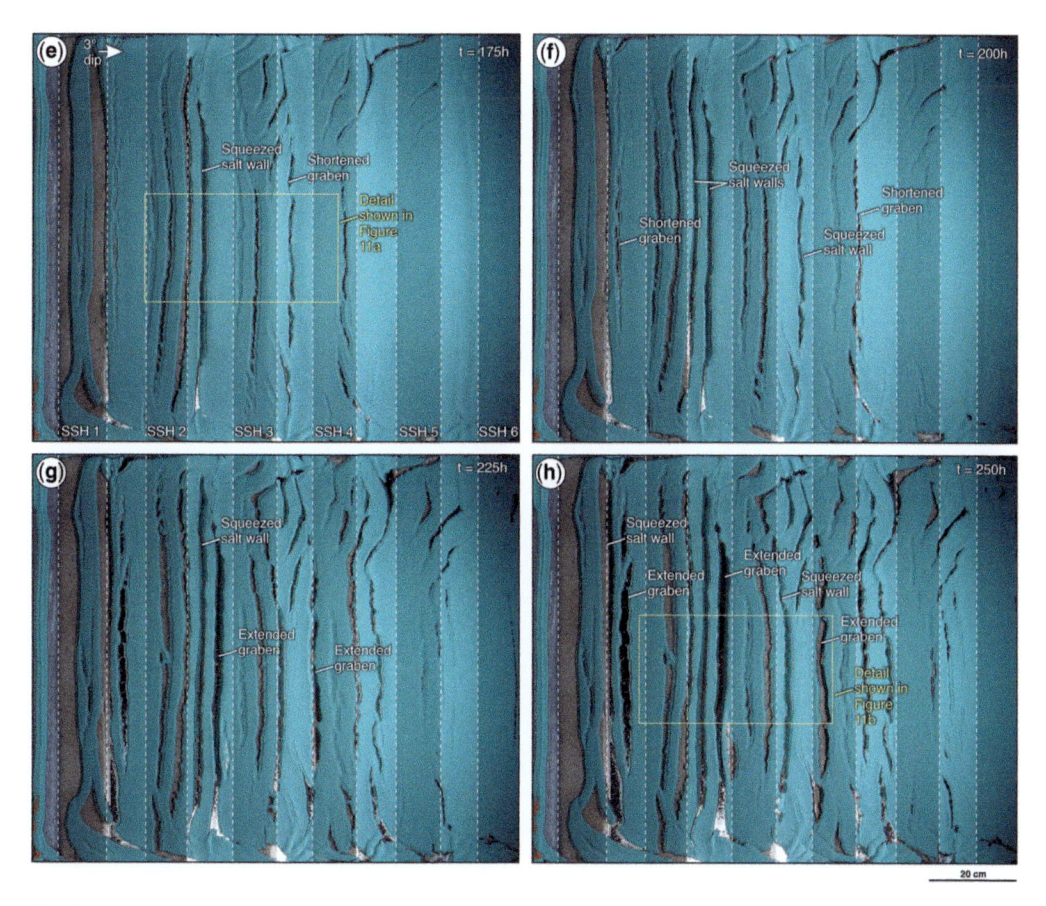

Fig. 9. *Continued.* (**e**) 175 h; (**f**) 200 h; (**g**) 225 h; and (**h**) 250 h. Note that after 250 h the entire slope of the model is dissected by numerous graben, squeezed graben and structures with multi-stage histories. See text for further details.

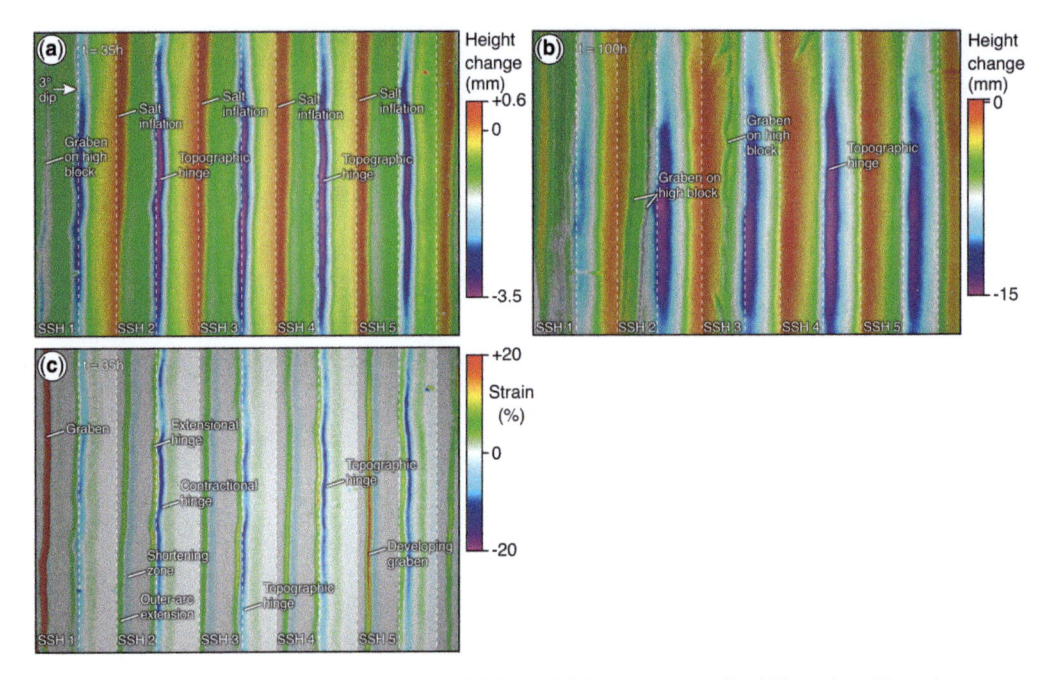

Fig. 10. 3D DIC data from selected runtimes of Model 2. (**a**) Height change map after 35 h runtime, illustrating repeated patterns of uplift and topographic lows at the updip and downdip edges of the sub-salt high blocks respectively. (**b**) Height change map after 100 h showing extensional collapse of the contractional salt-cored anticlines as a series of graben form above the high blocks. (**c**) Strain map of the model after 35 h. At the updip edges of the sub-salt high blocks there are linear zones of outer-arc extension above anticlines flanked by shortening strains. The topographic hinges (monocline) are picked out by paired zones of extension and contraction owing to the hinges developed in this location. Note that extensional values have been clipped to better illustrate shortening strains.

blocks that constitute the topographic monoclines (Fig. 12b). The patterns are identical to those seen in Models 1 and 2, although the number of structures are fewer owing to the thicker prekinematic section. Detailed views of the central portion of the model after the addition of syn-kinematic wedges (yellow and red granular mixtures) are shown in Figure 13. Graben G1 and G2 display significant extension and relief as they pass through the extensional hinge zone at the downdip edge of sub-salt highs 2 and 3, before being shortened as they move through the contractional hinge zones (Fig. 13a, b). Graben G2 expelled a small salt sheet, whereas graben G1 was buried under a second syn-kinematic wedge and shortening was manifest as a structural high in the inter-high block zone (Fig. 13b). Graben G1 began to extend again as it was translated downslope (Fig. 13c, d). A younger graben, G3, was cored by a major reactive diapir as it extended in the extensional hinge zone, before being shortened to expel a major salt sheet in the contractional hinge at the downdip edge of sub-salt high block 3 (Fig. 13b, c). Above sub-salt high 4, graben G4 also experienced enhanced stretching and subsequent contraction as it passed through the extension–contraction hinge pair at the downdip edge of this sub-salt high block (Fig. 13b–d). Once more, structures in the suprasalt section experienced multi-stage evolutionary histories, as they were translated downslope across the stepped base of salt.

The main structural styles of Model 3 are best illustrated by a series of cross-sections (Fig. 14). The prekinematic sections show variable raft separations and are yoked together by the syn-kinematic strata similar to the styles shown by Duval *et al.* (1992), and Pilcher *et al.* (2014). Syn-kinematic wedges form sedimentary thicks within the topographic monoclines developed above the downdip edges of the sub-salt high blocks, similar to the salt-detached ramp-syncline basins described from the Kwanza Basin, Angola (see Jackson & Hudec 2005, and references therein), the Campos Basin (Dooley *et al.* 2017) and the Santos Basin (Pichel *et al.* 2018). Some of these sedimentary thicks have been translated downdip, resulting in anomalous relationships whereby sedimentary thicks overlie sub-salt high blocks (e.g. Wedge 4 in Fig. 14). Graben G2, G3 and G4 discussed above, and associated salt sheets are shown on lines A–A′, B–B′ and C–C′ (Figs 13 & 14). In section A–A′ graben G2

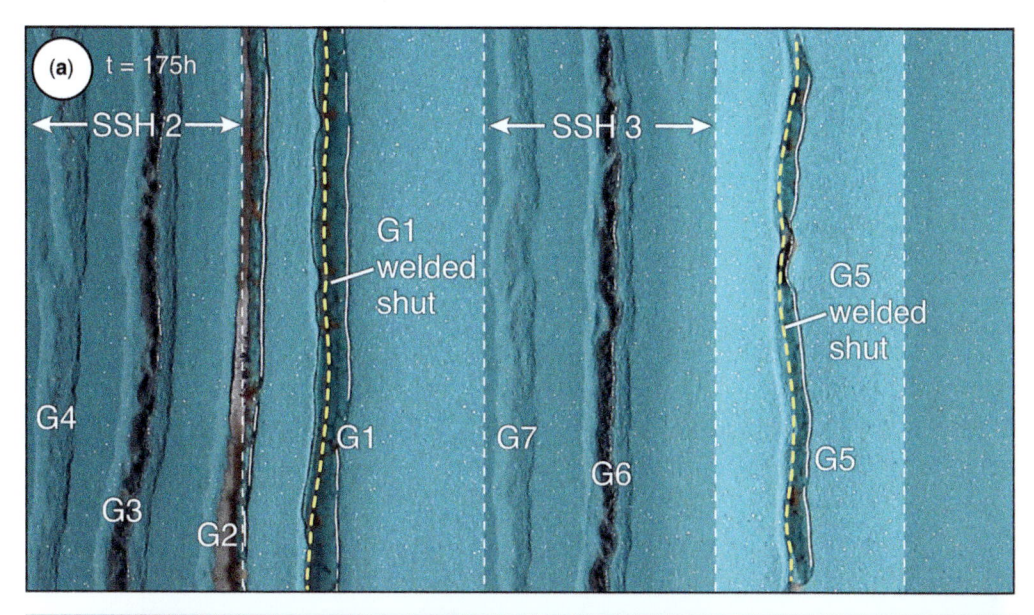

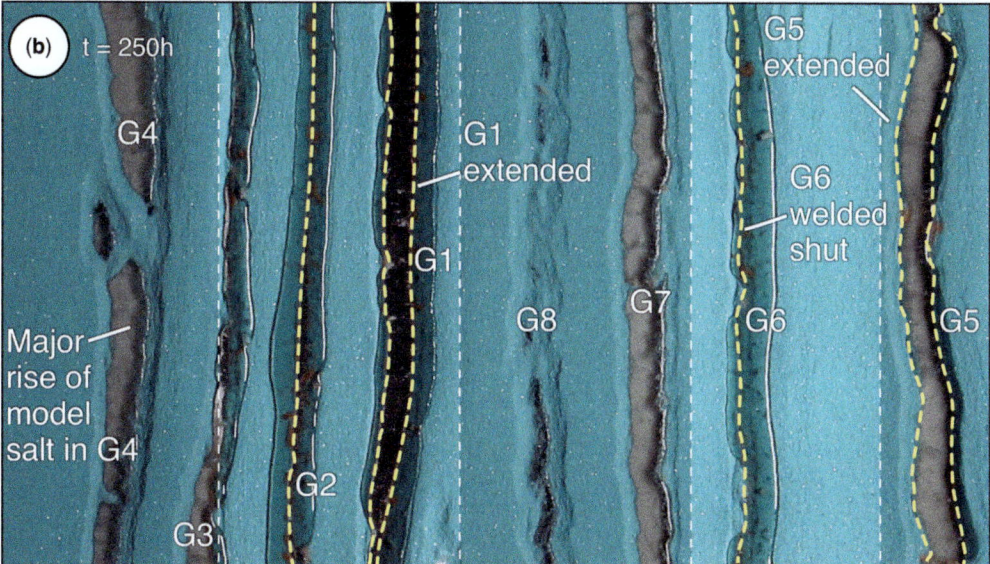

Fig. 11. Detailed views of surface deformation patterns in Model 2. Locations are shown on Figure 9e & h. (**a**) Detailed overhead view at 175 h runtime. Graben G1 and G5, formed above the sub-salt highs just updip of their current locations are welded shut as they pass through the contractional hinge. Graben G2 is beginning to expel a salt sheet as it starts to enter this contractional hinge. (**b**) Detailed overhead view at 250 h runtime. Graben G1 and G5 reopened as they entered a zone of faster moving salt (extension) close to the updip edges of the sub-salt high blocks. Graben G2 and G6 are welded shut in the contractional hinge, and graben G4 and G7 widen in the extensional hinge. G3 is being shortened as it enters the contractional hinge.

is shortened and the intervening diapir is welded shut along a secondary weld and isolated from the small salt sheet that overlies the prekinematic section below Wedge 4 (A–A′ in Fig. 14). Graben G3 is located in the contractional hinge associated with sub-salt high block 3, and has expelled a significant salt sheet onto the predominantly prekinematic strata (Fig. 14). In one section wedge 4 touches down onto the sub-canopy strata along a tertiary weld (B–B′, Fig. 14). Line D–D′ shows the earliest formed

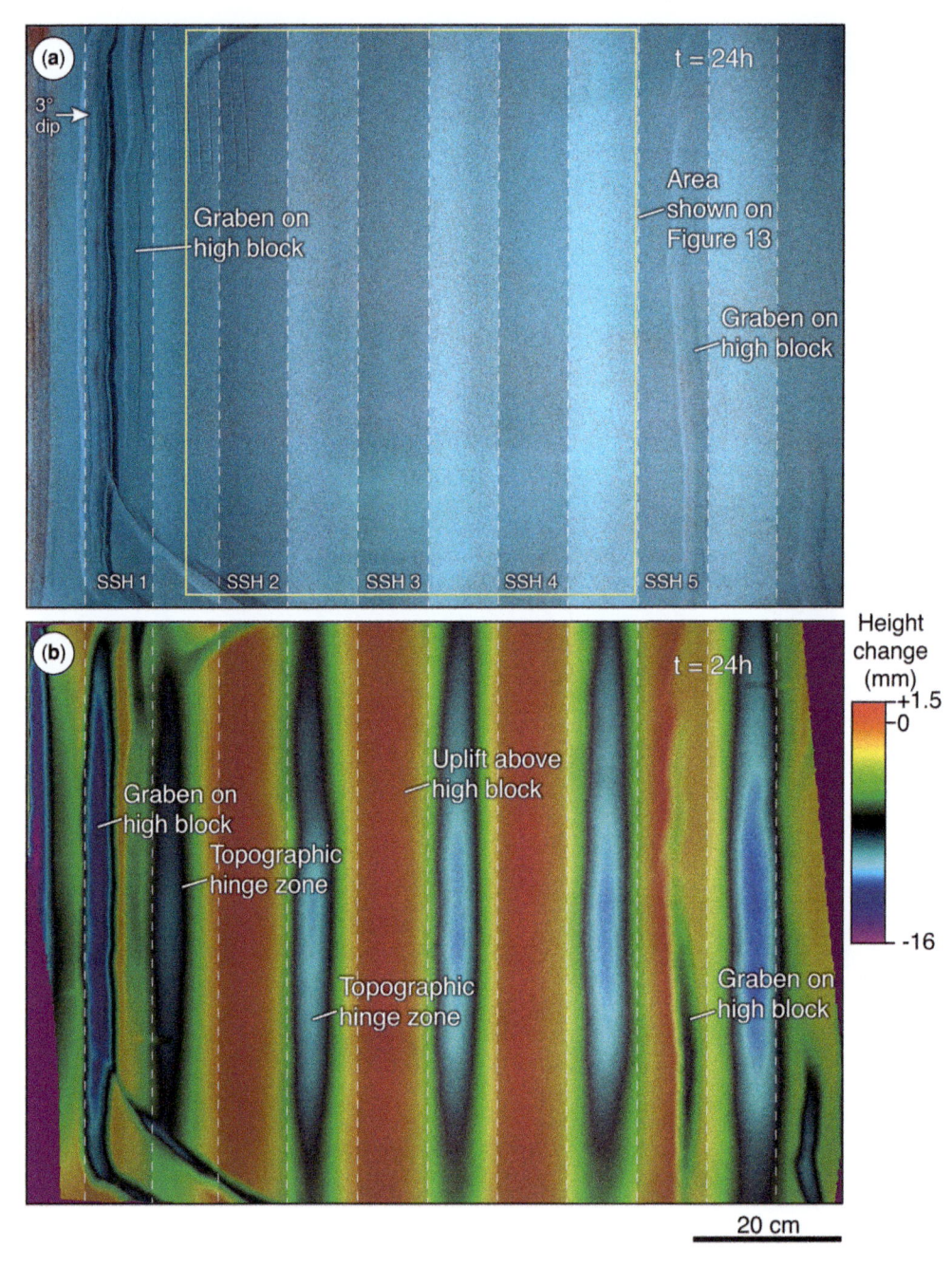

Fig. 12. (**a**) Overhead view of Model 3 after 24 h runtime. With a thicker and stronger roof in Model 3 (see Table 1), the most obvious features are extensional graben close to the updip breakaway and the open toe of the system. Yellow box indicates portion of Model 3 shown in Figure 13. (**b**) Height change map from DIC data after 24 h illustrates the same repeated patterns of rise and fall, as well as topographic hinge zones along the length of the model.

graben, G1, and associated diapir (Fig. 14). This diapir underwent initial extension and associated reactive rise of salt, followed by contraction that squeezed the diapir and caused thrusting of its roof, and finally back into extension as it moved out of the contractional hinge (Figs 13 & 14). The

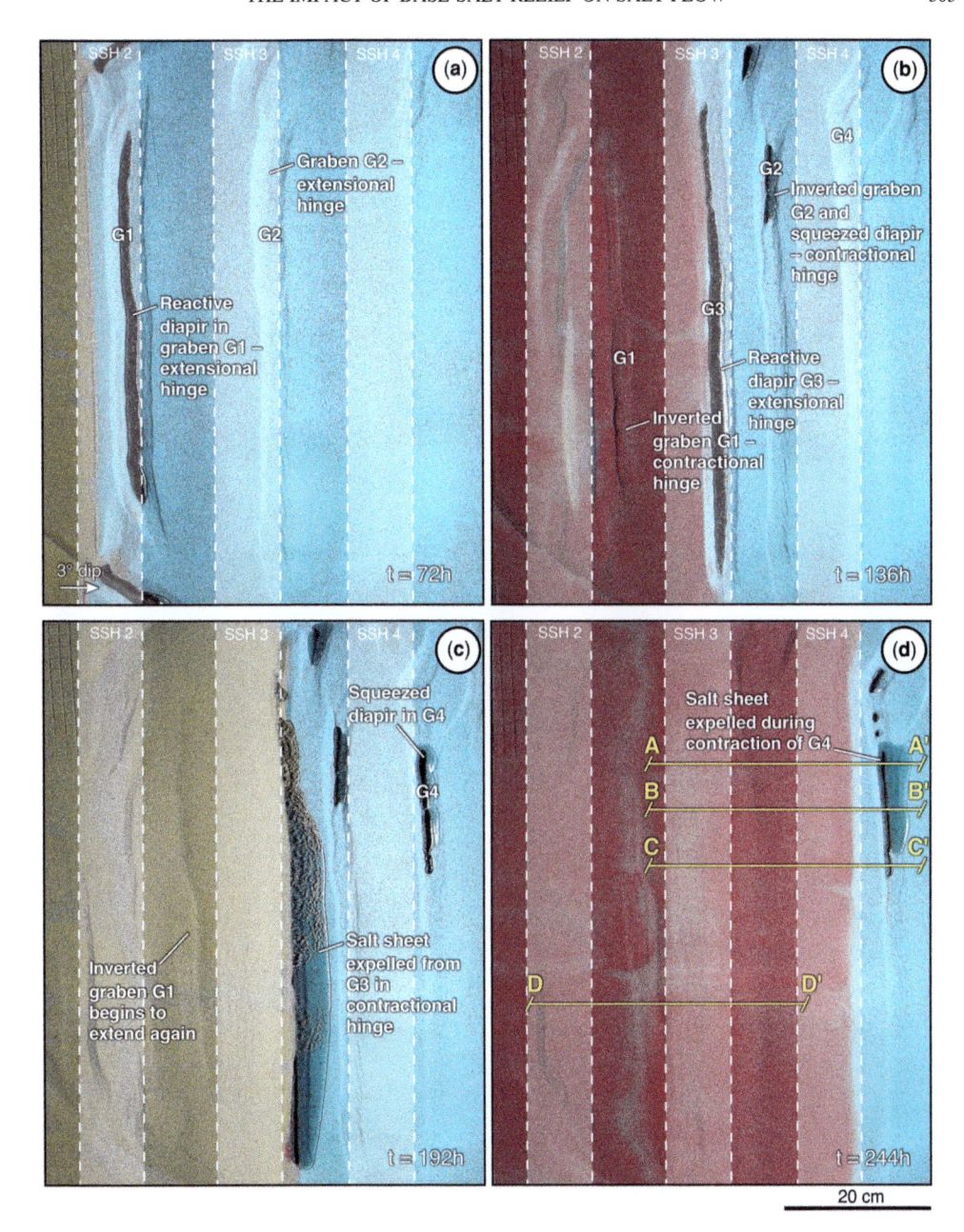

Fig. 13. Overhead views of a portion of Model 3 during the addition of syn-kinematic wedges. Model illuminated from the right. Syn-kinematic wedges with a 1° dip gradually prograded out across the system over time. Location shown on Figure 12a. (**a**) Toe of yellow Wedge 1 visible on the left side of this image. Model runtime was 72 h. Graben G1 and G2 widen in the extensional hinge of the monocline. (**b**) Overhead view after addition of Wedge 2 and model runtime increased to 136 h. Graben G1 was buried by this wedge but shows clear signs of shortening at surface. Graben G2 was also shortened in the contractional hinge and expelled a small salt sheet. Graben G3 broadens in extensional hinge. (**c**) Overhead view after addition of Wedge 3 and an increase in model runtime to 192 h. A major salt sheet was expelled from Graben G3 as it shortened in the contractional hinge. Graben G1 began to extend again as it approached the sub-salt high block with accelerated flow. (**d**) Overhead view of Model 3 after Wedge 4 and a runtime increase to 244 h. The more proximal systems have been buried and activity wanes as things begin to weld. Graben G4 was squeezed and expelled a small salt sheet. AA′, BB′, CC′ and DD′ are the locations of the model cross-sections shown in Figure 14.

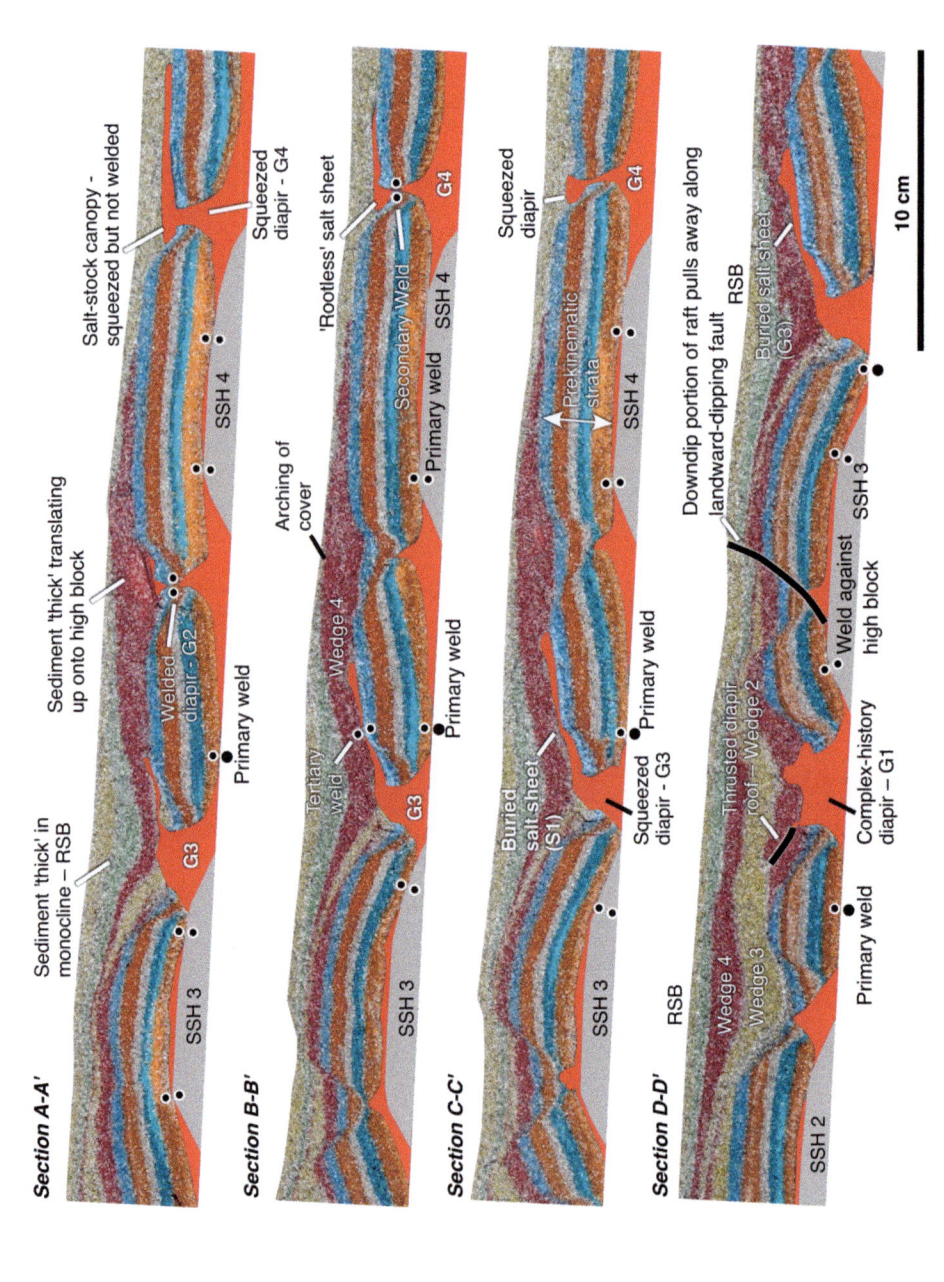

Fig. 14. Series of partial sections through Model 3. See Figure 13 for locations. Sections A–A′, B–B′ and C–C′ illustrate the salt sheets expelled from graben G3 and G4, and the welded neck of the diapir in graben G2. Section D–D′ (note different location) illustrates graben G1 which underwent a complex history of extension, contraction and later extension. A thrusted roof fragment of Wedge 2 is still present. The buried salt sheet expelled from the diapir in graben G3 is located further downslope. Note that where a portion of a raft is welded against a sub-salt high block, the more mobile portion of the raft pulls away along a landward-dipping fault. RSB, Ramp-syncline basin.

horned top-salt geometry seen in G1 is typical of a falling diapir, where extension outpaces salt supply (D–D′, Fig. 14; Vendeville & Jackson 1992). Primary welds are ubiquitous in all sections through Model 3, and are more common above and against the sub-salt high blocks (Fig. 14). Landward-dipping listric faults develop where a raft welds against the updip edge of a high block, but still retains mobility on its downdip side, and pulls away from the primary weld (D–D′, Fig. 14). Similar structural relationships are seen in Model 4, where a raft block, and associated ramp-basin sediments, is welded against a sub-salt high block and a landward-dipping fault develops downdip of this primary weld (Fig. 15). The only difference between these models is a much thinner prekinematic sequence in Model 4, resulting in much greater separation of the phase 1 rafts (rafts of prekinematic sediments – see Duval *et al.* 1992; Table 1; Fig. 15).

Effects of isolated, base-salt high blocks orthogonal to mean salt-flow direction

Models 1–4 all had sub-salt high blocks that were continuous along strike, resulting in salt flow and roof translation that was predominantly parallel to regional dip. Dooley & Hudec (2017) presented more complex scenarios whereby plunging sub-salt arches were somewhat discontinuous and the resultant flux perturbations resulted in significant rotations of early formed structures as the model evolved and structures were translated downdip. Model 5 investigates the impact of isolated sub-salt high blocks with triangular profiles (Fig. 2, Table 1). The two updip high blocks had the crest of the triangle on the downdip side, whereas the downdip block had the crest on its updip side (Figs 2 & 16). Figure 16 shows a series of photographs of the model surface as it evolved through 85 h of runtime. Initially, the model displays strong similarities to Models 1–4, with updip extension controlled by the geometry of the breakaway, a monocline at the downdip edge of the landward high blocks and local extension just downdip of the high blocks as thicker salt pulls away from the base-salt relief (Fig. 16a). 2D surface strain data illustrate the development of an extensional hinge and paired contractional hinges associated with the landward high blocks, as well as the localized extension zone downdip of these high blocks (Fig. 17a). At the updip edge of the seaward high block a zone of mild contraction is seen above its crest line, and a zone of outer-arc extension above a weakly developed salt-cored fold in this location (Fig. 17a). As the model evolved to 34 and 50 h runtime extensional graben and their cores of reactive diapirs were translated across the crests of the landward high blocks, and through the paired hinge zones, where they experienced enhanced extension and subsequent contraction to form local salt-cored anticlines (Figs 16b, c & 17b). Above the crest of the seaward high block the thickened salt in the former contraction zone began to collapse under extension (Figs 16b, c & 17b).

In the latter portion of Model 5 runtime, structural complexity increased (Fig. 16c, d). Reactive diapirs were squeezed at the contractional hinges just downdip of the two landward high blocks, and subsequently translated downdip with some being extended once more (Fig. 16c, d). Much of the slope is cut by salt-cored graben with variable orientations, except for the presence of an undeformed 'megaraft' to the north of the seaward high block (Fig. 16c). This 'megaraft' was separated from the basement high block to the south by a salt-detached tear fault (Fig. 16c). However, the main structural complexities stem from the major rotations observed around and between the two landward high blocks (Fig. 16d). Clockwise and counterclockwise rotations of raft blocks up to 67° were recorded by the passive marker grid, resulting in complex extensional and contractional deformation of the inter-raft diapirs with varying orientations (Fig. 16d). Displacement analyses illustrate salt flow being channelled around, and through, the basement high blocks, especially the two landward ones (Fig. 18). Immediately downdip of the landward high blocks are zones of anomalously low downdip displacements flanked by more rapidly moving salt and its overburden, imparting shear and associated deformation (Fig. 18a, c). Divergent and convergent flow on the updip and downdip sides respectively of the basement high blocks are documented by vector maps and north–south motions of the suprasalt cover (Fig. 18b, d). These flow perturbations, channelized and divergent/convergent flows, drove the complex rotations and resultant raft jostling and diapir deformation seen in the latter stages of Model 5.

Set 2 models: effects of regional dip, salt budget and base-salt relief on canopy coalescence and flow during subsequent shortening

Model 6 was constructed as shown in Figure 3. An array of vertical diapirs was initiated by differential loading of the deep source-layer salt. They grew upwards by downbuilding, began to spread laterally as salt rise outpaced our model sedimentation rate, and finally spread asymmetrically by tilting the base of the model to 1° (Fig. 3). Spreading salt sheets fused along frontal and lateral allosutures (see Dooley *et al.* 2012, for suture terminology) to form the canopy illustrated in Figures 19a & 20. The

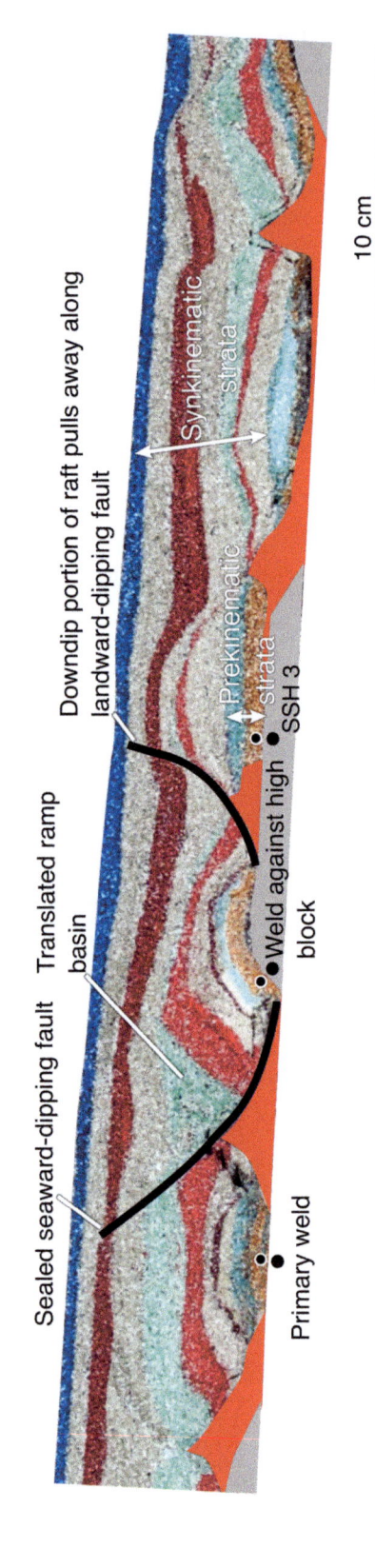

Fig. 15. Cross-section through Model 4 (see Table 1). Model setup was identical to that of Model 3 but the initial prekinematic thickness was thinner. In a similar fashion to Model 3, welding against the updip edge of a sub-salt high block resulted in the formation of a landward-dipping fault system as the more mobile downdip sequence continued to glide downslope.

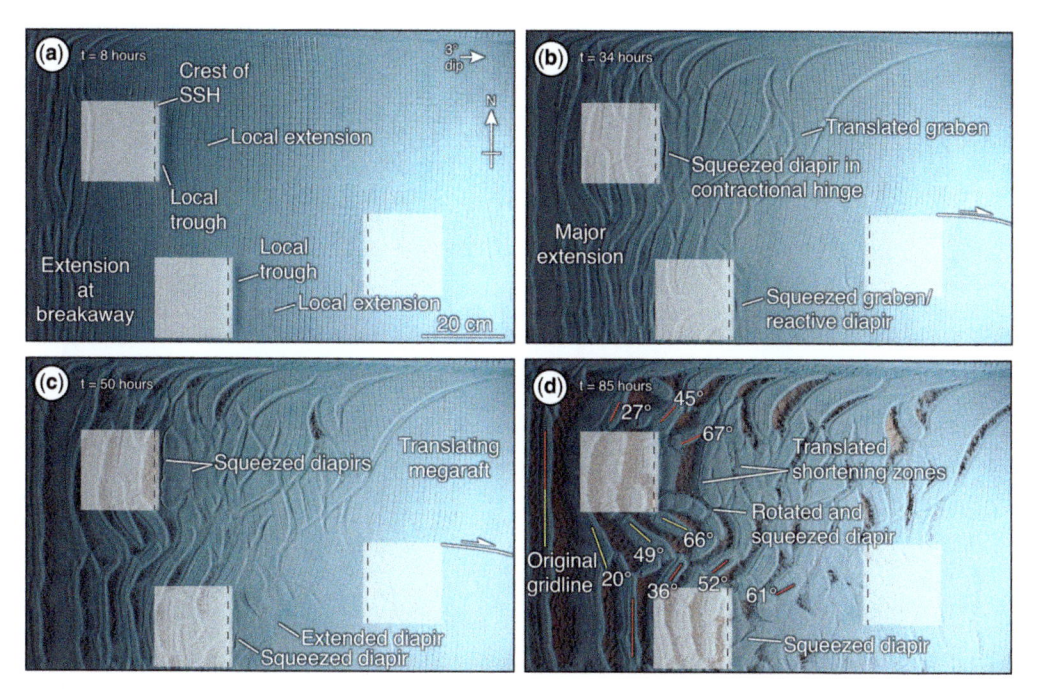

Fig. 16. Overhead views of Model 5 after 8 h (**a**), 34 h (**b**), 50 h (**c**) and 85 h (**d**) runtime. Sub-salt basement high blocks are indicated by white overlays. Crests of the high blocks are marked by black dashed lines. Clockwise and counterclockwise rotations of the passive marker grid are indicated by red and yellow lines respectively on (d).

introduction of a regional dip affected the spreading direction of individual salt stocks, skewing their streamlines downslope and causing asymmetric, or sub-radial, spreading (Figs 3a–c, 19b & 20a). Individual salt sheets continued to spread and climb up-section along a series of flats and ramps in the syn-kinematic strata (Fig. 3d). Surface views of a portion of Model 6 illustrate the suturing process and coalescence of a portion of the canopy (Fig. 19b, d). On initial collision the sutures between salt sheets were linear (e.g. suture between sheets 7 and 10, Fig. 19b), and gradually became bowed in the direction of override (e.g. suture between sheets 4 and 7, Fig. 19b). With continued coalescence the sutures in Model 6 became increasingly bowed and boudinaged by circumferential extension, forming fused salt sheets with heavily dismembered or unrecognizable sutures (Fig. 19c, d). The final canopy system had a highly irregular and cuspate boundary, with several small windows through the canopy

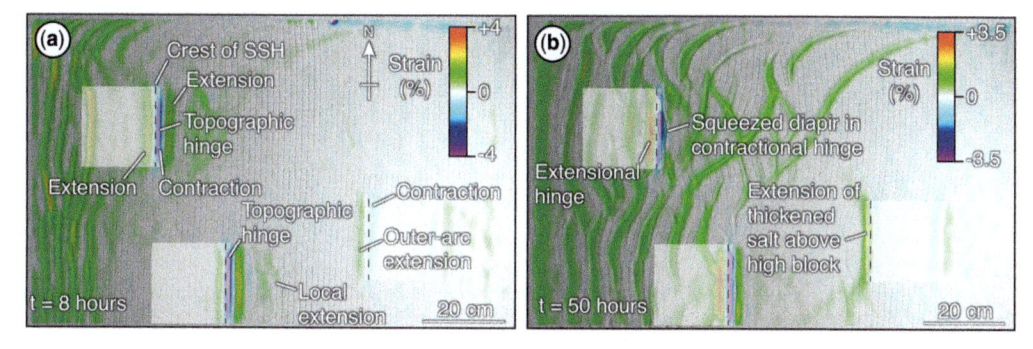

Fig. 17. 2D DIC data from Model 5 showing differential strain maps overlain on imagery from 8 h (**a**) and 50 h (**b**) runtime. These maps display the strains recorded over a 20 min period. See text for details.

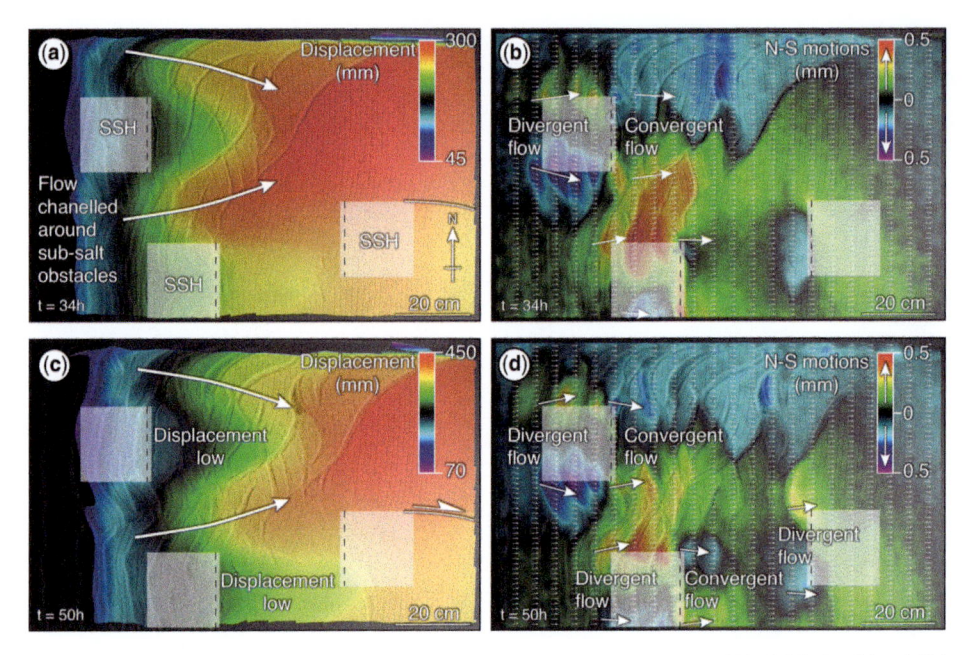

Fig. 18. Dip-parallel displacement (a, b) and y motions (c, d: north–south movements) of Model 5 after 34 and 50 h runtime. Note that flow is being channelled around and between the high blocks. Divergent and convergent flow cells are recorded on the updip and downdip sides of the basement high blocks, respectively.

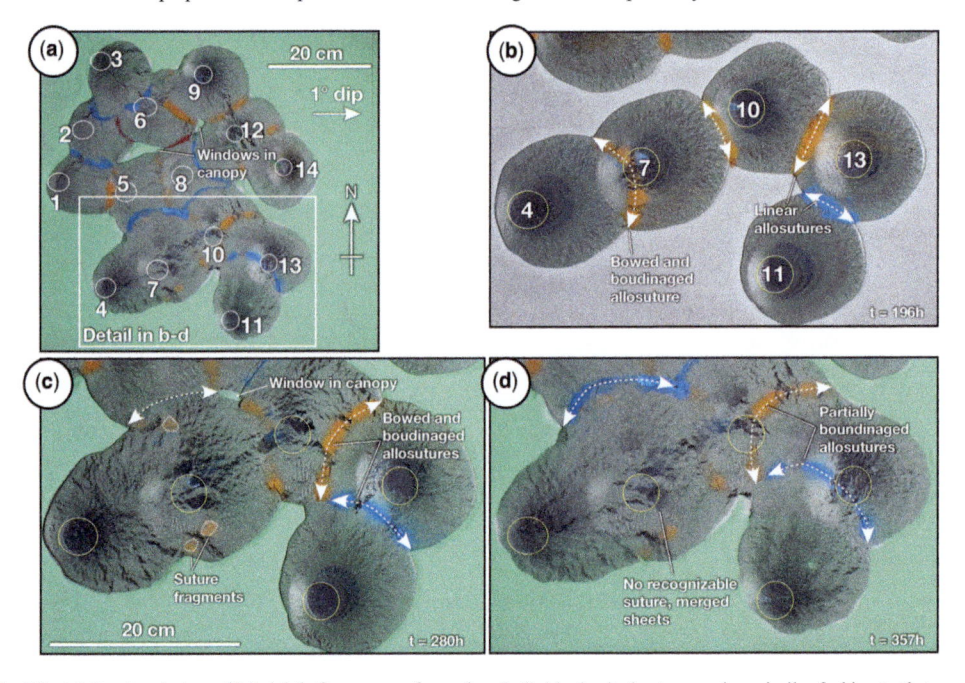

Fig. 19. (**a**) Overhead view of Model 6 after canopy formation. Individual salt sheets grew in a similar fashion to that illustrated in Figure 3. Diapir locations are numbered 1–14. Powder paints mark the location of frontal and lateral allosutures between the sheets (see Dooley *et al.* 2012, for details of suture terminology). Many of these sutures are dismembered during advanced coalescence. Rectangle marks the area shown in detail in (b)–(d). (**b**) Detailed view showing allosutures between five salt sheets. Sutures are initially linear and subsequently bow during override. (**c, d**) Detailed views showing advanced suturing between the five salt sheets and adjacent sheets. Sutures are gradually dismembered during this process. Note that there is a strong component of flow from canopy periphery toward its centre.

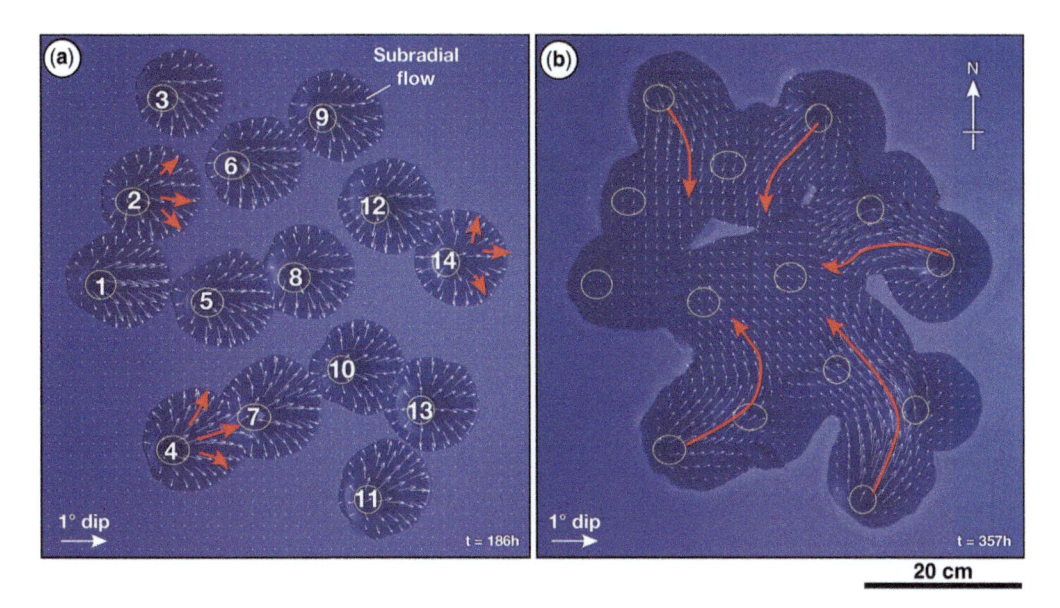

Fig. 20. Overhead views with superposed displacement vector maps of two stages of Model 6 after imposing a regional dip of 1°. (**a**) Introducing this gentle regional slope affected the direction of spreading of individual salt stocks, so that flow patterns of stocks were initially sub-radial, then their streamlines became slightly skewed down the regional slope as the diapirs widened and spread asymmetrically. (**b**) During the later stages of canopy coalescence, flowlines became dominated by inward flow from the canopy periphery, rather than roughly downslope. This was because outer feeders flowed more vigorously than inner feeders because there was less competition for source-layer salt in the peripheral regions.

salt (Fig. 19a). Detailed views of the suturing process along with a displacement vector map of the latter stages of canopy coalescence illustrate flow directions within the canopy that are highly oblique to regional dip (Figs 19c, d & 20b). In the latter stages of canopy coalescence flow was dominated by inward flow from the northern and southern margins, and backflow, against regional dip, from the eastern

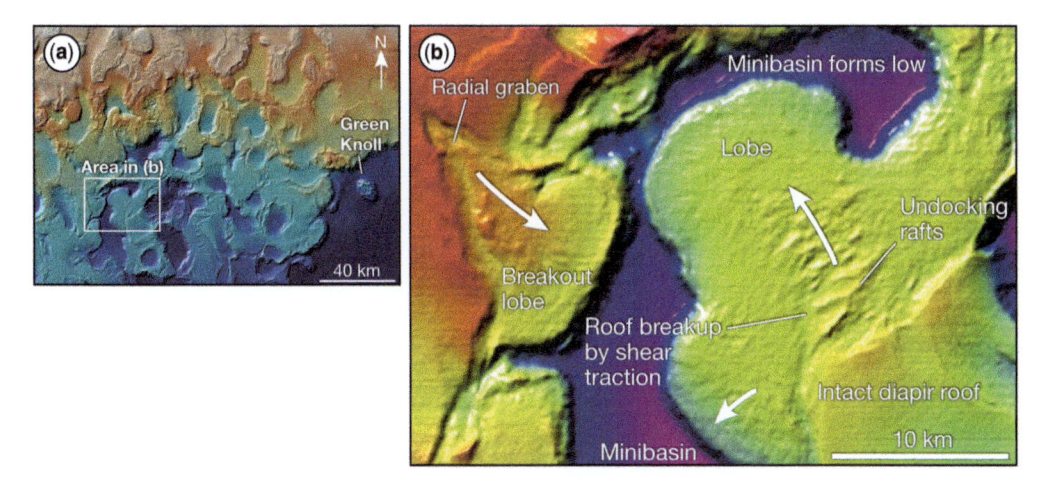

Fig. 21. (**a**) Bathymetric map of a portion of the Sigsbee Canopy, northern Gulf of Mexico. (**b**) Area of detail shown in (a) illustrating a spreading salt sheet flowing northward into a low formed by a minibasin. Diagram modified from Dooley *et al.* (2015). Bathymetry from GeoMapApp.

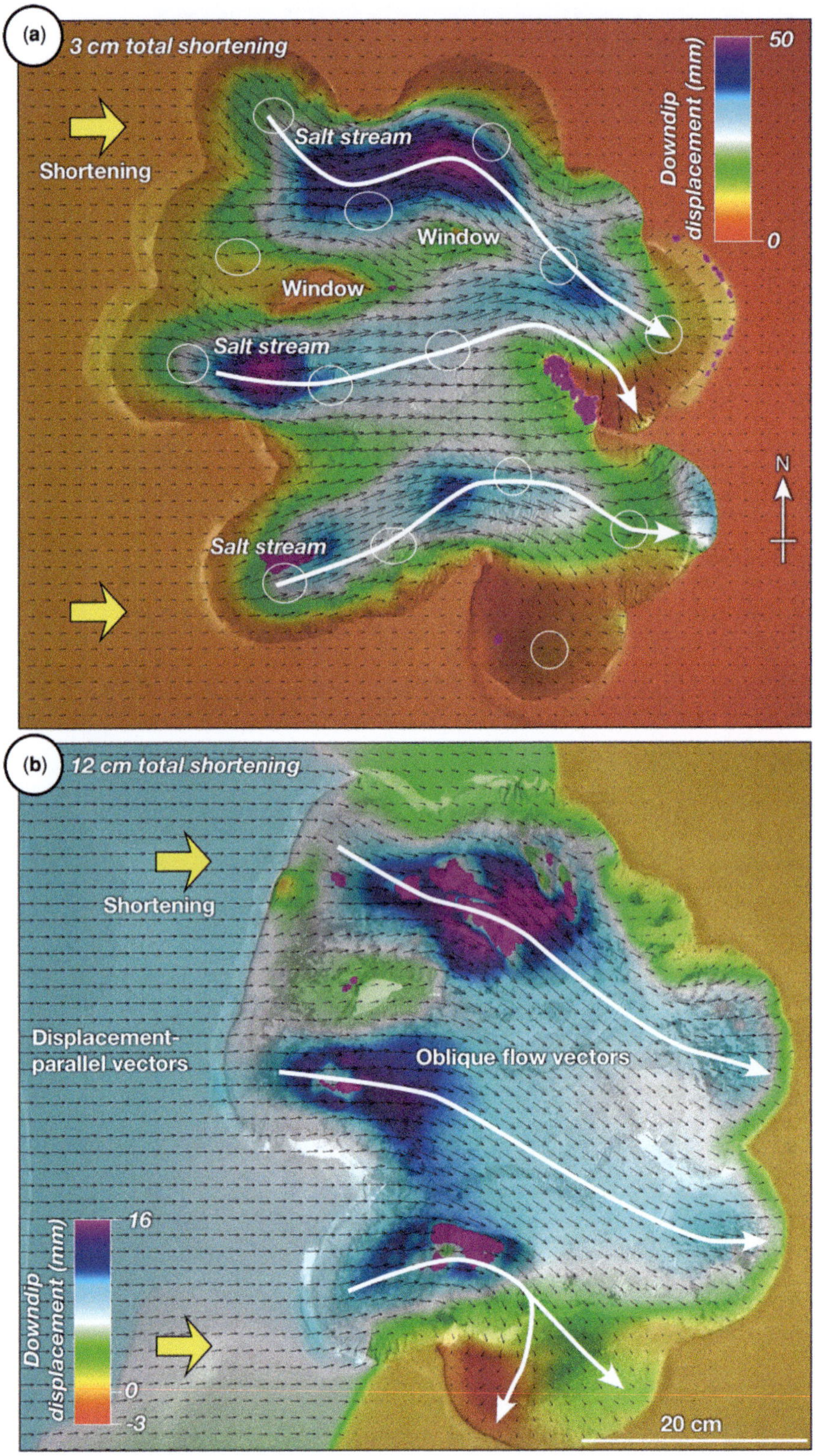
(a)
3 cm total shortening
Shortening
Salt stream
Window
Window
Salt stream
Salt stream
Downdip displacement (mm)
50
0
N
(b)
12 cm total shortening
Shortening
Displacement-parallel vectors
Oblique flow vectors
Downdip displacement (mm)
16
0
-3
20 cm

margin (Fig. 20b). The change in flow dynamics from predominantly downdip flow to cross- and backflows is a direct consequence of the salt budget, with feeders on the periphery of the diapir array facing less competition for source-layer salt than those in the centre of the canopy. These outer feeders flowed more vigorously, pumping salt through the canopy into the less active interior of the canopy (Fig. 20b). Because feeders in the centre of the canopy were less active, or inactive owing to the development of peripheral primary welds (Fig. 3), they also formed topographic lows within the canopy, facilitating backflows from the more actively growing downdip feeders (feeders 12 and 14, Fig. 20b). An example of canopy cross- and backflow from the Sigsbee salt canopy in the northern Gulf of Mexico is shown in Figure 21. Here a minibasin forms a topographic low in the canopy and vigorously rising diapirs flanking flow into this minibasin from a variety of directions.

After canopy coalescence Model 6 was subjected to 12 cm shortening by a computer-controlled moving endwall. The base-salt relief in a canopy system fed by numerous feeders is highly rugose, similar to an egg carton with irregular spacing, with feeders separated by base-salt highs and the salt thinning laterally to the canopy margins (see Fig. 2 of Dooley *et al.* 2012). Shortening-parallel displacement maps with superposed vectors are shown in Figure 22. During shortening, flow in the canopy returned to a dominant downdip direction, but significant deflections occurred around the local buttresses (windows in the canopy), and at the lobate lateral and downdip margins of the canopy (Fig. 22a). The fastest moving salt, a series of three salt streams (*sensu* Talbot & Pohjola 2009), links the feeder systems across the canopy (Fig. 22a). These salt streams formed where the salt was thickest and less impacted by the subcanopy topography. As the canopy thickened during advanced shortening, flow became more uniform as the effects of base-salt relief lessened (Fig. 22b). However, flow remained oblique to the shortening direction owing to the continued impact, albeit lessened, of sub-salt topography (ramps and windows in the canopy), the shape of the canopy margin and the location of the thickest salt above the more downdip and active feeders (Fig. 22b).

Concluding remarks

Physical models presented in this study document the major impact that base-salt relief has on salt flow and suprasalt deformation patterns. At the autochthonous or paraautochthonous level during early-stage margin tilt and downdip flow of the salt and its thin overburden across a rugose or stepped base of salt, our models illustrate the formation of localized zones of extension and contraction, where there are significant flux mismatches caused by this base-salt relief. Salt flowing up onto a high block is impacted by basal drag, resulting in local contraction owing to the mismatch between the flux that is accommodated through the thinner salt and the flux that is fed onto the base-salt high block from the updip region of thicker salt. This initially results in a zone of contractional uplift of thickened salt. Eventually the gradual thickening of salt above the high block allows the salt velocity to increase as the effects of basal drag become proportionally less, resulting in extensional collapse of the thickened salt mass and the formation of graben that migrate across the base-salt high block. At the downdip edge of the base-salt high block the salt flux exiting the high block is less than that in the thicker salt mass flowing away from the high block. The resultant flux imbalance generates a topographic monocline with an updip extensional hinge and downdip contractional hinge within the thicker salt. Extensional graben translated across and off the base-salt high block are initially rapidly extended through the extensional hinge and are then shortened and inverted as they pass through the contractional hinge. Models with multiple base-salt high blocks demonstrate the formation of these localized zones of extension and contraction down the entire slope, with suprasalt structures undergoing multiple phases of extension and contraction wherever base-salt relief is encountered. Reactive diapirs widen and grow in zones of extension and are then squeezed and expel salt sheets as they migrate through contractional hinges.

Topographic monoclines developed by the base-salt relief form salt-detached ramp basins in the presence of syn-kinematic sedimentation (e.g. Fig. 14). These structures were recognized in the Kwanza

---

Fig. 22. Dip-parallel displacement maps during early- (**a**), and late-stage (**b**) shortening of the canopy of Model 6. In general flow in the canopy returned to mostly downdip. The fastest avenues of moving salt (cool colours) are salt streams (after terminology in Talbot & Pohjola 2009). These three salt streams link feeder systems where salt was thickest, and less impacted by the sub-canopy topography. As the canopy thickened during advanced shortening (b), flow became more uniform as the impact of the sub-canopy topography lessened. However, flow remained oblique to the shortening direction, affected by: (1) sub-salt topography (ramps and windows in canopy); (2) shape of the canopy margin; and (3) location of thickest salt above downdip feeders. Note that these maps show flow values values for each increment of shortening, and not the total displacements recorded.

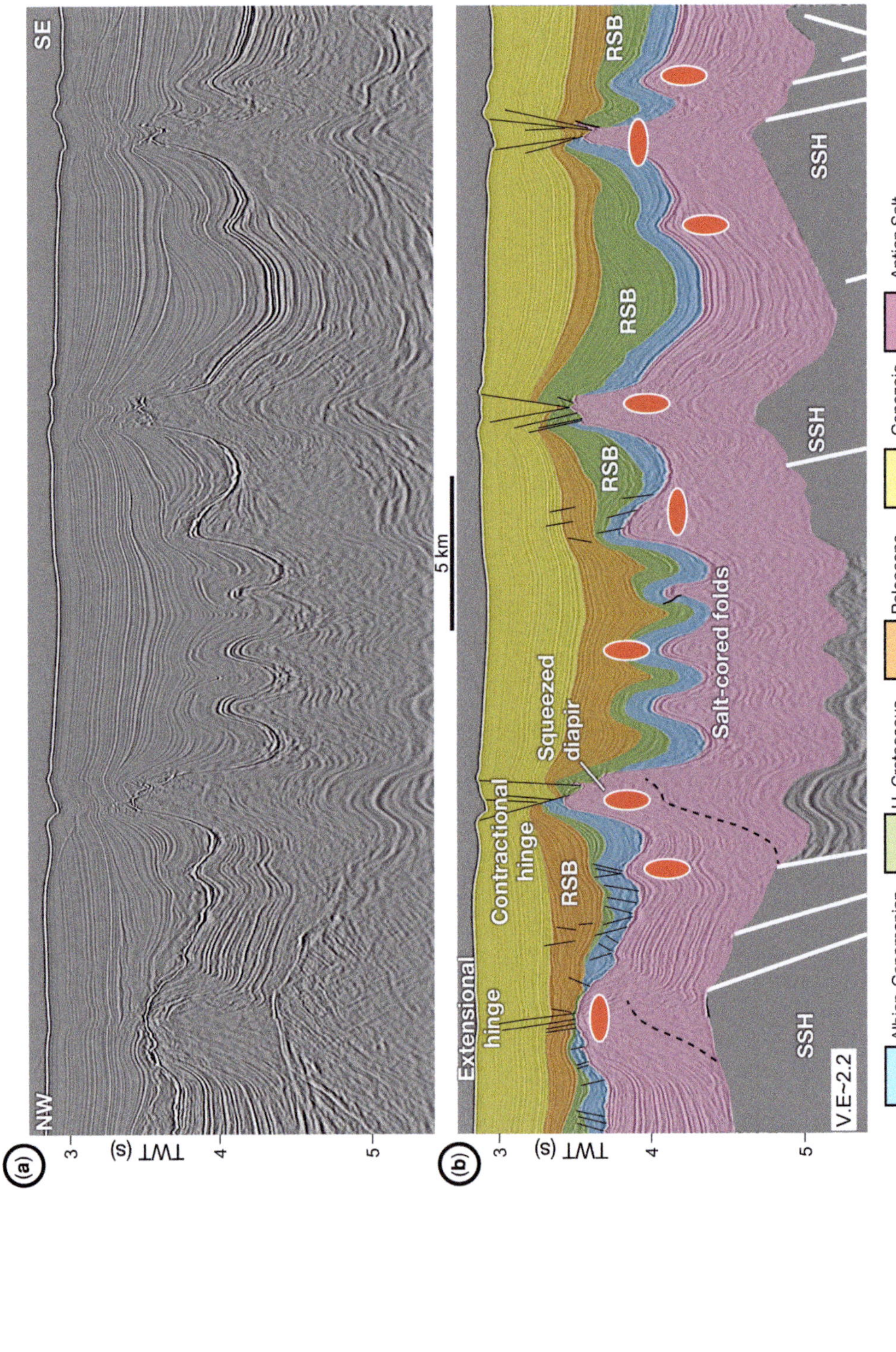
(a)
NW
SE
TWT (s)
3
4
5
5 km
(b)
Extensional hinge
Contractional hinge
Squeezed diapir
RSB
Salt-cored folds
SSH
V.E~2.2
Albian-Cenomanian
U. Cretaceous
Paleocene
Cenozoic
Aptian Salt

Basin (see Jackson & Hudec 2005, and references therein), and more recently by Dooley *et al.* (2017) and Pichel *et al.* (2018) from the Campos and Santos Basins, respectively. Seismic data from the São Paulo Plateau, Santos Basin, Brazil beautifully shows the complex deformation patterns associated with translation of salt and its overburden across a series of base-salt highs (Pichel *et al.* 2018; Fig. 23). The section shows a zone of monoclinal folding and subsidence with development of a large ramp-syncline basin (RSB), partially disrupted by a diapir, over the large base-salt step at the edge of the Tupi High (*c.* 2 km of structural relief). Basinward-dipping normal faults and minor reactive rise record extension at the top of the ramp (extensional hinge), and contraction at the bottom of the ramp is recorded by active diapirism and salt-cored folds (contractional hinge; Fig. 23). Further downdip, two more gentle and closely spaced, pre-salt, tilted fault-blocks result in additional contraction and minor extension (and associated diapirism), with subsidence and development of RSBs over their basinward-dipping edge (Fig. 23b). Our models also record the formation of landward-dipping or counter-regional faults when portions of the composite rafts are welded against a base-salt high block, and the more mobile downdip portions of these rafts continue to translate downslope (Fig. 15). Karlo *et al.* (2014) described a similar process from the Central North Sea, with counter-regional extension beginning as rafts began to ground out at base-salt level (see their Fig. 4). Base-salt relief is unlikely to be laterally continuous across a salt-bearing basin, and our models of isolated base-salt high blocks document even more complex deformation patterns and associated flow deviations (Figs 16–18). In the vicinity of the crests of these triangular-shaped sub-salt high blocks localized extension and contractional hinges related to the formation of topographic monoclines owing to flux mismatches occur as predicted by our other models with laterally continuous high blocks. However, these models also document major rotations with the concomitant formation of variably oriented salt walls as salt diverges and converges updip and downdip of these isolated high blocks. Unlike our well-behaved extension–translation–contraction model in Figure 1, our Series 1 models, like those of Dooley *et al.* (2017), Dooley & Hudec (2017) and Ferrer *et al.* (2017), demonstrate that extensional diapirs and shortening-related fold-thrust belts can form anywhere on a slope, and in variable orientations, as salt accelerates and decelerates when flowing across and between base-salt relief. Continued seaward flow translates the diapirs and associated suprasalt structures further downslope, obscuring the relationship between the base-salt relief that caused this deformation, as well as forming a structurally weakened salt-sediment package that may later be exploited in deep water as the margin evolves.

As a salt-bearing passive margin matures, processes such as differential loading may exploit an overburden already weakened by structures generated by the processes described above, driving salt rise upwards to the allochthonous level via feeders and associated salt sheets. Flow patterns at the allochthonous level can be perturbed by regional dip, salt budget and extreme base-salt relief associated with the coalescence of a canopy system. During the early stages of canopy formation the flow patterns of individual salt sheets were sub-radial with streamlines skewed down the regional slope, as expected. In the late stages of canopy coalescence, flowlines were dominated by inward flow from the canopy periphery. The reason for this was that outer feeders flowed more vigorously owing to less competition for source-layer salt. Sutures between salt sheets were bowed in the direction of override and became partially dismembered as they were deformed. Salt flow during a subsequent shortening phase was grossly parallel to the regional shortening direction. However, canopy salt flows fastest where it is thickest and thus the chains of feeders channelled the flow of salt streams within the canopy. This created intracanopy flow cells, and flow directions that were oblique to the regional slope and branched regularly. In addition to the above-listed effects, canopy flow will also be impacted by intrasalt inclusions (sutures and other intrasalt inclusions, intrasalt minibasins, etc.) as well as suprasalt minibasins (e.g. Duffy *et al.* 2017; Fernandez *et al.* 2017), creating additional complexities.

Fig. 23. (**a**) Uninterpreted and (**b**) interpreted seismic section from the São Paulo Plateau, Santos Basin, offshore Brazil, showing the effects associated with a major step (NW side of the seismic line; edge of the Tupi High) at base salt (*c.* 2 km) on suprasalt deformation patterns. Translation over this base-salt ramp results in a zone of monoclinal folding and development of an RSB at the top of basinward-dipping ramps delimited by updip extension at the top of the largest ramp (extensional hinge) and contraction at the bottom of the ramp (contractional hinge). Translation over narrow and more gentle base-salt steps formed by pre-salt tilted fault-blocks further downdip (to the SE) results in additional extensional and contractional deformation, similar to the physical models presented in this study. Note that the bulk of the translation and associated deformation occurred during Units II and III (Upper Cretaceous to Paleocene; see Pichel *et al.* 2018, for further details).

Acknowledgements T.D. thanks James Donnelly, Nathan Ivicic, Josh Lambert, Brandon Williamson, William Molthen and Rudy Lucero for logistical support in the modelling laboratories. This paper is dedicated to the memory of our friend and mentor, Martin Jackson. Seismic data in Figure 23 were provided to L. Pichel by CGG. Thanks to Thomas Hearon and Oriol Ferrer for constructive reviews of an earlier version of this manuscript, and to Ken McClay and Stephanie Jones for additional editing. Research was funded by the Applied Geodynamics Laboratory Industrial Associates program, comprising the following companies: Anadarko, Apache, Aramco Services, BHP Billiton, BP, CGG, Chevron, Cobalt, Condor, ConocoPhillips, EcoPetrol, ENI, ExxonMobil, Freeport-McMoRan, Fugro, Hess, Ion-GXT, LUKOIL Overseas Services, Maersk, Marathon, Murphy, Nexen USA, Noble, Pemex, Petrobras, PGS, Repsol, Rockfield, Samson, Shell, Spectrum, Statoil, Stone Energy, TGS, Total, Venari Resources and Woodside (http://www.beg.utexas.edu/agl/sponsors). The authors received additional support from the Jackson School of Geosciences, The University of Texas at Austin. Publication was authorized by the Director, Bureau of Economic Geology, The University of Texas at Austin.

References

ADAM, J. & KRÉZSEK, C. 2012. Basin-scale salt tectonic processes of the Laurentian Basin, Eastern Canada: insights from integrated regional 2D seismic interpretation and 4D physical experiments. *In*: ALSOP, G.I., ARCHER, S.G., HARTLEY, A.J., GRANT, N.T. & HODGKINSON, R. (eds) *Salt Tectonics, Sediments and Prospectivity*. Geological Society, London, Special Publications, **363**, 331–360, https://doi.org/10.1144/SP363.15

ADAM, J., URAI, J. *ET AL*. 2005. Shear localisation and strain distribution during tectonic faulting – new insights from granular-flow experiments and high-resolution optical image correlation techniques. *Journal of Structural Geology*, **27**, 283–301, https://doi.org/10.1016/j.jsg.2004.08.008

BRUN, J.P. & FORT, X. 2011. Salt tectonics at passive margins: geology v. models. *Marine and Petroleum Geology*, **28**, 1123–1145, https://doi.org/10.1016/j.marpetgeo.2011.03.004

COBBOLD, P.R. & SZATMARI, P. 1991. Radial gravitational gliding on passive margins. *Tectonophysics*, **188**, 249–289.

DEPTUCK, M.E. & KENDELL, K.L. 2017. A review of Mesozoic–Cenozoic salt tectonics along the Scotian Margin, Eastern Canada. *In*: SOTO, J.I., FLINCH, J.F. & TARI, G. (eds) *Permo-Triassic Salt Provinces of Europe, North Africa and the Atlantic Margins*. Elsevier, 287–312, https://doi.org/10.1016/B978-0-12-809417-4.00014-8

DOOLEY, T.P. & HUDEC, M.R. 2017. The effects of base-salt relief on salt flow and suprasalt deformation patterns – part 2: application to the eastern Gulf of Mexico. *Interpretation*, **5**, SD25–SD38, https://doi.org/10.1190/INT-2016-0088.1

DOOLEY, T.P., HUDEC, M.R. & JACKSON, M.P.A. 2012. The structure and evolution of sutures in allochthonous salt. *AAPG Bulletin*, **96**, 1045–1070.

DOOLEY, T.P., JACKSON, M.P.A. & HUDEC, M.R. 2013. Coeval extension and shortening above and below salt canopies on an uplifted, continental margin: application to the northern Gulf of Mexico. *AAPG Bulletin*, **97**, 1737–1764.

DOOLEY, T.P., JACKSON, M.P.A. & HUDEC, M.R. 2015. Breakout of squeezed stocks: dispersal of roof fragments, source of extrusive salt and interaction with regional thrust faults. *Basin Research*, **27**, 3–25, https://doi.org/10.1111/bre.12056

DOOLEY, T.P., HUDEC, M.R., CARRUTHERS, D., JACKSON, M.P.A. & LUO, G. 2017. The effects of base-salt relief on salt flow and suprasalt deformation patterns – part 1: flow across simple steps in the base of salt. *Interpretation*, **5**, SD1–SD23, https://doi.org/10.1190/INT-2016-0087.1

DUFFY, O.B., FERNANDEZ, N., HUDEC, M.R., JACKSON, M.P.A., BURG, G., DOOLEY, T.P. & JACKSON, C.A.-L. 2017. Lateral mobility of minibasins during shortening: insights from the SE Precaspian Basin, Kazakhstan. *Journal of Structural Geology*, **97**, 257–276, https://doi.org/10.1016/j.jsg.2017.02.002

DUVAL, B., CRAMEZ, C. & JACKSON, M.P.A. 1992. Raft tectonics in the Kwanza Basin, Angola. *Marine and Petroleum Geology*, **9**, 389–404.

FERNANDEZ, N., DUFFY, O.B., HUDEC, M.R., JACKSON, M.P.A., BURG, G., JACKSON, C.A.-L. & DOOLEY, T.P. 2017. The origin of salt-encased sediment packages: observations from the SE Precaspian Basin (Kazakhstan). *Journal of Structural Geology*, **97**, 237–256, https://doi.org/10.1016/j.jsg.2017.01.008

FERRER, O., GRATACÓS, O., ROCA, E. & MUÑOZ, J.A. 2017. Modeliing the interaction between presalt seamounts and gravitational failure in salt-bearing passive margins: the Messinian case in the northwestern Mediterranean Basin. *Interpretation*, **5**, SD99–SD117, https://doi.org/10.1190/INT-2016-0096.1

FORT, X., BRUN, J.-P. & CHAUVEL, F. 2004. Contraction induced by block rotation above salt (Angolan margin). *Marine and Petroleum Geology*, **21**, 1281–1294, https://doi.org/10.1016/j.marpetgeo.2004.09.006

GAULLIER, V., BRUN, J., GUERIN, G. & LECANU, H. 1993. Raft tectonics: the effects of residual topography below a salt décollement. *Tectonophysics*, **228**, 363–381.

HUDEC, M.R. & JACKSON, M.P.A. 2004. Regional restoration across the Kwanza Basin, Angola: salt tectonics triggered by repeated uplift of a metastable passive margin. *AAPG Bulletin*, **88**, 971–990, https://doi.org/10.1306/02050403061

HUDEC, M.R., NORTON, I.O., JACKSON, M.P.A. & PEEL, F.J. 2013. Jurassic evolution of the Gulf of Mexico salt basin. *AAPG Bulletin*, **97**, 1683–1710, https://doi.org/10.1306/04011312073

JACKSON, M.P.A. & HUDEC, M.R. 2005. Stratigraphic record of translation down ramps in a passive-margin salt detachment. *Journal of Structural Geology*, **27**, 889–911.

JACKSON, M.P.A. & HUDEC, M.R. 2017. *Salt Tectonics: Principles and Practice*. Cambridge University Press, Cambridge, https://doi.org/10.1017/9781139 003988

JACKSON, M.P.A., DOOLEY, T.P., HUDEC, M.R. & McDONNELL, A. 2011. The pillow fold belt: a key subsalt structural province in the Northern Gulf of Mexico. *AAPG Search and Discovery Article*, no. 10329.

KARLO, J.F., VAN BUCHEM, F.S.P., MOEN, J. & MILROY, K. 2014. Triassic-age salt tectonics of the Central North Sea. *Interpretation*, **2**, SM19–SM28.

Krantz, R.W. 1991. Measurements of friction coefficients and cohesion for faulting and fault reactivation in laboratory models using sand and sand mixtures. *Tectonophysics*, **188**, 203–207, https://doi.org/10.1016/0040-1951(91)90323-K

Maillard, A., Gaullier, V., Vendeville, B.C. & Odonne, F. 2003. Influence of differential compaction above basement steps on salt tectonics in the Ligurian-Provencal Basin, northwest Mediterranean. *Marine and Petroleum Geology*, **20**, 13–27.

McClay, K.R. 1990. Extensional fault systems in sedimentary basins: a review of analog model studie. *Marine and Petroleum Geology*, **7**, 206–233.

McClay, K.R., Dooley, T. & Lewis, G. 1998. Analog modeling of progradational delta systems. *Geology*, **26**, 771–774.

Peel, F.J. 2014. The engines of gravity-driven movement on passive margins: quantifying the relative contribution of spreading v. gravity sliding mechanisms. *Tectonophysics*, **633**, 126–142, https://doi.org/10.1016/j.tecto.2014.06.023

Pichel, L.M., Peel, F., Jackson, C.A.-L. & Huuse, M. 2018. Geometry and kinematics of salt-detached ramp syncline basins. *Journal of Structural Geology*, **115**, 208–230, https://doi.org/10.1016/ j.jsg.2018.07.016

Pilcher, R.S., Murphy, R.T. & McDonough Ciosek, J. 2014. Jurassic raft tectonics in the northeastern Gulf of Mexico. *Interpretation*, **2**, SM39–SM55, https://doi.org/10.1190/INT-2014-0058.1

Quirk, D.G., Schodt, N., Lassen, B., Ings, S.J., Hsu, D., Hirsch, K.K. & von Nicolai, C. 2012. Salt tectonics on passive margins: examples from Santos, Campos and Kwanza basins. Geological Society, London, Special Publications, **363**, 207–224, https://doi.org/10.1144/SP363.10

Rossi, D. & Storti, F. 2003. New artificial granular materials for analog laboratory experiments: aluminum and siliceous microspheres. *Journal of Structural Geology*, **25**, 1893–1899, https://doi.org/10.1016/S0191-8141(03)00041-5

Schellart, W.P. 2000. Shear test results for cohesion and friction coefficients for different granular materials: scaling implications for their usage in analog modeling. *Tectonophysics*, **324**, 1–16, https://doi.org/10.1016/S0040-1951(00)00111-6

Talbot, C.J. & Pohjola, V. 2009. Subaerial salt extrusions in Iran as analogues of ice sheets, streams and glaciers. *Earth-Science Reviews*, **97**, 155–183.

Tari, G. & Jabour, H. 2013. Salt tectonics along the Atlantic margin of Morocco. *In*: Mohriak, W.U., Danforth, A., Post, P.J., Brown, D.E., Tari, G.C., Nemčok, M. & Sinha, S.T. (eds) *Conjugate Divergent Margins*. Geological Society, London, Special Publications, **369**, 337–353, https://doi.org/10.1144/SP369.23

Vendeville, B.C. & Jackson, M.P.A. 1992. The fall of diapirs during thin-skinned extension. *Marine and Petroleum Geology*, **9**, 354–371.

Weijermars, R. 1986. Polydimethylsiloxane flow defined for experiments in fluid dynamics. *Applied Physics Letters*, **48**, 109–111, https://doi.org/10.1063/1.97008

Weijermars, R., Jackson, M.P.A. & Vendeville, B.C. 1993. Rheological and tectonic modeling of salt provinces. *Tectonophysics*, **217**, 143–174, https://doi.org/10.1016/0040-1951(93)90208-2

Role of outer marginal collapse on salt deposition in the eastern Gulf of Mexico, Campos and Santos basins

JAMES PINDELL[1,2]*, ROD GRAHAM[3] & BRIAN W. HORN[2]

[1]*Tectonic Analysis Ltd, Duncton, West Sussex GU28 0LH, UK*

[2]*ION E&P Advisors, 2501 CityWest Boulevard, Houston, TX 77042, USA*

[3]*Department of Earth Science and Engineering, Imperial College London, South Kensington Campus, London SW7 2AZ, UK*

J.P., 0000-0001-9098-782X

Correspondence: jim@tectonicanalysis.com

Abstract: Outer marginal collapse (OMC), a recently proposed process by which top-rift and base-salt unconformities formed near sea level may subside rapidly to 2.5–3 km at continental margins as mantle exhumation or seafloor spreading begins, needs further examination. We examine salt deposition at three margins and find that the differing positions and volumes of salt can be related to different durations of salt deposition as OMC and subsequent mantle exhumation proceed. Along NW Florida, salt is thin but deep and is interpreted as having formed at the start of OMC, before drowning further to abyssal depths. In the Campos Basin, salt is thick and extends across tens of kilometres of interpreted exhumed mantle, interpreted as having formed during the entire period of OMC before spreading onto mantle during exhumation. In the Santos Basin, salt is thick and extends across c. 100 km of interpreted exhumed mantle and/or oceanic crust, arguably requiring 'lateral tectonic accommodation', whereby salt deposition persists near global sea level across the conjugated salt basin during mantle exhumation beneath mobile salt. The supposition that OMC can account for salt deposition in three different basins without invoking problematic 1.5–2 km-deep subaerial depressions provides further support for the process.

As Rowan (2014) has pointed out, evaporites can be deposited as any stage of rifting and passive-margin formation. Our concern here is the creation of accommodation space for the thick sequences of evaporite in 'mega-salt basins', such as the Gulf of Mexico, and the Campos and the Santos basins, large parts of which have planar base-salt unconformities and thus post-date all or much of the rifting. The deposition of these salt sequences seems to have been rapid. Salvador (1987, 1991) and Davison *et al.* (2012) have suggested that up to 5 km of salt have been deposited over minimally faulted post-rift sag basins or above the post-rift unconformity itself, apparently in less than 5 myr. These rates are much faster than thermal subsidence can account for if deposition occurs within a few hundred metres of global sea level, which is commonly assumed. Even the youngest oceanic crust subsides thermally only about 1 km in 9 myr (Parsons & Sclater 1977).

Three theories for salt deposition in mega-salt basins are most commonly considered. The 'air-filled hole' model (Fig. 1a) presumes large subaerial depressions with the depositional surfaces of synrift or sag sections at *c.* 1.5 km below global sea level, produced by extension and thermal subsidence while largely starved of sediment. Accommodation for salt thus predates salt deposition, and the salt is progressively added over a sill by episodic marine flood and dessication cycles (Burke & Sengör 1988; Karner & Gambôa 2007), or more steadily through a semi-permeable sill (Warren 2006). A second theory presumes a similarly formed depression that then becomes marine but highly restricted, such that salt is initially a deep-water, hypersaline deposit whose deposition shallows upwards as salt is progressively deposited (not shown in Fig. 1, but consider a basin like that shown in Fig. 1a subsequently being inundated with hypersaline seawater and the resulting deep-water salt deposition). A third theory (Fig. 1b) presumes salt deposition near (<500 m) global sea level, above synrift and sag sections that keep the basin mostly filled, after which rapid basinwards tilting of the future rifted margins of the order of 2 km provides a rapid, tectonic driver for the majority of salt accommodation, behind a restrictive sill to the world's oceans (Pindell *et al.* 2014). In the case of the Gulf of Mexico, the sill was onshore Mexico and/or the Straits of Florida, and for the Campos and Santos basins, the sill was the Walvis–Río Grande magmatic ridge. Pindell *et al.* (2014) referred to this tilting process as 'outer marginal collapse'.

It is not our intention in this paper to try to prove one of these models over the others: no doubt, all of them, or combinations of them, have merit in

From: MCCLAY, K. R. & HAMMERSTEIN, J. A. (eds) 2020. *Passive Margins: Tectonics, Sedimentation and Magmatism*. Geological Society, London, Special Publications, **476**, 317–331.
First published online March 6, 2018, https://doi.org/10.1144/SP476.4
© 2018 The Author(s). Published by The Geological Society of London. All rights reserved.
For permissions: http://www.geolsoc.org.uk/permissions. Publishing disclaimer: www.geolsoc.org.uk/pub_ethics

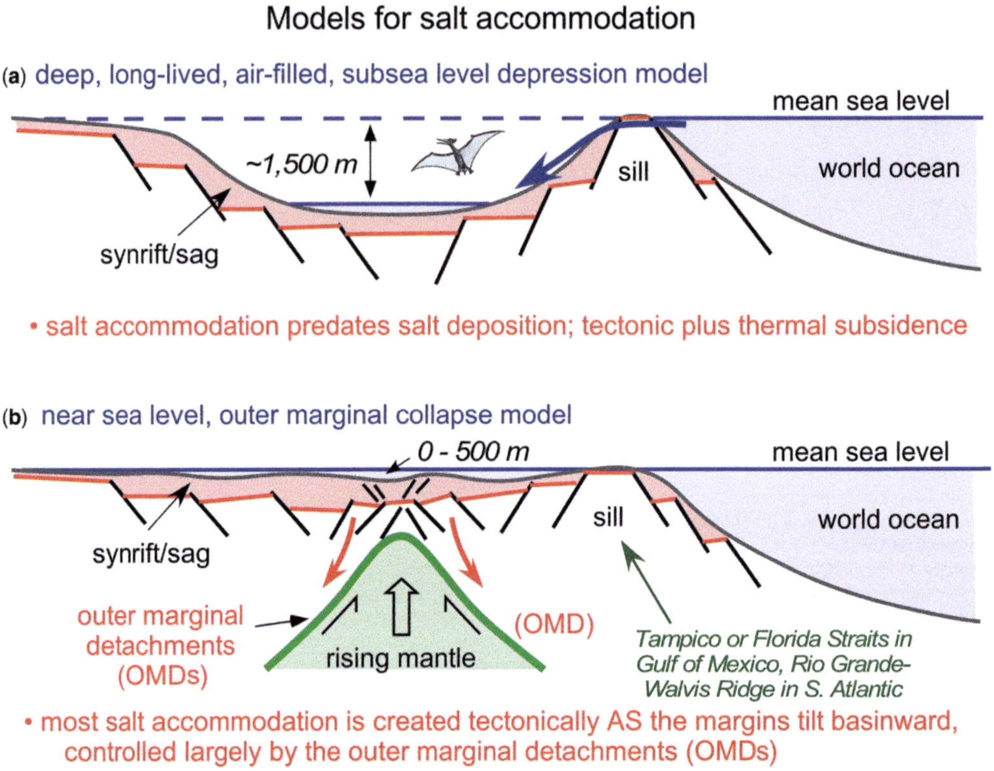

Models for salt accommodation

(a) deep, long-lived, air-filled, subsea level depression model

- salt accommodation predates salt deposition; tectonic plus thermal subsidence

(b) near sea level, outer marginal collapse model

- most salt accommodation is created tectonically AS the margins tilt basinward, controlled largely by the outer marginal detachments (OMDs)

Fig. 1. Two alternative models for the depositional setting of syn-exhumation mega-salt basins. (**a**) The 'air-filled, subsea-level hole' hypothesis. (**b**) The 'outer marginal collapse' hypothesis, in which rapid basinwards tilting of the nascent margins between the rifting and the drifting stages provides tectonically controlled accommodation for salt, in a basin setting behind an unrelated tectonic sill separating the basin from the world's oceans (Pindell *et al.* 2014). (b) is drawn to portray a time prior to, or at the initiation of, outer marginal collapse.

different basins around the world. Instead, we wish to demonstrate how the concept of outer marginal collapse (OMC) can be applied to salt occurrence in three important mega-salt basins (the eastern Gulf of Mexico, and the Campos and Santos basins), and to propose that salt thickness and position relative to crustal or basement structure in these basins can be explained simply by considering the timing and duration of salt deposition relative to the timing of OMC.

Concept of 'outer marginal collapse', as deduced from the NW Florida margin

Before we can relate salt deposition to models of accommodation, we must first define OMC. 'Outer marginal collapse' was first defined from interpretation of deeply penetrating 2D seismic reflection lines along the NW Florida margin of the NE Gulf of Mexico Basin (Pindell *et al.* 2014). To outline the concept for our purposes here, the main architectural elements of the NW Florida margin are shown in Figure 2a, and these elements and surfaces are shown schematically in Figure 2b for magma-poor margins in general, following the terminology of Pindell *et al.* (2014). The noted features are also applicable to most of the northern Yucatán salt-bearing margin along the southern side of the Gulf of Mexico Basin, which is the conjugate to NW Florida (Pindell *et al.* 2016). We acknowledge that there is a magmatic component within both the NW Florida (Eddy *et al.* 2014) and parts of the Yucatán margins (unpublished ION Geophysical data), but this occurs below the base-salt or, where present, the top-rift/sub-sag unconformity. The Gulf of Mexico had local centres of synrift magmatism, but its margins were magma-poor (i.e. magma-normal) by the time of final break-up and onset of seafloor spreading, as judged by outer marginal detachments (OMDs) rising to the top of the oceanic crust, which does not occur at margins like Argentina where break-up

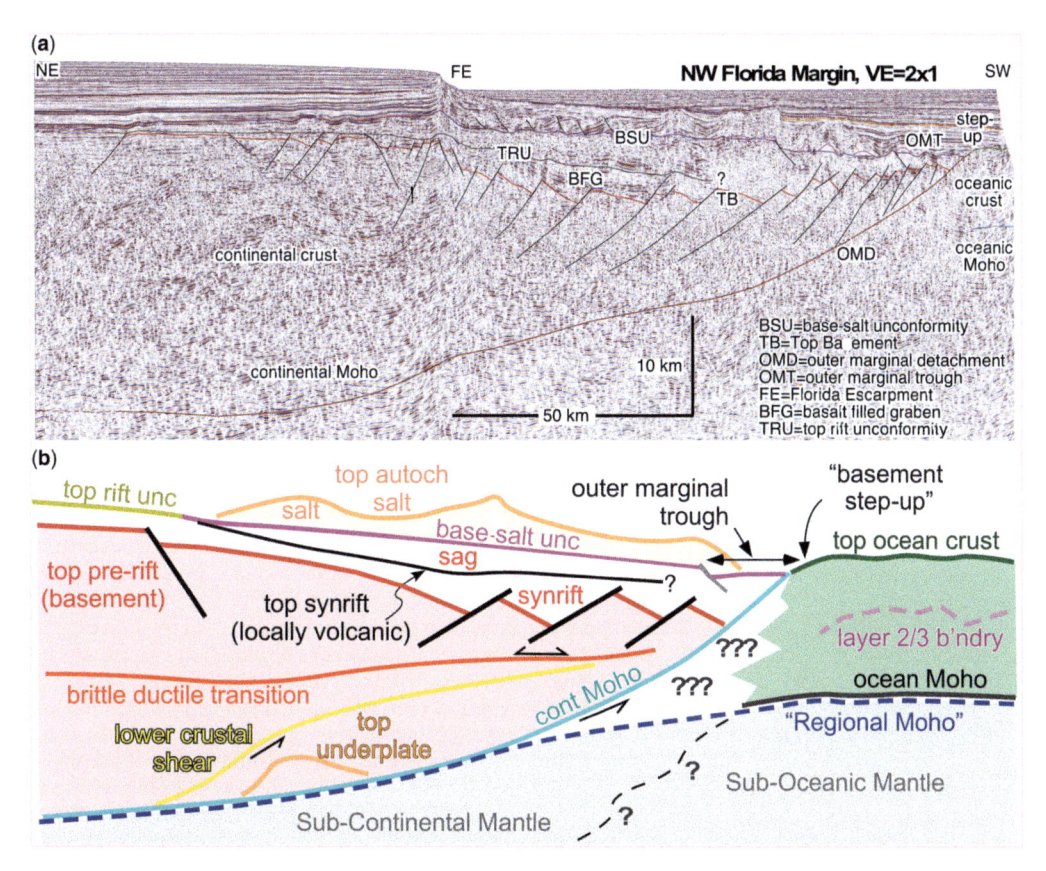

Fig. 2. (**a**) Sample seismic line from the NW Florida basin margin, partly interpreted for crustal and rift-related surfaces. The 'step-up' is the rise in the basement surface from the rifted continental crust or exhumed mantle to the top of the oceanic crust. The question mark denotes the point at which it becomes difficult to trace the top-rift unconformity (TRU) basinwards. (**b**) Schematic surface diagram for a typical magma-poor margin. 'Basement faults' (black) enter the top pre-rift surface; 'sedimentary section faults' (grey) do not. 'Regional Moho' (dashed) is the smoothed composite of the oceanic and continental Moho. 'Layer 2/3 boundary' refers to layers 2 and 3 of the oceanic crust (top gabbroic intrusive rock). '???' is probably serpentinized sub-continental mantle, suspected to be heavily intruded with gabbro, diabase during exhumation at this level. Boundary between sub-oceanic (harzburgite) and sub-continental (llerzolite) mantle is rarely imaged. 'Top underplate' probably marks gabbroic bodies chilled/ subcreted to the base of the crust. 'Lower crustal shear' may be detachments allowing lower-crustal necking (extension). VE, vertical exaggeration.

was magma-rich. Conversely, other margins (e.g. Argentina) had a magma-poor rift stage (note the Colorado and San Jorge basins), but a magma-rich break-up stage (Becker *et al.* 2014). This distinction is, in our opinion, worth making and can circumvent misunderstanding.

Figure 2 shows the 'outer marginal detachment' as an extensional shear zone along the Moho which rises from under thinned continental crust and breaks out (breaches the top of basement) where the mantle becomes exhumed at the palaeo-seafloor. We interpret only narrow zones of exhumed mantle along NW Florida, whereas western Iberia (Peréz-Gussinye 2012) and, locally, eastern India (Pindell *et al.* 2014) have wider areas of apparent exhumed mantle. 'Outer marginal troughs' (Fig. 2) at magma-poor margins are believed to be floored largely by exhumed mantle between the outermost rifted continental and the oceanic crusts (Pindell *et al.* 2014). Their width reflects the amount of mantle exhumed at the OMD. Where continent–ocean transitions are more magma-rich, the outer marginal troughs (OMTs) are filled with substantial thicknesses of magmatic rock, often expressed as seawards-dipping reflectors and their intrusive underpinnings (Paton *et al.* 2017). The OMDs do not rise and break-out to the seafloor; instead, the continental Moho tends to be continuous with the oceanic Moho beneath the magmatic fill of the OMTs. Further, magmatic underplating is common beneath the outermost continental crustal fragments, sandwiching the outer continental crust between intrusive (below) and extrusive magmatic rock (above) (e.g. see fig. 4 of Pindell *et al.* 2014).

Above the thinned continental crust in NW Florida (Fig. 2), we interpret a faulted and locally magmatic synrift section up to several kilometres thick, which is covered by a 'sag' section with minimal faulting comparable to those developed at the Brazilian and Angolan margins. The sag, in turn, is capped by a planar, minimally faulted, base-salt unconformity that deepens basinwards. Although partially welded, this surface spans the entire distance from the near-shore Late Jurassic (post-salt) open-marine onlap limit (Dobson & Buffler 1997) to the edge of the oceanic crust, suggesting that the basin was filled with salt to global sea level at the end of salt deposition (Hudec *et al.* 2013). At its oceanwards limit, the base-salt unconformity is commonly as deep as, or deeper than, the top of the adjacent oceanic crust, for which backstripping suggests a 2.5–3 km depth of subaqueous plate accretion (A.B. Watts pers. comm. in Pindell *et al.* 2014). If any significant faults cut the unconformity, they are usually limited to within 30 km of the oceanic crust and only rarely cut into the basement beneath the OMT's sedimentary section (Fig. 2). Thus, salt is effectively a postrift unit at this margin, as seems to be true for the entire Gulf of Mexico. Updip, salt along the base-salt unconformity has slipped basinwards to facilitate the formation of salt-cored folds in the OMT at the downdip end of the gravitational system. The salt has arguably flowed across any faults that cut the base-salt unconformity (see also Hudec *et al.* 2013; Pilcher *et al.* 2014).

Because they do not often enter the basement, and therefore do not thin the crust, we judge that the few faults cutting the base-salt unconformities control regional subsidence very little. We also judge that such faults, along with downslope slip, on the base-salt unconformity are subsidiary accommodation structures that form in response to the larger-scale process of extensional shearing and mantle exhumation along the OMD (Pindell *et al.* 2014). The association of mantle exhumation, OMC, and downslope salt migration and toe folding, all prior to the onset of seafloor spreading, is supported by the observation that the oldest abyssal plain strata onlap onto both the deformed salt and the adjacent oceanic crust (Fig. 2a). We believe that OMC can be considered as a distinct stage of continental margin formation, separating the traditional rifting and drifting stages. It is probably analogous to the exhumation stage of Perón-Pinvidic & Manatschal (2009), occurring when continental thinning by faulting has ceased and all displacement is concentrated on a major detachment fault, the OMD. This displacement is almost certainly facilitated by serpentinization (Reston 2007). We consider that the OMD effectively serves as an interim plate boundary between the rifting and drifting stages.

Salt in the NE Gulf of Mexico reaches the Upper Jurassic (post-salt) open-marine landwards onlap limit, and is interpreted as always having been thin (<300 m) based on the paucity of overlying stratal disruption during halokinesis (Fig. 2a). Pindell *et al.* (2014), therefore, suggested that the base-salt unconformity, all the way to its basinwards limit, was little more than 300 m below global sea level at the end of salt deposition. However, the basinwards limit of the base-salt unconformity lies roughly at the depth of the top of the oceanic crust, which is 'normal thickness' and was accreted by seafloor spreading at the commonly observed depth of 2.5–3 km subsea. Figure 3 highlights this apparent dilemma. Figure 3a shows the margin as deep-water sedimentation begins to bury the ocean–continent transition as the oceanic crust began to form. No observable faults cut upwards from basement into the deep-marine succession onlapping the OMT (e.g. Fig. 2a). Figure 3b shows the margin with a thin salt layer (<500 m) lying near sea level, with the base-salt unconformity always within 500 m of the global sea level, and with the future oceanic crust dashed at the depth of plate accretion. The question mark in Figure 3b represents the dilemma of how the shallow base-salt unconformity collapses (tilts basinwards) to the depth of 2.5 km prior to the formation of the oceanic crust. Thermal subsidence of the rifted margin in Figure 3b cannot be invoked because, first, more than 36 myr would be required to achieve 2 km of thermal subsidence, and, secondly, the rifted margin should not subside faster thermally than the oceanic crust, and thus should not be able to acquire the same depth by that means. Faults could, of course, generate the necessary rapid tectonic subsidence rates, but no basement-involved faults are apparent except for the OMD itself. Figure 2a shows that structure and similar structures in other (ION Geophysical) datasets from magma-poor margins are almost always imaged (e.g. fig. 1 of Pindell *et al.* 2014, from east India).

Inherent in this model for NW Florida is that salt deposition ceased in the Gulf of Mexico when salt was still thin along NW Florida (as depicted in Fig. 3b), and that OMC then proceeded to drown the thin salt to abyssal depths prior to the onset of seafloor spreading (Fig. 3a). Whether salt deposition, flanked by Norphlet Formation coastal/aeolian sand systems (Lovell & Weislogel 2010), occurred prior to the beginning of collapse, or records the start of the collapse, cannot be directly determined. However, the thin salt along NW Florida is depositionally

(a) Configuration since onset of seafloor spreading

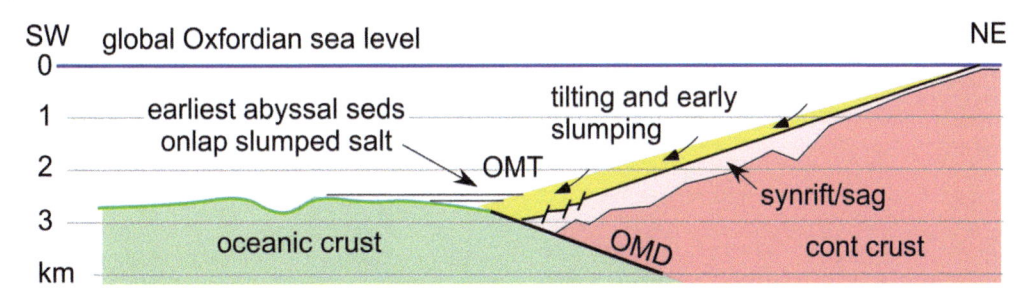

[Cross-section diagram showing: SW to NE profile, 0-3 km depth. Labels: "global Oxfordian sea level", "earliest abyssal seds onlap slumped salt", "tilting and early slumping", "OMT", "oceanic crust", "OMD", "synrift/sag", "cont crust"]

(b) Configuration during salt deposition (Yucatán conjugate not shown)

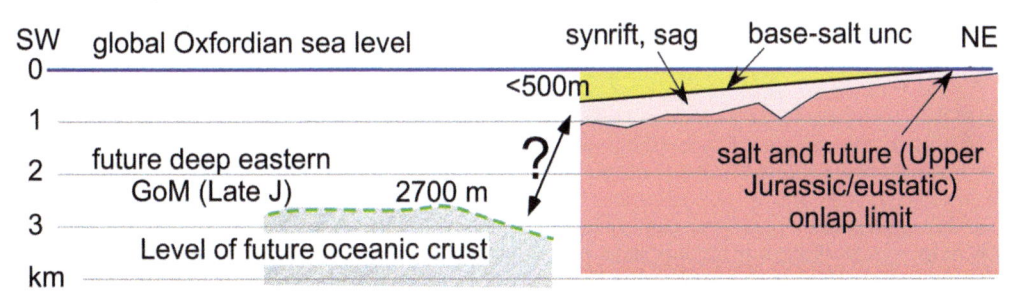

[Cross-section diagram showing: SW to NE profile, 0-3 km depth. Labels: "global Oxfordian sea level", "synrift, sag", "base-salt unc", "<500m", "future deep eastern GoM (Late J)", "2700 m", "Level of future oceanic crust", "salt and future (Upper Jurassic/eustatic) onlap limit"]

Fig. 3. (a) General margin architecture and salt relationships after seafloor spreading begins, simplified from Figure 2, for the eastern Gulf of Mexico (GoM) margin with thin salt. (b) Schematic reconstruction of (a) for the time of salt deposition, prior to seafloor spreading, as argued herein (shallow, thin salt near global sea level).

continuous with much thicker salt offshore Alabama, where the base-salt unconformity is also planar, so we suspect that the salt in Florida records an eastwardly diachronous onset of OMC (Pindell *et al.* 2014).

Using the NW Florida margin as a reference, we conclude that continental margins can experience rapid and large tectonic subsidence that is separate from and post-dates continental rifting, presumably as the margin sloughs off the sub-continental mantle as the latter rises along OMDs. Further, we believe OMDs serve as interim plate boundaries, after which further plate separation takes place by seafloor spreading whenever rising asthenosphere is able to penetrate the exhumed sub-continental mantle.

Alternatives to this 'near-sea-level' salt deposition model for margins with thin salt (Figs 1b & 3) are: (1) that salt might have been a deep-water deposit, forming a roughly isopachous blanket across the entire margin; or (2) that the salt in the eastern Gulf of Mexico represents only the onlapping fringe of a deepening water column that filled the Gulf of Mexico from bottom to top, such that the onlapping salt would be time equivalent to deeper-water (abyssal) facies outboard. In our opinion, neither of these deep-water scenarios seems likely, and the associated idea that the base-salt unconformity was situated far below (>1.5 km) global sea level prior to salt deposition does not appear to have been the case for the Kwanza Basin, which also has thick salt over a planar base-salt unconformity. There, it has been reported that recent drilling in Block 38 of the outer Kwanza Basin penetrated a thin, open-marine sedimentary section beneath the salt. These sediments lie above the post-rift unconformity, are very little faulted and were interpreted to have been deposited in less than 600 m of open-marine water (Sharp *et al.* 2016). This is evidence for a near-sea-level starting point for salt deposition; although no open-marine strata have been reported beneath the base-salt unconformity in the Gulf of Mexico with which to allow a similar analysis, the Kwanza results appear to fit best with the OMC model of Figures 1b & 3.

We will now explore the Campos and Santos basins to show how the duration of salt deposition relative to OMC can explain the great differences seen in the volume and position of salt relative to basement structure.

Architecture of the Campos and Santos rifted margins

Campos Basin margin

The basement architecture of the Campos margin (Fig. 4) is similar to that of the NW Florida margin and closely resembles the geometries of Figure 2b, with two primary differences. First, there is a significant gap between the interpreted oceanic crust and the continental crust with its planar base-salt unconformity. In this gap, the basement is deeper than the oceanic crust, and its surface expression on seismic is irregular ('rough'). Salt appears to rest directly on the basement, but there is no indication if this is a depositional or a structural (halokinetic) contact. The roughness of this basement, the apparent lack of a subsalt sedimentary section and unconformity, and the fact that basement sits 1–2 km deeper than in the adjacent oceanic crust suggest an interpretation of exhumed mantle. Zalán et al. (2011) also interpreted this approximate zone as exhumed mantle.

Second, the thickness of salt in the Campos Basin is much greater than in the NW Florida margin, with the most voluminous area of salt continuing basinwards across the area of interpreted exhumed mantle to the edge of the oceanic crust. If one assumes that the thickness of salt over the exhumed mantle is depositional, it directly follows that the 'air-filled hole' model (Fig. 1a) might apply, because the mantle was never likely to have been near sea level. However, if the salt was emplaced tectonically by gravitational processes (spreading), this does not necessarily apply. Following this latter view, we suggest, first, that the salt is thick over the outer continental crust because it was deposited during the entire period of OMC, which provides a c. 2.5 km driving subsidence at the position of the outer limit of continental crust. When the loading effect of salt is considered, this driving tectonic subsidence can lead to depositional salt thicknesses approaching 5 km, because salt has roughly twice the density of seawater. Secondly, the lack of subsalt sediment and a base-salt unconformity over the interpreted exhumed mantle can be interpreted to indicate that the salt there was emplaced by gravitational spreading. Considering continental break-up from the perspective of the Campos–Kwanza conjugate basin, the latter of which we also interpret as having a zone of exhumed mantle, the salt was likely to have spilled onto the widening zone of exhumed mantle from both sides, prior to final salt separation in the two margins.

Santos Basin margin

Interpretations of the basement structure, composition, and rift- and break-up-related magmatism in the more complicated Santos Basin vary considerably (e.g. Jackson et al. 2000; Cobbold et al. 2001; Meisling et al. 2001; Davison 2007; Zalán et al. 2011; Davison et al. 2012; Klingelhoefer et al. 2015). For our purposes here, we have chosen to present a seismic line that closely parallels the plate separation direction, thus simplifying the overall structure, and which crosses the central part of the sag-basin section (Fig. 5). Like the NW Florida and Campos margins, the Santos Basin has, beneath the shallow continental shelf in the west, a

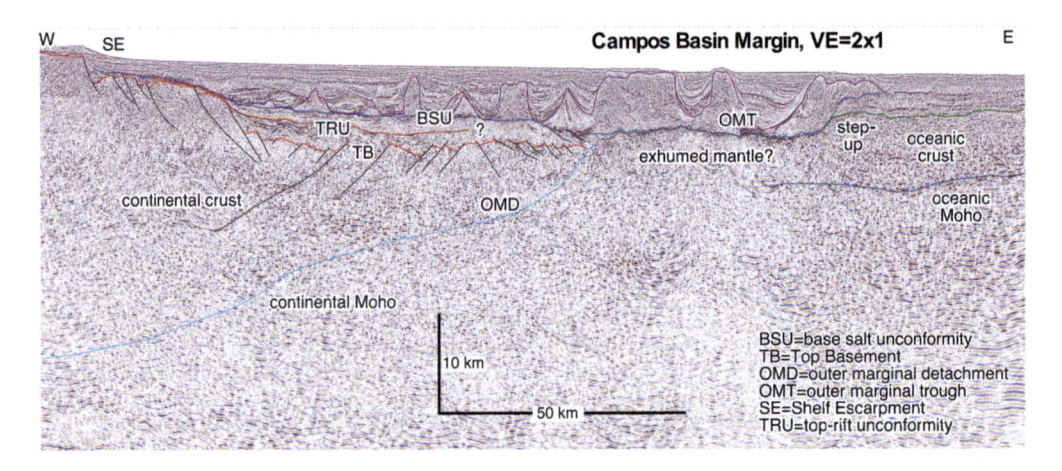

Fig. 4. Sample seismic line from the northern Campos Basin margin, showing crustal and rift-related surfaces crossing from the shelf to the deep basin. The 'step-up' is the rise in the basement surface from exhumed mantle to the top of the oceanic crust. The question mark denotes the point at which it becomes difficult to trace the top-rift unconformity (TRU) basinwards. VE, vertical exaggeration.

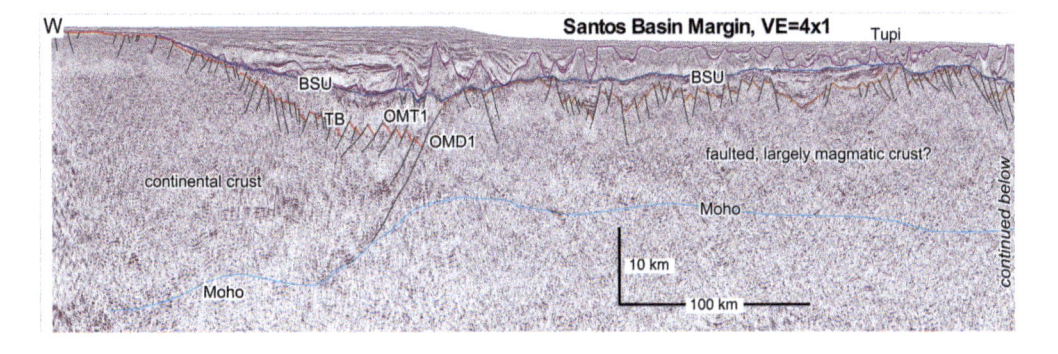

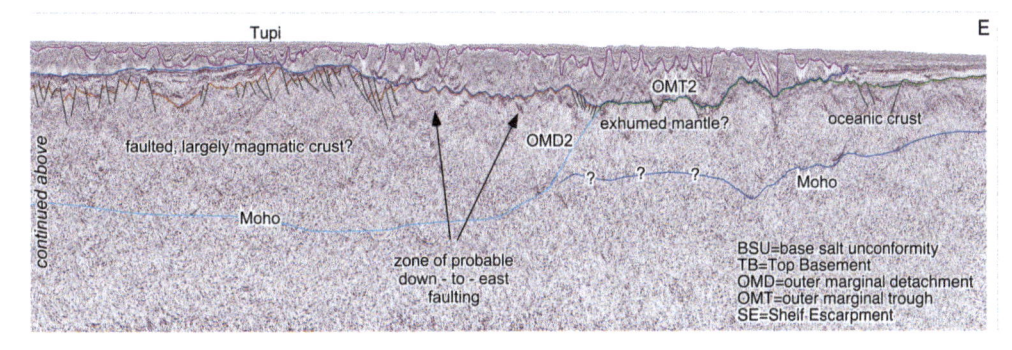

Fig. 5. Sample seismic line from the central Santos Basin margin, roughly parallel to the margin extension direction, showing crustal and rift-related surfaces crossing from the shelf to the deep basin. OMD1 and OMD2 are primary marginal detachments, and OMT1 and OMT2 are the associated outer marginal troughs. The Tupi producing area is shown for reference. VE, vertical exaggeration.

basinwards-tilting planar base-salt unconformity with a synrift section and thick, faulted continental crust beneath it; a well-defined outer marginal detachment (OMD1 in Fig. 5); and an abrupt change in the dip of the base-salt unconformity where this detachment breaks to the palaeosurface, forming an apparent OMT. As Davison et al. (2012) has already suggested, it is likely that salt has slipped down the base-salt unconformity towards this apparent OMT. However, the similarity with Florida and Campos ends there, because the 300–400 km-wide, drowned Sao Paulo Plateau lies oceanwards of this proximal OMD and its associated trough.

The Sao Paulo Plateau (between OMD1 and OMD2 in Fig. 5) has a crustal thickness of 10–20 km, a characteristic that has led to uncertainty over whether basement is continental or magmatic (e.g. Karner & Gambôa 2007). Extensional faults exist locally but portions of the plateau show little apparent sign of rotational block faulting and display continuous, strong undulating reflectors that seem more characteristic of magmatic plateaux. It is our understanding that through 2016, every well drilled to 'basement' encountered mafic volcanic rock rather than continental crust, and seismic refraction work suggests crustal velocities somewhere between those of continental and oceanic crust (Klingelhoefer et al. 2015). More regionally, if the Sao Paulo Plateau were entirely continental, then the southern South Atlantic Ocean (south of the Walvis and Río Grande fracture zones) would not be restorable to a pre-break-up configuration without invoking large, transcontinental transform offsets (Moulin et al. 2010), for which there is very little support. We view the Sao Paulo Plateau as having thick, mainly magmatic crust that shows internal extension (like Iceland: Jóhannesson & Sæmundsson 1998), emplaced during initial continental separation but with some dispersed continental blocks or ribbons within it, implying a disorganized rifting history involving multiple or shifting sites of crustal break-up. The extensional faulting seen in Figure 5 was accompanied by 'synrift' but highly mafic sedimentation in probable subaerial settings, in turn overlain by a minimally faulted, lacustrine or marginally marine sag basin, itself truncated to form the base-salt unconformity prior to salt deposition. Near and west of the Tupi Field in Figure 5, the base-salt unconformity is minimally faulted except for the primary inflection at the OMD1 breakout. The gentle westwards tilt of the base-salt unconformity in this western Sao Paulo Plateau zone is probably due to longer-term sediment loading caused by the overlying wedge of continental shelf strata, and is probably not an original feature of the margin (also noted by Davison et al. 2012), but the OMD1 inflection at the top of the crust probably always existed.

Along the outer Sao Paulo Plateau (east of Tupi in Fig. 5), southwards to about 26°S (this discussion of the plateau should not be assigned to the area south of 26°S), basement descends eastwards to what can be interpreted as a second outer marginal detachment (OMD2). In the depth image of Figure 5, the base-salt unconformity, although undulating, dips regionally eastwards towards OMD2, and there is little or no apparent synrift or sag section beneath the unconformity. Although the interpretation in Figure 5 suggests faulting as the cause for the undulation, this is not entirely clear: the undulating surface could be due to incomplete velocity correction for the overlying salt in the depth imaging, or to an igneous basement with significant relief. Thus, it is difficult to judge how far east an originally planar base-salt unconformity might have continued. Nevertheless, because of the great volume of salt basinwards from this portion of the basement surface, we judge that salt was probably deposited directly upon much, if not all, of it.

At the eastern limit of this second eastwards-deepening basement zone, beyond the OMD2 breakout, there is a zone of deep, rough basement with no apparent subsalt section, and farther east lying slightly shallower is typical oceanic crust with an imaged Moho. We cannot be certain if the deep zone between the oceanic crust and the OMD2 break is exhumed mantle (as in Zalán et al. 2011) or oceanic crust. If the former, the exhumed mantle may have local mound-forming extrusive igneous centres. Very thick (up to 7 km), recently diapiric, highly deformed evaporite with associated mini-basins overlies the basement here, with the thickest salt in Figure 5 overlying 'outer marginal trough 2'. The salt continues east from the zone of interpreted exhumed mantle over the first 40–50 km of the oceanic crust before ramping up over younger sedimentary section. Large extensional gaps in the original salt layer occur updip over much of the plateau, with strong shortening downdip in the toe of the gravitational system (OMT2 in Fig. 5).

The area of thickest salt corresponds to the areas of probable downfaulted crust (east of Tupi) and possible exhumed mantle in OMT2, the depth of which was likely to never to have been shallower than 2.5–3 km below sea level. The pre-salt depositional surface in the conjugate outer Kwanza Basin was relatively shallow, apparently within 600 m of global sea level (Cowie et al. 2015; Sharp et al. 2016, see above). From these observations, we conclude that the outer Santos Basin (east of Tupi) underwent a subsidence similar to that suggested for the NW Florida and Campos margins (i.e. Fig. 3). However, we are uncertain as to what degree the basement faulting contributed to the subsidence.

Regionally, we view the Sao Paulo Plateau as a remnant of a more regional complex of strong magmatism that initially occupied a large part of the growing gap between the continental crusts of South America and Africa, with an eventual waning of magmatism and break-up of the plateau that mimicked a second rifting episode which produced OMD2 and OMT2 in Figure 5. This second break-up event appears to have been accompanied by local mantle exhumation, forming the deepest areas of basement along the margin, even though the margin had previously been magma-rich. If so, and as with NW Florida and Campos, mantle exhumation eventually gave way to more normal seafloor spreading. At the latitude of the Sao Paulo Plateau, the line of this second break-up was off basin centre, much closer to the African margin, leaving most of the magmatic plateau on the South American side (Sao Paulo Plateau). Integrating the margin history with the age constraints discussed by Karner & Gambôa (2007) and Davison et al. (2012), salt deposition (116–111 Ma) post-dated continental break-up and the Hauterivian–Barremian formation of the largely magmatic Sao Paulo Plateau, and deposition of the Barremian–Early Aptian sag section. The salt was thus deposited contemporaneously with the break-up of the plateau from Africa, accomplished by motion on OMD2 in the Late Aptian, and prior to our inferred age of Early Albian for the onset of seafloor spreading (c. 112–110 Ma). Limestones immediately overlying the salt in the Santos and Campos basins, at least where drilled, are of a shallow-water, open-marine nature of earliest Albian–Cenomanian age, with no Aptian section overlying salt (Davison et al. 2012; unpublished well data).

Depositional model for mega-salt basins, integrating outer marginal collapse

If accommodation space for salt in our basin examples can be created or augmented by OMC, then the salt section is the sedimentary record of the collapse. Potentially, only salt can provide such a record because salt can be deposited so much faster than other types of sediment in most situations. In contrast, OMC at most salt-free margins (i.e. those that are not isolated from the world oceans as they form) leaves little geological record except for rapid drowning and stratal condensation. Outer marginal collapse was suggested by Pindell et al. (2014) to account for the rapid, apparently multi-kilometric Middle Jurassic drowning of the Briançonnais carbonate platform in the western French Alps (Lemoine 1975, 1988; Rudkiewicz 1988; Lemoine et al. 2000; Manatschal et al. 2009; Mohn et al. 2010; Graciansky et al. 2011).

Figure 6 shows eight sequential stages of margin development that can be applied to the Santos Basin (north of 26°S), certain stages of which also apply to the Florida and Campos margins. Figure 6a–e portrays the two rifting events in which the depositional surface in the basin remains within 500 m of global sea level, involving the OMD1 and the OMD2 and associated structures. The thinner ruled crust denotes the dominantly magmatic crust of the Sao Paulo Plateau, but equally represents thinned continental crust for the Campos and NW Florida margins. Figure 6e–h can be related to the cessation of salt deposition, due to unspecified oceanographic/climatic factors, in our three examples: NW Florida (Fig. 6e), and the Campos (Fig. 6f) and Santos (Fig. 6g) basins, as discussed below. Figure 6h shows the Santos Basin after salt deposition had ceased and the original salt mass had spread gravitationally onto the young oceanic crust as seafloor spreading ensued (see Fig. 5). Primary differences in salt occurrence (i.e. salt thickness and position relative to crustal or basement features) in the three basins can be explained simply by considering the timing and duration of salt deposition relative to the timing of OMC.

For the NW Florida margin, we have already made the case that salt deposition ceased early in the period of OMC. This resulted in a thin salt layer (<300 m thick) tilting quickly basinwards to the 2–3 km depth of accretion for the adjacent oceanic crust. Thin salt deposited near sea level was rapidly drowned; a drowning that presumably also affected other depositional units associated with the salt, such as the more basinwards parts of the Norphlet sandstone. In addition, the zone of exhumed mantle in NW Florida is so narrow and the salt was so thin that the downslope movement of the sedimentary section on salt merely led to the early creation of some open salt-cored folds that were buttressed from further migration onto the oceanic crust by the 'basement step-up' (Fig. 2a).

In the Campos Basin, the original salt was much thicker than in NW Florida, up to 3–4 km judging by the enveloping surfaces of salt-cored anticlines and the large volumes of salt remaining today (Figs 4 & 6f). We suggest this relatively greater thickness is due to salt deposition persisting for the entire period of OMC, rather than just the start of it, thereby providing a relatively larger tectonic driver for salt accommodation (c. 2 km). The larger tectonic driver is required given the great thickness and apparent time interval of deposition (116–111 Ma: Davison et al. 2012). The only apparent imaged structure that could be responsible is the OMD.

There does not seem to be any sediment separating the interpreted zone of exhumed mantle and the base of salt, and, as has been shown, the salt is most voluminous in the area of interpreted exhumed mantle. However, the salt is still not as voluminous as would be expected if exhumed mantle had been buried by salt all the way to the former sea level (which would suggest a 5.3 km original thickness, if 3 km of water were replaced by salt). This suggests that salt flowed early onto the exhumed mantle, prior to any significant sedimentation there, such that the salt–mantle contact is halokinetic, not depositional. When considered in conjunction with the conjugate Angolan margin, we suggest that salt deposition ceased in the Campos–Angolan conjugate basin close to the end of OMC (the stage shown in Fig. 4f) but probably before mantle exhumation began. Subsequently, as mantle exhumation continued between the two future margins, salt spread gravitationally and laterally to match the basement extension without being replenished at shallow levels. Since the base-salt unconformities on both margins dipped towards the basin centre (Fig. 4f), salt was able to migrate downslope and form the thickest part of its isopach directly over the area of exhumed mantle. However, with further continental separation and mantle exhumation, the salt would eventually thin and become separated into the two conjugate salt basins; we speculate that salt separation occurred roughly when seafloor spreading began. The Red Sea today is at an analogous stage to the early salt separation in the Campos–Angola conjugate basin; Augustin et al. (2014) showed the breakout of basement between gliding salt glaciers into basement deeps, with imminent separation of salt from the two sides. However, the youngest gravity gliding in the Campos Basin (Neogene–Present) has emplaced salt over the fringe of the oceanic crust and some of its autochthonous sedimentary cover (Fig. 4).

In the Santos Basin, the mappable area of the planar subsalt unconformity lies quite far from the area of interpreted oceanic crust. Between them lies: (1) a zone of possible exhumed mantle about 60 km wide on our sample seismic line (typically wider than in Campos) that was likely to have never been shallow, and that has no imaged sedimentary section between the mantle and salt; and (2) a zone of inclined, probably magmatic crust that may be incrementally faulted down to the east (Fig. 5). Although it is difficult to demonstrate with our dataset, we consider that much of this area of possible faulted crust was originally an eastwards extension of the more westerly parts of the Sao Paulo Plateau, with a base-salt unconformity that was analogous to that at Tupi. As in the Campos Basin, the thick salt over the area of possible exhumed mantle is interpreted to be allochthonous, with a halokinetic contact between the salt and the mantle. However, in the Santos Basin,

(a) Pre-rift stage, 35 km crust, 90 km mantle.
  • shown as passive rift, no thermal uplift, but some could have occurred.

(b) First synrift stage (Neocomian-Barremian).
  • synrift magmatism, creation of Sao Paulo Plateau (highly magmatic).
  • synrift deposits, deposition near sea level and filling most graben, except for basement highs.
  • extended zone comprises variable proportions of magmatic and continental crust.
  • first outer marginal detachment (OMD1) formed in Santos Basin.

(c) Sag basin stage (Barremian-Early Aptian).
  • deposition of sag section on top-rift unconformity, subsidence mechanism probably largely thermal.
  • extension possibly migrating to central rift axis.
  • depositional setting fluvial-lacustrine, isolated from sea, but depositional surface near sea level (<500 m).

(d) Incipient breakup stage (mid-Aptian).
  • outermost sag sections and some underlying section uplifted on both margins, local volcanism.
  • uplift probably formed a discontinuous 'central' outer high.
  • associated with asthenospheric rise and possible dyking.
  • can be viewed as onset of second rifting event in Santos.

(e) Outer marginal collapse and initial salt stage (Late Aptian).
  • extension focuses on outer marginal detachments, OMD2 in Santos.
  • outer marginal detachments break through synrift, sag, to salt.
  • tilts margins and separates sag basin into two halves.
  • provides rapid accommodation for salt.
  • salt precipitation matches space created by outer marginal collapse, maintains depositional surface near sea level over entire basin.
  • In NW Florida, salt deposition ends early in the history of collapse, putting thin salt in deep water.

(f) Break-up stage, with salt deposition keeping pace with collapse.
  • salt deposition continues at sea level, concurrent with salt flow into central zone of extension as base salt becomes tilted.
  • plate divergence occurring on 'outer marginal detachments'.
  • Rapid subsidence and salt accumulation is vertical, but beginning to become lateral (arrows in salt) as mantle exhumation begins.

(g) Mantle exhumation stage.
  • irregular mantle/local igneous basement widens.
  • salt deposition continues, but moves concurrently onto exhuming mantle by 'lateral tectonic accommodation', to keep basin filled with salt.
  • exhumed mantle is loaded by salt, up to 6 km thick, but this is not a depositional thickness.
  • salt deposition ceases during mantle exhumation.
  • continuous spreading ridge will nucleate as mantle exhumation wanes, asthenosphere approaches surface.
  • salt deposition ending but lateral flow continues.

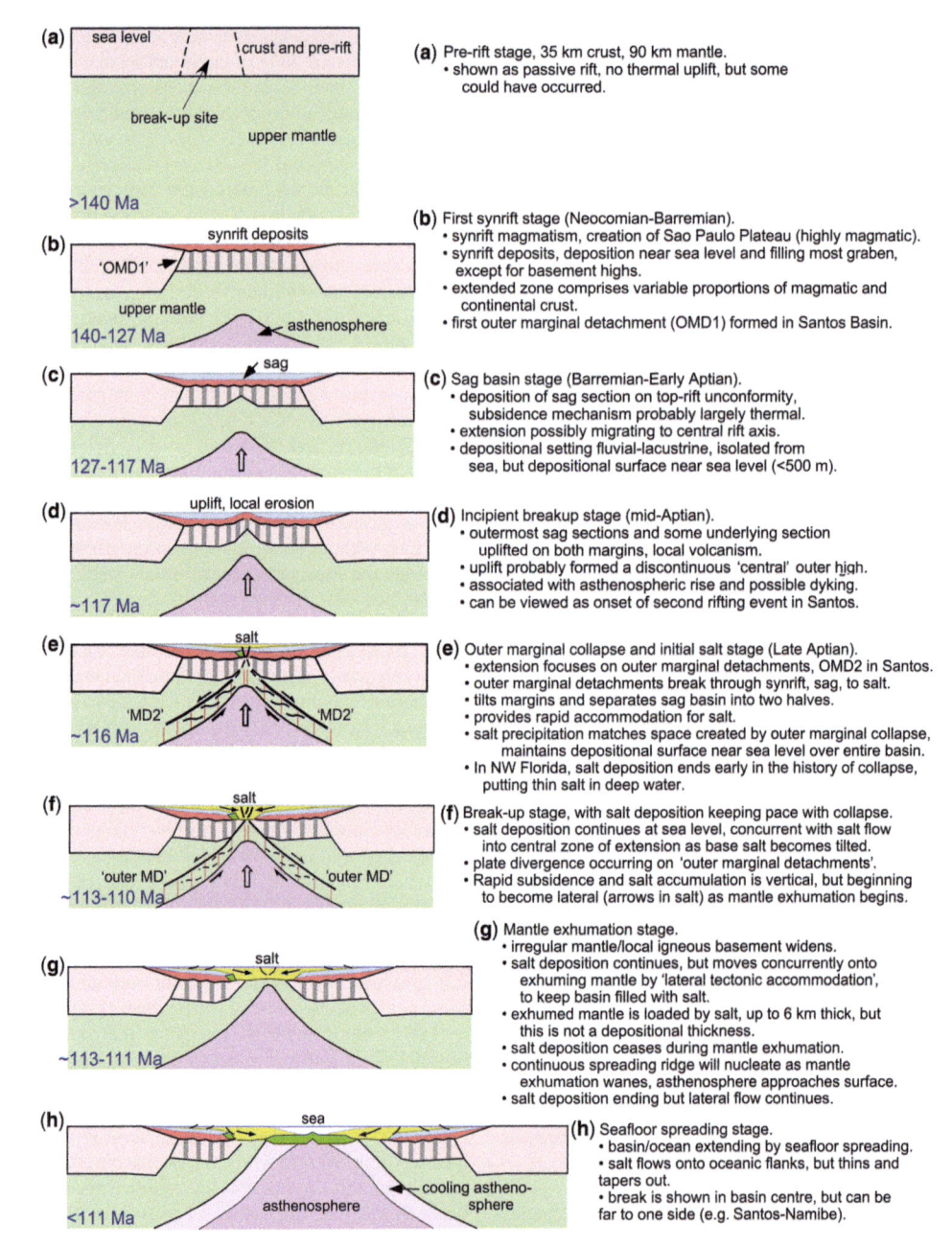

(h) Seafloor spreading stage.
  • basin/ocean extending by seafloor spreading.
  • salt flows onto oceanic flanks, but thins and tapers out.
  • break is shown in basin centre, but can be far to one side (e.g. Santos-Namibe).

Fig. 6. Eight-stage cartoon for rifting and break-up between South America and Africa at the Santos Basin. Key elements of each frame are outlined in points to the right. (**a**)–(**d**) Generic, simple synrift stages for margins; the nature of the thinned crust is not critical here. (**e**) Salt deposition in an early stage of OMC. If salt deposition ceased at this stage, we argue that the thin salt would continue to collapse into deep water (as exemplified by NW Florida: Fig. 2a). (**f**) Salt deposition has accompanied the entire period of OMC, and thus is thick (as exemplified by the Campos Basin (collapse plus loading can produce salt thicknesses up to $c.$ 5 km). (**g**) Salt deposition has outlasted collapse and persisted into the period of mantle exhumation, filling laterally created tectonic accommodation space, as exemplified by the Santos Basin. (**h**) Post-salt, syn-seafloor spreading stage, showing salt spreading and tapering onto the seafloor crust shortly after it has formed. The diagram is drawn with incipient plate break-up in the basin centre, which was not the case in the Santos segment of the margin. The vertical scale for basin elements is exaggerated for clarity; ages are estimates as judged from the literature and unpublished wells.

thick salt continues for several tens of kilometres over the more obvious oceanic crust before thinning basinwards, with no visible pre-salt sedimentary section (Fig. 5). In the Santos Basin, we judge that post-depositional downslope flow of salt onto the interpreted exhumed mantle from areas of updip thinning and welding could not have supplied as much salt onto the mantle as is observed. We suggest that in the Santos Basin, salt production near sea level persisted longer than the period of OMC on OMD2, so that salt production near sea level and lateral flow into the areas of deepest basement was sustained while the salt basin continued to widen (Fig. 6g). We envisage a conjugate salt basin whose depositional surface was near sea level in which new salt was produced as mantle exhumation continued at depth. Extension in the salt section must have matched basement extension, probably manifested as a double conveyor belt for salt migrating downdip from both basin flanks, while the depositional surface of the entire basin remained near sea level. We call this mechanism for creating overall salt volume horizontally 'lateral tectonic accommodation', as opposed to vertical accommodation. In order to comprehend the volume and position of salt in the Santos Basin, salt deposition probably ended after mantle exhumation had begun (Fig. 6g), so that the ensuing post-salt gravitational tapering of originally thick salt, as plate divergence continued, overran the western fringe of the more obvious oceanic crust as, or very soon after, it was formed by seafloor spreading (Fig. 6h).

Considering the Campos and Santos basins together, all indications suggest that salt deposition ended concurrently in both basins. If so, our model suggests that in the Santos Basin, both break-up (on OMD2) and a portion of the mantle exhumation had been achieved by the time of continental break-up in the Campos Basin, when salt deposition ceased and basinwards gravitational spreading began in both basins. This difference is expected, given that the Campos Basin was situated closer to the South Atlantic early pole of rotation (Heine et al. 2013), with northwards propagation of break-up and mantle exhumation.

Salt deposition in the context of the South Atlantic break-up history

The interpreted existence of exhumed mantle and oceanic crust beneath allochthonous salt along the Campos and Santos basins suggests that the outer limits of the South American and African salt cannot define any particular plate reconstruction. Norton et al. (2016) pointed out that better salt-age plate reconstructions can be made by aligning the actual continent–ocean boundaries as mapped from seismic. These authors also suggested an early stage of subsalt seafloor spreading to account for the occurrence of salt on the oceanic crust. While we certainly agree that the line of final continental crust break-up lies well inboard of the current limits of salt, we prefer to view the occurrence of salt on the oceanic crust as a halokinetic relationship (i.e. salt is allochthonous, but was emplaced on the oceanic crust very early). We also judge the exhumed mantle–salt relationship in the Campos Basin as largely halokinetic, given the large amounts of updip extension in the salt (Fig. 4). For the Santos Basin, the mantle–salt relationship is also largely halokinetic, even though, we argue, salt deposition near sea level across the entire conjugate basin persisted well into the period of mantle exhumation. In both basins, getting away from the idea that salt was deposited directly onto exhumed mantle (or oceanic crust) is an important strike against the 'air-filled hole' hypothesis of Figure 1a.

Invoking OMC as the means of creating salt accommodation suggests that plate reconstructions depicting the start of salt deposition in the conjugate Campos and Kwanza basins should attempt to realign the lines where the OMDs break to surface. These structures control vertical motions during OMC, and allow the base-salt unconformities to subside to the level of the exhumed mantle. In the Santos Basin, the same should be true, but the possible syn-salt crustal extension observed in the outer part of the Sao Paulo Plateau (east of the Tupi Field: Fig. 5) would tend to shift the Brazilian line of OMD2 breakout farther east than it would have been had this extension not occurred.

We reviewed G-Plates (2016) reconstructions in 1 myr increments from 116 to 111 Ma (possible span of salt deposition) in order to determine which reconstruction best aligned the proposed OMD breakouts that we mapped along parts of the Brazilian and African conjugate margins. We found that the 116 Ma G-Plates reconstruction provided the best overall realignment, at least southward to 26°S latitude (Fig. 7). The alignment is good for the Campos segment (enlarged in Fig. 7), and about 50 km too tight at the position of our sample line in the Santos Basin, implying perhaps that the extension seen as faulting in the outer Sao Paulo Plateau sums to about 50 km. Accepting this, the G-Plates reconstruction for 116 Ma is very good. However, if the true age of the salt is closer to, or persists to, the younger end of the permitted age span (i.e. 111 Ma), then the reconstruction may be a bit too loose for 116 Ma, and might be better considered for later in the Aptian.

Summary and discussion

We have shown how the concept of outer marginal collapse (OMC) (Pindell et al. 2014) can be applied successfully to three salt basins with significant

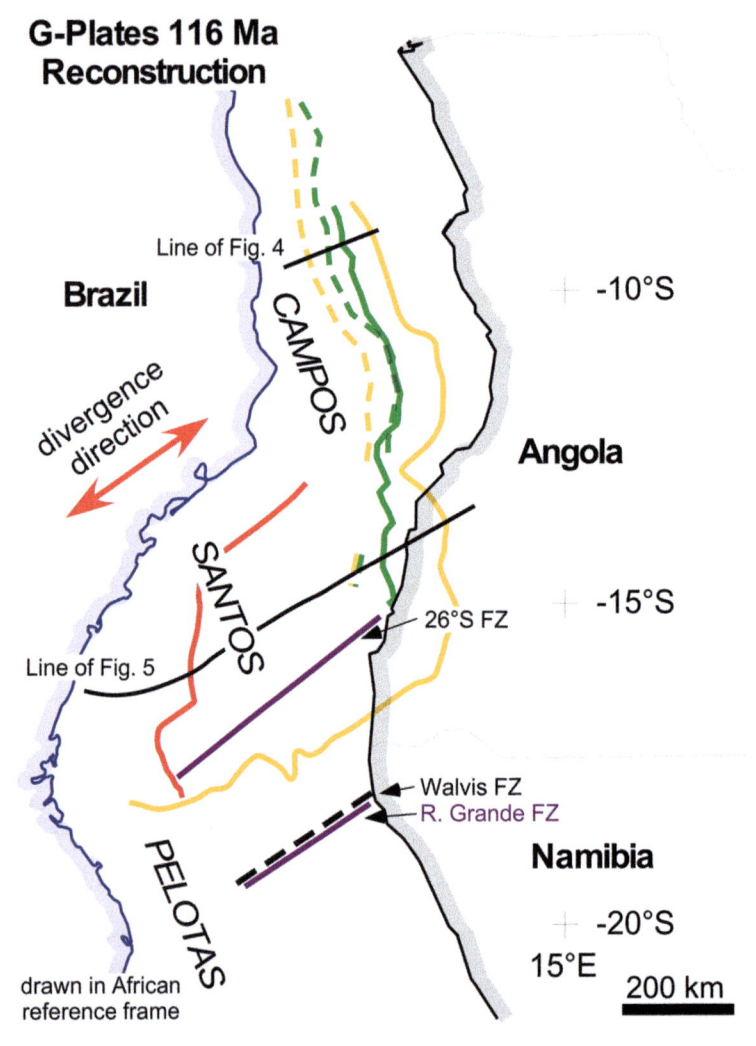

Fig. 7. Traces of outer marginal detachment (OMD) breakouts from both margins, and today's basinwards limits of salt, plotted on the 2016 G-Plates 116 Ma reconstruction of the central South Atlantic. Solid lines denote South American features; dashed lines denote African features. Green, breakout lines of outer marginal detachments, OMD2 in the Santos Basin; orange, today's outer limits of salt; red, in the Santos Basin only, outer marginal detachment OMD1. The South American coastline is in blue; the African coastline (reference frame) is in black. The red double-headed arrow is the plate separation azimuth in the G-Plates model, with which we agree. The heavy dashed black line is the trace of the Walvis Fracture Zone off Namibia; heavy solid purple lines are the traces of the Río Grande Fracture Zone off Pelotas Basin (more southerly fault zone), and the '26°S fracture zone' (our informal usage, more northerly line). Note the good alignment and the similar azimuths of the fracture zones with the G-Plate-modelled separation direction.

differences. We have argued that in the eastern Gulf of Mexico, OMC outlasted salt deposition so that a thin salt deposited near sea level was rapidly drowned to abyssal depths as OMC continued. In the Campos Basin, we suggest that salt production and collapse ended at about the same time, creating a thick salt layer above a planar, basinwards-deepening base-salt unconformity, much of which subsequently flowed basinwards onto a zone of exhumed mantle which widened between the two sides of the Campos–Kwanza conjugate basin. In the Santos Basin, salt deposition outlasted OMC and continued well into the mantle exhumation stage, taking advantage of 'lateral tectonic accommodation'. The potential significance of lateral tectonic accommodation can be appreciated by considering that, in any given cross-section, if (1) salt was 5 km thick over the mantle prior to the end of salt deposition, and (2) the full tectonic exhumation rate was, say, 60 km Ma^{-1} across the conjugate salt basin, then 300 km^2 of salt can be added to the cross section for each 1 myr of salt deposition. Thus, primary differences in salt occurrence (i.e. salt thickness and position relative to crustal or basement features) can be explained simply by considering the timing and duration of salt deposition relative to the timing of outer marginal collapse.

Because salt can be deposited so quickly relative to other types of sediment, OMC is probably best recorded by mega-salt basins. At salt-free margins, the most obvious effect of OMC is that it causes extremely rapid drowning of once-shallow features (e.g. top rift unconformities) to abyssal depths. Although we have explored salt relationships to show viability in the model at margins which have undergone magma-poor break-up, we presently can describe this process only qualitatively in terms of 'already-rifted continental margins sloughing off rising sub-continental mantle along outer marginal detachments'. We look forward to more quantitative efforts that can test links between the observations for rapid drowning of margins and the processes that may be responsible. One such mechanism may be depth-dependent stretching (Karner & Gambôa 2007), whereby lower crust stretches independently to drive subsidence, but it is not clear how this happens where OMC seems to occur at both conjugate margins. Another may relate to the rise of the asthenosphere between the two conjugate flanks of exhumed sub-continental mantle as they separate to allow the onset of seafloor spreading. This could potentially be viewed as an independent rifting event within the exhumed (and embrittled) upper mantle.

But, perhaps, the most significant observation is that the continental margins were already thinned prior to collapse; hence, the base-salt unconformities are not faulted. This implies that we should not be looking for rapid subsidence mechanisms per se, but, rather, for ways of temporarily supporting thinned pairs of conjugate continental margins near global sea level at the end of rifting, which can be turned off rapidly in order to allow for OMC to proceed. The fact that outer marginal detachments (OMDs) are the structures that are used may be simply due to the fact that they are the only structures available that can be used. Obvious potential ways for supporting thinned crust at non-isostatic elevations are transient heat pulses or short-lived mantle convection cells (dynamic topography). Curiously, break-up at the NW Florida margin was clearly magma-poor, although rifting did involve magmatic extrusion (Fig. 4) and probable underplating (Eddy et al. 2014). The synrift of the Campos and Santos basins also possessed an igneous component.

Our estimates for the original salt thicknesses in the three basin examples are necessarily qualitative. Due to the possibility of salt dissolution through time, they are minima at best, and it may be that attempting more quantitative assessments of original salt thickness is of limited value. However, if loss of salt through time has been substantial, then the depositional surfaces within the basins were, if anything, closer to sea level than we have suggested.

We have included the NW Florida margin, and the Campos and the Santos basin margins in our discussion. We have omitted the northern and SW Gulf of Mexico margins because the relatively greater amount of Cenozoic clastic input drives a correspondingly greater amount of salt deformation in both settings, and this, in turn, greatly hinders the ability to gauge salt deposition in relation to basement structure and crustal boundaries.

We consider that the occurrence of thick salt bodies over the deepest parts of basins, and over basement that should never have been situated near sea level (e.g. exhumed mantle), is partly responsible for the common belief that an 'air-filled hole' model, with salt deposition directly on basement, is a necessary stage in the development of mega-salt basins. We, instead, have tried to show that downslope movement of salt into areas with the deepest basement (e.g. areas of exhumed mantle), prior to salt separation between the conjugate margins, provides an alternative perspective that is more in keeping with the regional geology of the basins concerned. In the Santos–Namibe conjugate basin, we suggest that salt deposition near global sea level across the entire basin, prior to salt separation, continued well into the period of mantle exhumation (or earliest seafloor spreading, depending on one's interpretation of basement beneath OMT2 in Fig. 5), whereby a great percentage of the overall accommodation was formed laterally, by 'lateral tectonic accommodation'. Other questions that remain unanswered by the 'air-filled hole' model include: (1) How does a basin remain so starved of sediment for several tens of millions of years to form the deep depressions required to accommodate so much salt? (2) Why do we not see more interfingering of fans and fluvial deposits in salt, and why do we not see eroded channels at the base-salt unconformities along basin margin slopes, if salt had, in fact, filled pre-existing subaerial depressions?

Acknowledgements The authors are grateful to ION Geophysical for granting permission to publish Figures 2a, 4 & 5, and for access to the seismic reflection and well datasets on which much of our discussion is based. We thank Marek Kaminski for providing the top-salt reflector in Figures 4 & 5, and Ken McDermott, Paul Bellingham, Doug Paton, David Lewis, Lorcan Kennan, Alan Roberts and

Nick Kusznir for deliberations on outer marginal collapse. Thomas Hearon and an anonymous reviewer provided insightful reviews, for which we are grateful.

References

AUGUSTIN, N., DEVEY, C., VAN DER ZWAN, F.M., FELDENS, P., TOMINAGA, M., BANTAN, R.A. & KWASNITSCHKA, T. 2014. The rifting to spreading transition in the Red Sea. *Earth and Planetary Science Letters*, **395**, 217–230.

BECKER, K., FRANKE, D. ET AL. 2014. Asymmetry of high-velocity lower crust on the South Atlantic rifted margins and implications for the interplay of magmatism and tectonics in continental breakup. *Solid Earth*, **5**, 1011–1026, https://doi.org/10.5194/se-5-1011-2014

BURKE, K. & SENGÖR, C. 1988. Ten-meter global sea-level change associated with South Atlantic Aptian salt deposition. *Marine Geology*, **83**, 309–312.

COBBOLD, P.R., MEISLING, K.E. & MOUNT, V.S. 2001. Reactivation of an obliquely rifted margin, Campos and Santos basins, southeastern Brazil. *AAPG Bulletin*, **85**, 1925–1944.

COWIE, L., ANGELO, R.M., KUSZNIR, N., MANATSCHAL, G. & HORN, B.W. 2015. The palaeo-bathymetry of base Aptian salt deposition on the northern Angolan rifted margin: constraints from flexural back-stripping and reverse post-break-up thermal subsidence modeling. *Petroleum Geoscience*, **22**, 59–70, https://doi.org/10.1144/petgeo2014-087

DAVISON, I. 2007. Geology and tectonics of the South Atlantic Brazilian salt basins. *In*: REIS, A.C., BUTLER, R.W.H. & GRAHAM, R.H. (eds) *Deformation of the Continental Crust: The Legacy of Mike Coward*. Geological Society, London, Special Publications, **272**, 345–359, https://doi.org/10.1144/GSL.SP.2007.272.01.18

DAVISON, I., ANDERSON, L. & NUTTALL, P. 2012. Salt deposition, loading and gravity drainage in the Campos and Santos salt basins. *In*: ALSOP, G.I., ARCHER, S.G., HARTLEY, A.J., GRANT, N.T. & HODGKINSON, R. (eds) *Salt Tectonics, Sediments and Prospectivity*. Geological Society, London, Special Publications, **363**, 159–174, https://doi.org/10.1144/SP363.8

DOBSON, L.M. & BUFFLER, R.T. 1997. Seismic stratigraphy and geologic history of Jurassic rocks, northeastern Gulf of Mexico. *AAPG Bulletin*, **81**, 100–120.

EDDY, D.R., VAN AVENDONK, H.J.A., CHRISTESON, G.L., NORTON, I.O., KARNER, G.D., JOHNSON, C.A. & SNEDDEN, J.W. 2014. Deep crustal structure of the northeastern Gulf of Mexico: implications for rift evolution and seafloor spreading. *Journal of Geophysical Research*, **119**, 6802–6822, https://doi.org/10.1002/2014JB011311

GRACIANSKY, P.C., ROBERTS, D.G. & TRICART, P. 2011. *The Western Alps, from Passive Margin to Orogenic Belt*. Developments in Earth Surface Processes, **14**. Elsevier, Amsterdam.

HEINE, C., ZOETHOUT, J. & MÜLLER, R.D. 2013. Kinematics of the South Atlantic rift. *Solid Earth*, **4**, 215–253.

HUDEC, M.R., NORTON, I.O., JACKSON, M.P.A. & PEEL, F.J. 2013. Jurassic evolution of the Gulf of Mexico salt basin. *AAPG Bulletin*, **97**, 1683–1710.

JACKSON, M.P.A., CRAMEZ, C. & FONCK, J.-M. 2000. Role of sub-aerial volcanic rocks and mantle plumes in creation of South Atlantic margins: implications for salt tectonics and source rocks. *Marine and Petroleum Geology*, **17**, 477–498.

JÓHANNESSON, H. & SÆMUNDSSON, K. 1998. *Geological Map of Iceland, 1:500 000, Bedrock Geology*. Icelandic Institute of Natural History and Iceland Geodetic Survey, Reykjavík, Iceland.

KARNER, G.D. & GAMBÔA, L.A.P. 2007. Timing and origin of the South Atlantic pre-salt sag basins and their capping evaporates. *In*: SCHREIBER, B.C., LUGLI, S. & BABEL, M. (eds) *Evaporites through Space and Time*. Geological Society, London, Special Publications, **285**, 15–35, https://doi.org/10.1144/SP285.2

KLINGELHOEFER, F., EVAIN, M. ET AL. 2015. Imaging proto-oceanic crust off the Brazilian Continental Margin. *Geophysical Journal International*, **200**, 471–488.

LEMOINE, M. 1975. Mesozoic sedimentation and tectonic evolution of the Briançonnais Zone in the Western Alps – possible evidence for an Atlantic-type margin between the European Craton and the Tethys. *International Sedimentology Congress, Nice, Theme*, **4**, 211–216.

LEMOINE, M. 1988. Des nappes embryonnaires aux blocs bascules: evolution de idée's et des modeles sur l'histoire Mesozoique des Alpes Occidentales. *Bulletin Société Géologique de France, Paris, 8ème serie*, **4**, 787–797 [From embryonic nappes to tilted blocks: Evolution of Ideas and Models on the Mesozoic History of the Western Alps. Bulletin Geological Society of France, Paris, 8th series].

LEMOINE, M., DE GRACIANSKY, P.C. & TRICART, P. 2000. *De l'Ocean a la Chaine de Montagnes: Tectonique des plaques dons les Alpes*. Gordon & Breach, Paris.

LOVELL, T. & WEISLOGEL, A. 2010. Detrital zircon U–Pb age constraints on the provenance of the Upper Jurassic Norphlet Formation, eastern Gulf of Mexico: implications for paleogeography. *Gulf Coast Association of Geological Societies Transactions*, **60**, 443–460.

MANATSCHAL, G., MOHN, G., MASINI, E., PERON-PINVIDIC, G., UNTERNEHR, P. & KARNER, G.D. 2009. Mapping structures of ancient exposed hyperextended margins in the Alps: A key to understand the evolution of ultradeep water passive continental margins? AAPG Search and Discovery Article #50172, *AAPG International Conference and Exhibition*, October 26–29, 2008, Cape Town, South Africa.

MEISLING, K.E., COBBOLD, P.R. & MOUNT, V.S. 2001. Segmentation of an obliquely rifted margin, Campos and Santos Basins, southeastern Brazil. *AAPG Bulletin*, **85**, 1903–1924.

MOHN, G., MANATSCHAL, G., MUNTENER, O., BELTRANDO, M. & MASINI, E. 2010. Unravelling the interaction between tectonic and sedimentary processes during lithospheric thinning in the Alpine Tethys margins. *International Journal of Earth Sciences*, **99**, 75–101, https://doi.org/10.1007/s00531-010-0566-6

MOULIN, M., ASLANIAN, D. & UNTERNEHR, P. 2010. A new starting point for the South and Equatorial Atlantic Ocean. *Earth-Science Review*, **97**(1–4), 59–95, https://doi.org/10.1016/j.earscirev.2009.08.001

NORTON, I.O., CARRUTHERS, D. & HUDEC, M.R. 2016. Rift to drift transition in the South Atlantic salt basins: a new flavor of oceanic crust. *Geology*, **44**, 55–58.

PARSONS, B. & SCLATER, J.G. 1977. An analysis of the variation of ocean floor bathymetry and heat flow with age. *Journal of Geophysical Research*, **82**, 803–827.

PATON, D.A., PINDELL, J., MCDERMOTT, K., BELLINGHAM, P. & HORN, B. 2017. Evolution of seaward-dipping reflectors at the onset of oceanic crust formation at volcanic passive margins: Insights from the South Atlantic. *Geology*, **45**, 439–442, https://doi.org/10.1130/G38706.1

PERÉZ-GUSSINYE, M. 2012. A tectonic model for hyperextension at magma-poor rifted margins: an example from the West Iberia–Newfoundland conjugate margins. *In*: MOHRIAK, W.U., DANFORTH, A., POST, P.J., BROWN, D.E., TARI, G.C., NEMCOK, M. & SINHA, S.T. (eds) *Conjugate Divergent Margins*. Geological Society, London, Special Publications, **369**, 403–427, https://doi.org/10.1144/SP369.19

PERÓN-PINVIDIC, G. & MANATSCHAL, G. 2009. The final rifting evolution at deep magma-poor passive margins from Iberia–Newfoundland: a new point of view. *International Journal of Earth Sciences*, **98**, 1581–1597.

PILCHER, R.S., MURPHY, R.T. & MCDONOUGH CIOSEK, J. 2014. Jurassic raft tectonics in the northeastern Gulf of Mexico. *Interpretation*, **2**, SM39–SM55.

PINDELL, J., GRAHAM, R. & HORN, B.W. 2014. Rapid outer marginal collapse at the rift to drift transition of passive margin evolution, with a Gulf of Mexico case study. *Journal of Basin Research*, **26**, 1–25.

PINDELL, J., MIRANDA, E., CERÓN, A. & HERNANDEZ, L. 2016. Aeromagnetic map constrains Jurassic–Early Cretaceous synrift, break up, and rotational seafloor spreading history in the Gulf of Mexico. *In*: LOWERY, C., SNEDDEN, J. *ET AL.* (eds) *35th Annual GCSSEPM Foundation Bob F Perkins Research Conference 2016: Mesozoic of the Gulf Rim and Beyond: New Progress in Science and Exploration of the Gulf of Mexico Basin*. Gulf Coast Section SEPM (GCSSEPM), Houston, Texas, USA, 89–112.

RESTON, T. 2007. The formation of non-volcanic rifted margins by the progressive extension of the lithosphere: the example of the West Iberian margin. *In*: KARNER, G.D., MANATSCHAL, G. & PINHEIRO, L.M. (eds) *Imaging, Mapping and Modelling Continental Lithosphere Extension and Breakup*. Geological Society, London, Special Publications, **282**, 77–110, https://doi.org/10.1144/SP282.5

ROWAN, M.G. 2014. Passive-margin salt basins: hyperextension, evaporite deposition, and salt tectonics. *Journal of Basin Research*, **26**, 154–182.

RUDKIEWICZ, J.-L. 1988. *Structure et subsidence de la Marge Tethysienne entre Grenoble et Briançon au Lias et au Dogger*. PhD thesis, Ecole de Mines Superieur, Paris.

SALVADOR, A. 1987. Late Triassic–Jurassic paleogeography and origin of Gulf of Mexico Basin. *AAPG Bulletin*, **71**, 419–451.

SALVADOR, A. 1991. Triassic–Jurassic. *In*: SALVADOR, A. (ed.) *The Gulf of Mexico Basin*. Decade of North American Geology, J. Geological Society of America, Boulder, Colorado, USA, 131–180.

SHARP, I., HIGGINS, S. *ET AL.* 2016. Tectono-stratigraphic evolution of the onshore Namibe–Benguela–Kwanza basins, Angola – Implications for margin evolution models. *Abstract presented at the RIFTS III Conference*, 22–24 April 2016, London.

WARREN, J.K. 2006. *Evaporites: Sediments, Resources, and Hydrocarbons*. Springer, Berlin.

ZALÁN, P., SEVERINO, M.C., RIGOTI, C.A., MAGNAVITA, L.P., DE OLIVEIRA, J.A.B. & VIANA, A.R. 2011. An entirely new 3-D view of the crustal and mantle structure of a South Atlantic passive margin; Santos, Campos and Espirito Santos basins, Brazil. *Extended abstract presented at the AAPG Annual Convention and Exhibition*, April 10–13, 2011, Houston, Texas, USA.

The South Atlantic and Gulf of Mexico salt basins: crustal thinning, subsidence and accommodation for salt and presalt strata

MARK G. ROWAN

Rowan Consulting, Inc., 850 8th St., Boulder, CO 80302, USA

mgrowan@frii.com

Abstract: The South Atlantic and Gulf of Mexico conjugate-margin salt basins display similar relationships between crustal architecture and presalt and salt sequences. 3D and 2D depth-migrated seismic data reveal that: (1) the base salt is mostly smooth but drops into an outer marginal trough just landward of and deeper than oceanic crust; (2) the Moho climbs basinward to shallow depths and is largely absent below the trough; (3) faults are low-angle and asymmetrical beneath the smooth base salt but steeper and symmetrical in the trough; (4) the trough contains triangular highs between faulted lows; and (5) the smooth base salt is underlain by sag sequences that often dip and thicken basinward. The observations suggest that these salt basins shared a common evolution. Crustal faulting gradually shifted basinward. Consequent thermal/loading subsidence plus lower-crustal thinning generated basinward-shifting accommodation for sag sequences, but slow sedimentation relative to subsidence resulted in deep depressions. A switch to symmetrical boudinage of thinned crust created the troughs and possible reactive mantle diapirs. Evaporites formed during this stage, with deposition near sea-level in proximal positions but 2–3 km deep in basin centres. Oceanic spreading separated the salt into conjugate basins, with allochthonous flow out over oceanic crust.

Many conjugate divergent margins have segments containing salt (Hudec & Jackson 2007), for example the Gulf of Mexico (Mexico and the USA), the South Atlantic (Brazil and Angola, Democratic Republic of Congo, Republic of Congo and Gabon), the Central Atlantic (Nova Scotia and Morocco), the southern North Atlantic (Portugal and Newfoundland), the Bay of Biscay (Spain and France), the northern North Atlantic (Norwegian Barents Sea and northeastern Greenland) and the Red Sea (Saudi Arabia and Egypt, Sudan and Eritrea). The salt may be deposited prior to rifting or at any time during rifting, needing only a combination of an arid climate, an isolated basin and an adequate supply of water. The relative timing impacts the spatial and thickness distribution of the evaporites as well as the triggers and evolution of salt-related deformation (Rowan 2014).

The South Atlantic and Gulf of Mexico conjugate salt basins (Fig. 1) were, according to some researchers, deposited after or during the onset of oceanic spreading, either as individual basins separated by the proto-spreading centre (e.g. Jackson *et al.* 2000; Marton *et al.* 2000; Quirk *et al.* 2013) or as a single basin deposited partly above normal oceanic crust (Imbert 2005; Montaron & Tapponnier 2010; Norton *et al.* 2016) or early, intrusive oceanic crust (Norton *et al.* 2015a). Others, however, have suggested that the salt was deposited late in the rift history as a contiguous basin just prior to oceanic spreading (e.g. Evans 1978; Davison 1999; Pindell & Kennan 2007; Karner & Gambôa 2007; Mohriak *et al.* 2008; Torsvik *et al.* 2009; Lentini *et al.* 2010; Kneller & Johnson 2011; Davison *et al.* 2012; Hudec *et al.* 2013; Rowan 2014; Pindell *et al.* 2014, 2016; Nguyen & Mann 2016; Pascoe *et al.* 2016). Moreover, although some interpretations identify these as magma-rich margins (e.g. Imbert 2005; Imbert & Philippe 2005; Post 2005; Pindell & Kennan 2007; Mickus *et al.* 2009; Mohriak & Leroy 2013; Eddy *et al.* 2014; Mohriak 2015; Nguyen & Mann 2016; Norton *et al.* 2016; Lundin & Doré 2017), others interpret these margins as relatively magma poor, without significant magmatic contribution until final break-up and steady-state oceanic spreading (e.g. Contrucci *et al.* 2004; Unternehr *et al.* 2010; Stern *et al.* 2011; Zalán *et al.* 2011; Peron-Pinvidic *et al.* 2013; Rowan 2014; Pindell *et al.* 2014, 2016). Rowan (2014) termed these syn-exhumation salt basins based on the crustal model of Péron-Pinvidic & Manatschal (2009). The base salt is largely unfaulted and underlain by similarly unfaulted 'sag' sequences (e.g. Karner & Gambôa 2007; Unternehr *et al.* 2010; Miranda *et al.* 2013; Rowan 2014; Pindell *et al.* 2014).

The processes that generated the accommodation for the salt and presalt sag sequences are controversial, as is the associated depth of evaporite deposition relative to global sea-level. In some interpretations, the salt and sag sequences were deposited in deep depressions that were created during earlier and ongoing rifting (e.g. Warren 2006; Karner & Gambôa 2007; Reston 2010; Davison *et al.* 2012; Garcia

From: McCLAY, K. R. & HAMMERSTEIN, J. A. (eds) 2020. *Passive Margins: Tectonics, Sedimentation and Magmatism.* Geological Society, London, Special Publications, **476**, 333–363.
First published online April 17, 2018, https://doi.org/10.1144/SP476.6
© 2018 The Author(s). Published by The Geological Society of London. All rights reserved.
For permissions: http://www.geolsoc.org.uk/permissions. Publishing disclaimer: www.geolsoc.org.uk/pub_ethics

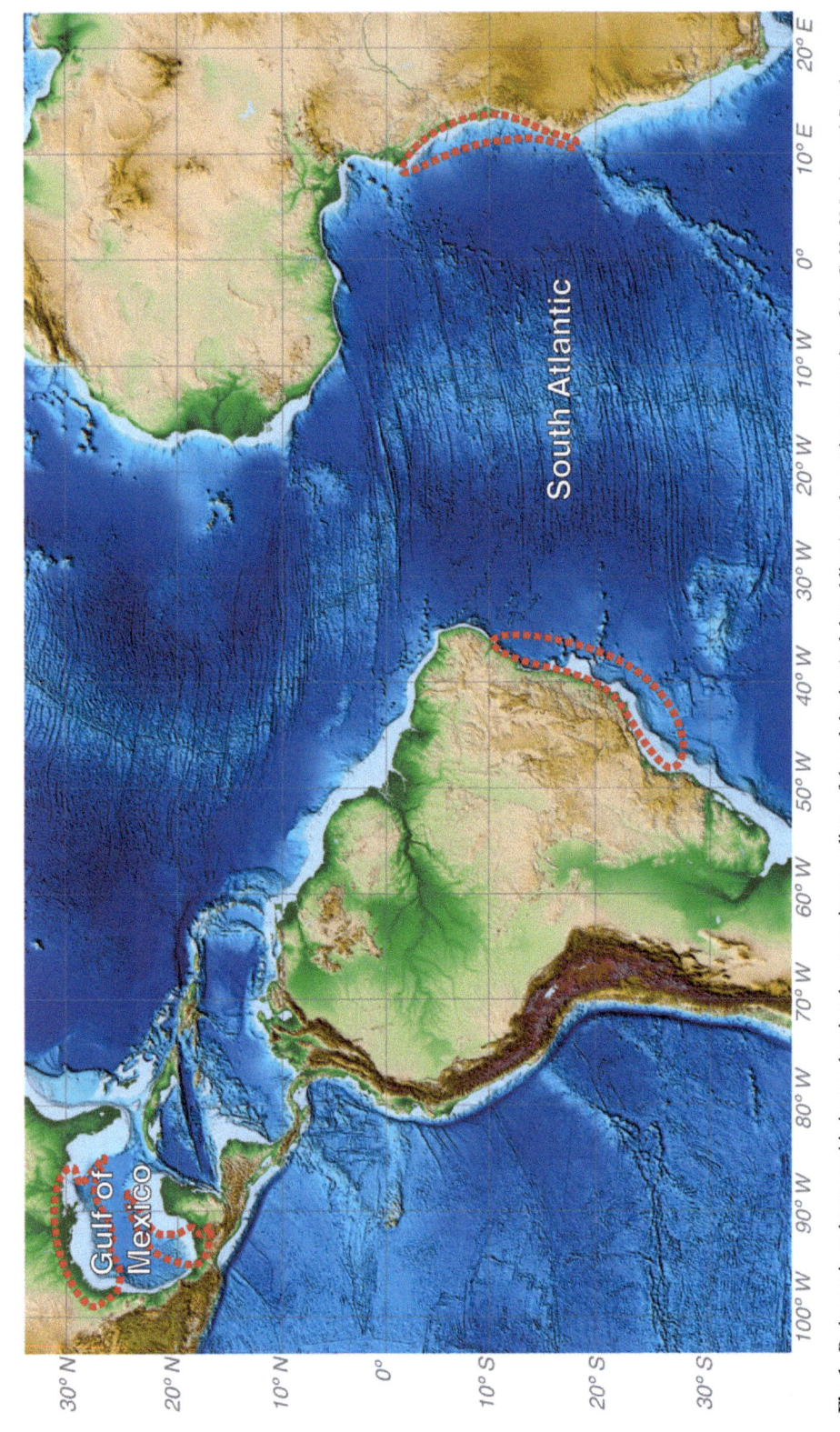

Fig. 1. Bathymetric and topographic image showing the approximate outlines of the salt basins (red dotted lines) on the conjugate margins of the Gulf of Mexico and South Atlantic. Modified from https://maps.ngdc.noaa.gov/viewers/wcs-client/

et al. 2012; Hudec et al. 2013; Rowan 2014; Lundin & Doré 2017) or during early oceanic spreading (Montaron & Tapponnier 2010). Others invoked at least the end of salt deposition as occurring effectively at sea-level (e.g. Moulin et al. 2005; Aslanian et al. 2009; Hudec et al. 2013; Pindell et al. 2014; Norton et al. 2015a). In the model by Hudec et al. (2013), the top salt then collapsed up to 3 km vertically owing to ongoing extension of underlying transitional crust and consequent lateral stretching of the salt prior to the onset of oceanic spreading. The top salt thus ended up approximately at the level of the top oceanic crust. Similarly, the South Atlantic was also inferred to have been deposited at sea-level and then dropped owing to underlying extension, but in this case by early, subsalt oceanic spreading (Norton et al. 2015a). In contrast, Rowan (2015) inferred that the top salt in both basins had topographic relief during deposition, occurring just beneath sea-level in proximal positions but as deep as several kilometres in the basin centres where the spreading centres subsequently developed.

A different model for the Gulf of Mexico was proposed by Pindell et al. (2014). They suggested that there was no existing depression prior to evaporite deposition, i.e. that synrift deposition kept pace with subsidence. Thus, salt deposition occurred near global sea-level throughout its history. Accommodation for the salt, as well as subsidence of the top salt by 2–3 km immediately following evaporite deposition, was generated by so-called outer marginal collapse in as little as 3 myr. Effectively, the outer continental margins became crustal-scale half graben in the hanging walls of landward-dipping shear zones detached near the Moho, with subcontinental mantle being exhumed in the footwalls.

On the Angolan margin, Cowie et al. (2016, 2017) used flexural backstripping and reverse thermal subsidence modelling constrained by gravity-anomaly inversion to calculate the palaeobathymetry at the onset of evaporite deposition. The results indicated shallow water (200–600 m) in proximal positions but palaeobathymetries of 2–3 km, and as much as 4.5 km, in distal domains. This matches the water-loaded depths of 2–3 km derived from flexural backstripping of the conjugate South Atlantic margins by Norton et al. (2015b); values would be less if it was an air-filled depression. It also matches the results of cross-section restoration in the Santos Basin of Brazil (Garcia et al. 2012). However, the preferred interpretation of Cowie et al. (2016, 2017) was that all the salt was deposited at shallow depths, that the proximal salt subsided solely by thermal and loading subsidence, but that the distal salt experienced additional, late synrift tectonic subsidence.

In this paper, I take another look at these issues using 2D and 3D depth-migrated seismic data from the northeastern and Yucatan segments of the Gulf of Mexico conjugate margins and the Kwanza Basin of Angola (presented in abbreviated form in Rowan 2015). I first describe and interpret the seismic data and then present a model based on the observed crustal, sag and salt geometries. This model is discussed with reference to the interpretations and models cited above, interpretations from other magma-poor but non-evaporite margins and numerical models of passive-margin formation. I argue that accommodation for the salt and sag sequences in the South Atlantic and Gulf of Mexico was generated over most of the margin extents by a basinward progression of lower-crustal thinning and thermal and loading subsidence, except in the outermost domain. There, ongoing rifting switched from asymmetrical, low-angle faulting to more symmetrical, high-angle faulting and possible coeval exhumation of subcontinental mantle and/or lower continental crust. This last stage of rifting was a form of large-scale boudinage that ultimately facilitated the rise of asthenospheric mantle and steady-state oceanic spreading. Evaporites were deposited during the final stage of rifting, with palaeobathymetry of the top salt ranging from relatively shallow water at the basin margins to as much as 2–3 km in the basin centres.

Seismic observations and interpretations

In the following sections, the seismic data are described and interpreted. Given that these are depth-migrated data, thicknesses and dips are dependent on the velocity models applied. Nevertheless, any errors are likely to be minor and thus the geometries are considerably more realistic than those obtained from time-migrated data.

I first address the large, margin-scale features such as the base-salt geometry and the Moho. I then focus on the details of several aspects: the crustal extensional structures; the presalt sag sequences; and the geometry of the ocean–continent transition zone. Note that these are all 2D interpretations, even those of 3D data; structures have been examined but not mapped out in three dimensions. I conclude this section by summarizing the observations and presenting a schematic cross-section that incorporates the key elements of the crustal, presalt and base-salt geometries.

Margin-scale geometries

Possibly the best location to observe the key large-scale features of late synrift (syn-exhumation) salt basins is the eastern Gulf of Mexico. Basinward of the Yucatan Platform, the base salt or its equivalent weld is a mostly smooth and undeformed surface

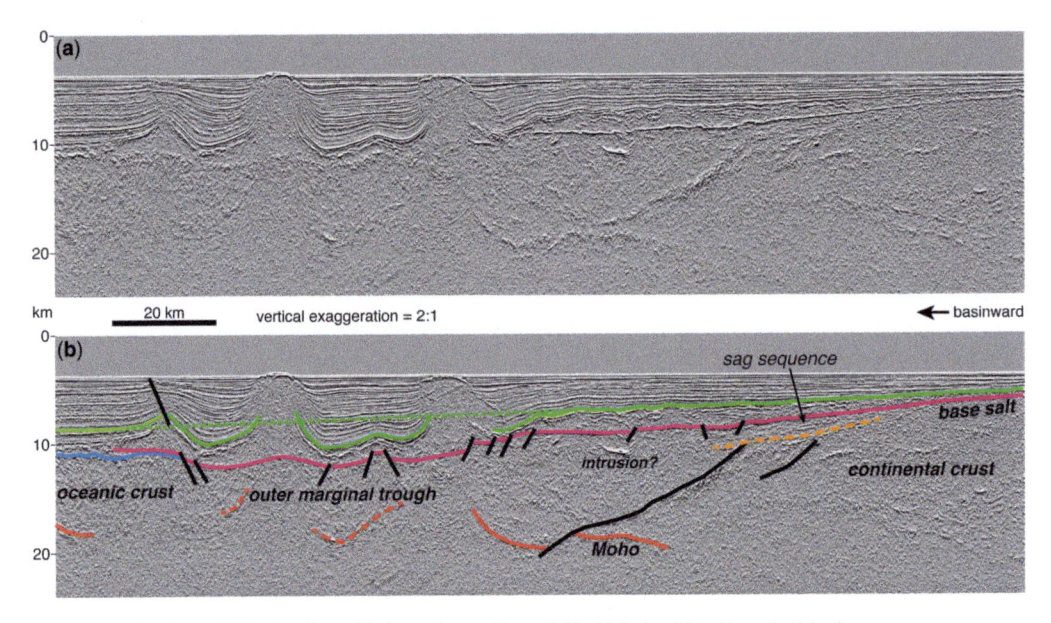

Fig. 2. Depth-migrated 2D seismic profile from the southern Gulf of Mexico (data from the Maximus survey courtesy of MultiClient Geophysical ASA): (**a**) uninterpreted; and (**b**) interpreted. Black lines are faults, red is the Moho (dashed where uncertain), blue is the top oceanic crust, orange is the base of the sag sequence, warm pink is the base salt or equivalent weld and green is the top Cretaceous. The dotted green line marks the regional elevation for the top Cretaceous, demonstrating that strata are below regional in the outer marginal trough.

for much of its extent (Fig. 2). However, there is a distal zone where the base salt drops abruptly into an outer marginal trough (Pindell *et al.* 2014; salt trough of Imbert 2005) and then climbs up onto oceanic crust. The landward-dipping basement step (Barker & Mukherjee 2011) at the basinward edge of the trough was termed the inner ramp by Hudec *et al.* (2013) and interpreted as the limit of oceanic crust and thus the break-up edge of salt. Salt basinward of the inner ramp was considered to represent early lateral flow over newly formed oceanic crust and overlying abyssal strata (e.g. Peel 2001; Imbert & Philippe 2005; Fiduk *et al.* 2007; Hudec *et al.* 2013).

Several aspects of the suprasalt deformation warrant mention. First, the outer marginal trough is the site of several diapirs that are still growing (Fig. 2). Second, the deepest (Upper Jurassic to Paleogene) strata are more than 2 km below their regional elevation, probably owing to subsidence into deeper and originally thicker salt in the trough. Third, there is significant proximal extension (not seen at this scale) that is not balanced by an equivalent amount of distal contraction (Imbert & Philippe 2005; Hudec *et al.* 2013).

Beneath the smooth salt is a prominent wedge of basinward-dipping and -thickening, largely unfaulted strata (Fig. 2), a geometry that I term an asymmetrical sag sequence and that will be addressed below. This overlies crustal extensional geometries, including a major basinward-dipping fault that cuts the entire crust. At the base of the section, a bright series of reflectors probably represents the continental Moho. At the basinward end of the section is the oceanic Moho at the base of 7 km-thick oceanic crust. Beneath the smooth base salt, the Moho is at about 20 km but climbs sharply to 16 km or less at the landward edge of the outer marginal trough. Beneath the trough, a possible Moho has significant relief.

The Moho is imaged more clearly in the northeastern Gulf of Mexico (Fig. 3). There, the Moho gradually climbs from about 34 km beneath the Florida Platform to a depth of 14 km just 5 km below the basinward termination of a smooth and unfaulted base salt or equivalent weld. The interpreted depth of the Moho is roughly supported by refraction seismic data (Christeson *et al.* 2014) and more exactly by gravity inversion modelling (N. Kusznir, pers. comm. 2013) The Moho is either absent or poorly imaged beneath the outer marginal trough, where the base salt is faulted, but then reappears beneath oceanic crust. The oceanic crust is abnormally thick (up to 9 km), which may represent an excess magma event related to break-up at magma-poor margins (Bronner *et al.* 2011). Potential field data acquired along the SuperCache seismic lines show that the outer marginal trough in this area is

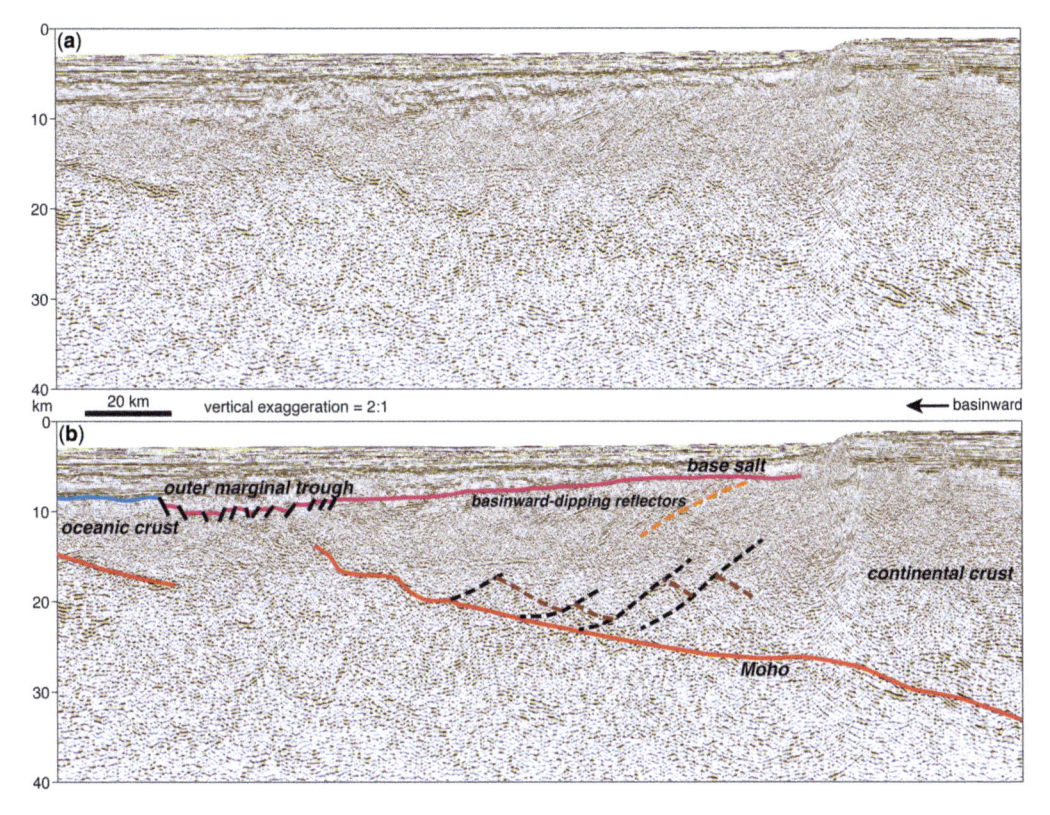

Fig. 3. Depth-migrated 2D seismic profile from the northeastern Gulf of Mexico (data from the SuperCache survey courtesy of Dynamic Data Services): (**a**) uninterpreted; and (**b**) interpreted. Black lines are faults (dashed where approximate), red is the Moho, brown depicts possible crustal fault blocks, blue is the top oceanic crust, orange is the base of prominent basinward-dipping reflectors and warm pink is the base salt.

characterized by a prominent linear low on RTP (reduction to pole) magnetic data but a less obvious signature on gravity data.

A more detailed image is provided by 3D seismic data in the same area (Fig. 4), although the Moho cannot be seen. Again, the base salt/weld is effectively unfaulted over a distance of at least 45 km but drops abruptly by 3.5 km at the landward edge of the outer marginal trough. The base is then irregular, with highs and lows, but gradually climbs to the level of the top oceanic crust, which is at approximately the same depth as the basinward termination of the smooth base salt. Suprasalt strata also drop into the outer marginal trough, which is c. 15 km wide.

Remarkably similar features can be observed in the Kwanza Basin of Angola (Fig. 5). Again, the base salt is undeformed for a distance of at least 60 km but abruptly becomes irregular and indistinct, forming an outer marginal trough c. 40 km wide and at least 1.5 km deep. The seismically interpreted Moho has some relief where it is offset by crustal faults but is otherwise subhorizontal beneath the smooth base of salt. Published data show that it deepens to the east to a depth of 25–30 km beneath normal-thickness continental crust (e.g. Contrucci et al. 2004; Unternehr et al. 2010). Moving basinward, the interpreted Moho climbs to a level about 2.5 km below the distal end of the unfaulted base of salt, and is not apparent beneath the outer marginal trough (Fig. 5).

Extensional structures

The area of smooth base salt in Angola is underlain by largely unfaulted strata which in turn overlie rift geometries (Figs 5 & 6). The extended crust is dominated by half graben and associated low-angle, basinward-dipping faults (with dips as low as 20°), some of which appear to offset the Moho (Fig. 5). Interpreted throw and heave on the faults can be as much as 4.5 and 5.5 km, respectively. The hanging walls sometimes contain wedge geometries, interpreted as syntectonic strata, that are separated from the overlying sag sequence by a locally distinct unconformity (Fig. 6).

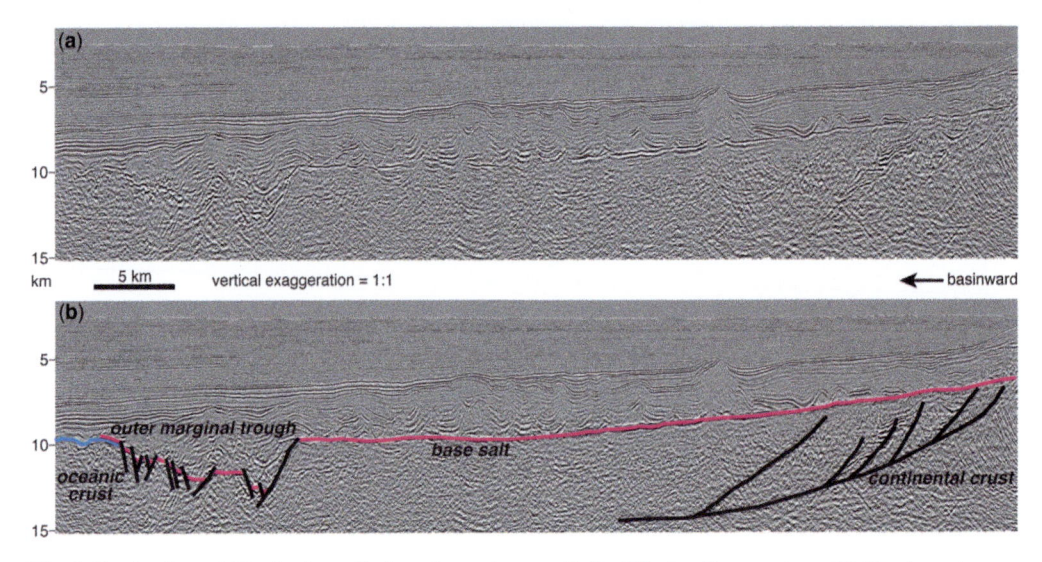

Fig. 4. Depth-migrated 3D seismic profile from the northeastern Gulf of Mexico (data courtesy of Schlumberger): (**a**) uninterpreted; and (**b**) interpreted. Black lines are faults, blue is the top oceanic crust and warm pink is the base salt or equivalent weld.

Although the rift geometry is generally not as well imaged in the Gulf of Mexico, similar structures can locally be observed (Figs 2 & 3). In Figure 7, from the northeastern Gulf of Mexico, the smooth base salt is underlain by a series of mostly basinward-dipping faults but is offset slightly by

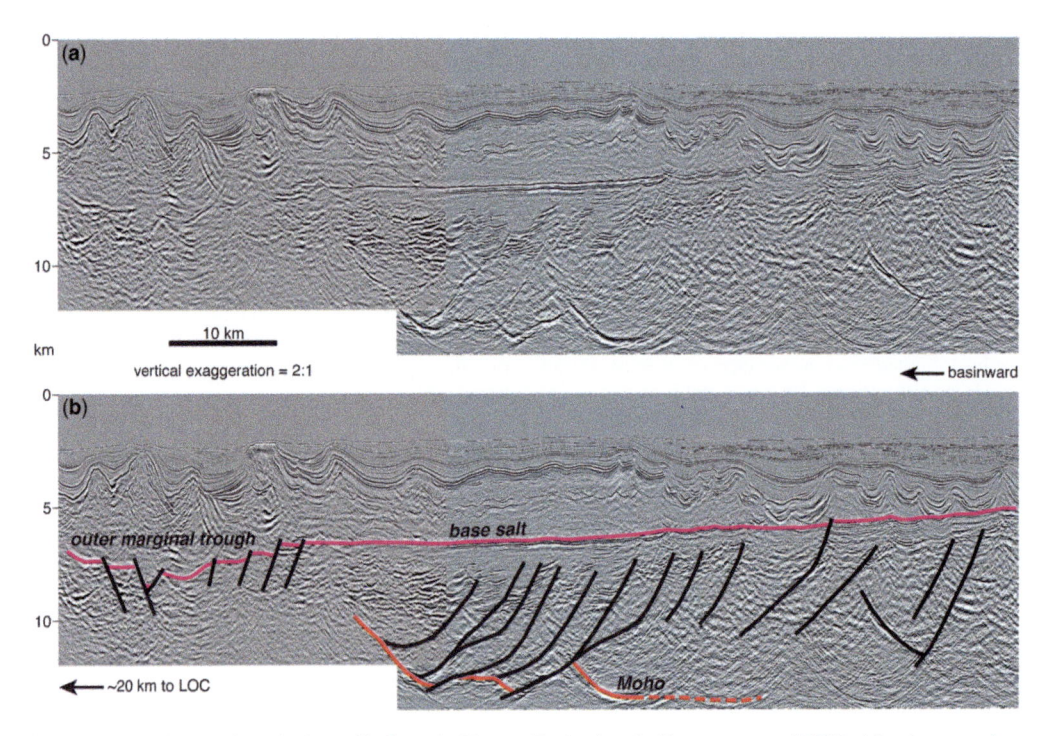

Fig. 5. Depth-migrated 3D seismic profile from the Kwanza Basin, Angola (data courtesy of TGS): (**a**) uninterpreted; and (**b**) interpreted. Black lines are faults, red is the Moho (dashed where uncertain) and warm pink is the base salt or equivalent weld. LOC is the limit of oceanic crust picked on longer 2D profiles.

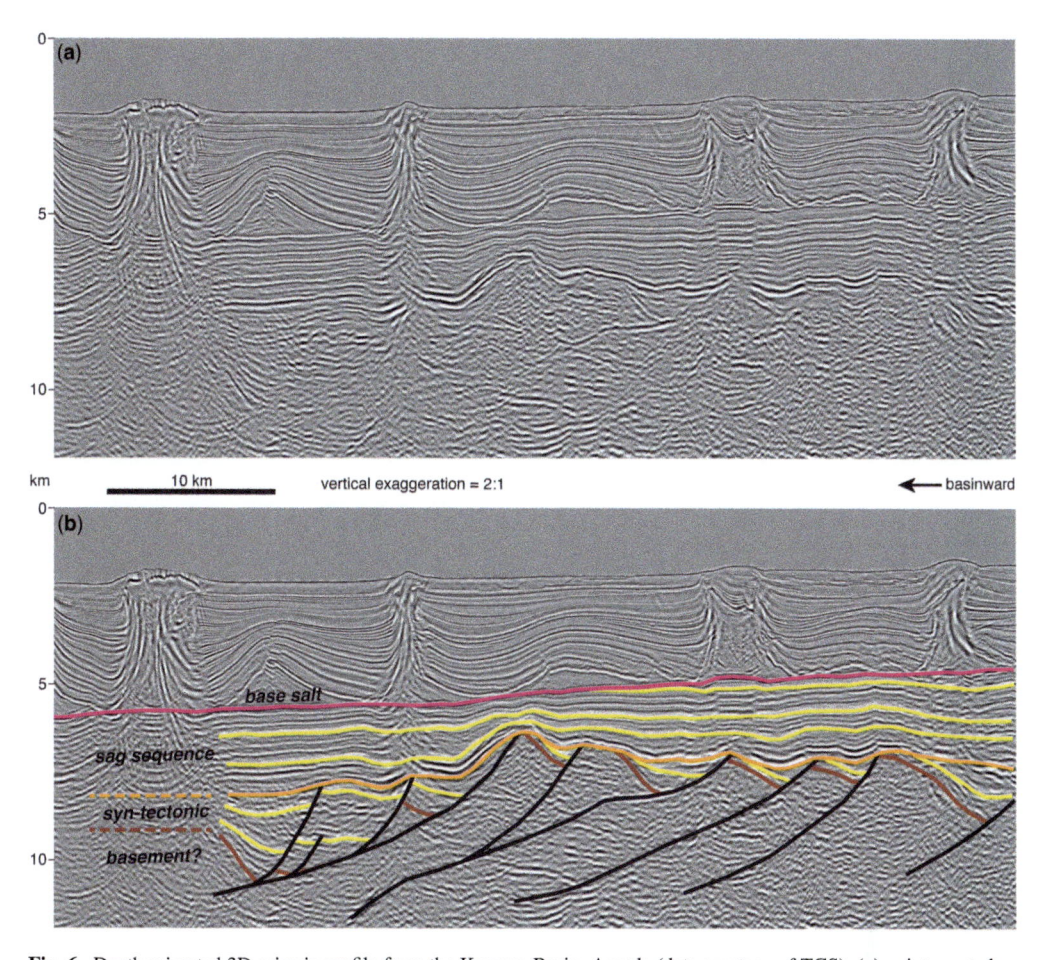

Fig. 6. Depth-migrated 3D seismic profile from the Kwanza Basin, Angola (data courtesy of TGS): (**a**) uninterpreted; and (**b**) interpreted. Black lines are faults, brown is top basement, orange is top syntectonic section (base sag sequence), warm pink is the base salt or equivalent weld, and yellow lines highlight stratal geometries.

only two of them. Fault dips are as low as 20°, but the amount of displacement is unknown. These faults have a WNW–ESE trend, i.e. approximately parallel to the limit of oceanic crust in this part of the basin. Thus the extension direction associated with the thinning stage was broadly coaxial with oceanic spreading but highly oblique to the NW–SE extension direction of the older stretching stage (as shown in, e.g. Kneller & Johnson 2011; Hudec *et al.* 2013).

Sag sequence

'Sag sequence' is a somewhat contentious name given to strata immediately beneath the salt in the South Atlantic (e.g. Henry *et al.* 2004; Karner & Gambôa 2007). Here I use it in a geometric sense: it is characterized by mostly undeformed strata, although it can have minor faults and there may be compactional drape over underlying basement highs (Fig. 6). It is separated from syntectonic growth strata in any given location by a horizon that is generally unconformable above footwall highs but conformable above the hanging walls of the half graben. The base salt forms the top of the sag sequence and appears locally to be a low-angle unconformity. Alternatively, the boundary may be conformable above convergent strata that are below the limits of seismic resolution.

Many interpretations illustrate the sag sequence in the South Atlantic basins as broadly symmetrical, thinning both landward and basinward (e.g. Marton *et al.* 2000; Lentini *et al.* 2010; Unternehr *et al.* 2010). However, 3D seismic data often display basinward-dipping and -thickening strata, even in the distal domains. A subtle example from the Kwanza Basin is shown in Figure 8, but this type of asymmetrical sag sequence is better seen in the deepwater Gabon margin (pers. obs. 2016).

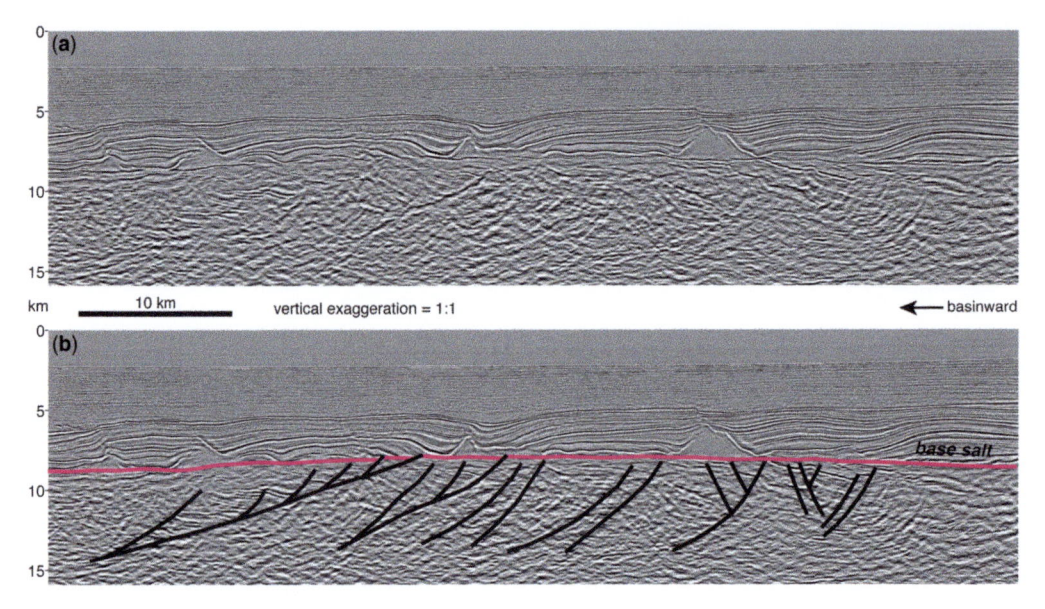

Fig. 7. Depth-migrated 3D seismic profile from the northeastern Gulf of Mexico (data courtesy of Schlumberger): (**a**) uninterpreted; and (**b**) interpreted. Black lines are faults and warm pink is the base salt or equivalent weld.

Additionally, a line from the Campos Basin in Brazil (Unternehr *et al.* 2010, fig. 4) appears to show a similar pattern.

Sag sequences have generally not been recognized in the Northern Gulf of Mexico because of the deep depths to the base salt and the difficult seismic imaging owing to shallow salt canopies. However, both of these problems largely disappear in the eastern Gulf of Mexico, and modern 2D and 3D seismic data do show the presence of a largely unfaulted sequence beneath the salt (Figs 2, 3 & 9). In the northeastern Gulf of Mexico, a symmetrical sequence thins slightly over minor presalt structural highs (Fig. 9) and also thins eastward and onlaps the Southern Platform, a major basement high with thin to absent salt. To the south of the high, the section beneath the salt comprises an unfaulted, asymmetrical package of basinward-dipping and -thickening strata that is up to 9 km thick according to the applied velocity model and thus fills more than 50% of the distance between the base salt and the Moho (Figs 3 & 10). A similar geometry is observed off the Yucatan Platform in the Southern Gulf of Mexico (Fig. 11a; Miranda *et al.* 2013).

The nature of these dipping strata is unknown, and there are several possible interpretations. First, they might represent folded and eroded Palaeozoic strata such as those penetrated on the Florida Platform (e.g. Dobson & Buffler 1997). However, they have a different seismic character than where Palaeozoic strata have been penetrated and they strike parallel to the edge of oceanic crust, suggesting a genetic relationship to late stages of rifting. Second, the geometry is similar to that of progradational siliciclastic or carbonate sequences, but the total relief in the Gulf of Mexico examples is greater than even the relief on continental slopes (e.g. Patruno *et al.* 2015) and would require progradation into a depression 6–9 km deep. Third, because these are basinward-dipping reflectors, they may represent SDRs (seaward-dipping reflectors) in the genetic sense, i.e. subaerial basalts derived from asthenospheric mantle (e.g. Imbert 2005; Imbert & Philippe 2005; Post 2005; Rowan 2014; Norton *et al.* 2015*a*, 2016; Pascoe *et al.* 2016). However, SDRs typically have concave-downward geometries and thicken basinward (e.g. Paton *et al.* 2017), whereas these are sigmoidal (Figs 10 & 11). Moreover, there is no prominent magnetic anomaly as is common on margins with SDRs (e.g. Talwani *et al.* 1995; Hinz *et al.* 1999; Blaich *et al.* 2009). Although a large magnetic anomaly is illustrated in Pindell *et al.* (2016), the EMAG2 data (Maus *et al.* 2009) and proprietary magnetic data from the SuperCache seismic survey both show that high values extend from the landward portion of the basinward-dipping reflectors onto the Florida Platform, suggesting that the magnetic high represents continental crust, not volcanic rocks. Also, SDRs typically extend right up to the edge of normal oceanic crust, but these reflectors are usually separated from oceanic crust by some tens of kilometres. Fourth, the dipping presalt reflectors may represent a series of basinward-shifting depocentres; although the lithologies are unknown, interpretations

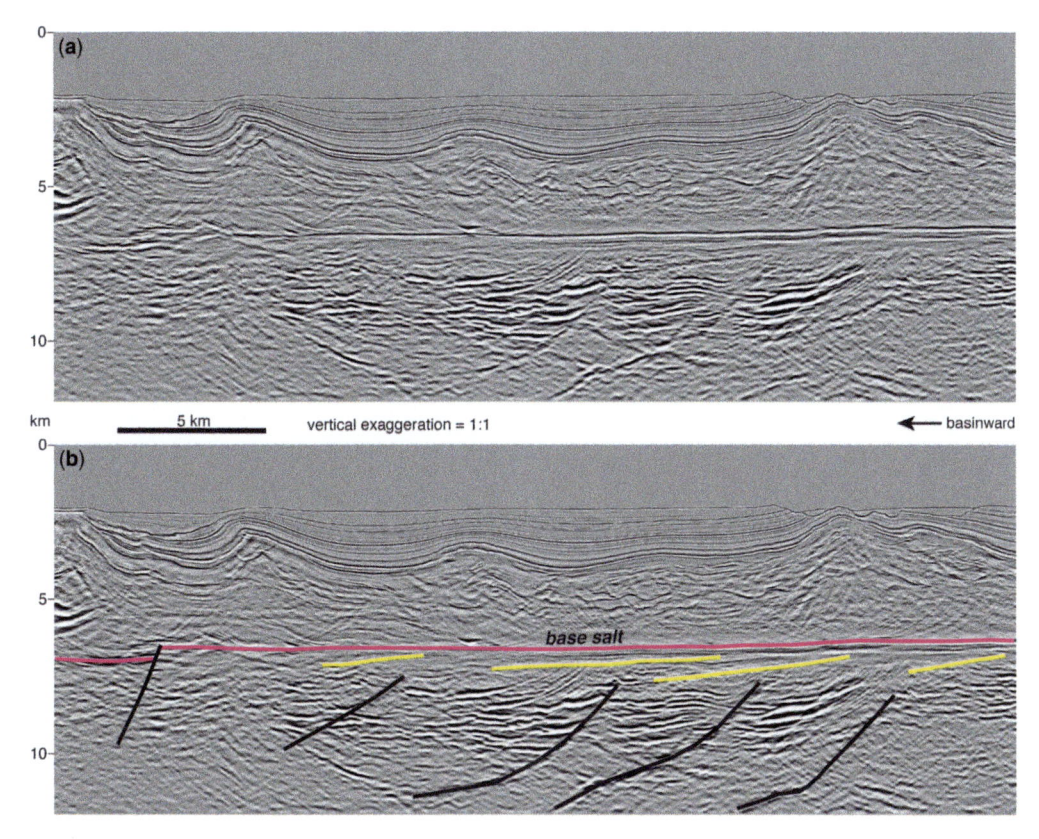

Fig. 8. Depth-migrated 3D seismic profile from the Kwanza Basin, Angola (data courtesy of TGS): (**a**) uninterpreted; and (**b**) interpreted. Black lines are faults, warm pink is the base salt or equivalent weld and yellow lines highlight stratal geometries.

of the Yucatan examples (Fig. 11a) suggest that the strata comprise primarily sedimentary rocks, with some local development of synrift volcanic rocks (Miranda *et al.* 2013; Goswami *et al.* 2016). Numerical models have generated 4–7 km-thick packages of basinward-shifting depocentres, similar to those described here, simply by aggradation during crustal thinning (Allen & Beaumont 2016, figs 4 & 11); others have produced basinward-younging sag sequences above prominent unconformities in cases of strong lower crust (Andrés-Martínez *et al.* 2017).

I interpret these basinward-dipping presalt strata as asymmetrical sag sequences based on the arguments above. Admittedly, an SDR interpretation cannot be ruled out for this easternmost portion of the Gulf of Mexico. In any case, the geometry requires that subsidence progressively shifted basinward late in the rift history, prior to the onset of evaporite deposition (Fig. 11). Broadly similar late synrift patterns on the conjugate south Australian and Antarctic margins have been interpreted as sediment onlapping newly created exhumed subcontinental mantle (Gillard *et al.* 2015*a*). However, most of the basinward-dipping strata in the northeastern Gulf of Mexico occur above a clear Moho at 20–25 km (Fig. 3; see also Allen & Beaumont 2016), so do not onlap exhumed mantle. Thus, the shifting subsidence would have been accompanied by shifting areas of relative uplift in more distal positions (Fig. 11), as invoked for other magma-poor margins (Masini *et al.* 2013; Haupert *et al.* 2016).

Outer marginal trough

The outer marginal troughs in the South Atlantic and Gulf of Mexico salt basins are pronounced lows bounded updip by steep basinward-dipping faults and downdip by the edge of oceanic crust (Figs 2–5). Somewhat chaotic reflectors probably indicate the presence of intrusive or even extrusive magmatic rocks. In a detailed example from the Kwanza Basin, the trough comprises an alternating pattern of apparent lows and triangular structural highs (Fig. 12a). Two of the highs are asymmetrical, contain basinward-dipping reflectors and are bounded by landward-dipping faults (Fig. 12a, location 2),

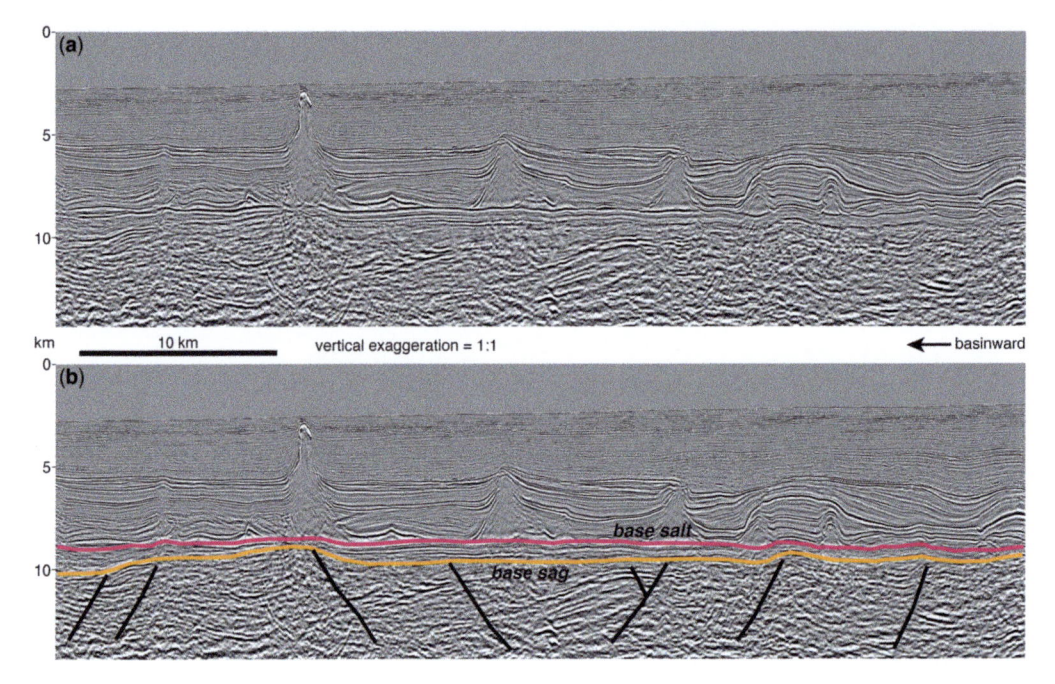

Fig. 9. Depth-migrated 3D seismic profile from the northeastern Gulf of Mexico (data courtesy of Schlumberger): (**a**) uninterpreted; and (**b**) interpreted. Black lines are faults, orange is the base sag sequence and warm pink is the base salt or equivalent weld.

and are thus interpreted as rotated fault blocks with half graben filled with salt (Fig. 12b). However, two of the highs are symmetrical and transparent (Fig. 12a, locations 3), with offset reflectors that appear to dip away on both sides of the highs. The geometry is analogous to that of reactive salt diapirism (Vendeville & Jackson 1992), which is effectively a form of large-scale boudinage. Thus, the symmetrical, transparent highs may represent reactive diapirs of relatively ductile rock such as serpentinized upper lithospheric mantle, weak lower lithospheric mantle or even asthenospheric mantle. Alternative interpretations are certainly possible.

Somewhat different geometries are seen in a detailed image of the outer marginal trough in the northeastern Gulf of Mexico (Fig. 13). The base salt has structural relief formed by a series of fault blocks, with both basinward-dipping, low-angle faults and more symmetrical high-angle faults that appear to cross-cut the former. The low-angle faults detach on a seismic reflector that is interpreted as the Moho, which again rises to a very shallow level at the landward margin of the trough (Fig. 3). This interpreted Moho has several triangular highs just beneath conjugate sets of the steeper faults, again a geometry analogous to that of reactive salt diapirs. Alternative interpretations are possible, but note also the presence of broadly symmetrical triangular geometries of the interpreted Moho in the outer marginal trough of the conjugate Yucatan margin (Fig. 2).

Summary of the observations

The seismic data illustrated here demonstrate that many features of the South Atlantic and Gulf of Mexico salt basins are remarkably similar. These are combined into a schematic cross-section representing these syn-exhumation (late synrift) salt basins (Fig. 14). The model adopts the Peron-Pinvidic et al. (2013) terminology of proximal, necking, distal, outer and oceanic domains. Key elements of the geometry are summarized below.

(1) The Moho climbs basinward from beneath the proximal domain, through the necking domain, to the distal domain. It then shallows more gently beneath the distal domain before ramping up again toward the outer domain, where it either has significant relief or is not observed, before appearing again in the oceanic domain. The distal domain, with the lower-angle Moho, may be relatively wide (Kwanza Basin, Unternehr et al. 2010) or relatively narrow (northeastern Gulf of Mexico, Fig. 3). Whether this is a function of upper-plate or lower-plate position (e.g. Péron-Pinvidic & Manatschal 2009) or Type I or

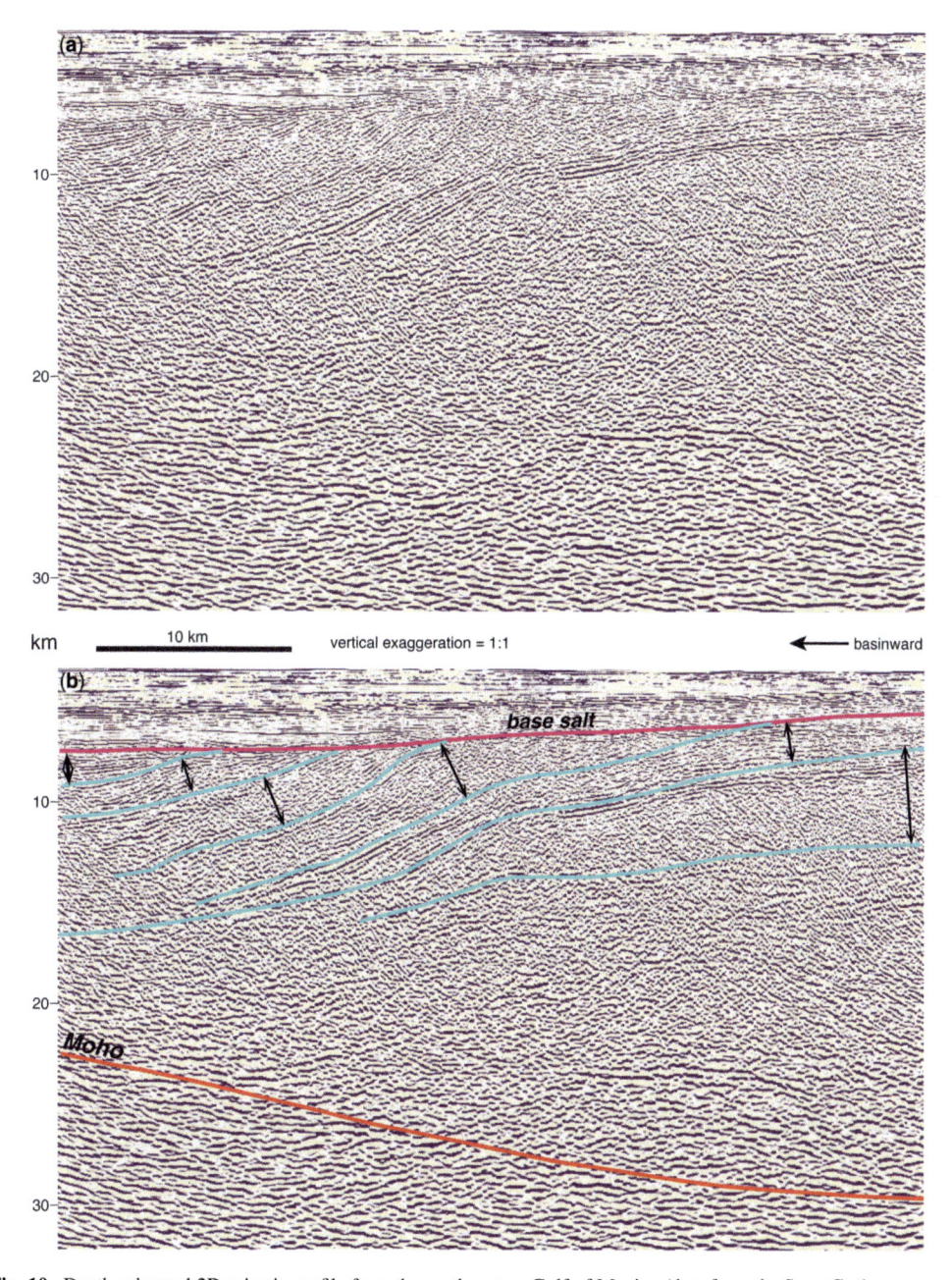

Fig. 10. Depth-migrated 2D seismic profile from the northeastern Gulf of Mexico (data from the SuperCache survey courtesy of Dynamic Data Services): (**a**) uninterpreted; and (**b**) interpreted. Red line is the Moho, warm pink is the base salt and blue lines highlight stratal geometries. Double-headed black arrows indicate thickest area of each interval.

Type II style deformation (Huismans & Beaumont 2011) is uncertain.

(2) The base salt or its equivalent weld is largely unfaulted in the proximal to distal domains. This is not to say that it has no structural relief, as it may be offset by small faults (Fig. 2) or draped over larger presalt crustal faults owing to differential compaction, for example in the Santos Basin of Brazil (e.g. Kumar *et al.* 2013, fig. 5b; Quirk *et al.* 2013, fig. 2). The

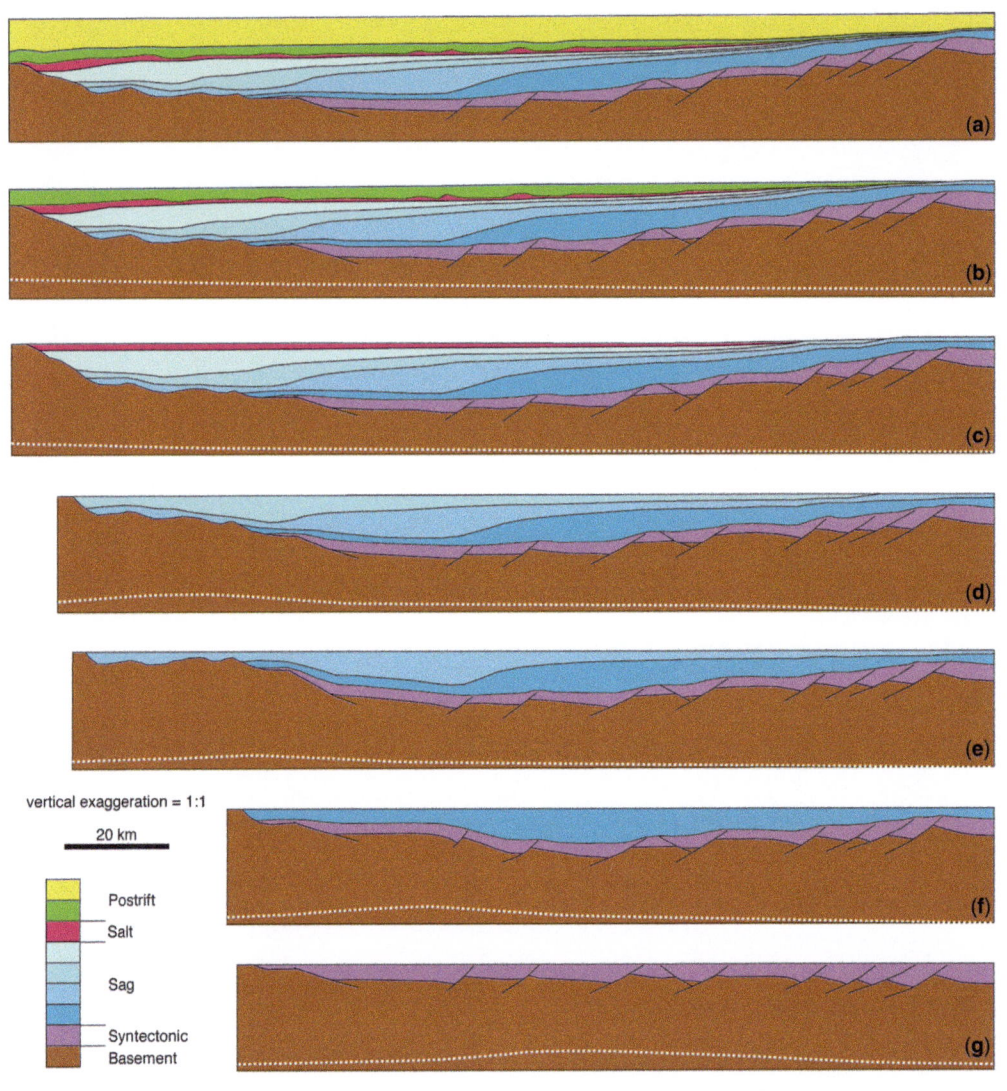

Fig. 11. (**a**) Depth conversion of a 2D seismic profile from the southern Gulf of Mexico (from Miranda *et al.* 2013) using assumed average velocities. (**b–g**) Sequential restorations generated using Move™ from Midland Valley Exploration. Restorations were by vertical simple shear and incorporated decompaction and flexural isostasy using an effective elastic thickness of 5 km. Each horizon was restored to horizontal and the water depths are not shown. The white dotted lines in (b–g) indicate the amount of crustal subsidence, measured from the baseline up, calculated for deposition of the next stratigraphic interval.

mostly smooth base salt ends abruptly, however, at the landward edge of the outer marginal trough (outer domain), where it drops basinward by as much as 3.5 km (Figs 2–5, 12 & 13). This is the same location where the Moho climbs to its shallowest level. The base salt in the outer marginal trough appears to have significant structural relief, although the image is often complicated by probable intrusive and/or extrusive magmatic rocks.

The basinward margin of the outer marginal trough is marked by the basement step (Barker & Mukherjee 2011) or inner ramp (Hudec *et al.* 2013), where the base salt often ramps up onto and over oceanic crust (Figs 2 & 13). Very similar geometries to all these described features were obtained in a numerical model for an intermediate-width margin with late synrift salt (Allen & Beaumont 2016, fig. 11).

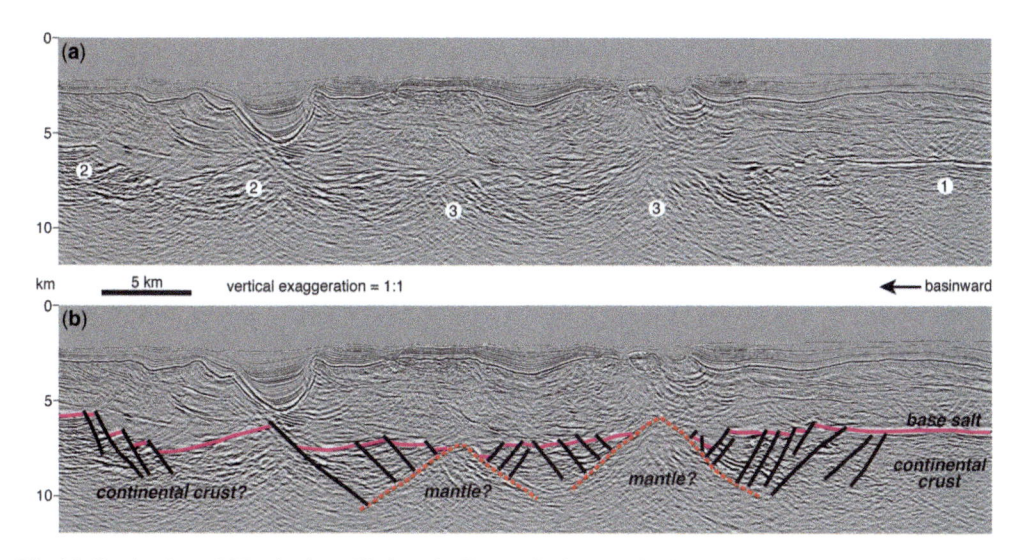

Fig. 12. Depth-migrated 3D seismic profile from the Kwanza Basin, Angola (data courtesy of TGS): (**a**) uninterpreted; and (**b**) interpreted (alternative interpretations are possible). Black lines are faults, warm pink is the base salt or equivalent weld and red dashed line is possible Moho over reactive mantle diapirs. The numbers in white circles mark features discussed in the text.

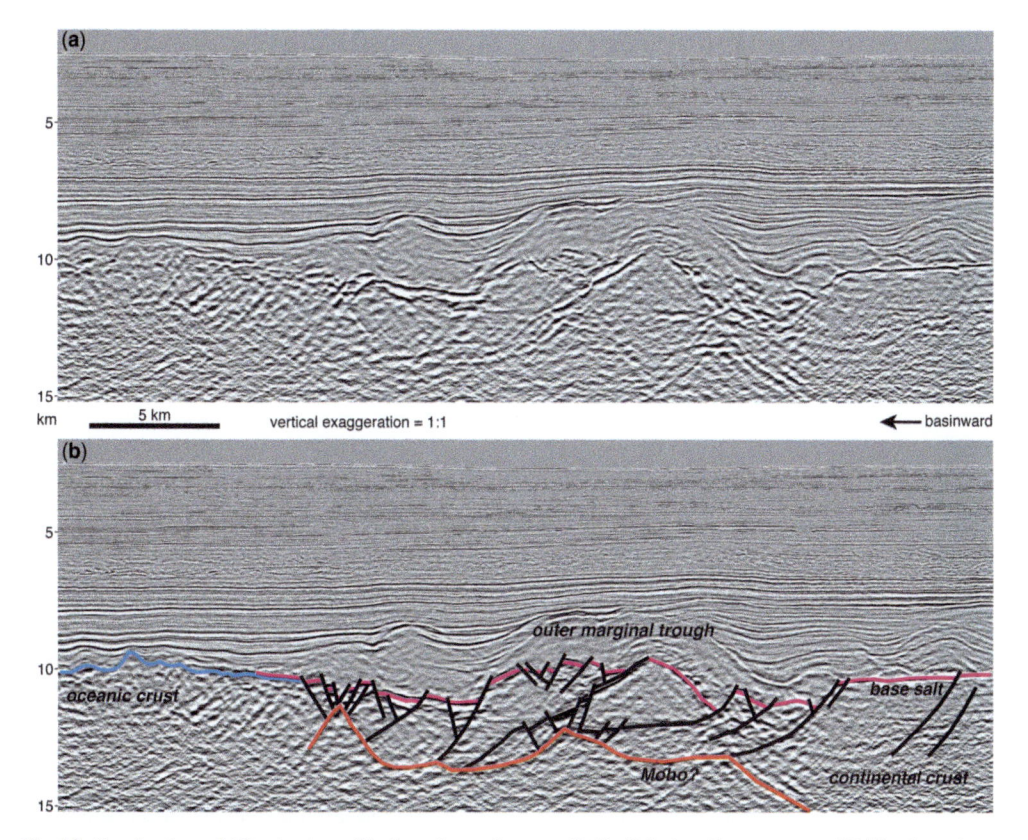

Fig. 13. Depth-migrated 3D seismic profile from the northeastern Gulf of Mexico (data courtesy of Schlumberger): (**a**) uninterpreted; and (**b**) interpreted (alternative interpretations are possible). Black lines are faults, red is the interpreted Moho, blue is the top oceanic crust and warm pink is the base salt or equivalent weld.

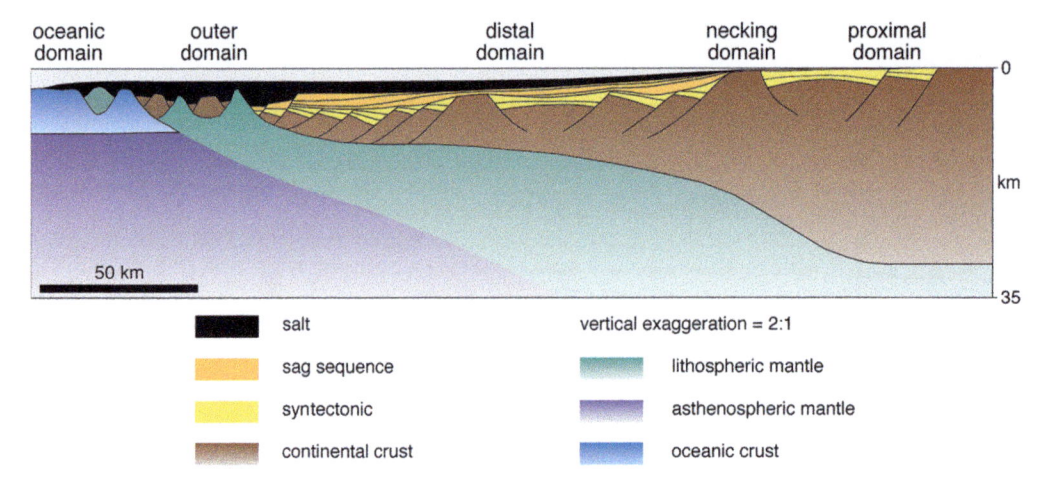

Fig 14. Schematic cross-section illustrating the various features observed and interpreted on seismic data from the Gulf of Mexico and South Atlantic salt basins (originally shown in Rowan 2015). Domain names are taken from Peron-Pinvidic *et al.* (2013). The geometries are simplified and there can be many natural variations in crustal and stratigraphic thicknesses, domain widths, fault patterns and makeup of the outer domain.

(3) Beneath the smooth base salt, which locally may be a minor angular unconformity, is a largely unfaulted sag sequence that may be as thick as 6–9 km (Figs 10 & 11; see also Henry *et al.* 2004; Karner & Gambôa 2007). Although broadly symmetrical in some areas and thus typical of presalt sag sequences (Figs 6 & 9), the strata dip basinward in other areas, often with a pseudo-clinoform geometry (Figs 10 & 11). The sag sequence is generally not apparent in the outer marginal trough, although parallel reflectors in some fault blocks (Fig. 12) probably represent the same stratigraphy. Interestingly, although the landward margin of the outer marginal trough is a fault with up to 3.5 km of relief (Fig. 4), there is no apparent footwall uplift as is typical with large rift faults (Figs 12 & 13).

(4) Underlying the sag sequence are syntectonic growth strata, with the boundary between the two marked by a local unconformity over basement highs. Although the boundary may be a single horizon over distances of at least 50 km (Fig. 6), it may be diachronous over the entire length of the necking and distal domains, with both the syntectonic and local postrift strata becoming increasingly younger basinward (e.g. Ranero & Pérez-Gussinyé 2010; Brune *et al.* 2014; Gillard *et al.* 2015a; Haupert *et al.* 2016; Andrés-Martínez *et al.* 2017). The rift geometry is generally asymmetrical, dominated by low-angle, basinward-dipping faults (Figs 2 & 4–8). Some of these faults appear to cut only the upper crust; others cut the entire crust and offset the Moho.

(5) The fault geometry changes markedly in the outer domain. Faults tend to be steeper and more symmetrical (Figs 2–5 & 12), although such faults may offset older, more asymmetrical faults (Fig. 13). The base salt is offset by the faults, forming a series of highs and lows; although the overall geometry is that of a trough situated outboard of the area of smooth base salt and inboard of oceanic crust, some highs are above the regional level of the base salt (Figs 12 & 13). Some of the highs appear to be rotated fault blocks of probable continental crust, but others have transparent, triangular shapes and overlying symmetrical fault arrays. These are similar to reactive salt diapirs, which form by boudinage of competent rocks between ductile salt below and air or water above. In this case, if the analogy is warranted, the salt forms the upper ductile layer and the lower ductile layer would represent weak mantle. Strong upper lithospheric mantle may be weakened by serpentinization; although the presence of thick overlying salt may inhibit downward percolation of water, reflux (downward seeping of brines) is a common process in salt basins (Warren 2006). Alternatively, the lower ductile layer might be the weak lower lithospheric mantle or even asthenospheric mantle.

The idealized cross-section of Figure 14 displays most of the features summarized above. However, it is highly schematic and there are many possible variations. For example: (1) the crustal geometries, including the relative widths of the various domains, can differ considerably; (2) the timing of faulting in the necking and distal domains may progressively young basinward, with significant overlap in ages, or there may be discrete jumps, with active extension in one area largely ceasing once rifting has shifted to more distal positions; (3) the sag sequence can range in thickness from less than 1 km to as much as 9 km (if the applied velocity models are correct) and can display varying degrees of asymmetry; (4) the base salt in the necking and distal domains may have variable amounts of relief owing to minor fault offsets or drape over deeper structures; and (5) the outer marginal trough (the outer domain) can be highly variable in terms of its width, the amount of relief on the base salt, the nature of the highs (fault blocks or diapirs), the presence and geometry of any Moho reflector, the relative proportion of upper continental crust, lower continental crust, lithospheric mantle and asthenospheric mantle and the degree of serpentinization and magmatic infiltration.

Discussion

In the following sections, I first summarize why I consider the Gulf of Mexico and South Atlantic salt basins to be situated along magma-poor margin segments. I then use the seismic observations, along with observations from other magma-poor margins and several published models, to address various aspects of the interactions between rifting and evaporite deposition: (1) the nature of lithospheric thinning; (2) the generation of accommodation for the sag and evaporite sequences; (3) the timing of salt deposition relative to development of the outer marginal trough; and (4) the palaeobathymetry during salt deposition. This leads to a general model for the evolution of syn-exhumation salt basins, and finally a discussion of some three-dimensional aspects of the Gulf of Mexico.

Magma-rich or magma-poor?

As summarized in the introduction, there is significant disagreement concerning whether the Gulf of Mexico and South Atlantic salt basins are along magma-rich or magma-poor margin segments. Some of us have even gone both ways, interpreting the northeastern Gulf of Mexico first as magma rich, containing SDRs (Pindell & Kennan 2007; Rowan 2014) but subsequently as magma poor (Pindell *et al.* 2014, 2016; and this paper). For the purposes of this discussion, I define magma-rich margins as containing large volumes of magmatic intrusive and extrusive rocks (SDRs) emplaced late in the rift history, as is typical for magma-rich margins around the world such as the southernmost South Atlantic (e.g. Franke 2013).

In the Gulf of Mexico, arguments supporting a magma-rich character centre on two lines of evidence. The first is a large-magnitude magnetic high along the Texas coastline (e.g. Mickus *et al.* 2009). However, this anomaly is located roughly 300 km from the limit of oceanic crust and is associated with uplift in the Late Triassic, c. 50 myr prior to break-up. Thus, it is more likely to represent a series of igneous plutons emplaced early in the rift history. Alternatively, it may mark a suture within the late Palaeozoic orogen of southeastern North America (Pascoe *et al.* 2016). The second line of evidence is the system of basinward-dipping reflectors observed on seismic data in the northeastern Gulf of Mexico (Fig. 10). Although these have been interpreted as volcanic SDRs (e.g. Imbert 2005; Imbert & Philippe 2005; Post 2005; Pascoe *et al.* 2016), I argued above that these reflectors do not have the geometry or potential-field signature of SDRs. Nevertheless, it is possible that the easternmost portion of the Gulf of Mexico is magma rich but becomes magma poor along-strike to the west (see Rowan 2014).

In the South Atlantic salt basins, there are again two primary lines of evidence used to invoke a magma-rich designation, both in offshore Brazil. The first is that many wells have penetrated volcanic rocks in the presalt section (e.g. Winter *et al.* 2007; Alvarenga *et al.* 2016). However, these are found mostly in early syntectonic growth strata, c. 20 myr prior to break-up; the presalt sag sequence is dominated by carbonates and siliciclastics. Second, an area of basinward-dipping reflectors in the Campos Basin has been interpreted as volcanic SDRs (e.g. Norton *et al.* 2015a; see also Torsvik *et al.* 2009). However, this geometry is rare on the Brazilian and conjugate west African margins, and thus might simply represent a sediment-filled half graben bounded by a landward-dipping fault or basinward-dipping sag-sequence strata.

I conclude that the evidence for large volumes of magma emplaced late in the rift history is rare in the outer domains of both the Gulf of Mexico and South Atlantic salt basins. Instead, both 2D and 3D seismic data show geometries characteristic of hyperextension and possible mantle exhumation (Figs 2–9 & 12–13). Thus, although both intrusive and extrusive magmatic rocks may be present locally, especially during early stages of rifting or in the outer marginal troughs, I classify both conjugate sets of salt basins as occurring on magma-poor margin segments.

Rifting and evaporite deposition

Lithospheric thinning. Much previous work has focused on the issue of symmetrical v. asymmetrical styles of crustal extension on magma-poor, conjugate divergent margins (e.g. McKenzie 1978; Le Pichon & Sibuet 1981; Lister *et al.* 1986; Brun

& Beslier 1996; Huismans & Beaumont 2002, 2003). In recent years, there has been an increasing tendency to invoke a transition from early symmetrical extension to late-stage asymmetrical rifting, ultimately leading to exhumation of subcontinental mantle in the footwalls of low-angle detachment faults (e.g. Péron-Pinvidic & Manatschal 2009; Reston 2009b, 2010; Peron-Pinvidic et al. 2013; Pérez-Gussinyé 2013). Yet the observations reported here suggest, at least for the Gulf of Mexico and South Atlantic salt-bearing margins, that there was a final switch back to more symmetrical extension. The common existence of prominent lows in the base salt on both sides of oceanic crust on conjugate margins and the internal block faulting of these lows (e.g. northeastern Gulf of Mexico and Yucatan margin, Figs 2–4) shows that the last stage of rifting prior to final break-up created a broadly symmetrical low that was subsequently divided by oceanic spreading. Moreover, the shallow Moho or lack of Moho in these outer domains suggests that lithospheric mantle was exhumed to very shallow levels beneath these troughs. Certainly extension becomes symmetrical by the time normal oceanic spreading begins, with the development of spreading centres; I am merely suggesting that the switch from asymmetrical to symmetrical deformation can begin sooner, during the exhumation stage.

Other recent work is broadly compatible with the results presented here. First, the geometry of synexhumation deposition on the Iberia–Newfoundland and Alpine conjugate margins was used to suggest that mantle exhumation was more symmetrical than the asymmetrical hyperextension of the thinning stage (Haupert et al. 2016). Second, a study of the Australia–Antarctica conjugate margins proposed that symmetrical, higher-angle faults formed after initial mantle exhumation in the footwalls of detachment faults (Gillard et al. 2015b, 2016). They suggested that the deformation then cycles back to the development of a new exhumation detachment, but that is not observed in the data illustrated here. Third, although in a different tectonic setting (back-arc basin), field data from the western Betics record the development of late-stage, high-angle faults during mantle exhumation (Frasca et al. 2016). Once the ductile middle crust was thinned enough, the crust and upper, high-strength mantle became mechanically coupled, with steeper brittle faults cutting older, low-angle shear zones. Finally, in one of their recent compilations, Péron-Pinvidic and co-workers invoke late-stage high-angle normal faults in the outer domain (Péron-Pinvidic et al. 2017, fig. 2).

I have postulated that the outer marginal trough may contain a complex assemblage of upper continental crust, lower continental crust, lithospheric mantle, magmatic intrusive and extrusive rocks and possibly asthenospheric mantle. Whatever the mixture is, the geometry is broadly that of large-scale boudinage. Boudinage has been produced in analogue models of rifted margins in which the deeper ductile layer is the weak lower lithospheric mantle (Brun & Beslier 1996) and in numerical models in which the deeper layer is the ductile lower continental crust (Huismans & Beaumont 2014). Boudinage leads to necking of the competent beam, which in turn generates triangular diapirs of the deeper ductile level (Vendeville & Jackson 1992). Mantle diapirs have been proposed in the past (e.g. Ramberg 1972) but were generally assumed to rise actively through buoyancy, separating and pulling apart the thinned crustal blocks (e.g. Pérez-Gussinyé 2013). However, reactive diapirs form by the passive rise of the ductile material into the space created by local necking of competent overlying rocks during extension (Vendeville & Jackson 1992). One possible difference between reactive diapirism of salt and mantle concerns the strength contrasts of the different layers. The competent overburden above salt has pronounced contrasts between it and the air or water above and the salt below. In contrast, the competent layer of thinned crust and possible upper lithospheric mantle has a sharp strength contrast with the overlying salt but a more gradational lower boundary owing to gradual weakening with depth (e.g. see the strength profiles for coupled deformation from Pérez-Gussinyé et al. 2001; Huismans & Beaumont 2002).

The cause of the switch from asymmetrical to symmetrical deformation is uncertain. Gillard et al. (2015b) suggested that it is caused by magmatic infiltration in the footwall of an exhumation detachment and consequent strain hardening. Alternatively, the cause may be geometric: normal faults tend to dip in the same direction as the underlying detachment when it has a slope but tend to dip in both directions as conjugate sets when the detachment is subhorizontal (e.g. McClay & Ellis 1987; Stewart 1999). It may be that during asymmetrical hyperextension, low-angle faults detach near the top of basinward-dipping ductile middle crust, but once the ductile crust is thinned enough and crustal and upper lithospheric mantle extension are coupled, more symmetrical faults may detach near the top of subhorizontal lower lithospheric mantle. Another possibility is that the evolving mechanical stratigraphy during depth-dependent stretching somehow favours the transition from more asymmetrical to more symmetrical deformation.

The observed geometries have implications for the relative timing of evaporite deposition and oceanic spreading. Active faulting over broad areas during salt deposition would suggest that deposition occurred prior to spreading, in contrast to some interpretations (e.g. Jackson et al. 2000; Marton et al.

2000; Imbert 2005; Montaron & Tapponnier 2010; Quirk et al. 2013; Norton et al. 2016); if the onset of spreading predated evaporite deposition, ongoing separation would be localized at the spreading centre. Norton et al. (2015a) proposed an innovative model in which the material beneath the salt in the outer marginal trough of the South Atlantic basins is intrusive oceanic crust and that deposition occurred as spreading commenced. However, the geometries beneath the salt in the outer marginal trough observed on 3D seismic data are not compatible with this interpretation. Specifically, the layered character in some areas (Fig. 12) and the common bright amplitudes beneath the salt in the outer marginal trough (Figs 5, 12 & 13), which indicate strong internal acoustic–impedance contrasts and thus significant variations in lithology, are both incompatible with intrusive asthenospheric magma. In addition, the apparent lack of Moho beneath outer marginal troughs in most cases, if not owing to seismic illumination issues, suggests that oceanic crust is not present. Thus, I conclude that, for both the Gulf of Mexico and South Atlantic, single pre-existing salt basins were separated into conjugate pairs by post-depositional oceanic spreading.

Generation of accommodation for sag and salt sequences. Accommodation for sediment in rift environments is provided by some combination of upper-crust brittle faulting, lower-crust ductile thinning, thermal subsidence owing to cooling of deeper lithospheric levels brought up to relatively shallow depths and flexural subsidence beneath added depositional loads. The first provides only local accommodation in half graben, whereas the other three generate more regional subsidence. One of the ongoing discussions in the literature regards just how and when the space for sag and salt sequences was generated (e.g. Karner & Driscoll 1999; Karner et al. 2003; Karner & Gambôa 2007; Reston 2010; Hudec et al. 2013; Huismans & Beaumont 2014; Rowan 2014; Pindell et al. 2014). Although some of these papers deal with just the salt or presalt, it is important to consider both the salt and the sag sequences since they are broadly parallel and thus probably shared a common tectonic origin.

A key observation is that the base salt in the outer marginal trough is below the level of oceanic crust in the northeastern Gulf of Mexico (Hudec et al. 2013; Pindell et al. 2014; Rowan 2014). Even the basinward limit of the smooth base salt, elevated with respect to the outer marginal trough, is roughly level with the top of oceanic crust. There are two end-member explanations. First, there could have been a deep, basinwide depression at the onset of evaporite deposition, i.e. sag-sequence strata did not fill the basin to sea-level (e.g. Warren 2006; Karner & Gambôa 2007; Reston 2009a, 2010; Davison et al. 2012; Garcia et al. 2012; Hudec et al. 2013; Rowan 2014; Lundin & Doré 2017). In contrast, the top of the presalt strata could have been essentially at sea-level when evaporite deposition commenced (e.g. Pindell et al. 2014; Cowie et al. 2016, 2017), in which case there had to be rapid subsidence of at least 2–3 km during and immediately after salt deposition. Specifically, Pindell et al. (2014) invoked tectonic collapse on conjugate landward-dipping 'outer marginal detachments' connecting the basinward margins of the outer marginal troughs to shear zones near the base of the crust.

Although presalt penetrations are absent in the Gulf of Mexico margins, except in the most proximal portions, well data in the South Atlantic basins show that shallow-water facies are locally present just beneath the salt (e.g. Bertani & Carozzi 1985; Bate 1999; Karner & Gambôa 2007; Saller et al. 2016), which might be interpreted to support the idea that the basin was filled and that early salt was deposited close to global sea-level. However, deeper-water facies are also known. For example, Karner et al. (2003) pointed out that water depths were at least 500 m in the Argilles Vertes to the east of the Atlantic Hinge on the Congo margin and should have been even deeper in more distal positions. More importantly, shallow water does not necessarily mean near sea-level when the strata are lacustrine, as lake surfaces in isolated depressions may be significantly deeper than sea-level. Saller et al. (2016) identified the upper synrift to sag sequences of the Kwanza Basin as shallow- to deepwater, exclusively lacustrine strata deposited in an increasingly saline and alkaline environment that eventually led to the precipitation of evaporites, an interpretation indicative of deposition in an isolated depression with one or more large lakes.

How deep the depressions might have been depends on the interplay between subsidence rate and sediment accumulation rate during the presalt interval. Flexural backstripping, with corrections for thermal subsidence, has yielded maximum depths at the onset of evaporite deposition in the South Atlantic of 2–3 km, and even up to 4.5 km (Karner & Driscoll 1999; Norton et al. 2015b; Cowie et al. 2016, 2017), and a value of 2 km has been derived for the northeastern Gulf of Mexico (A. Roberts, pers. comm. 2013). Reconstructions of the northern Bay of Biscay margin have yielded palaeobathymetry of 3–4 km at the onset of spreading, with the deepest water depths in the transition from continental to oceanic crust, and outcrops in the Alps and French Pyrenees show that strong subsidence and rapid creation of accommodation began during the onset of hyperextension (Masini et al. 2013; Tugend et al. 2015). Numerical forward modelling of magma-poor margins has generated basin depths (variously air-loaded, water-loaded or partly sediment-loaded) of 3–5 km just prior to mantle exhumation or oceanic spreading (Lavier & Manatschal 2006; Huismans & Beaumont 2014; Brune et al. 2014; Mohn et al. 2015; Allen & Beaumont 2016). In most cases these depths were reached after 16–26 myr of total rifting. Time-span estimates for sag and evaporite strata in the South Atlantic are 5–10 and <1–5 myr, respectively (e.g. Freitas 2006; Karner & Gambôa 2007; Unternehr et al. 2010; Davison et al. 2012; Mohriak & Leroy 2013; Quirk et al. 2013; Saller et al. 2016; A. Pulham, pers. comm. 2016). Thus there was no problem in generating enough subsidence for sag and salt deposition through normal rift-related processes.

Sedimentation may have been rapid enough that the basin floor remained near sea-level during crustal subsidence and deposition of the sag sequence (e.g. Pindell et al. 2014). However, underfilled basins are known in extensional environments ranging from relatively narrow rift basins such as the northern North Sea, where there was up to 1 km of topographic relief at the end of rifting (e.g. Barr 1991; Gabrielsen et al. 2001), to passive margins such as the Alps and offshore Portugal, Norway or east India, where observed onlap geometries on seismic data show that the rift geometry was only gradually infilled (e.g. Péron-Pinvidic et al. 2007; Osmundsen & Redfield 2011; Masini et al. 2013; Haupert et al. 2016). Furthermore, radiolarian cherts deposited directly above exhumed subcontinental mantle in the Alps and offshore Portugal document deep, underfilled settings near the end of rifting (Péron-Pinvidic & Manatschal 2009). In numerical models, there is still 2–3 km of bathymetry after deposition of 7 and 4 km-thick sag sequences, respectively (Allen & Beaumont 2016). The South Atlantic and Gulf of Mexico were enormous salt basins, with maximum dimensions of about 2500 by 650 km and 1500 by 800 km, respectively (using reconstructions from Lentini et al. 2010; Hudec et al. 2013). To have completely filled that much rift-created accommodation with sediment during rifting is unlikely, an argument supported by the almost complete lack of interbedded siliciclastics in the evaporite sequences (e.g. Salvador 1987; Gambôa et al. 2008).

Why is the sag sequence sometimes more symmetrical (e.g. Figs 6 & 9) and sometimes more asymmetrical (e.g. Figs 2, 3, 8, 10 & 11)? Similar variations in late synrift stratal geometries have been observed on magma-poor margins without evaporites, but conflicting patterns have been noted. Whereas Decarlis et al. (2015) showed the asymmetrical geometry on the upper plate, Masini et al. (2013) and Gillard et al. (2015a, 2016) showed it on the lower plate and Haupert et al. (2016) suggested that it occurs on both plates in the zone of exhumation, with symmetrical geometries dominating in the hyperextension province. It may be that asymmetrical sag strata form where there is significant and rapid basinward younging of crustal extension, which may not fit simple upper plate–lower plate models. The distribution of symmetrical and asymmetrical sag sequences may also reflect variations in the relative importance of loading and thermal subsidence vs that owing to lower-crustal thinning. Alternatively, recent numerical models show that sag sequences are more symmetrical for weak lower crust and more asymmetrical for strong lower crust (Andrés-Martínez et al. 2017; Andrés-Martínez, pers. comm. 2017).

A final point is that evaporite deposition itself contributes to the accommodation for further evaporites owing to loading subsidence. Davison et al. (2012) determined that 3 km of salt creates another 1.5–2 km of subsidence, and Hudec et al. (2013) stated that a final salt thickness of 3–4 km requires only a depression slightly over 1 km deep. These calculations assumed that salt filled the depression to sea-level, but that is not a requirement.

Generation of the outer marginal trough and timing of salt deposition. Where the base salt is smooth, the salt was deposited above the sag sequence and thus post-dated most extensional faulting in those areas. Yet what was the relative timing of salt deposition and formation of the outer marginal trough? One of the difficulties in determining this is that there are no growth geometries within salt, so that relative timing is unclear. Thus there are three simple possibilities: (1) the salt was deposited prior to development of the outer marginal trough, so that it was locally prerift with the base salt subsequently offset by later faulting; (2) the trough formed during evaporite deposition, making the salt locally synrift; and (3) development of the trough occurred prior to salt deposition, meaning the salt was locally postrift as in more proximal positions. Of course, combinations are possible in any given location and the relative timing might vary along-strike (see below). In this discussion, and for the reasons cited above, I assume that the outer marginal trough formed prior to oceanic crust and thus discount the model in which it represents a form of intrusive oceanic crust (Norton et al. 2015a).

Could evaporite deposition have predated, at least to some degree, formation of the outer marginal trough? Yes, but in this case there would have been matching suprasalt extension. In other words, as rifting was accommodated by lithospheric extension beneath the salt in the outer domain, the suprasalt cover would also have been extended. However, it would not have had to occur in the same location; instead, it could have been offset laterally owing to decoupling of deformation at the evaporite level, as is typical in thick-skinned extension involving salt (e.g. Jackson & Vendeville 1994; Coward & Stewart 1995). In fact, this is observed in the northeastern Gulf of Mexico and Yucatan margins (see below).

Could evaporite deposition have been entirely synchronous with outer marginal trough development? In other words, could salt represent the entire time recorded by formation of the outer marginal trough? This is considered implausible for two reasons. First, evaporite deposition is generally rapid, typically occurring in 2–3 myr at most (and maybe much less), so is unlikely to span the entire time of formation of the outer trough. Second, this model would require a gap in observed extension during sag-sequence deposition, which would have post-dated crustal extension in the distal domain but pre-dated extension in the outer marginal trough.

Could at least some of the development of the outer marginal trough have predated evaporite deposition? This would solve the problem of the apparent time gap in that deposition of the sag sequence in the distal domain would have been synchronous with extension in the outer marginal trough. In other words, strata making up the sag sequence, which is locally postrift, would represent syntectonic growth strata in the outer marginal trough. There is some evidence for this in offshore west Africa at least. There, the inner boundary of the outer marginal trough has a kinked geometry, with an upper, lower-angle portion and a lower, steeper portion (Fig. 15). One possible explanation is that the upper segment represents an early fault that was rotated during deformation and then cut by a younger, steeper fault (e.g. Proffett 1977; Reston 2005). However, the lack of stratal rotation in the footwall precludes this interpretation. Alternatively, the upper segment represents an eroded fault scarp, suggesting that there was relict topography at the time of evaporite deposition in the outer marginal trough. Thus, the salt in this case would have post-dated at least the early development of the outer marginal trough.

Evaporite deposition. I have argued that the onset of evaporite deposition occurred in an arid, isolated depression formed during the preceding rifting and associated subsidence. The seismic geometries suggest that ongoing rifting during salt deposition, prior to oceanic spreading, was localized in the outer marginal trough. An important observation is that where the salt is thin, for example the northeastern Gulf of Mexico, the top salt (in addition to the base salt) is also below the level of oceanic crust and is onlapped by strata that abut against the edge of oceanic crust (Fig. 3; Pindell *et al.* 2014; Rowan 2014). As pointed out by Pindell *et al.* (2014), there are two possible solutions: either the top salt, like the base salt, was originally at depth in a depression; or the top salt was near sea-level and subsided rapidly to oceanic depths.

Although Hudec *et al.* (2013) invoked the onset of salt deposition occurring in a depression, they claim that salt eventually filled the basin to sea-level. The evaporites then stretched and thinned as some form of extension (hyperextension or mantle exhumation) occurred beneath the salt, thereby dropping the top salt to near the level of oceanic crust and the base salt even deeper. However, whereas the zone of extension that occurred after (or during) salt deposition is over 400 km wide in their model, the data shown here demonstrate that such extension was confined to the outer marginal trough, a much narrower zone. A quick calculation shows that creation of a a trough 50 km wide and 2 km deep would drop the top of an originally 200 km wide and 3 km thick salt basin by only 1 km, not enough to get to the level of oceanic crust. Moreover, applying this model to areas of thin distal salt (e.g. northeastern Gulf of Mexico and Yucatan margins) is more problematic. For example, if the salt was originally 500 m thick and the top was approximately at sea-level, then the base would have been at 500 m below sea-level. It would then have to drop very rapidly by 2 or more km, but with no obvious mechanism in the area landward of the outer marginal trough where there was no ongoing extension beneath the salt. This argument also applies to the variation on this model by Norton *et al.* (2015a), in which the extra space required to drop the salt is created only in the outer marginal trough.

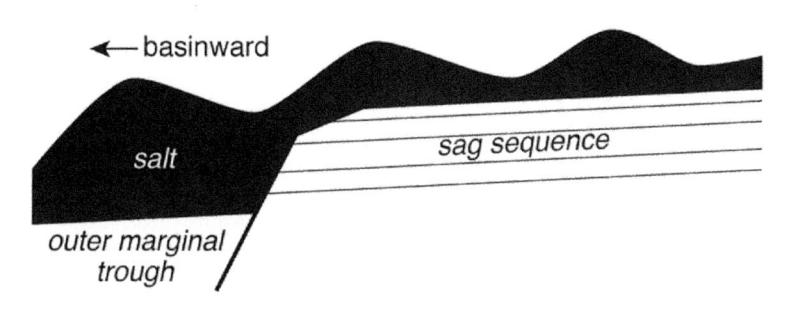

Fig. 15. Simplified sketch of geometry observed on 3D seismic data from a West African South Atlantic salt basin.

One possible solution to this problem is outer marginal collapse, in which both base and top salt would indeed have dropped significantly during mantle exhumation, but over a wider zone including updip of the outer marginal trough (Pindell et al. 2014). This model requires a drop of the top salt from sea-level to c. 2.5 km depth in the very short time between the cessation of evaporite deposition and the onset of oceanic spreading (Pindell et al. 2014), which is, in my opinion, problematic. Moreover, one of the arguments underlying outer marginal collapse is based on a seismic profile from the northeastern Gulf of Mexico on which thin salt can be traced directly from its proximal onlap limit in the Apalachicola Embayment to its distal termination below the level of oceanic crust (Pindell et al. 2014, fig. 13). Since the updip limit of salt was supposedly at sea-level, the distal salt must supposedly also have been at sea-level, a conclusion presumably based on the assumption that the top salt should have been horizontal at the time of deposition. Supporting evidence would be provided by data showing that strata directly above the salt were deposited in the same water depths in both locations. In fact, well data show that the salt in proximal areas is overlain by and laterally equivalent to eolian sands, which could have been deposited above sea-level or in a subaerial depression below sea-level, and then a sequence of shales, thin dolostone and anhydrite and micritic limestone (Brand 2016). More distal wells, however, encountered shallow- to deeper-water siliciclastics and carbonates above the salt (S. Krueger, pers. comm. 2017), suggesting there was topographic relief on the top salt at the end of evaporite deposition. Moreover, dune preservation suggests rapid drowning, which excludes progressive transgression of the global ocean and consequent reworking of the dunes (Imbert & Philippe 2005). Such drowning is more compatible with gradual filling of an air-filled depression at the end of salt and dune deposition than with any kind of tectonic or thermal/loading subsidence and consequent transgression.

Wells in offshore Brazil also provide critical data, at least for that margin. Proximal wells have encountered up to several kilometres of Albian shallow-water marine carbonates directly above the salt (e.g. Williams & Hubbard 1984; Modica & Brush 2004; Okubo et al. 2015). In contrast, the Albian in the first Tupi well, in the deepwater Santos Basin, penetrated only 140 m of shales and marls, with minor micritic limestone and sandstone, highly suggestive of deepwater sedimentation (Mohriak, this volume, in review). Again, the conclusion is that the top salt was not horizontal.

The belief that the top of a salt layer must have been horizontal at the time of deposition is a common misconception. Although salt (in the general sense) behaves as a viscous material, its flow rate, and thus its ability to level out, depends in part on its viscosity (which is to a large degree dependent on the mechanical stratigraphy of the layered evaporite sequence) and its thickness. For example, increasing the viscosity of modelled salt from 10^{18} to 10^{19} Pa s significantly reduces salt flow during margin tilt (Goteti et al. 2013). Thus, the top salt can maintain a slope, especially over a short time interval and/or where the salt is thin. Numerical models of passive margins show that, even where the halite is continuous, the top salt can maintain up to 2.5 km of relief for up to 5 my (Goteti et al. 2013; Allen & Beaumont 2016). Sandbox models of tilted half graben show that the top silicone polymer attained a subhorizontal attitude where the salt was thickened by lateral flow but maintained a slope of close to 10° in the areas thinned by downdip flow (B. Vendeville, pers. comm. 2016). Submarine namakiers (salt glaciers) in the Red Sea have relief on the top salt of up to 600 m over a horizontal distance of 28 km (Augustin et al. 2014), giving an average slope of over 1.2°. For passive margins with salt 100, 200 and 300 km wide, 2.5 km of relief on the top salt requires slopes of 1.4°, 0.72° and 0.48°, respectively (the angles decrease if oceanic spreading occurred at depths shallower than 2.5 km). Thus there is no mechanical reason why the top salt could not have been near sea-level in proximal positions and near the level of oceanic crust in distal domains on the Gulf of Mexico and South Atlantic margins.

Established models of evaporite deposition in deep depressions are based on the Messinian evaporites of the Mediterranean. Basin isolation in an arid climate led to evaporation of seawater, drawdown of the brine level by up to 1.5 km, and cyclical precipitation of evaporites in shallow water but a deep basin (e.g. Warren 2006; Ryan 2009). In contrast, a new model invokes linked erosion and deposition in a non-desiccated deepwater basin (Roveri et al. 2014). Others have argued that the drawdown model is not needed if the precursor setting was lacustrine at levels below sea-level (e.g. Reston 2010), which is consistent with the facies encountered beneath the salt in offshore Angola (Saller et al. 2016). In the Gulf of Mexico, unpublished work suggests that the evaporite geochemistry requires a constant-recharge model of precipitation in water depths ranging from shallow in proximal areas to over 2 km in distal areas (F. Peel, pers. comm. 2014). In any case, whether the Gulf of Mexico and South Atlantic evaporites were deposited in mostly air-filled depressions or mostly water-filled depressions is immaterial to the focus of this paper. In either case, the difference in palaeobathymetry in strata just above the salt, between shallow-water to nonmarine facies in proximal settings and deepwater facies in distal settings, shows that there was topographic relief on the top of salt and that the basins were thus not filled to sea-level. Moreover, this gets around the significant problems associated with published models (Hudec et al. 2013; Pindell et al. 2014; Norton et al. 2015a) that would require the entire thickness (top and base) of originally thin salt in the northeastern Gulf of Mexico to drop multiple kilometres in a very short time span.

One other aspect of evaporite deposition is that its areal distribution is more widespread than that of the sag sequence. Whereas sag strata are confined mostly to the necking and distal domains, relatively thin salt is found also in the proximal domain, where it occurs above a major unconformity (incorrectly termed the break-up unconformity; see Rowan 2014) underlain by basement or early synrift redbeds. This is analogous to the Bay of Biscay–Pyrenean system, where thin post-hyperextension (syn-exhumation) shallow-water facies were deposited above eroded basement in the proximal domains, suggesting the onset of regional thermal subsidence (Tugend et al. 2015).

Evolutionary model

The various arguments cited above are combined into a model that illustrates the simplified genesis and evolution of the geometric features observed and interpreted on the seismic data and shown schematically in Figure 14. This model borrows elements from other models (including Péron-Pinvidic & Manatschal 2009; Peron-Pinvidic et al. 2013; Brune et al. 2014; Allen & Beaumont 2016), and I emphasize that many variations are of course possible. The progressive development is depicted in Figure 16 and summarized below.

(1) Rifting begins by roughly symmetrical brittle faulting of the upper crust in areas that will become the proximal parts of the margins (Fig. 16a), although more distributed early extension is also possible. Depth-dependent stretching leads to ductile thinning of the lower crust in areas that will become the distal portions. This is equivalent to the stretching stage of Péron-Pinvidic & Manatschal (2009).

(2) In the thinning stage, upper-crustal extension becomes more asymmetrical and shifts laterally into what will become the necking domain on one margin (Fig. 16b) and eventually into the distal domain (Fig. 16c, d). Ductile thinning of the lower crust eventually leads to coupled deformation, with low-angle faults cutting the entire crust (Fig. 16d).

(3) Subsidence occurs in the abandoned portions of the rifted margin, driven by some combination of thermal and loading subsidence and ongoing lower-crustal extension. The locus of subsidence shifts laterally, following behind the active upper-crustal rifting, thereby generating basinward-shifting sag sequences (Fig. 16c–e).

(4) Minor relative uplift occurs in areas of active upper-crustal rifting, possibly owing to rising asthenospheric mantle beneath the area of greatest lithospheric thinning, which also shifts laterally (Fig. 16c–e). This pattern has been interpreted in late synrift strata of the Alps (Masini et al. 2013). Proximal uplift and erosion also occur (Fig. 16d, e), producing the so-called break-up unconformity, which actually correlates temporally with the thinning and exhumation stages (see Rowan 2014 for a discussion of the processes that might contribute to both proximal and distal uplift).

(5) The deformation becomes more symmetrical again during the exhumation stage in the outer domain (Fig. 16e, f), taking the form of large-scale boudinage and leading to the development of the outer marginal trough and possible reactive mantle diapirs.

(6) Evaporites are deposited over a short time interval during the final stage of rifting, with the greatest thicknesses occurring in the outer marginal trough owing to ongoing extension (Fig. 16f). Subsidence is now margin-wide, similar to that of the postrift (post-salt) section, so that the salt has a wider extent than the sag sequence, including over the proximal unconformity in places. The salt is deposited in a pre-existing depression and the top salt has topographic relief, near sea-level in proximal positions and near the level of oceanic crust in distal positions. Some of the relief on top salt may be caused by syn-depositional lengthening and thinning of the salt owing to ongoing extension in the outer marginal trough (Hudec et al. 2013; Norton et al. 2015a).

(7) Eventually, continued symmetrical extension and thinning (boudinage) leads to breakthrough of asthenospheric mantle, possibly as reactive diapirs, and the development of steady-state oceanic spreading (Fig. 16g). Margin-wide subsidence continues and salt may flow out over newly created oceanic crust.

Three-dimensional aspects

The interpretations and models in this paper are all two-dimensional, but of course there can be significant along-strike variations in crustal geometries, timing, synrift strata and original salt thickness and distribution. To take just one example, the thickness and asymmetry of the sag sequence can vary considerably. This may be caused by such factors as differences in crustal rheology, pre-existing structural fabric, differences in the amount of extension on

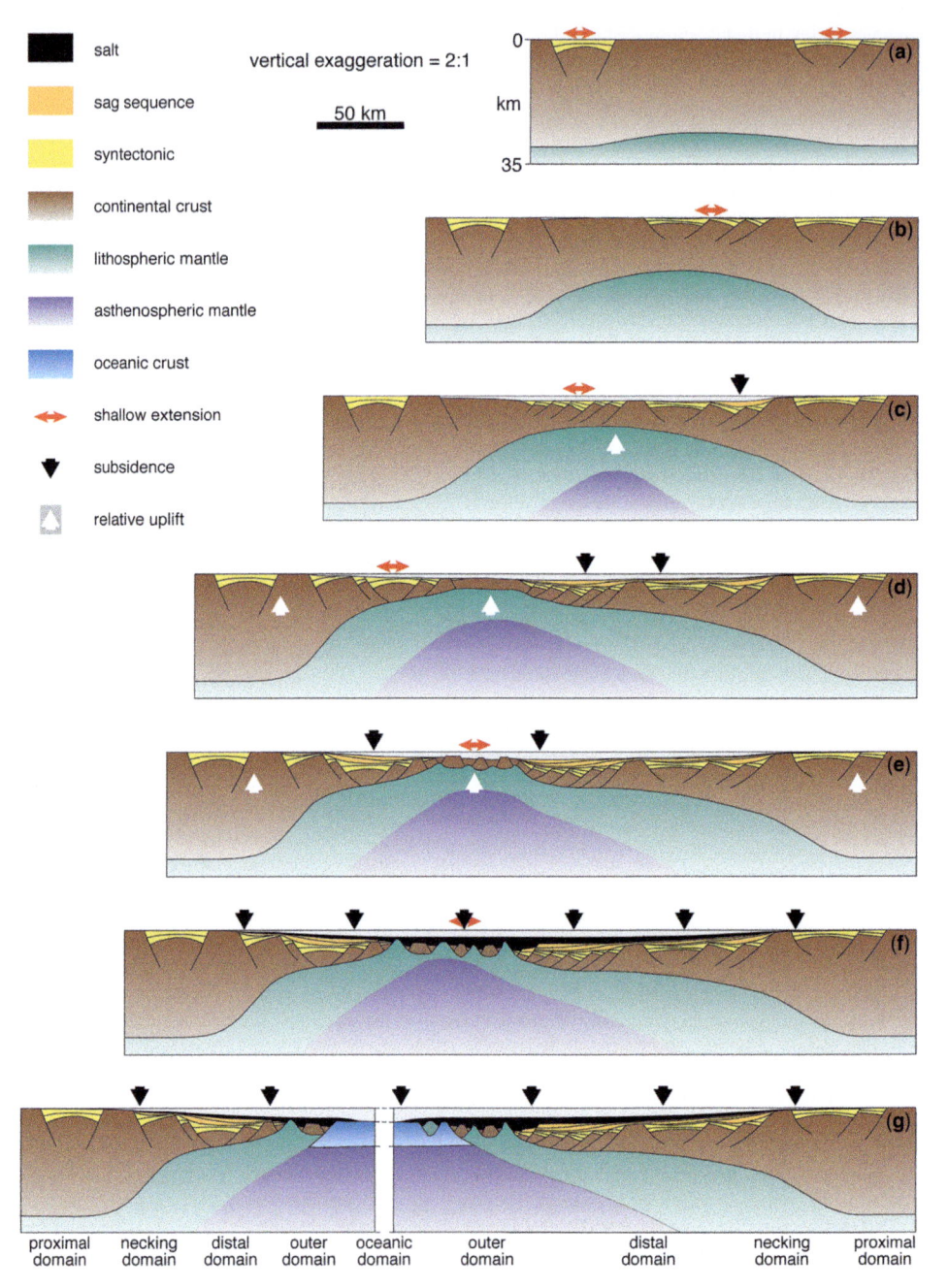

Fig. 16. Schematic evolution of syn-exhumation (late synrift) salt basins (originally shown in Rowan 2015) that draws on the seismic observations made here as well as models by Péron-Pinvidic & Manatschal (2009), Peron-Pinvidic *et al.* (2013), Brune *et al.* (2014) and Allen & Beaumont (2016). (**a**) Stretching stage with symmetrical upper-crustal rifting on steep faults and ductile thinning of the lower continental crust. (**b–d**) Thinning stage with basinward-shifting extension on asymmetrical low-angle faults, associated syntectonic deposition, continued ductile thinning of the lower crust and consequent basinward-shifting subsidence enhanced by thermal and loading effects. (**e, f**) Shift to symmetrical extension during mantle exhumation stage, with boudinage of the thinned crust and formation of reactive mantle diapirs; evaporites are deposited during exhumation, with thickest salt in the outer marginal trough (outer domain). (**g**) Breakthrough of asthenospheric mantle and development of normal oceanic crust, separating the salt basin into two conjugate basins.

underlying faults, complex three-dimensional flow of ductile crust and mantle layers, the evolving thermal regime, and so on.

Here I examine, first the effects of variable timing of rifting along the length of the northern Gulf of Mexico and, second, the differences in salt flow between ridge- and transform-dominated segments of oceanic break-up. Most interpretations show that the oceanic basin opened by anticlockwise rotation of Yucatan about a pole located near the western tip of Cuba (e.g. Pindell 1985; Marton & Buffler 1994; Imbert & Philippe 2005; Pindell & Kennan 2007, 2009; Pindell et al. 2016). Thus the oceanic crust is wider and, if the onset of spreading propagated to the east (Pindell et al. 2014, 2016; Pascoe et al. 2016), older in the western Gulf of Mexico. By inference, the formation of the outer marginal trough was also older to the west (Pindell et al. 2014). The trough is narrow in the northeastern Gulf of Mexico, but it is unknown whether it widens to the west. Kneller & Johnson (2011) showed an interpretation in which a zone of ultra-slow spreading lithosphere (possible exhumed subcontinental mantle) was wider to the west and narrowed eastward.

Several other changes occur along-strike. First, there are fewer diapirs and salt sheets to the east, suggesting that salt was originally thicker in the centre and west. This would have been enhanced in the central outer domain by basinward flow of salt both during evaporite deposition (Davison et al. 2012; Quirk et al. 2013) and subsequently beneath a thin overburden. Such flow would have been convergent in the highly arcuate northern Gulf of Mexico margin (Fort & Brun 2008). Second, the height of the basement step, from the base of salt up onto the top of the oceanic crust, increases to the west (Barker & Mukherjee 2011; Hudec et al. 2013). Seismic profiles show relief of 1–2 km in the east and 3–5 km in the west (Hudec et al. 2013; Rowan 2014). Third, the amount of Upper Jurassic suprasalt extension in the east is greater than the amount of distal contraction (Imbert & Philippe 2005; Hudec et al. 2013). Unpublished interpretations suggest that this discrepancy is reduced or absent to the west, but whether there is a gradual change or an abrupt decrease across a transform fault is unclear. A similar pattern is observed in the southern Gulf of Mexico, where there was relatively little Upper Jurassic extension above the salt in the Bay of Campeche but significant amounts along the NW Yucatan margin (Hudec 2017).

The observations and interpretations cited above lead to a simple model of the variable evolution along the strike of the northern Gulf of Mexico (Fig. 17). Whereas the stage of extension at any given time differed from west to east (earlier in the west), evaporite deposition across the Gulf is assumed to have been broadly synchronous. Because the salt is thought to have been deposited in a pre-existing depression, both the onset and cessation of evaporite deposition are assumed to have been coeval throughout the basin because they would have been controlled largely by climate. This is in contrast to Pindell et al. (2014), who depicted salt deposition

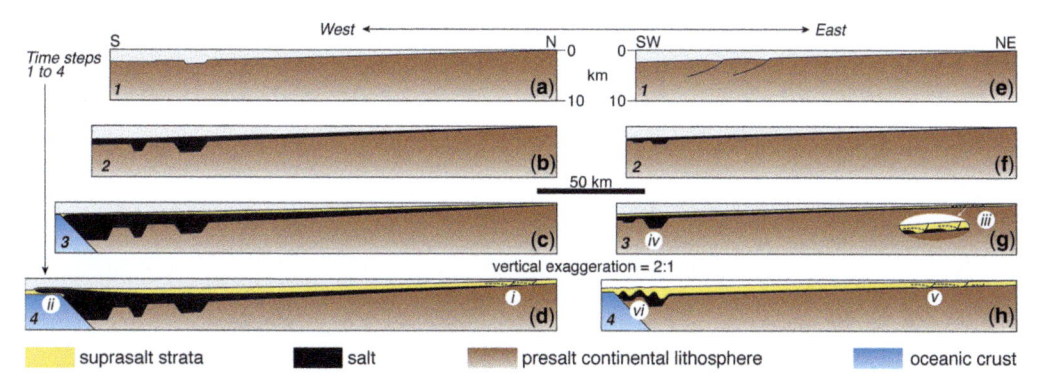

Fig. 17. Schematic depiction of along-strike variations in the evolution of the northern Gulf of Mexico, with time steps 1–4 meant to be coeval in the West and East. The geometries are highly simplified; for example, no early diapirism is depicted and the presalt continental lithosphere includes sag and syntectonic strata, continental crust and lithospheric mantle that are not indicated. (a–d) West-central: (a) after onset of symmetrical extension and initiation of outer marginal trough; (b) continued extension and early salt deposition; (c) after further extension and evaporite deposition followed by emplacement of oceanic crust, with minor distal inflation of the now-covered salt; and (d) linked gravity-driven proximal extension (location *i*) and distal salt breakout and nappe advance (location ii). (e–h) Northeastern: (e) still in hyperextension stage with low-angle faults; (f) initiation of outer marginal trough and early salt deposition; (g) ongoing evaporite deposition and then burial during decoupled extension, with proximal suprasalt faulting (location iii, see inset) and distal presalt extension in the outer marginal trough (location iv); and (h) linked gravity-driven proximal extension (location v) and contraction buttressed by edge of oceanic crust (location vi).

beginning earlier in the west owing to earlier creation of accommodation. The model also assumes that the evaporites were deposited over a period of several million years, as indicated by strontium isotope ratios in onshore salt domes (A. Pulham, pers. comm. 2016). It is possible that evaporation took place over a much shorter time frame (as little as several tens of thousands of years; F. Peel, pers. comm. 2014), but this would introduce only minor changes to the depicted model.

At some point near the end of the rifting history, the western Gulf was already starting to form an outer marginal trough, with more symmetrical deformation, while the eastern Gulf was still undergoing asymmetrical hyperextension (Time 1, Fig. 17a, e). The outer marginal trough in the west continued to widen while it was just getting started in the east (Time 2, Fig. 17b, f). Evaporite formation began during this stage, with thicker salt where active extension was occurring in the outer marginal trough. Crustal extension (and possible mantle exhumation) was still ongoing in the east by the time oceanic spreading started in the west (Time 3, Fig. 17c, g). Furthermore, loading of the thicker salt in the west initiated distal inflation of the salt beneath a thin overburden (Fig. 17c). In contrast, presalt extension in the outer marginal trough in the east (Fig. 17g, location iv) was accommodated by suprasalt extension in more proximal positions (Fig. 17g, location iii), a form of decoupled thick-skinned extension. By the time spreading commenced in the eastern Gulf of Mexico (Time 4, Fig. 17d, h), gravitational failure was well under way in the west, with proximal extension (Fig. 17d, location i) linked to distal inflation, salt breakout and nappe emplacement over the oceanic crust and abyssal strata (Fig. 17d, location ii). Inflation was minimal to the east owing to the thin salt and consequent relative lack of Poiseuille flow. Instead, proximal gravity-driven extension (Fig. 17h, location v) was balanced by contraction buttressed by the edge of oceanic crust (Fig. 17h, location vi).

Another aspect of varying interaction between rifting and salt geometry around the Gulf of Mexico is the distribution of early allochthonous salt nappes emplaced out over oceanic crust. The maximum distance from the break-up edge to the toe of the nappe differs significantly (Hudec *et al.* 2013, figs 8 & 13). First, nappes are non-existent in the far eastern Gulf of Mexico, probably owing to the originally thin salt as argued above. Second, salt moved at most 20 km along the NW margin of Yucatan, again because of relatively thin salt. Third, it flowed up to 100 km in the north-central Gulf of Mexico owing to much thicker depositional salt and convergent flow from the NW, north and NE (Fort & Brun 2008). Fourth, and most enigmatically, it generally flowed less in the NW Gulf of Mexico and especially

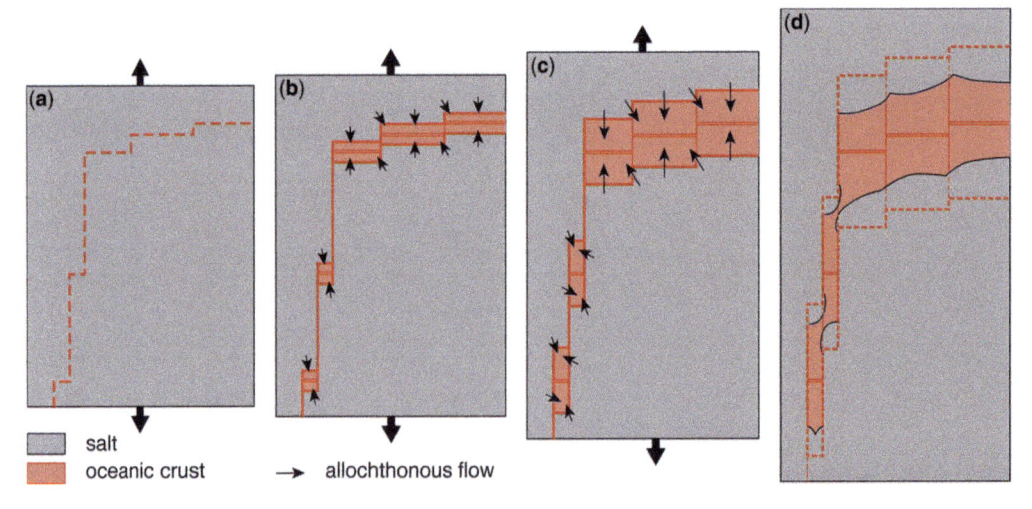

Fig. 18. Schematic illustration of salt basin separated by oceanic spreading. (**a**) Portion of salt basin with dashed line indicating geometry of imminent break-up, which is characterized by transform segments to the left and ridge segments to the right. (**b**) Early spreading creates small isolated windows of oceanic crust along transform-dominated portion and continuous oceanic crust along ridge-dominated portion; salt starts flowing out over oceanic crust (not shown but indicated by black arrows). (**c**) Continued spreading with salt assumed to preferentially flow over the oldest (deepest) oceanic crust. (**d**) Further spreading and final salt nappe geometry, with only minor flows along transform-dominated segment and larger, cuspate-lobate emplacement along ridge-dominated segment. Note that neither diachroneity of spreading nor differences between the northern and southern Gulf of Mexico are included.

the western edge of the Bay of Campeche, even though there was plenty of salt and it was the area of earliest spreading.

The relative lack of allochthonous salt flow in the western Gulf of Mexico is related to the spreading-centre geometry in the oceanic basin, which is dominated by transform segments in the west and ridge segments in the centre and east (e.g. Sandwell et al. 2014; Pindell et al. 2016). A simple cartoon (Fig. 18) shows how only isolated windows of oceanic crust form during early spreading in the west. Although these eventually link up, salt flow is limited to small areas in the corners of transform-ridge segment transitions where the oceanic crust is oldest and thus deepest. In contrast, laterally continuous nappes can form where ridge segments dominate owing to flow out over larger areas of older oceanic crust.

Conclusions

The Gulf of Mexico and South Atlantic salt basins are remarkably similar in many ways, not so much in the styles of salt tectonics but more in the crustal architecture and its relationship to evaporite deposition. In this paper, I have used 3D and 2D depth-migrated seismic data, primarily from the northeastern Gulf of Mexico and the deepwater Kwanza Basin of Angola, to document and argue the following points concerning the modes of crustal thinning, the generation of accommodation for the salt and presalt strata and the actual deposition of salt.

(1) The base salt is smooth and mostly unfaulted over much of its length but drops abruptly into an outer marginal trough several tens of kilometres or more wide just landward of oceanic crust. The base salt is highly irregular within the trough, with both highs and lows, and often climbs up and over oceanic crust.

(2) The seismic Moho shallows basinward from beneath normal-thickness continental crust to a level only 5 km or less beneath the base salt at the landward edge of the outer marginal trough. It is absent rugose or poorly imaged beneath the trough but reappears beneath oceanic crust.

(3) Crustal deformation was dominated by low-angle, basinward-dipping faults in the necking and distal domains but was more symmetrical in the outer marginal trough. There, extension was accommodated by steeper faults in the form of large-scale boudinage, with salt filling the lows and possible reactive diapirs of weak lithospheric mantle forming triangular highs. The seismic character suggests an increased magmatic (intrusive and/or extrusive) contribution in this area.

(4) Beneath the smooth base salt is a largely unfaulted sag sequence in the necking and distal domains that may be as much as 8–9 km thick. Although sometimes symmetrical, the strata often dip basinward, in some cases forming basinward-shifting depocentres.

(5) Accommodation for the sag and salt sequences was provided mostly by a combination of lower-crustal thinning and basinward-younging thermal and loading subsidence. The exception was in the outer marginal trough, where active extension generated space for the thickest evaporites.

(6) Sedimentation during hyperextension could not keep pace with subsidence, resulting in enormous, underfilled depressions 1500–2500 km long and up to 600–800 km wide.

(7) Evaporites formed in these isolated depressions, with deposition near sea-level in proximal positions but as much as 2–3 km deep in the basin centres. By the time oceanic spreading commenced, the base salt was 1–5 km below the level of oceanic crust, or c. 2.5–7.5 km beneath sea-level depending on the depth of formation of the oceanic crust. The depth of the base salt was enhanced owing to crustal loading by the salt itself. The top salt was roughly at the same depth as the top oceanic crust, i.e. well below sea-level, but was inflated in places owing to basinward flow of salt, resulting in emplacement of allochthonous salt out over oceanic crust.

(8) In the Gulf of Mexico (and probably in the South Atlantic), along-strike variations in the timing of rifting and spreading and the distribution of ridge- and transform-dominated segments of oceanic break-up explain differences in the relationship between crustal geometries and salt distribution.

Acknowledgments Because this paper is ultimately based on wonderful seismic data, I first want to thank the various seismic vendors and the individuals that facilitated my getting permission to publish the data: Dynamic Data Services and S. Venkatraman and P. Nuttall; Multi-Client Geophysical ASA and K. Mohn; Schlumberger and V. Robertson; and TGS and C. Sanders and A. Herrá at Repsol. I also want to acknowledge the many people with whom I have had discussions on these topics over the years or have helped in some way on this paper, including (but not limited to) N. Kusznir, G. Manatschal, W. Mohriak, J.A. Muñoz, R. Pascoe, F. Peel, J. Pindell, A. Pulham and C. Sanders. I also thank M. Hudec, S. Krueger and K. McClay for reviews that helped me refine my story and strengthen my arguments. Finally, I thank Midland Valley for access to their restoration software Move™. Partial funding was provided by the members of the Salt–Sediment Interaction Research Consortium at the University of Texas, El Paso:

BHP-Billiton, BP, Chevron, ConocoPhillips, ExxonMobil, Hess, Kosmos and Shell.

References

ALLEN, J. & BEAUMONT, C. 2016. Continental margin syn-rift salt tectonics at intermediate width margins. *Basin Research*, **28**, 598–633, https://doi.org/10.1111/bre.12123

ALVARENGA, R.S., IACOPINI, D., KUCHLE, J., SCHERER, C.M.S. & GOLDBERG, K. 2016. Seismic characteristics and distribution of hydrothermal vent complexes in the Cretaceous offshore rift section of the Campos Basin, offshore Brazil. *Marine and Petroleum Geology*, **74**, 12–25, https://doi.org/10.1016/j.marpetgeo.2016.03.030

ANDRÉS-MARTÍNEZ, M., PÉREZ-GUSSINYÉ, M., ARMITAGE, J.J. & MORGAN, J.P. 2017. Approaching the meaning of regional unconformities of tectonic origin present in passive margins from a numerical-modelling perspective. 79th EAGE Conference and Exhibition Extended Abstract, https://doi.org/10.3997/2214-4609.201701308

ASLANIAN, D., MOULIN, M. *ET AL.* 2009. Brazilian and African passive margins of the Central Segment of the South Atlantic Ocean: kinematic constraints. *Tectonophysics*, **468**, 98–112.

AUGUSTIN, N., DEVEY, C.W., VAN DER ZWAN, F.M., FELDENS, P., TOMINAGA, M., BANTAN, R.A. & KWASNITSCHKA, T. 2014. The rifting to spreading transition in the Red Sea. *Earth and Planetary Science Letters*, **395**, 217–230, https://doi.org/10.1016/j.epsl.2014.03.047

BARKER, S. & MUKHERJEE, S.S. 2011. Interpretation of the basement step – some observations and implications in the Gulf of Mexico. *AAPG Search and Discovery Article*, 90124, http://www.searchanddiscovery.com/abstracts/html/2011/annual/abstracts/Barker.html

BARR, D. 1991. Subsidence and sedimentation in semi-starved half-graben: a model based on North Sea data. *In*: ROBERTS, A.M., YIELDING, G. & FREEMAN, B. (eds) *The Geometry of Normal Faults*. Geological Society, London, Special Publications, **56**, 17–28, https://doi.org/10.1144/GSL.SP.1991.056.01.02

BATE, R.H. 1999. Non-marine ostracod assemblages of the pre-salt rift basins of West Africa and their role in sequence stratigraphy. *In*: CAMERON, N.R., BATE, R.H. & CLURE, V.S. (eds) *The Oil and Gas Habitats of the South Atlantic*. Geological Society, London, Special Publications, **153**, 283–292, https://doi.org/10.1144/GSL.SP.1999.153.01.17

BERTANI, R.T. & CAROZZI, A.V. 1985. Lagoa Feia Formation (Lower Cretaceous), Campos Basin, offshore Brazil: rift valley stage carbonate reservoirs I & II. *Journal of Petroleum Geology*, **8**, 37–58.

BLAICH, O.A., FALEIDE, J.I., TSIKALAS, F., FRANKE, D. & LEÓN, E. 2009. Crustal-scale architecture and segmentation of the Argentine margin and its conjugate off South Africa. *Geophysical Journal International*, **178**, 85–105, https://doi.org/10.1111/j.1365-246X.2009.04171.x

BRAND, J.H. 2016. Stratigraphy and mineralogy of the Oxfordian Lower Smackover Formation in the eastern Gulf of Mexico. *In*: LOWERY, C.M., SNEDDEN, J.W. & ROSEN, N.C. (eds) *Mesozoic of the Gulf Rim and Beyond: New Progress in Science and Exploration of the Gulf of Mexico Basin*. 35th Annual Gulf Coast Section SEPM Foundation Perkins-Rosen Research Conference, GCSSEPM Foundation, Houston, TX, USA, 14–35.

BRONNER, A., SAUTER, D., MANATSCHAL, G., PÉRON-PINVIDIC, G. & MUNSCHY, M. 2011. Magmatic breakup as an explanation for magnetic anomalies at magma-poor rifted margins. *Nature Geoscience*, **4**, 549–553, https://doi.org/10.1038/ngeo1201

BRUN, J.-P. & BESLIER, M.O. 1996. Mantle exhumation at passive margins. *Earth and Planetary Science Letters*, **142**, 161–173, https://doi.org/10.1016/0012-821X(96)00080-5

BRUNE, S., HEINE, C., PÉREZ-GUSSINYÉ, M. & SOBOLEV, S.V. 2014. Rift migration explains continental margin asymmetry and crustal hyper-extension. *Nature Communications*, **5**, 1–9, https://doi.org/10.1038/ncomms5014

CHRISTESON, G.L., VAN AVENDONK, H.J.A., NORTON, I.O., SNEDDEN, J.W., EDDY, D.R., KARNER, G.D. & JOHNSON, C.A. 2014. Deep crustal structure in the eastern Gulf of Mexico. *Journal of Geophysical Research: Solid Earth*, **119**, 6782–6801, https://doi.org/10.1002/2014JB011045

CONTRUCCI, I., KLINGELHÖFER, F. *ET AL.* 2004. The crustal structure of the NW Moroccan continental margin from wide-angle and reflection seismic data. *Geophysical Journal International*, **159**, 117–128, https://doi.org/10.1111/j.1365-246X.2004.02391.x

COWARD, M.P. & STEWART, S. 1995. Salt-influenced structures in the Mesozoic-Tertiary cover of the southern North Sea, U.K. *In*: JACKSON, M.P.A., ROBERTS, D.G. & SNELSON, S. (eds) *Salt Tectonics: A Global Perspective*. AAPG, Memoirs, Tulsa, OK, USA, **65**, 229–250.

COWIE, L., KUSZNIR, N., ANGELO, R., MANATSCHAL, G. & HORN, B. 2016. The palaeo-bathymetry of base Aptian salt deposition and the composition of underlying basement on the northern Angolan rifted margin. *Petroleum Geoscience*, **22**, 59–70, https://doi.org/10.1190/ice2016-6524491.1

COWIE, L., ANGELO, R., KUSZNIR, N., MANATSCHAL, G. & HORN, B. 2017. Structure of the ocean-continent transition, location of the continent-ocean boundary and magmatic type of the northern Angolan margin from integrated quantitative analysis of deep seismic reflection and gravity anomaly data. *In*: SABATO CERALDI, T., HODGKINSON, R.A. & BACKÉ, G. (eds) *Petroleum Geoscience of the West African Margin*. Geological Society, London, Special Publications, **438**, 159–176, https://doi.org/10.1144/SP438.6

DAVISON, I., ANDERSON, L. & NUTTALL, P. 2012. Salt deposition, loading and gravity drainage in the Campos and Santos salt basins. *In*: ALSOP, G.I., ARCHER, S.G., HARTLEY, A.J., GRANT, N.T. & HODGKINSON, R. (eds) *Salt Tectonics, Sediments and Prospectivity*. Geological Society, London, Special Publications, **363**, 159–173.

DAVISON, I. 1999. Tectonics and hydrocarbon distribution along the Brazilian South Atlantic margin. *In*: CAMERON, N.R., BATE, R.H. & CLURE, V.S. (eds) *The Oil and Gas Habitats of the South Atlantic*. Geological Society, London, Special Publications, **153**, 133–151, https://doi.org/10.1144/GSL.SP.1999.153.01.09

DECARLIS, A., MANATSCHAL, G., HAUPERT, I. & MASINI, E. 2015. The tectono-stratigraphic evolution of distal, hyper-extended magma-poor conjugate rifted margins: examples from the Alpine Tethys and Newfoundland–Iberia. *Marine and Petroleum Geology*, **68**, 54–72, https://doi.org/10.1016/j.marpetgeo.2015.08.005

DOBSON, L.M. & BUFFLER, R.T. 1997. Seismic stratigraphy and geologic history of Jurassic rocks, northeastern Gulf of Mexico. *AAPG Bulletin*, **81**, 100–120.

EDDY, D.R., VAN AVENDONK, H.J.A., CHRISTESON, G.L., NORTON, I.O., KARNER, G.D., JOHNSON, C.A. & SNEDDEN, J.W. 2014. Deep crustal structure of the northeastern Gulf of Mexico: implications for rift evolution and sea-floor spreading. *Journal of Geophysical Research: Solid Earth*, **119**, 6802–6822, https://doi.org/10.1002/2014JB011311

EVANS, R. 1978. Origin and significance of evaporites in basins around Atlantic margin. *AAPG Bulletin*, **62**, 223–234.

FIDUK, J.C., ANDERSON, L.E., SCHULTZ, T.R. & PULHAM, A.J. 2007. Deep-water depositional trends of Mesozoic and Paleogene strata in the central northern Gulf of Mexico. *In*: KENNAN, L., PINDELL, J. & ROSEN, N.C. (eds) *The Paleogene of the Gulf of Mexico and Caribbean Basins: Processes, Events, and Petroleum Systems*. 27th Annual Gulf Coast Section SEPM Foundation Bob F. Perkins Research Conference, GCSSEPM Foundation, Houston, TX, USA, 45–53.

FORT, X. & BRUN, J.-P. 2008. Salt flow from basin-scale in the Gulf of Mexico. *AAPG Search and Discovery Article*, 90078, http://www.searchanddiscovery.com/abstracts/html/2008/annual/abstracts/409435.htm

FRANKE, D. 2013. Rifting, lithosphere breakup and volcanism: comparison of magma-poor and volcanic rifted margins. *Marine and Petroleum Geology*, **43**, 63–87, https://doi.org/10.1016/j.marpetgeo.2012.11.003

FRASCA, G., GUEYDAN, F., BRUN, J.-P. & MONIÉ, P. 2016. Deformation mechanisms in a continental rift up to mantle exhumation. Field evidence from the western Betics, Spain. *Marine and Petroleum Geology*, **76**, 310–328, https://doi.org/10.1016/j.marpetgeo.2016.04.020

FREITAS, J.T.R. 2006. *Ciclos deposicionais evaporíticos da Bacia de Santos: uma análise cicloestratigráfica a partir de dados de 2 poços e da traços de sismica*. MSc thesis, Universidade Federal do Rio Grande do Sul, Porto Alegre, Brazil.

GABRIELSEN, R.H., KYRKJEBØ, R. & FALEIDE, J.I. 2001. The Cretaceous post-rift basin configuration of the northern North Sea. *Petroleum Geoscience*, **7**, 137–154, https://doi.org/10.1144/petgeo.7.2.137

GAMBÔA, L.A.P., MACHADO, M.A.P., DA SILVEIRA, D.P., DE FREITAS, J.T.R. & DA SILVA, S.R.P. 2008. Evaporitos estratificados no Atlântico Sul: interpretação sísmica e controle tectono-estratigráfico na Bacia de Santos. *In*: MOHRIAK, W., SZATMARI, P. & COUTO ANJOS, S.M. (eds) *Sal: Geologia e Tectônica*. Beca Edições Ltda, São Paulo, 340–359.

GARCIA, S.F.DE M., LETOUZEY, J., RUDKIEWICZ, J.-L., DANDERFER FILHO, A. & FRIZON DE LAMOTTE, D. 2012. Structural modeling based on sequential restoration of gravitational salt deformation in the Santos Basin (Brazil). *Marine and Petroleum Geology*, **35**, 337–353, https://doi.org/10.1016/j.marpetgeo.2012.02.009

GILLARD, M., AUTIN, J., MANATSCHAL, G., SAUTER, D., MUNSCHY, M. & SCHAMING, M. 2015a. Tectonomagmatic evolution of the final stages of rifting along the deep conjugate Australian–Antarctic magma-poor rifted margins: constraints from seismic observations. *Tectonics*, **34**, 1–31, https://doi.org/10.1002/(ISSN)1944-9194

GILLARD, M., MANATSCHAL, G. & AUTIN, J. 2015b. How can asymmetric detachment faults generate symmetric Ocean Continent Transitions? *Terra Nova*, **28**, 27–34, https://doi.org/10.1111/ter.12183

GILLARD, M., AUTIN, J. & MANATSCHAL, G. 2016. Fault systems at hyper-extended rifted margins and embryonic oceanic crust: structural style, evolution and relation to magma. *Marine and Petroleum Geology*, **76**, 51–67, https://doi.org/10.1016/j.marpetgeo.2016.05.013

GOSWAMI, A., HARTWIG, A., REUBER, K. & PINDELL, J. 2016. Imaging the pre-salt Gulf of Mexico: basin classification and comparison with other world class petroleum basins. *AAPG Search and Discovery Article*, 90260, http://www.searchanddiscovery.com/abstracts/html/2016/90260ice/abstracts/2474591.html

GOTETI, R., BEAUMONT, C. & INGS, S.J. 2013. Factors controlling early-stage salt tectonics at rifted continental margins and their thermal consequences. *Journal of Geophysical Research: Solid Earth*, **118**, 3190–3220.

HAUPERT, I., MANATSCHAL, G., DECARLIS, A. & UNTERNEHR, P. 2016. Upper-plate magma-poor rifted margins: stratigraphic architecture and structural evolution. *Marine and Petroleum Geology*, **69**, 241–261, https://doi.org/10.1016/j.marpetgeo.2015.10.020

HENRY, S., DANFORTH, A., VENKATRAMAN, S. & WILLACY, C. 2004. PSDM subsalt imaging reveals new insights into petroleum systems and plays in Angola-Congo-Gabon. *Petroleum Exploration Society of Great Britain–Houston Geological Society Joint Africa Symposium*, 7–8th September 2004, London.

HINZ, K., NEBEN, S., SCHRECKENBERGER, B., ROESER, H.A., BLOCK, M., DE SOUZA, G.K. & MEYER, H. 1999. The Argentine continental margin north of 48°S: sedimentary successions, volcanic activity during breakup. *Marine and Petroleum Geology*, **16**, 1–25, https://doi.org/10.1016/S0264-8172(98)00060-9

HUDEC, M.R. 2017. Basement structure and Jurassic evolution of the southern Gulf of Mexico salt province. *AAPG Search and Discovery Article*, 90291.

HUDEC, M.R. & JACKSON, M.P.A. 2007. Terra infirma: understanding salt tectonics. *Earth-Science Reviews*, **82**, 1–28, https://doi.org/10.1016/j.earscirev.2007.01.001

HUDEC, M.R., NORTON, I.O., JACKSON, M.P.A. & PEEL, F.J. 2013. Jurassic evolution of the Gulf of Mexico salt basin. *AAPG Bulletin*, **97**, 1683–1710, https://doi.org/10.1306/04011312073

HUISMANS, R. & BEAUMONT, C. 2011. Depth-dependent extension, two-stage breakup and cratonic underplating at rifted margins. *Nature*, **473**, 74–78, https://doi.org/10.1038/nature09988

HUISMANS, R.S. & BEAUMONT, C. 2002. Asymmetric lithospheric extension: the role of frictional plastic strain softening inferred from numerical experiments. *Geology*, **30**, 211–214, https://doi.org/10.1130/0091-7613(2002)0302.0.CO;2

HUISMANS, R.S. & BEAUMONT, C. 2003. Symmetric and asymmetric lithospheric extension: relative effects of frictional-plastic and viscous strain softening. *Journal of Geophysical Research: Solid Earth*, **108**, 1–22, https://doi.org/10.1029/2002JB002026

HUISMANS, R.S. & BEAUMONT, C. 2014. Rifted continental margins: the case for depth-dependent extension. *Earth and Planetary Science Letters*, **407**, 148–162, https://doi.org/10.1016/j.epsl.2014.09.032

IMBERT, P. 2005. The Mesozoic Opening of the Gulf of Mexico: Part 1, Evidence for oceanic accretion during and after salt deposition. *In*: POST, P.J., ROSEN, N.C., OLSON, D.L., PALMES, S.L., LYONS, K.T. & NEWTON, G.B. (eds) *Petroleum Systems of Divergent Continental Margin Basins*. 25th Annual Gulf Coast Section SEPM Foundation Bob F. Perkins Research Conference, GCSSEPM Foundation, Houston, TX, USA, 1119–1150.

IMBERT, P. & PHILIPPE, Y. 2005. The Mesozoic opening of the Gulf of Mexico: Part 2. Integrating seismic and magnetic data into a general opening model. *In*: POST, P.J., ROSEN, N.C., OLSON, D.L., PALMES, S.L., LYONS, K.T. & NEWTON, G.B. (eds) *Petroleum Systems of Divergent Continental Margin Basins*. 25th Annual Gulf Coast Section SEPM Foundation Bob F. Perkins Research Conference, GCSSEPM Foundation, Houston, TX, USA, 1151–1190, https://doi.org/10.5724/gcs.05.25.1151

JACKSON, M.P.A. & VENDEVILLE, B.C. 1994. Regional extension as a geologic trigger for diapirism. *Geological Society of America Bulletin*, **106**, 57–73.

JACKSON, M.P.A., CRAMEZ, C. & FONCK, J.-M. 2000. Role of subaerial volcanic rocks and mantle plumes in creation of South Atlantic margins: implications for salt tectonics and source rocks. *Marine and Petroleum Geology*, **17**, 477–498, https://doi.org/10.1016/S0264-8172(00)00006-4

KARNER, G.D. & GAMBÔA, L.A.P. 2007. Timing and origin of the South Atlantic pre-salt sag basins and their capping evaporites. *In*: SCHREIBER, B.C., LUGLI, S. & BABEL, M. (eds) *Evaporites Through Space and Time*. Geological Society, London, Special Publications, **285**, 15–35, https://doi.org/10.1144/SP285.2

KARNER, G.D. & DRISCOLL, N.W. 1999. Tectonic and stratigraphic development of the West African and eastern Brazilian Margins: insights from quantitative basin modelling. *In*: CAMERON, N.R., BATE, R.H. & CLURE, V.S. (eds) *The Oil and Gas Habitats of the South Atlantic*. Geological Society, London, Special Publications, **153**, 11–40.

KARNER, G.D., DRISCOLL, N.W. & BARKER, D.H.N. 2003. Syn-rift regional subsidence across the West African continental margin: the role of lower plate ductile extension. *In*: ARTHUR, T.J., MACGREGOR, D.S. & CAMERON, N.R. (eds) *Petroleum Geology of Africa: New Themes and Developing Technologies*. Geological Society, London, Special Publications, **207**, 105–129, https://doi.org/10.1144/GSL.SP.2003.207.6

KNELLER, E.A. & JOHNSON, C.A. 2011. Plate kinematics of the Gulf of Mexico based on integrated observations from the Central and South Atlantic. *Gulf Coast Association of Geological Societies Transactions*, **61**, 283–299.

KUMAR, N., DANFORTH, A., NUTTALL, P., HELWIG, J., BIRD, D.E. & VENKATRAMAN, S. 2013. From oceanic crust to exhumed mantle: a 40 year (1970–2010) perspective on the nature of crust under the Santos Basin, SE Brazil. *In*: MOHRIAK, W.U., DANFORTH, A., POST, P.J., BROWN, D.E., TARI, G.C., NEMČOK, M. & SINHA, S.T. (eds) *Conjugate Divergent Margins*. Geological Society, London, Special Publications, **369**, 147–165, https://doi.org/10.1144/SP369.16

LAVIER, L.L. & MANATSCHAL, G. 2006. A mechanism to thin the continental lithosphere at magma-poor margins. *Nature*, **440**, 324–328, https://doi.org/10.1038/nature04608

LENTINI, M.R., FRASER, S.I., SUMNER, H.S. & DAVIES, R.J. 2010. Geodynamics of the central South Atlantic conjugate margins: implications for hydrocarbon potential. *Petroleum Geoscience*, **16**, 217–229, https://doi.org/10.1144/1354-079309-909

LE PICHON, X. & SIBUET, J.-C. 1981. Passive margins: a model of formation. *Journal of Geophysical Research: Solid Earth*, **86**, 3708–3720, https://doi.org/10.1029/JB086iB05p03708

LISTER, G.S., ETHERIDGE, M.A. & SYMONDS, P.A. 1986. Detachment faulting and the evolution of passive continental margins. *Geology*, **14**, 246, https://doi.org/10.1130/0091-7613(1986)142.0.CO;2

LUNDIN, E.R. & DORÉ, A.G. 2017. The Gulf of Mexico and Canada Basin: genetic siblings on either side of North America. *GSA Today*, **27**, 4–11, https://doi.org/10.1130/GSATG274A.1

MARTON, G. & BUFFLER, R.T. 1994. Jurassic reconstruction of the Gulf of Mexico Basin. *International Geology Review*, **36**, 545–586, https://doi.org/10.1080/00206819409465475

MARTON, G.L., TARI, G.C. & LEHMANN, A.C.T. 2000. Evolution of the Angolan passive margin, West Africa, with emphasis on post-salt structural styles. *In*: MOHRIAK, W. & TALWANI, M. (eds) *Atlantic Rifts and Continental Margins*. AGU Geophysical Monographs, American Geophysical Union, Washington, D.C., **115**, 129–150.

MASINI, E., MANATSCHAL, G. & MOHN, G. 2013. The Alpine Tethys rifted margins: reconciling old and new ideas to understand the stratigraphic architecture of magma-poor rifted margins. *Sedimentology*, **60**, 174–196. https://doi.org/10.1111/sed.12017

MAUS, S., BARCKHAUSEN, U. ET AL. 2009. EMAG2: a 2-arc min resolution Earth Magnetic Anomaly Grid compiled from satellite, airbornce, and marine magnetic measurements. *Geochemistry, Geophysics, Geosystems*, **10**, Q08005, https://doi.org/10.1029/2009GC002471

MCCLAY, K.R. & ELLIS, P.G. 1987. Analogue models of extensional fault geometries. *In*: COWARD, M.P., DEWEY, J.F. & HANCOCK, P.L. (eds) *Continental Extensional Tectonics*. Geological Society, London, Special Publications, **28**, 109–125, https://doi.org/10.1144/GSL.SP.1987.028.01.09

MCKENZIE, D. 1978. Some remarks on the development of sedimentary basins. *Earth and Planetary Science Letters*, **40**, 25–32, https://doi.org/10.1016/0012-821X(78)90071-7

MICKUS, K., STERN, R.J., KELLER, G.R. & ANTHONY, E.Y. 2009. Potential field evidence for a volcanic rifted margin along the Texas Gulf Coast. *Geology*, **37**, 387–390, https://doi.org/10.1130/G25465A.1

MIRANDA, L.R., CARDENAS, A., MALDONADO, R., REYES, E., RUIZ, J. & WILLIAMS, C. 2013. Play hipotético pre-sal en aguas profundas del Golfo de México. *Congreso Mexicano del Petróleo*, Riviera Maya, Mexico.

MODICA, C.J. & BRUSH, E.R. 2004. Postrift sequence stratigraphy, paleogeography, and fill history of the deep-water Santos Basin, offshore southeast Brazil. *AAPG Bulletin*, **88**, 923–945, https://doi.org/10.1306/01220403043

MOHN, G., KARNER, G.D., MANATSCHAL, G. & JOHNSON, C.A. 2015. Structural and stratigraphic evolution of the Iberia–Newfoundland hyper-extended rifted margin: a quantitative modelling approach. *In*: GIBSON, G.M., ROURE, F. & MANATSCHAL, G. (eds) *Sedimentary Basins and Crustal Processes at Continental Margins: From Modern Hyperextended Margins to Deformed Ancient Analogues*. Geological Society, London, Special Publications, **413**, 53–89, https://doi.org/10.1144/SP413.9

MOHRIAK, W., NEMČOK, M. & ENCISO, G. 2008. South Atlantic divergent margin evolution: rift-border uplift and salt tectonics in the basins of SE Brazil. *In*: PANKHURST, R.J., TROUW, R.A., NEVES, B.B. & DE WIT, M.J. (eds) *West Gondwana: Pre-Cenozoic Correlations Across the South Atlantic Region*. Geological Society, London, Special Publications, **294**, 365–398, https://doi.org/10.1144/SP294.19

MOHRIAK, W.U. 2015. Rift basins in the Red Sea and Gulf of Aden: analogies with the southern South Atlantic. *In*: POST, P.J., COLEMAN, J.L., ROSEN, N.C., BROWN, D.E., ROBERTS-ASHBY, T., KAHN, P. & ROWAN, M.G. (eds) *Petroleum Systems in Rift Basins*. 31st Annual Gulf Coast Section SEPM Foundation Perkins-Rosen Research Conference, GCSSEPM Foundation, Houston, TX, USA, 789–826.

MOHRIAK, W.U. In review. Rifting and salt deposition in passive continental margin basins: differences and similarities between the Red Sea and the South Atlantic. *In*: MCCLAY, K.R. (eds) *Passive Margins: Tectonics, Sedimentation and Magmatism*. Geological Society, London, Special Publications **476**.

MOHRIAK, W.U. & LEROY, S. 2013. Architecture of rifted continental margins and break-up evolution: insights from the South Atlantic, North Atlantic and Red Sea–Gulf of Aden conjugate margins. *In*: MOHRIAK, W.U., DANFORTH, A., POST, P.J., BROWN, D.E., TARI, G.C., NEMČOK, M. & SINHA, S.T. (eds) *Conjugate Divergent Margins*. Geological Society, London, Special Publications, **369**, 497–535, https://doi.org/10.1144/SP369.17

MONTARON, B. & TAPPONNIER, P. 2010. A quantitative model for salt deposition in actively spreading basins. *AAPG Search and Discovery Article*, **30117**, 1–8, http://www.searchanddiscovery.com/documents/2010/30117montaron/ndx_montaron.pdf

MOULIN, M., ASLANIAN, D. *ET AL.* 2005. Geological constraints on the evolution of the Angolan Margin based on reflection and refraction seismic data (Zalango project). *Geophysical Journal International*, **162**, 793–810.

NGUYEN, L.C. & MANN, P. 2016. Gravity and magnetic constraints on the Jurassic opening of the oceanic Gulf of Mexico and the location and tectonic history of the Western Main transform fault along the eastern continental margin of Mexico. *Interpretation*, **4**, SC23–SC33, https://doi.org/10.1190/INT-2015-0110.1

NORTON, I.O., CARRUTHERS, D.T. & HUDEC, M.R. 2015*a*. Rift to drift transition in the South Atlantic salt basins: a new flavor of oceanic crust. *Geology*, **44**, 55–58, https://doi.org/10.1130/G37265.1

NORTON, I.O., CARRUTHERS, D.T. & HUDEC, M.R. 2015*b*. Rift to drift transition in the South Atlantic salt basins: a new flavor of oceanic crust, supplementary information. *GSA Data Repository*, **2016013**, 1–15, http://www.geosociety.org/pubs/ft2016.htm

NORTON, I.O., LAWVER, L.A. & SNEDDEN, J.W. 2016. Gulf of Mexico tectonic evolution from Mexico deformation to oceanic crust. *In*: LOWERY, C.M., SNEDDEN, J.W. & ROSEN, N.C. (eds) *Mesozoic of the Gulf Rim and Beyond: New Progress in Science and Exploration of the Gulf of Mexico Basin*. 35th Annual Gulf Coast Section SEPM Foundation Perkins-Rosen Research Conference, GCSSEPM Foundation, Houston, TX, USA, 1–12.

OKUBO, J., LYKAWKA, R., WARREN, L.V., FAVORETO, J. & DIAS-BRITO, D. 2015. Depositional, diagenetic and stratigraphic aspects of Macaé Group carbonates (Albian): example from an oilfield from Campos Basin. *Brazilian Journal of Geology*, **45**, 243–258, https://doi.org/10.1590/23174889201500020005

OSMUNDSEN, P.T. & REDFIELD, T.F. 2011. Crustal taper and topography at passive continental margins. *Terra Nova*, **23**, 349–361, https://doi.org/10.1111/j.1365-3121.2011.01014.x

PASCOE, R., NUTTALL, P., DUNBAR, D. & BIRD, D. 2016. Constraints on the timing of continental rifting and oceanic spreading for the Mesozoic Gulf of Mexico Basin. *In*: LOWERY, C.M., SNEDDEN, J.W. & ROSEN, N.C. (eds) *Mesozoic of the Gulf Rim and Beyond: New Progress in Science and Exploration of the Gulf of Mexico Basin*. 35th Annual Gulf Coast Section SEPM Foundation Perkins-Rosen Research Conference, GCSSEPM Foundation, Houston, TX, USA, 81–122.

PATON, D.A., PINDELL, J., MCDERMOTT, K., BELLINGHAM, P. & HORN, B. 2017. Evolution of seaward-dipping reflectors at the onset of oceanic crust formation at volcanic passive margins: insights from the South Atlantic. *Geology*, **45**, 439–442, https://doi.org/10.1130/G38706.1

PATRUNO, S., HAMPSON, G.J. & JACKSON, C.A.-L. 2015. Quantitative characterisation of deltaic and subaqueous clinoforms. *Earth-Science Reviews*, **142**, 79–119, https://doi.org/10.1016/j.earscirev.2015.01.004

PEEL, F. 2001. Emplacement, inflation and folding of an extensive allochthonous salt sheet in the Late Mesozoic (ultra-deepwater Gulf of Mexico). *AAPG Search and Discovery Article*, 90906, http://www.searchanddiscovery.com/abstracts/html/2001/annual/abstracts/0614.htm

PÉREZ-GUSSINYÉ, M. 2013. A tectonic model for hyperextension at magma-poor rifted margins: an example from the West Iberia–Newfoundland conjugate margins. *In*: MOHRIAK, W.U., DANFORTH, A., POST, P.J., BROWN, D.E., TARI, G.C. & SINHA, S.T. (eds) *Conjugate Divergent Margins*. Geological Society, London, Special Publications, **369**, 403–427, https://doi.org/10.1144/SP369.19

PÉREZ-GUSSINYÉ, M., RESTON, T.J. & PHIPPS MORGAN, J. 2001. Serpentinization and magmatism during extension at non-volcanic margins: the effect of initial lithospheric structure. *In*: WILSON, R.C.L., WHITMARSH, R.B., TAYLOR, B. & FROITZHEIM, N. (eds) *Non-Volcanic Rifting of Volcanic Margins: A Comparison of Evidence from Land and Sea*. Geological Society, London, Special Publications, **187**, 551–576, https://doi.org/10.1144/GSL.SP.2001.187.01.27

PÉRON-PINVIDIC, G. & MANATSCHAL, G. 2009. The final rifting evolution at deep magma-poor passive margins from Iberia–Newfoundland: a new point of view. *International Journal of Earth Sciences (Geologisches Rundschau)*, **98**, 1581–1597, https://doi.org/10.1007/s00531-008-0337-9

PÉRON-PINVIDIC, G., MANATSCHAL, G., MINSHULL, T.A. & SAWYER, D.S. 2007. Tectonosedimentary evolution of the deep Iberia–Newfoundland margins: evidence for a complex breakup history. *Tectonics*, **26**, TC2011, https://doi.org/10.1029/2006TC001970

PERON-PINVIDIC, G., MANATSCHAL, G. & OSMUNDSEN, P.T. 2013. Structural comparison of archetypal Atlantic rifted margins: a review of observations and concepts. *Marine and Petroleum Geology*, **43**, 21–47, https://doi.org/10.1016/j.marpetgeo.2013.02.002

PÉRON-PINVIDIC, G., MANATSCHAL, G., MASINI, E., SUTRA, E., FLAMENT, J.M., HAUPERT, I. & UNTERNEHR, P. 2017. Unravelling the along-strike variability of the Angola–Gabon rifted margin: a mapping approach. *In*: SABATO CERALDI, T., HODGKINSON, R.A. & BACKÉ, G. (eds) *Petroleum Geoscience of the West African Margin*. Geological Society, London, Special Publications, **438**, 49–76, https://doi.org/10.1144/SP438.1

PINDELL, J. 1985. *Plate tectonic evolution of the Gulf of Mexico and Caribbean Region*. PhD thesis, University of Durham.

PINDELL, J. & KENNAN, L. 2007. Rift models and the salt-cored marginal wedge in the Northern Gulf of Mexico: implications for deep-water Paleogene Wilcox deposition and basinwide maturation. *In*: KENNAN, L., PINDELL, J. & ROSEN, N.C. (eds) *The Paleogene of the Gulf of Mexico and Caribbean Basins: Processes, Events, and Petroleum Systems*. 27th Annual Gulf Coast Section SEPM Foundation Bob F. Perkins Research Conference, GCSSEPM Foundation, Houston, TX, USA, 146–186, https://doi.org/10.5724/gcs.07.27.0146

PINDELL, J. & KENNAN, L. 2009. Tectonic evolution of the Gulf of Mexico, Caribbean and northern South America in the mantle reference frame: an update. *In*: JAMES, K.H., LORENTE, M.A. & PINDELL, J.L. (eds) *The Origin and Evolution of the Caribbean Plate*. Geological Society, London, Special Publications, **328**, 1–55, https://doi.org/10.1144/SP328.1

PINDELL, J., GRAHAM, R. & HORN, B. 2014. Rapid outer marginal collapse at the rift to drift transition of passive margin evolution, with a Gulf of Mexico case study. *Basin Research*, **26**, 701–725, https://doi.org/10.1111/bre.12059

PINDELL, J., MIRANDA, C.E., CERÓN, A. & HERNANDEZ, L. 2016. Aeromagnetic map constrains Jurassic–Early Cretaceous synrift, break up, and rotational seafloor spreading history in the Gulf of Mexico. *In*: LOWERY, C.M., SNEDDEN, J.W. & ROSEN, N.C. (eds) *Mesozoic of the Gulf Rim and Beyond: New Progress in Science and Exploration of the Gulf of Mexico Basin*. 35th Annual Gulf Coast Section SEPM Foundation Perkins-Rosen Research Conference, GCSSEPM Foundation, Houston, TX, USA, 123–153.

POST, P.J. 2005. Constraints on the interpretation of the origin and early development of the Gulf of Mexico Basin. *In*: POST, P.J., ROSEN, N.C., OLSON, D.L., PALMES, S.L., LYONS, K.T. & NEWTON, G.B. (eds) *Petroleum Systems of Divergent Continental Margin Basins*. 25th Annual Gulf Coast Section SEPM Foundation Bob F. Perkins Research Conference, GCSSEPM Foundation, Houston, TX, USA, 1016–1061, https://doi.org/10.5724/gcs.05.25.1016

PROFFETT, J.M. 1977. Cenozoic geology of the Yerrington district, Nevada, and implications for the nature of Basin and Range faulting. *Bulletin of the Geological Society of America*, **88**, 247–266.

QUIRK, D.G., HERTLE, M. *ET AL*. 2013. Rifting, subsidence and continental break-up above a mantle plume in the central South Atlantic. *In*: MOHRIAK, W.U., DANFORTH, A., POST, P.J., BROWN, D.E., TARI, G.C., NEMČOK, M. & SINHA, S.T. (eds) *Conjugate Divergent Margins*. Geological Society, London, Special Publications, **369**, 185–214, https://doi.org/10.1144/SP369.20

RAMBERG, H. 1972. Theoretical models of density stratification and diapirism in the Earth. *Journal of Geophysical Research: Solid Earth*, **77**, 877–889, https://doi.org/10.1029/JB077i005p00877

RANERO, C.R. & PÉREZ-GUSSINYÉ, M. 2010. Sequential faulting explains the asymmetry and extension discrepancy of conjugate margins. *Nature*, **468**, 294–299, https://doi.org/10.1038/nature09520

RESTON, T.J. 2005. Polyphase faulting during the development of the west Galicia rifted margin. *Earth and Planetary Science Letters*, **237**, 561–576.

RESTON, T.J. 2009*a*. The extension discrepancy and synrift subsidence deficit at rifted margins. *Petroleum Geoscience*, **15**, 217–237, https://doi.org/10.1144/1354-079309-845

RESTON, T.J. 2009*b*. The structure, evolution and symmetry of the magma-poor rifted margins of the North and Central Atlantic: a synthesis. *Tectonophysics*, **468**, 6–27, https://doi.org/10.1016/j.tecto.2008.09.002

RESTON, T.J. 2010. The opening of the central segment of the South Atlantic: symmetry and the extension discrepancy. *Petroleum Geoscience*, **16**, 199–206, https://doi.org/10.1144/1354-079309-907

ROVERI, M., MANZI, V., BERGAMASCO, A., FALCIERI, F.M., GENNARI, R., LUGLI, S. & SCHREIBER, B.C. 2014. Dense shelf water cascading and Messinian Canyons: a new scenario for the Mediterranean salinity crisis. *American Journal of Science*, **314**, 751–784, https://doi.org/10.2475/05.2014.03

ROWAN, M.G. 2014. Passive-margin salt basins: hyperextension, evaporite deposition, and salt tectonics. *Basin Research*, **26**, 154–182, https://doi.org/10.1111/bre.12043

ROWAN, M.G. 2015. Synexhumation salt basins: crustal thinning, subsidence, and accommodation for salt and presalt strata. *In*: POST, P.J., COLEMAN, J.L., ROSEN, N.C., BROWN, D.E., ROBERTS-ASHBY, T., KAHN, P. & ROWAN, M.G. (eds) *Petroleum Systems in Rift Basins*. 31st Annual Gulf Coast Section SEPM Foundation Perkins-Rosen Research Conference, 827–835.

Ryan, W.B.F. 2009. Decoding the Mediterranean salinity crisis. *Sedimentology*, **56**, 95–136, https://doi.org/10.1111/j.1365-3091.2008.01031.x

Saller, A., Rushton, S., Buambua, L., Inman, K., McNeil, R. & Dickson, J.A.D.T. 2016. Presalt stratigraphy and depositional systems in the Kwanza Basin, offshore Angola. *AAPG Bulletin*, **100**, 1135–1164, https://doi.org/10.1306/02111615216

Salvador, A. 1987. Late Triassic-Jurassic Paleogeography and Origin of Gulf of Mexico Basin. *AAPG Bulletin*, **71**, 419–451.

Sandwell, D.T., Muller, R.D., Smith, W.H.F., Garcia, E. & Francis, R. 2014. New global marine gravity model from CryoSat-2 and Jason-1 reveals buried tectonic structure. *Science*, **346**, 65–67, https://doi.org/10.1126/science.1258213

Stern, R.J., Anthony, E.Y. *et al.* 2011. Southern Louisiana salt dome xenoliths: first glimpse of Jurassic (c. 160 Ma) Gulf of Mexico crust. *Geology*, **39**, 315–318, https://doi.org/10.1130/G31635.1

Stewart, S.A. 1999. Geometry of thin-skinned tectonic systems in relation to detachment layer thickness in sedimentary basins. *Tectonics*, **18**, 719–732, https://doi.org/10.1029/1999TC900018

Talwani, M., Ewing, J., Sheridan, R.E., Holbrook, W.S. & Glover, L. 1995. The edge experiment and the U.S. East Coast Magnetic Anomaly. *In*: Banda, E., Talwani, M. & Torne, M. (eds) *Rifted Ocean-Continent Boundaries*. Kluwer Academic, Dordrecht, 155–181.

Torsvik, T.H., Rousse, S., Labails, C. & Smethurst, M.A. 2009. A new scheme for the opening of the South Atlantic Ocean and the dissection of an Aptian salt basin. *Geophysical Journal International*, **177**, 1315–1333, https://doi.org/10.1111/j.1365-246X.2009.04137.x

Tugend, J., Manatschal, G., Kusznir, N.J. & Masini, E. 2015. Characterizing and identifying structural domains at rifted continental margins: application to the Bay of Biscay margins and its Western Pyrenean fossil remnants. *In*: Gibson, G.M., Roure, F. & Manatschal, G. (eds) *Sedimentary Basins and Crustal Processes at Continental Margins: From Modern Hyper-extended Margins to Deformed Ancient Analogues*. Geological Society, London, Special Publications, **413**, 171–203, https://doi.org/10.1144/SP413.3

Unternehr, P., Peron-Pinvidic, G., Manatschal, G. & Sutra, E. 2010. Hyper-extended crust in the South Atlantic: in search of a model. *Petroleum Geoscience*, **16**, 207–215, https://doi.org/10.1144/1354-079309-904

Vendeville, B.C. & Jackson, M.P.A. 1992. The rise of diapirs during thin-skinned extension. *Marine and Petroleum Geology*, **9**, 331–354, https://doi.org/10.1016/0264-8172(92)90047-I

Warren, J. 2006. *Evaporites*. 2nd edn. Springer-Verlag, Berlin.

Williams, B.G. & Hubbard, R.J. 1984. Seismic stratigraphic framework and depositional sequences in the Santos Basin, Brazil. *Marine and Petroleum Geology*, **1**, 90–104.

Winter, W.R., Jahnert, R.J. & Franca, A.B. 2007. Bacia de Campos. *Boletim de Geociências da Petrobras*, **15**, 511–529.

Zalán, P.V., Severino, M. & Rigoti, C.A. 2011. An entirely new 3D-view of the crustal and mantle structure of a South Atlantic passive margin – Santos, Campos and Espírito Santo basins, Brazil. *AAPG Search and Discovery Article*, 30117, http://www.searchanddiscovery.com/documents/2011/30177zalan/ndx_zalan.pdf

Tectono-stratigraphic evolution of the SE Mediterranean passive margin, offshore Egypt and Libya

LYDIA J. JAGGER[1]*, TIM G. BEVAN[2] & KEN R. McCLAY[1]

[1]*Fault Dynamics Research Group, Royal Holloway University of London, Egham, TW20 0EX, UK*

[2]*BP Egypt, N Road 90, 5th Settlement, New Cairo, Egypt*

L.J.J., 0000-0002-3254-2053
Correspondence: lydia.jagger@gmail.com

Abstract: The regional tectono-stratigraphic evolution of the offshore SE Mediterranean passive margin is evaluated using detailed 2D seismic interpretations. New models for the development of the margin are proposed in the context of the break-up of northern Gondwana and the subsequent evolution of the southern Neotethys Ocean. The SE Mediterranean margin is segmented into distinct rift and transform-dominated tectonic domains as a consequence of multiple phases of rifting and continental break-up during the Middle Triassic–Middle Jurassic (c. 240–170 Ma) and the Late Jurassic–Mid-Cretaceous (c. 145–93 Ma), controlled by reactivation of pre-existing Pan-African basement fabrics and shear zones in varying regional stress fields. The pre-existing basement-involved extensional fault systems were repeatedly reactivated during major phases of inversion in the late Santonian–Maastrichtian (c. 84–65 Ma) and Middle–Late Eocene (c. 49–37 Ma) and episodes of mild inversion during the Oligocene–Early Pleistocene, as a consequence of the convergence of the African–Arabian and Eurasian plates and closure of the Neotethys oceans. The inversion history was fundamentally controlled by the structure and along-strike segmentation of the margin inherited from Neotethyan rifting.

The Eastern Mediterranean is now a significant hydrocarbon province; however, the Mesozoic–Early Cenozoic basin evolution remains the subject of much debate, despite rapid advances in the understanding of tectonic-related processes and settings in the region (Belopolsky *et al.* 2012; Cowie & Kusznir 2012*a*; Robertson *et al.* 2012; Xypolias *et al.* 2016). The SE Mediterranean margin comprises the Libyan, Egyptian and Levant continental margins and provides important insights into the evolution of the East Mediterranean Basin as a whole (Fig. 1).

The East Mediterranean Basin is considered to be a remnant of the Mesozoic Neotethys Ocean (Figs 2 & 3), which developed along the northern margin of Gondwana as Pangaea broke apart (Garfunkel 1998, 2004). There is near consensus that several Neotethyan ocean basins existed in the Eastern Mediterranean, separated by elongate continental fragments (Robertson 2007; Robertson *et al.* 2012). However, the timing, geometry and processes of rifting, drifting and later amalgamation of the continental fragments involved are not well understood (Robertson *et al.* 2012; Xypolias *et al.* 2016), and many contrasting models have been developed to explain the rift history of the Eastern Mediterranean region (discussed in detail by Robertson *et al.* 1996; Robertson 2007). These can be categorized into three main groups that infer: (1) Late Paleozoic opening of a single Neotethys oceanic domain along the northern margin of Gondwana (Fig. 2a, b; e.g. Stampfli & Borel 2002, 2004); (2) Late Triassic–Early Jurassic development of a southern Neotethys spreading ridge in the Eastern Mediterranean (Fig. 2c, d; e.g. Garfunkel 1998; Gardosh *et al.* 2010; Robertson *et al.* 2012); and (3) Early–Middle Cretaceous opening of a southern Neotethys ocean basin in the Eastern Mediterranean (Fig. 2e, f; e.g. Dercourt *et al.* 1986; Hallett 2002).

In addition, the origin and evolution of the Sirt Basin (Fig. 1), which formed during the Early Cretaceous, within this complex Eastern Mediterranean tectonic framework, remains uncertain and is widely debated (e.g. Hallett 2002; Guiraud *et al.* 2005; Abadi *et al.* 2008; Capitanio *et al.* 2009; Frizon de Lamotte *et al.* 2011; Cowie & Kusznir 2012*a, b*).

Seafloor spreading in the Neotethys ended in the late Santonian (c. 84 Ma) owing to a major change in the relative motion between the African and Eurasian plates (Bosworth *et al.* 2008) and the onset of plate convergence (Fig. 3). The complex Late Cretaceous–Present-day collision history and subduction of the Neotethys oceanic domain resulted in widespread reactivation or overprinting of the pre-existing Mesozoic rift structures (Fig. 1). As a result, key uncertainties remain regarding the crustal structure and development of the Eastern Mediterranean basins, and the extent to which the deepwater southern margins were subsequently affected by collision

From: McCLAY, K. R. & HAMMERSTEIN, J. A. (eds) 2020. *Passive Margins: Tectonics, Sedimentation and Magmatism*.
Geological Society, London, Special Publications, **476**, 365–401.
First published online June 12, 2018, https://doi.org/10.1144/SP476.10
© 2018 The Author(s). Published by The Geological Society of London. All rights reserved.
For permissions: http://www.geolsoc.org.uk/permissions. Publishing disclaimer: www.geolsoc.org.uk/pub_ethics

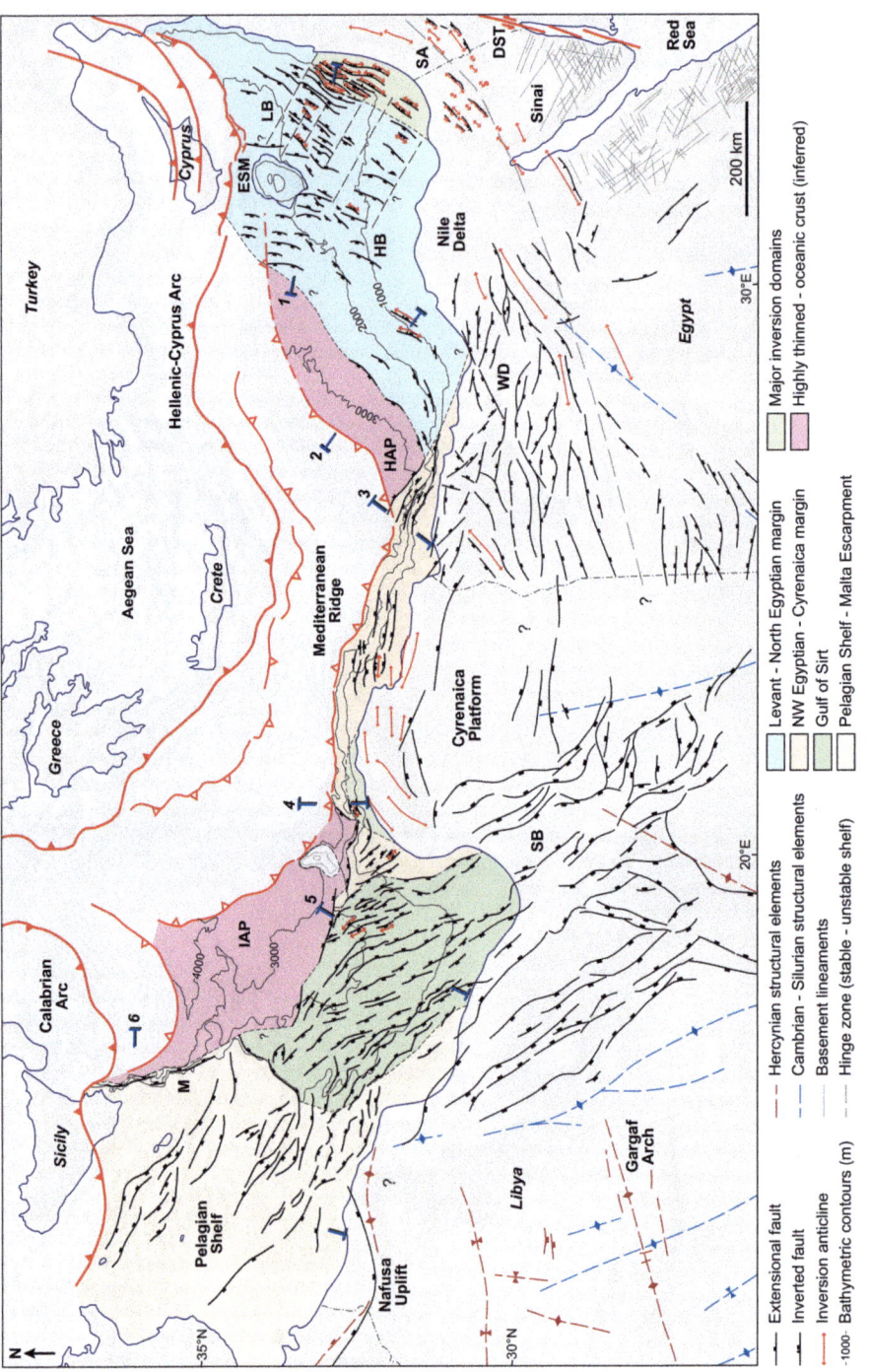

Fig. 1. Tectonic setting of the East Mediterranean Basin with overlain structure map of the SE Mediterranean margin, showing the offshore sub-salt tectonic domains, major Tethyan structural elements, and major Pan-African and Hercynian basement structural trends and lineaments. Offshore faults and tectonic provinces mapped from the 2D seismic dataset used in this study. Onshore faults compiled from Bosworth et al. (1999), Abadi et al. (2008), Bosworth et al. (2008) and Fiduk (2009). Basement lineaments interpreted from DEM and SRTM30 elevation models, integrated with magnetic and residual gravity data (courtesy of BP, 2009). Pan-African and Hercynian basement fabrics compiled from Bosworth & McClay (2001), Hallett (2002), Coward & Ries (2003), and Craig et al. (2008). Numbered cross-sections are the locations of interpreted regional seismic profiles 1–6 (Figs 5, 7–10 & 12). DST, Dead Sea Transform; ESM, Eratosthenes Seamount; HAP, Herodotus Abyssal Plain; HB, Herodotus Basin; IAP, Ionian Abyssal Plain; LB, Levant Basin; M, Malta Escarpment; SA, Syrian Arc fold belt; SB, Sirt Basin; WD, Western Desert.

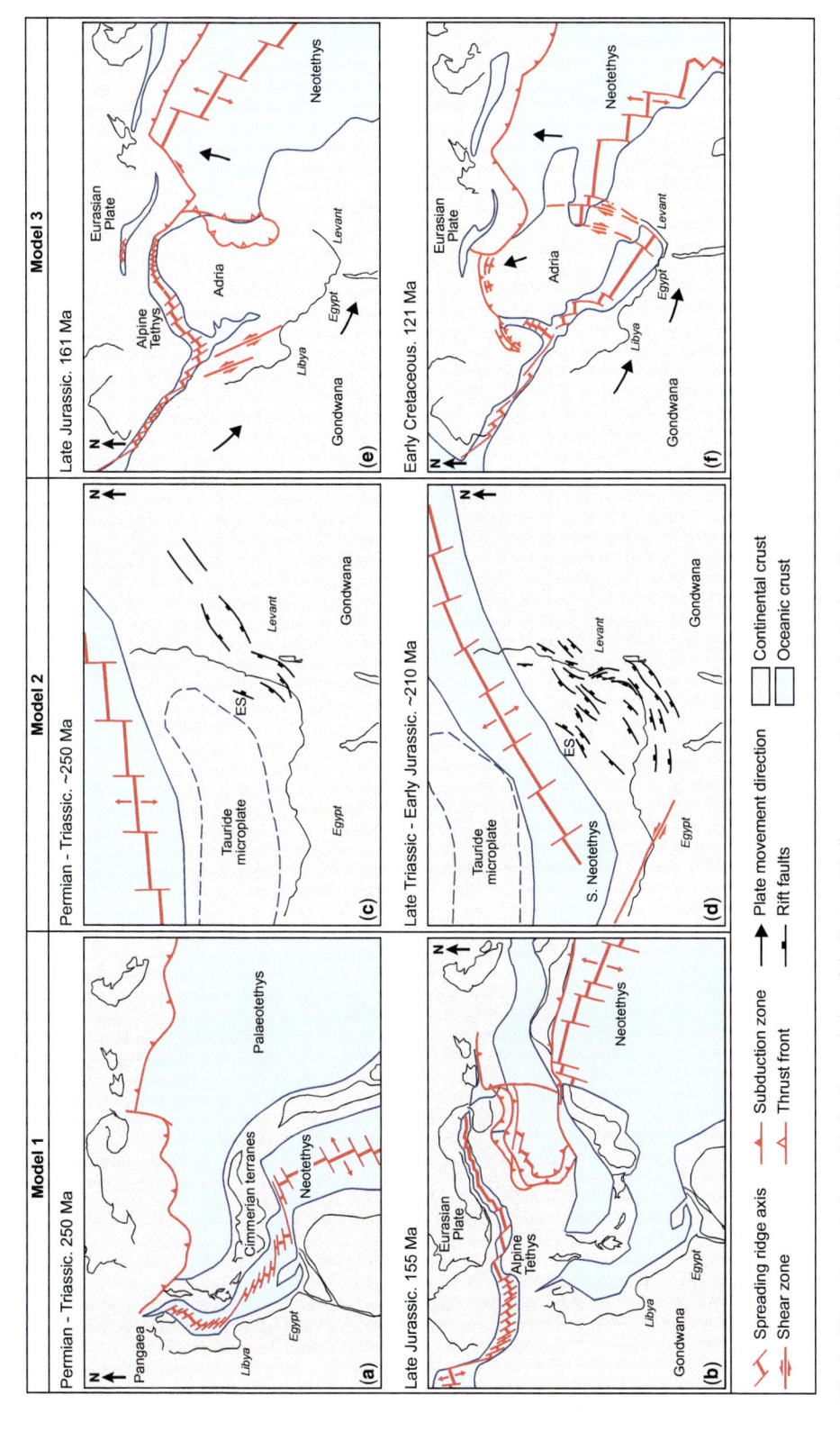

Fig. 2. Alternative tectonic models for the evolution of the Eastern Mediterranean region, showing the key rift phases of each model. (**a, b**) Model 1: Middle–Late Permian–Triassic single Neotethys model (modified after Stampfli & Borel 2002, 2004). (**c, d**) Model 2: Late Triassic–Early Jurassic southern Neotethys model (modified after Garfunkel 1998; Gardosh et al. 2010). (**e, f**) Model 3: Early–Middle Cretaceous southern Neotethys model (modified after Dercourt et al. 1986; Hallett 2002). ES, Eratosthenes Seamount.

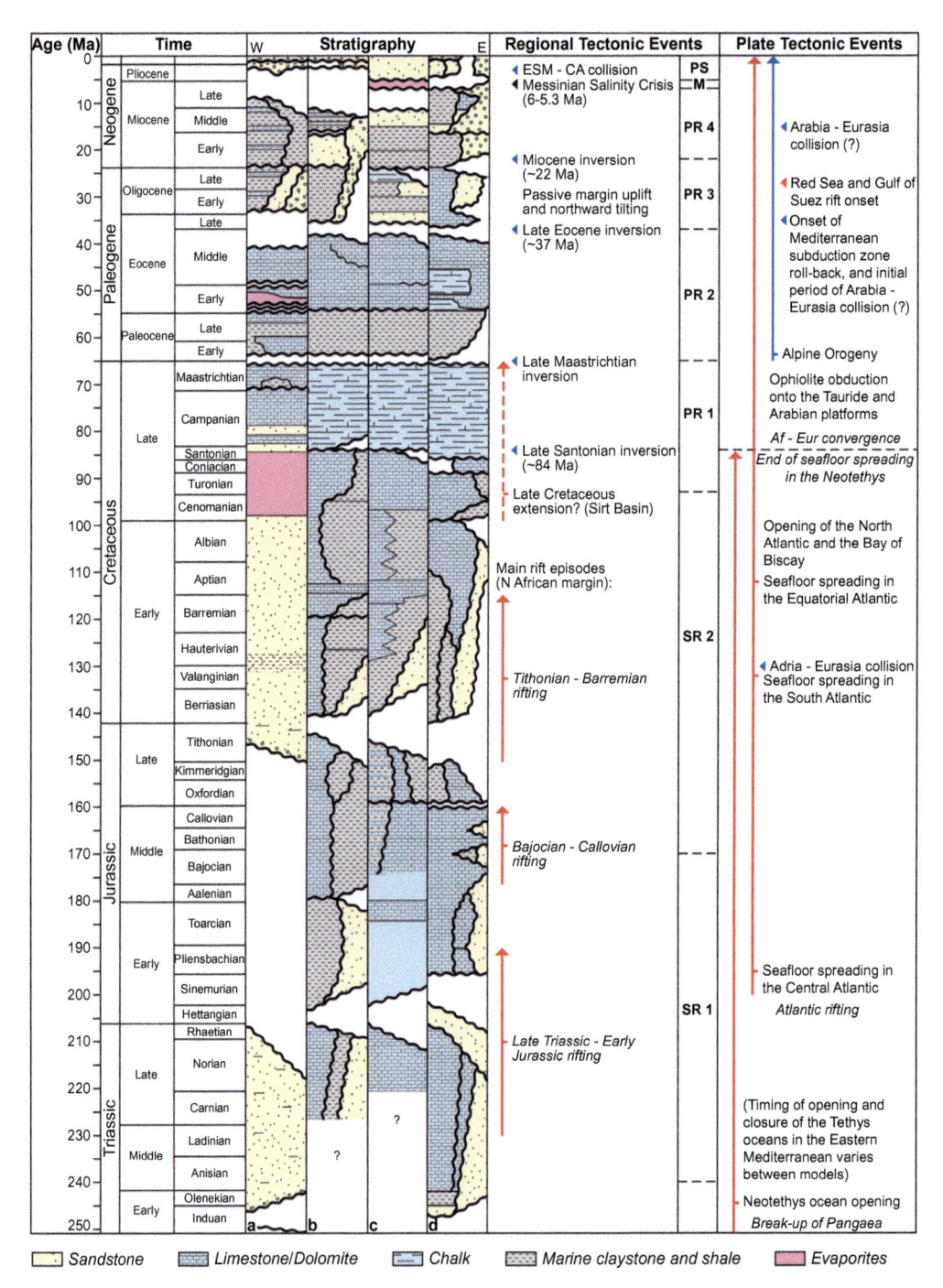

Fig. 3. Regional tectono-stratigraphic chart for the North African margin, showing the correlation with the offshore megasequences interpreted in this study (SR1 to PS). (**a**) Sirt Basin, (**b**) Western Desert, (**c**) Nile Delta, and (**d**) Syrian Arc, southern Israel. Stratigraphy after Guiraud *et al.* (2005). Key tectonic events compiled from Guiraud & Bosworth (1997), Hallett (2002), Dolson *et al.* (2005), Guiraud *et al.* (2005), Rouchy & Caruso (2006), Bosworth *et al.* (2008), Jolivet *et al.* (2008), Moulin *et al.* (2010), and Robertson *et al.* (2012). Interpreted offshore megasequences: SR1, Syn-rift 1; SR2, Syn-rift 2; PR1, Post-rift 1; PR2, Post-rift 2; PR3, Post-rift 3; PR4, Post-rift 4; M, Messinian evaporites; PS, Post-salt. CA, Cyprus Arc; ESM, Eratosthenes Seamount.

tectonics (Briand 2000; Cowie & Kusznir 2012a, b; Robertson et al. 2012). These aspects of the regional geology directly control hydrocarbon prospectivity as they determine key petroleum system elements such as heat flow, source rock and reservoir distribution, and trap formation and integrity (Belopolsky et al. 2012). The SE Mediterranean passive margin remained several hundred kilometres south of the Africa–Eurasia collision front (Fig. 1) during the Alpine Orogeny, and was only mildly deformed (Garfunkel 2004; Gardosh et al. 2011). As a result, the margin preserves the record of the main tectonic events associated with the opening and closure of the Neotethys oceans, and provides significant insights into the evolution of the East Mediterranean Basin and the Tethyan passive margins (Ben-Avraham et al. 2006; Gardosh et al. 2011), as well as the inboard northern margin of the African plate.

This paper presents an integrated regional study of the tectono-stratigraphy and structure of the offshore SE Mediterranean passive margin, based on the results of detailed 2D seismic interpretations and analysis from the Libyan, Egyptian and Levant continental margins. The various hypotheses for the evolution of the Neotethys in the Eastern Mediterranean are tested and we present a new model for the geodynamic evolution of this part of the East Mediterranean Basin.

Geological setting

The SE Mediterranean margin is located on the northeastern margin of the African plate and covers an area of c. 2 208 000 km^2 between longitudes 12–37° E and latitudes 26–37° N (Fig. 1). The margin extends for a distance of roughly 3300 km from the Pelagian Shelf, offshore NW Libya, to the Cyprus Arc, offshore Syria, and varies in orientation along strike from NW–SE in the west to NNE–SSW in the east.

During the Late Paleozoic–Mesozoic the SE Mediterranean margin formed part of the southern continental margin of the Neotethys Ocean and records a complex polyphase rift history (Figs 1 & 3) associated with the episodic break-up of northern Gondwana from the Middle Triassic–Mid-Cretaceous (c. 240–93 Ma) (Guiraud et al. 2005; Bosworth et al. 2008; Gardosh et al. 2011). The onset of plate convergence in the late Santonian and progressive closure of the Neotethys resulted in selective reactivation and inversion of the Mesozoic rift fault systems along the margin (Figs 1 & 3) during episodes of compression in the late Santonian (c. 84 Ma), latest Maastrichtian, Bartonian–Priabonian (c. 37 Ma) and Aquitanian (c. 22 Ma)–Early Pleistocene (Guiraud & Bosworth 1997; Guiraud et al. 2005; Bosworth et al. 2008). The Mediterranean Basin was progressively isolated as the African–Arabian and Eurasian plates collided, which resulted in widespread evaporite deposition and significant basin margin erosion during the Messinian Salinity Crisis (6–5.3 Ma) (Jolivet et al. 2006; Rouchy & Caruso 2006; Roveri et al. 2016) (Fig. 3).

The present-day tectonic setting of the Eastern Mediterranean (Fig. 1) is dominated by slab retreat and back-arc extension as remnant Neotethys oceanic crust, preserved in the Herodotus and Ionian abyssal plains (de Voogd et al. 1992; Cernobori et al. 1996; Catalano et al. 2001; Longacre et al. 2007; Cowie & Kusznir 2012a, b), continues to be subducted along the Hellenic–Cyprus and Calabrian arcs (Ben-Avraham et al. 2006; Jolivet et al. 2008; Nocquet 2012). Locally, the SE Mediterranean margin is in the early stages of continental collision along the southern margins of Turkey and the Aegean Sea (Bosworth et al. 2008). The irregular geometries of the margin and the active convergent plate boundaries have resulted in diachronous collision and strain partitioning (Ben-Avraham et al. 2006), with the northern Egyptian and Libyan margins, including the offshore Sirt Basin, occupying a more passive position (Hallett 2002).

Dataset and methodology

The dataset used in this study consisted of regional 2D seismic reflection profiles extending across the offshore Libyan, Egyptian and Levant continental margins, totalling 83 463 line km. Seismic interpretation was constrained by well picks from 27 nearshore wells as well as potential field data covering the entire Eastern Mediterranean region.

The seismic interpretation workflow comprised regional megasequence and fault mapping to construct regional cross-sections, structural maps and tectono-chronostratigraphic charts documenting the variations in structural style and tectonic evolution along the margin.

The seismic dataset includes several 2D surveys of different vintages, acquired during 1999–2005 and processed using pre-stack time migration, that cover most of the margin in a seismic grid with spacing of c. 8–30 km. 2D seismic data quality are generally good, showing coherent reflection events to c. 9–10 s TWT. Key limitations were the orientations of the dip lines, generally trending oblique to the dominant Tethyan structural trend, which resulted in poor imaging of steeply dipping fault planes, syn-rift stratal geometries and inversion anticlines. In some areas, strong vertical acoustic impedance contrasts associated with the Messinian evaporite sequences resulted in poor sub-salt seismic imaging and loss of reflection continuity, particularly across the outer continental margins where the continental–oceanic crust transition was typically not clearly imaged.

Depositional megasequences bounded by regional unconformities were identified and key horizons defining the major tectono-stratigraphic boundaries were mapped and correlated across the dataset where possible. Horizon and megasequence interpretation was dominantly based on seismic character owing to the limited well constraint across the deepwater regions of the margin. Depositional sequences were distinguished and subdivided using their seismic character and internal stratal geometries, and the morphologies of their bounding surfaces. Sequence boundaries were identified based on reflection amplitude and continuity, and termination of stratal reflections. Continuous and prominent horizons were mapped outwards from areas of high confidence in a grid across the 2D surveys. Regional correlation of well picks, which reach the Late Cretaceous to Paleozoic sequences, and published well data (Gardosh & Druckman 2006; Gardosh et al. 2010; Lie et al. 2011; Steinberg et al. 2011) from the basin margins to the deepwater regions provided additional constraint on horizon interpretation and the identification and ages of the post-rift and syn-inversion megasequences.

Major basin-controlling faults with strike lengths greater than the survey dip line spacing (c. 8–15 km) were mapped to define the offshore structural framework. Fault interpretation in areas of poor data quality was aided by the integration of regional gravity and magnetic data into the seismic project. The potential field maps also provided additional detail on the basement and crustal structure, the boundaries between different crustal types, and the regional structural framework in the SE Mediterranean.

Offshore SE Mediterranean margin architecture and Tethyan structural framework

The bathymetry of the SE Mediterranean margin defines four distinct morphological domains (Fig. 1). These are: (1) the Levant–north Egyptian margin; (2) the NW Egyptian–Cyrenaica margin; (3) the Gulf of Sirt; and (4) the Pelagian Shelf–Malta Escarpment.

Levant–north Egyptian margin

The Levant–north Egyptian domain extends from the Cyprus Arc, offshore Syria, to the north Egyptian continental shelf underlying the West Nile Delta. The Nile Delta is prograding over a broad, NE–SW-oriented, continental shelf and slope 150–250 km wide that deepens gradually towards the NW to a maximum depth of 3000–3200 m in the Herodotus Abyssal Plain. To the NNE, the Eratosthenes Seamount forms a prominent bathymetric feature, elongated NE–SW, and acts as a buttress to the gravity-driven system (e.g. Loncke et al. 2010; Sellier et al. 2013). At the toe of the system, the Mediterranean Ridge accretionary complex forms a major arc-shaped feature in the Eastern Mediterranean, associated with the Hellenic–Cyprus subduction zone.

The deep marine Levant and Herodotus basins form the major depocentres along the Levant–north Egyptian margin. The offshore structural framework consists of a NE–SW- to NNE–SSW-trending basement-involved rift system, which extends across the continental margin to the Herodotus Abyssal Plain. Along the southeastern margins of the Levant and Herodotus basins, the major fault systems were subsequently inverted, producing a series of NE–SW to NNE–SSW doubly plunging subsurface anticlines that form the offshore deformation front of the Syrian Arc (e.g. Shahar 1994; Bosworth et al. 1999) fold belt. In particular, inversion was focused on the southern Levant Basin margin and offshore North Sinai.

NW Egyptian–Cyrenaica margin

The NW Egyptian–Cyrenaica domain extends from the offshore Western Desert, west of the Nile Delta, to the offshore Cyrenaica Platform, NE Libya. The bathymetry of this part of the margin is characterized by a much narrower continental shelf and steep slope that rapidly deepens to abyssal plain depths of 3200 m over a distance of 50 km. To the north, this part of the margin is colliding with the outer domain of the Mediterranean Ridge accretionary complex. The offshore structural framework is characterized by a significant change in the dominant extensional fault trend to WNW–ESE to NW–SE. Offshore NW Cyrenaica, NE–SW- to east–west-trending basins were subsequently inverted during Syrian Arc compression.

Gulf of Sirt

The Gulf of Sirt, offshore northern Libya, is characterized by a broad, NW–SE-oriented, continental shelf and slope up to 300 km wide that gradually deepens northwards over hundreds of kilometres to a maximum depth of 4000 m in the Ionian Abyssal Plain. The offshore Sirt Basin forms the major depocentre along the north Libyan margin. The dominant Tethyan structural framework consists of a NNW–SSE- to NW–SE-trending rift system that extends northwards from the onshore Sirt Basin to the Ionian Abyssal Plain. Mild inversion occurred locally within the offshore Sirt Basin during its post-rift evolution.

Pelagian Shelf–Malta Escarpment

The Pelagian Shelf–Malta Escarpment domain extends from the western margin of the Gulf of Sirt and the Ionian Abyssal Plain to the Calabrian Arc, offshore Tunisia. The bathymetry of the Pelagian Shelf is characterized by a broad shallow shelf with water depths generally up to 500 m. The steep NNW–SSE-trending Malta Escarpment, bounding the Pelagian Shelf to the east, shows a similar present-day morphology to the NW Egyptian–Cyrenaica margin, characterized by a rapid increase in water depth from the shelf to the Ionian Abyssal Plain. Across the Pelagian Shelf, the offshore Tethyan structural framework consists of a NW–SE- to WNW–ESE-trending extensional system, continuous with the NW Egyptian–Cyrenaica margin.

2D seismic stratigraphy and megasequences

Eight regional depositional megasequences have been distinguished across the SE Mediterranean margin (Figs 3 & 4).

Basement

The syn-rift megasequences unconformably overlie the Precambrian–Paleozoic crystalline basement, and pre-rift units in some areas, along the margin (Fig. 3). The basement is characterized by discontinuous and chaotic seismic reflections (Fig. 4). The top of the basement is defined by an angular unconformity (U1), onlapped by fanning syn-rift growth strata, and a significant change in seismic character at the base of the sedimentary rift sequence. The depth to the continental basement (U1) ranges from c. 2–4 s TWT on the north Libyan continental shelf to c. 8–9 s TWT in the central Levant Basin.

In the Herodotus and Ionian abyssal plains, the top of the oceanic crust was interpreted to occur at a depth of c. 9–10 s TWT (Fig. 4), characterized by a relatively flat-lying, high-amplitude reflection with high continuity, and a change in seismic character at the base of the post-rift sediment sequence. In contrast to the continental basement, the oceanic crust is generally undeformed and characterized by low–moderate-amplitude, discontinuous, sub-parallel, subhorizontal and diffractive seismic reflections.

Syn-rift 1 (SR1) megasequence. Middle Triassic–Middle Jurassic

The SR1 megasequence extends from the Levant Basin to NW Cyrenaica (Fig. 4a) but is not present west of the Cyrenaica Platform (Fig. 4b), where the main Sirt rift sequence (SR2) is younger than the Levant–north Egyptian rift sequence.

The SR1 megasequence is approximately mid-Anisian to Bajocian in age (Garfunkel 2004; Gardosh & Druckman 2006; Gardosh et al. 2011) and is defined by the top of the basement (U1) and the Middle Jurassic unconformity (U2). The stratigraphy (Fig. 3) is thought to consist of terrestrial to shallow marine shelf siliciclastic and carbonate sediments (Tawadros 2001; Hallett 2002; Gardosh & Druckman 2006; Moustafa 2010). The seismic character (Fig. 4a) consists of moderate- to high-amplitude continuous reflections that diverge and thicken towards the hanging walls of extensional faults. These wedge-shaped growth geometries and fault-related thickness variations indicate syn-tectonic deposition and reflect the main phase of rifting and basin development along this part of the margin, the end of which is marked by U2.

U2 is an angular unconformity that truncates the top of the SR1 sequence across the continental shelf regions and defines a change in seismic character and stratal geometries between the SR1 and SR2 megasequences (Fig. 4a). In the Levant Basin, U2 was interpreted as the top of a series of high-amplitude reflections within the rift sequence (e.g. Gardosh & Druckman 2006).

Syn-rift 2 (SR2) megasequence. Middle–Late Jurassic–Mid-Cretaceous

The rift sequence in the offshore Sirt Basin was interpreted to be Late Jurassic to Mid-Cretaceous (c. 145–93 Ma) in age based on the available well data, and the geology and subsidence history of the onshore Sirt Basin (e.g. Abadi et al. 2008).

The SR2 megasequence offshore northern Libya (Fig. 4b) is defined by the Late Jurassic (?) base syn-rift unconformity (U3) and the Mid-Cretaceous (Cenomanian, c. 93 Ma) top syn-rift unconformity (U4). The stratigraphy (Fig. 3) consists of fluvial–marine to shallow marine carbonates and clastic sediments (Tawadros 2001; Hallett 2002; Abadi et al. 2008). Deposition was controlled by growth on domino-style extensional faults on the continental shelf and in the offshore Sirt Basin, and defines the main rift phase offshore northern Libya. The seismic character consists of moderate- to high-amplitude, continuous reflections that thicken and diverge into hanging wall depocentres.

To the east, the SR2 rift geometries could not be identified across the Cyrenaica shelf (Fig. 4b). Late Jurassic and Early Cretaceous sediments were undeformed by the rift fault systems and show progradational passive margin geometries. This indicates that the main phase of extension on the Cyrenaica Platform, correlated to the Middle Triassic–Middle Jurassic rifting along the Levant and north Egyptian margins (SR1), had ended by the time of rifting in the Sirt Basin.

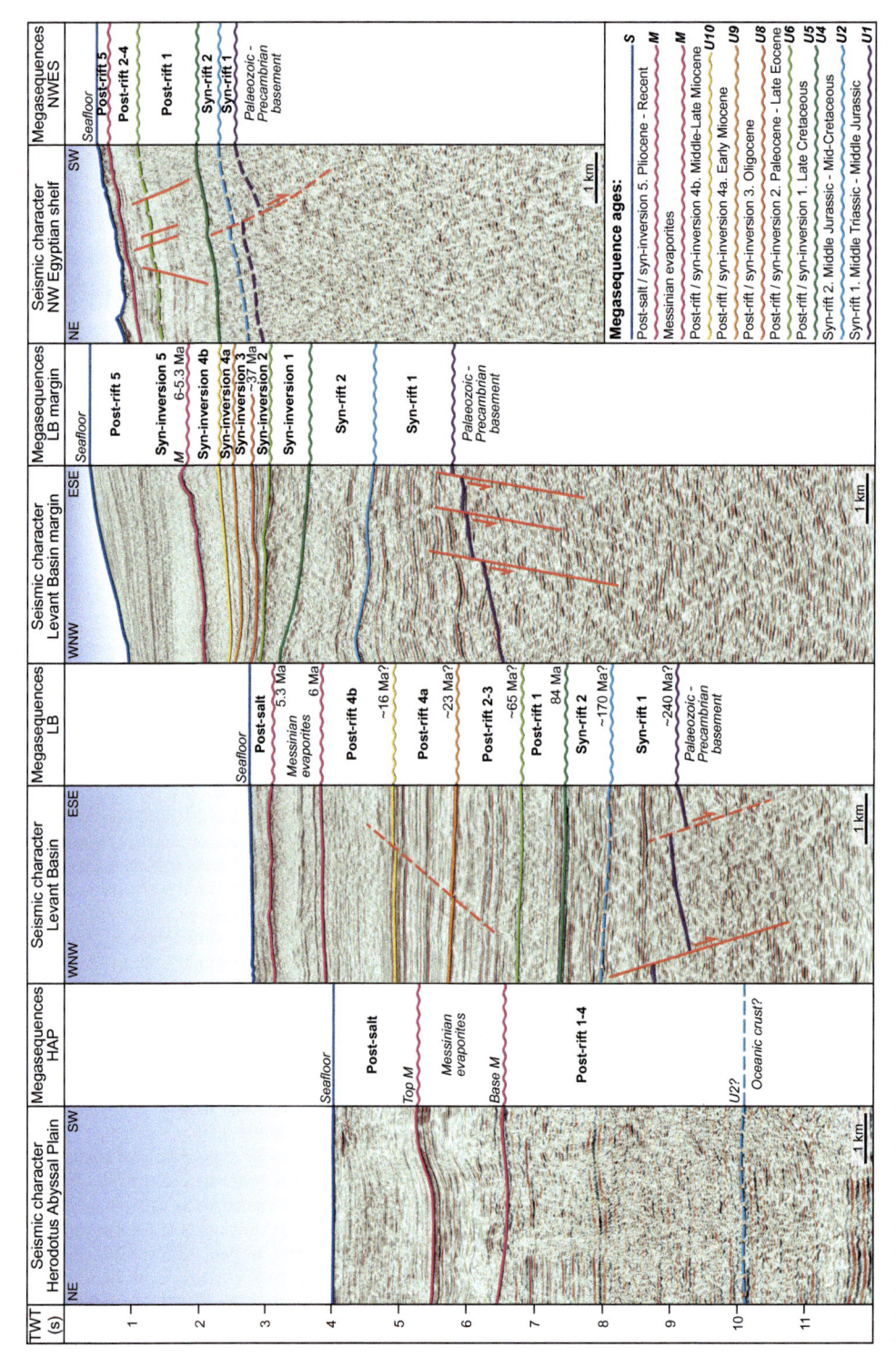

Fig. 4. Regional seismic stratigraphy of the offshore SE Mediterranean margin, showing the seismic character and key regional unconformities interpreted across the (**a**) Levant–north Egyptian margin.

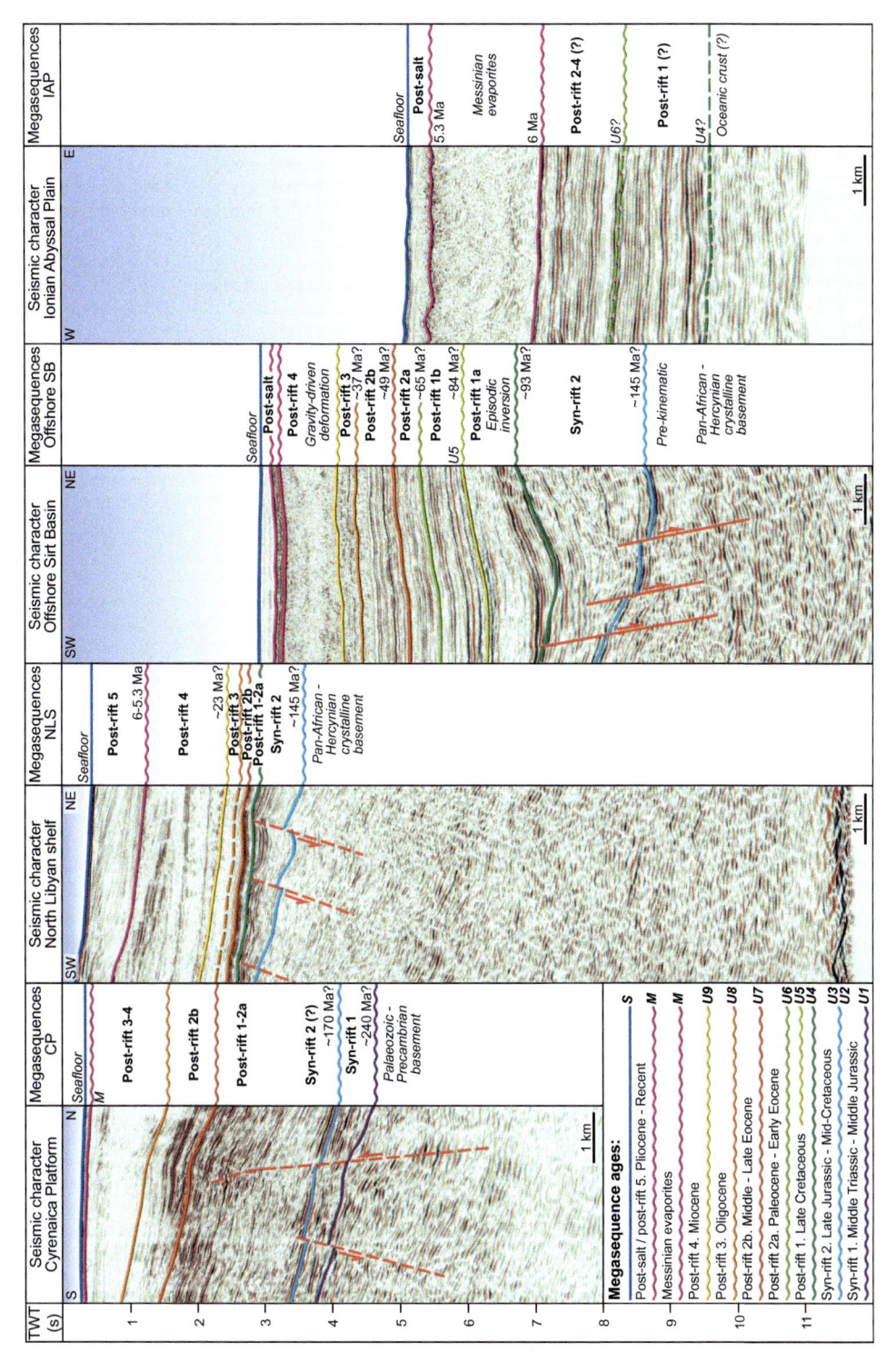

Fig. 4. (b) north Libyan margin. Note that several unconformities could not be correlated regionally across the entire SE Mediterranean margin. CP, Cyrenaica Platform; HAP, Herodotus Abyssal Plain; IAP, Ionian Abyssal Plain; LB, Levant Basin; NLS, North Libyan shelf; NWES, NW Egyptian shelf; SB, Sirt Basin.

Along the Levant, north Egyptian, and Cyrenaica margins, the SR2 megasequence (Fig. 4a) is approximately Bajocian to Turonian–Coniacian in age (e.g. Gardosh & Druckman 2006; Yousef et al. 2010; Bevan & Moustafa 2012) and is defined by the Middle Jurassic unconformity (U2) and the Mid-Cretaceous top syn-rift unconformity (U4). The stratigraphy (Fig. 3) is characterized by mixed carbonate–siliciclastic platform sedimentation, comprising shallow marine shelf carbonates and shales, on the margin of a deep marine basin, comprising slope and deep marine shales and marls basinward of the shelf (Tawadros 2001; Hallett 2002; Garfunkel 2004; Gardosh & Druckman 2006; Moustafa 2010; Tassy et al. 2015). Aptian–Albian shallow-water carbonates form the Zohr hydrocarbon reservoir. The seismic character at the basin margins consists of variable low- to high-amplitude, discontinuous seismic reflections. Reflection geometries are generally parallel to sub-parallel and prograding. However, slight fault-related thickness variations, characterized by diverging and wedge-shaped growth geometries in the hanging walls of major rift fault systems, and fault offsets, indicate extensional reactivation of basin-bounding faults during the deposition of this sequence. In contrast, the sequence in the central Levant Basin is typically characterized by sub-parallel, moderate- to high-amplitude, continuous reflections, indicating the dominance of thermal subsidence within the basin.

U4 is an angular unconformity (Fig. 4), overlain and onlapped or downlapped by post-rift–syn-inversion units, and which truncates rotated syn-rift strata offshore northern Libya (Fig. 4b). Depth to the top of the rift sequence ranges from c. 2 s TWT across the north Libyan and NW Egyptian shelf regions to c. 7.5 s TWT in the central Levant Basin.

Post-rift–syn-inversion 1 (PR–SI1) megasequence. Late Cretaceous

In the Gulf of Sirt, the Late Cretaceous (Cenomanian–Maastrichtian, c. 93–65 Ma?) PR1 megasequence (Fig. 4b) is defined by the Cenomanian top syn-rift unconformity (U4) and the latest Cretaceous (U6) unconformity. The stratigraphy (Fig. 3) consists of marine shales and shallow marine platform carbonates (Hallett 2002; Abadi et al. 2008).

Along the north Libyan shelf (Fig. 4b), the Late Cretaceous was characterized by condensed sedimentation and erosion on the uplifted basin margin, where the seismic character consists of low- to high-amplitude, continuous and closely spaced parallel reflections. The megasequence thickens significantly into the deepwater Sirt Basin, where the Late Cretaceous basin fill can be subdivided into two semi-regional sequences – PR1a and PR1b.

Based on available well data, the PR1a sequence was interpreted to be approximately Cenomanian–Coniacian/early Santonian (?) in age. Sediment deposition was controlled by thermal subsidence, characterized by low-amplitude, discontinuous seismic reflections with parallel to sub-parallel geometries. The PR1a strata locally show fanning syn-inversion growth geometries, associated with mild inversion of half-graben bounding extensional fault systems, which thin and onlap onto uplifted and folded SR2 strata in the hanging walls of the inverted faults. The top of the PR1a sequence is marked by an erosional unconformity (U5), interpreted to be Santonian (c. 84 Ma?) in age, which recorded a phase of mild regional uplift across the basin and truncates the PR1a sequence towards the basin margins.

The PR1b sequence correlates to the main PR(–SI)1 megasequence identified across the Levant and north Egyptian margins (Fig. 4a). This megasequence is approximately Coniacian–Santonian (c. 84 Ma?) to Maastrichtian (c. 65 Ma?) in age (Moustafa 2010; Yousef et al. 2010; Gardosh et al. 2011) and is defined by the Coniacian–Santonian unconformity (U5) and the Late Cretaceous unconformity (U6). The stratigraphy (Fig. 3) consists of shallow marine shelf carbonates, and deep marine shelf and slope mudstone, chalk and marl deposition (Gardosh & Druckman 2006; Moustafa 2010; Yousef et al. 2010; Lie et al. 2011). Along the shelf regions, particularly NW Egypt and Cyrenaica, sedimentation was dominated by continued carbonate platform build-up. This produced thick sequences characterized by low- to moderate-amplitude, relatively continuous and parallel to progradational internal reflections. On the Levant and Herodotus basin margins, SI1 sediment deposition was controlled by reverse reactivation of major extensional faults and the growth of hanging wall inversion anticlines. The seismic character consists of low-amplitude discontinuous reflections that define fanning syn-inversion growth strata, which thin and onlap towards the fold crests. Within the Levant, Herodotus and Sirt basins (Fig. 4), PR1 sediment deposition was controlled by continued post-rift thermal subsidence, which resulted in drowning of the Zohr structure, and episodic mild reactivation of the inverted faults. The seismic character consists of variably low- to high-amplitude, relatively continuous, flat-lying and parallel to sub-parallel reflections, with some areas of discontinuous reflections. The basin fill sequences have a more uniform thickness, and thin away from the centre of the basins and onlap onto the basin margins and carbonate platforms.

In the Herodotus and Ionian abyssal plains, the post-rift sediment cover overlying the oceanic crust could not be subdivided into individual megasequences (Fig. 4). The post-rift sequences are characterized by high-amplitude, continuous and flat-lying reflections with parallel to sub-parallel geometries.

Post-rift–syn-inversion 2 (PR–SI2) megasequence. Paleocene (?)–Late Eocene

The Danian (c. 65 Ma?) to Priabonian (c. 37 Ma?) PR(–SI)2 megasequence is defined by the Late Cretaceous unconformity (U6) and the Late Eocene unconformity (U8). The stratigraphy (Fig. 3) consists of shallow marine shelf carbonate deposition at the basin margins, and deep marine shales and pelagic sedimentation within the basins (Gardosh & Druckman 2006; Moustafa 2010; Yousef et al. 2010; Lie et al. 2011). Carbonate platform build-up continued along the NW Egyptian and Cyrenaica shelf regions (Fig. 4), characterized by low- to high-amplitude reflections with progradational geometries that downlap onto the Cretaceous platform sequences. Along the Levant and Herodotus basin margins (Fig. 4a), further inversion of the major fault systems resulted in amplification of the hanging wall folds. The SI2 seismic character consists of high-amplitude, continuous reflections defining fanning growth sequences that thin and onlap onto the fold limbs. Within the Levant and Herodotus basins, sediment deposition was controlled by continued post-rift thermal subsidence, and characterized by moderate- to high-amplitude, continuous and parallel reflections that onlap onto the basin margins and Jurassic–Cretaceous carbonate platform.

In the Gulf of Sirt (Fig. 4b), the Paleocene–Early Eocene (c. 65–49 Ma) PR2a megasequence is defined by the latest Cretaceous unconformity (U6) and the Early Eocene unconformity (U7, late Ypresian), and the Middle–Late Eocene PR2b megasequence (c. 49–37 Ma?) is defined by the Early Eocene unconformity (U7, late Ypresian) and the Late Eocene unconformity (U8, Priabonian).

The stratigraphy (Fig. 3) consists of shallow water shelf and open-marine carbonates and marine shales (Tawadros 2001; Hallett 2002; Abadi et al. 2008). Along the north Libyan shelf, deposition was controlled by basin margin uplift and erosion, and carbonate shelf progradation. The PR2a shelf sequence was subsequently eroded, but, where present, seismic reflections show parallel to slight prograding geometries, with downlap onto U6 towards the NE. U7, characterized by a high-amplitude reflection throughout the basin, is a strongly erosional unconformity along the Sirt Basin margin where it commonly removed the underlying post-rift sediments, such that Middle–Late Eocene sediments unconformably overlie the Paleocene and Late Cretaceous sequences. The seismic character consists of low-amplitude continuous reflections, and progradational clinoform geometries, downlapping towards the NE onto U7, define a well-developed Middle–Late Eocene carbonate shelf.

In the deepwater Sirt Basin, deposition was dominated by continued post-rift thermal subsidence and further mild inversion of basin-controlling faults. The seismic character of the sediment fill consists of moderate- to high-amplitude continuous and widely spaced reflections with parallel to sub-parallel geometries. Mass transport complexes in the upper part of the sequence are characterized by chaotic and discontinuous reflections. U8 is an erosional unconformity that truncates the underlying sequence.

Post-rift–syn-inversion 3 (PR–SI3) megasequence. Oligocene–Early Miocene

The Oligocene–Early Miocene (c. 37–23 Ma?) PR(–SI)3 megasequence is defined by the Late Eocene unconformity (U8) and the Base Miocene unconformity (U9). The stratigraphy (Fig. 3) consists of fluvio-deltaic and marine clastic-dominated sequences, shallow marine shelf carbonates, and deep marine distal turbidite sequences, pelagic shales and marls (Tawadros 2001; Hallett 2002; Dolson et al. 2005; Lie et al. 2011; Bevan & Moustafa 2012). Sediment deposition was controlled by uplift and basinward-tilting of the passive margin. As a result, high volumes of clastic sediments were input into the basin offshore northern Egypt (e.g. Dolson et al. 2005); however, the NW Egyptian margin and platform areas were characterized by much lower clastic sediment supply (e.g. Tari et al. 2012).

Across the shelf regions, this megasequence has an erosional base (U8) related to uplift and tilting of the margin (Fig. 4). The seismic character consists of low-amplitude, discontinuous to chaotic reflections with parallel, sub-parallel and progradational clinoform geometries that onlap and downlap onto U8 and the platform margin. Along the Levant and Herodotus basin margins (Fig. 4a), hanging wall fold growth continued to control sediment deposition. The SI3 seismic character consists of variably low- to high-amplitude, continuous reflections that show slight thinning and convergence towards the crests of the anticlines. The sequence thickens basinward, where the seismic character in the Levant and Sirt basins (Fig. 4) consists of low- to moderate-amplitude, variably continuous and parallel reflections that thin and onlap onto the basin margins in the Levant Basin (Fig. 4a), to discontinuous and chaotic reflections with occasional higher-amplitude continuous reflections defining episodic mass transport complex deposition in the Sirt Basin (Fig. 4b). U9 is an erosional unconformity that truncates the Oligocene shelf sequence across the shelf areas.

Post-rift–syn-inversion 4 (PR–SI4) megasequence. Miocene

The Miocene (c. 23–6 Ma?) PR(–SI)4 megasequence is defined by the Base Miocene unconformity (U9) and the Base Messinian unconformity (M). Continued passive margin uplift and tilting was the dominant tectonic control on sediment deposition. The stratigraphy (Fig. 3) is characterized by clastic-dominated sedimentation.

Offshore northern Libya (Fig. 4b), a major delta system occupied the north Libyan shelf and prograded into the Gulf of Sirt during the Miocene (Fiduk 2009). This produced a thick sequence, >1 s TWT, with highly variable seismic character. Continuous moderate- to high-amplitude reflections define multiple clastic prograding systems, and mass transport complexes, characterized by discontinuous and chaotic to transparent reflections, onlap the Eocene–Oligocene shelf and thicken into the Sirt Basin. In the deepwater Sirt Basin, the seismic character consists of highly disrupted, chaotic and transparent reflection sequences, indicating the development of multiple gravity-driven slumps and mass transport complexes within the basin.

Along the Levant and north Egyptian margins (Fig. 4a), the Early Miocene (c. 23–16 Ma?) PR(–SI)4a megasequence is defined by the Base Miocene unconformity (U9) and the Base Middle Miocene unconformity (U10), and the Middle–Late Miocene (c. 16–6 Ma) PR(–SI)4b megasequence is defined by the Base Middle Miocene unconformity (U10) and the Base Messinian unconformity (M) or its correlative conformity.

Along the shelf regions, the stratigraphy (Fig. 3) consists of clastic-dominated sequences and shallow marine shelf carbonates on platform areas (Tawadros 2001; Hallett 2002). Sediment deposition along the north Egyptian margin was characterized by continued high volumes of fluvial–deltaic clastic sediments entering the basin through the Nile Delta system (e.g. Dolson et al. 2005). In contrast, the NW Egyptian margin and the Cyrenaica Platform were characterized by continued low clastic sediment supply. On the platform areas (Fig. 4), low- to moderate-amplitude reflections with discontinuous to progradational geometries define shelf-edge sediment sequences.

The offshore basin stratigraphy (Fig. 3) is characterized by deep marine siliciclastic-dominated sedimentation in a distal turbidite setting (Lie et al. 2011). On the Herodotus and Levant Basin margins (Fig. 4a), and locally within the Levant Basin, sediment deposition was dominantly controlled by continued inversion of major extensional faults and hanging wall fold amplification. The SI4 seismic character consists of low- to moderate-amplitude, typically discontinuous and chaotic, reflections that thin and onlap onto U9 and U10 towards the crests of the folds and diverge away from the fold crests. The sequence thickens dramatically basinward where PR4 sediment deposition was controlled by subsidence. In the central Levant Basin, the PR4a seismic character is dominated by a high-amplitude, parallel to sub-parallel reflection series that onlaps onto the basin margins. The reflections are generally laterally continuous but are offset by minor post-rift faults that die out within the underlying and overlying post-rift sequences. In the overlying PR4b sequence, the reflections are low- to moderate-amplitude, with zones of chaotic and discontinuous reflections indicating mass transport and turbidite deposition.

The Base Messinian unconformity is an erosional unconformity that truncates the underlying sequence at a depth of c. 0.5–1 s TWT across the NW Egyptian and north Libyan shelf regions, and its correlative conformity occurs at a maximum depth of c. 4 s TWT in the central Levant Basin. The unconformity is characterized by a continuous high-amplitude reflection, produced owing to the seismic velocity contrast between the siliciclastic sequence and the overlying evaporites. In the Herodotus and Ionian abyssal plains, the base of the Messinian salt occurs at a depth of c. 6.5–7 s TWT.

Messinian evaporite sequence

The Messinian (6–5.3 Ma) evaporite sequence can be correlated throughout the Eastern Mediterranean and is defined by the Base and Top Messinian unconformities (M).

At the basin margins and across the continental shelf–platform regions (Fig. 4), the Messinian sequence is characterized by a major, regionally continuous, high-amplitude erosional unconformity. Within the Levant Basin (Fig. 4a), the evaporite sequence is up to 1 s TWT thick and pinches out towards the fold systems along the eastern margin of the basin. The sequence generally shows a chaotic and discontinuous seismic character with areas of transparent reflections. Continuous moderate- to high-amplitude reflections are likely to represent anhydrite and shale units within the massive salt unit (Gardosh & Druckman 2006). In places these reflection geometries indicate internal deformation and thrust fault detachments within the evaporite sequence. In the deepwater Sirt Basin (Fig. 4b), the interbedded evaporite–clastic sequence (Fiduk 2009; Bowman 2012) is <0.5 s TWT thick and is characterized by a series of high-amplitude continuous and chaotic reflections.

The Messinian sequence thickens basinward into the Herodotus and Ionian abyssal plains (Fig. 4), where mobile halite units (e.g. Fiduk 2009; Bowman 2012) are up to 2 s TWT thick. The salt is characterized by a transparent zone, with higher-amplitude discontinuous internal reflections, bounded by continuous high-amplitude reflections produced by the seismic velocity contrast between the evaporite units and the Miocene–Pliocene clastic sediments.

The Top Messinian unconformity is characterized by a high-amplitude reflection at a minimum depth of c. 0.5–1 s TWT along the continental shelf–platform regions, where it merges with the Base Messinian unconformity, to a maximum depth of c. 4.5–5.5 s TWT in the Ionian and Herodotus abyssal plains.

Post-salt–syn-inversion 5 (PS–SI5) megasequence. Pliocene–Recent

The Pliocene–Recent (5.3–0 Ma) PS(–SI5) megasequence is defined by the Top Messinian unconformity (M) and the present-day seafloor reflection (S). The stratigraphy (Fig. 3) consists of Nile-derived clastic-dominated sediments and siliciclastic shelf-edge delta deposits (Tawadros 2001; Hallett 2002; Gardosh & Druckman 2006; Yousef *et al.* 2010), characterized by variably low- to high-amplitude continuous reflections.

On the Levant and Herodotus basin margins (Fig. 4a), the thick post-Messinian sequence has a progradational character, with downlapping clinoform geometries. Seismic reflections also show extensional growth geometries related to updip listric fault development. Interbedded discontinuous and chaotic to transparent reflection units indicate deposition of mass transport complexes. Along some parts of the continental shelf, the Messinian unconformity was uplifted and folded and onlapping growth sequences were developed (SI5), related to continued inversion of the sub-salt basement fault systems during the Late Miocene–Early Pleistocene. In the deep marine Levant Basin, the PS sequence was deformed by thin-skinned tectonics related to the uplift- and gravity-driven internal deformation and basinward movement of the underlying evaporite sequence (e.g. Cartwright & Jackson 2008; Cartwright *et al.* 2012).

To the west, on the NW Egyptian and Cyrenaica platforms (Fig. 4), the sequence is much thinner and is characterized by downlapping progradational reflection geometries, with chaotic and discontinuous reflections developed in places, associated with gravity-driven deformation and shelf collapse.

In the Gulf of Sirt (Fig. 4b), a second major delta system (Fiduk 2009) is characterized by progradational clinoform geometries downlapping onto the Messinian unconformity and eroded Miocene delta system. In the deepwater Sirt Basin, seismic reflection geometries are generally parallel to sub-parallel and onlap onto the pre-existing Messinian topography. Troughs eroded within the section indicate incision and infill of submarine canyons and post-Messinian slumping.

In the Herodotus Abyssal Plain (Fig. 4a), the sequence is c. 1–1.5 s TWT thick. The gravity-driven contractional domain is characterized by parallel and fanning reflection geometries related to salt-cored folding and thrust fault development. In the Ionian Abyssal Plain (Fig. 4b), seismic reflections onlap the Top Messinian unconformity, indicating salt movement and associated deformation of the overlying cover sequence. The post-salt section is thinner and limited salt-related deformation occurred along the Libyan margin in comparison to the north Egyptian–Levant margins.

Regional structure of the offshore SE Mediterranean margin

Seismic profiles 1–6 illustrate the characteristic structural styles and continent–ocean transition structure of the different morphological domains of the SE Mediterranean margin (Fig. 1).

Levant margin

The Levant margin (Fig. 5) is characterized by a wide zone of rifted and thinned continental crust, which extends over a distance of at least 400 km to the oceanic crust in the Herodotus Abyssal Plain.

The inner continental margin, where the Levant Basin (Fig. 6) forms the major depocentre, is c. 220 km wide. The basin is underlain by thinned Paleozoic–Precambrian basement, which was cut by NE–SW-trending, planar, high-angle extensional fault systems bounding a series of horst and graben structures infilled by wedge-shaped SR1 growth strata (Fig. 4a), during the main phase of rifting in the Middle Triassic–Middle Jurassic. Extensional reactivation of the major faults, and deposition of the SR2 sequence (Fig. 4a), occurred during the Middle Jurassic–Mid-Cretaceous.

Following the end of the rift phases in the Turonian–Coniacian, the basin was infilled with a thick post-rift sediment sequence (PI1–4) during Late Cretaceous–Late Miocene thermal subsidence. Major NE–SW-trending extensional fault systems on the southeastern margin of the southern Levant Basin, as well as depocentre-bounding faults within the basin, were moderately inverted during its post-rift evolution (Figs 5 & 6). This produced asymmetric hanging wall anticlines onlapped by fanning syn-inversion growth stratal packages (SI1–4; Fig. 4a), which define episodic phases of inversion that occurred from the Late Cretaceous (Santonian?) to the Middle–Late Miocene.

Following the Messinian Salinity Crisis, uplift and tilting of the margin, and associated basinward

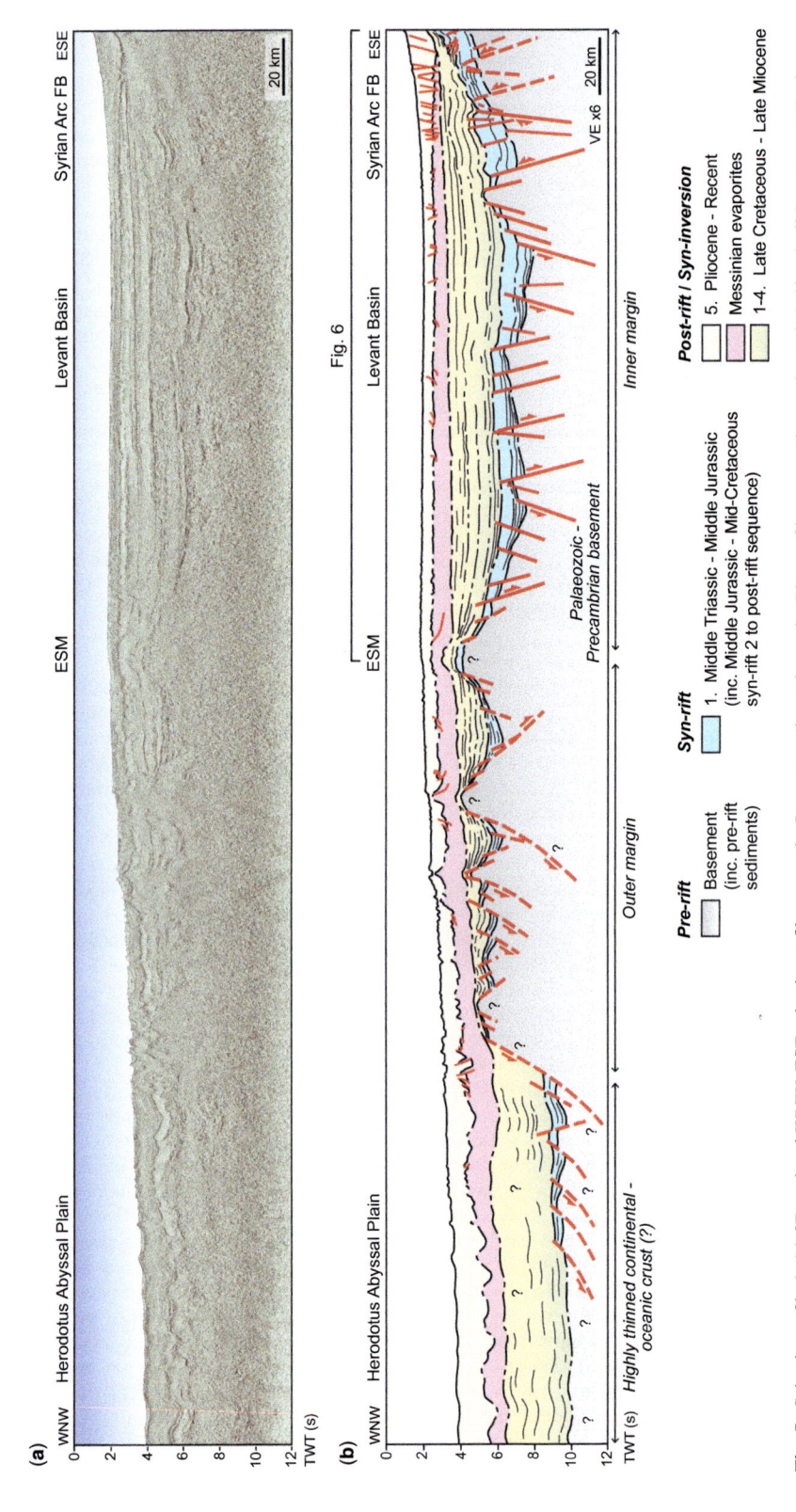

Fig. 5. Seismic profile 1. (**a**) 2D regional WNW–ESE seismic profile across the Levant continental margin. The profile crosses the continental shelf and offshore Levant Basin and extends to the Herodotus Abyssal Plain. (**b**) Line diagram interpretation of (**a**). Vertical exaggeration is c. 6:1. Numbers in key refer to the regional megasequences identified in this study (Fig. 4). Figure 1 shows the location of regional seismic profile 1. ESM, Eratosthenes Seamount; FB, fold belt.

Messinian salt flow, occurred in the Mid–Late Pliocene (e.g. Cartwright & Jackson 2008; Cartwright *et al.* 2012). This produced a salt-detached gravity-driven system, decoupled from the underlying basement-involved structures, in the Pliocene–Recent post-salt section (Fig. 5). Along the basin margin, the sub-salt inversion anticlines mark the updip limit of the post-salt extensional domain (Figs 5 & 6). Continued sub-salt fold amplification during the Pliocene–Early Pleistocene (SI5) resulted in uplift of the Messinian unconformity and development of crestal collapse faults within the post-salt section, indicating that the basement-involved structures influenced post-salt tectonics at the basin margin.

Basinward of the Eratosthenes Seamount, the outer continental margin is *c.* 145 km wide (Fig. 5). The structure of the outer margin consists of a series of tilted basement fault blocks bounded by dominantly seaward-dipping faults. Isolated half-graben basins were developed in the fault hanging walls. There is a relatively abrupt transition from large-scale rotated fault blocks to more listric and increasingly low-angle faults bounding the Herodotus Abyssal Plain, across which the continental basement deepens towards the oceanic domain. The transition to oceanic crust in the Herodotus Abyssal Plain was interpreted to occur where there was no longer any evidence for block faulting or deformation of the basement.

Levant Basin

The internal structure of the Levant Basin (Fig. 6a) is characterized by NE–SW-trending extensional faults, with very little variation in orientation between the rift border fault systems and the intra-rift faults. The extensional fault systems are segmented and offset along strike by a series of NW–SE-trending linear features that were interpreted as transfer or accommodation zones.

The basin is subdivided into a number of distinct NE–SW-trending depocentres and its structure and segmentation change significantly from south to north (Fig. 6b). The southern Levant Basin (cross-sections C and D) consists of segmented asymmetric syn-rift depocentres separated by intra-rift basement horst blocks. In contrast, the northern Levant Basin (cross-sections A and B) is characterized by a broadly symmetric basin geometry as the basin deepens and segmentation decreases towards the north.

To the west, the Levant Basin is bounded by the Eratosthenes Seamount (Figs 5 & 6). The structure of the basin suggests that the seamount was rifted from the Levant margin in a NW direction during the main SR1 phase of basin development, and subsequently formed a structural high at the northwestern margin of the basin during its post-rift evolution, onlapped by the post-rift megasequences. The seismic sections show that the entire Levant Basin is underlain by rifted continental basement as no evidence has been found for the existence of oceanic crust within the basin.

North Egyptian margin

The Pliocene–Recent evolution of the north Egyptian margin (Fig. 7) was characterized by high deltaic sediment supply, basinward Messinian salt flow and post-salt thin-skinned gravity-driven deformation (e.g. Loncke *et al.* 2006).

In the sub-salt section, rifted and thinned continental crust extends for a distance of at least 150 km from the shelf region to the Herodotus Abyssal Plain. Middle Triassic–Middle Jurassic rifting resulted in the development of NE–SW-trending basement-involved extensional fault systems across the margin, with fault-controlled hanging wall depocentres infilled by wedge-shaped SR1 strata. Extensional reactivation of the major faults occurred during the Middle Jurassic–Mid-Cretaceous (SR2). Along the continental shelf, the rift fault systems on the southeastern margin of the Herodotus Basin, with a dominant counter-regional dip towards the SE, were inverted during episodic phases of compression from the Late Cretaceous to the Late Miocene–Early Pleistocene. Inversion produced a series of broad asymmetric anticlines, onlapped by syn-inversion growth sequences (SI1–5) in the shelf region. In the proximal part of the Nile Delta system, pinchout of the salt layer is located basinward of the sub-salt inverted fault systems.

Basinward of the continental shelf, the transition to oceanic crust across the outer part of the margin is characterized by a series of tilted fault blocks bounded by listric seaward-dipping faults. The continent–ocean transition was inferred to be present below the thick Messinian salt and highly deformed sediments in the Herodotus Abyssal Plain.

NW Egyptian margin

The morphology of the NW Egyptian margin (Fig. 8) is characterized by a narrow continental shelf, *c.* 50 km wide, and a steep narrow slope, *c.* 15 km wide, across which there is an abrupt transition to the Herodotus Abyssal Plain.

The main Middle Triassic–Middle Jurassic rift phase (SR1) resulted in the development of high-angle, planar extensional fault systems across the margin. Below the continental shelf break, the top of the basement deepens abruptly across a major seaward-dipping extensional fault zone with an offset of *c.* 4 s TWT. This fault zone defines a transition from rifted continental crust in the footwall to highly thinned crust in the hanging wall, and subsequently to typical undeformed oceanic crust below 8 s

(a)

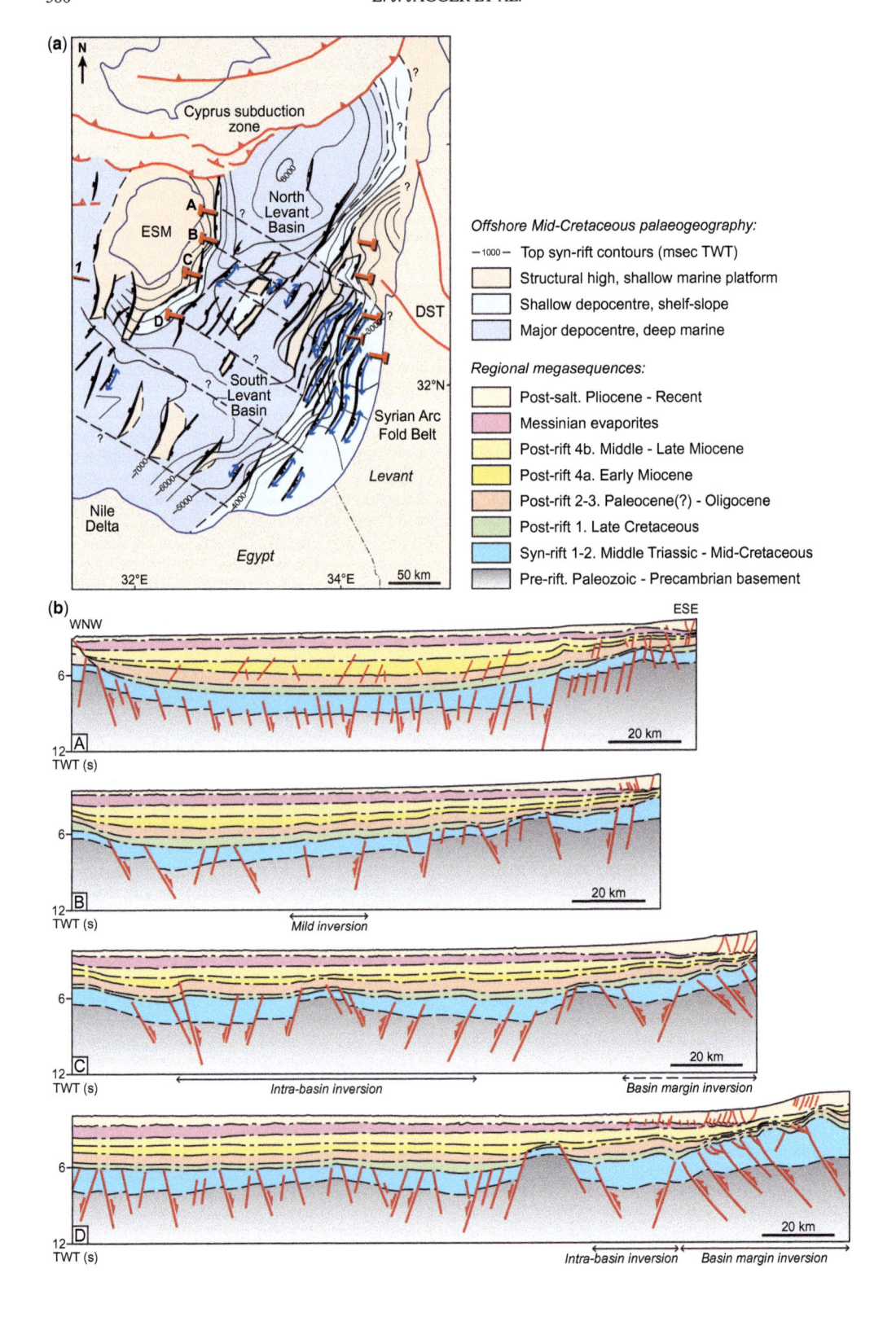

Offshore Mid-Cretaceous palaeogeography:
- —1000— Top syn-rift contours (msec TWT)
- Structural high, shallow marine platform
- Shallow depocentre, shelf-slope
- Major depocentre, deep marine

Regional megasequences:
- Post-salt. Pliocene - Recent
- Messinian evaporites
- Post-rift 4b. Middle - Late Miocene
- Post-rift 4a. Early Miocene
- Post-rift 2-3. Paleocene(?) - Oligocene
- Post-rift 1. Late Cretaceous
- Syn-rift 1-2. Middle Triassic - Mid-Cretaceous
- Pre-rift. Paleozoic - Precambrian basement

(b)

TWT in the Herodotus Abyssal Plain. In the hanging wall, the continent–ocean transition zone is c. 40 km wide and is characterized by a series of highly rotated basement fault blocks cut by steeply dipping faults. Narrow, deep half-graben basins were developed in the hanging walls of the rotated fault blocks, and infilled with antithetic-dipping wedges of SR1 strata (Fig. 4a). To the NE, the transition to undeformed oceanic crust, which was interpreted based on its seismic stratigraphic characteristics (Fig. 4a) as well as a lack of evidence for basement deformation, occurs across seaward-dipping extensional faults.

The SR2 strata define a carbonate platform sequence built up overlying the rifted basement on the continental shelf. Stratal package geometries are mainly sub-parallel and suggest that platform build-up was dominated by aggradation, which continued throughout the Cretaceous. This produced a flat-topped, attached carbonate platform, characterized by a basinward-thickening sediment wedge bounded by a steep seaward-facing slope above the major basement fault zone. The aggradational shallow marine platform geometry suggests that it was built up on a narrow shelf bounding the adjacent Herodotus deep marine basin, which was already developed by the Middle Jurassic–Mid-Cretaceous.

In contrast to the north Egyptian and Levant margins, there is no evidence for inversion of any of the major fault systems on the NW Egyptian shelf. The Cretaceous–Recent post-rift (PR1–PS) evolution of the margin was dominated by carbonate platform aggradation and clastic shelf progradation. The thin post-rift megasequences (<1 s TWT) indicate that subsidence and accommodation space were limited, potentially hidden by thermal uplift of the margin. This produced a narrow, steep and unstable shelf margin above the fault-bounded basement platform, onlapped by basinward-thickening post-rift sediments.

Cyrenaica Platform

The morphology of the Cyrenaica margin (Fig. 9) is similarly characterized by a narrow continental shelf, c. 50 km wide. Middle Triassic–Middle Jurassic rifting resulted in the development of narrow NE–SW- to WNW–ESE-trending half-grabens, infilled by wedge-shaped SR1 stratal packages, along the continental shelf. Extensional reactivation of the basin-bounding faults occurred during the Late Jurassic–Mid-Cretaceous (SR2). The outer part of the shelf is bounded by a WSW–ENE- to WNW–ESE-trending structurally high basement ridge and major seaward-dipping extensional fault system, across which the top of the basement is offset from a depth of c. 3 s TWT to c. 7–8 s TWT, that define an abrupt transition from rifted continental crust in the footwall to inferred highly thinned or oceanic crust underlying the outer domain of the Mediterranean Ridge.

The post-rift evolution of the Cyrenaica margin was characterized by carbonate platform progradation on the continental shelf throughout the Cretaceous–Late Eocene (PR1–2; Fig. 4b). The well-developed Cretaceous shelf margin suggests that the Cyrenaica Platform remained structurally high throughout the Middle Jurassic–Mid-Cretaceous rifting in the adjacent Sirt Basin (SR2). Multi-phase inversion occurred along the western Cyrenaica margin during the Late Cretaceous and Middle–Late Eocene (Fig. 9). Episodes of compression were accommodated by reverse reactivation of NE–SW to ENE–WSW half-graben bounding fault systems, which resulted in uplift of the asymmetric basin fill above regional level, folding of the post-rift carbonate platform sequences and deposition of syn-inversion growth stratal packages (SI1 and SI2) onlapping the inversion anticlines. Underlying the Mediterranean Ridge thrust front, significant buckling, uplift and inversion occurred in the hanging walls of the major seaward-dipping fault systems on the outer part of the Cyrenaica margin.

Gulf of Sirt

The Gulf of Sirt margin (Fig. 10) is characterized by a broad zone of rifted and thinned continental crust, which extends over a distance of at least 350–400 km to highly thinned continental crust and oceanic crust in the Ionian Abyssal Plain. The offshore Sirt Basin (Fig. 11) is up to 210 km wide and shows a typical rift structure, characterized by high-angle rotated basement fault systems overlain by thick post-rift sequences that were deposited during thermal subsidence. The Sirt Basin is bounded to the NE by major NW–SE- to WNW–ESE-oriented seaward-dipping extensional fault systems that define the transition to the Ionian Abyssal Plain.

The SR2 megasequence (Fig. 4b), characterized by fault-controlled wedge-shaped stratal packages,

---

Fig. 6. Structure of the Levant Basin. (**a**) Tethyan structural elements and summary Mid-Cretaceous palaeogeography of the Levant Basin, mapped from the 2D seismic dataset. Contours shown are the top syn-rift horizon, highlighting the structural architecture of the basin. Cross-section 1 shows the location of seismic profile 1 (Fig. 5). (**b**) Series of cross-sections across the Levant Basin, interpreted from the 2D seismic dataset, illustrating the variations in regional structural style along strike from north (A) to south (D). Vertical exaggeration is c. 3:1. Cross-section locations are shown in (a). DST, Dead Sea Transform; ESM, Eratosthenes Seamount.

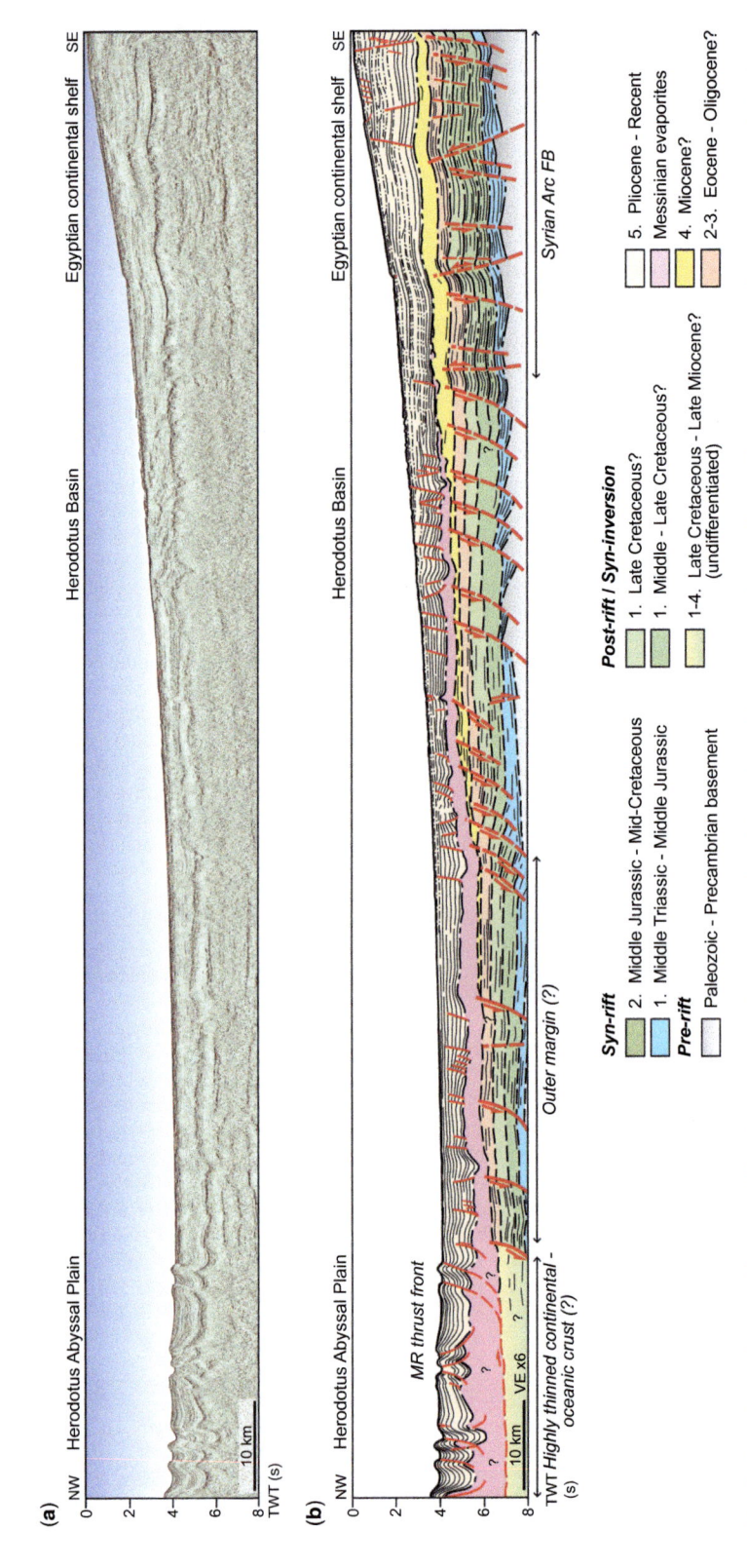

Fig. 7. Seismic profile 2. (**a**) 2D regional NW–SE seismic profile across the West Nile Delta, offshore northern Egypt. The profile crosses the north Egyptian continental shelf and Herodotus Basin, and extends to the Herodotus Abyssal Plain and the outer domain of the Mediterranean Ridge. (**b**) Line diagram interpretation of (**a**). Vertical exaggeration is c. 6:1. Numbers in key refer to the regional megasequences identified in this study (Fig. 4). Figure 1 shows the location of regional seismic profile 2. FB, Fold belt; MR, Mediterranean Ridge.

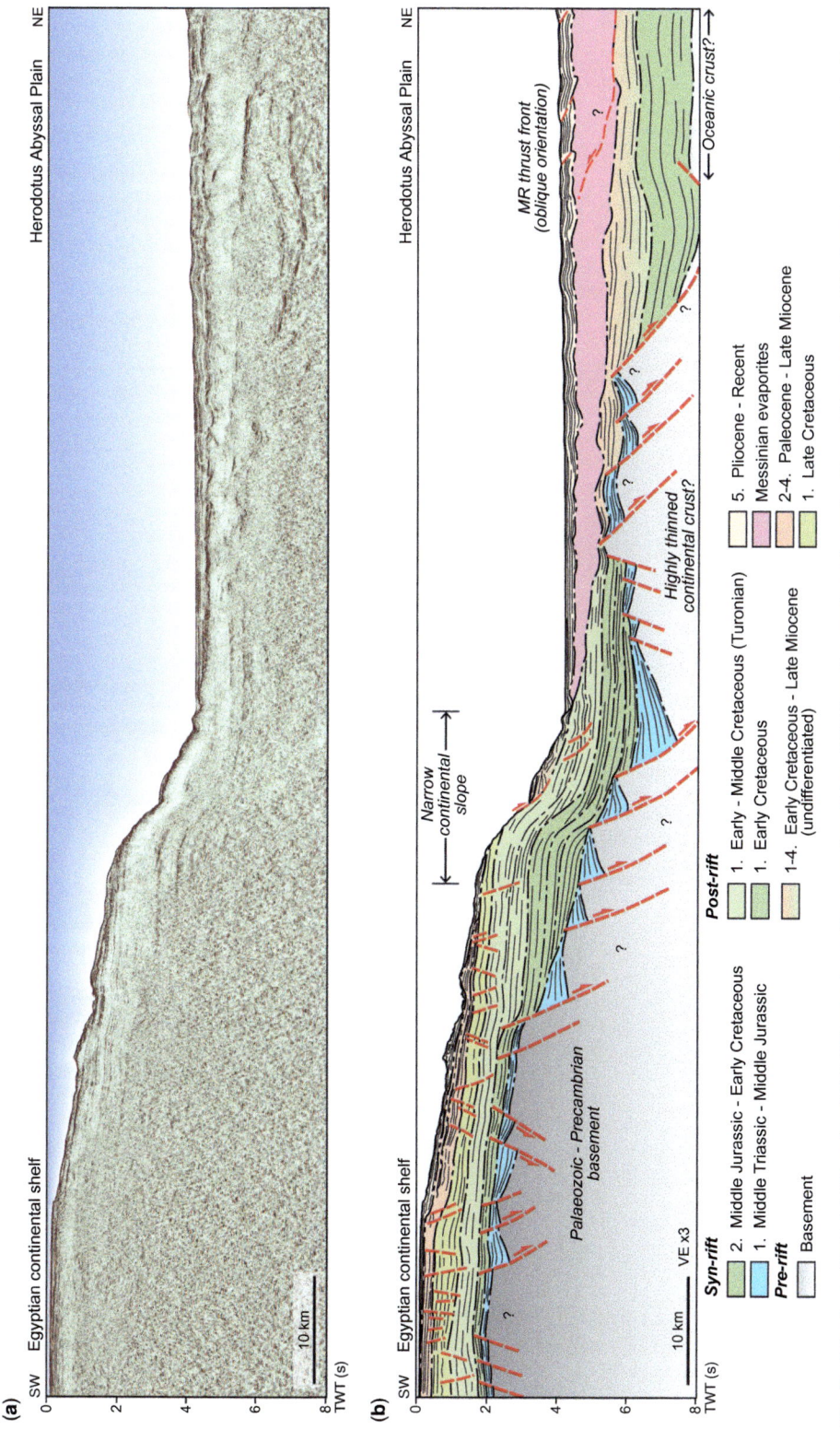

Fig. 8. Seismic profile 3. (**a**) 2D regional NE–SW seismic profile across the offshore NW Egyptian continental margin. The profile crosses the NW Egyptian continental shelf and extends to the Herodotus Abyssal Plain. (**b**) Line diagram interpretation of (a). Note that the Early Cretaceous syn-rift to post-rift sequences could not be differentiated. Vertical exaggeration is c. 3:1. Numbers in key refer to the regional megasequences identified in this study (Fig. 4). Figure 1 shows the location of regional seismic profile 3. MR, Mediterranean Ridge.

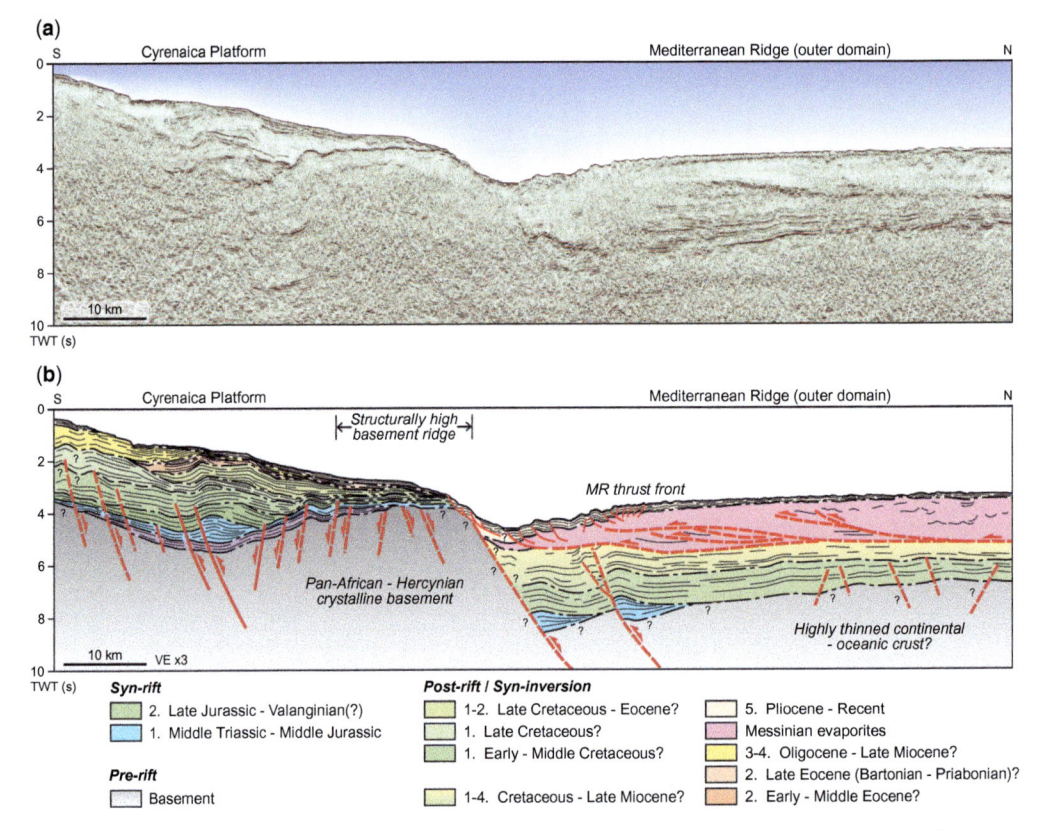

Fig. 9. Seismic profile 4. (**a**) 2D regional north–south seismic profile across the western Cyrenaica margin, offshore NE Libya. The profile crosses the Cyrenaica Platform and extends to the outer domain of the Mediterranean Ridge. (**b**) Line diagram interpretation of (a). Vertical exaggeration at the seafloor is c. 3:1. Numbers in key refer to the regional megasequences identified in this study (Fig. 4). Figure 1 shows the location of regional seismic profile 4. MR, Mediterranean Ridge.

defines the main phase of rifting and basin development during the Late Jurassic–Mid-Cretaceous. The end of the rift phases in the offshore Sirt Basin in the Cenomanian was followed by thermal subsidence and deposition of the Late Cretaceous–Late Eocene post-rift sequences (PR1–2; Fig. 4b). Middle–Late Eocene carbonate platforms were well developed on the basin margins (Fig. 10). In contrast to the Cyrenaica margin, where the carbonate shelf developed during the Late Jurassic–Cretaceous, this suggests a later break-up age for the Ionian Abyssal Plain and formation of the Gulf of Sirt margin.

Thermal subsidence was interrupted by episodic phases of compression in the Late Cretaceous (Santonian–Maastrichtian) and Middle–Late Eocene (late Ypresian–Priabonian) (Fig. 4b). This resulted in mild uplift of the basin and mild inversion of NNW–SSE-trending half-graben bounding faults within the basin. Inversion produced a series of gentle hanging wall folds in the post-rift basin fill, although fault reactivation was limited and faults remained in net extension. Inversion in the Sirt Basin was restricted to the main depocentre and no reactivation occurred on the basin margins.

Late Eocene (Priabonian)–Late Miocene episodic passive margin uplift and tilting resulted in delta progradation across the continental shelf and gravity-driven slumping within the basin (PR2b–4; Fig. 4b). In contrast to the Levant and Herodotus basins, the absence of mobile Messinian halite within the Sirt Basin resulted in limited post-salt deformation.

Offshore Sirt Basin

The internal structure of the offshore Sirt Basin (Fig. 11a) is characterized by two distinct basement-involved fault trends, which show considerable variation between the orientations of the rift border faults and the intra-rift faults. NW–SE- to WNW–ESE-oriented extensional faults extend across the offshore north Libyan margin from the Cyrenaica

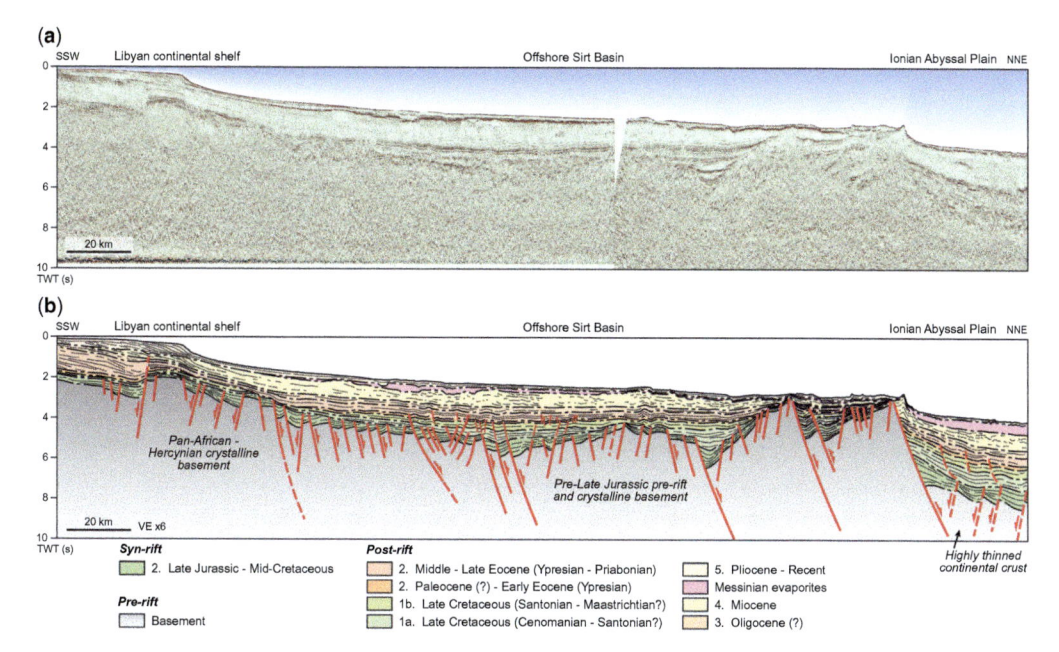

Fig. 10. Seismic profile 5. (**a**) 2D regional NNE–SSW seismic profile across the offshore north Libyan continental margin. The profile crosses the north Libyan continental shelf and the offshore Sirt Basin, and extends to the Ionian Abyssal Plain. (**b**) Line diagram interpretation of (a). Vertical exaggeration is *c.* 6:1. Numbers in key refer to the regional megasequences identified in this study (Fig. 4). Figure 1 shows the location of regional seismic profile 5.

Platform to the Pelagian Shelf. The major seaward-dipping fault systems bounding the Ionian Abyssal Plain also have a WNW–ESE to NW–SE trend, whereas, in the Gulf of Sirt, the fault systems dominantly strike NNW–SSE, parallel to the rift axis, and define the main depocentre of the offshore Sirt Basin. The rift border fault systems and half-graben bounding faults are segmented along strike with relay ramps formed between the overlapping tips of the en echelon fault segments.

The basin architecture is strongly asymmetric, and shows significant variations in depocentre segmentation and polarity along strike (Fig. 11b). The northwestern Sirt Basin (cross-sections A–C) is dominated by major NNW–SSE-trending rotated basement fault blocks that define segmented, strongly asymmetric half-graben depocentres. These fault-controlled depocentres change polarity across NE–SW- to NNE–SSW-trending, high-relief accommodation zones oblique to the rift trend (Fig. 11a). Towards the SE (cross-sections D and E), the basin is characterized by a major asymmetric to broadly symmetric NNW–SSE-trending depocentre, which similarly changes polarity along strike. The characteristics of the continent–ocean transition zone on the northeastern boundary of the Sirt Basin also change along strike. In the northwestern Sirt Basin (cross-sections A–C), the transition to the Ionian Abyssal Plain occurs across a series of seaward-dipping tilted fault blocks, whereas towards the SE (cross-sections D and E), the continent–ocean transition is defined by a structurally high basement ridge, which extends from the Cyrenaica Platform, bounded by major high-angle seaward-dipping extensional fault systems.

Pelagian Shelf and Malta Escarpment

The Pelagian Shelf (Fig. 12) Tethyan basins are characterized by a series of NW–SE-trending horst and graben structures. Late Jurassic–Mid-Cretaceous SR2 strata show fault-related thickness variations, which indicate that extension on the shelf was active during rifting in the Sirt Basin to the SE. In contrast, rifting on the Pelagian Shelf did not progress to seafloor spreading and the region did not subside substantially following continental break-up in the Ionian Abyssal Plain. The NW–SE basin-bounding fault systems were repeatedly reactivated during the Cenozoic (Fig. 12). In particular, there is strong evidence for extensive Late Miocene–Recent tectonics. Deep extensional and rhomboidal pull-apart basins are well developed across the continental shelf (e.g. Jongsma *et al.* 1985), preferentially controlled by the pre-existing Tethyan structures, and the Oligocene–Miocene post-rift section is highly faulted.

The northeastern part of the Pelagian continental shelf is bounded by the NNW–SSE-trending

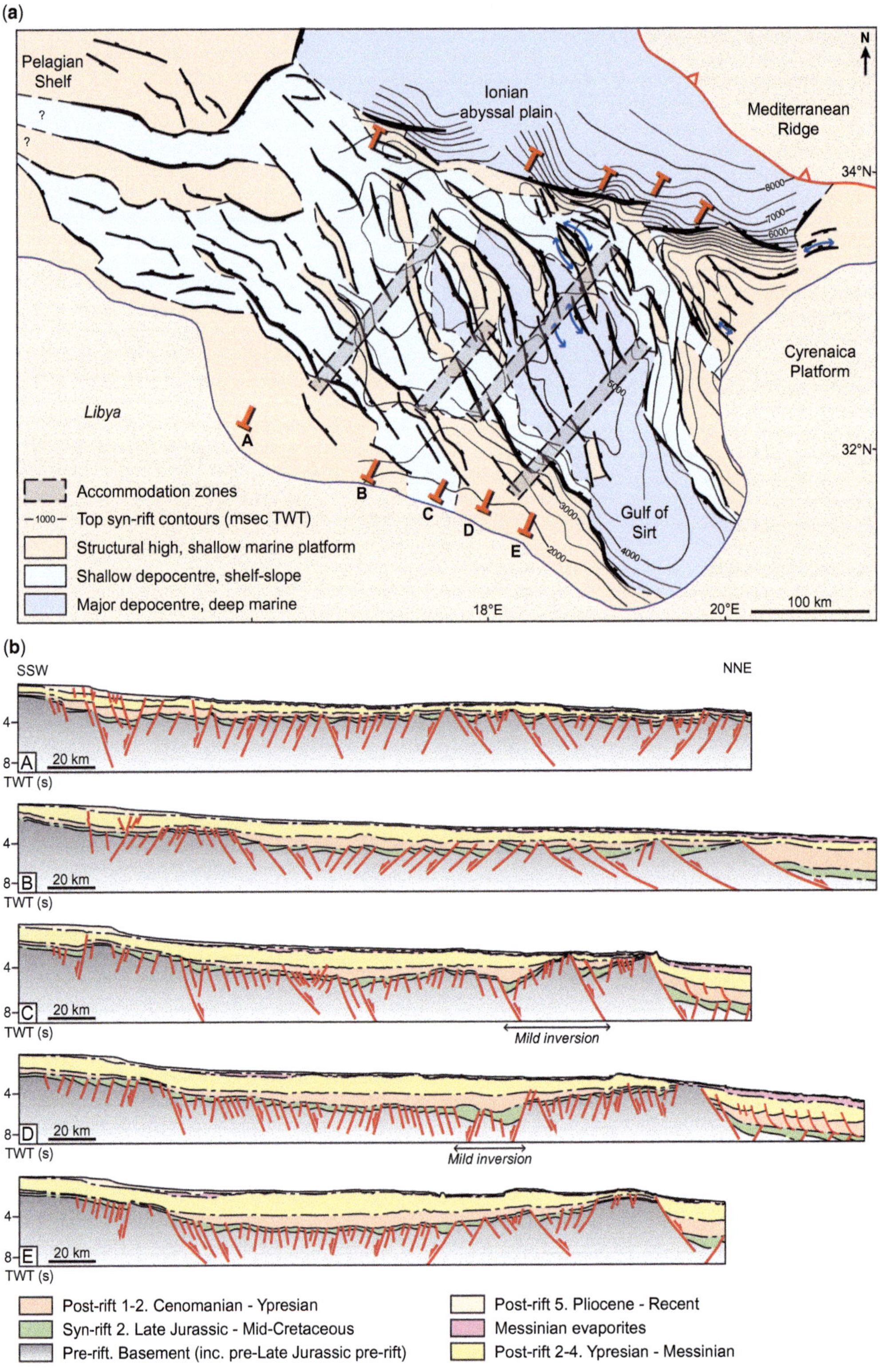
(a)
Pelagian Shelf
Ionian abyssal plain
Mediterranean Ridge
N
34°N
8000
7000
6000
Cyrenaica Platform
Libya
32°N
5000
A
B
C
D
E
3000
2000
4000
Gulf of Sirt
Accommodation zones
1000 Top syn-rift contours (msec TWT)
Structural high, shallow marine platform
Shallow depocentre, shelf-slope
Major depocentre, deep marine
18°E
20°E
100 km
(b)
SSW
NNE
4
8 A 20 km
TWT (s)
4
8 B 20 km
TWT (s)
4
8 C 20 km
TWT (s)
Mild inversion
4
8 D 20 km
TWT (s)
Mild inversion
4
8 E 20 km
TWT (s)
Post-rift 1-2. Cenomanian - Ypresian
Syn-rift 2. Late Jurassic - Mid-Cretaceous
Pre-rift. Basement (inc. pre-Late Jurassic pre-rift)
Post-rift 5. Pliocene - Recent
Messinian evaporites
Post-rift 2-4. Ypresian - Messinian

Malta Escarpment (Fig. 12), which defines a steep and abrupt shelf break across which the transition to the Ionian Abyssal Plain occurs. The continental shelf area consists of a structurally high basement platform, bounded by a major seaward-dipping extensional fault system that marks the abrupt transition from rifted continental basement in the footwall to interpreted oceanic crust in the hanging wall. There is no evidence for compression or inversion of the major NNW–SSE-trending fault systems during the post-rift evolution of the Malta Escarpment. Slumping and collapse of the steep, unstable shelf edge dominated the post-rift morphology of the margin. In the Ionian Abyssal Plain, the thickening Pliocene–Recent sediment wedge above the Messinian salt indicates that reactivation of the main seaward-dipping extensional fault system, and fault-controlled subsidence relative to the Pelagian Shelf, is continuing to the present day.

Discussion

Our study shows that the offshore SE Mediterranean margin was segmented into discrete tectonic provinces along strike (Fig. 1) during the episodic break-up of northern Gondwana from the Middle Triassic–Mid-Cretaceous (c. 240–93 Ma). These regional tectonic provinces are characterized by distinct morphologies, structural styles and tectonostratigraphic histories (Figs 1 & 13). We attribute this segmentation of the margin to multiple phases of rifting and continental break-up in the Eastern Mediterranean Neotethys, and the reactivation of pre-existing basement fabrics in varying regional stress fields (Fig. 14). The along-strike segmentation of the margin subsequently exerted a regional-scale control on the episodic late Santonian (c. 84 Ma)–Early Pleistocene inversion history, associated with the convergence of the African–Arabian and Eurasian plates and the polyphase closure of the Neotethys oceans from the late Santonian to the Present day (Fig. 15).

Segmentation of the SE Mediterranean rifted margin

In common with rifted continental margins worldwide, the SE Mediterranean margin is composed of rifted and sheared margin segments.

Rifted margin segments. The crustal structure of the Levant–north Egyptian (Figs 5 & 7) and the Gulf of Sirt (Fig. 10) margin segments is typical of rifted continental margins. They are characterized by wide regions of rifted and highly thinned continental crust, less than 25 km thick, that extend at least 400 km from the continental shelf regions, where the Moho lies at a depth of 30–35 km, to the 5–9 km-thick oceanic crust in the Herodotus and Ionian abyssal plains, where the Moho depth is c. 17–23 km (e.g. de Voogd et al. 1992; Ben-Avraham et al. 2002; Marone et al. 2003; Cowie & Kusznir 2012a, b).

Extension and subsidence were focused along the inner margins, where the Levant and offshore Sirt basins (Figs 6 & 11) form the major depocentres c. 200 km wide, containing post-rift sequences up to c. 10 km thick. The rift basins are underlain by highly thinned continental crust c. 8–20 km thick, with β stretching factors of up to 2.3–3 below the Levant Basin (e.g. Ben-Avraham et al. 2002; Gardosh et al. 2010; Cowie & Kusznir 2012a, b). The outer margins are characterized by a series of high-angle seaward-dipping tilted basement fault blocks (Figs 5, 7 & 10). The transition from rifted continental basement to highly thinned continental–transitional and oceanic crust in the Herodotus and Ionian abyssal plains is interpreted to be relatively abrupt and occurs across major seaward-dipping extensional fault systems. This indicates much higher β stretching factors on the outermost distal parts of the margins.

In comparison, Robertson (2007) observed that the structure of the Eastern Tethyan rifted margins does not correspond to ideal volcanic or magma-poor rifted margins; instead, they correspond to an 'intermediate' type, characterized by pulsed rifting, limited rift volcanism and a narrow continent–ocean transition zone. Other intermediate-type margins, such as the North West Shelf, Western Australia (e.g. Symonds et al. 1998; Dore & Stewart 2002; Direen et al. 2008) and the South China Sea (e.g. Franke et al. 2011; Ding et al. 2013; Franke 2013), where break-up was similarly achieved by multiphase rifting and episodic terrane detachment from pre-existing continental margins, show structural features comparable with the SE Mediterranean rifted margins summarized above. Seismic and numerical modelling studies have further shown that pulsed rifting and tectonic inheritance could explain the development of rift basins offset from

Fig. 11. Structure of the offshore Sirt Basin. (**a**) Tethyan structural elements and summary Mid-Cretaceous palaeogeography of the offshore Sirt Basin mapped from the 2D seismic dataset. Contours shown are the top syn-rift horizon, highlighting the structural architecture of the basin. (**b**) Series of cross-sections across the offshore Sirt Basin, which illustrate the variations in basin structure along strike from NW (A) to SE (E). Vertical exaggeration is c. 4:1. Cross-section locations are shown in (a).

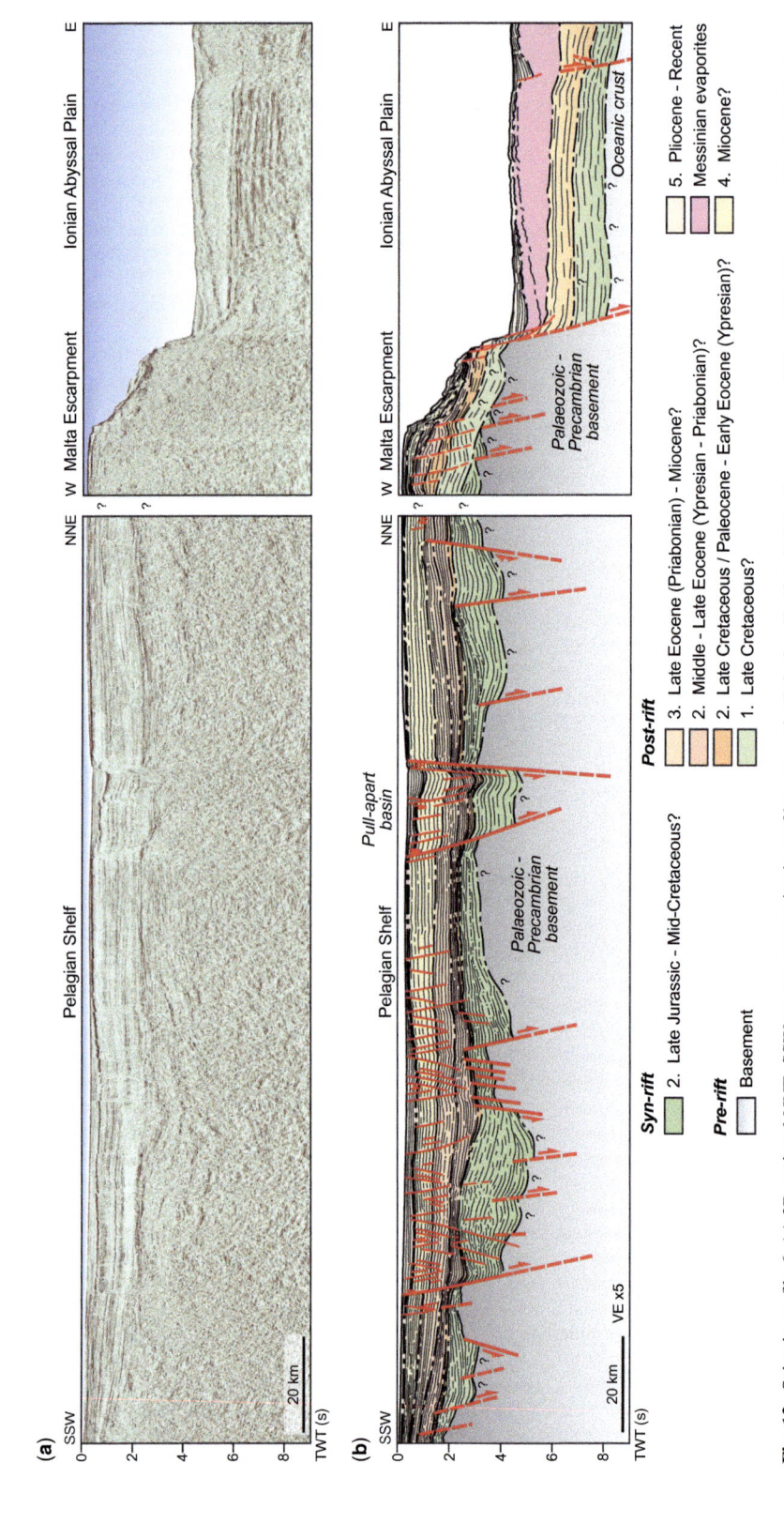

Fig. 12. Seismic profile 6. (**a**) 2D regional NNE–SSW to east–west seismic profiles across the Pelagian Shelf and Malta Escarpment, offshore NW Libya, extending to the Ionian Abyssal Plain. (**b**) Line diagram interpretation of (**a**). Vertical exaggeration is c. 5:1. Numbers in key refer to the regional megasequences identified in this study (Fig. 4). Figure 1 shows the location of regional seismic profile 6.

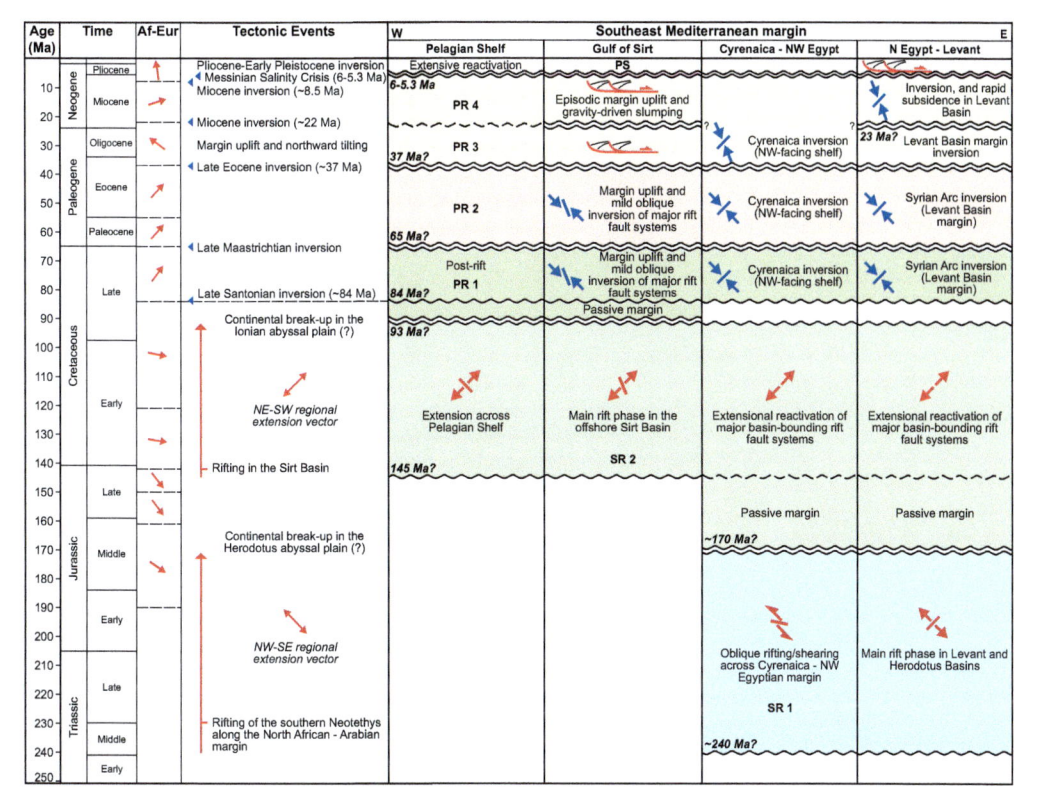

Fig. 13. Summary of the tectono-stratigraphic histories of the different domains of the offshore SE Mediterranean margin, the key regional tectonic events and their correlation across the margin, and their relationship to the motion history of the African–Arabian and Eurasian plates. Interpreted megasequences: SR1, syn-rift 1; SR2, syn-rift 2; PR1, post-rift 1; PR2, post-rift 2; PR3, post-rift 3; PR4, post-rift 4; PS, post-salt. Motion history of the African plate relative to Eurasia after Savostin *et al.* (1986), Guiraud & Bosworth (1997), and Guiraud *et al.* (2005). Regional tectonic events based on this study, with ages after Guiraud & Bosworth (1997), Guiraud (1998), Guiraud *et al.* (2005), and Bosworth *et al.* (2008).

the main locus of rifting and continental break-up (e.g. Gartrell 2000; Chenin & Beaumont 2013). This suggests that inherited Pan-African–Paleozoic weak zones and terrane fragmentation processes along the north Gondwana margin exerted a major control on the structure and evolution of the south Neotethyan rifted margins (Fig. 14).

Transform-dominated margin segments. The structure of the NW Egyptian–Cyrenaica (Figs 8 & 9) and Malta Escarpment (Fig. 12) margin segments is characteristic of transform margins (e.g. Bird 2001; Nemčok *et al.* 2016). The narrow continental shelf regions, where the Moho lies at a depth of c. 30 km, are c. 50 km wide and are underlain by continental crust that thins from c. 30 to 22 km thick towards the continent–ocean transition (e.g. Bronner & Makris 2000; Marone *et al.* 2003). Below the steep continental shelf break, major seaward-dipping extensional fault zones with offsets of c. 4 s TWT define an abrupt transition from rifted continental basement in the footwall to highly thinned and faulted continental–transitional crust interpreted in the hanging wall, and subsequently to typical undeformed oceanic crust, c. 5–9 km thick, underlying the Herodotus and Ionian abyssal plains. Our seismic observations along the NW Egyptian–Cyrenaica margin correspond closely to the gravity models of Longacre *et al.* (2007) and Cowie & Kusznir (2012a, b), which show an abrupt change in the geometry of the Moho and decrease in crustal thickness across a narrow zone of extreme crustal thinning below the outer continental shelf. Tassy *et al.* (2015) mapped the 3D seismo-stratigraphic architecture of the Jurassic–Cretaceous carbonate platform–slope–basin system along the Egyptian continental margin, and similarly concluded that the structure of the abrupt NW–SE-trending palaeo-shelf edge corresponds to a transform margin hinge zone and was strongly controlled by deep crustal Tethyan transform faults.

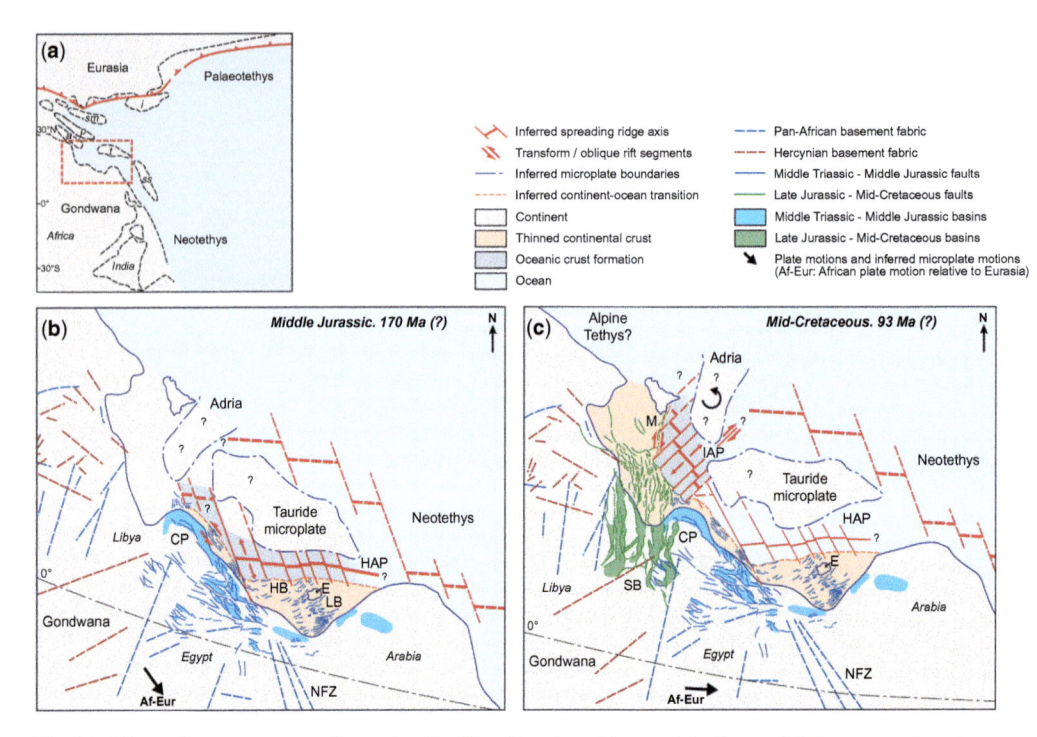

Fig. 14. Schematic summary maps illustrating the rift and break-up history of the Eastern Mediterranean, based on the results of this study. (**a**) Simplified tectonic setting and palaeogeography of the eastern Tethys realm during the Late Triassic–Early Jurassic, modified after Robertson (2007). a, Adria; i, Iranian units; p, Pelagonian; sm, Serbo-Macedonia; ss, Sirandaj–Sirjan; t, Tauride units. Red box highlights the study area shown in detail in (b) and (c). (**b**) Middle Jurassic–Middle Triassic (*c.* 240–170 Ma), opening of the Herodotus Abyssal Plain. (**c**) Late Jurassic–Mid-Cretaceous (*c.* 145–93 Ma), opening of the Ionian Abyssal Plain. Positions of Gondwana-derived microplates are approximate. Northern rifted microplates and Eurasian margin not shown. Offshore faults mapped from the 2D seismic dataset used in this study. Onshore faults after Abadi *et al.* (2008), Bosworth *et al.* (2008), and Fiduk (2009). Relative Af-Eur plate motions after Savostin *et al.* (1986) and Guiraud *et al.* (2005). Basement fabrics after Bosworth & McClay (2001), Hallett (2002), Coward & Ries (2003), and Craig *et al.* (2008). CP, Cyrenaica Platform; E, Eratosthenes Seamount; HAP, Herodotus Abyssal Plain; HB, Herodotus Basin; IAP, Ionian Abyssal Plain; LB, Levant Basin; M, Malta Escarpment; NFZ, Najd Fault Zone; SB, Sirt Basin.

The morphological and structural characteristics of the SE Mediterranean transform-dominated margin segments described above compare well with the Atlantic transform margins, offshore West Africa, which developed during oblique rifting of the Equatorial Atlantic (e.g. Turner *et al.* 2003; Wilson *et al.* 2003; Antobreh *et al.* 2009; Turner & Wilson 2009; Rupke *et al.* 2010; Nemčok *et al.* 2016). However, characteristic marginal ridge structures, such as that bounding the Romanche Fracture Zone (Antobreh *et al.* 2009; Rupke *et al.* 2010; Nemčok *et al.* 2016), are not observed along the SE Mediterranean transform margin segments, which are typically characterized by less pronounced fault-bounded basement horsts at the continental shelf boundaries. This was similarly observed along the obliquely divergent margin offshore Equatorial Guinea (Turner & Wilson 2009), which suggests that there was also an oblique component of extension during the development of the SE Mediterranean transform margins (Fig. 14).

Offshore NW Egypt, the continent–ocean transition, characterized by a narrow zone of highly faulted, rotated and thinned continental–transitional crust (Fig. 8), is comparable with the structures observed offshore Equatorial Guinea, where the Ascension Fracture Zone accommodated highly oblique rifting and break-up along the Gulf of Guinea margin (Wilson *et al.* 2003; Turner & Wilson 2009). Similar fault rotation during shearing may therefore be related to oblique transtensile strain and the development of a major fracture zone offshore NW Egypt. Gravity and magnetic maps (e.g. Longacre *et al.* 2007) indicate a major NW–SE boundary offshore NW Egypt, oriented *c.* 10–20° oblique to the margin segment. This may define the fracture zone location, supporting the interpretation of the NW Egyptian–Cyrenaica margin as an obliquely divergent margin.

Fig. 15. Schematic summary maps illustrating the post-rift inversion and uplift history of the SE Mediterranean margin, based on the results of this study. (a) Late Santonian (84 Ma); (b) Priabonian (37 Ma); (c) Aquitanian (23 Ma); and (d) Late Messinian–Early Pleistocene. Positions of rifted microplates and Eurasian margin not shown. Offshore faults and inversion domains mapped from the 2D seismic dataset used in this study. Onshore faults after Abadi et al. (2008), Bosworth et al. (2008), and Fiduk (2009). Relative Af-Eur plate motions after Savostin et al. (1986) and Guiraud et al. (2005). CP, Cyrenaica Platform; ESM, Eratosthenes Seamount; HAP, Herodotus Abyssal Plain; HB, Herodotus Basin; IAP, Ionian Abyssal Plain; LB, Levant Basin; M, Malta Escarpment; SA, Syrian Arc fold belt; SB, Sirt Basin.

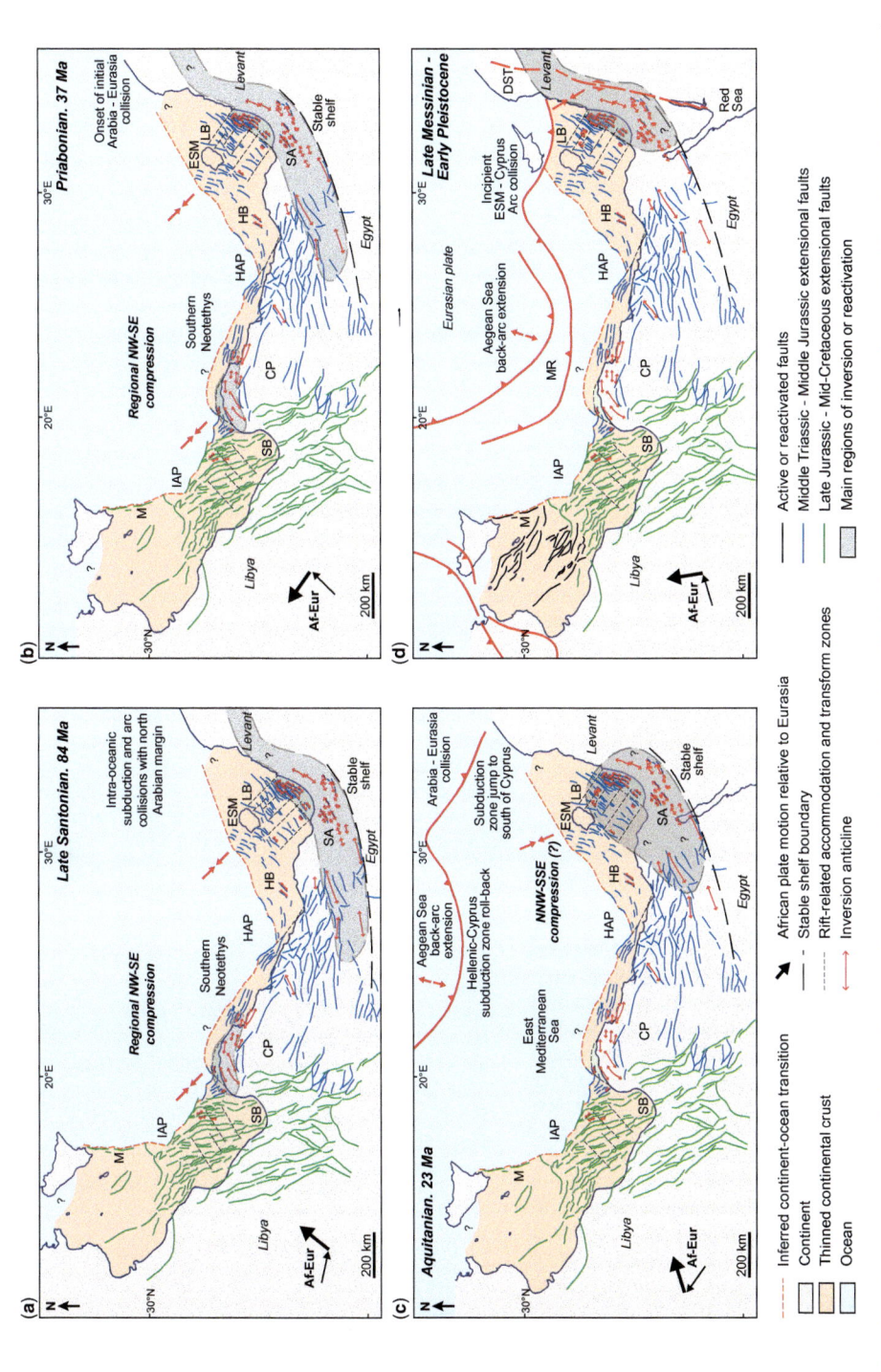

Pan-African and Hercynian tectonic inheritance

Tectonic inheritance is known to fundamentally influence the geometry of rifting and continental break-up, as well as the locations and orientations of transform faults and oceanic fracture zones (e.g. Clemson *et al.* 1997; Meisling *et al.* 2001; Miller *et al.* 2002; Thomas 2006; Antobreh *et al.* 2009; Ebbing & Olesen 2010; Gibson *et al.* 2013; Tasrianto & Escalona 2015). As such, the inherited Pan-African basement fabrics and Hercynian structures are key to understanding the break-up history in the Eastern Mediterranean (Figs 1 & 14).

Levant–north Egyptian margin. Along the Levant–north Egyptian rifted margin segment, the Precambrian crystalline basement has a dominant NW-trending Pan-African fabric (Fig. 1), which is present across NE Africa and the Arabian–Nubian Shield (e.g. Younes & McClay 2002; Coward & Ries 2003; Craig *et al.* 2008). The Tethyan fault pattern and structure of the Levant Basin (Figs 1 & 6) are characteristic of an orthogonal rift (e.g. McClay *et al.* 2002; Guiraud *et al.* 2010; Zwaan *et al.* 2016). The dominant NE–SW rift fault trend indicates that the regional extension vector was oriented NW–SE, at a high angle to the structural trend, and that the fault orientations were not strongly influenced by the pre-existing basement fabrics. The rift border fault systems show curved geometries and fault linkage across NW–SE-trending accommodation zones, and the intra-rift faults and depocentres are also segmented and offset across these zones. This suggests that the pre-existing NW–SE Pan-African basement fabrics were reactivated during rifting, and controlled the basin segmentation and rift evolution (Fig. 14).

NW Egyptian–Cyrenaica margin. The continent–ocean transition structure of the NW Egyptian–Cyrenaica margin segment (Figs 8 & 9), in comparison with the West African transform margins, indicates that a major NW–SE-oriented transform–oceanic fracture zone was developed offshore NW Egypt. Studies of the southern margin of Australia (e.g. Miller *et al.* 2002; Gibson *et al.* 2013) have suggested that the positions and orientations of fracture zones along rifted margins are controlled by pre-existing basement structures. Onshore northern Egypt, NW- to NNW-trending Pan-African fabrics are well developed in the basement of Sinai and the Eastern Desert (Fig. 1). Major potential controlling structures include the Pan-African Najd shear zone systems of the Arabian plate (Fig. 14). The Najd fault systems define a major NW–SE crustal-scale shear zone, *c.* 250 km wide and *c.* 1100 km long with a cumulative left-lateral offset of 240–300 km, which extends to the Egyptian side of the Red Sea (Younes & McClay 2002; Craig *et al.* 2008). Reactivation of this left-lateral shear zone during Neotethys rifting is likely to have exerted a strong influence on the development of a transform–oceanic fracture zone offshore NW Egypt.

Gulf of Sirt. The Precambrian crystalline basement in the Gulf of Sirt has a dominant north–south to NW–SE Pan-African fabric (Fig. 1; Hallett 2002). The Tethyan fault pattern, characterized by two distinct rift fault populations, and architecture of the offshore Sirt Basin (Fig. 11) are characteristic of an oblique rift (e.g. Bosworth & McClay 2001; McClay *et al.* 2002; Mortimer *et al.* 2016; Zwaan *et al.* 2016). The basin is segmented by relatively high-relief accommodation zones (e.g. Bosworth & McClay 2001; McClay *et al.* 2002), oriented parallel to the extension direction, which typically developed above offsets in the pre-existing zones of weakness. Across these accommodation zones, segmented asymmetric offset depocentres flip polarity along strike associated with the change in dominant fault dip. Therefore, we infer that rifting in the Gulf of Sirt was strongly controlled by inherited basement fabrics. There is a strong correlation between the NNW–SSE basement-involved fault trend in the Sirt Basin and the north–south to NW–SE Pan-African fabric that was developed across Libya (Fig. 1). In contrast, the dominant NW–SE fault trend across the north Libyan margin, and the development of extension-parallel NE–SW-trending accommodation zones in the offshore Sirt Basin, indicate a regional NE–SW extension vector oblique (*c.* 60°) to the trend of the major basement fault systems and pre-existing structures (Fig. 14).

Pelagian Shelf–Malta Escarpment. Extensional fault systems across the Pelagian Shelf (Fig. 1) trend NW–SE, consistent with the dominant fault trend across the north Libyan margin and perpendicular to the inferred regional extension vector (Fig. 14), which indicates limited basement control on rifting along this part of the margin. However, the NNW–SSE-trending Malta Escarpment bounding the Pelagian Shelf is oriented parallel to the Pan-African trend (Fig. 1) and is suggested to be a reactivated Pan-African basement lineament (Fig. 14), which played an active role in segmentation of the North African margin during Neotethys rifting (e.g. Briand 2000).

Eastern Mediterranean Tethys rift history.
Middle Triassic–Mid-Cretaceous (*c.* 240–93 Ma?)

Based on the along-strike segmentation of the SE Mediterranean margin and the tectono-stratigraphic histories of the margin segments (Figs 1 & 13), we present a new tectonic evolutionary model for the development of the Neotethys margin in the context of the break-up of northern Gondwana from the Middle Triassic–Mid-Cretaceous (c. 240–93 Ma?) (Fig. 14).

Middle Triassic–Middle Jurassic (c. 240–170 Ma?).
Seafloor spreading in the easternmost Mediterranean began during the Late Triassic–Early Jurassic, as the result of Middle Triassic–Middle Jurassic multiphase rifting (SR1) and break-up along the Levant–north Egyptian and the NW Egyptian–Cyrenaica margin segments (Figs 13 & 14b).

The timing of rifting is consistent with previous models that inferred detachment of the Tauride microplate from the Levant–north Egyptian margin and the development of a southern Neotethys spreading ridge in the Herodotus Abyssal Plain during the Late Triassic–Early Jurassic (Fig. 2c, d; e.g. Garfunkel 1998; Robertson 2007; Gardosh et al. 2010; Robertson et al. 2012). Rifting also occurred between the Eratosthenes Seamount and the Levant shelf during the main rift phase (Fig. 14b). However, although the failed rifting of the Eratosthenes Seamount resulted in significant crustal thinning underlying the Levant Basin (e.g. Gardosh et al. 2010), it did not progress to seafloor spreading and as a result the seamount remained stranded at the northwestern margin of the basin (Fig. 6).

Continental break-up in the easternmost Mediterranean was the result of NW–SE regional extension, perpendicular to the orthogonally rifted Levant–north Egyptian margin segment (Fig. 14b). The southern Neotethys spreading ridge axis was oriented NE–SW, parallel to the Levant margin but outboard of the Eratosthenes Seamount, consistent with the majority of the Late Triassic–Early Jurassic southern Neotethys models (Fig. 2d; e.g. Garfunkel 1998; Gardosh et al. 2010). Similarly, the highly oblique orientation of the regional extension vector relative to the NW Egyptian–Cyrenaica margin segment resulted in the development of a southern sinistral transform margin. The transform margin orientation is consistent with the southeastward divergence of the African plate relative to Eurasia during the Middle Triassic–Late Jurassic, associated with the opening of the Central Atlantic Ocean (e.g. Savostin et al. 1986; Guiraud et al. 2005); NW–SE transform faults were similarly developed in the Alpine Tethys, which opened during the Middle Jurassic c. 165 Ma following Late Triassic–Early–Middle Jurassic rifting (de Graciansky et al. 2010).

The westward extent of the Late Triassic–Early Jurassic southern Neotethys is difficult to determine owing to subsequent overprinting by the Sirt rift system. Based on the ages of the regional syn-rift megasequences in this study (Figs 4 & 13), we suggest that the ocean extended as far as the northwestern margin of Cyrenaica (Fig. 14b). The NW–SE break-up kinematics restricted the space available for the Tauride microplate to move once it detached from Gondwana (Garfunkel 2004), resulting in the development of a narrow Red Sea-scale ocean basin (Morris et al. 2006; Robertson 2007; Robertson et al. 2012) in the easternmost Mediterranean. The limited westward extent of the southern Neotethys, in contrast with models inferring that NW–SE break-up extended to the Alpine Tethys (e.g. Frizon de Lamotte et al. 2011), is consistent with the majority of studies that suggest that the oceanic crust in the Ionian Abyssal Plain is younger than the Herodotus Abyssal Plain (e.g. de Voogd et al. 1992; Catalano et al. 2001).

Models inferring north–south to NE–SW break-up and the development of a major transform fault along the Levant margin (Fig. 2a, b & e, f; e.g. Dercourt et al. 1986; Hallett 2002; Stampfli & Borel 2002, 2004) are not supported by the structural styles or crustal structure of the offshore Levant Basin. One of the major arguments for a Levant transform is the inferred presence of oceanic crust in the northern Levant Basin and the steep geometry of the northeastern basin margin (Ben-Avraham et al. 2002; Schattner & Ben-Avraham 2007). However, the offshore seismic interpretation in this study, as well as seismic refraction (e.g. Netzeband et al. 2006) and regional gravity data (e.g. Longacre et al. 2007; Cowie & Kusznir 2012a, b), show that the entire basin is underlain by rifted and highly thinned continental basement, supporting the interpretation of the Levant segment as a rifted margin.

Late Jurassic–Mid-Cretaceous (145–93 Ma?).
Further west, offshore northern Libya, seafloor spreading occurred during the Cenomanian–late Santonian, as the result of Late Jurassic–Cenomanian rifting (SR2) and break-up along the Gulf of Sirt and the Malta Escarpment margin segments (Figs 13 & 14c).

This second rift phase resulted in widespread extensional reactivation across the south Tethyan margin (Fig. 13). Along the offshore Levant–Cyrenaica margin segments, major basin-bounding fault systems were reactivated. Onshore, Late Jurassic–Early Cretaceous regional uplift and widespread alkaline basalt magmatism, as well as ENE–WSW to ESE–WNW extensional growth faults, are documented along the Levant margin, which could not previously be well explained (Brew et al. 1999, 2001; Walley 2001; Homberg et al. 2009; Hardy et al. 2010).

Rift flank uplift and erosion at the margins of the offshore Sirt Basin, and the development of a thermally subsiding post-rift sag basin (Fig. 4b), suggest that continental break-up in the Ionian Abyssal Plain occurred during the Mid–Late Cretaceous

(Cenomanian?). This corresponds to the onset of a major period of subsidence in the onshore Sirt Basin c. 99 Ma (Abadi et al. 2008) and is consistent with observations from the Malta Escarpment and the Cyrenaica Platform where Albian–Cenomanian interstratified pillow lavas and shallow marine platform carbonates are overlain by Turonian–Maastrichtian pelagic carbonates (Dercourt et al. 1986; Biju-Duval et al. 1987).

The offshore Sirt margin structure and fault pattern (Fig. 11) indicate a NE–SW extension direction, oblique to the pre-existing Pan-African basement fabrics (Fig. 14c). Late Jurassic rotation of the regional extension vector from NW–SE towards NE–SW can be partly attributed to the onset of seafloor spreading in the South and Equatorial Atlantic oceans, which resulted in a change in the relative motion of the African plate (Fig. 13) from southeastward to dominantly eastward divergence from the Eurasian plate (Janssen et al. 1995; Bumby & Guiraud 2005; Moulin et al. 2010).

Mid-Cretaceous opening of the Ionian Abyssal Plain in our model (Fig. 14c) is similar to previous Early–Middle Cretaceous southern Neotethys models (Fig. 2e, f; Dercourt et al. 1986; Hallett 2002), although these models indicate that the Herodotus Abyssal Plain also opened at the same time. They proposed that the Ionian Abyssal Plain opened as the result of an anticlockwise rotation of Adria relative to Africa (Fig. 2f), owing to the onset of collision of the northern Adriatic margin with Eurasia c. 130 Ma (Dercourt et al. 1986; Janssen et al. 1995; Channell 1996; Hallett 2002). We consider this to be possible assuming an original position of Adria in the Gulf of Sirt (Fig. 14c; e.g. Smith 2006).

Break-up in the Ionian Abyssal Plain extended NW–SE along the northern margin of the offshore Sirt Basin from NW Cyrenaica to the Malta Escarpment, and focused rifting in the Sirt Basin (Fig. 14c). On the southern margin of Adria, the Apulia Escarpment is considered to be the conjugate passive margin to the Malta Escarpment (Catalano et al. 2001; Rosenbaum et al. 2004).

However, the origin and evolution of the Sirt Basin and its relationship to the Neotethys–Atlantic oceanic systems remain widely debated. Alternative models include: (1) rifting as part of the Early Cretaceous (c. 142–101 Ma) West and Central African rift system (e.g. Guiraud & Maurin 1992; Guiraud et al. 2005; Moulin et al. 2010; Fairhead et al. 2013); (2) southeastward continuation of Early Cretaceous rifting to the NW, which led to Albian–Cenomanian continental break-up and seafloor spreading within the Bay of Biscay (Cowie & Kusznir 2012a); and (3) rifting driven by the slab pull exerted by the Hellenic subduction c. 70 Ma (Capitanio et al. 2009).

Based on the age of the offshore rift sequence in this study, the timing of subsidence in the onshore Sirt Basin (Abadi et al. 2008), and the basin structure, we suggest that the Late Jurassic–Mid-Cretaceous rifting in the Sirt Basin and break-up in the Ionian Abyssal Plain (c. Cenomanian) were the result of an anticlockwise rotation of Adria following the collision of its northern margin with Eurasia (Fig. 14c). Potential controls on the mechanics of rifting and style of extension in the Sirt Basin are the pre-existing south Tethyan margin architecture, coinciding with the westernmost extent of the crustal-scale transform–fracture zone to the north of the NW Egyptian–Cyrenaica margin, and possible anticlockwise rotation of Adria during rifting, which may have resulted in greater extension and crustal thinning towards the east and produced the asymmetric basin structure (Fig. 11).

Passive margin uplift and inversion in the SE Mediterranean. Late Cretaceous (c. 84 Ma)– Present day

Following the end of Neotethys seafloor spreading in the late Santonian (84 Ma), the post break-up evolution of the SE Mediterranean margin was characterized by short-lived pulses of compression that propagated across the African plate, associated with the northward convergence of Africa–Arabia relative to Eurasia during the Alpine Orogeny (Guiraud & Bosworth 1997; Bosworth et al. 2008). This resulted in episodic phases of uplift and selective inversion of the Neotethyan rift fault systems along the margin, during the: (1) Late Cretaceous, (2) Middle–Late Eocene, (3) Oligocene–Early Miocene, (4) Early–Middle Miocene, and (5) Pliocene–Early Pleistocene (Figs 4 & 13). The distribution, structural styles and mechanisms of inversion were fundamentally controlled by the structure and along-strike segmentation of the margin inherited from Neotethyan rifting (Figs 1 & 15).

Phases 1 and 2. Santonian–Maastrichtian (c. 84– 65 Ma?) and late Ypresian–Priabonian (c. 49– 37 Ma?). Inversion was focused on the NW-facing Levant–north Egyptian (Figs 5–7) and NW Cyrenaica (Fig. 9) margin segments, and dominantly occurred around the proximal margins of the offshore basins and the inboard rift basins adjacent to the stable Egyptian and Cyrenaica basement platforms (Fig. 15a, b). The inverted faults trend NE–SW and are associated with gentle fault-parallel hanging wall anticlines, which indicates coaxial shortening in a NW–SE regional compressional stress field. Localized, highly oblique reactivation of major NNW–SSE- and WNW–ESE-trending faults, and mild uplift, also occurred within the offshore Sirt Basin and offshore NE Cyrenaica.

In comparison with the Cyrenaica 'shock absorber' concept proposed by Bosworth et al. (2008), we suggest that this inhomogeneous shortening strain pattern was controlled by the Tethyan margin structure and the orientations of the pre-existing rift fault systems relative to the regional compressional stress fields (Fig. 15). The NNW–SSE to NW–SE faults and margin segments were not optimally oriented for reactivation in the regional NW–SE compressional stress field, which resulted in strain partitioning and focusing of the shortening strain on the NW-facing margin segments. The magnitudes of inversion were strongly influenced by the proximity of the inverted basins to the stable north Egyptian and Cyrenaica basement platforms (Figs 1 & 15), which Bevan & Moustafa (2012) suggested acted as a buttress, to the north of which there was significant inversion.

Late Santonian inversion (Fig. 15a) is widely documented along the onshore North African margin, and was linked to the abrupt change in the motion of the African plate relative to Eurasia from left-lateral divergence to northeastward convergence c. 84 Ma, which produced a major change in the trans-African stress field to NNW–SSE to WNW–ESE shortening (Janssen et al. 1995; Guiraud & Bosworth 1997; Guiraud et al. 2005; Bosworth et al. 2008). Inversion continued through the Maastrichtian along the offshore SE Mediterranean margin (Fig. 4), which corresponds to a second compressional event in the latest Maastrichtian–early Paleocene, oriented NNW–SSE to WNW–ESE, recorded along the onshore margin (Guiraud & Bosworth 1997; Guiraud 1998; Guiraud et al. 2005). No major relative plate motion changes occurred during the Maastrichtian; however, this inversion event may be related to trench–margin collision and widespread ophiolite obduction along the north and northeastern margins of Arabia (Brew et al. 2001).

The Middle–Late Eocene inversion phase (Fig. 15b) corresponds to a compressional event, characterized by inversion, folding and local thrusting, recorded in NW Africa and along the NE African–Arabian margin, where this was the main phase of uplift along the Syrian Arc of Israel, Lebanon and Syria, associated with NW–SE- to NNW–SSE-directed shortening (Walley 1998; Guiraud & Bosworth 1999; Brew et al. 2001; Guiraud et al. 2005; El Amawy et al. 2010; Moustafa 2010; Bevan & Moustafa 2012; Arsenikos et al. 2013). This event can generally be attributed to a second major anticlockwise rotation in the relative convergent motion of the African plate, c. 37 Ma, which resulted from changes in the rates and directions of opening of the Central, South and North Atlantic oceans (e.g. Savostin et al. 1986; Guiraud & Bosworth 1999; Guiraud et al. 2005).

Phases 3–5. Oligocene–Early Miocene (c. 37–23 Ma); Early–Middle Miocene (c. 23–15 Ma); and Pliocene–Early Pleistocene (c. 5.3–3 Ma). Mild broad-wavelength fold amplification and uplift of the pre-existing fold systems occurred through the Oligocene–Early Miocene along the Levant–north Egyptian and NW Cyrenaica margin segments (Figs 4 & 13). In the Gulf of Sirt, following northward tilting of the Mediterranean margin in the Late Eocene–Early Oligocene (Hallett 2002; Dolson et al. 2005), the margin evolution was dominated by episodic uplift and tilting, which produced multiple gravity-driven slumps and mass transport complexes (Figs 4 & 13). Synchronous compressional deformation occurred along the onshore margin where folding and uplift of the Syrian Arc and growth on the inversion structures in North Sinai, the Western Desert and Cyrenaica similarly continued through the Oligocene and into the Early Miocene (Walley 1998, 2001; El Amawy et al. 2010; Moustafa 2010; Bevan & Moustafa 2012; Abd El-Wahed & Kamh 2013; Arsenikos et al. 2013). The African plate rotated strongly anticlockwise towards the NW with respect to Eurasia during the Oligocene–Early Miocene (Guiraud et al. 2005), which probably resulted in a rotation of the regional compressional stress field towards NNW–SSE to north–south (Figs 13 & 15), and produced the observed change in inversion styles and magnitude through oblique reactivation of the pre-existing inversion structures.

The Early–Middle Miocene inversion phase (Fig. 15c) was focused on the Levant–north Egyptian margin. Fold amplification continued along the margins of the Levant and Herodotus basins; however, inversion also occurred in the deepwater Levant Basin (Fig. 6), where intra-rift fault systems within the segmented depocentres in the southern part of the basin were reactivated. The distribution of inversion structures suggests that the Eratosthenes Seamount and the structurally high basement horst blocks may have acted as buttresses in the c. NNW–SSE compressional stress field, focusing inversion in the southern Levant Basin. Along the onshore margin, uplift of the Syrian Arc along the Arabian margin persisted into the Early Miocene and possibly later (Walley 1998; Guiraud et al. 2005). A new change in relative plate motions occurred c. 22 Ma, potentially caused by a temporary plate reorganization associated with Early Miocene Arabia–Eurasia collision along the Bitlis–Zagros suture zone (e.g. Okay et al. 2010; Robertson et al. 2012; McQuarrie & van Hinsbergen 2013), and Africa began to move northeastward (Figs 13 & 15), which is thought to have generated major folding and shearing along the Syrian Arc (Guiraud et al. 2005). Intra-plate compressional stresses associated with the collision may also have been significant in causing inversion along the Levant–north Egyptian margin. During the Miocene, the Hellenic–Cyprus subduction zone jumped to its present location south of Cyprus following the cessation of subduction further north, associated with the westward escape of the Anatolian plate (Kempler 1998; Robertson 1998). This produced rapid subsidence in the northern Levant Basin (Fig. 6) and is thought to have resulted in tectonic uplift of the Eratosthenes Seamount by at least 1 km (Robertson 1998), which may have caused the inversion in the Levant Basin to the SE.

Following the Messinian Salinity Crisis, the Pliocene–Early Pleistocene inversion phase (Fig. 15d) was similarly focused on the Levant–north Egyptian margin segment. Approximately NNW–SSE compression resulted in further mild oblique reactivation and uplift of the major faults on the southeastern margins of the Levant and Herodotus basins; however, inversion within the Levant Basin did not continue (Fig. 6). To the west, the margin segments from NW Egypt to the Malta Escarpment record Pliocene–Pleistocene uplift, erosion and gravity-driven slumping (Figs 8–10 & 12). Along the onshore margin, many of the Syrian Arc structures were reactivated or overprinted in a NNW–SSE compressional stress field associated with left-lateral transpression on the Dead Sea Transform (e.g. Walley 1998; Joseph-Hai et al. 2010; Moustafa 2010). Early Pleistocene NW–SE to north–south shortening is well documented across the onshore North African–Arabian margin and throughout the Eastern Mediterranean (Guiraud et al. 2005; Schattner 2010). A potential cause for this inversion event is the progressive collision of the Eratosthenes Seamount with the Cyprus Arc. Initial collision significantly disrupted subduction beneath the Cyprus Arc c. 3 Ma; this resulted in rapid uplift of southern Cyprus and may have temporarily interrupted the anticlockwise plate motion, which altered the regional stress regime and induced synchronous shortening throughout the Eastern Mediterranean (Kempler 1998; Robertson 1998; Schattner 2010; Kinnaird & Robertson 2013).

Conclusions

- The SE Mediterranean margin was segmented into distinct rift and transform-dominated tectonic domains during the polyphase break-up of the northern margin of Gondwana and opening of the southern Neotethys Ocean during the Middle Triassic–Mid-Cretaceous (c. 240–93 Ma).
- Reactivation of pre-existing Pan-African basement fabrics and shear zones in NW–SE and NE–SW regional extensional stress fields exerted a major control on the segmentation of the margin, rift basin architectures and evolution, and transform–fracture zone development.
- Middle Triassic–Middle Jurassic (c. 240–170 Ma) NW–SE regional extension resulted in the development of the Levant–north Egyptian orthogonally rifted margin and the NW Egyptian–Cyrenaica obliquely divergent transform margin. Continental break-up in the Herodotus Abyssal Plain occurred during the Late Triassic–Early Jurassic as a southern branch of the Neotethys opened in the Eastern Mediterranean.
- Late Jurassic–Mid-Cretaceous (c. 145–93 Ma) NE–SW regional extension resulted in the development of the Gulf of Sirt obliquely rifted margin and the Malta Escarpment transform margin. Continental break-up in the Ionian Abyssal Plain occurred during the Mid-Cretaceous, possibly Cenomanian.
- The convergence history of the African–Arabian and Eurasian plates from c. 84 Ma to the present day resulted in episodic uplift and inversion of NE–SW-trending Neotethyan rift fault systems along the margin in response to NW–SE to NNW–SSE regional compression during the late Santonian–Maastrichtian (c. 84–65 Ma), Middle–Late Eocene (c. 49–37 Ma) and Oligocene–Early Pleistocene.
- The distribution, structural styles and mechanisms of inversion were fundamentally controlled by the pre-existing structure and along-strike segmentation of the margin inherited from Neotethyan rifting.

Acknowledgements The authors wish to thank BP for permission to publish this paper. BP Exploration kindly provided the seismic data and funding for this research programme. The 2D seismic interpretation was carried out using GeoFrame 4.4, provided by Schlumberger. This study was conducted as part of the PhD research of L. Jagger, which was undertaken with the Fault Dynamics Research Group at Royal Holloway University of London. Martin Insley and Neil Hodgson are thanked for their thorough and constructive reviews, which greatly improved the manuscript.

References

ABADI, A.M., VAN WEES, J.-D., VAN DIJK, P.M. & CLOETINGH, S.A.P.L. 2008. Tectonics and subsidence evolution of the Sirt Basin, Libya. *AAPG Bulletin*, **92**, 993–1027. https://doi.org/10.1306/03310806070

ABD EL-WAHED, M.A. & KAMH, S.Z. 2013. Evolution of strike-slip duplexes and wrench-related folding in the central part of Al Jabal Al Akhdar, NE Libya. *The Journal of Geology*, **121**, 173–195. https://doi.org/10.1086/669249

ANTOBREH, A.A., FALEIDE, J.I., TSIKALAS, F. & PLANKE, S. 2009. Rift-shear architecture and tectonic development of the Ghana margin deduced from multichannel seismic reflection and potential field data. *Marine and Petroleum Geology*, **26**, 345–368. https://doi.org/10.1016/j.marpetgeo.2008.04.005

ARSENIKOS, S., FRIZON DE LAMOTTE, D., CHAMOT-ROOKE, N., MOHN, G., BONNEAU, M.-C. & BLANPIED, C. 2013. Mechanism and timing of tectonic inversion in Cyrenaica (Libya): integration in the geodynamics of the East Mediterranean. *Tectonophysics*, **608**, 319–329. https://doi.org/10.1016/j.tecto.2013.09.025

BELOPOLSKY, A., TARI, G., CRAIG, J. & ILIFFE, J. 2012. New and emerging plays in the Eastern Mediterranean: an introduction. *Petroleum Geoscience*, **18**, 371–372. https://doi.org/10.1144/petgeo2011-096

BEN-AVRAHAM, Z., GINZBURG, A., MAKRIS, J. & EPPELBAUM, L. 2002. Crustal structure of the Levant Basin, eastern Mediterranean. *Tectonophysics*, **346**, 23–43. https://doi.org/10.1016/S0040-1951(01)00226-8

BEN-AVRAHAM, Z., WOODSIDE, J., LODOLO, E., GARDOSH, M.A., GRASSO, M., CAMERLENGHI, A. & VAI, G.B. 2006. Eastern Mediterranean basin systems. *In*: GEE, D.G. & STEPHENSON, R.A. (eds) *European Lithosphere Dynamics*. Geological Society, London, Memoirs, **32**, 263–276. https://doi.org/10.1144/GSL.MEM.2006.032.01.15

BEVAN, T.G. & MOUSTAFA, A.R. 2012. Inverted rift basins of northern Egypt. *In*: ROBERTS, D.G. & BALLY, A.W. (eds) *Regional Geology and Tectonics: Phanerozoic Rift Systems and Sedimentary Basins: 1B*. Elsevier, Amsterdam, 482–507.

BIJU-DUVAL, B., MOREL, Y., BUROLLET, P.F., WINNOCK, E., RAVENNE, C., MASCLE, G. & CHARIER, S. 1987. Le Plateau Cyrénien: promontoire Africain sur la Marge Ionienne. *Revue de L'Institut Français du Pétrole*, **42**, 419–447.

BIRD, D. 2001. Shear margins: continent–ocean transform and fracture zone boundaries. *The Leading Edge*, **20**, 150–159. https://doi.org/10.1190/1.1438894

BOSWORTH, W. & MCCLAY, K. 2001. Structural and stratigraphic evolution of the Gulf of Suez rift, Egypt: A synthesis. *In*: ZIEGLER, P.A., CAVAZZA, W., ROBERTSON, A.H.F. & CRASQUIN-SOLEAU, S. (eds) *Peri-Tethys Memoir 6: Peri-Tethyan Rift/Wrench Basins and Passive Margins*. Memoires du Museum National d'Histoire Naturelle, Paris, **186**, 567–606.

BOSWORTH, W., GUIRAUD, R. & KESSLER,, II, L.G. 1999. Late Cretaceous (ca. 84 Ma) compressive deformation of the stable platform of northeast Africa (Egypt): far-field stress effects of the 'Santonian event' and origin of the Syrian Arc deformation belt. *Geology*, **27**, 633–636. https://doi.org/10.1130/0091-7613(1999)027<0633:LCCMCD>2.3.CO;2

BOSWORTH, W., EL-HAWAT, A.S., HELGESON, D.E. & BURKE, K. 2008. Cyrenaican 'shock absorber' and associated inversion strain shadow in the collision zone of northeast Africa. *Geology*, **36**, 695–698. https://doi.org/10.1130/G24909A.1

BOWMAN, S.A. 2012. A comprehensive review of the MSC facies and their origins in the offshore Sirt Basin, Libya. *Petroleum Geoscience*, **18**, 457–469. https://doi.org/10.1144/petgeo2011-070

BREW, G., LITAK, R., BARAZANGI, M. & SAWAF, T. 1999. Tectonic evolution of northeast Syria: regional implications and hydrocarbon prospects. *GeoArabia*, **4**, 289–318.

BREW, G., BARAZANGI, M., AL-MALEH, A.K. & SAWAF, T. 2001. Tectonic and geologic evolution of Syria. *GeoArabia*, **6**, 573–616.

BRIAND, F. 2000. *African Continental Margins of the Mediterranean Sea*, Djerba (Tunisia), 22–25 November. CIESM Workshop Series, CIESM, Monaco, **13**.

BRONNER, M. & MAKRIS, J. 2000. Crustal structure of the Libyan margin. *In*: BRIAND, F. (ed.) *African Continental Margins of the Mediterranean Sea*. CIESM Workshop Series, CIESM, Monaco, **13**, 63–66.

BUMBY, A.J. & GUIRAUD, R. 2005. The geodynamic setting of the Phanerozoic basins of Africa. *Journal of African Earth Sciences*, **43**, 1–12. https://doi.org/10.1016/j.jafrearsci.2005.07.016

CAPITANIO, F.A., FACCENNA, C. & FUNICIELLO, R. 2009. The opening of Sirte basin: result of slab avalanching? *Earth and Planetary Science Letters*, **285**, 210–216. https://doi.org/10.1016/j.epsl.2009.06.019

CARTWRIGHT, J.A. & JACKSON, M.P.A. 2008. Initiation of gravitational collapse of an evaporite basin margin: the Messinian saline giant, Levant Basin, eastern Mediterranean. *Geological Society of America Bulletin*, **120**, 399–413. https://doi.org/10.1130/B26081X.1

CARTWRIGHT, J.A., JACKSON, M.P.A., DOOLEY, T.P. & HIGGINS, S. 2012. Strain partitioning in gravity-driven shortening of a thick, multilayered evaporite sequence. *In*: ALSOP, G.I., ARCHER, S.G., HARTLEY, A.J., GRANT, N.T. & HODGKINSON, R. (eds) *Salt Tectonics, Sediments and Prospectivity*. Geological Society, London, Special Publications, **363**, 449–470. https://doi.org/10.1144/SP363.21

CATALANO, R., DOGLIONI, C. & MERLINI, S. 2001. On the Mesozoic Ionian Basin. *Geophysical Journal International*, **144**, 49–64. https://doi.org/10.1046/j.0956-540X.2000.01287.x

CERNOBORI, L., HIRN, A., MCBRIDE, J.H., NICOLICH, R., PETRONIO, L. & ROMANELLI, M. 1996. Crustal image of the Ionian basin and its Calabrian margins. *Tectonophysics*, **264**, 175–189. https://doi.org/10.1016/S0040-1951(96)00125-4

CHANNELL, J.E.T. 1996. Palaeomagnetism and palaeogeography of Adria. *In*: MORRIS, A. & TARLING, D.H. (eds) *Palaeomagnetism and Tectonics of the Mediterranean Region*. Geological Society, London, Special Publications, **105**, 119–132. https://doi.org/10.1144/GSL.SP.1996.105.01.11

CHENIN, P. & BEAUMONT, C. 2013. Influence of offset weak zones on the development of rift basins: activation and abandonment during continental extension and breakup. *Journal of Geophysical Research: Solid Earth*, **118**, 1698–1720. https://doi.org/10.1002/jgrb.50138

CLEMSON, J., CARTWRIGHT, J.A. & BOOTH, J. 1997. Structural segmentation and the influence of basement structure on the Namibian passive margin. *Journal of the Geological Society, London*, **154**, 477–482. https://doi.org/10.1144/gsjgs.154.3.0477

COWARD, M.P. & RIES, A.C. 2003. Tectonic development of North African basins. *In*: ARTHUR, T.J., MACGREGOR, D.S. & CAMERON, N.R. (eds) *Petroleum Geology of Africa: New Themes and Developing Technologies*. Geological Society, London, Special Publications, **207**, 61–83. https://doi.org/10.1144/GSL.SP.2003.207.4

Cowie, L. & Kusznir, N. 2012a. Mapping crustal thickness and oceanic lithosphere distribution in the Eastern Mediterranean using gravity inversion. *Petroleum Geoscience*, **18**, 373–380. https://doi.org/10.1144/petgeo2011-071

Cowie, L. & Kusznir, N. 2012b. Gravity inversion mapping of crustal thickness and lithosphere thinning for the eastern Mediterranean. *The Leading Edge*, **31**, 810–814. https://doi.org/10.1190/tle31070810.1

Craig, J., Rizzi, C. et al. 2008. Structural styles and prospectivity in the Precambrian and Palaeozoic hydrocarbon systems of North Africa. In: Salem, M.J., Oun, K.M. & Essed, A.S. (eds) *The Geology of East Libya 4*. Earth Science Society of Libya, Tripoli, 51–122.

Dercourt, J., Zonenshain, L.P. et al. 1986. Geological evolution of the Tethys belt from the Atlantic to the Pamirs since the Lias. *Tectonophysics*, **123**, 241–315. https://doi.org/10.1016/0040-1951(86)90199-X

Ding, W., Franke, D., Li, J. & Steuer, S. 2013. Seismic stratigraphy and tectonic structure from a composite multi-channel seismic profile across the entire Dangerous Grounds, South China Sea. *Tectonophysics*, **582**, 162–176. https://doi.org/10.1016/j.tecto.2012.09.026

Direen, N.G., Stagg, H.M.J., Symonds, P.A. & Colwell, J.B. 2008. Architecture of volcanic rifted margins: new insights from the Exmouth–Gascoyne margin, Western Australia. *Australian Journal of Earth Sciences*, **55**, 341–363. https://doi.org/10.1080/08120090701769472

Dolson, J.C., Boucher, P.J., Siok, J. & Heppard, P.D. 2005. Key challenges to realizing full potential in an emerging giant gas province: Nile Delta/Mediterranean offshore, deep water, Egypt. In: Dore, A.G. & Vining, B.A. (eds) *Petroleum Geology: North-West Europe and Global Perspectives – Proceedings of the 6th Petroleum Geology Conference*. Geological Society, London, Petroleum Geology Conference Series, **6**, 607–624. https://doi.org/10.1144/0060607

Dore, A.G. & Stewart, I.C. 2002. Similarities and differences in the tectonics of two passive margins: the Northeast Atlantic Margin and the Australian North West Shelf. In: Keep, M. & Moss, S.J. (eds) *The Sedimentary Basins of Western Australia 3. Proceedings of the Petroleum Exploration Society of Australia Symposium*, Perth, 89–117.

Ebbing, J. & Olesen, O. 2010. New compilation of top basement and basement thickness for the Norwegian continental shelf reveals the segmentation of the passive margin system. In: Vining, B.A. & Pickering, S.C. (eds) *Petroleum Geology: From Mature Basins to New Frontiers – Proceedings of the 7th Petroleum Geology Conference*. Geological Society, London, Petroleum Geology Conference Series, **7**, 885–897. https://doi.org/10.1144/0070885

El Amawy, M.A., Muftah, A.M., Abd El-Wahed, M. & Nassar, A. 2010. Wrench structural deformation in Ras Al Hilal-Al Athrun area, NE Libya: a new contribution in Northern Al Jabal Al Akhdar Belt. *Arabian Journal of Geosciences*, **4**, 1067–1085. https://doi.org/10.1007/s12517-009-0114-5

Fairhead, J.D., Green, C.M., Masterton, S.M. & Guiraud, R. 2013. The role that plate tectonics, inferred stress changes and stratigraphic unconformities have on the evolution of the West and Central African Rift System and the Atlantic continental margins. *Tectonophysics*, **594**, 118–127. https://doi.org/10.1016/j.tecto.2013.03.021

Fiduk, J.C. 2009. Evaporites, petroleum exploration, and the Cenozoic evolution of the Libyan shelf margin, central North Africa. *Marine and Petroleum Geology*, **26**, 1513–1527. https://doi.org/10.1016/j.marpetgeo.2009.04.006

Franke, D. 2013. Rifting, lithosphere breakup and volcanism: comparison of magma-poor and volcanic rifted margins. *Marine and Petroleum Geology*, **43**, 63–87. https://doi.org/10.1016/j.marpetgeo.2012.11.003

Franke, D., Barckhausen, U. et al. 2011. The continent–ocean transition at the southeastern margin of the South China Sea. *Marine and Petroleum Geology*, **28**, 1187–1204. https://doi.org/10.1016/j.marpetgeo.2011.01.004

Frizon de Lamotte, D., Raulin, C., Mouchot, N., Wrobel-Daveau, J.-C., Blanpied, C. & Ringenbach, J.-C. 2011. The southernmost margin of the Tethys realm during the Mesozoic and Cenozoic: initial geometry and timing of the inversion processes. *Tectonics*, **30**. TC3002, https://doi.org/10.1029/2010TC002691

Gardosh, M.A. & Druckman, Y. 2006. Seismic stratigraphy, structure and tectonic evolution of the Levantine Basin, offshore Israel. In: Robertson, A.H.F. & Mountrakis, D. (eds) *Tectonic Development of the Eastern Mediterranean Region*. Geological Society, London, Special Publications, **260**, 201–227. https://doi.org/10.1144/GSL.SP.2006.260.01.09

Gardosh, M.A., Garfunkel, Z., Druckman, Y. & Buchbinder, B. 2010. Tethyan rifting in the Levant Region and its role in Early Mesozoic crustal evolution. In: Homberg, C. & Bachmann, M. (eds) *Evolution of the Levant Margin and Western Arabia Platform since the Mesozoic*. Geological Society, London, Special Publications, **341**, 9–36. https://doi.org/10.1144/SP341.2

Gardosh, M.A., Weimer, P. & Flexer, A. 2011. The sequence stratigraphy of Mesozoic successions in the Levant margin, southwestern Israel: a model for the evolution of southern Tethys margins. *AAPG Bulletin*, **95**, 1763–1793. https://doi.org/10.1306/02081109135

Garfunkel, Z. 1998. Constrains on the origin and history of the Eastern Mediterranean basin. *Tectonophysics*, **298**, 5–35. https://doi.org/10.1016/S0040-1951(98)00176-0

Garfunkel, Z. 2004. Origin of the Eastern Mediterranean basin: a reevaluation. *Tectonophysics*, **391**, 11–34. https://doi.org/10.1016/j.tecto.2004.07.006

Gartrell, A.P. 2000. Rheological controls on extensional styles and the structural evolution of the Northern Carnarvon Basin, North West Shelf, Australia. *Australian Journal of Earth Sciences*, **47**, 231–244. https://doi.org/10.1046/j.1440-0952.2000.00776.x

Gibson, G.M., Totterdell, J.M., White, L.T., Mitchell, C.H., Stacey, A.R., Morse, M.P. & Whitaker, A. 2013. Pre-existing basement structure and its influence on continental rifting and fracture zone development along Australia's southern rifted margin. *Journal of the Geological Society, London*, **170**, 365–377. https://doi.org/10.1144/jgs2012-040

de Graciansky, P.C., Roberts, D.G. & Tricart, P. 2010. *The Western Alps, from Rift to Passive Margin to*

Orogenic Belt: An Integrated Geoscience Overview. Elsevier, Amsterdam.

GUIRAUD, R. 1998. Mesozoic rifting and basin inversion along the northern African Tethyan margin: An overview. *In*: MACGREGOR, D.S., MOODY, R.T.J. & CLARK-LOWES, D.D. (eds) *Petroleum Geology of North Africa*. Geological Society, London, Special Publications, **132**, 217–229. https://doi.org/10.1144/GSL.SP.1998.132.01.13

GUIRAUD, R. & BOSWORTH, W. 1997. Senonian basin inversion and rejuvenation of rifting in Africa and Arabia: synthesis and implications to plate-scale tectonics. *Tectonophysics*, **282**, 39–82. https://doi.org/10.1016/S0040-1951(97)00212-6

GUIRAUD, R. & BOSWORTH, W. 1999. Phanerozoic geodynamic evolution of northeastern Africa and the northwestern Arabian platform. *Tectonophysics*, **315**, 73–104. https://doi.org/10.1016/S0040-1951(99)00293-0

GUIRAUD, R. & MAURIN, J.C. 1992. Early Cretaceous rifts of Western and Central Africa: an overview. *Tectonophysics*, **213**, 153–168. https://doi.org/10.1016/0040-1951(92)90256-6

GUIRAUD, R., BOSWORTH, W., THIERRY, J. & DELPLANQUE, A. 2005. Phanerozoic geological evolution of Northern and Central Africa: an overview. *Journal of African Earth Sciences*, **43**, 83–143. https://doi.org/10.1016/j.jafrearsci.2005.07.017

GUIRAUD, M., BUTA-NETO, A. & QUESNE, D. 2010. Segmentation and differential post-rift uplift at the Angola margin as recorded by the transform-rifted Benguela and oblique-to-orthogonal-rifted Kwanza basins. *Marine and Petroleum Geology*, **27**, 1040–1068. https://doi.org/10.1016/j.marpetgeo.2010.01.017

HALLETT, D. 2002. *Petroleum Geology of Libya*. Elsevier Science, Amsterdam.

HARDY, C., HOMBERG, C., EYAL, Y., BARRIER, É. & MÜLLER, C. 2010. Tectonic evolution of the southern Levant margin since Mesozoic. *Tectonophysics*, **494**, 211–225. https://doi.org/10.1016/j.tecto.2010.09.007

HOMBERG, C., BARRIER, E., MROUEH, M., HANDMAN, W. & HIGAZI, F. 2009. Basin tectonics during the Early Cretaceous in the Levant margin, Lebanon. *Journal of Geodynamics*, **47**, 218–223. https://doi.org/10.1016/j.jog.2008.09.002

JANSSEN, M.E., STEPHENSON, R.A. & CLOETINGH, S.A.P.L. 1995. Temporal and spatial correlations between changes in plate motions and the evolution of rifted basins in Africa. *Geological Society of America Bulletin*, **107**, 1317–1332. https://doi.org/10.1130/0016-7606(1995)107<1317:TASCBC>2.3.CO;2

JOLIVET, L., AUGIER, R., ROBIN, C., SUC, J.-P. & ROUCHY, J.M. 2006. Lithospheric-scale geodynamic context of the Messinian salinity crisis. *Sedimentary Geology*, **188–189**, 9–33. https://doi.org/10.1016/j.sedgeo.2006.02.004

JOLIVET, L., AUGIER, R. ET AL. 2008. Subduction, convergence and the mode of backarc extension in the Mediterranean region. *Bulletin de la Société Géologique de France*, **179**, 525–550. https://doi.org/10.2113/gssgfbull.179.6.525

JONGSMA, D., VAN HINTE, J.E. & WOODSIDE, J.M. 1985. Geologic structure and neotectonics of the North African Continental Margin south of Sicily. *Marine and Petroleum Geology*, **2**, 156–179. https://doi.org/10.1016/0264-8172(85)90005-4

JOSEPH-HAI, N., EYAL, Y. & WEINBERGER, R. 2010. Mesoscale folds and faults along a flank of a Syrian Arc monocline, discordant to the monocline trend. *In*: HOMBERG, C. & BACHMANN, M. (eds) *Evolution of the Levant Margin and Western Arabia Platform since the Mesozoic*. Geological Society, London, Special Publications, **341**, 211–226. https://doi.org/10.1144/SP341.10

KEMPLER, D. 1998. Eratosthenes Seamount: The possible spearhead of incipient continental collision in the eastern Mediterranean. *In*: ROBERTSON, A.H.F., EMEIS, K.-C., RICHTER, C. & CAMERLENGHI, A. (eds) *Proceedings of the Ocean Drilling Program, Scientific Results*, College Station, TX (Ocean Drilling Program), **160**, 709–721.

KINNAIRD, T. & ROBERTSON, A.H.F. 2013. Tectonic and sedimentary response to subduction and incipient continental collision in southern Cyprus, easternmost Mediterranean region. *In*: ROBERTSON, A.H.F., PARLAK, O. & ÜNLÜGENC, U.C. (eds) *Geological Development of Anatolia and the Easternmost Mediterranean Region*. Geological Society, London, Special Publications, **372**, 585–614. https://doi.org/10.1144/SP372.10

LIE, O., SKIPLE, C. & LOWREY, C. 2011. New insights into the Levantine Basin. *GeoExpro*, **8**, 24–27.

LONCKE, L., GAULLIER, V., MASCLE, J., VENDEVILLE, B.C. & CAMERA, L. 2006. The Nile deep-sea fan: an example of interacting sedimentation, salt tectonics, and inherited subsalt paleotopographic features. *Marine and Petroleum Geology*, **23**, 297–315. https://doi.org/10.1016/j.marpetgeo.2006.01.001

LONCKE, L., VENDEVILLE, B.C., GAULLIER, V. & MASCLE, J. 2010. Respective contributions of tectonic and gravity-driven processes on the structural pattern in the Eastern Nile deep-sea fan: insights from physical experiments. *Basin Research*, **22**, 765–782. https://doi.org/10.1111/j.1365-2117.2009.00436.x

LONGACRE, M., BENTHAM, P., HANBAL, I., COTTON, J. & EDWARDS, R. 2007. New crustal structure of the Eastern Mediterranean Basin: Detailed integration and modelling of gravity, magnetic, seismic refraction, and seismic reflection data. EGM 2007 International Workshop, Innovation in EM, Grav and Mag Methods: a new Perspective for Exploration, Capri, Italy, April 15–18, 2007.

MARONE, F., VAN DER MEIJDE, M., VAN DER LEE, S. & GIARDINI, D. 2003. Joint inversion of local, regional and teleseismic data for crustal thickness in the Eurasia–Africa plate boundary region. *Geophysical Journal International*, **154**, 499–514. https://doi.org/10.1046/j.1365-246X.2003.01973.x

MCCLAY, K.R., DOOLEY, T.P., WHITEHOUSE, P. & MILLS, M. 2002. 4-D evolution of rift systems: insights from scaled physical models. *AAPG Bulletin*, **86**, 935–959. https://doi.org/10.1306/61EEDBF2-173E-11D7-8645000102C1865D

MCQUARRIE, N. & VAN HINSBERGEN, D.J.J. 2013. Retrodeforming the Arabia–Eurasia collision zone: age of collision v. magnitude of continental subduction. *Geology*, **41**, 315–318. https://doi.org/10.1130/G33591.1

MEISLING, K.E., COBBOLD, P.R. & MOUNT, V.S. 2001. Segmentation of an obliquely rifted margin, Campos and Santos basins, southeastern Brazil. *AAPG Bulletin*, **85**, 1903–1924.

MILLER, J.M., NORVICK, M.S. & WILSON, C.J. 2002. Basement controls on rifting and the associated formation of ocean transform faults – Cretaceous continental extension of the southern margin of Australia. *Tectonophysics*, **359**, 131–155. https://doi.org/10.1016/S0040-1951(02)00508-5

MORRIS, A., ANDERSON, M.W., INWOOD, J. & ROBERTSON, A.H.F. 2006. Palaeomagnetic insights into the evolution of Neotethyan oceanic crust in the eastern Mediterranean. *In*: ROBERTSON, A.H.F. & MOUNTRAKIS, D. (eds) *Tectonic Development of the Eastern Mediterranean Region*. Geological Society, London, Special Publications, **260**, 351–372. https://doi.org/10.1144/GSL.SP.2006.260.01.15

MORTIMER, E.J., PATON, D.A., SCHOLZ, C.A. & STRECKER, M.R. 2016. Implications of structural inheritance in oblique rift zones for basin compartmentalization: Nkhata Basin, Malawi Rift (EARS). *Marine and Petroleum Geology*, **72**, 110–121. https://doi.org/10.1016/j.marpetgeo.2015.12.018

MOULIN, M., ASLANIAN, D. & UNTERNEHR, P. 2010. A new starting point for the South and Equatorial Atlantic Ocean. *Earth-Science Reviews*, **98**, 1–37. https://doi.org/10.1016/j.earscirev.2009.08.001

MOUSTAFA, A.R. 2010. Structural setting and tectonic evolution of North Sinai folds, Egypt. *In*: HOMBERG, C. & BACHMANN, M. (eds) *Evolution of the Levant Margin and Western Arabia Platform Since the Mesozoic*. Geological Society, London, Special Publications, **341**, 37–63. https://doi.org/10.1144/SP341.3

NEMČOK, M., RYBÁR, S., SINHA, S.T., HERMESTON, S.A. & LEDVÉNYIOVÁ, L. (eds) 2016. *Transform Margins: Development, Controls and Petroleum Systems*. Geological Society, London, Special Publications, **431**, https://doi.org/10.1144/SP431

NETZEBAND, G.L., GOHL, K., HÜBSCHER, C.P., BEN-AVRAHAM, Z., DEHGHANI, G.A., GAJEWSKI, D. & LIERSCH, P. 2006. The Levantine Basin – crustal structure and origin. *Tectonophysics*, **418**, 167–188. https://doi.org/10.1016/j.tecto.2006.01.001

NOCQUET, J.-M. 2012. Present-day kinematics of the Mediterranean: a comprehensive overview of GPS results. *Tectonophysics*, **579**, 220–242. https://doi.org/10.1016/j.tecto.2012.03.037

OKAY, A.I., ZATTIN, M. & CAVAZZA, W. 2010. Apatite fission-track data for the Miocene Arabia–Eurasia collision. *Geology*, **38**, 35–38. https://doi.org/10.1130/G30234.1

ROBERTSON, A.H.F. 1998. Tectonic significance of the Eratosthenes Seamount: a continental fragment in the process of collision with a subduction zone in the eastern Mediterranean (Ocean Drilling Program Leg 160). *Tectonophysics*, **298**, 63–82. https://doi.org/10.1016/S0040-1951(98)00178-4

ROBERTSON, A.H.F. 2007. Overview of tectonic settings related to the rifting and opening of Mesozoic ocean basins in the Eastern Tethys: Oman, Himalayas and Eastern Mediterranean regions. *In*: KARNER, G.D., MANATSCHAL, G. & PINHEIRO, L.M. (eds) *Imaging, Mapping and Modelling Continental Lithosphere Extension and Breakup*. Geological Society, London, Special Publications, **282**, 325–388. https://doi.org/10.1144/SP282.15

ROBERTSON, A.H.F., DIXON, J.E. *ET AL*. 1996. Alternative tectonic models for the Late Palaeozoic–Early Tertiary development of Tethys in the Eastern Mediterranean region. *In*: MORRIS, A. & TARLING, D.H. (eds) *Palaeomagnetism and Tectonics of the Mediterranean Region*. Geological Society, London, Special Publications, **105**, 239–263. https://doi.org/10.1144/GSL.SP.1996.105.01.22

ROBERTSON, A.H.F., PARLAK, O. & USTAOMER, T. 2012. Overview of the Palaeozoic–Neogene evolution of Neotethys in the Eastern Mediterranean region (southern Turkey, Cyprus, Syria). *Petroleum Geoscience*, **18**, 381–404. https://doi.org/10.1144/petgeo2011-091

ROSENBAUM, G., LISTER, G.S. & DUBOZ, C. 2004. The Mesozoic and Cenozoic motion of Adria (central Mediterranean): a review of constraints and limitations. *Geodinamica Acta*, **17**, 125–139. https://doi.org/10.3166/ga.17.125-139

ROUCHY, J.M. & CARUSO, A. 2006. The Messinian salinity crisis in the Mediterranean basin: a reassessment of the data and an integrated scenario. *Sedimentary Geology*, **188–189**, 35–67. https://doi.org/10.1016/j.sedgeo.2006.02.005

ROVERI, M., GENNARI, R. *ET AL*. 2016. The Messinian salinity crisis: open problems and possible implications for Mediterranean petroleum systems. *Petroleum Geoscience*, **22**, 283–290. https://doi.org/10.1144/petgeo2015-089

RUPKE, L.H., SCHMID, D.W., HARTZ, E.H. & MARTINSEN, B. 2010. Basin modelling of a transform margin setting: structural, thermal and hydrocarbon evolution of the Tano Basin, Ghana. *Petroleum Geoscience*, **16**, 283–298. https://doi.org/10.1144/1354-079309-905

SAVOSTIN, L.A., SIBUET, J.-C., ZONENSHAIN, L.P., LE PICHON, X. & ROULET, M-J. 1986. Kinematic evolution of the Tethys belt from the Atlantic Ocean to the Pamirs since the Triassic. *Tectonophysics*, **123**, 1–35. https://doi.org/10.1016/0040-1951(86)90192-7

SCHATTNER, U. 2010. What triggered the early-to-mid Pleistocene tectonic transition across the entire eastern Mediterranean? *Earth and Planetary Science Letters*, **289**, 539–548. https://doi.org/10.1016/j.epsl.2009.11.048

SCHATTNER, U. & BEN-AVRAHAM, Z. 2007. Transform margin of the northern Levant, eastern Mediterranean: from formation to reactivation. *Tectonics*, **26**, TC5020. https://doi.org/10.1029/2007TC002112

SELLIER, N.C., VENDEVILLE, B.C. & LONCKE, L. 2013. Post-Messinian evolution of the Florence Rise area (Western Cyprus Arc) Part II: experimental modeling. *Tectonophysics*, **591**, 143–151. https://doi.org/10.1016/j.tecto.2011.07.003

SHAHAR, J. 1994. The Syrian arc system: an overview. *Palaeogeography, Palaeoclimatology, Palaeoecology*, **112**, 125–142. https://doi.org/10.1016/0031-0182(94)90137-6

SMITH, A.G. 2006. Tethyan ophiolite emplacement, Africa to Europe motions, and Atlantic spreading. *In*: ROBERTSON, A.H.F. & MOUNTRAKIS, D. (eds) *Tectonic Development of the Eastern Mediterranean Region*. Geological Society, London, Special Publications, **260**, 11–34. https://doi.org/10.1144/GSL.SP.2006.260.01.02

STAMPFLI, G.M. & BOREL, G.D. 2002. A plate tectonic model for the Paleozoic and Mesozoic constrained by dynamic plate boundaries and restored synthetic oceanic isochrons. *Earth and Planetary Science Letters*,

196, 17–33. https://doi.org/10.1016/S0012-821X(01) 00588-X

STAMPFLI, G.M. & BOREL, G.D. 2004. The TRANSMED transects in space and time: Constraints on the palaeotectonic evolution of the Mediterranean domain. *In*: CAVAZZA, W., ROURE, F.M., SPAKMAN, W., STAMPFLI, G.M. & ZIEGLER, P.A. (eds) *The TRANSMED Atlas: The Mediterranean Region from Crust to Mantle*. Springer, Berlin; CD-ROM.

STEINBERG, J., GVIRTZMAN, Z., FOLKMAN, Y. & GARFUNKEL, Z. 2011. Origin and nature of the rapid late Tertiary filling of the Levant Basin. *Geology*, **39**, 355–358. https://doi.org/10.1130/G31615.1

SYMONDS, P.A., PLANKE, S., FREY, O. & SKOGSEID, J. 1998. Volcanic evolution of the western Australian continental margin and its implications for basin development. *In*: PURCELL, P.G. & PURCELL, R.R. (eds) *The Sedimentary Basins of Western Australia 2*. Proceedings of the Petroleum Exploration Society of Australia Symposium, Perth, 33–54.

TARI, G., HUSSEIN, H., NOVOTNY, B., HANNKE, K. & KOHAZY, R. 2012. Play types of the deep-water Matruh and Herodotus basins, NW Egypt. *Petroleum Geoscience*, **18**, 443–455. https://doi.org/10.1144/petgeo2012-011

TASRIANTO, R. & ESCALONA, A. 2015. Rift architecture of the Lofoten–Vesterålen margin, offshore Norway. *Marine and Petroleum Geology*, **64**, 1–16. https://doi.org/10.1016/j.marpetgeo.2015.02.036

TASSY, A., CROUZY, E., GORINI, C., RUBINO, J.-L., BOUROULLEC, J.-L. & SAPIN, F. 2015. Egyptian Tethyan margin in the Mesozoic: evolution of a mixed carbonate–siliciclastic shelf edge (from Western Desert to Sinai). *Marine and Petroleum Geology*, **68**(Part A), 565–581. https://doi.org/10.1016/j.marpetgeo.2015.10.011

TAWADROS, E. 2001. *Geology of Egypt and Libya*. A.A. Balkema, Brookfield, Rotterdam.

THOMAS, W.A. 2006. Tectonic inheritance at a continental margin. *GSA Today*, **16**, 4–11.

TURNER, J.P. & WILSON, P.G. 2009. Structure and composition of the ocean-continent transition at an obliquely divergent transform margin, Gulf of Guinea, West Africa. *Petroleum Geoscience*, **15**, 305–311. https://doi.org/10.1144/1354-079309-846

TURNER, J.P., ROSENDAHL, B.R. & WILSON, P.G. 2003. Structure and evolution of an obliquely sheared continental margin: Rio Muni, West Africa. *Tectonophysics*, **374**, 41–55. https://doi.org/10.1016/S0040-1951(03) 00325-1

DE VOOGD, B., TRUFFERT, C., CHAMOT-ROOKE, N., HUCHON, P., LALLEMANT, S. & PICHON, X. 1992. Two-ship deep seismic soundings in the basins of the eastern Mediterranean Sea (Pasiphae cruise). *Geophysical Journal International*, **109**, 536–552. https://doi.org/10.1111/j.1365-246X.1992.tb00116.x

WALLEY, C.D. 1998. Some outstanding issues in the geology of Lebanon and their importance in the tectonic evolution of the Levantine region. *Tectonophysics*, **298**, 37–62. https://doi.org/10.1016/S0040-1951(98) 00177-2

WALLEY, C.D. 2001. The Lebanon passive margin and the evolution of the Levantine Neo-Tethys. *In*: CAVAZZA, W., ROBERTSON, A.H.F. & ZIEGLER, P. (eds) *Peri-Tethyan Rift/Wrench Basins and Passive Margins*. Memoires du Museum National d'Histoire Naturelle, Paris, **186**, 407–439.

WILSON, P.G., TURNER, J.P. & WESTBROOK, G.K. 2003. Structural architecture of the ocean–continent boundary at an oblique transform margin through deep-imaging seismic interpretation and gravity modelling: equatorial Guinea, West Africa. *Tectonophysics*, **374**, 19–40. https://doi.org/10.1016/S0040-1951(03)00326-3

XYPOLIAS, P., USTAÖMER, T. & ZULAUF, G. 2016. Eastern mediterranean tectonics. *International Journal of Earth Sciences*, **105**, 1437–3262. https://doi.org/10.1007/s00531-016-1392-2

YOUNES, A.I. & MCCLAY, K.R. 2002. Development of accommodation zones in the Gulf of Suez–Red Sea rift, Egypt. *AAPG Bulletin*, **86**, 1003–1026. https://doi.org/10.1306/61EEDC10-173E-11D7-8645000102 C1865D

YOUSEF, M., MOUSTAFA, A.R. & SHANN, M. 2010. Structural setting and tectonic evolution of offshore North Sinai, Egypt. *In*: HOMBERG, C. & BACHMANN, M. (eds) *Evolution of the Levant Margin and Western Arabia Platform since the Mesozoic*. Geological Society, London, Special Publications, **341**, 65–84. https://doi.org/10.1144/SP341.4

ZWAAN, F., SCHREURS, G., NALIBOFF, J. & BUITER, S.J.H. 2016. Insights into the effects of oblique extension on continental rift interaction from 3D analogue and numerical models. *Tectonophysics*, first published online March 8, 2016, https://doi.org/10.1016/j.tecto.2016.02.036

Contourites along the Iberian continental margins: conceptual and economic implications

ESTEFANÍA LLAVE[1]*, F. JAVIER HERNÁNDEZ-MOLINA[2], MARGA GARCÍA[3], GEMMA ERCILLA[4], CRISTINA ROQUE[5], CARMEN JUAN[4,6], ANXO MENA[2,5,7], BENEDICT PREU[8], DAVID VAN ROOIJ[6], MICHELE REBESCO[9], RACHEL BRACKENRIDGE[10], GLORIA JANÉ[1], MARÍA GÓMEZ-BALLESTEROS[11] & DORRIK STOW[10]

[1]*Instituto Geológico y Minero de España (IGME), Rios Rosas 23, 28003 Madrid, Spain*

[2]*Department Earth Sciences, Royal Holloway University of London (RHUL), Egham Hill, Egham, Surrey TW20 0EX, UK*

[3]*Instituto Andaluz de Ciencias de la Tierra (IACT–CSIC), Universidad de Granada, Avenida de las Palmeras 4, 18100 Armilla, Granada, Spain*

[4]*Centro de Investigaciones Marinas y Ambientales (CMIMA-CSIC), Paseo Marítimo Barceloneta, 37–49, 08003 Barcelona, Spain*

[5]*Divisão de Geologia e Georecursos Marinhos, Instituto Português do Mar e da Atmosfera (IPMA), Rua C do Aeroporto, 1749-077 Lisboa, Portugal*

[6]*Department of Geology and Soil Science, Ghent University, Krijgslaan 281 S8, B-9000 Gent, Belgium*

[7]*Departamento de Xeociencias Mariñas e Ordenación do Territorio, Universidad de Vigo, E-36310 Vigo, Pontevedra, Spain*

[8]*Chevron Corporation, 6001 Bollinger Canyon Road, San Ramon, CA 94583, USA*

[9]*OGS – Istituto Nazionale di Oceanografia e di Geofisica Sperimentale, Borgo Grotta Gigante 42/C, 34010 Sgonico, TS, Italy*

[10]*Institute of Petroleum Engineering, Heriot-Watt University, Third Gait, Edinburgh EH14 4AS, UK*

[11]*Instituto Español de Oceanografía (IEO), C/Corazón de María 8, 28002 Madrid, Spain*

EL, 0000-0001-7873-7211; FJH-M, 0000-0002-7483-144;
MG, 0000-0003-2632-6132; GE, 0000-0002-9709-2938;
CR, 0000-0002-2583-594X; CJ, 0000-0003-3752-4919;
DVR, 0000-0003-3633-3344; MR, 0000-0002-9492-4081;
RB, 0000-0002-0572-314X; MG-B, 0000-0002-0772-9211;
DS, 0000-0001-5082-7848
*Correspondence: e.llave@igme.es

Abstract: This work uses seismic records to document and classify contourite features around the Iberian continental margin to determine their implications for depositional systems and petroleum exploration. Contourites

From: McCLAY, K. R. & HAMMERSTEIN, J. A. (eds) 2020. *Passive Margins: Tectonics, Sedimentation and Magmatism.* Geological Society, London, Special Publications, **476**, 403–436.
First published online December 3, 2019, https://doi.org/10.1144/SP476-2017-46
© 2019 The Author(s). Published by The Geological Society of London. All rights reserved.
For permissions: http://www.geolsoc.org.uk/permissions. Publishing disclaimer: www.geolsoc.org.uk/pub_ethics

include depositional features (separated, sheeted, plastered and confined drifts), erosional features (abraded surfaces, channels, furrows and moats) and mixed features (contourite terraces). Drifts generally show high- to moderate-amplitude reflectors, which are cyclically intercalated with transparent layers. Transparent layers may represent finer-grained deposits, which can serve as seal rocks. High-amplitude reflectors (HARs) are likely to represent sandier layers, which could form hydrocarbon reservoirs. HARs occur on erosive features (moats and channels), and are clearly developed on contourite terraces and overflow features. Most of the contourite features described here are influenced by Mediterranean water masses throughout their Pliocene and Quaternary history. They specifically record Mediterranean Outflow Water, following its exit through the Gibraltar Strait. This work gives a detailed report on the variation of modern contourite deposits, which can help inform ancient contourite reservoir interpretation. Further research correlating 2D and 3D seismic anomalies with core and well-logging data is needed to develop better diagnostic criteria for contourites. This can help to clarify the role of contourites in petroleum systems.

The characterization of contourite features has been a critical area of research in marine geology over the past three decades due to their implications for stratigraphy, sedimentology, palaeoceanography, palaeoclimatology, sedimentary instability processes and energy resources (Rebesco et al. 2014). Studies have shown that contourites are common in deep marine environments, but remain poorly understood in terms of their composition, sedimentary processes, origin, sequence, lithology, seismic facies, petrophysical characteristics (porosity, permeability, etc.) and role in petroleum systems (Viana 2008; Stow et al. 2011a, b; Shanmugam 2012; Brackenridge et al. 2013; among others). Advances in understanding of contourites have yielded a clearer picture of their lateral and temporal variability, as well as their relationships with alongslope processes. Previous work has helped to formalize the terms 'contourite depositional system' (CDS) and 'contourite depositional complex' (CDC) (Hernández-Molina et al. 2003, 2008a, b; Rebesco & Camerlenghi 2008). Contourites, however, frequently occur interbedded or simultaneously deposited with sedimentary facies resulting from down-slope processes. These deposits represent mixed turbidite–contourite systems (Faugères et al. 1999; Rebesco & Camerlenghi 2008; Creaser et al. 2017). Mixed systems are common along continental margins where bottom currents rework and/or redistribute pre-existing gravitational deposits (Marchès et al. 2010; Brackenridge et al. 2013; Mulder et al. 2013). When down-slope processes dominate alongslope processes, gravitational deposits (such as turbidites) may overprint or inhibit the development of contourites. When strong alongslope currents dominate, turbidity currents may deviate and feed contourite drifts (e.g. Faugères et al. 1999; Mulder et al. 2003, 2006; Viana et al. 2007).

This paper follows the contourite drift classification criteria of Faugères et al. (1999) and Rebesco (2005). Sediment drifts are commonly bounded by and/or associated with erosional contourite features (such as contourite channels or moats). Compared to depositional features, these latter erosional features have generally received less attention and their genetic relationships with oceanographic processes remain unclear (Nelson et al. 1993, 1999; Stow & Mayall 2000; Hernández-Molina et al. 2006, 2008a, 2015; Hernández-Molina et al. 2016c; García et al. 2009). Erosional contourite features are clearly identifiable in seismic reflection profiles, allowing large-scale interpretation (Faugères et al. 1999; Stow et al. 2002a, b, c; Hernández-Molina et al. 2008a, b, 2016a, b, c; Nielsen et al. 2008; Brackenridge et al. 2013; Llave et al. 2015; Calvin Campbell & Mosher 2016; Delivet et al. 2016; Gruetzner & Uenzelmann-Neben 2016; Kuijpers & Nielsen 2016; among others). In the past, some contourite deposits have been considered as potential (hydrocarbon) source rocks, but ones less likely to form reservoirs than turbidites (Pickering et al. 1989; Pickering & Hiscott 2016). Sediments that make up such contourite drifts are typically muddy but can reach substantial local thickness, and include well-sorted sands that themselves are thick and laterally extensive enough to play a role in deep-water petroleum systems. Recent studies have interpreted bottom currents as a crucial factor in hydrocarbon reservoir development because weak flows enable the accumulation of mud-rich deposits (such as contourites). These can serve as both cap rock (seals) or, with deeper burial, as potential source rocks or shale-gas reservoirs when adequately enriched in organic matter (up to 2 wt%). On the other hand, high-velocity flows may represent a mechanism for 'mature sand' accumulation in deep-water environments. These sands form excellent reservoir units (Enjorlas et al. 1986; Colella 1990; Mutti 1992; Shanmugam et al. 1993; Viana et al. 1998; Shanmugam 2006, 2012, 2013a, b; Stow & Faugères 2008; Viana 2008; Stow et al. 2011a, 2013b; Mutti & Carminatti 2012). Sandy contourites and related deposits with good lateral continuity and exposed to long-term effects of current winnowing may have greater textural maturity and better developed primary interstices than turbidites. Sandy contourites can thus present good petrophysical characteristics, including high values for porosity, permeability, and lateral and vertical transmissivity of fluids (Shanmugam 2008; Viana 2008).

In spite of the potential role they play in deep-water petroleum systems and, by extension, their economic significance, contourites and mixed-drift depositional systems are not well understood. In particular, the scientific literature on sandy contourites, including ancient analogues in outcrop, is sparse. Mixed-drift systems do not yet enjoy the benefit of well-defined interpretive models. Both of these could help refine and direct turbidite exploration.

Published seismic and sedimentological data from the Iberian continental margin have reported on the interrelationships of turbidite and contourite depositional systems, especially contourites within mixed turbidite–contourite systems (Mulder et al. 2003, 2006, 2008; Llave et al. 2006; Marchès et al. 2007; García et al. 2016). The Iberian margin is affected by several vigorous water masses interacting along upper and middle continental slopes, and by weaker water masses moving along the lower slope and abyssal plains (Hernández-Molina et al. 2011, 2016c) (Fig. 1). Each of these domains hosts extensive and complex contourite features of variable dimensions and sedimentary thicknesses that are often poorly understood in terms of their oceanographic/depositional contexts (Maestro et al. 2013; Llave et al. 2015). Along the Iberian margin, large volumes of sand have been efficiently transported, re-deposited or reworked by the persistent hydrodynamic regimes of the Gulf of Cadiz (Buitrago et al. 2001; Habgood et al. 2003; Llave et al. 2005; Hernández-Molina et al. 2006; Akhmetzhanov et al. 2007; Brackenridge et al. 2011, 2013). IODP Expedition 339 and other cores from a number of cruises furthered understanding of sand-rich contourite deposits by presenting a facies model for sandy contourites (Expedition 339 Scientists 2012; Hernández-Molina et al. 2013; Stow et al. 2013a; Brackenridge et al. 2018). This research also considered contourites' potential role in deep-water petroleum systems. A thorough record of these features as they occur around the Iberian margin can further understanding of both their scientific significance and their economic potential (e.g. Rebesco & Camerlenghi 2008; Hernández-Molina et al. 2011 and references therein).

This study provides a regional review of alongslope processes and their sedimentary features around the Iberian continental margin based on 2D seismic profiles from published and unpublished sources. This work also discusses sandy contourites and how these features can be used to interpret ancient contourite deposits and explore for petroleum in deep-water settings.

Data and methodology

This study interprets geophysical surveys of the Iberian continental margin. Surveys used acoustic techniques to develop a 2D vertical profile of the structure underlying the seafloor at water depths of 200–5000 m. This study primarily reviews contourite features as characterized by methods and nomenclature described in Faugères et al. (1999), and further developed by Rebesco & Stow (2001) and Nielsen et al. (2008). Contourite morphologies developed at different depths and under the influence of multiple water masses are defined as 'contourite features' (CFs). Thickness is expressed in two-way travel time in seconds (TWT (s)).

The dataset was collected by several Spanish and international research projects, as well as by commercial petroleum exploration projects. Data were made available through the Geophysical Information System (Sistema de Información Geofísico – SIGEOF (http://cuarzo.igme.es/sigeco/default.htm), the Institut de Ciències del Mar–Consejo Superior de Investigaciones Científicas (CSIC) (http://www.icm.csic.es/geo/gma/SurveyMaps/) and the Instituto Portugues do Mar e da Atmosfera – IPMA (http://www.ipma.pt). Data were obtained by various seismic reflection methods described here. Low-resolution methods are commonly used for hydrocarbon exploration purposes. These methods record multichannel seismic profiles penetrating several kilometres but provide comparatively low-resolution (>50 m) images. Moderate-resolution/-penetration methods use air guns (Uniboom and Sparker) that penetrate from 100 m to 2 km with a resolution of 1 and 10 m (respectively). High-resolution/low-penetration seismic methods use 3.5 kHz echo sounders, TOPAS (topographical parametric sonar) and Parasound, which penetrate the upper few to tens of metres of the subsurface and record it at centimetre resolution. The various penetration depths and resolutions of these seismic survey systems record contourite features on different scales. Morphology and boundaries of the deposit are recorded at larger scales. The architecture of discrete internal depositional units is recorded at medium scales, and seismic facies are recorded at smaller scales. These features are referred to as first-, second- and third-order seismic elements in Nielsen et al. (2008).

A digital bathymetric model obtained from GEBCO (2003) and Zitellini et al. (2009) served as the base map used in the study. The *World Ocean Atlas 2012* (National Oceanographic Data Center 2012) provided the source data for the selected vertical hydrographic profiles.

Contourite features along the Iberian margin

The water masses around Iberia control alongslope sedimentation, and thus shape intermediate- and deep-water bathymetric features (Hernández-Molina et al. 2011 and references therein) (Fig. 1; Table 1). In

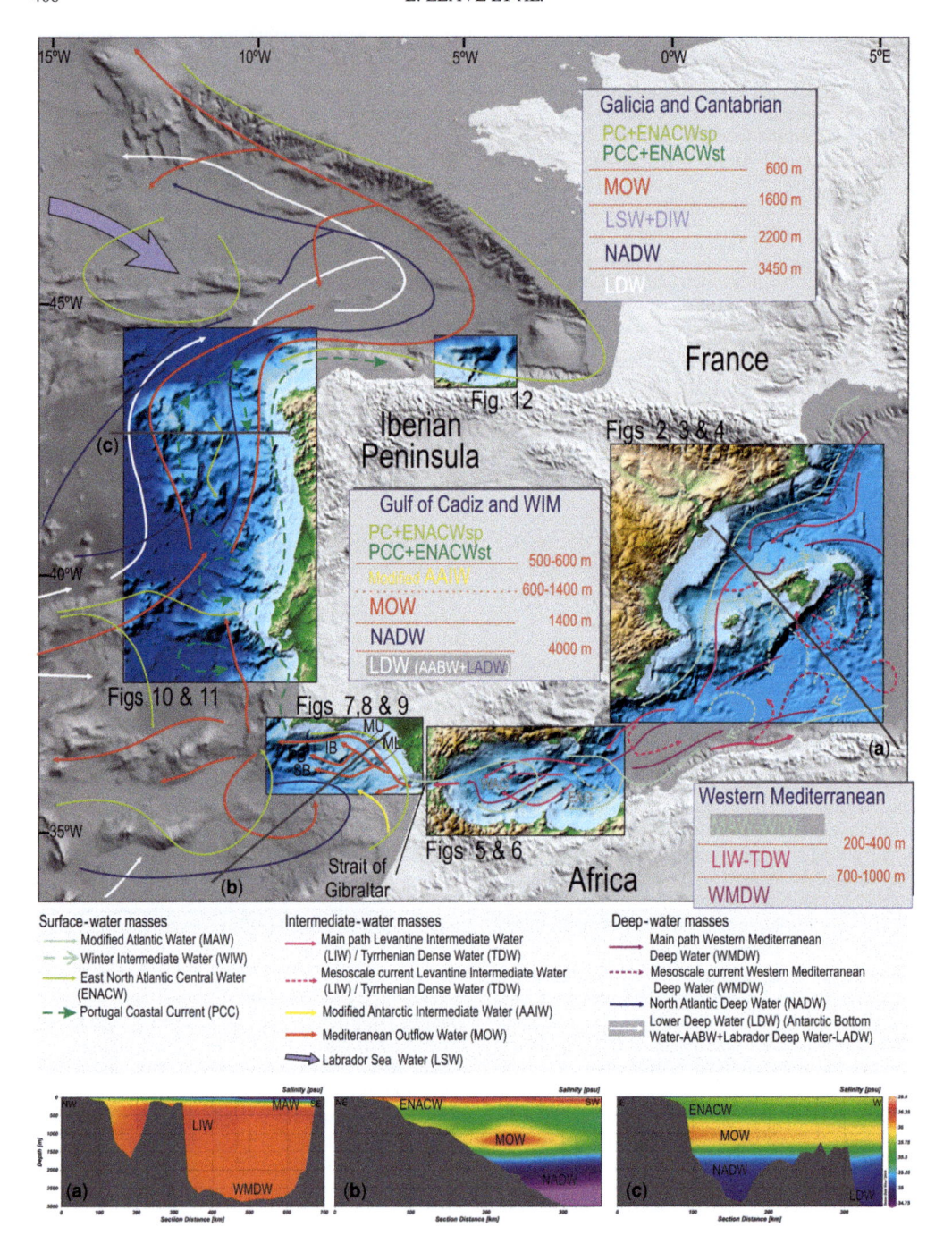

Fig. 1. Surficial-, intermediate- and deep-water circulation around the Iberian continental margin (modified from Hernández-Molina *et al.* 2011); digital bathymetric model obtained from GEBCO (2003) and Zitellini *et al.* (2009). Vertical hydrographic profiles: (**a**) east Iberian margin, (**b**) Gulf of Cadiz and (**c**) Galician margins (source data from the *World Ocean Atlas 2012* (National Oceanographic Data Center 2012). Locations in the study areas are also shown.

Table 1. *Acronyms of the main water masses present along the Iberian margin*

Mediterranean Sea
AW	Atlantic Water
MAW	Modified Atlantic Water
EAG	Eastern Alboran Gyres
WAG	Western Atlantic Gyre
LIW	Levantine Intermediate Water
WMDW	Western Mediterranean Deep Water
WIW	Western Intermediate Water
TDW	Tyrrhenian Dense Water
LMW	Light Mediterranean Waters
DMW	Dense Mediterranean Waters

Gulf of Cadiz and West Iberia
AIW	Inflow of the Atlantic Water
PC	Portugal Current
PCCC	Portugal Coastal Counter Current
ENACW	Eastern North Atlantic Central Water
AAIW	Modified Antarctic Intermediate Water
MOW	Mediterranean Outflow Water
MU	Mediterranean Upper Core
ML	Mediterranean Lower Core
NADW	North Atlantic Deep Water
LDW	Lower Deep Water
LADW	Labrador Deep Water
AABW	Antarctic Bottom Water

Galicia and Cantabrian
LSW	Labrador Sea Water

spite of many bottom-current measurements, the Iberian margin lacks long-term hydrodynamic records and has many areas with few or no measurements. Its mean recorded velocities are commonly low (5–15 cm s^{-1}), but certain water masses can travel across the seafloor at relatively high velocities, exceeding 80 cm s^{-1} and occasionally reaching almost 300 cm s^{-1}: for example, within the Strait of Gibraltar (Madelain 1970; Mélières et al. 1970; Ambar & Howe 1979; Iorga & Lozier 1999; Candela 2001). These water masses interact along the upper and middle continental slopes, and, with less intensity, along lower slope areas and the abyssal plain (Fig. 1).

Contourite features with large-scale dimensions and a range of sedimentary thicknesses occur at the following locations: the NE Iberian margin, Alboran Sea, Gulf of Cadiz CDS (well studied), western Iberian margin, Galician margin, Ortegal Spur, and Le Danois Bank or 'Cachucho'. These features are complex and poorly understood. The following subsections summarize the main water masses shaping the Iberian margin and describe contourite features formed by their interaction with the seafloor.

Mediterranean Sea

Oceanographic setting. According to traditional definitions, the Mediterranean Sea hosts three main water masses (Fig. 1; Table 1): Atlantic Water, Levantine Intermediate Water and Western Mediterranean Deep Water.

Atlantic Water (AW) or Modified Atlantic Water (MAW) forms due to the mixing of North Atlantic Surface Water, which enters through the Strait of Gibraltar at a velocity of approximately 1 m s^{-1} (Salat & Cruzado 1981; Gascard & Richez 1985). The MAW flows eastwards at a depth of 100–200 m and forms two anticyclonic gyres (Fig. 1): a quasi-permanent gyre in the western Alboran Basin (Western Alboran Gyre: WAG) and a semi-permanent gyre in the eastern Alboran Basin (Eastern Alboran Gyre: EAG) (e.g. Perkins et al. 1990; Millot 1999; Robinson et al. 2001). The energy of the WAG may exert an effect down to water depths of approximately 500–700 m (Cheney & Doblar 1982; Heburn & La Violette 1990; Perkins et al. 1990; Viúdez et al. 1998).

Levantine Intermediate Water (LIW) originates in the Strait of Sicily and flows westwards into the Western Mediterranean at depths of 200–600 m. After emerging from the Strait of Sicily, the LIW flows along the Iberian margin and into the central part of the Alboran Sea at velocities of up to 14 cm s^{-1} (Fig. 1).

Western Mediterranean Deep Water (WMDW) forms locally in the Alboran Basin and its surroundings during winter months, and generally flows westwards at water depths down to 600 m, although it can also reach depths of 2000 m (Fig. 1). Certain WMDW pulses can travel at velocities as high as 22 cm s^{-1} (Gascard & Richez 1985; Parrilla & Kinder 1987; Millot 1999; Fabres et al. 2002).

In recent studies of the Strait of Gibraltar and the nearby Alboran Sea, Millot (2009, 2014) proposed a more complex oceanographic structure and defined two additional water masses for this area: Western Intermediate Water (WIW: 100–300 m), situated between the AW and LIW; and Tyrrhenian Dense Water (TDW: variable water depth), situated between the LIW and WMDW (Fig. 1). Ercilla et al. (2016) categorized these five Alboran Sea water masses as light or intermediate-water masses (Light Mediterranean Waters (LMW): 100–600 m), and dense or deep-water masses (Dense Mediterranean Waters (DMW): >275 m).

Contourite features (CFs). The best-studied contourites in the Mediterranean Sea occur in the Alboran Sea. Evidence also indicates contourite deposition elsewhere along eastern margins of Iberia. Along the continental slope between Cap de Creus and Blanes Canyons (Fig. 2a; Table 2), two mounded, upward-prograding stratified drifts reach thicknesses of approximately 1 s and widths of 5 km at water depths of 1000–2300 m (Barcelona CFs). One of these occurs just south of La Fonera Canyon at

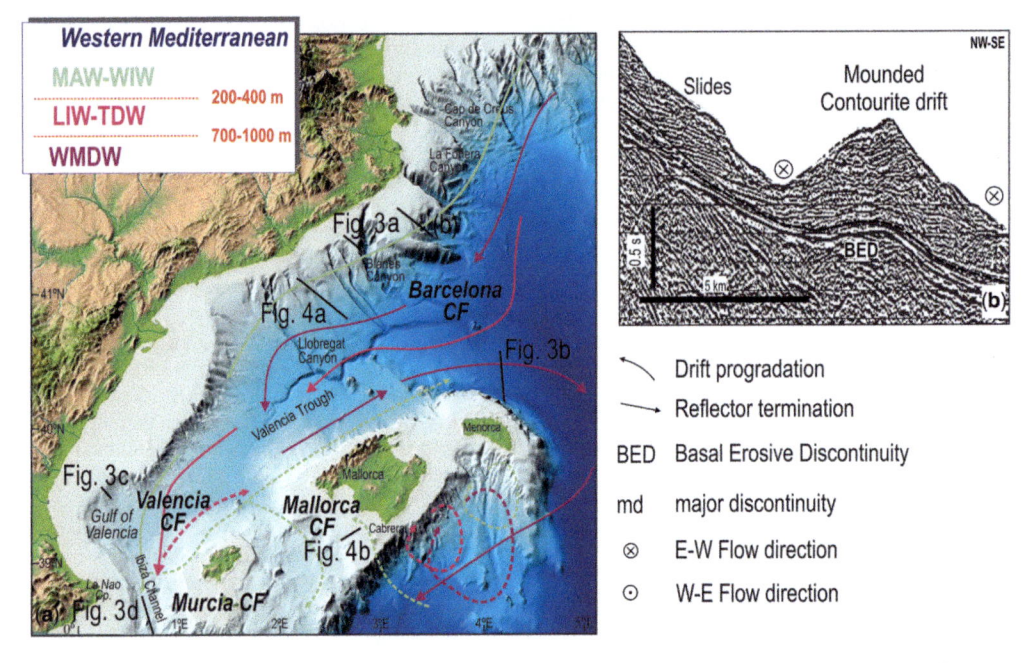

Fig. 2. (**a**) Digital bathymetric model of the NE Iberian margin, and the locations of the main water masses and seismic profiles. (**b**) Multichannel seismic profile showing a mounded drift example, as well as erosive features.

water depths of 1200–2300 m (Canals 1985), and another is located between the La Fonera and Blanes canyons at depths of 1000–1300 m (Fig. 2b). This second drift has not been analysed for specific contourite features but is similar to the one studied by Canals (1985).

Upslope, prograding, mounded, elongated and separated drifts (hereafter referred to simply as separated drifts) have been described from several localities. An isolated separated drift south Blanes Canyon reaches about 0.8 s in thickness and 6 km in width (Fig. 3a). A broad separated drift north of Menorca occurs on the lower slope at depths down to 2000 m. This drift is approximately 150 km long, 0.5 s thick, 25 km wide and reaches 100 m elevation above the seafloor (Menorca CFs) (Fig. 3b). This drift also exhibits sediment waves and a contourite moat (200 m deep and 5 km wide) along its seaward margin (Mauffret 1979; Velasco et al. 1996). Sediment waves also occur along the Gulf of Valencia continental margin (Valencia CFs). One of these situated along the outer continental shelf exhibits wavelengths of 400–800 m and heights of 2–4 m. A second example at 250–850 m water depth exhibits wavelengths of 500–1000 m and wave amplitudes of 2–50 m (Ribó et al. 2016). This latter field of sediment waves is developed along a possible prograding plastered drift approximately 0.5 s thick (Fig. 3c; Table 2). Several separated drifts have also been identified developed locally around seamounts in this sector (Fig. 3d).

Several plastered drifts with an upslope-prograding stacking pattern have also been reported. These include 10 km-wide plastered drifts with thicknesses ranging from 1 to 0.5 s, and developed at water depths of 400–800 and 1100–1400 m (Barcelona CFs) (Fig. 4a), and mounded, plastered, shallow-water contourites (Mallorca CFs) in a slope canyon near western Cabrera Island (SW of Mallorca) at water depths of 250–600 m (Fig. 4b; Table 2). A field of sediment waves is also present at the surface (Vandorpe et al. 2011; Lüdmann et al. 2012). Several plastered drifts of approximate 0.5 s thickness also occur around seamounts in the Valencia Trough at depths of between 500 and >1100 m (Valencia CFs), in the southern Ibiza Channel, and along the SE Iberian margins between depths of 600 and 900 m (Murcia CFs) (Fig. 3d; Table 2).

Contourites in the Alboran Sea display a great variety of both depositional and erosional features that range from a few to several tens of kilometres in length (Alboran CDS) (Fig. 5a). Their morphological and sedimentary characteristics have recently been described by Ercilla et al. (2016) and Juan et al. (2016). Depositional features predominate and are categorized as different types of drifts (Ercilla et al. 2002, 2016; Palomino et al. 2011; Juan et al. 2012, 2013, 2014, 2016). The largest drifts are plastered and sheeted types (Figs 5b, c & 6a; Table 2).

Plastered drifts occur along the continental slopes of both the Iberian and African margins between water depths of 235 and 575–1000 m, respectively. These drifts appear in seismic images as upslope onlapping stratified facies that pinch out in both upslope and downslope directions (Fig. 5b, c). The sheeted drifts occur along the base of the central Iberian slope, and in the western and southern Alboran basins at water depths below approximately 500 m. These appear as parallel-stratified seismic layers with a sub-tabular geometry beneath a relatively flat seafloor (Fig. 6a). Several small plastered and sheeted drifts occur locally along the flanks and tops of seamounts (Fig. 5c). Smaller drifts include separated (Figs 5b & 6b), channel-related (Fig. 6c) and mounded, confined drifts (Fig. 6d) (Palomino et al. 2010, 2011; Ercilla et al. 2016; Juan et al. 2012, 2013, 2014). Separated drifts occur along the westernmost upper and lower African slopes, and at the foot of structural seamounts and diapirs (Fig. 6b). In seismic images, they show onlapping stratified facies with internal discontinuities that display a mounded geometry. Channel-related drifts make up part of the seafloor within the Alboran Trough, appearing in the corridor between the Xauen Bank and the African slope. Channel-related drifts consist of isolated irregular, stratified layers scattered within the channel floor (Fig. 6c). Confined drifts, situated between structural highs, display a mounded morphology, which consists of internally stratified layers (Fig. 6d).

The erosional contourite features are located primarily along the Alboran margin, and consist of moats, scarps and furrows (Ercilla et al. 2016; Juan et al. 2016). The moats are mostly associated with separated and confined drifts. A few incipient moats also occur along the walls of the structural highs, where the slope exhibits changes in bathymetric patterns (Figs 5b & 6b). Scarps are narrow, steep erosional surfaces roughly parallel to the margin, which occur in association with landward flanks of terraces and other basal features along the slope. Scarps mark the transition between the basin's physiographical/oceanographic provinces. Furrows occur as linear features incised into steep, distal erosional escarpments of the African margin, near the Strait of Gibraltar. Extensive terraces represent mixed depositional and erosional features. These form the tops of plastered drifts along the continental slope. Terraces along the African margin are more pronounced (Fig. 5b). Terraces transition seawards into an onlapping, concordant surface.

Gulf of Cadiz

Oceanographic setting. The modern hydrodynamic setting of the Gulf of Cadiz is dominated by the exchange of water between the Atlantic Ocean and Mediterranean Sea through the Strait of Gibraltar. This gateway allows egress of warm, saline Mediterranean Outflow Water (MOW), which flows into the Atlantic Ocean, and the overlying inflow of the Atlantic Water into the Mediterranean Sea (Lacombe & Lizeray 1959; Ochoa & Bray 1991; Baringer & Price 1999; Iorga & Lozier 1999; Nelson et al. 1999; Potter & Lozier 2004; Lozier & Sindlinger 2009) (Fig. 1; Table 1). The MOW is an intermediate-water mass composed of waters originating in the Mediterranean Basin (Ambar & Howe 1979; Bryden & Stommel 1984; Bryden et al. 1994; Millot et al. 2006). The water mass accelerates through the narrow Strait of Gibraltar, reaching local velocities as high as 300 cm s^{-1} (Ambar & Howe 1979; Mulder et al. 2003), and moves northwestwards along the middle continental slope of the Gulf of Cadiz. This MOW flow occurs under the AIW and above the North Atlantic Deep Water (NADW). The AIW consists of the North Atlantic Superficial Water (NASW), from the surface down to a depth of approximately 100 m, and the Eastern North Atlantic Central Water (ENACW), which flows between depths of 100 and 600 m. In the Gulf of Cadiz, the Modified Antarctic Intermediate Water (AAIW) (Louarn & Morin 2011) circulates above the MOW (Hernández-Molina et al. 2014b). The underlying NADW flows southwards from the Greenland–Norwegian Sea region at depths greater than 1500 m (Ambar et al. 1999; Baringer & Price 1999; Serra et al. 2005).

In the Gulf of Cadiz, the MOW flow is controlled by the complex morphology of the continental slope. Flow is locally enhanced where salt tectonics have created diapiric ridges oblique to the MOW flow direction (Fig. 1). These ridges are partially responsible for splitting the MOW into numerous distinctive cores (Fig. 1), although vertical layering within the main MOW core has also been proposed as an alternative controlling mechanism (Sannino et al. 2007; Millot 2009; Copard et al. 2011). The main water cores are the Mediterranean Upper Core (MU) and the Mediterranean Lower Core (ML), each of whose branches displays unique salinity, temperature and average velocity (Madelain 1970; Zenk 1975; Ambar & Howe 1979; Gründlingh 1981; Borenäs et al. 2002; Serra et al. 2005) (Fig. 1). The MU flows alongslope along the southwestern Iberian margin at depths of 500–800 m, and part of its flow is captured by the Portimão Canyon along the Algarve margin (Marchès et al. 2007). The ML generally flows northwestwards at an average velocity of 20–30 cm s^{-1} (Llave et al. 2007), with the major part of flow concentrated west of 7° W (Madelain 1970). At this longitude, a branch detaches from the south side of the ML and flows SW. At approximately 7° 20′ W, the ML divides into three distinct branches that generally flow NW: the Southern

Table 2. *Acoustic facies, morphology and location of the main contourite features along the Iberian margin*

Main depositional features	Acoustic facies	Shape/dimensions	Location
Mediterranean Sea: Barcelona CF			
Mounded drifts	Prograding upslope	High mound shape, c. 5 km wide, a few hundreds of metres of relief	Middle–lower slope
Separated drifts	Prograding upslope	High mound shape, c. 6 km wide, a few tens of metres of relief	Middle–lower slope
Plastered drifts	Prograding upslope–downslope	Low mound shape, 10 km wide, a few tens of metres of relief	Upper–Middle slope
Mediterranean Sea: Valencia and Balearic Islands CF			
Plastered	Prograding upslope–downslope	Low mound shape, 10 km wide, a few tens of metres of relief	Middle slope
Mediterranean Sea: Murcia CF			
Plastered	Prograding upslope–downslope	Low mound shape, 10 km wide, a few tens of metres of relief	Middle slope
Mediterranean Sea: Alboran CDS			
Plastered drifts	Prograding upslope–downslope	Low to high mound shape, up to a few hundreds of kilometres long (<300 km), 5.5–40 km wide, tens to a few hundreds of metres of relief	Large drifts: Spanish and Moroccan slopes Small drifts: seamounts flanks, Spanish base-of-slope
Sheeted drifts	Aggrading sub-parallel stratified facies	Subtabular geometry; < 100 km long, 15 to 50 km wide	Large drifts: Spanish base-of-slope and sub-basins Small drifts: Alboran Ridge, seamounts tops
Channel-related drifts	Aggrading and prograding downwards	Low mound shape. c. 10 km long, <5 km wide	Alboran Trough
Mounded confined drifts	Prograding downwards	High mound shape; a few to tens of kilometres long and wide, 100–300 m of relief	Marginal shelf banks

Separated drifts	Prograding upslope	Low to high mound shape; <40 km long, 20 km wide, a few tens of metres of relief	Moroccan slope and shelf-break scarp; locally at the foot of seamounts and diapirs
Atlantic: Gulf of Cadiz CDS			
Plastered drifts	Prograding and aggrading downwards	Low mound shape; c. 10 km long, 5 km wide, a few metres of relief	Upper–middle slope
Separated drifts	Prograding upslope	High mound shape; 90 km long, 10–20 km wide, 150–200 m of relief	Middle and lower slope
Sheeted drifts	Aggrading sub-parallel stratified facies	Subtabular geometry; hundreds of kilometres long, tens of kilometres wide	Middle slope
Atlantic: Western Iberian Margin CDS			
Mounded drift	Prograding upslope	Low mound shape; c. 10 km long, c. 10 km wide, a few metres of relief	Middle–lower slope
Separated drifts	Prograding upslope	High mound shape; 5 km long, 5 km wide, a few hundreds of metres of relief	Middle–lower slope; around structural high continental rise
Atlantic: Galicia and Ortegal CDS			
Plastered drifts	Prograding–aggrading upslope–downslope	Low mound shape, a few kilometres long, tens to a few hundreds of metres of relief	Middle–lower slope; at the base if structural highs slope – abyssal plain
Separated drifts	Prograding upslope	Low to high mound shape; 5–22 km long, 1–10 km wide, a few tens to a few hundreds of metres of relief	Middle–lower slope
Atlantic: Le Danois CDS			
Plastered drifts	Prograding upslope–downslope	Low mound shape, 10–20 km long, c. 6 km wide, a few metres of relief	Upper southern slope of Le Danois Bank
Separated drifts	Prograding upslope	High mound shape; 10–45 km long, 3–10 km wide, a few tens of metres of relief	Middle slope

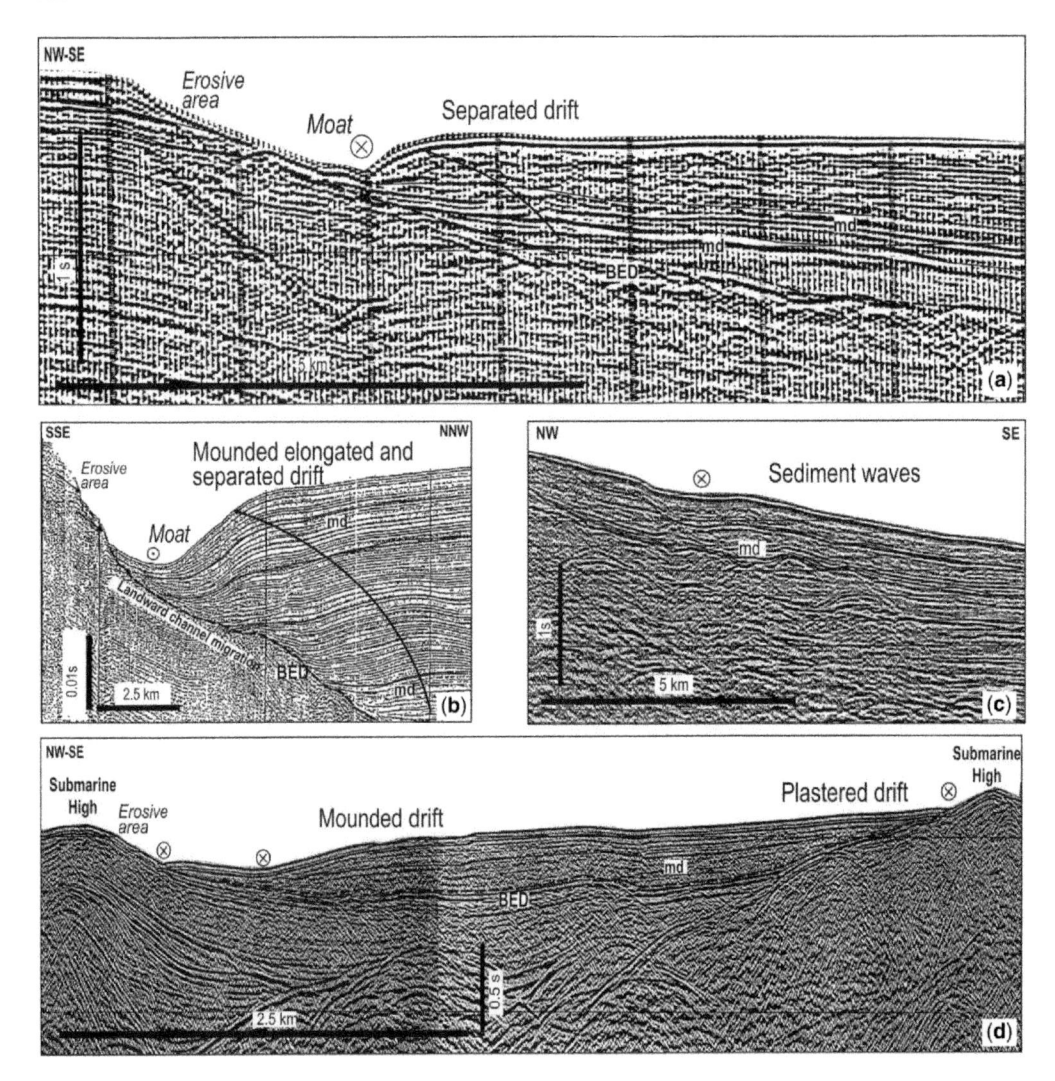

Fig. 3. (a) Multichannel and (b) sparker seismic profile (modified from Velasco *et al.* 1996) showing mounded separated drifts and moats. (c) Multichannel seismic profile where sediment waves are developed on a plastered drift. (d) Multichannel seismic profile showing mounded and plastered drifts, as well as erosional features around submarine highs. For the locations of the seismic profiles see Figure 2. BED, Basal Erosive Discontinuity; md, major discontinuity.

Branch (SB), the Principal Branch (PB) and the Intermediate Branch (IB) (Fig. 1). Portimão Canyon and Cape St Vincent also create a series of eddies, referred to as meddies due to their MOW origins (Ambar *et al.* 2002, 2008; Serra *et al.* 2005).

Contourite features. Erosional features are common in the Strait of Gibraltar, where high bottom-current velocities prevent deposition (Kelling & Stanley 1972; Stanley *et al.* 1975; Serrano *et al.* 2005). Esteras *et al.* (2000) identified several large channels likely to be associated with MOW circulation as the dominant contourite features. A few isolated plastered drifts also occur along the northern continental slope (Fig. 7a, b).

Beyond where the MOW exits the Strait of Gibraltar, its interaction with the middle slope of the southwestern Iberian margin has produced one of the most extensive and complex contourite depositional systems ever described. This feature is referred to as the Gulf of Cadiz CDS (Fig. 7; Table 2) (e.g. Madelain 1970; Kenyon & Belderson 1973; Gonthier *et al.* 1984; Nelson *et al.* 1999; Stow *et al.* 2002a, b, c; Habgood *et al.* 2003;

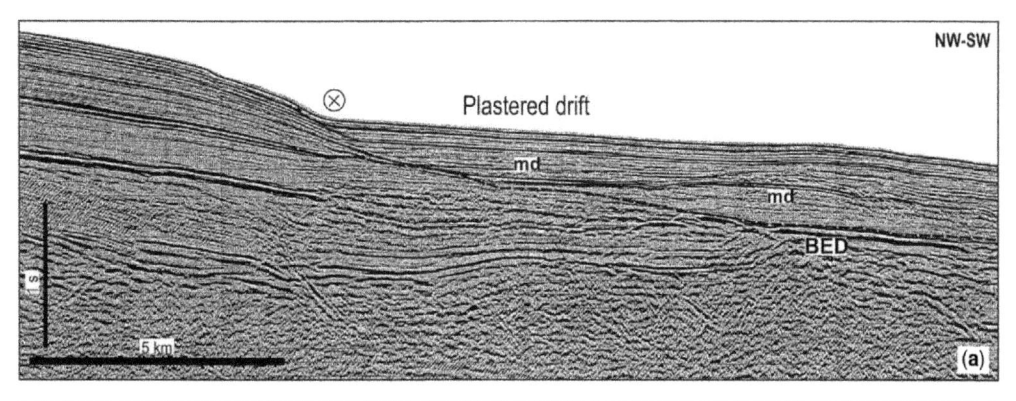

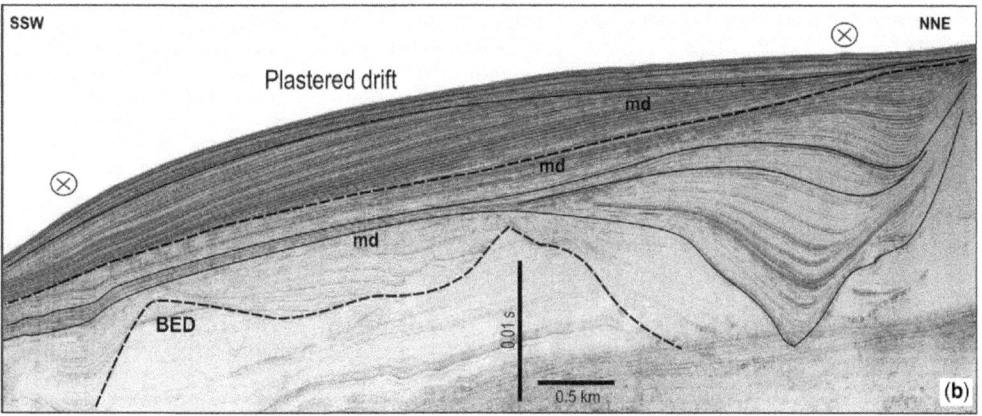

Fig. 4. (a) Multichannel and (b) sparker seismic profile (modified from Vandorpe *et al.* 2011) showing plastered drifts. For the locations of the seismic profiles see Figure 2. BED, Basal Erosive Discontinuity; md, major discontinuity.

Hernández-Molina *et al.* 2006; Mulder *et al.* 2006; Hanquiez *et al.* 2007; Marchès *et al.* 2007; Roque *et al.* 2012; Brackenridge *et al.* 2013; Llave *et al.* 2015 and references therein). The main depositional features are sediment wave fields, sediment lobes, mixed drifts, plastered drifts, separated drifts and sheeted drifts. The major erosional features are contourite channels, furrows, marginal valleys and moats (García *et al.* 2009). All of these features occur at specific locations along the margin, and their distributions correspond to five morphosedimentary sectors within the CDS: Sector 1 includes proximal scours and ribbons; Sector 2 includes overflow sediment lobes; Sector 3 includes channels and ridges; Sector 4 includes contourite deposition; and Sector 5 includes submarine canyons (Hernández-Molina *et al.* 2003, 2006; Llave *et al.* 2007). The development of each of these five sectors through time is related to an overall systematic deceleration of the MOW along the margin, with localized acceleration due to its interaction with irregularities along the seafloor. The middle slope of the Gulf of Cadiz is a bathymetrically complex area composed of mixed contourite features. This slope hosts four relatively flat contourite terraces with gradients of less than 0.5° at average depths of 500, 675, 750 and 850 m (García *et al.* 2009; Hernández-Molina *et al.* 2014*b*). These terraces are bounded by relatively steep risers with gradients of 1.5°–3°. Terraces form by both depositional and erosional processes.

Four sets of separated drifts have been described at water depths of 500–700 m along the Algarve margin: the Faro-Albufeira, Portimão, Lagos and Sagres drifts (Table 2). The Faro-Albufeira drifts are the most extensive and best developed. The Álvarez Cabral moat restricts these features to the upper slope. The drifts display asymmetrical shapes and sigmoidal-oblique, prograding stacking patterns (Fig. 7c). These drifts rise 150–200 m above the adjacent moat axis and surrounding deposits, and grade southwards into the Faro and the Bartolomeu Dias sheeted drifts, where the seafloor is smooth and almost flat (Fig. 7d). The Portimão and Lagos

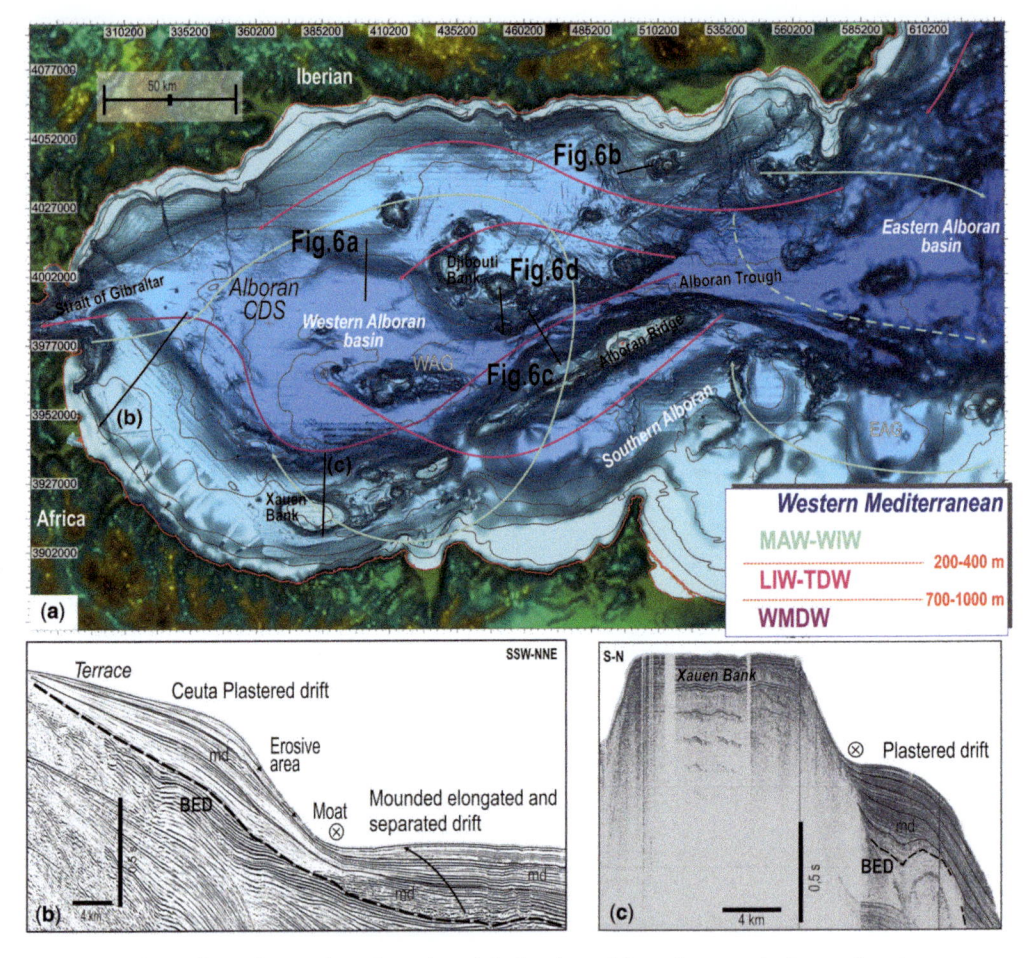

Fig. 5. Map of the Alboran Sea continental margin and the locations of the main contourite features based on (**a**) digital bathymetric data, and (**b**) airgun and (**c**) sparker seismic profiles (modified from Juan *et al.* 2012, 2013; Ercilla *et al.* 2016) showing examples of plastered and mounded and separated drifts, as well as erosive features. BED, Basal Erosive Discontinuity; md, major discontinuity.

drifts extend approximately 50 km from the Portimão Canyon but are separated by the Lagos Canyon. These display a smooth mounded morphology and are categorized as sheeted drifts (Hernández-Molina *et al.* 2003; Marchès *et al.* 2007, 2010). However, Roque *et al.* (2012) reported that the Lagos Drift grades from a sheeted drift near the Lagos Canyon to a plastered drift near the San Vicente high and finally to a separated drift. The Sagres Drift occurs at the transition between the southern and western Portuguese margins.

Sheeted drifts, with an aggrading stacking pattern and gentle morphology, occur between water depths of 600 and 1600 m in the contourite depositional and submarine canyons sectors (Vanney & Mougenot 1981). These were deposited across hundreds of kilometres, over which they are crossed by diapiric ridges, deformed and eroded by large contouritic channels of Sector 3 (channels and ridges sector) (Fig. 8a–c) (García 2002; Mulder *et al.* 2002, 2003; Habgood *et al.* 2003; Hernández-Molina *et al.* 2003, 2006; Llave 2003; Llave *et al.* 2007).

Extensive bedforms include ripple marks, sand ribbons and sediment waves (Fig. 8d). Longitudinal mounded drifts (Fig. 9a) are well developed in a NW–SE direction in Sector 1 (proximal scours and ribbons sector) and in Sector 2 (overflow sediment lobes sector) (Hernández-Molina *et al.* 2006; Llave *et al.* 2007).

Composite erosional features occur along the Guadalquivir Bank. Along with numerous diapiric highs (isolated and ridge-form), these include four types of submarine valleys (moats, channels, marginal valleys and furrows) in Sector 3 (García

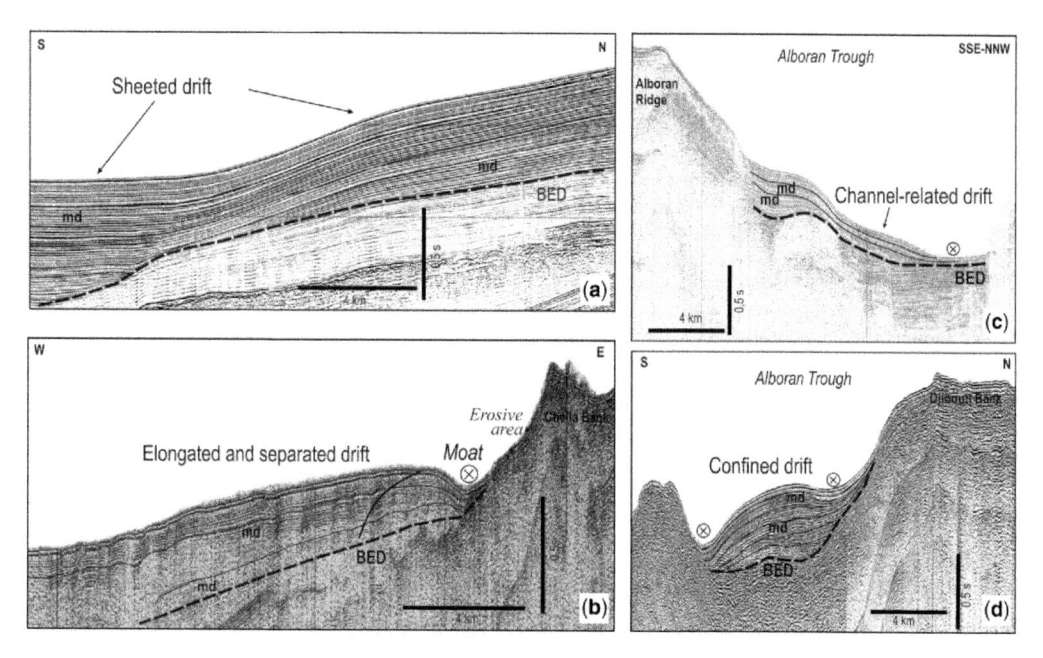

Fig. 6. (**a**) Airgun seismic profile showing an example of sheeted drift. (**b**) Airgun seismic profile showing an example of a channel-related drift. (**c**) Sparker seismic profile showing a mounded separated drift and moat. (**d**) Airgun seismic profile showing a confined drift (modified from Juan *et al.* 2012, 2013; Ercilla *et al.* 2016). For the locations of the seismic profiles see Figure 5. BED, Basal Erosive Discontinuity; md, major discontinuity.

2002; Hernández-Molina *et al.* 2003, 2006; García *et al.* 2009) (Fig. 9). The main erosional features along the middle slope of the central Gulf of Cadiz (Sector 3) include five major contourite channels (Fig. 8a–c). These occur along the southern flanks of the diapiric ridges and the Guadalquivir Bank. They include (from north to south) the Diego Cao, Huelva, Gusano, Guadalquivir and Cadiz Contourite channels (García *et al.* 2009). Different channels assume different dimensions but all display sinuous trends in the downslope to alongslope directions. The Cadiz and Guadalquivir channels (Fig. 8a) represent the largest features (1 and 12 km wide, respectively, more than 100 km long, and up to 130 m deep). Marginal valleys are unique erosional features located along NW sides of the diapiric ridges and isolated diapirs within the channels and ridges sector (Fig. 9e). These features exhibit irregular orientations but a locally sinuous morphology with a predominantly NE–SW trend (García *et al.* 2009). Their incision depths can exceed 250 m at some localities. The most prominent example of these features in the Gulf of Cadiz is the Alvarez Cabral contourite moat (Fig. 7c), which is associated with the Faro-Albufeira mounded drifts (Sector 4) (Gonthier *et al.* 1984; Llave *et al.* 2001; García 2002; Stow *et al.* 2002b; Hernández-Molina *et al.* 2003, 2006; Marchès *et al.* 2007). This moat (80 km long and 3.5–11 km wide) incises the base of the Algarve margin's upper slope and trends WSW, parallel to the slope.

Sector 1, located at water depths of between 500 and 1000 m, is where the majority of the erosional features occur. This locality hosts abrasive surfaces (Fig. 9b) and several NW–SE-trending erosional scours (Fig. 9c) formed across an extensive area (90 km long and 30 km wide) (Kenyon & Belderson 1973; Habgood 2002; Hernández-Molina *et al.* 2003, 2006). This sector also hosts two main channels (Fig. 9a) (Hernández-Molina *et al.* 2014b). The southern channel forms due to WSW-trending erosion from the Camarinal Sill and, at 3–4 km in width, represents the most significant erosional feature near the Strait of Gibraltar. The northern channel is obscured by infill near the Strait but becomes more distinct towards the NW, where it joins the Cadiz and Guadalquivir contourite channels (Hernández-Molina *et al.* 2014b). Both southern and northern channels host an associated mounded drift along the seaward side of the channel and numerous small oblique furrows (Fig. 9a) (Hernández-Molina *et al.* 2014b). Several reports have interpreted furrows in the overflow sedimentary lobe sector (Sector 2) as erosional features related to MOW dynamics and gravitational processes (Habgood *et al.* 2003; Mulder *et al.* 2003; Hanquiez *et al.* 2007; García *et al.* 2009). The best-developed furrow is the Gil Eanes furrow (Fig. 9d), situated at water depths of

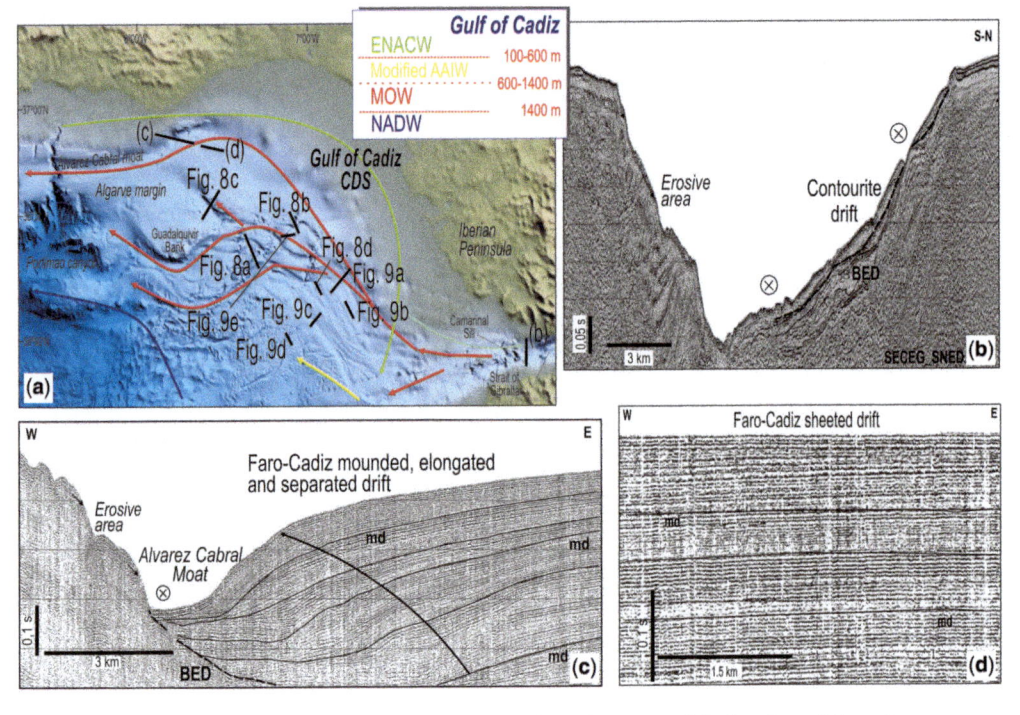

Fig. 7. (a) Digital bathymetric model of the Gulf of Cadiz continental margin and the locations of the main water masses. (b) Airgun seismic profile showing contourite drifts close to the Strait of Gibraltar. (c) & (d) Sparker seismic profiles showing mounded elongated and separated drift, and its basinward prolongation as sheeted drift (modified from Llave *et al.* 2001). BED, Basal Erosive Discontinuity; md, major discontinuity.

between 900 and 1200 m. Kenyon & Belderson (1973) first described this feature, while Habgood *et al.* (2003), Hanquiez *et al.* (2007) and García *et al.* (2009) published subsequent descriptions. The furrow is approximately 50 km long and has a width of 0.8–1.7 km, a sinuous trend, and erosional incision of up 90 m.

West Iberia

Oceanographic setting. Four main water masses flow along the western Iberian margin at different depths (Fig. 1; Table 1). As the main shallow currents, (1) the Portugal Current (PC) flows southwards, while the Portugal Coastal Current (PCC) flows towards the north (Fiúza *et al.* 1998; Pérez *et al.* 2001; Martins *et al.* 2002; Peliz *et al.* 2005; Varela *et al.* 2005). The ENACW (2) of subtropical origin extends down to water depths of 600 m (McCartney & Talley 1982; Fiúza 1984; Pollard & Pu 1985; Ambar & Fiúza 1994; Fiúza *et al.* 1998; Pérez *et al.* 2001) and moves north from about 200–300 m. A component of ENACW of subpolar origin moves south from about 300–400 m (Ambar & Fiúza 1994; Fiúza *et al.* 1998). After exiting the Gulf of Cadiz, the MOW (3) splits into three principal branches: a main branch flowing northwards, a second branch flowing westwards and a third branch flowing southwards towards the Canary Islands before veering westwards (Ambar & Howe 1979; Iorga & Lozier 1999; Slater 2003). The northern branch flows along the middle slope of the Portuguese margin towards the Galician margin and the Bay of Biscay. This branch includes two distinct cores centred at water depths of 800 m and nearly 1200 m. Along the Oporto continental slope, the MOW mixes with the ENACW flowing at depths of between 250 and 540 m. At approximately 42° N, the MOW bifurcates intermittently into two branches (Mazé *et al.* 1997). Of these, one branch flows west of the Galicia Bank plateau and the other flows north along the continental slope of the Iberian Peninsula (Iorga & Lozier 1999; González-Pola 2006). The deep-water masses (below 2000 m water depth) of the western Iberian margin consist of the southwards-flowing NADW and the northwards-flowing Lower Deep Water (LDW) (4) (van Aken 2000). This LDW forms primarily from the mixing of the deep Antarctic Bottom Water (AABW) and the Labrador Deep Water (LADW) (Le Floch 1969; Botas *et al.* 1989; Haynes & Barton 1990; Pingree & Le Cann 1990; McCartney 1992;

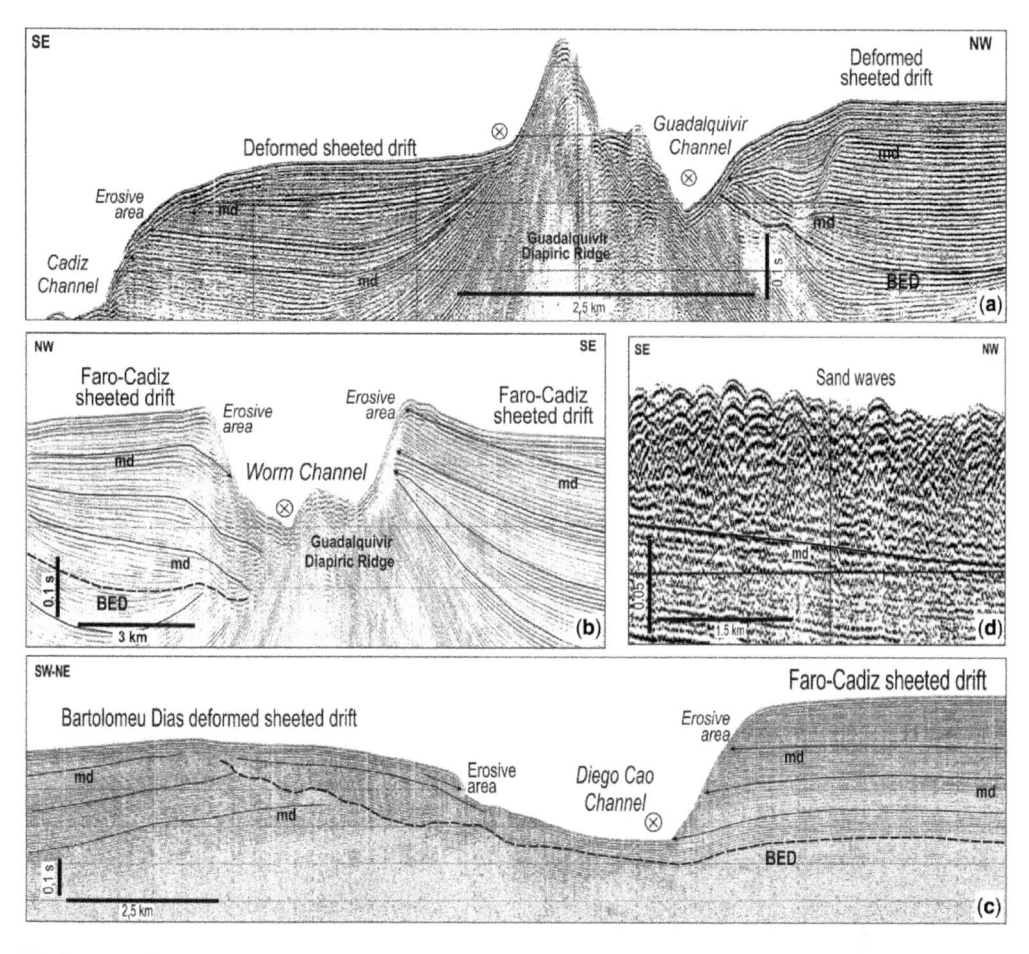

Fig. 8. (a)–(c) Sparker seismic profiles showing deformed sheeted drifts, as well as contourite channels around diapiric ridges. (d) Sparker seismic profile showing sand waves (modified from Llave *et al.* 2001; Hernández-Molina *et al.* 2006, 2014*b*; García *et al.* 2009). For the locations of the seismic profiles see Figure 7. BED, Basal Erosive Discontinuity; md, major discontinuity.

Van Aken 2000; McCave *et al.* 2001; Valencia *et al.* 2004).

Contourite features. The Sines Drift (a separated drift) represents the main contourite depositional feature of western Iberia. Mougenot (1989) first identified this feature as a contourite based on its general mounded morphology and wavy seismic patterns (Fig. 10a, b; Table 2). The Sines Drift is bounded by two of the Portuguese margin's major canyons: the Setúbal Canyon to the north and the San Vicente Canyon to the south (Fig. 10a). Formed by MOW circulation, the drift is an elongated, plastered sedimentary body formed below 750 m water depth along the gentle (*c.* 0.5°) north–south-trending continental slope of the Alentejo margin (AM) (Fig. 10b). Roque *et al.,* (2015) recently discovered an extensive area (approximately 52 km long and 34 km wide) of the Sines Drift affected by slope failure and mass wasting.

Recent reports have documented additional local contourite features occurring around structural highs and topographical irregularities associated with the circulation of either the MOW (300–2000 m water depth) or LDW (>2000 m water depth). Roque *et al.* (2015) recently identified a new separated drift offshore of Aveiro on the continental rise at a water depth of approximately 2500 m (Fig. 10c; Table 2). Neves *et al.* (2009) reported evidence of a contourite between water depths of 2300 and 3000 m reaching approximately 1.5 s thickness and 2.5 km width. This contourite drift is separated from a structural high by a moat (Fig. 10c). Large-scale sediment waves also appear in high-resolution seismic reflection data (Fig. 10d), along with localized synsedimentary deformation and primary

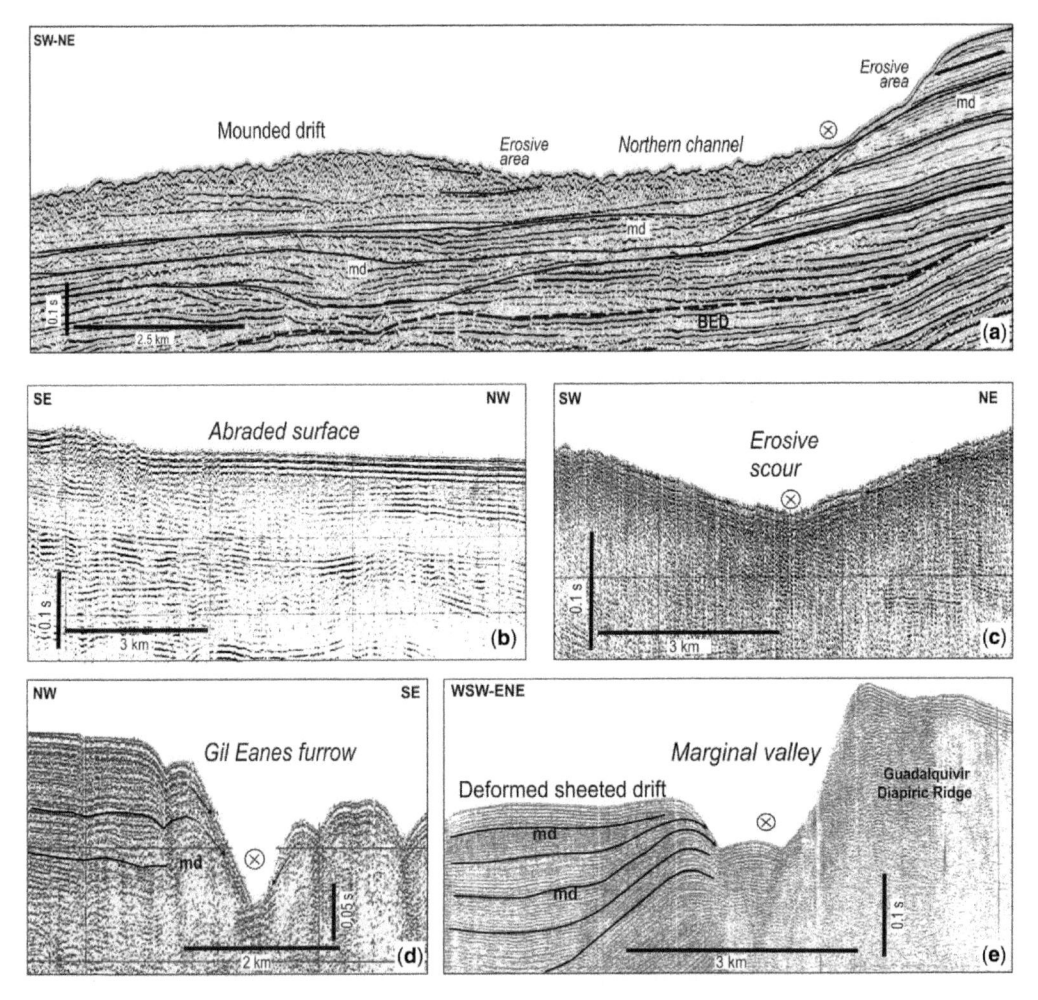

Fig. 9. (a) Airgun seismic profile where mounded drifts and contourite channels are described. (b)–(e) Sparker seismic profiles showing examples of contourite erosional features such as an abraded surface, an erosive scour, a furrow and a marginal valley (modified from Llave *et al.* 2001; Hernández-Molina *et al.* 2006, 2014*b*; García *et al.* 2009). For the locations of the seismic profiles see Figure 7. BED, Basal Erosive Discontinuity; md, major discontinuity.

faulting within contourite drift strata (Alves *et al.* 2000, 2003).

Clear examples of both depositional (plastered drift, separated drift, sediment waves) and erosional (terraces, moats, furrows) features occur along the Galician continental margin and rise (Galicia Bank CFs) (Ercilla *et al.* 2006, 2008*b*, 2009, 2011; Bender *et al.* 2012; Hanebuth *et al.* 2015; Llave *et al.* 2018; Collart *et al.* 2018). The plastered drifts occur as low-relief mounds of a few kilometres in length, and tens to a few hundred metres in thickness. Their internal structure as revealed by high- to very-high-amplitude acoustic reflections highlights well-stratified, aggradational–progradational seismic features of good lateral continuity (Fig. 10e; Table 2). Plastered drifts along the continental slope occur: (i) on the northwestern flank (2100 m water depth); (ii) along the northern scarp (1600 m water depth); (iii) at the base of certain structural highs (between 1500 and 4980 m deep) within the Galicia Bank plateau (Ercilla *et al.* 2011); and (iv) along the distal part of the Ortegal marginal platform (OMP) between water depths of 700 and 1100 m (Fig. 10e) (Jané *et al.* 2012; Llave *et al.* 2013, 2018). These drifts display smooth and terraced morphology (Ortegal CDS), a morphology which contrasts the numerous adjacent submarine canyons incised into the seafloor.

Major separated drifts occur at four locations. The first location lies along the lower continental

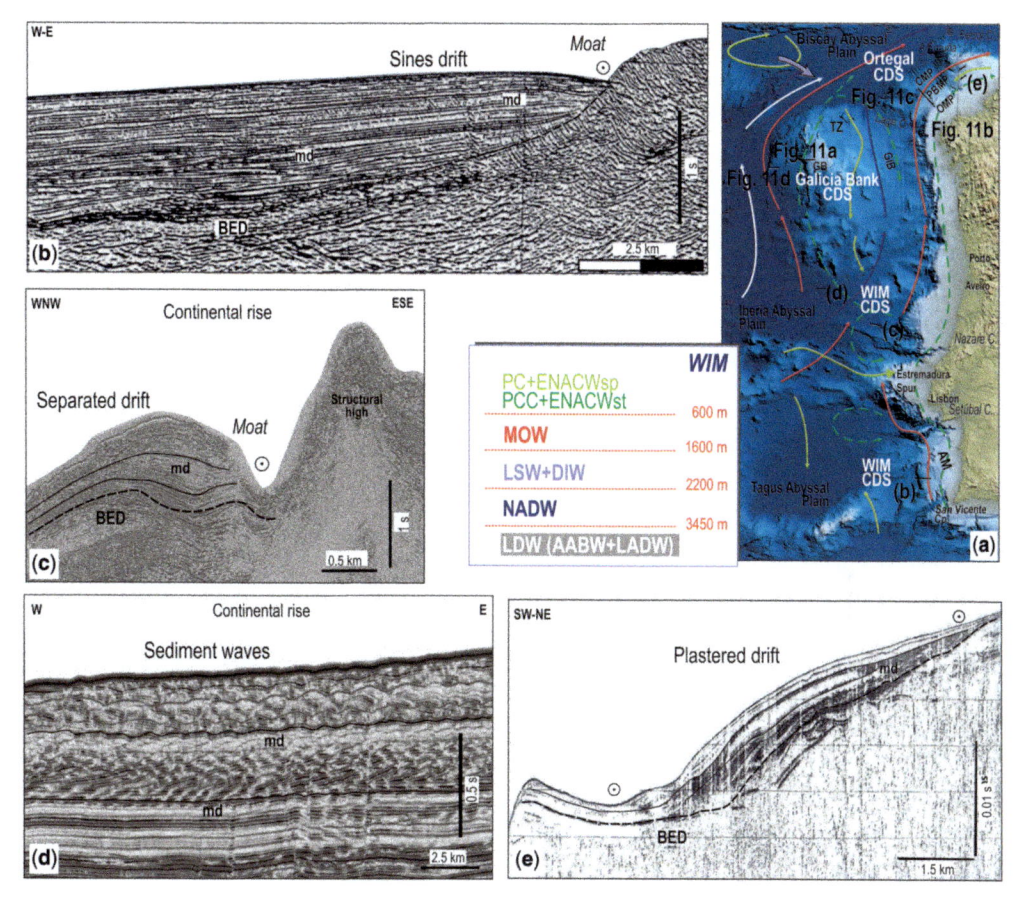

Fig. 10. (**a**) Digital bathymetric model of the western Iberian continental margin and continental rise, and the locations of the main water masses and seismic profiles. (**b**) & (**c**) Airgun and sparker seismic profiles showing examples of mounded separated drifts and moats, and (**d**) sediment waves (modified from Alves *et al.* 2003, 2006; Pereira & Alves 2011; Roque *et al.* 2012). (**e**) TOPAS seismic profile where an example of a plastered drift is shown (modified from Llave *et al.* 2013). AM, Alentejo Margin; CMP, Castro Marginal Platform; GB, Galicia Bank; GIB, Galicia Interior Basin; OMP, Ortegal Marginal Platform; PBMP, Pardo Bazán Marginal Platform; TZ, Transitional Zone; BED, Basal Erosive Discontinuity; md, major discontinuity.

slope of the western Galician continental margin, specifically at the foot of and near highs reaching approximately 2000 m water depth (Ercilla *et al.* 2006; Bender *et al.* 2012). Sediment waves contour the surface of this drift. The second drift occurs in the Transitional Zone (TZ: 1600–2500 m) (Ercilla *et al.* 2011). A third occurs along the Galicia Bank plateau (GB: 700–800 m water depth), where several moats and associated drifts (15–250 m tall from the moat axis to the top of the drift crest and 1–5 km wide) are developed at its foot and around the numerous highs (Fig. 11a). Elongated separated drifts have formed on one flank adjacent to three of the aforementioned highs, while local plastered drifts form on the other flank (Ercilla *et al.* 2008b, 2011). As such, these separated drifts form part of the Galicia Bank CFs. The fourth separated drift locality occurs at the heads of the Ferrol and A Coruña canyons at water depths of 500 and 700 m, and as part of the Pardo Bazán marginal platform at a depth of 1600 m in the Ortegal CFs (Fig. 11b). These drifts exhibit mounded shapes, and are 5–22 km long, 2–10 km wide and average 50 m in thickness (Jané *et al.* 2012; Llave *et al.* 2013, 2018; Collart *et al.* 2018). A separated drift occurs at the foot of the highs across the lower slope of the western Galician continental margin (Bender *et al.* 2012). Several of these drifts host sediment waves, which are also developed at the heads of the Ferrol and A Coruña canyons (Jané *et al.* 2012; Collart *et al.* 2018; Llave *et al.* 2018).

Abraded surfaces are tens of metres in relief and several hundreds of metres long. Seismic images

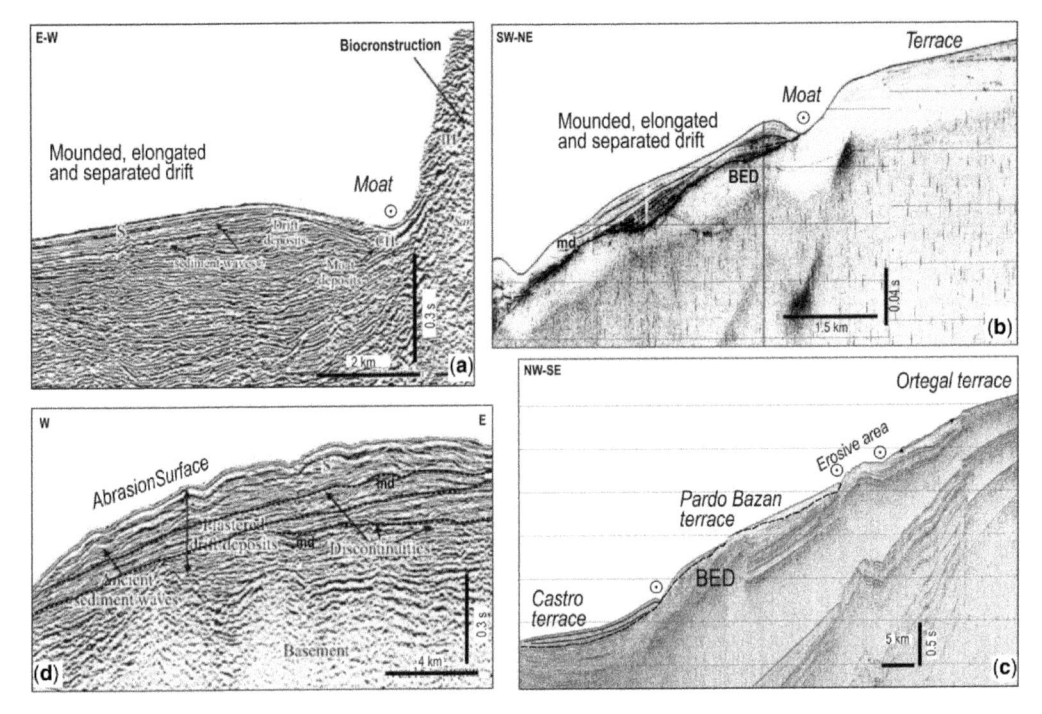

Fig. 11. (a) & (b) TOPAS seismic profiles showing examples of mounded separated drifts and moats (modified from Ercilla *et al.* 2011; Llave *et al.* 2013). (c) Airgun seismic profile showing the distribution of three contourite terraces (modified from Llave *et al.* 2013). (d) TOPAS seismic profile where an abrasion surface is shown on a plastered drift (modified from Ercilla *et al.* 2011). For the locations of the seismic profiles see Figure 10. BED, Basal Erosive Discontinuity; md, major discontinuity.

show reflectors for these features terminating against eroded seafloor surfaces that form large-scale contourite terraces. The main abraded surfaces occur along the Ortegal, Pardo Bazán and Castro marginal platforms (OMP, PBMP and CMP, respectively) where they form three contourite terraces at water depths of 200–600, 900–1800 and 2000–3500 m, respectively (Ortegal CDS) (Fig. 11c) (Jané *et al.* 2012; Llave *et al.* 2018). These represent structural highs (Fig. 11d) (Ercilla *et al.* 2008b, 2011). The Ortegal terrace is the most extensive of these highs, spanning approximately 150 km in length, up to 70 km in width to the north and approximately 20 km width in the south (Llave *et al.* 2013, 2018). Moats are erosional features that typically occur at the foot of structural scarps and highs (Fig. 11a). These can represent tens of metres (depth) and hundreds of metres (width) of erosional incision. Their asymmetrical, V-shaped cross-sections exhibit steeper eastern margins (Ercilla *et al.* 2011). Several moats occur on the Ortegal terrace and at the heads of the Ferrol and A Coruña canyons (Fig. 11b) (Jané *et al.* 2012; Llave *et al.* 2013, 2018; Collart *et al.* 2018).

Older contourite deposits may also occur along the western Iberian margin, particularly within the Iberian abyssal plain (southern Galicia Bank CFs). Sediment waves occur within the upper rise according to Miocene–Quaternary sediments drilled at ODP Site 1069 (Shipboard Scientific Party 1998) and according to middle Eocene deposits offshore of Oporto, drilled by ODP Leg 149 and in the DSDP Leg 48B area (Wilson *et al.* 1996, 2001). Soares *et al.* (2014) have described several Cretaceous drifts consisting of elongated mounded drifts along the outer proximal margin. These workers inferred the presence of sheeted drifts along the distal margin offshore of NW Portugal deposited after the Aptian–Albian lithospheric break-up. Near the continental margin of the Galicia Bank, deposits drilled by ODP Leg 103 (Boillot *et al.* 1987; Comas & Maldonado 1988), DSDP Leg 47B Site 398 along the southern margin of the Vigo Seamount (Maldonado 1979) and ODP Leg 149 sites 897–901 (Alonso *et al.* 1996; Milkert *et al.* 1996) were also interpreted as contourites deposited in the late Miocene.

Cantabrian margin

Oceanographic setting. Along the Cantabrian continental margin, most of the water masses originate in the North Atlantic or result from interactions between waters originating in the Atlantic and Mediterranean. The uppermost water mass is the ENACW, which extends to depths of approximately 400–600 m and flows westwards along the continental margin (Fig. 1; Table 1) (González-Pola 2006). It generally flows at a velocity of 1 cm s^{-1}, although it can occasionally reach velocities of up to 10 cm s^{-1} (Pingree & Le Cann 1990). Between water depths of 400–500 and 1500 m, the MOW follows the continental slope as a contour current (Fig. 1). Seafloor irregularities and the Coriolis effect control the local MOW circulation. Although detailed information is lacking on MOW circulation in the Bay of Biscay, it appears to split into two branches around the Galicia Bank. One of these continues northwards, while the other flows eastwards along the Cantabrian margin slope (Iorga & Lozier 1999; González-Pola 2006). From Ortegal Spur to Santander, the MOW propagates along the slope at reduced velocities. Its minimum velocity is around 2–3 cm s^{-1} at 6° W and 8° W (Pingree & Le Cann 1990; Díaz del Río et al. 1998). The NADW occurs between water depths of 1500 and 3000 m (Fig. 1), and includes a core of LSW at a depth of approximately 1800 m (Vangriesheim & Khripounoff 1990; McCartney 1992). The LDW forms beneath the NADW primarily due to mixing of the deep AABW and the LSW (Botas et al. 1989; Haynes & Barton 1990; McCartney 1992) (Fig. 1). A cyclonic recirculation cell develops over the Biscay abyssal plain. This feature exhibits a characteristic polewards velocity near the continental margin of 1.2 (± 1.0) cm s^{-1} (Dickson et al. 1985; Paillet & Mercier 1997).

Contourite features. The Le Danois CDS has been identified along the Cantabrian margin (Fig. 12; Table 2) in an area surrounded predominantly by downslope processes (Ercilla et al. 2008a; Iglesias 2009; Van Rooij et al. 2010). This CDS includes both depositional and erosional features (separated drifts, plastered drifts, moats and scours) generated by the MOW circulation and controlled by seafloor irregularities including two topographical highs: the large Le Danois Bank and the smaller Vizco High (Fig. 12a) (Van Rooij et al. 2010).

The Le Danois CDS includes two separated drifts: the Gijón and Le Danois drifts. The Gijon Drift occurs along the upper slope at water depths of approximately 400–850 m and has a maximum thickness of 0.25 s (Fig. 12b; Table 2). The Le Danois Drift occurs at the foot of the southern side of the Le Danois Bank at water depths of between 800 and 1500–1600 m. This feature is approximately 0.3 s thick and varies in width, reaching as much as 10 km width in its central part but only spanning 3.5–4 km to the west and 4.7 km to the east (Fig. 12b). Three plastered drifts of about 1 s thickness occur along the upper, southern slope of the Le Danois Bank (Van Rooij et al. 2010). These appear as mounded relief along the western edge between water depths of 600 and 750 m, and on the eastern edge at depths of 750–1100 and 1100–1550 m (Fig. 12c; Table 2).

Moats and scours are the main erosional features of the Le Danois CDS. The two moats identified are referred to as the Gijón and Le Danois moats. As the upslope continuation of Gijón Canyon, the Gijón moat trends NW–SE, and spans approximately 45 km length and 1–4 km width. This feature begins at 1100 m water depth, incises an additional 400 m in the west and disappears to the east (Fig. 12b). The Le Danois moat trends in a WNW–ESE direction and extends from depths of 800–1500 m towards the east. It spans 48 km in length, varies from 0.8 to 2.8 km in width and incises 75–105 m of surface (Fig. 12b). Scour alignments trending NE and ENE occur on one of the plastered drifts. These span 5.5–28 km in length, 1250 m in width and run c. 5 m deep (Van Rooij et al. 2010).

Sedimentology of contourite features

Contourite deposits along the Iberian margin range in grain size from clay to sand but primarily are categorized as mud-rich. Deposits include thin, interbedded layers of fine-grained sand and silt of terrigenous and biogenic origin. The fine-grained beds are predominantly poorly sorted, intensely bioturbated and typically display broad rhythmic bedding (Stow et al. 2002a, c).

Faugères et al. (1984) and Gonthier et al. (1984) proposed the original contourite facies model sequence based on research across the Faro Drift along the middle slope of the Gulf of Cadiz. Their model consists of two superposed units, including a basal coarsening-upward unit grading from homogeneous mud (clay, fine silt) to mottled coarser silt and finally to sandy silt/silty sand. This is followed by a fining-upward unit with an inverse facies succession. This general 'bigradational' model theoretically represents an increase bottom-current flow followed by a decline in current strength (Stow & Holbrook 1984; Stow et al. 2002b; Huneke & Stow 2008). Stow & Faugères (2008) extended this model to include five sedimentary divisions applied to the standard bigradational contourite sequence (C1–C5) (Fig. 13). The model can be applied to contourites of siliciclastic, volcanic, bioclastic or mixed composition. Bioturbation limits preservation of sedimentary structures but indistinct to discontinuous parallel laminations, coarser sand layers or rare cross-laminations may occur. The facies model interprets the bigradational sequence as two shifts in the strength of the bottom-current flow: from weak to

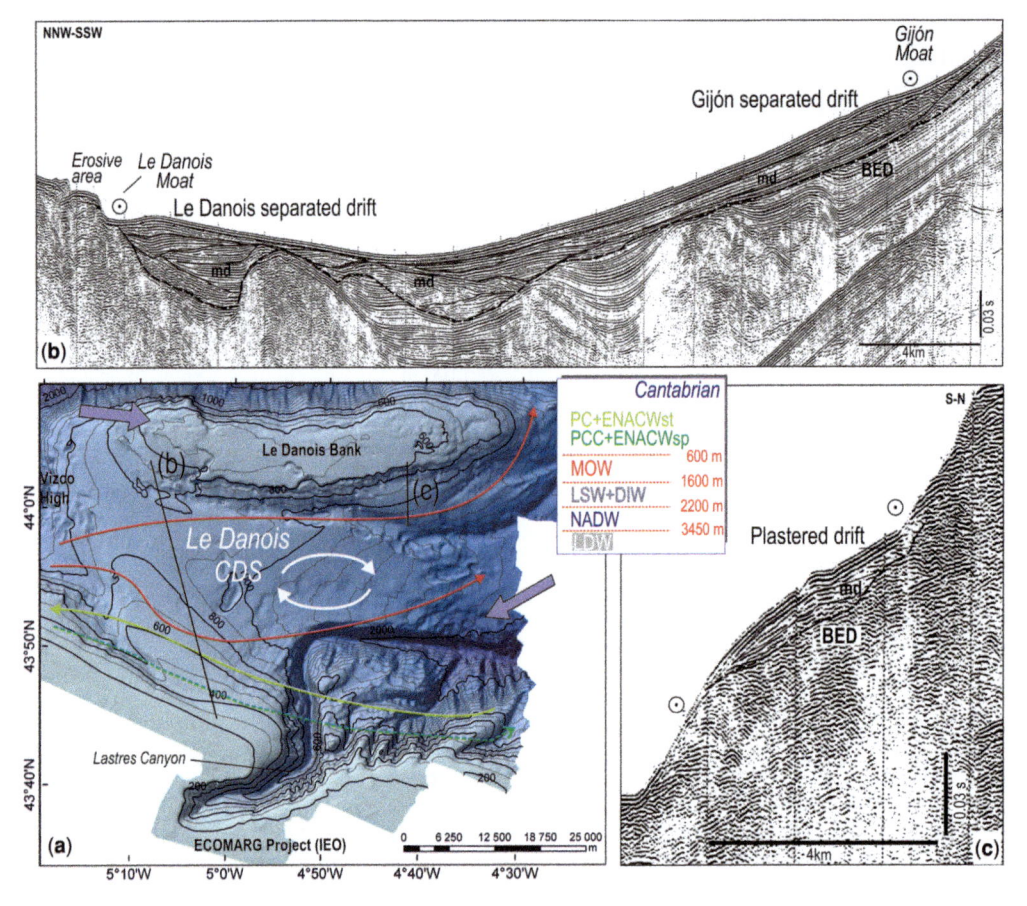

Fig. 12. (a) Digital bathymetric model of the Le Danois continental margin, and the description of the main contourite features based on the location of the main water masses and seismic profiles. (b) & (c) Airgun seismic profiles where mounded separated drifts and moats, as well as a plastered drift are shown (modified from Van Rooij *et al.* 2010). BED, Basal Erosive Discontinuity; md, major discontinuity.

strong, and then back to weak. The continued presence of coarsening-/fining-upward trends in grain size along with strong bioturbation and mottling indicates continuous, gradual (relative to gravity flow deposits) deposition of sediments through bottom-current processes. As the velocity increases, coarser sediment predominates as the removal of the fine fraction produces coarser, better-sorted sedimentary packages (Nelson *et al.* 1993). Stow & Faugères (2008) also described sandy contourites that contain an inversely graded lower sub-sequence (mud + mottled silt and mud units), a middle sandy silt and an upper normally graded sub-sequence (mottled silt and mud + mud units). Most of these deposits were interpreted as bottom-current-modified turbidites (Faugères & Stow 1993, 2008; Stow *et al.* 2002*a*, *b*; Stow & Faugères 2008). Mulder *et al.* (2013) concluded that contourite sequences record changes in bottom-current velocity and flow competency but may also depend on sediment supply. These workers hypothesized that increased erosion of mud along the flanks of confined contourite channels and moats or increases in sediment supply by rivers and downslope mass transport along the continental shelf and upper slope provided the coarse, terrigenous sediment observed.

Coarse terrigenous sediment is significant in the Gulf of Cadiz. Research from IODP Expedition 339 and cores from other cruises have detected voluminous, mature and well- to moderately sorted contourite sands that form laterally extensive sheeted drifts, channel-floor cover or patch drifts in contourite channels (Buitrago *et al.* 2001; Viana *et al.* 2007; Hanquiez 2006; Hanquiez *et al.* 2007; Viana 2008; Stow *et al.* 2011*a*, *b*, 2013*b*; Brackenridge *et al.* 2013, 2018; Hernández-Molina *et al.* 2014*a*, *b*).

The Ortegal Spur of the Ortegal contourite terrace (Llave *et al.* 2013) also hosts compositionally

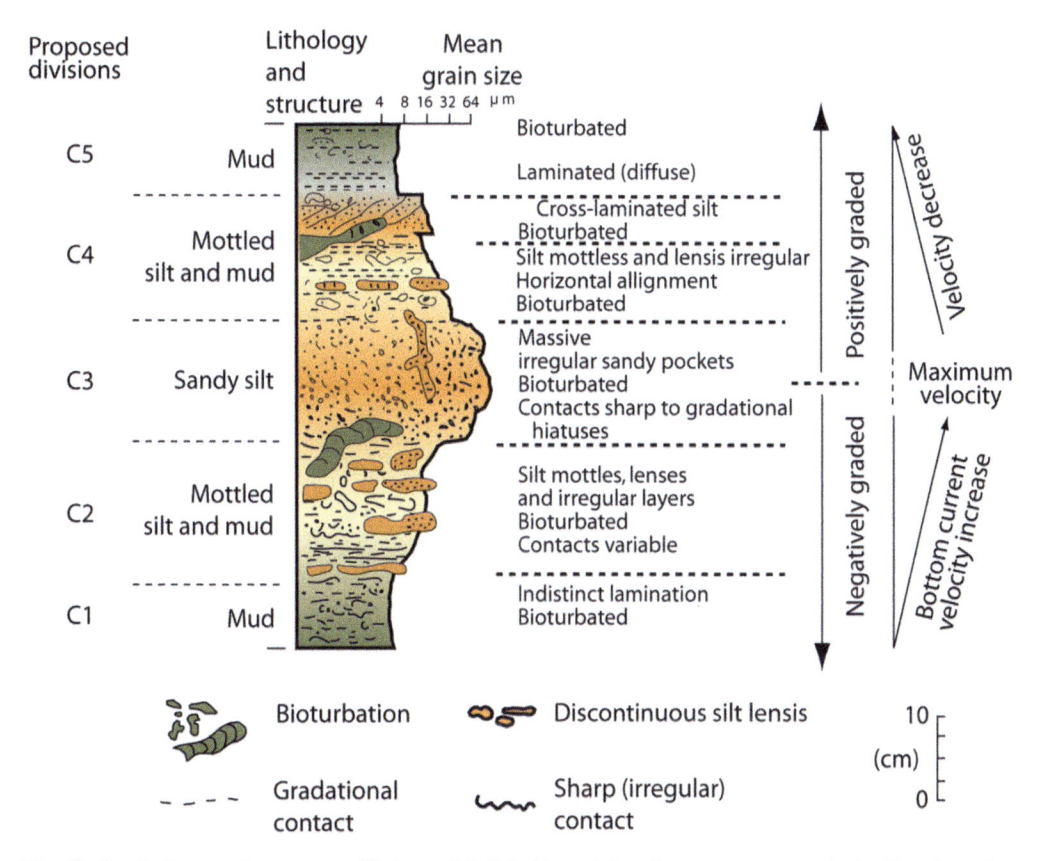

Fig. 13. Standard contourite sequence of facies models linked to variations in contour–current velocity (from Stow & Faugères 2008; based on the original figure from Gonthier et al. 1984).

mature, coarse and silty sands. Sands specifically consist of subrounded and well-sorted quartz, and glauconite grains with abundant bioclastic fragments. The bioclasts include fragments of foraminifera, gastropods, bivalves and pteropods (Alejo et al. 2012).

Translating seismic facies to sedimentary facies

Contourite features are described and systematized here according to their seismic characteristics. Interpretations therefore depend heavily on the external drift morphology, internal stacking pattern of depositional units and other aspects apparent in seismic images. Local and regional water-circulation patterns and palaeoceanographical models support these interpretations.

Drift morphologies (i.e. separated, sheeted, plastered and confined drifts) are recognized at the scale of the drifts themselves (Fig. 14). Slope-plastered contourite drifts predominate upper slope settings. These often occur in the presence of elongated erosional surfaces along the uppermost slope and in downslope areas, accompanied by a mounded alongslope elongated drift, with frequent sediment waves and internal erosional surfaces (Fig. 14). In middle and lower slope areas, complex seafloor physiography caused by tectonics creates obstacles for current flow. These features induce local acceleration, and create a considerable variety of drifts and erosional features such as moats, channels and furrows (Fig. 14). Middle and lower slope areas also host contourite terraces, which form from both depositional and erosional processes (Fig. 14).

Depositional units within contourites reported around Iberia are generally lenticular in shape, and have well-layered, convex-up seismic units of good lateral continuity along both strike and dip (Figs 2–12). Stacking patterns show downlapping (onlapping on steep slopes) and sigmoidal progradational reflector patterns where downstream and upslope migration has occurred (Figs 2–12).

Table 3 lists seismic facies most commonly recognized from Iberian contourite drifts. Among these are: (a) transparent layers of variable thickness intercalated with zones of high- to moderate-

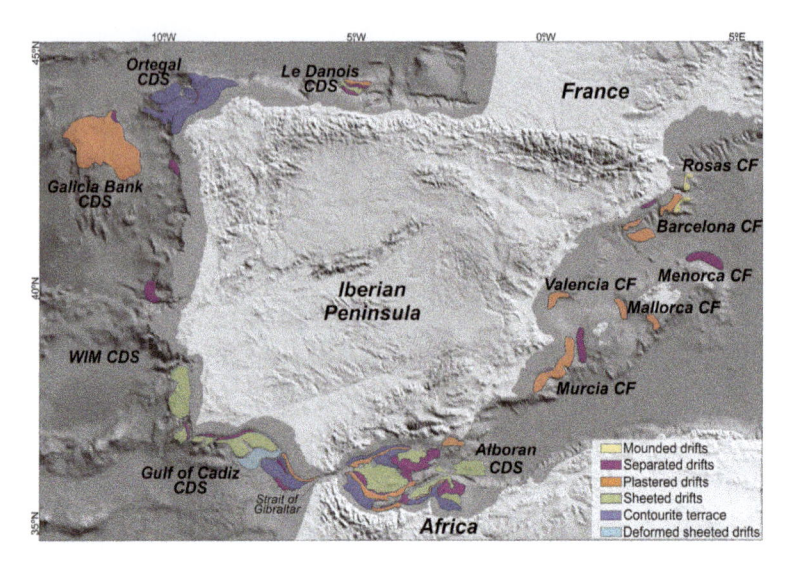

Fig. 14. Locations and depths (in metres) of the main types of drifts along the Iberian margin: C, confined; Ch, channel related; M, mound; P, plastered; S, separated; Sh, sheeted; Sw, sediment waves. The image also shows contourite features (red), erosive features (purple) and unpublished features (orange). See Table 1 for more detail regarding their characteristics.

amplitude seismic reflectors, particularly in sheeted drifts (Figs 6a, 7d, 8a–c); and (b) smooth, parallel, moderate- to high-amplitude reflectors typically interbedded with transparent zones in plastered drifts and common throughout mounded, confined and channel-related drifts (Figs 2b, 3a–d, 4a, b, 5b, c, 6b–d, 7c, 9a, 10b, c, 11a, b & 12b, c). Discontinuous seismic facies include short, discontinuous to chaotic reflectors (c) of moderate to high amplitude occurring in most drifts, particularly in mounded drifts and moats (Figs 2a, 3a & 3d, 6b, c, 7c, 9a, 10c, 11a & 12b). Sigmoid progradational reflectors (d) occur in mounded drifts where strong downstream and/or oblique migration has occurred. They are also common in separated drifts (Figs 3b, 7c & 12b). Gently wavy reflectors (e) are common over parts of several drifts (Figs 3c, 8d & 10d). Contourite terraces and practically all contourite depositional systems host horizontal and low-inclination (f) reflectors truncated at the seafloor or by an internal erosional surfaces (e.g. Figs 4a, b, 5b, 9a, b & 11b, c).

Relatively uniform deposits (fine or coarse sediments) tend to exhibit transparent or weak seismic facies, whereas extensive sheets of interbedded coarse- and fine-grained sediments exhibit higher-amplitude seismic reflectors with good lateral continuity. The aforementioned laterally extensive progradational to aggradational seismic units with sub-parallel, variable amplitude reflectors indicate muddy compositions. By contrast, moats within contourite channels and many contourite terraces exhibit high-amplitude reflections (HARs). These high-amplitude seismic reflections span a few kilometres in width and are interpreted as coarse-grained sediments (Deptuck et al. 2003; Posamentier & Kolla 2003). These facies occur in erosional features such as moats and along channel axes (Table 3), which develop in channels around seafloor irregularities. Contourite terraces (e.g. Alboran, Gulf of Cadiz, WIM and Ortegal) (Fig. 14; Table 3) and overflow features can also appear as HAR seismic facies. The Gulf of Cadiz' proximal Sector 1 is an example of an overflow feature where an exceptionally thick sandy sheeted drift appears as an extensive HAR unit (Fig. 14).

Contourite features described here occur at depths corresponding to the principal interfaces between MAW–LIW–WMDW in the Western Mediterranean Sea, and between the ENACW–MOW–NADW in the Atlantic. The HARs, however, do not appear to correspond with enhanced bottom currents, development of terraces on top of plastered drifts or bottom-current modification of sand layers exposed at the seafloor surface. Their origin and interpretation remain unresolved in spite of their economic potential as hydrocarbon reservoirs.

Discussion

Factors influencing contourite seismic/sedimentary facies changes

Contourite drifts along the Iberian margin exhibit certain stratigraphic patterns in their reflection

Table 3. *Main high-amplitude reflectors (HARs) for contourite features along the Iberian margins*

Seismic section	Seismic facies characteristics	Depositional setting	Location
	Transparent layers intercalated with moderate- to high-amplitude reflectors	Sheeted drifts	Proximal sector Gulf of Cádiz CDS
	Smooth, parallel moderate- to high-amplitude reflectors typically interbedded with transparent zones	Plastered drifts and throughout mounded, confined and channel-related drifts	Around Iberia middle continental slopes
	Short, discontinuous to chaotic reflectors of moderate- to high-amplitude reflectors	Moats, channels, furrows	Around Iberian continental slope bathymetric irregularities
	Sigmoid and/or oblique progradational reflectors with strong downstream migration	Mounded drifts	Around Iberia middle continental slopes
	Gently wavy reflectors	Over parts of several drifts	Around Iberia middle continental slopes and rise
	Horizontal or low-inclination moderate- to high-amplitude reflectors truncated by HAR erosional surfaces	Contourite terraces	Alboran Sea, Gulf of Cadiz, Galician margin (marginal platforms)

amplitudes (Figs 2–12). These patterns include a more transparent facies (T) at the base, higher amplitude reflections (R) in the upper part and a subtle but continuous high-amplitude erosional surface at the top. In the case of the Faro Drift (Gulf of Cadiz), these variations have been attributed to long-term changes in bottom-current strength at different scales through the latest Pliocene and Quaternary (Llave *et al.* 2001; Stow *et al.* 2002*a*, *b*, *c*; Hernández-Molina *et al.* 2016*a*). This mechanism creates repeated coarsening-upward sequences bounded by erosional surfaces at unit and sub-unit scales.

In the case of the Gulf of Cadiz CDS, several studies interpreted these facies as reflecting changes in MOW bottom-current strength (e.g. Gonthier *et al.* 1984; Faugères *et al.* 1985*a*, *b*; Stow *et al.* 1986, 2002*b*; Sierro *et al.* 1999; Roque *et al.* 2012) linked to eustatic/climatic drivers and MOW variability (Voelker *et al.* 2006; Toucanne *et al.* 2007; García *et al.* 2009; Rogerson *et al.* 2012; Bahr *et al.* 2014; Hernández-Molina *et al.* 2014*a*). IODP Expedition 339 sampled some of the Gulf of Cadiz CDS seismic facies and found that transparent seismic facies corresponded to fine-grained contourites, while HARs corresponded to mature, well-sorted contourite sands that reached up to 10 m in thickness (Stow *et al.* 2013*b*; Hernández-Molina *et al.* 2014*b*, 2016*a*).

Interacting factors determine the degree to which bottom currents can influence the morphology of the Iberian margin. At longer timescales, tectonic factors determine the role that downslope sediment transport plays in margin development. Tectonics also influence contourite sedimentation as the mechanism producing remnant marginal platforms, structural highs and the opening of the Gibraltar Strait itself (i.e. northwestern Iberian margins: Maestro *et al.* 2015; Llave *et al.* 2018). On shorter timescales, climate and sea-level changes cause deepening or shoaling of water masses. This in turn controls the vertical distribution of sand and mud deposits, associated acoustic facies, stratigraphic stacking patterns and thicknesses. Climate and sea-level changes also drive the general behaviour of the water masses and influence depositional styles along the margins. In the Gulf of Cadiz and west of Iberia, Mediterranean water masses flowed more swiftly and at greater depths during glacial periods (Schönfeld & Zahn 2000; Rogerson *et al.* 2005; Llave *et al.* 2006; Voelker *et al.* 2006; Ercilla *et al.* 2016). This would favour local development of sandy contourites at intermediate to deeper water depths along the continental slope. These factors explain the Iberian margin's overall sedimentary evolution from dominantly downslope or mixed alongslope–downslope sedimentary processes at the beginning of the Pliocene to more alongslope sedimentation with the Quaternary opening of the Strait of Gibraltar (Roque *et al.* 2012; Brackenridge *et al.* 2013; Hernández-Molina *et al.* 2016*a*; Ercilla *et al.* 2016).

The relationship between oceanographic processes and the depositional, erosional and mixed features observed among Pliocene and Quaternary contourites allows for interpretation of similar features in the ancient record. Ancient oceans consisted of different water masses and circulation regimes than those observed today (e.g. Hay 2009). The most discernible patterns belong to extreme glacial maxima and greenhouse conditions (Pickering & Hiscott 2016). Records left by ancient water masses should, nevertheless, resemble sandy and muddy deposits described here.

Implications for petroleum exploration

Thick and widespread progradational to aggradational depositional units characterized by sub-parallel reflectors of varying amplitude in seismic images are interpreted as representing fine-grained drifts occurring throughout the Iberian margin. Continuous HARs, indicative of sandier contourites, occur in moats and channels, in contourite terraces, and in sectors with sheeted drifts affected by overflows. IODP Expedition 339 drilled several of these drifts around the Gulf of Cadiz and western Iberian margin. These features exhibit seismic facies similar to those observed in other sectors of the Iberian margin. Sampling of these drifts showed that sub-parallel reflectors of variable amplitude corresponded to muddy contourites that were sometimes enriched in organic carbon (up to 2 wt%). These units could therefore serve as both seals and potential source rocks (Hernández-Molina *et al.* 2013; Stow *et al.* 2013*b*). Sampling of seismic features associated with sandy contourites also suggests extensive distribution of mature, well-sorted Pliocene–Quaternary sands (Expedition 339 Scientists 2012; Hernández-Molina *et al.* 2013; Stow *et al.* 2013*b*). HARs indicate sandy contourites: for example, in southeasterly areas of the Gulf of Cadiz (affected by the overflow processes), where Brackenridge *et al.* (2013, 2018) identified a tabular, aggradational sedimentary stacking pattern associated with a buried, mixed contourite–turbidite succession. This feature includes a sand- and clay-rich interval between 925 and 1740 m which reaches 815 m in thickness. With 600 m of sand alone, this contourite could serve as a potential reservoir unit (Buitrago *et al.* 2001; García-Mojonero & Martínez del Olmo 2001). Cakebread-Brown *et al.* (2003) have also interpreted HARs and AVO (amplitude v. offset) anomalies from this unit as high-porosity (30%), gas-bearing contourite sands.

The Santos Drift in the northern Santos Basin includes more than 600 m of fine-grained Neogene-aged sand and silt, and thus acts as seal rock for

Paleogene oil-bearing sandstones (Duarte & Viana 2007). Similar sandy contourites have been described along the Uruguayan margin (Hernández-Molina *et al.* 2016*b*), within the Pliocene sedimentary record of the Gulf of Mexico (Shanmugam *et al.* 1993), in Eocene deposits of the Campos Basin, Brazil (Mutti *et al.* 1980; Viana & Rebesco 2007) and in the northeastern Atlantic (Nelson *et al.* 1993; Howe *et al.* 1994; García-Mojonero & Martínez del Olmo 2001; Stow *et al.* 2002*a, c*, 2013*a, b*; Habgood *et al.* 2003; Akhmetzhanov *et al.* 2007; Hernández-Molina *et al.* 2014*a, b*). In these cases, HARs record contourite deposits mainly composed of medium- to fine-grained sand with common bedforms, such as mega-ripples and sand waves (Stoker *et al.* 1998; Viana *et al.* 1998, 2002; Masson *et al.* 2004; Shanmugam 2006, 2012, 2013*b*; Mutti & Carminatti 2012).

Similarities between these features and those described from the Gulf of Cadiz (Hernández-Molina *et al.* 2016*a*) and Iberian continental margin indicate the economic potential of sandy contourites from these areas. The occurrence of fine- to coarse-grained sedimentary materials associated with contourites around Iberia could inform interpretation of deeper-water sedimentary facies from the ancient record. Emerging information on contourites could also facilitate innovations in deep-water petroleum exploration strategies.

Conclusions and future areas of study

Contourite features of the Iberian continental margin include extensive depositional, erosional and mixed (depositional and erosional) features developed along the continental slope due to bottom-current dynamics. Depositional features include mounded, elongated and separated, sheeted, plastered, confined, and channel-related drifts. Erosional features include moats, channels and furrows, while terraces are interpreted as mixed erosional–depositional features. Large mud-dominated contourite drifts with good, alongslope continuity could serve as petroleum source rocks. Moats, channels and sheeted drifts proximal to overflows and contourite terraces exhibit high-amplitude reflections (HARs) in seismic images. These are interpreted as extensive sandy contourites that could also serve as potential hydrocarbon reservoirs.

The muddy/sandy contourites described here, along with the overall deep-water morphology of the Iberian margin generally record intermediate- to deep-water masses and their interfaces from the Pliocene to Quaternary. The connection between observed features and mechanisms makes these features good analogues for interpreting recent or ancient deep-water environments, and useful in identifying potential seals and reservoir rocks. Similar contourite features in deeper or older sediments could represent future petroleum exploration opportunities. Advances in geophysics, offshore 3D seismic imaging and robust correlation with well-log and borehole data will help further understanding of contourites and their role in petroleum systems.

Acknowledgements The research was conducted in the framework of 'The Drifters Research Group' of the Royal Holloway University of London (UK). Thanks to SECEG and SNED for permission to include a seismic profile obtained during an oceanographic cruise in the Strait of Gibraltar (Fig. 7b), and to the ECOMARG Project for providing the multibeam bathymetry data obtained in Le Danois or 'El Cachucho' (Fig. 11a). We thank the editor, Dr Ken McClay, and the two reviewers, Dr Domenico Chiarella and anonymous, for their positive and useful comments, which have improved the original submitted version.

Funding This contribution is a product of the IGCP-619 and INQUA-1204 projects, and is partially supported through the CTM 2008-06399-C04/MAR (CONTOURIBER), CGL2011-16057-E (MOW), CTM 2012-39599-C03 (MOWER), CGL2016-80445-R (SCORE), FCT-PTDC/GEO-GEO/4430/2012 (CONDRIBER), CTM2016-75129-C3-1-R and CGL2015-74216-JIN projects.

Author contributions EL: Investigation (Lead), Writing – Original Draft (Lead); FJH-L: Investigation (Equal), Writing – Review & Editing (Equal); MG: Investigation (Supporting), Writing – Review & Editing (Supporting); GE: Investigation (Supporting), Writing – Review & Editing (Supporting); CR: Investigation (Supporting), Writing – Review & Editing (Supporting); CJ: Investigation (Supporting), Writing – Review & Editing (Supporting); AM: Investigation (Supporting), Writing – Review & Editing (Supporting); BP: Investigation (Supporting), Writing – Review & Editing (Supporting); DVR: Investigation (Supporting), Writing – Review & Editing (Supporting); MR: Investigation (Supporting), Writing – Review & Editing (Supporting); RB: Investigation (Supporting), Writing – Review & Editing (Supporting); GJ: Investigation (Supporting), Writing – Review & Editing (Supporting); MG-B: Investigation (Supporting), Writing – Review & Editing (Supporting); DS: Investigation (Supporting), Writing – Review & Editing (Supporting).

References

AKHMETZHANOV, A., KENYON, N.H., HABGOOD, E., VAN DER MOLLEN, A.S., NEILSEN, T., IVANOV, M. & SHASHKIN, P. 2007. North Atlantic contourite sand channels. *In*: VIANA, A.R. & REBESCO, M. (eds) 2007. *Economic and Palaeoceanographic Significance of Contourite Deposits*. Geological Society, London, Special

Publications, **276**, 25–47, https://doi.org/10.1144/GSL.SP.2007.276.01.02

ALEJO, I., NOMBELA, M.A. *ET AL.* 2012. Caracterización de los sedimentos superficiales en tres sistemas deposicionales contorníticos (Golfo de Cádiz, Cabo Ortegal y El Cachucho): implicaciones conceptuales. Resúmenes extendidos del VIII, Congreso Geológico de España. *Geo-Temas*, **13**, 1781–1784.

ALONSO, B., COMAS, M.C., ERCILLA, G. & PALANQUES, A. 1996. Data report: textural and mineral composition of Cenozoic sedimentary facies off the western Iberian Peninsula, Sites 897, 898, 899 and 900. *In*: WHITMARSH, R.B., SAWYER, D.S., KLAUS, A. & MASSON, D.G. (eds) *Proceedings of Ocean Drilling Program, Scientific Results, Volume 149*. Ocean Drilling Program, College Station, TX, 741–754.

ALVES, T.M., GAWTHORPE, R.L., HUNT, D. & MONTEIRO, J.H. 2000. Tertiary evolution of the São Vicente and Setúbal submarine canyons (southwest Portugal): insights from seismic stratigraphy. *Cenozoico de Portugal, Ciências da Terra*, **14**, 243–256, http://hdl.handle.net/10362/4716

ALVES, T.M., GAWTHORPE, R.L., HUNT, D.W. & MONTEIRO, J.H. 2003. Cenozoic tectono-sedimentary evolution of the western Iberian margin. *Marine Geology*, **195**, 75–108, https://doi.org/10.1016/S0025-3227(02)00683-7

ALVES, T.M., MOITA, C., SANDNES, F., CUNHA, T., MONTEIRO, J.H. & PINHEIRO, L.M. 2006. Mesozoic–Cenozoic evolution of North Atlantic continental-slope basins: the Peniche basin, western Iberian margin. *American Association of Petroleum Geologists Bulletin*, **90**, 31–60, https://doi.org/10.1306/08110504138

AMBAR, I. & FIÚZA, A. 1994. Some features of the Portugal Current System: A poleward slope undercurrent, an upwelling related southward flow and an autumn-winter poleward coastal surface current. *In*: *Proceedings of the 2nd International Conference on Air–Sea Interaction and on Meteorology and Oceanography of the Coastal Zone*. American Meterological Society, Boston, MA, 286–287.

AMBAR, I. & HOWE, M.R. 1979. Observations of the Mediterranean outflow – II. The deep circulation in the vicinity of the Gulf of Cadiz. *Deep Sea Research*, **26A**, 555–568, https://doi.org/10.1016/0198-0149(79)90096-7

AMBAR, I., ARMI, L., BOWER, A. & FERREIRA, T. 1999. Some aspects of time variability of the Mediterranean water off south Portugal. *Deep Sea Research, I*, **46**, 1109–1136, https://doi.org/10.1016/S0967-0637(99)00006-0

AMBAR, I., SERRA, N. *ET AL.* 2002. Physical, chemical and sedimentological aspects of the Mediterranean outflow off Iberia. *Deep Sea Research II*, **49**, 4163–4177, https://doi.org/10.1016/S0967-0645(02)00148-0

AMBAR, I., SERRA, N., NEVES, F. & FERREIRA, T. 2008. Observations of the Mediterranean Undercurrent and eddies in the Gulf of Cadiz during 2001. *Journal of Marine Systems*, **71**, 195–220, https://doi.org/10.1016/j.jmarsys.2007.07.003

BAHR, A., JIMÉNEZ-ESPEJO, F.J. *ET AL.* 2014. Deciphering bottom current velocity and paleoclimate signals from contourite deposits in the Gulf of Cádiz during the last 140 kyr: An inorganic geochemical approach. *Geochemistry, Geophysics, Geosystems*, **15**, 3145–3160, https://doi.org/10.1002/2014GC005356

BARINGER, M.O. & PRICE, J.F. 1999. A review of the physical oceanography of the Mediterranean Outflow. *Marine Geology*, **155**, 63–82, https://doi.org/10.1016/S0025-3227(98)00141-8

BENDER, V.B., HANEBUTH, T.J.J., MENA, A., BAUMANN, K.H., FRANCÉS, G. & VON DOBENECK, T. 2012. Control of sediment supply, palaeoceanography and morphology on late Quaternary sediment dynamics at the Galician continental slope. *Geo-Marine Letters*, **32**, 313–335, https://doi.org/10.1007/s00367-012-0282-2

BOILLOT, G., MALOD, J.-A., DUPEUBLE, P.-A. & CYBERE GROUP 1987. Mesozoic evolution of Ortegal Spur, North Galicia margin: Comparison with adjacent margins. *In*: BOILLOT, G., WINTERER, E.L. *ET AL.* (eds) *Proceedings of the Ocean Drilling Program Initial Reports (Part A), Volume 103*. Ocean Drilling Program, College Station, TX, 107–119.

BORENÄS, K.M., WAHLIN, A.K., AMBAR, I. & SERRA, N. 2002. The Mediterranean outflow splitting- a comparison between theoretical models and CANIGO data. *Deep Sea Research II*, **49**, 4195–4205, https://doi.org/10.1016/S0967-0645(02)00150-9

BOTAS, J.A., FÉRNANDEZ, E., BODE, A. & ANADÓN, R. 1989. Water masses off central Cantabrian coast. *Scientia Marina*, **53**, 755–761.

BRACKENRIDGE, R., STOW, D.A.V. & HERNÁNDEZ-MOLINA, F.J. 2011. Contourites within a deep-water sequence stratigraphic framework. *Geo-Marine Letters*, **31**, 343–360, https://doi.org/10.1007/s00367-011-0256-9

BRACKENRIDGE, R.A., HERNÁNDEZ-MOLINA, F.J., STOW, D.A.V. & LLAVE, E. 2013. A Pliocene mixed contourite–turbidite system offshore the Algarve Margin, Gulf of Cadiz: Seismic response, margin evolution and reservoir implications. *Marine and Petroleum Geology*, **46**, 36–50, https://doi.org/10.1016/j.marpetgeo.2013.05.015

BRACKENRIDGE, R.E., STOW, D.A.V. *ET AL.* 2018. Textural characteristics and facies of sand-rich contourite depositional systems. *Sedimentology*, **65**, 2223–2252, https://doi.org/10.1111/sed.12463

BRYDEN, H.L. & STOMMEL, H.M. 1984. Limiting processes that determine basic features of the circulation in the Mediterranean Sea. *Oceanological Acta*, **7**, 289–296.

BRYDEN, H.L., CANDELA, J. & KINDER, T.H. 1994. Exchange through the Strait of Gibraltar. *Progress in Oceanography*, **33**, 201–248, https://doi.org/10.1016/0079-6611(94)90028-0

BUITRAGO, J., GARCÍA, C., CAJEBREAD-BROW, J., JIMÉNEZ, A. & MARTÍNEZ DEL OLMO, W. 2001. Contouritas: Un Excelente Almacén Casi Desconocido (Golfo de Cádiz, SO de España). Paper presented at the Congreso Técnico Exploración y Producción REPSOL-YPF, 24–27 September 2001, Madrid, Spain.

CAKEBREAD-BROWN, J., GARCIA-MOJONERO, C., BORTZ, R., SCHWARZHANS, W., CORTES, L. & OLMO, W.M. 2003. Offshore Algarve, Portugal: A prospective extension of the Spanish Gulf of Cadiz Miocene play. AAPG Intl Conf., AAPG Bull., Barcelona, Spain.

CALVIN CAMPBELL, D. & MOSHER, D.C. 2016. Geophysical evidence for widespread Cenozoic bottom current activity from the continental margin of Nova Scotia, Canada. *In*: VAN ROOIJ, D., CALVIN CAMPBELL, D., RUEGGEBERG, A. & WÅHLIN, A. (eds) *The Contourite Log-Book: Significance for Palaeoceanography, Ecosystems and Slope*

Instability. Marine Geology, **378**, 237–260, https://doi.org/10.1016/j.margeo.2015.10.005

CANALS, M. 1985. *Estructura sedimentaria y evolución morfológico del talud y el glacis continentales del Golfo de León: fenómenos de desestabilización de la cobertera sedimentaria plio-cuaternaria*. PhD thesis, University of Barcelona.

CANDELA, J. 2001. Mediterranean Water and global circulation. *In*: SIEDLER, G., CHURCH, J. & GOULD, J. (eds) *Ocean Circulation and Climate: Observing and Modelling the Global Ocean*. International Geophysics, **77**. Academic Press, San Diego, CA, 419–429.

CHENEY, R.E. & DOBLAR, R.A. 1982. Structure and variability of the Alborán Sea frontal system. *Journal of Geophysical Research*, **87**, 585–594, https://doi.org/10.1029/JC087iC01p00585

COLLART, T., VERREYDT, W. ET AL. 2018. Sedimentary processes and cold-water coral mini-mounds at the Ferrol canyon head, NW Iberian margin. *Progress in Oceanography*, **169**, 48–65, https://doi.org/10.1016/j.pocean.2018.02.027

COLELLA, A. 1990. Active tidal sand waves at bathyal depths observed from submersible and bathysphere (Messina Strait, southern Italy). *In: Proceedings of the Thirteenth International Sedimentological Congress, 26–31 August 1990, Nottingham, England. Abstracts*. International Association of Sedimentologists, 98–99.

COMAS, M.C. & MALDONADO, A. 1988. Late Cenozoic sedimentary facies and processes in the Iberian Abyssal Plain, Site 637, ODP Leg 103. *In*: BOILLOT, G., WINTERER, E.L. ET AL. (eds) *Proceedings of the ODP Scientific Results, Volume 103*. Ocean Drilling Program, College Station, TX, 635–655.

COPARD, K., COLIN, C., FRANK, N., JEANDEL, C., MONTERO-SERRANO, J.-C., REVERDIN, G. & FERRON, B. 2011. Nd isotopic composition of water masses and dilution of the Mediterranean outflow along the South-West European margin. *Geochemistry, Geophysics, Geosystems*, **12**, Q06020, https://doi.org/10.1029/2011GC003529

CREASER, A., HERNÁNDEZ-MOLINA, F.J., BADALINI, G., THOMPSON, P., WALKER, R., SOTO, M. & CONTI, B. 2017. A Late Cretaceous mixed (turbidite–contourite) system along the Uruguayan Margin: Sedimentary and palaeoceanographic implications. *Marine Geology*, **390**, 234–253, https://doi.org/10.1016/j.margeo.2017.07.004

DELIVET, S., VAN EETVELT, B., MONTEYS, X., RIBÓ, M. & VAN ROOIJ, D. 2016. Seismic geomorphological reconstructions of Plio-Pleistocene bottom current variability at Goban Spur. *In*: VAN ROOIJ, D., CALVIN CAMPBELL, D., RUEGGEBERG, A. & WÅHLIN, A. (eds) *The Contourite Log-Book: Significance for Palaeoceanography, Ecosystems and Slope Instability*. Marine Geology, **378**, 261–275, https://doi.org/10.1016/j.margeo.2016.01.001

DEPTUCK, M.E., STEFFENS, G.S., BARTON, M., PIRMEZ, C. 2003. Architecture and evolution of upper fan channel-belts on the Niger Delta slope and in the Arabian Sea. *Marine and Petroleum Geology*, **20**, 649–676.

DÍAZ DEL RÍO, G., GONZÁLEZ, N. & MARCOTE, D. 1998. The intermediate Mediterranean water inflow along the northern slope of the Iberian Peninsula. *Oceanologica Acta*, **21**, 157–163, https://doi.org/10.1016/S0399-1784(98)80005-4

DICKSON, R.R., GOULD, W.J., MULLER, T.J. & MAILLARD, C. 1985. Estimates of the mean circulation in the deep (>2000 m) layer of the eastern North Atlantic. *Progress in Oceanography*, **14**, 103–127, https://doi.org/10.1016/0079-6611(85)90008-4

DUARTE, C.S. & VIANA, A. 2007. Santos Drift System: stratigraphic organization and implications for late Cenozoic palaeocirculation in the Santos Basin, SW Atlantic Ocean. *In*: VIANA, A. & REBESCO, M. (eds) *2007. Economic and Palaeoceanographic Significance of Contourite Deposits*. Geological Society, London, Special Publications, **276**, 171–198, https://doi.org/10.1144/GSL.SP.2007.276.01.09

ENJORLAS, J.M., GOUADAIN, J., MUTTI, E. & PIZON, J. 1986. New turbiditic model for the Lower Tertiary sands in the South Viking Graben. *In*: SPENCER, A.M. (ed.) *Habitat of Hydrocarbons on the Norwegian Continental Shelf*. Norwegian Petroleum Society, Oslo, 171–178.

ERCILLA, ET AL. 2016. Significance of bottom currents in deep-sea morphodynamics: An example from the Alboran Sea. *Marine Geology*, **378**, 157–170.

ERCILLA, G., BARAZA, J., ALONSO, B., ESTRADA, F., CASAS, D. & FARRÁN, M. 2002. The Ceuta Drift, Alboran Sea (southwestern Mediterranean). *In*: STOW, D.A.V., PUDSEY, C.J., HOWE, J.A., FAUGÈRES, J.C. & VIANA, A.R. (eds) 2002. *Deep-Water Contourite Systems: Modern Drifts and Ancient Series, Seismic and Sedimentary Characteristics*. Geological Society, London, Memoirs, **22**, 155–170, https://doi.org/10.1144/GSL.MEM.2002.022.01.12

ERCILLA, G., CÓRDOBA, D. ET AL. 2006. Geological characterization of the *Prestige* sinking area. *Marine Pollution Bulletin*, **53**, 208–219, https://doi.org/10.1016/j.marpolbul.2006.03.016

ERCILLA, G., CASAS, D. ET AL. 2008a. Morphosedimentary features and recent depositional architectural model of the Cantabrian continental margin. *Marine Geology*, **247**, 61–83, https://doi.org/10.1016/j.margeo.2007.08.007

ERCILLA, G., GARCÍA-GIL, S. ET AL. 2008b. High resolution seismic stratigraphy of the Galicia Bank Region and neighbouring abyssal plains (NW Iberian continental margin). *Marine Geology*, **249**, 108–127, https://doi.org/10.1016/j.margeo.2007.09.009

ERCILLA, G., CASAS, D. ET AL. 2009. Cartografiando la dinámica sedimentaria de la región del Banco de Galicia. *In: Nuevas Contribuciones al Margen Ibérico: 6th Symposium on the Atlantic Iberian Margin (MIA 09) Abstract Volume*. ASG, Oviedo, Spain, 201–204.

ERCILLA, G., CASAS, D. ET AL. 2011. Imaging the recent sediment dynamics of the Galicia Bank region (Atlantic, NW Iberian Peninsula). *Marine Geophysical Researches*, **32**, 99–126, https://doi.org/10.1007/s11001-011-9129-x

ERCILLA, G., JUAN, C. ET AL. 2016. Significance of bottom currents in deep-sea morphodynamics: An example from the Alboran Sea. *In*: VAN ROOIJ, D., CALVIN CAMPBELL, D., RUEGGEBERG, A. & WÅHLIN, A. (eds) *The Contourite Log-Book: Significance for Palaeoceanography, Ecosystems and Slope Instability*. Marine Geology, **378**, 157–170, https://doi.org/10.1016/j.margeo.2015.09.007

ESTERAS, M., IZQUIERDO, J., SANDOVAL, N.G. & MAMAD, A. 2000. Evolución morfológica y estratigráfica plio-cuaternaria del Umbral de Camarinal (Estrecho de

Gibraltar) basada en sondeos marinos. *Revista de la Sociedad Geológica de España*, **13**, 539–550.

EXPEDITION 339 SCIENTISTS 2012. *Mediterranean Outflow: Environmental Significance of the Mediterranean Outflow Water and its Global Implications*. IODP Preliminary Report 339, https://doi.org/10.2204/iodp.pr.339.2012

FABRES, J., CALAFAT, A., SANCHEZ-VIDAL, A., CANALS, M. & HEUSSNER, S. 2002. Composition and spatio-temporal variability of particle fluxes in the Western Alboran Gyre, Mediterranean Sea. *Journal of Marine Systems*, **33-34**, 431–456, https://doi.org/10.1016/S0924-7963(02)00070-2

FAUGÈRES, J.-C. & STOW, D.A.V. 1993. Bottom-current-controlled sedimentation: a synthesis of the contourite problem. *Sedimentary Geology*, **82**, 287–297, https://doi.org/10.1016/0037-0738(93)90127-Q

FAUGÈRES, J.-C. & STOW, D.A.V. 2008. Sediment drift drifts: nature, evolution and controls. *In*: REBESCO, M. & CAMERLENGHI, A. (eds) *Sediment Drifts*. Developments in Sedimentology, **60**. Elsevier, Amsterdam, 259–288.

FAUGÈRES, J.C., GONTHIER, E. & STOW, D.A.V. 1984. Contourite drift molded by deep Mediterranean outflow. *Geology*, **12**, 296–300, https://doi.org/10.1130/0091-7613(1984)12<296:CDMBDM>2.0.CO;2

FAUGÈRES, J.C., CREMER, M., MONTEIRO, H. & GASPAR, L. 1985a. Essai de reconstitution des processus d`edification de la ride sedimentaire de Faro (Marge Sud-Portugaise). *Bulletin de l'Institut de géologie du Bassin d'Aquitaine*, **37**, 229–258.

FAUGÈRES, J.C., FRAPPA, M., GONTHIER, E., DE RESSEGUIER, A. & STOW, D.A.V. 1985b. Modelé et facies de type contourite a la surface d'une ride sédimentaire édifiée par des courants issus de la veine d'eau méditerranéenne (ride du Faro, Golfe de Cadix). *Bulletin de la Société Géologique de France I*, **8**, 35–47.

FAUGÈRES, J.C., STOW, D.A.V., IMBERT, P. & VIANA, A. 1999. Seismic features diagnostic of contourite drifts. *Marine Geology*, **162**, 1–38, https://doi.org/10.1016/S0025-3227(99)00068-7

FIÚZA, A.F.G. 1984. *Hidrologia e dinamica das aguas costeiras de Portugal*. PhD thesis, University of Lisbon.

FIÚZA, A.F.G., HAMANN, M., AMBAR, I., DÍAZ DEL RÍO, G.D., GONZÁLEZ, N. & CABANAS, J.M. 1998. Water masses and their circulation off western Iberia during May 1993. *Deep Sea Research*, **45**, 1127–1160, https://doi.org/10.1016/S0967-0637(98)00008-9

GARCÍA, M. 2002. *Caracterización Morfológica del Sistema de Canales y Valles Submarinos del Talud Medio del Golfo de Cádiz (SO de la Península Ibérica): Implicaciones Oceanográficas*. Degree thesis, University of Cádiz.

GARCÍA, M., HERNÁNDEZ-MOLINA, F.J. *ET AL.* 2009. Contourite erosive features caused by the Mediterranean Outflow Water in the Gulf of Cadiz: Quaternary tectonic and oceanographic implications. *Marine Geology*, **257**, 24–40, https://doi.org/10.1016/j.margeo.2008.10.009

GARCÍA, M., HERNÁNDEZ-MOLINA, F.J., ALONSO, B., VÁZQUEZ, J.T., ERCILLA, G., LLAVE, E. & CASAS, D. 2016. Erosive sub-circular depressions on the Guadalquivir Bank (Gulf of Cadiz): Interaction between bottom current, mass-wasting and tectonic processes.

In: VAN ROOIJ, D., CALVIN CAMPBELL, D., RUEGGEBERG, A. & WÅHLIN, A. (eds) *The Contourite Log-Book: Significance for Palaeoceanography, Ecosystems and Slope Instability*. Marine Geology, **378**, 5–19, https://doi.org/10.1016/j.margeo.2015.10.004

GARCÍA-MOJONERO, C. & MARTÍNEZ DEL OLMO, W. 2001. One sea level fall and four different gas plays: the Gulf of Cadiz Basin, SW Spain. *In*: FILLON, R.H. & ROSEN, N.C. (eds) *Petroleum Systems of Deep-Water Basins: Global and Gulf of Mexico Experience*. Gulf Coast Section SEPM Foundation 21st Annual Bob F. Perkins Research Conference 2001. Program and Abstracts. Gulf Coast Section SEPM (GCSSEPM), Houston, TX, 357–367.

GASCARD, J.C. & RICHEZ, C. 1985. Water masses and circulation in the Western Alboran Sea and in the Straits of Gibraltar. *Progress in Oceanography*, **15**, 157–216, https://doi.org/10.1016/0079-6611(85)90031-X

GEBCO 2003. *One Minute Global Bathymetric Grid from the General Bathymetric Chart of the Oceans*, https://www.gebco.net

GONTHIER, E.G., FAUGÈRES, J.C. & STOW, D.A.V. 1984. Contourite facies of the Faro Drift, Gulf of Cadiz. *In*: STOW, D.A.V. & PIPER, D.J.W. (eds) 1984. *Fine-Grained Sediments, Deep-Water Processes and Facies*. Geological Society, London, Special Publications, **15**, 275–292, https://doi.org/10.1144/GSL.SP.1984.015.01.18

GONZÁLEZ-POLA, C. 2006. *Variabilidad climática en la región sureste del Golfo de Vizcaya*. Master's thesis, University of Oviedo.

GRUETZNER, J. & UENZELMANN-NEBEN, G. 2016. Contourite drifts as indicators of Cenozoic bottom water intensity in the eastern Agulhas Ridge area, South Atlantic. *In*: VAN ROOIJ, D., CALVIN CAMPBELL, D., RUEGGEBERG, A. & WÅHLIN, A. (eds) *The Contourite Log-Book: Significance for Palaeoceanography, Ecosystems and Slope Instability*. Marine Geology, **378**, 350–360, https://doi.org/10.1016/j.margeo.2015.12.003

GRÜNDLINGH, M. 1981. On the observation of a solitary event in the Mediterranean Outflow west of Gibraltar. *Meteor-Forschungsergebnisse A/B*, **23**, 15–46.

HABGOOD, E.L. 2002. *Alongslope and Downslope Sediment Transport Processes in the Gulf of Cádiz*. PhD thesis, University of Southampton.

HABGOOD, E.L., KENYON, N.H., MASSON, D.G., AKHMETZHANOV, A., WEAVER, P.P.E., GARDNER, J. & MULDER, T. 2003. Deep-water sediment wave fields, bottom current sand channels and gravity flow channel–lobe systems: Gulf of Cádiz, NE Atlantic. *Sedimentology*, **50**, 483–510, https://doi.org/10.1046/j.1365-3091.2003.00561.x

HANEBUTH, T.J.J., ZHANG, W., HOFMANN, A.L., LOEWEMARK, L.A. & SCHWENK, T. 2015. Oceanic density fronts steering bottom-current induced sedimentation deduced from a 50 ka contourite drift record and numerical modelling (off NW Spain). *Quaternary Science Reviews*, **112**, 207–225, https://doi.org/10.1016/j.quascirev.2015.01.027

HANQUIEZ, V. 2006. *Processus sédimentaires et activité de la Veine d'Eau Méditerranéenne au cours du Quaternaire terminal dans le Golfe de Cadix*. Doctoral thesis, University of Bordeaux 1.

Hanquiez, V., Mulder, T., Lecroart, P., Gonthier, E., Marchès, E. & Voisset, M. 2007. High resolution seafloor images in the Gulf of Cadiz, Iberian margin. *Marine Geology*, **28**, 42–59, https://doi.org/10.1016/j.margeo.2007.08.002

Hay, W.W. 2009. Cretaceous oceans and ocean modeling. *In*: Hu, X., Wang, C., Scott, R.W., Wagreich, M. & Jansa, L. (eds) *Cretaceous Oceanic Red Beds: Stratigraphy, Composition, Origins, and Paleoceanographic and Paleoclimatic Significance*. SEPM Special Publications, **91**, 243–271.

Haynes, R. & Barton, E.D. 1990. A poleward flow along the Atlantic coast of the Iberian Peninsula. *Journal of Geophysical Research*, **95**, 11425–11441, https://doi.org/10.1029/JC095iC07p11425

Heburn, G.W., La Violette, P.E. 1990. Variations in the structure of the anticyclonic gyres found in the Alboran Sea. *Journal of Geophysical Researches*, **95**, 1599–1613.

Hernández-Molina, F.J., Llave, E. et al. 2003. Looking for clues to paleoceanographic imprints: a diagnosis of the gulf of Cadiz contourite depositional systems. *Geology*, **31**, 19–22, https://doi.org/10.1130/0091-7613(2003)031<0019:LFCTPI>2.0.CO;2

Hernández-Molina, F.J., Llave, E. et al. 2006. The contourite depositional system of the Gulf of Cádiz: A sedimentary model related to the bottom current activity of the Mediterranean outflow water and its interaction with the continental margin. *Deep Sea Research II*, **53**, 1420–1463, https://doi.org/10.1016/j.dsr2.2006.04.016

Hernández-Molina, F.J., Llave, E. & Stow, D.A.V. 2008a. Continental slope contourites. *In*: Rebesco, M. & Camerlenghi, A. (eds) *Contourites*. Developments in Sedimentology, **60**. Elsevier, Amsterdam, 379–408.

Hernández-Molina, F.J., Maldonado, A. & Stow, D.A.V. 2008b. Sediment drift of Abyssal plains and oceanic basins. *In*: Rebesco, M. & Camerlenghi, A. (eds) *Contourites*. Developments in Sedimentology, **60**. Elsevier, Amsterdam, 347–378.

Hernández-Molina, F.J., Serra, N., Stow, D.A.V., Ercilla, G., Llave, E. & Van Rooij, D. 2011. Along-slope oceanographic processes and sedimentary products around Iberia. *In*: Hernández-Molina, F.J., Stow, D.A.V. et al. (eds) *Deep Water Circulation: Processes and Products*. Geo-Marine Letters, **31**, 315–341.

Hernández-Molina, F.J., Stow, D.A.V. & Alvarez-Zarikian, C., Expedition IODP 339 Scientists 2013. IODP Expedition 339 in the Gulf of Cadiz and off West Iberia: decoding the environmental significance of the Mediterranean Outflow Water and its global influence. *Scientific Drilling*, **16**, 1–11.

Hernández-Molina, F.J., Llave, E. et al. 2014a. Contourite processes associated with the Mediterranean Outflow Water after its exit from the Strait of Gibraltar: Global and conceptual implications. *Geology*, **42**, 227–230, https://doi.org/10.1130/G35083.1

Hernández-Molina, F.J., Stow, D.A.V., Alvarez-Zarikian, C. & Expedition IODP 339 Scientists 2014b. Onset of Mediterranean outflow into the North Atlantic. *Science*, **344**, 1244–1249, https://doi.org/10.1126/science.1251306

Hernández-Molina, F.J., Wåhlin, A. et al. 2015. Procesos oceanográficos y sus productos alrededor del margen de Iberia: Una nueva aproximación multidisciplinar. *In*: *Boletin Geologico y Minero*. **126**, 2–3, 279–326.

Hernández-Molina, F.J., Sierro, F.J. et al. 2016a. Evolution of the Gulf of Cadiz margin and southwest Portugal contourite depositional system: Tectonic, sedimentary and paleoceanographic implications from IODP expedition 339. *Marine Geology*, **377**, 7–39, https://doi.org/10.1016/j.margeo.2015.09.013

Hernández-Molina, F.J., Soto, M. et al. 2016b. A contourite depositional system along the Uruguayan continental margin: Sedimentary, oceanographic and paleoceanographic implications. *In*: Van Rooij, D., Calvin Campbell, D., Rueggeberg, A. & Wåhlin, A. (eds) *The Contourite Log-Book: Significance for Palaeoceanography, Ecosystems and Slope Instability*. Marine Geology, **378**, 333–349.

Hernández-Molina, F.J., Wåhlin, A. et al. 2016c. Oceanographic processes and products around the Iberian margin: a new multidisciplinary approach. *In*: Van Rooij, D., Calvin Campbell, D., Rueggeberg, A. & Wåhlin, A. (eds) *The Contourite Log-Book: Significance for Palaeoceanography, Ecosystems and Slope Instability*. Marine Geology, **378**, 127–156.

Howe, J.A., Stoker, M.S. & Stow, D.A.V. 1994. Late Cenozoic sediment drift complex, northeast Rockall trough, North Atlantic. *Paleoceanography and Paleoclimatology*, **9**, 989–999, https://doi.org/10.1029/94PA01440

Huneke, N.V. & Stow, D. 2008. Identification of ancient contourites: problems and palaeoceanographic significance. *In*: Rebesco, M. & Camerlenghi, A. (eds) *Contourites*. Developments in Sedimentology, **60**. Elsevier, Amsterdam, 323–344.

Iglesias, J. 2009. *Sedimentation on the Cantabrian Continental Margin from Late Oligocene to Quaternary*. PhD thesis, University of Vigo.

Iorga, M. & Lozier, M.S. 1999. Signatures of the Mediterranean outflow from a North Atlantic climatology. 1. Salinity and density fields. *Journal of Geophysical Research*, **194**, 25985–26029, https://doi.org/10.1029/1999JC900115

Jané, G., Llave, E. et al. 2012. Contourite features along the northwestern Iberian continental margin and abyssal plain: the local influence of regional water masses in a down-slope dominate margin. *In*: *34th International Geological Congress (IGC)*, 5–10 August, Brisbane, Australia, CD-ROM.

Juan, C., Ercilla, G. et al. 2012. Contourite sedimentation in the Alboran Sea: Plio-Quaternary evolution. *Geo-Temas*, **13**, 1817–1820.

Juan, C., Ercilla, G. et al. 2013. Paleo-circulation patterns in the Alboran Sea inferred from the contourite register since the Messinian. *In*: *30th International Association of Sedimentologists Meeting of Sedimentology*, September 2013, Manchester, UK, Abstract Volume T3S4. International Association of Sedimentologists.

Juan, C., Ercilla, G. et al. 2014. (Paleo) circulation models in the Alboran seas during the Pliocene and Quaternary. *In*: Van Rooij, D. & Rüggeberg, A. (eds) *Book of Abstracts. 2nd Deep-Water Circulation Congress: The Contourite Log-Book, Ghent, Belgium, 10–12 September 2014*. Vilzz Special Publications, **69**, 91–92.

JUAN, C., ERCILLA, G. ET AL. 2016. Current-controlled sedimentation in the Alboran Sea during the Pliocene and Quaternay. In: VAN ROOIJ, D., CALVIN CAMPBELL, D., RUEGGEBERG, A. & WÅHLIN, A. (eds) The Contourite Log-Book: Significance for Palaeoceanography, Ecosystems and Slope Instability. Marine Geology, 378, 292–311, https://doi.org/10.1016/j.margeo.2016.01.006

KELLING, G. & STANLEY, D.J. 1972. Sedimentary evidence of bottom current activity, Strait of Gibraltar region. Marine Geology, 13, M51–M60, https://doi.org/10.1016/0025-3227(72)90087-4

KENYON, N.H. & BELDERSON, R.H. 1973. Bed forms of the Mediterranean undercurrent observed with side-scan sonar. Sedimentary Geology, 9, 77–99, https://doi.org/10.1016/0037-0738(73)90027-4

KUIJPERS, A. & NIELSEN, T. 2016. Near-bottom current speed maxima in North Atlantic contourite environments inferred from current-induced bedforms and other seabed evidence. In: VAN ROOIJ, D., CALVIN CAMPBELL, D., RUEGGEBERG, A. & WÅHLIN, A. (eds) The Contourite Log-Book: Significance for Palaeoceanography, Ecosystems and Slope Instability. Marine Geology, 378, 230–236, https://doi.org/10.1016/j.margeo.2015.11.003

LACOMBE, H. & LIZERAY, J.C. 1959. Sur le régime des courants dans le Détroit de Gibraltar. Comptes Rendus de l'Académie des Sciences de Paris, 248, 2502–2504.

LE FLOCH, J. 1969. Evolution rapide de regimes de circulation non permanents des couches d'eaux super-cielles dans le secteur sud-est du Golfe de Gascogne. Cahiers Oceanographiques, XXII, 269–276.

LLAVE, E. 2003. Análisis morfosedimentario y estratigráfico de los depósitos contorníticos del Golfo de Cádiz: Implicaciones paleoceanográficas. Doctoral thesis, University of Cádiz

LLAVE, E., HERNÁNDEZ-MOLINA, F.J., SOMOZA, L., DÍAZ-DEL-RÍO, V., STOW, D.A.V., MAESTRO, A. & ALVEIRINHO DIAS, J.M. 2001. Seismic stacking pattern of the Faro-Albufeira contourite system (Gulf of Cadiz): A Quaternary record of paleoceanographic and tectonic influences. Marine Geophysical Researches, 22, 487–508, https://doi.org/10.1023/A:1016355801344

LLAVE, E., HERNÁNDEZ-MOLINA, F.J., SOMOZA, L., STOW, D.A.V. & DÍAZ-DEL-RÍO, V. 2005. The contourite depositional system in the Gulf of Cadiz: a unique Quaternary example of different drifts with reservoir potential characteristics. In: WENCESLAO, M.O. (ed.) 5 Aniversario de la Asociación de Geólogos y Geofísicos Españoles del Petróleo, Asociación de Geólogos y Geofísicos Españoles del Petróleo. (ed) Aggep. W. Martínez del Olmo, 53–74.

LLAVE, E., SCHÖNFELD, J., HERNÁNDEZ-MOLINA, F.J., MULDER, T., SOMOZA, L., DÍAZ DEL RÍO, V. & SÁNCHEZ-ALMAZO, I. 2006. High-resolution stratigraphy of the Mediterranean outflow contourite system in the Gulf of Cadiz during the late Pleistocene: The impact of Heinrich events. Marine Geology, 227, 241–262, https://doi.org/10.1016/j.margeo.2005.11.015

LLAVE, E., HERNÁNDEZ-MOLINA, F.J., SOMOZA, L., STOW, D. & DÍAZ DEL RÍO, V. 2007. Quaternary evolution of the contourite depositional system in the Gulf of Cadiz. In: VIANA, A. & REBESCO, M. (eds) 2007. Economic and Palaeoceanographic Importance of Contourites. Geological Society, London, Special Publications, 276, 49–79, https://doi.org/10.1144/GSL.SP.2007.276.01.03

LLAVE, E., JANÉ, G. ET AL. 2013. Sandy contourites along the continental margin of the northwestern Iberian Peninsula. In: 30th International Association of Sedimentologists Meeting of Sedimentology, September 2013, Manchester, UK, Abstract Volume T3S4. International Association of Sedimentologists.

LLAVE, E., HERNÁNDEZ-MOLINA, F.J. ET AL. 2015. Bottom current processes along the Iberian continental margin. In: MAESTRO, A., ERCILLA, G. & HERNÁNDEZ-MOLINA, F.J. (eds) Procesos geológicos en el Margen Continental Ibérico. Boletín Geológico y Minero, 126, 219–256.

LLAVE, E., JANÉ, G., MAESTRO, A., LÓPEZ-MARTÍNEZ, J., HERNÁNDEZ-MOLINA, F.J., MINK, S. 2018. Geomorphological and sedimentary processes of the glacially influenced northwestern Iberian margin and abyssal plains. Geomorphology, 312, 60–85, https://doi.org/10.1016/j.geomorph.2018.03.022

LOUARN, E. & MORIN, P. 2011. Antarctic Intermediate Water influence on Mediterranean Sea Water outflow. Deep Sea Research Part I: Oceanographic Research Papers, 58, 932–942, https://doi.org/10.1016/j.dsr.2011.05.009

LOZIER, S. & SINDLINGER, L. 2009. On the source of Mediterranean Overflow Water property changes. Journal of Physical Oceanography, 39, 1800–1819, https://doi.org/10.1175/2009JPO4109.1

LÜDMANN, T., WIGGERSHAUS, S., BETZLER, C. & HÜBSCHER, C. 2012. Southwest Mallorca Island: A cool-water carbonate margin dominated by drift deposition associated with giant mass wasting. Marine Geology, 289, 1–15, https://doi.org/10.1016/j.margeo.2011.09.008

MADELAIN, F. 1970. Influence de la topographie du fond sur l'ecoulement méditerranéen entre le Detroit de Gibraltar et le Cap Saint-Vincent. Cahiers Océanographiques, 22, 43–61.

MAESTRO, A., LÓPEZ-MARTÍNEZ, J. ET AL. 2013. Morphology of the Iberian continental margin. Geomorphology, 196, 13–35, https://doi.org/10.1016/j.geomorph.2012.08.022

MALDONADO, A. 1979. Upper Cretaceous and Cenozoic depositional processes and facies in the distal North Atlantic continental margin off Portugal, DSDP Site 398. In: SIBUET, J.-C., RYAN, W.B.F. ET AL. (eds) Initial Reports of the Deep Sea Drilling Project, Volume 47. United States Government Printing Office, Washington, DC, 373–402.

MAESTRO, A., BOHOYO, F. ET AL. 2015. Influencia de los procesos tectónicos y volcánicos en la morfología de los márgenes continentales ibéricos. Boletín Geológico y Minero, 126, 427–482

MARCHÈS, E., MULDER, T., CREMER, M., BONNEL, C., HANQUIEZ, V., GONTHIER, E. & LECROART, P. 2007. Contourite drift construction influenced by capture of Mediterranean Outflow Water deep-sea current by the Portimão submarine canyon (Gulf of Cadiz, South Portugal). Marine Geology, 242, 247–260, https://doi.org/10.1016/j.margeo.2007.03.013

Marchès, E., Mulder, T., Gonthier, E., Cremer, M., Hanquiez, V., Garlan, T. & Lecroart, P. 2010. Perched lobe formation in the Gulf of Cadiz: Interactions between gravity processes and contour currents (Algarve Margin, Southern Portugal). *Sedimentary Geology*, **229**, 81–94, https://doi.org/10.1016/j.sedgeo.2009.03.008

Martins, C.S., Hamann, M. & Fiuza, A.F.G. 2002. Surface circulation in the eastern North Atlantic from drifters and altimetry. *Journal of Geophysical Research*, **107**, 3217, https://doi.org/10.1029/2000JC000345

Masson, D.G., Wynn, R.B. & Bett, B.J. 2004. Sedimentary environment of the Faeroe–Shetland Channel and Faeroe Bank channels, NE Atlantic, and the use of bedforms as indicators of bottom current velocity in the deep ocean. *Sedimentology*, **51**, 1–35, https://doi.org/10.1111/j.1365-3091.2004.00668.x

Mauffret, A. 1979. Etude géodynamique de la marge des Illes Baléares. *Mémoires de la Société Géologique de France*, **LVI**, 1–96.

Mazé, J.P., Arhan, M. & Mercier, H. 1997. Volume budget of the eastern boundary layer off the Iberian Peninsula. *Deep Sea Research*, **44**, 1543–1574, https://doi.org/10.1016/S0967-0637(97)00038-1

McCartney, M.S. 1992. Recirculating components to the deep boundary current of the northern North Atlantic. *Progress in Oceanography*, **29**, 283–383, https://doi.org/10.1016/0079-6611(92)90006-L

McCartney, M.S. & Talley, L.D. 1982. The Subpolar Mode Water of the North Atlantic Ocean. *Journal of Physical Oceanography*, **12**, 1169–1188, https://doi.org/10.1175/1520-0485(1982)012<1169:TSMWOT>2.0.CO;2

McCave, I.N., Hall, I.R. *et al.* 2001. Distribution, composition and flux of particulate material over the European margin at 47°–50° N. *Deep Sea Research I*, **48**, 3107–3139, https://doi.org/10.1016/S0967-0645(01)00034-0

Mélières, F., Nesteroff, W.D. & Lancelot, Y. 1970. Etude photographique des fonds du Golfe de Cádiz. *Cahiers Oceanographiques*, **22**, 63–72.

Milkert, D., Alonso, B., Liu, L., Zhao, X., Comas, M. & de Kaenel, E. 1996. Sedimentary facies and depositional history of the Iberia abyssal plain. *In*: Whitmarsh, R.B., Sawyer, D.S., Klaus, A. & Masson, D.G. (eds) *Proceedings of the Ocean Drilling Program Scientific Results, Volume 149*. Ocean Drilling Program, College Station, TX, 685–704.

Millot, C. 1999. Circulation in the Western Mediterranean Sea. *Journal of Marine Systems*, **20**, 423–442, https://doi.org/10.1016/S0924-7963(98)00078-5

Millot, C. 2009. Another description of the Mediterranean outflow. *Progress in Oceanography*, **82**, 101–124, https://doi.org/10.1016/j.pocean.2009.04.016

Millot, C. 2014. Heterogeneities of in- and out-flows in the Mediterranean Sea. *Progress in Oceanography*, **120**, 254–278, https://doi.org/10.1016/j.pocean.2013.09.007

Millot, C., Candela, J., Fuda, J.L. & Tber, Y. 2006. Large warming and salinification of the Mediterranean outflow due to changes in its composition. *Deep Sea Research I*, **53**, 656–666, https://doi.org/10.1016/j.dsr.2005.12.017

Mougenot, D. 1989. *Geologia da Margem Portuguesa*. Instituto Hidrográfico, Lisbon.

Mulder, T., Lecroart, P. *et al.* 2002. Past deep-ocean circulation and the palaeoclimate record-Gulf of Cadiz. *EOS, Transations of the American Geophysical Union*, **83**, 481–488, https://doi.org/10.1029/2002EO000337

Mulder, T., Voisset, M. *et al.* 2003. The Gulf of Cadiz: an unstable giant contouritic levee. *Geo-Marine Letters*, **23**, 7–18, https://doi.org/10.1007/s00367-003-0119-0

Mulder, T., Lecroart, P. *et al.* 2006. The western part of the Gulf of Cadiz: contour currents and turbidity currents interactions. *Geo-Marine Letters*, **26**, 31–41, https://doi.org/10.1007/s00367-005-0013-z

Mulder, T., Faugères, J.C. & Gonthier, E. 2008. Mixed turbidite–contourite systems. *In*: Rebesco, M. & Camerlenghi, A. (eds) *Contourites*. Developments in sedimentology, **60**. Elsevier, Amsterdam, 435–456.

Mulder, T., Hassan, R. *et al.* 2013. Contourites in the Gulf of Cadiz: a cautionary note on potentially ambiguous indicators of bottom current velocity. *Geo-Marine Letters*, **33**, 357–367, https://doi.org/10.1007/s00367-013-0332-4

Mutti, E. 1992. *Turbidite Sandstones : San Donato*. Milanese, Universitá di Parma, Agip, 275 p.

Mutti, E. & Carminatti, M. 2012. Deep-water sands of the Brazilian offshore basins. Search and Discovery Article 30219 presented at the American Association of Petroleum Geologists International Conference and Exhibition, October 23–26, 2011, Milan, Italy.

Mutti, E., Barros, M., Possato, S. & Rumenos, L. 1980. Deep-sea fan turbidite sediments winnowed by bottom-currents in the Eocene of the Campos Basin, Brazilian Offshore. *In*: *1st European Meeting of the International Association of Sedimentologists, Bochum, Abstracts*. International Association of Sedimentologists, 114.

National Oceanographic Data Center 2012. *World Ocean Atlas 2012*. National Oceanographic Data Center, http://www.nodc.noaa.gov/

Nelson, C.H., Baraza, J. & Maldonado, A. 1993. Mediterranean undercurrent sandy contourites, Gulf of Cadiz, Spain. *Sedimentary Geology*, **82**, 103–131, https://doi.org/10.1016/0037-0738(93)90116-M

Nelson, C.H., Baraza, J., Rodero, J., Maldonado, A., Escutia, C. & Barber, J.H., Jr. 1999. Influence of the Atlantic inflow and Mediterranean outflow currents on Late Pleistocene and Holocene sedimentary facies of Gulf of Cadiz continental margin. *Marine Geology*, **155**, 99–129, https://doi.org/10.1016/S0025-3227(98)00143-1

Neves, M.C., Terrinha, P., Afilhado, A., Moulin, M., Matias, L. & Rosas, F. 2009. Response of a multi-domain continental margin to compression: Study from seismic reflection–refraction and numerical modelling in the Tagus Abyssal Plain. *Tectonophysics*, **468**, 113–130, https://doi.org/10.1016/j.tecto.2008.05.008

Nielsen, T., Kuijpers, A. & Knutz, P. 2008. Seismic expression of contourite depositional systems. *In*: Rebesco, M. & Camerlenghi, A. (eds) *Contourites*. Developments in Sedimentology, **60**. Elsevier, Amsterdam, 301–322.

OCHOA, J. & BRAY, N.A. 1991. Water mass exchange in the Gulf of Cadiz. *Deep Sea Research*, **38**, S465–S503, https://doi.org/10.1016/S0198-0149(12)80021-5

PAILLET, J. & MERCIER, H. 1997. An inverse model of the eastern North Atlantic general circulation and thermocline ventilation. *Deep Sea Research*, **44**, 1293–1328, https://doi.org/10.1016/S0967-0637(97)00019-8

PALOMINO, D., VÁZQUEZ, J.T., ERCILLA, G., ALONSO, B., LÓPEZ-GONZÁLEZ, N., DÍAZ DEL RÍO, V. & FERNÁNDEZ-SALAS, L.M. 2010. Influence of water masses on the seabed morphology around the Seamounts of the Motril Marginal Shelf (Alboran Sea, Western Mediterranean). *Geo-Temas*, **11**, 131–132.

PALOMINO, D., VÁZQUEZ, J.T., ERCILLA, G., ALONSO, B., LÓPEZ-GONZÁLEZ, N., DÍAZ DEL RÍO, V. & FERNÁNDEZ-SALAS, L.M. 2011. Interaction between seabed morphology and water masses around the seamounts on the Motril Marginal Plateau (Alboran Sea, Western Mediterranean). *Geo-Marine Letters*, **31**, 465–479, https://doi.org/10.1007/s00367-011-0246-y

PARRILLA, G. & KINDER, T.H. 1987. Oceanografía física del mar de Alborán. *Boletín Instituto Español de Oceanografía*, **4**, 133–165.

PELIZ, A., DUBERT, J., SANTOS, A., OLIVEIRA, P. & LE CANN, B. 2005. Winter upper ocean circulation in the western Iberian Basin – fronts, eddies and poleward flows: An overview. *Deep Sea Research I*, **52**, 621–646, https://doi.org/10.1016/j.dsr.2004.11.005

PEREIRA, R. & ALVES, T. 2011. Margin segmentation prior to continental break-up: A seismic-stratigraphic record of multiphased rifting in the North Atlantic (Southwest Iberia). *Tectonophysics*, **505**, 17–34, https://doi.org/10.1016/j.tecto.2011.03.011

PÉREZ, F.F., CASTRO, C.G., ÁLVAREZ–SALGADO, X.A. & RÍOS, A.F. 2001. Coupling between the Iberian basin-scale circulation and the Portugal boundary current system. A chemical study. *Deep Sea Research I*, **48**, 1519–1533, https://doi.org/10.1016/S0967-0637(00)00101-1

PERKINS, H., KINDERS, T. & VIOLETTE, P. 1990. The Atlantic inflowing the Western Alboran Sea. *Journal of Physical Oceanography*, **20**, 242–263, https://doi.org/10.1175/1520-0485(1990)020<0242:TAIITW>2.0.CO;2

PICKERING, K.T. & HISCOTT, R.N. 2016. *Deep Marine Systems. Processes, Deposits, Environments, Tectonics and Sedimentation*. John Wiley, Chichester, UK.

PICKERING, K.T., HISCOT, R.N. & HEIN, F.J. 1989. *Marine Environments. Clastic Sedimentation and Tectonics*. Unwin Hyman, London.

PINGREE, R.D. & LE CANN, B. 1990. Structure, strength and seasonality of the slope currents in the Bay of Biscay region. *Journal of the Marine Biological Association of the United Kingdom*, **70**, 857–885, https://doi.org/10.1017/S0025315400059117

POLLARD, R. & PU, S. 1985. Structure and circulation of the upper Atlantic Ocean northeast of the Azores. *Progress in Oceanography*, **14**, 443–462, https://doi.org/10.1016/0079-6611(85)90022-9

POSAMENTIER, H.W., KOLLA, V. 2003. Seismic geomorphology and stratigraphy of depositional elements in deep-water settings. *Journal of Sedimentary Research* **73**, 367–388.

POTTER, R. & LOZIER, S. 2004. On the warming and salinification of the Mediterranean outflow waters in the North Atlantic. *Geophysical Research Letter*, **31**, L01202, https://doi.org/10.1029/2003GL018161

REBESCO, M. 2005. Contourites. *In*: RICHARD, C., SELLEY, R.C., COCKS, L.R.M. & PLIMER, I.R. (eds) *Encyclopedia of Geology, Volume 4*. Elsevier, London, 513–527.

REBESCO, M. & CAMERLENGHI, A. (eds). 2008. *Contourites*. Developments in Sedimentology, **60**. Elsevier, Amsterdam.

REBESCO, M. & STOW, D.A.V. 2001. Seismic expression of contourites and related deposits: a preface. *Marine and Geophysical Research*, **22**, 303–308, https://doi.org/10.1023/A:1016316913639

REBESCO, M., HERNÁNDEZ-MOLINA, F.J., VAN ROOIJ, D. & WAHLIN, A. 2014. Contourites and associated sediments controlled by deep-water circulation processes: State-of-the-art and future considerations. *Marine Geology*, **352**, 111–154, https://doi.org/10.1016/j.margeo.2014.03.011

RIBÓ, M., PUIG, P., URGELES, R., VAN ROOIJ, D., MUÑOZ, A. 2016. Spatio-temporal evolution of sediment waves developed on the Gulf of Valencia margin (NW Mediterranean) during the Plio-Quaternary. *In*: VAN ROOIJ, D., CALVIN CAMPBELL, D., RUEGGEBERG, A. & WÄHLIN, A. (eds) *The Contourite Log-Book: Significance for Palaeoceanography, Ecosystems and Slope Instability*. Marine Geology, **378**, 276–291, https://doi.org/10.1016/j.margeo.2015.11.011

ROBINSON, A., LESLIE, W., THEOCHARIS, A. & LASCARATOS, A. 2001. Mediterranean Sea circulation. *In: Encyclopedia of Ocean Sciences*. Academic Press, London, 1689–1706.

ROGERSON, M., ROHLING, E.J., WEAVER, P.P.E. & MURRAY, J.W. 2005. Glacial to interglacial changes in the settling depth of the Mediterranean Outflow plume. *Paleoceanography and Paleoclimatology*, **20**, PA3007, https://doi.org/10.1029/2004PA001106

ROGERSON, M., ROHLING, E.J., BIGG, G.R. & RAMIREZ, J. 2012. Paleoceanography of the Atlantic-Mediterranean exchange: Overview and first quantitative assessment of climatic forcing. *Reviews of Geophysics*, **50**, RG2003, https://doi.org/10.1029/2011RG000376

ROQUE, C., DUARTE, H. *ET AL*. 2012. Pliocene and Quaternary depositional model of the Algarve margin contourite drifts (Gulf of Cadiz, SW Iberia): seismic architecture, tectonic control and paleoceanographic insights. *Marine Geology*, **303–306**, 42–62, https://doi.org/10.1016/j.margeo.2011.11.001

ROQUE, C., HERNÁNDEZ-MOLINA, F.J. *ET AL*. 2015. Slope failure and mass movements in the Sines Contourite Drift (West Portuguese Margin): preliminary results. *In*: DÍAZ DEL RÍO, V., BÁRCENAS, P. *ET AL*. (eds) *Volumen de Comunicaciones presentadas en el VIII Simposio sobre el Margen Ibérico Atlántico, MIA15*. Ediciones Sia Graf, Málaga, Spain, 579–582.

SALAT, J. & CRUZADO, A. 1981. Masses d'eau dans la Méditerranée Occidentale. Mer Catalane et eaux adjacentes. *Rapports et Proces-verbaux des Réunions. Commission Internationale pour l'Exploration Scientifique de la Mer Mediterranée, CIESM*, **27**, 201–209.

SANNINO, G., CARILLO, A. & ARTALE, V. 2007. Three-layer view of transports and hydraulics in the Strait of

Gibraltar: A three-dimensional model study. *Journal of Geophysical Research*, **112**, C03010, https://doi.org/10.1029/2006JC003717

SCHÖNFELD, J. & ZAHN, R. 2000. Late Glacial to Holocene history of the Mediterranean Outflow. Evidence from benthic foraminiferal assemblages and stable isotopes at the Portuguese margin. *Palaeogeography, Palaeoclimatology, Palaeoecology*, **159**, 85–111, https://doi.org/10.1016/S0031-0182(00)00035-3

SERRA, N., AMBAR, I. & KÄSE, R. 2005. Numerical modelling of the Mediterranean Water splitting and eddy generation. *Deep Sea Research II*, **52**, 383–408, https://doi.org/10.1016/j.dsr2.2004.05.025

SERRANO, A., MAESTRO, A. & NOZAL, F. (co-ords). 2005. *Memoria del mapa Geomorfológico de España a escala 1:1 000 000: Geomorfología de la zona submarina española*. Instituto Geológico y Minero de España (IGME), Madrid.

SHANMUGAM, G. 2006. *Deep-Water Processes and Facies Models: Implications for Sandstone Petroleum Reservoirs*. Elsevier, Amsterdam.

SHANMUGAM, G. 2008. Deep-water bottom currents and their deposits. *In*: REBESCO, M. & CAMERLENGHI, A. (eds) *Contourites*. Developments in Sedimentology, **60**. Elsevier, Amsterdam, 59–81.

SHANMUGAM, G. 2012. New perspectives on deep-water sandstones: origin, recognition, initiation, and reservoir quality. *Handbook of Petroleum Exploration and Production 9*. Elsevier, Amsterdam, pp. 524.

SHANMUGAM, G. 2013a. Modern internal waves and internal tides along oceanic pycnoclines: challenges and implications for ancient deep-marine baroclinic sands. *AAPG Bull.* **97**, 767–811.

SHANMUGAM, G. 2013b. New perspectives on deep-water sandstones: implications. *Pet. Explor. Dev.* **40**, 316–324.

SHANMUGAM, G., SPALDING, T.D. & ROFHEART, D.H. 1993. Processes sedimentology and reservoir quality of deep-marine bottom-current reworked sands (sandy contourites): An example from the Gulf of Mexico. *AAPG Bulletin*, **77**, 1241–1259.

SHIPBOARD SCIENTIFIC PARTY 1998. Site 1067. *In*: WHITMARSH, R.B., BESLIER, M.-O. & WALLACE, P.J. (eds) *Proceedings of the Ocean Drilling Program Initial Reports, Volume 173*. Ocean Drilling Program, College Station, TX, 107–161.

SIERRO, F.J., FLORES, J.A. *ET AL*. 1999. Messinian climatic oscilations, astronomic cyclicity and reef growth in the western Mediterranean. *Marine Geology*, **153**, 137–146, https://doi.org/10.1016/S0025-3227(98)00085-1

SLATER, D.R. 2003. *The Transport of Mediterranean Water in the North Atlantic Ocean*. PhD thesis, University of Southampton.

SOARES, D.M., ALVES, T.M. & TERRINHA, P. 2014. Contourite drifts on distal margins as indicators of established lithospheric breakup. *Earth and Planetary Science Letters*, **401**, 116–131, https://doi.org/10.1016/j.epsl.2014.06.001

STANLEY, D.J., KELLING, G., VERA, J.A. & SLIENG, H. 1975. Sands in the Alborán Sea: A Model of Input in a Deep Marine Basin. Smithsonian Contributions to the Earth Sciences, **15**.

STOKER, M.S., AKHURST, M.C., HOWE, J.A. & STOW, D.A.V. 1998. Sediment drifts and contourites on the continental margin off northwest Britain. *Sedimentary Geology*, **115**, 33–51, https://doi.org/10.1016/S0037-0738(97)00086-9

STOW, D.A.V. & FAUGÈRES, J.-C. 2008. Contourite facies and the facies model. *In*: REBESCO, M. & CAMERLENGHI, A. (eds) *Contourites*. Developments in Sedimentology, **60**. Elsevier, Amsterdam, 223–256.

STOW, D.A.V. & HOLBROOK, J.A. 1984. North Atlantic contourites: an overview. *In*: STOW, D.A.V. & PIPER, D.J.W. (eds) 1984. *Fine-Grained Sediments: Deep-Water Processes and Facies*. Geological Society, London, Special Publications, **15**, 245–256, https://doi.org/10.1144/GSL.SP.1984.015.01.16

STOW, D.A.V. & MAYALL, M. 2000. Deep-water sedimentary systems: New models for the 21st century. *Marine and Petroleum Geology*, **17**, 125–135, https://doi.org/10.1016/S0264-8172(99)00064-1

STOW, D.A.V., FAUGÈRES, J.-C. & GONTHIER, E. 1986. Facies distribution and textural variation in Faro Drift contourites: velocity fluctuation and drift growth. *Marine Geology*, **72**, 71–100, https://doi.org/10.1016/0025-3227(86)90100-3

STOW, D.A.V., FAUGÈRES, J.C., HOWE, J.A., PUDSEY, C.J. & VIANA, A. 2002a. Bottom currents, contourites and deep-sea sediment drifts: current state-of-the-art. *In*: STOW, D.A.V., PUDSEY, C.J., HOWE, J.A., FAUGÈRES, J.C., VIANA, A. (eds) 2002. *Deep-water Contourite Systems: Modern Drifts and Ancient Series, Seismic and Sedimentary Characteristics*. Geological Society, London, Memoirs, **22**, 7–20, https://doi.org/10.1144/GSL.MEM.2002.022.01.02

STOW, D.A.V., FAUGÈRES, J.C. *ET AL*. 2002b. Faro-Albufeira drift complex, northern Gulf of Cadiz. *In*: STOW, D.A.V., PUDSEY, C.J., HOWE, J.A., FAUGÈRES, J.C., VIANA, A. (eds) 2002. *Deep-water Contourite Systems: Modern Drifts and Ancient Series, Seismic and Sedimentary Characteristics*. Geological Society, London, Memoirs, **22**, 137–154, https://doi.org/10.1144/GSL.MEM.2002.022.01.11

STOW, D.A.V., PUDSEY, C.J., HOWE, J.A., FAUGÈRES, J.C., VIANA, A. (eds). 2002c. *Deep-water Contourite Systems: Modern Drifts and Ancient Series, Seismic and Sedimentary Characteristics*. Geological Society, London, Memoirs, **22**, https://doi.org/10.1144/GSL.MEM.2002.022.01.32

STOW, D., BRACKENRIDGE, R. & HERNÁNDEZ-MOLINA, F.J. 2011a. Contourite sheet sands: New deepwater exploration target. Paper presented at the AAPG 2011 Annual Convention and Exhibition, Houston, Texas, 10–13 April, 2011.

STOW, D.A.V., HERNÁNDEZ-MOLINA, F.J., HODELL, D. & ALVAREZ ZARIKIAN, C.A. 2011b. *Mediterranean Outflow: Environmental Significance of the Mediterranean Outflow Water and its Global Implications. IODP Scientist Prospectus, Volume 339*. Integrated Ocean Drilling Program Management International, Tokyo, https://doi.org/10.2204/iodp.sp.339.2011

STOW, D.A.V., HERNÁNDEZ-MOLINA, F.J., ALVAREZ ZARIKIAN, C.A. & THE EXPEDITION 339 SCIENTISTS 2013a. *Proceedings of the Integrated Ocean Drilling Program, Volume 339*. Integrated Ocean Drilling Program

Management International, Tokyo, https://doi.org/10.2204/iodp.proc.339.2013

STOW, D.A.V., HERNÁNDEZ-MOLINA, F.J., LLAVE, E., GARCÍA, M., DÍAZ DEL RIO, V., SOMOZA, L. & BRUNO, M. 2013b. The Cadiz Contourite Channel: Sandy contourites, bedforms and dynamic current interaction. *Marine Geology*, **343**, 99–114, https://doi.org/10.1016/j.margeo.2013.06.013

TOUCANNE, S., MULDER, T. ET AL. 2007. Contourites of the Gulf of Cadiz: A high-resolution record of the paleocirculation of the Mediterranean outflow water during the last 50 000 years. *Palaeogeography, Palaeoclimatology, Palaeoecology*, **246**, 354–366, https://doi.org/10.1016/j.palaeo.2006.10.007

VALENCIA, V., FRANCO, J., BORJA, A. & FONTAN, A. 2004. Hydrography of the southeastern Bay of Biscay. *In*: BORJA, A. & COLLINS, M.B. (eds) *Oceanography and Marine Environment of the Basque Country*. Elsevier Oceanography Series, **70**. Elsevier, Amsterdam, 159–193.

VAN AKEN, H.M. 2000. The hydrography of the mid-latitude Northeast Atlantic Ocean II: The intermediate water masses. *Deep Sea Research I*, **47**, 789–824, https://doi.org/10.1016/S0967-0637(99)00112-0

VANDORPE, T., VAN ROOIJ, D., STOW, D.A.V. & HENRIET, J.-P. 2011. Pliocene to recent shallow-water contourite deposits on the shelf and shelf edge off south-western Mallorca, Spain. *Geo-Marine Letters*, **31**, 391–403, https://doi.org/10.1007/s00367-011-0248-9

VANGRIESHEIM, A. & KHRIPOUNOFF, A. 1990. Near-bottom particle concentration and flux: temporal variations observed with sediment traps and nephelometer on the Meriadzek Terrace, Bay of Biscay. *Progress in Oceanography*, **24**, 103–116, https://doi.org/10.1016/0079-6611(90)90023-U

VANNEY, J.-R. & MOUGENOT, D. 1981. *La plate-forme continentale du Portugal et les provinces adjacentes: analyse geomorphologique*. Memorias dos Servicos Geologicos de Portugal, **28**.

VAN ROOIJ, D., IGLESIAS, J. ET AL. 2010. The Le Danois Contourite Depositional System: Interactions between the Mediterranean Outflow Water and the upper Cantabrian slope (North Iberian margin). *Marine Geology*, **274**, 1–20, https://doi.org/10.1016/j.margeo.2010.03.001

VARELA, R.A., ROSON, G., HERRERA, J.L., TORRES-LOPEZ, S. & FERNANDEZ-ROMERO, A. 2005. A general view of the hydrographic and dynamical patterns of the Rias Baixas adjacent sea area. *Journal of Marine Systems*, **54**, 97–113, https://doi.org/10.1016/j.jmarsys.2004.07.006

VELASCO, J.P.B., BARAZA, J., CANALS, M. & BALÓN, J. 1996. La depresión periférica y el lomo contourítico de Menorca: evidencias de la actividad de corrientes de fondo al N del talud Balear. *Geogaceta*, **20**, 359–362.

VIANA, A. 2008. Economic relevance of contourites. *In*: REBESCO, M. & CAMERLENGHI, A. (eds) *Contourites*. Developments in Sedimentology, **60**. Elsevier, Amsterdam, 493–510.

VIANA, A. & REBESCO, M. (eds). 2007. *Economic and Palaeoceanopraphic Significance of Contourite Deposits*. Geological Society, London, Special Publications, **276**, https://doi.org/10.1144/GSL.SP.2007.276.01.17

VIANA, A.R., FAUGÈRES, J.C. & STOW, D.A.V. 1998. Bottom current-controlled sand deposits – a review of modern shallow- to deep-water environments. *Sedimentary Geology*, **115**, 53–80, https://doi.org/10.1016/S0037-0738(97)00087-0

VIANA, A.R., HERCOS, C.M., ALMEIDA, W., JR, MAGALHÃES, J.L.C. & ANDRADE, S.B. 2002. Evidence of bottom current influence on the Neogene to Quaternary sedimentation along the northern Campos Slope, SW Atlantic Margin. *In*: STOW, D.A.V., PUDSEY, C.J., HOWE, J.A., FAUGÈRES, J.C. & VIANA, A.R. (eds) 2002. *Deep-Water Contourite Systems: Modern Drifts and Ancient Series, Seismic and Sedimentary Characteristics*. Geological Society, London, Memoirs, **22**, 249–259, https://doi.org/10.1144/GSL.MEM.2002.022.01.18

VIANA, A.R., ALMEIDA, W., JR, NUNES, M.C.V. & BULHÕES, E.M. 2007. The economic importance of contourites. *In*: VIANA, A.R. & REBESCO, M. (eds) 2007. *Economic and Palaeoceanographic Significance of Contourite Deposits*. Geological Society, London, Special Publications, **276**, 1–23, https://doi.org/10.1144/GSL.SP.2007.276.01.01

VIÚDEZ, A., PINOT, J.M. & HANEY, R.L. 1998. On the upper layer circulation of the Alboran Sea. *Journal of Geophysical Research: Oceans*, **103**, 21653–21666, https://doi.org/10.1029/98JC01082

VOELKER, A., LEBREIRO, S., SCHÖNFELD, J., CACHO, I., EXLENKENSER, H. & ABRANTES, F. 2006. Mediterranean outflow strengthenings during Northern Hemisphere Coolings: a salt sources for the glacial Atlantic? *Earth and Planetary Science Letters*, **245**, 39–55, https://doi.org/10.1016/j.epsl.2006.03.014

WILSON, R.C.L., SAWYER, D.S., WHITMARSH, R.B., ZERONG, J. & CARBONELL, J. 1996. Seismic stratigraphy and tectonic history of the Iberia Abyssal Plain. *In*: WHITMARSH, R.B., SAWYER, D.S., KLAUS, A. & MASSON, D.G. (eds) *Proceedings of the Ocean Drilling Program Scientific Results, Volume 149*. Ocean Drilling Program, College Station, TX, 617–633.

WILSON, R.C.L., MANATSCHAL, G. & WISE, S. 2001. Rifting along non-volcanic passive margins: stratigraphic and seismic evidence from the Mesozoic successions of the Alps and western Iberia. *In*: WILSON, R.C.L., WHITMARSH, R.B., TAYLOR, B. & FROITZHEIM, N. (eds) *Non-Volcanic Rifting of Continental Margins: A Comparison of Evidence from Land and Sea*. Geological Society, London, Special Publications, **187**, 429–452, https://doi.org/10.1144/GSL.SP.2001.187.01.21

ZENK, W. 1975. On the Mediterranean outflow west of Gibraltar. *Meteor-Forschungsergebnisse*, **16**, 23–34.

ZITELLINI, N., GRÀCIA, E. ET AL. 2009. The Quest for the Africa–Eurasia plate boundary west of the Strait of Gibraltar. *Earth and Planetary Science Letters*, **280**, 13–50, https://doi.org/10.1016/j.epsl.2008.12.005

Index

Page numbers in *italics* refer to Figures. Page numbers in **bold** refer to Tables.

Abu Dabbab Formation 51, *54*
accommodation space
　generation for sag and salt sequences 317, *326*, 327, 328, 329, 333, 335, 349
　SE Indian margin *29*, 30
　Uruguay rifted margin 35
Adria, collision with Eurasia *367*, *368*, 394
African Plate *136*, 150
　convergence with Eurasian Plate 365, *368*, 369
　divergence from Arabian Plate 50
Agbada Formation 193, *195*
Agulhas Basin *136*
Agulhas Plateau *136*
Agulhas Rise *136*, 143, 150
Akata Formation 193, *195*, *199*
Alboran Basin 407
Alboran Sea, contourite features 407, 408, **410**, *414*, *415*
Alboran Trough, channel-related contourite drifts 409, **410**, *415*
Alentejo margin 417, *419*
Alpine Orogeny *368*, 369
Álvarez Cabral moat 413, 415, *416*
Alwyn Field, fault-scarp degradation 249, *252*
Amazon Craton 166, 184
Amazon divergent segment 163, *164*, *165*, 185
　geological setting 166
　seismic stratigraphy 169, *170*, 171, *173*, *174*, 176, 177, 178, *179*, *180*
Amazon Rift 87, 89
Amazon River *164*, 184
Angel Formation *263*
Angolan margin see Kwanza Basin
Antarctic Bottom Water (AABW) *406*, **407**, 416, 421
Antarctic Plate *136*
Anz fault segment 52, *53*
Aptian, Exmouth Plateau fault-scarp degradation *244*, *246*, *247*, 249, *250*, *251*, 253
Aptian Unconformity 215
Apulia Escarpment 394
Aqaba-Dead Sea transform 50
Arabella-1 well 262, *273*
Arabian Plate, divergence from African Plate 50
Argentine Basin *136*, 147, 150
　bathymetric anomaly 141, *148*
argillokinesis 194, *195*, *196*, 197, 201
Argo Abyssal Plain 206, *207*, 232, 233, 262
Argoland 209, 233
Ascension Hotspot 148
Athol Formation 233, *235*, 238, 245, 249, *263*
Atlantic Ocean see central Atlantic; Equatorial Atlantic; NE Atlantic margin; South Atlantic
Atlantic Water (AW) 407, **407**
Atlantic Water Inflow (AIW) **407**, 409
Australia, NW passive margin 2, 5, 232–253
Australian plate, collision with SE Asia microplates 233

Barcelona contourite features 407–408, *408*, **410**, *413*
Bare Formation *263*

Barreirinhas divergent segment 163, *164*, *165*, 167, 185
　geological setting 166–167
　seismic stratigraphy 169, *170*, 171, *172*, 176
Barreirinhas Fold Belt *167*
Barrow Group *235*, *236*, 238
Barrow Sub-Basin *207*, 209, 232, 233, *260*
Bartolomeu Dias sheeted drift 413, *417*
Base Cenozoic unconformity 215, 218
Base Cretaceous unconformity 215, 218
Base Messinian unconformity *372*, *373*, 376, 377
Base Middle Miocene unconformity 376
Base Miocene unconformity *372*, *373*, 375
base-salt 343–344, 346, 347
　Gulf of Mexico 335–336, *343*
　Kwanza Basin 337–338, *339*, *341*
base-salt relief 287–313
base-salt unconformity *321*, 339, *340*, *341*, *342*, *343*, 346
　Campos Basin 322
　Florida basin 318, *319*, 320
　Santos Basin 323–324
bathymetry
　present day, South Atlantic *158*
　residual anomalies 141–143
　　effect of crustal thickness 141–143, *145*
　　effect of dynamic topography 143, *146*, 147
　　effect of sediment thickness 141, *144*
　　as predictive tool 147–150, *151*, *152*
Bay of Bengal, gravity anomaly map *25*
Bay of Biscay, DSDP Leg 48 13–14, *15*
Beagle Platform 206, *207*, *208*, 209
Beagle Sub-Basin 5, *208*, 232
　Canning TQ3D seismic survey *207*, *208*, 209–211, 212–228
　Cenozoic-present day syn-inversion *208*, 209, *217*
　　faults *217*, 218
　　seismic stratigraphy 215
　　structural evolution 226, *227*
　Cretaceous post-extension *208*, 209, *216*
　　faults 218
　　seismic stratigraphy 215
　　structural evolution 226, *227*
　evolution 207, *208*, 209
　faults
　　architecture *214*, 215, *216*, *217*, 218, *219*
　　displacement analyses *221*, *222*, *223*, *225*
　　geometry and kinematic evolution 220–225
　　orthorhombic symmetry *219*, *224*, *225*, 226, 228
　　polygonal systems 226
　　vertical linkage 226
　geological setting 206–209
　horsts and graben *213*, *214*, 215, *216*
　hydrocarbons 207, 209
　Jurassic-Cretaceous syn-extension *208*, 209, *214*, *223*
　　seismic stratigraphy 215
　　structural evolution 226, *227*
　paraconformities 215
　seismic stratigraphy *210*, 212–215
　structural evolution 226, *227*

Beagle Sub-Basin (*Continued*)
  Triassic pre-extension *208*, 209, *213*
    seismic stratigraphy 212
  Triassic-Jurassic syn-extension *208*, 209
    faults *214*, 218
    seismic stratigraphy 212, 215
    structural evolution 226, *227*
  unconformities 212, 215, 218
Bengal Basin *25*, 27
Benue Trough *87*, 89, 193
Berriasian, Exmouth Plateau fault-scarp degradation *244*, 245, *246*, *247*, 249, *250*, *251*
Blanes Canyon, contourite features 407–408, *412*
Bonaparte Basin 207, 226, 232, 259, *260*
Borborema Province *164*
Border fault system 51, *52*, *55*, 56
boudinage 307, 335, 348, 353, *354*
Bouvet Hotspot *136*, 148
Brae fields, fault-scarp degradation 249
Bragge fault zone 72, *73*
Brazil, central equatorial margin *2*, *165*
  Cretaceous evolution 167–169, 185–186
  crustal domain architecture *168*, 171–177, 180–183
    exhumed domain *182*, 183, 186
    hyperextended domain *168*, *172*, 176, 181, 186
    Moho 176, 181
    necking domain *168*, 171, 176, 181, 186
    oceanic domain *168*, *174*, 177, 183, 186
    proximal domain *168*, 171, 181, 186
    transitional domain *168*, 176–177, 181, 183, 186
  crustal structure 5, 163–187
    comparative *181*, *182*, 186
  fracture zones *168*
  geological setting 165–169
  magmatism 184, 185
  mass-transport complexes 178
  seismic stratigraphy 169–171
  slide complexes 178, *180*, *181*, 184, 186
  structural highs *168*
  tectonostratigraphy *166*
  uplift and margin collapse 184–185, 187
  volcanic complexes *168*, 177–178, *179*, *180*, *181*, 184, 185
Brazilian-African conjugate margins 327, *328*, 329
breakaway zone *4*
breakup, crustal/lithospheric 27
breakup unconformity 353
breakup zone *4*
Bremstein fault complex 72, *73*
Brent fault block, fault-scarp degradation 249, *252*
Brigadier Formation 233, *235*, 238, 245, 249, *263*
Browse Basin 207, 232, 259, *260*
Byro Group *263*

Cabrera Island, contourite features 408, *413*
Cadiz contourite channel 415
Calabrian Arc *366*, 369, 371
Calypso Formation *263*
Cameroon volcanic line *93*, 95, *136*, 143
Campos Basin *328*
  margin architecture 322
  salt deposition
    OMC 6, 317, 322, 325, *326*, 327, 328, 329
    thickness 322, 325

Campos-Angola conjugate basin 322, 328
Canning Basin 207, 232
Canning TQ3D seismic survey *207*, 209–211, 212–228
Cantabrian margin
  contourite features 421
  oceanographic setting 420–421
canyons
  Exmouth Plateau 245
  Gulf of Cadiz 413
Cape Range Fracture Zone *232*
Cape Range Group 218
carbonate mounds, Beagle Sub-Basin 215
carbonate platforms, NW Egyptian-Cyrenaica margin 374, 375, 381, 384
Cauvery Basin *25*, 27
Cazadores 3D seismic survey *232*, *235*, 237
Ceará transform segment 163, *164*, *165*, 168
central Atlantic, opening 3, 393
  Jurassic V. Cretaceous *88*, 89, *92*, 93–94, 95, 104–105
  see also Equatorial Atlantic
Central North Sea, oblique rifting 278
Ceuta plastered drift *414*
Chain fracture zone *87*, 91, *92*, *93*, 94, 102, *110*
Charcot fracture zone *110*, 201
Chenaillet ophiolite 32
coal, Norian 237, 238, 245, *246*
Coastal fault system 51, 52, 54, *55*, 56, *57–58*
Colorado Basin *26*, 32
conjugate margins 3, *32*, 167
  East Greenland-Norway *91*
  Equatorial Atlantic 83–106
continent-ocean transition zone 24
  Côte d'Ivoire margin *110*
  magma-rich rifted margin 35
    Equatorial Atlantic *88*, 90
    West African margin 95, *96*, *101*
  subsidence
    South Atlantic 135, 150, 152–153
      depth through time model 137–141
  see also ocean-continent transition
contourites 6, 404
  Amazon divergent segment *173*, 185
  Beagle Sub-Basin *207*, 215
  bedforms 414
  channel-related drifts 409, **410**, *415*, *424*, **425**
  channels 414, 415, *417*, 418, **425**
  confined 409, **410**, *415*, *424*, **425**
  depositional features 404
  diapiric ridges 409, 414, 415, *417*
  erosional features 404, 409, 415, *416*, *424*
  facies change 424, 426
  furrows 409, 413, 414, 415, 418, **425**
  hydrocarbon systems 404, 405, 426–427
  Iberian continental margin 405, 407–427
    high amplitude reflectors 424, **425**, 426
    sedimentology 421–423
    seismic/sedimentary facies 423–424, **425**, 426
  moats 409, *412*, 413, *414*, 415, *415*, *416*, 417, 418, 419, 420, 421, **425**
  mounded drifts 408, 409, **410–411**, *412*, 414, *415*, *416*, *418*, 419, 420, *424*, **425**
  plastered drifts 408, 409, **410–411**, *412*, *413*, *414*, 418, *419*, *422*, *424*, **425**
  sand-rich 404, 405, 426–427

sand ribbons 414
scarps 409, 420
scours 421
sediment lobes 413, 414, 415
sediment waves 408, 414, *417*, 418, 419, 420, *424*
separated drifts 408, 409, **410–411**, *412*, 413, *415*, *416*, 417, 418, 419, *422*, *424*
sheeted drifts 409, **410–411**, 413, 414, *415*, *416*, *417*, 420, *424*, **425**
terraces 409, 413, *414*, 418, 420, 424, **425**
turbidite mixed drift deposits 404, 405
valleys 413, 414, *417*, *418*
corner folds
　NW Red Sea 72–73, *75*, 76
　Duwi half-graben 60
Cossigny Member 209, *263*
Côte d'Ivoire
　offshore structure *2*, *3*
　Romanche fracture zone 101–102
　transform margin
　　breakup 127–128
　　Cenomanian extensional faults 130
　　domino fault blocks *116*, 117
　　extension 128
　　faults
　　　Cenomanian extensional 130
　　　first generation 116–117, 119
　　　second generation 119, *120*, *121–122*, *123*, *127*
　　geological setting 109–111
　　half-graben *116*, 117
　　horst and graben *116*, 117, *118*
　　magmatic system 119, 122, *124–125*, *126*, 127
　　magmatism, mid-ocean *127*, 128
　　margin development 128–129
　　ridges *115*, *116*, 119, *121–122*, *127*, 128, *129*
　　rift-drift evolution 127
　　seafloor spreading 111, *112*, 128
　　seismic data 111–112, 114
　　seismic stratigraphy 111, *112*, *113*, 114–116
　　sills *112*, *115*, *117*, 122, *124–125*, *126*, 127
　　volcanoes 119, *120*, 122, *124–125*, *126*, *127*, 128
Côte d'Ivoire-Ghana Ridge 109, *110*, 111, 128
Coulomb-Mohr failure criterion 205
crust
　breakup 27
　thickness 3
　　effect on bathymetry 141–143, *145*
　　Equatorial Atlantic 87–89, 90
　　global 85–87
　　gravity inversion 85
　　prediction from residual bathymetry 147, *149*
　　SE Indian rifted margin *29*, *30*, 38
　　South American equatorial margin 90, *91*, *92*
　　South Atlantic, uncertainty 153–154, **155**
　　Uruguay rifted margin 35, *36*, *37*, 38
　　West African equatorial margin *93*, 94–96
CRUST1.0 crustal thickness grid 141, 143, *145*, 147, *149*
Cuvier Abyssal Plain *232*, *233*, 237, 262
Cyprus Arc *see* Hellenic-Cyprus Arc
Cyrenaica Platform *366*, 370
　megasequences 371, *373*, 374
　structure 381, *384*

DAKHLA experiment 102
Dakhla Formation 51, *54*, *58–59*, 60, *61*, *64*, 67
Dalia South Fault, footwall degradation scarps 241, 242
Dalia South-1 well *236*, 237, 238
Dampier Sub-Basin 5, 206, *207*, 209, 232, 233, 259–281, *260*
　Carboniferous-Permian syn-rift *267*, *268*, 271–272, 275, *280*
　　megasequence *265*, 266–268
　Cretaceous-Holocene passive margin, megasequence *265*, 269, *280*, 281
　depocentres 275, 278, 281
　depth structure *267*
　fault reactivation 271, *272*, *273*, 275, 279
　geological background 259, 262
　hydrocarbons 259
　Jurassic-Cretaceous post-rift *267*, *268*, *280*, 281
　　megasequence *265*, 269, 281
　Late Palaeozoic *265*, 266–269, 271–272, *273*, 275, 279, *280*
　Precambrian-Carboniferous pre-rift, megasequence 264, *265*, 266, *280*
　ramp syncline 275
　rifting 266–281
　　inherited weakness zone 275–276, *277*, 278
　　oblique 278
　　post-rift and passive margin 279, *280*, 281
　　rift basin expansion 278–279
　　rift basin localization 275–276, *277*, 278
　　Triassic-Jurassic 269, *274*, 275, 279, *280*
　seismic stratigraphy *265*
　seismic surveys 262, *264*, 266
　stratigraphy 262
　structural domains 264, *270*
　structural evolution 279–281
　structure
　　basin bounding faults 270–271, 275, 279
　　eastern margin *260*, *261*, *263*, 271, 275
　　　see also Enderby Terrace; Mermaid Nose
　　western margin *261*, *263*, 271
　　　see also Rankin Platform
　tectonics 259, 262
　tectonostratigraphy *263*, 264
　thickness 268
　Triassic post-rift *268*, *280*
　　megasequence *265*, 268–269
　Triassic-Jurassic syn-rift *267*, *268*, 275, *280*
　　megasequence *265*, 269
debris flows, Beagle Sub-Basin 215
decompression melting 24, 38, 40–41, 85, 86
　equatorial Atlantic *88*, 90
Deep Ivorian Basin *101*, 102
Delambre 1 well 209, *210*, 211, *212*, *213*, 215
Delambre Formation 218, *235*, *263*
deltas, progradational 5, 193–203
　see also Niger Delta; Nile Delta
Demerara margin 168
Demerara Plateau *87*, *91*, *92*, 95, *97*, *103*, 105
Dense Mediterranean Waters (DMW) 407, **407**
Diego Cao contourite channel 415, *417*
Dingo Claystone *263*
Discovery Seamounts *136*, 148, 150
distal domain *4*, 346, 347, 353, *354*
divergent margins, Equatorial Atlantic *164*

Dockrell Formation *263*
domino extensional faults
　Dampier Sub-Basin 271, 281
　Exmouth Plateau 233
　northern Libyan margin 371
domino fault blocks, Côte d'Ivoire margin *116*, 117
DSDP Leg 48, work of David Roberts 13–14, *15*
Duwi accommodation zone 50, 52, *55*, 56, *57*
Duwi Formation 51, *54*, *58–59*, *61*, *64*, 67
Duwi half-graben 56, *57–58*, 59–60, *61*, *62*, *63*, 72
　relay ramps 56, *57*, 60, *74–75*
Duwi Sub-Basin 51, *53*, 70
dynamic topography
　effect on bathymetry 143, *146*, 147
　prediction from bathymetric anomalies 150, *152*
　uncertainty in palaeobathymetric modelling 154–155

Eagle Graben 269
Early Cretaceous unconformity 215
Early Eocene unconformity *372*, *373*, 375
East Greenland-Norway conjugate margin *91*
East Mediterranean Basin
　evolution, tectonic models 365, *366*
　see also SE Mediterranean passive margin
East North Atlantic Central Water (ENACW) *406*, 416, 421
Eastern Alboran Gyre (EAG) 407, **407**
Eastern Margin *260*, *261*, *263*, 264, 271
Eastern North Atlantic Central Water (ENACW) **407**, 409, 424
Egyptian margin *see* NW Egyptian margin
El Quwyh shear zone *52*, *55*, *58*
Elan Bank microcontinent 27
Eliassen Formation *263*
Eliassen Terrace 264, *266*, *270*, 278
Enderby Terrace *260*, *263*, 264, *266*, 267, 269, *270*, 275
　faulting 270–271, *272*
Eocene unconformity 215, 218
equatorial African margin 109
　see also Côte d'Ivoire
Equatorial Atlantic
　crustal age *164*
　crustal thickness 87–89, *90*
　gravity inversion 83–85, *88*, 89–90
　margins 2, 3, 5, 90–102, 105–106, *164*
　ocean isochrons *88*, 89
　plate reconstruction 102–105
　rifting 167–169
　sediment thickness *88*, 89
Eratosthenes Seamount *366*, *368*, 370, *378*, 379, *380*
　rifting 393
Esna Formation 51, *54*, *58–59*, 60, *61*, *64*, 67
Etendeka LIP *26*
Eurasian Plate, convergence with African Plate 365, *368*, 369
evaporite deposition *see* salt deposition
exhumation faults, SE Indian rifted margin 29, 30, 31
exhumed mantle domain *see* mantle exhumation
Exmouth Plateau 5, 206, *207*, *208*, 232, 233, 237
　Cazadores seismic survey 237
　domino-style faults 259, 262
　fault reactivation 249, *250*, *251*
　fault-scarp degradation 232, 242–253
　　comparative examples 249, 252

　controls 245
　models 249, *250*, *251*
　Rhaetian *239*, 241–249, *250*, *251*, 253
　geology 232–233
　hanging-wall talus wedges 243, *244*, 245, *247*, 249, *250*, *251*, 253, 275
　megasequences 238
　structure *239*, *240*, 241–242, *261*
Exmouth Sub-Basin *207*, 209, *232*, 275

failed breakup basin, Guinea Plateau 95, *96*, *103*, 105
Falkland Plateau *136*
Falkland-Agulhas fracture zone 32, *136*
Faro-Albufeira mounded drifts 413, 415
Faro-Cadiz sheeted drift 413, *416*, *417*, 426
fault reactivation
　Dampier Sub-Basin 271, *272*, 275, 279
　Exmouth Plateau 249, *250*, *251*
fault systems
　Côte d'Ivoire margin *115*, 116–119, *120*, *121–122*, *123*, *127*, 130
　extensional folds 49–50, 245
　Red Sea 50–76
fault-propagation folding 231, 249
fault-scarp degradation 231
　Exmouth Plateau 5, 232, *234*, *236*, 242–253, *244*
　　controls 245
　　models 249, *250*, *251*
　Viking Graben 249, *252*, 253
faulting
　Beagle Sub-Basin 215–228
　　displacement analyses *221*, *222*, 225
　　geometry and kinematic evolution 220–225
　　orthorhombic symmetry *219*, *224*, 225
　　polygonal systems 226
　　vertical linkage 226
　Dampier Sub-Basin 270–271, 275–281
　development models 205–206
　orthorhombic, Beagle Platform 206
　ramp-flat-ramp geometry 271, 279
　Red Sea 3, 50–51, 56–76
　SE Indian margin 29, 30–31
　SE Mediterranean passive margin *366*, 370, 374–376, 377–387
　Uruguay margin 37
Florida passive margin, salt basins 6, 318–321
Florida Platform 336, 340
flowlines, Equatorial Atlantic *87*, *88*, 89
folds
　extensional fault-related 49–50, 72, *73–75*
　compactional drape 50
　drag 49
　fault-bend 49
　fault-propagation 50
　hanging wall 50, 56, 71, *73*
　longitudinal 49–50, 72
　role of mechanical stratigraphy 50
　transverse 49–50
fault-propagation 231, 249
footwall anticline 231, 249, 275
footwall collapse structures 231, *234*
footwall degradation *234*, *239*, *240*, *244*, 245
Forestier Claystone *235*, *236*, 238, 242, *244*, 249, *250*, *263*
fracture zones, Equatorial Atlantic *87*, 89, *91*

G-Plate reconstruction 3, 103, 104, 105, 327, *328*
Galicia Bank *15*, *419*
Galician margin
  contourite features 407, **411**, 418–419
  water masses *406*
Gascoyne Abyssal Plain 206, *207*, 233, 237, 262
GDH1 thermal model 137, *138*, *139*, *140*, 153, 157
Gearle Siltstone *263*
Gebel Duwi *61*, *62*, *63*
Gebel Hamrawin granite body 60
Gijón Drift 421, *422*
Gijón moat 421, *422*
Gil Eanes furrow 415–416
Giralia Calcarenite 218
Gondwana, Neotethys Ocean 365, *367*, 369
Goodwyn horst 271
Gough Hotspot 148
graben
  Dampier Sub Basin 267, 269, 271, *274*, 279
  salt flow experiments 293–294, *295*, *297*, 298–302, *303*, *304*
  SE Indian margin 30
  Uruguay margin 33, 35
  see also horst and graben
Graciansky, Professor Pierre Charles de *16*
  work with David Roberts 1, 16, 19
gravity anomaly, free-air *84*, 85, *87*, *88*, 89
gravity anomaly inversion 3, 83–84
  3D method 84–85
    thermal correction 85
  Equatorial Atlantic 83, *88*
    conditioning 89–90
    plate reconstruction 102–105
  South American equatorial margin 91, *92*, 96–100
  West African equatorial margin 100–102
gravity anomaly maps
  Bay of Bengal *25*
  Equatorial Atlantic *84*
  Uruguay rifted margin *26*
Griffith criterion 205
Guadalquivir Bank 414, 415
Guadalquivir contourite channel 415, *417*
Guadalquivir diapiric ridge *417*
Guinea Plateau *87*, *93*
  failed breakup basin 95, *96*, *103*, 105
Gulf of Cadiz 426
  contourite features 407, **411**, 412–416, 421–422
  oceanographic setting *406*, 409, 412, *416*
Gulf of Mexico, salt basins 333, *334*, 335–357
  accommodation generation 349–350
  base salt 335–336, *343*
  basinward-dipping strata 340–341, *343*, 346
  evolution 355–357
  magma-rich v. magma-poor margin 347
  margin-scale geometries 335–337
  outer marginal collapse 6, 317, 318–321, 328, 329, 335
  rifting 355
  sag-sequence 336, 340–341, *342*
  seismic surveys 335–337, *338*, 340, *342*, *344*, *345*
  suprasalt deformation 336
Gulf of Sirt 370, *389*
  megasequences *373*, 374, 375, 376, 377
  Pan-African and Hercynian tectonic inheritance 392

  structure 381, 384
  uplift and inversion 395
Gulf of Suez rift system 50, 51, *52*
  October fault system 72, *73*
Gurupi Belt 166–167
Gurupi-São Luis graben system *164*
Gusano contourite channel 415

half-graben
  Côte d'Ivoire margin *116*, 117
  Dampier Sub-Basin 267
  North West Shelf 262
  NW Red Sea *52*, *55*, 56, 60, *65*, 66–67
  SE Indian margin 30
half-space cooling models 137, 152–153
Hamadat fault segment *52*, *55*, 56, 60
  displacement *70*, 71
  hanging wall folds *65–67*
Hamadat half-graben 60, *65*, 66–67
Hamadat Sub-Basin 51, *53*
Hamrawin relay 60, *62*, *70*
Hamrawin shear zone 52, *53*, *55*, 56, *57–58*, *61*
hanging-wall syncline 231, 249, 279
hanging-wall talus wedges, Exmouth Plateau 243, *244*, *245*, *247*, 249, *250*, *251*, 253, 275
Haycock Marl *263*
Hellenic-Cyprus Arc *366*, *368*, 369, 370
Herodotus Abyssal Plain *366*, 369, 370, 377, *378*, 379, 381
  megasequences 371, *372*, 374, 377, *382*, *383*
  oceanic crust 393
  spreading ridge 393
Herodotus Basin *366*, 370, 374
  megasequences *372*, 375, 376, 377
  structure 379, *382*
horst and graben
  Beagle Sub-Basin 206, *213*, *214*, 215
  Côte d'Ivoire margin *116*, 117, *118*
  Dampier Sub-Basin 271, 281
  see also graben; half-graben
hotspots
  and crustal thickening 143
  crustal thickness variation 147–150, *151*, 160
  and rifting 26, 27, 32–33
Huelva contourite channel 415
hydrocarbons
  Beagle Sub-Basin 207
  contourites 404, 405, 426–427
  Dampier Sub-Basin 259
  Eastern Mediterranean 365, 369
  Equatorial Atlantic margins 83
  Ivorian Coastal Basin 109, *110*, 111, 130
  Niger Delta 193, 194, *199*
hyper-extension 'thinned' domain 4, 26–27, 348
  Brazilian equatorial margin *168*, *172*, 176, 181

Iberia-Newfoundland rifted margin 32, 38, *91*
Iberian continental margin
  contourite/turbidite mixed drift deposits 405, 407, **410–411**
  water masses 405, *406*, 407
IndiaSPAN deep seismic profile 24, *25*, 28, *29*
  PSDM 24, 25, 28, *29*
  PSTM 24, 25, 28, *29*

inversion *see* SE Mediterranean passive margin, uplift and inversion
ION Span deep seismic profiles 1, 24
see also IndiaSPAN deep seismic profile; UruguaySPAN deep seismic profile
Ionian Abyssal Plain *366*, 369, 370, 371, *386*
megasequences 371, *373*, 374, 377
oceanic crust 393
opening 394
structure *385*, 387, *388*
Islas Orcadas Rise *136*, 143
isochrons, ocean *88*, 89
Ivorian Basin *101*, 102
Ivorian Coastal Basin *110*, 111
Ivorian Tano Basin 109, *110*, 111, *113*, 128, *129*
hydrocarbon prospectivity *110*, 130
tectono-stratigraphy *112*
Ivory Coast *see* Côte d'Ivoire
Ivory Coast Basin, divergent margin *164*, 167–169
Ivory Coast-Ghana transform margin *164*, 168–169

Kallahin fault segment *52*, *55*, 56, *57–58*
displacement *70*, 71
Kangaroo Syncline *232*, 233
Kendrew Trough 264, 269, 278–279
Kennedy Group *263*, 267
Kerguelen hotspot 27
Krishna-Godavari Basin *25*, 27
Kwanza Basin
outer marginal trough 341–342, *345*
sag sequence 339, *341*
salt deposition 321, 324, 335
base-salt 337–338, *339*, *341*
extensional structures 337–339
margin-scale geometries 337, *338*
Kybra-1 well 262, 266

Labrador Deep Water (LADW) *406*, **407**, 416
Labrador Sea Water (LSW) *406*, **407**, 421
Lagos Canyon 414
Lagos drift 413–414
Lambert Formation *263*
Lambert Shelf 207, *208*
large igneous provinces
crustal thickness variation 150, 157
and rifting 32–33
uncertainty in palaeobathymetric models **155**, 156–157
Late Cretaceous unconformity *372*, *373*, 374, 375
Late Eocene unconformity *372*, *373*, 375
Late Jurassic unconformity 371, *372*
lateral tectonic accommodation *326*, 327, 328, *329*
Latest Cretaceous U6 unconformity *372*, *373*, 374, 375
Le Danois Bank, contourite features 407, **411**, 421, *422*
Le Danois Drift 421, *422*
Le Danois moat 421, *422*
Legendre Formation *263*
Legendre Trend *263*
Levant Basin *366*, 370
depocentres 377, 379
megasequences *372*, 374, 375, 376, 377
rifting 379, 387, 393
structure 377, *378*, 379, *380*
Levant margin

regional structure 377, *378*, 379, *380*
transform or rifted margin 393
Levant-North Egyptian margin *366*, 370, *389*
megasequence 371, *372*, 374, 376
Pan-African and Hercynian tectonic inheritance 392
rifting 387, 393
uplift and inversion 394–396
Levantine Intermediate Water (LIW) *406*, 407, **407**, 424
Lewis Trough 264, *265*, *266*, 269, 275, 279
Liberia divergent segment *164*
Libyan margin *see* northern Libyan margin
Light Mediterranean Waters (LMW) 407, **407**
lithosphere
breakup 27
oceanic
crustal thickness estimation 153–154
thermal subsidence 135, 137
half-space model 137, 152–153
plate cooling models 137, 152–153, 157
uncertainty in calculation 150, 152–153
transitional 24
Locker Shale
Beagle Platform 209
Dampier Sub-Basin 262, *263*, 268, 272, 275, 279
Exmouth Plateau 233, *235*, 237
Louann Salt 287
Lower Crustal Body (LCB) 37, 38
Lower Deep Water (LDW) *406*, **407**, 416, 417, 421
Lüderitz Basin *26*, 32
Lyons Group *263*

Madeleine Trend 264, *266*, *270*, 278–279, 281
magma-poor (magma-starved) margins 1, 23, 27
versus magma-rich 31, 32, 39–41
salt basins 333, 347
magma-rich (magma-dominated) margins 1, 23
SDRs 33
versus magma-poor 31, 39–41, 333
magmatic addition 85, 86
magma-rich v. 'normal'
Equatorial Atlantic *88*, 90
South American equatorial margin 91, *92*, 94
West African equatorial margin *93*, 94–95
SE Indian margin 28, 30, 38
Uruguay margin 33, 35, 38
magmatic system, Côte d'Ivoire margin 119, 122, *124–125*, *126*, 127
magmatism
at rift-to-drift transition 23–24, 37–38
budget 38–41
control of melt production 40–41
Brazilian equatorial margin 184, 185
see also volcanoes, Brazilian equatorial margin
mid-ocean ridge *127*, 128
Côte d'Ivoire 128
SE Indian rifted margin 28, *29*, 30, 31, 32
emplacement *29*, 37
Uruguay rifted margin 33, *34*, 35–37
timing and budget 37–38
Mahanadi Basin *25*, 27
Main Ethiopian Rift 37, 39, 278
Mallorca contourite features 408, **410**, *413*
Malta Escarpment *366*, 371
Pan-African and Hercynian tectonic inheritance 392

structure 387, *388*
transform margin 389, 394
Mandu Limestone *263*
mantle
 composition 41
 dynamic topography 143, 147
mantle diapirs 348, 353, *354*
mantle exhumation *4*, 26, 348
 Brazilian equatorial margin *182*, 183, 186
 Campos Basin 322, 327
 Gulf of Mexico 319, 320, 347
 salt basins 353, *354*
 Santos Basin 324, *326*, 327, 328
 SE Indian rifted margin 26, *29*, 30, 32, 38–39, 40
 South Atlantic salt basins 347
mantle plume, southern Exmouth Plateau 233
Marajó graben *164*
Maranhão Fold Belt *167*
Maranhão seamounts *165*
Marathon fracture zone *136*
Marginal Ridge, Romanche fracture zone 101–102
Martin Cas Hotspot 148
mass-transport complexes
 Beagle Sub-Basin 215
 Brazilian equatorial margin 178
Mediterranean Lower Core (ML) *406*, **407**, 409
Mediterranean Outflow Water (MOW) *406*, **407**, 409, 412, 413, 416, 417, 421, 424, 426
Mediterranean Ridge accretionary complex *366*, 370, *382*, *384*
Mediterranean Sea
 contourite features 407–409, **410**
 oceanographic setting 407
Mediterranean Upper Core (MU) *406*, **407**, 409
Menorca contourite features 408, **410**, *412*
Mermaid Fault System *261*, *263*, 266, 267, 271, *273*, *276*
Mermaid Nose *260*, 262, *263*, 264, *265*, *266*, 267, 269, 271, 272, 275
Mersa Alam Formation 51, *54*
Messinian evaporites 352, *368*, 369, *372*, *373*, 376–377, *378*
Meteor Rise *136*, 143, 150
Microbaculaspora tentula 267
Mid-Atlantic Rift *87*, 89
Middle Cretaceous unconformity 371, *372*, 374
Middle Jurassic unconformity 371, *372*, 374
Miria Formation *263*
Modified Antarctic Intermediate Water (AAIW) *406*, **407**, 409
Modified Atlantic Water (MAW) *406*, 407, **407**, 424
Moho *4*
 Brazilian equatorial margin *174*, 176, *182*, 183
 depth, gravity inversion 85, *88*, 89–90
 Gulf of Mexico 336–337, 342, 348
 Kwanza Basin 337, *338*, 342, 348
 outer marginal detachment 319–320
 reflective packages 32
 seismic
  SE Indian rifted margin 28, *29*, 30, 32
  Uruguay rifted margin 33, *34*, 35
 South American equatorial margin 91, *92*
Montadert, Professor Lucien, work with David Roberts 1, 13
Muderong Shale
 Beagle Platform 209, 215
 Dampier Sub Basin *263*
 Exmouth Plateau 233, *235*, *236*, 238, 242, 243, *244*, 249
Mungaroo Formation
 Beagle Platform fault system 209
 Dampier Sub-Basin 262, *263*, 268, 272
 Exmouth Plateau fault system 233, *235*, 237, 238, 241, 242–243, 245, 249, *250*, 253
Murat Formation 233, *235*, 238, 245, 249, *263*
Murcia contourite features 408, **410**, *412*

Nakheil fault system *52*, *53*, *55*, 56, *57–59*
 displacement *70*, 71
 hanging wall folds 56–67
 relay ramps and corner folds 60, *61*, *62*, *63*, *64*
Nakheil Formation 51, *54*, 56, *58–59*, *63*, *64*
Nakheil longitudinal syncline 56, *57*, 59, *60*, *62*, *63*, *64*
Nakheil relay 60, *62*, *63*, *64*, 70
Namibian rifted margin 32
NE Atlantic margins, work of David Roberts 12–13
NE Egyptian margin, structure 379, *382*, 389
NE Georgia Rise *136*, 143, 150
NE Iberian margin, contourites 407–409
necking domain *4*, 26–27, *346*, 347, 353, *354*
 Brazilian equatorial margin *168*, 171, 176, 181
 see also thinned domain
Neotethys Ocean
 opening 209, 233, 365, *367*, *368*, 369, 393
 subduction 365, 369
Niger Delta 2, 5, *87*, 193–203
 along-strike variability 201–202
 brittle deformation model *195*, 197–201, 203
 clastic wedge deformation 193, *199*
 compressional (toe-thrust) domain 193, 194, *195*, 197, *198*, 199, 200, 201
 depobelts 201
 ductile deformation model 194, *195*, *196*, 197, *200*, 202–203
 extensional domain 193, *195*, 197, 199, 200, 201
 hydrocarbons 193, 194, *199*, 202
 jump-tie correlation 200
 mobile shale 194, *195*, *196*, 197, 201
  shale diapirism 201, 202, 203
 mud escape features *196*, 197, 202
 overpressure 194, *199*, 201, 203
 sediment thickness 89, 193
 translational domain/inner deformation front 193–194, *195*, 197, *198*, 199, 201
 wedge taper 201, *202*
Nile Delta *366*, *368*, 370, *382*
Ninetyeast Ridge *25*, 27
Norian, Exmouth Plateau fault-scarp degradation 243, *244*, 245, *246*, *250*, 253
North African margin, tectonostratigraphy *368*
North Atlantic Deep Water (NADW) *406*, **407**, 409, 416, 421, 424
North Atlantic Surface Water (NASW) 407, 409
North Rankin Formation *263*
North West Shelf, passive margin 5, 206, 207, 232, 259, *260*
 Carboniferous-Permian initiation 233, 275
 collision with Eurasian plate 262
 Cretaceous extension 233, *234*, *235*
 Cretaceous-present day passive margin 233, *234*, *235*

North West Shelf, passive margin (*Continued*)
  fault reactivation 262
  Jurassic-Cretaceous post-extension 262
    megasequence 238
  Triassic-Jurassic syn-extension 233, *234, 235,* 259
    megasequence 238, 259
  unconformity 233, 238, *240*
  see also Beagle Sub-Basin; Exmouth Plateau; Northern Carnarvon Basin
Northern Carnarvon Basin 207, 209, *232,* 259, 262
  geology 232–233
  rift system *261,* 275
northern Libyan margin
  megasequences 371, *373,* 375, 376
  rifting 393
  structure 384, *385*
Nubia Formation 51, *54, 58–59,* 59, *61, 63, 64,* 67
NW Egyptian margin
  megasequences *372*
  structure 379, 381, *383*
  transform margin 390
NW Egyptian-Cyrenaica margin *366,* 370, 374, *389*
  megasequences *372, 373,* 376, 377
  Pan-African and Hercynian tectonic inheritance 392
  rifting 393
  transform margin 389, 390
  uplift and inversion 394–395
NW Florida margin
  mantle exhumation 319
  outer marginal collapse 318–321, 325, *326,* 328, 329
  outer marginal detachment 318–320

ocean-continent transition (OCT) 85
  South American equatorial margin 91, 94
  West African equatorial margin 95
  see also continent-ocean transition zone
oceanic domain *4,* 26, *346, 354*
  Brazilian equatorial margin *168, 174,* 177, 183
  SE Indian rifted margin 26, *29,* 30
  Uruguay rifted margin *34,* 35
oceanic lithosphere *see* lithosphere, oceanic
OCTek Gravity Inversion 85, *86*
  South American equatorial margin 97–100
  West African equatorial margin 100–102
October fault system 72, *73*
Orange Basin *26,* 32
Ortegal Spur, contourite features 407, **411**, 418, *419,* 420, 422
outer domain *4, 346, 354*
outer marginal collapse
  rift-drift transition 320
  salt deposition 6, 317, 335, 352
    Campos Basin 322, 325, *326,* 328
    model 324–329
    NW Florida margin 318–321, 325, *326,* 328
    Santos Basin 323–324, 325, *326,* 328
  outer marginal detachments 318–320, *321,* 323–324, 325, *326,* 329, 349
  outer marginal troughs 319–320, *321,* 323, 324, 329, 341–342, 344, *345,* 347, 348, 349
  generation, and timing of salt deposition 350–351
overburden deformation, salt flow evolution 295–296, *297*
Oxfordian, Exmouth Plateau fault-scarp degradation *244, 246, 247,* 249, *250, 251,* 253

Oxfordian unconformity 209, 215, 233, 238, *240,* 241–242, 243, 271

palaeobathymetry
  Early Cretaceous, South Atlantic margins 3, 135–160
  grid model uncertainty 150, 152–157, *159*
    crustal thickness estimation 153–154
    dynamic topography models 154–155
    sediment calculations 154
    thermal subsidence 150, 152–153
  and plate tectonics 135
Palar-Penmar Basin *25,* 27
Pará-Maranhão Basin *167*
Pará-Maranhão slide complex 178, *180*
Pará-Maranhão transform segment 163, *164, 165,* 183–184, 185
  geological setting 166, *167*
  seismic stratigraphy *166,* 169, *170,* 171, *172, 173,* 176, *177,* 178
Paraná-Etendeka LIP *26,* 32, 33
Parnaiba Basin *164*
passive margins
  classification 1, 23
  distribution 1, *2*
  formation 1
  history of research 1
  structure 1, 3, *4,* 6–7
  see also magma-poor margins; magma-rich margins; rifted margins
Pelagian Shelf *366,* 369, 371, *389*
  Pan-African/Hercynian tectonic inheritance 392
  structure 385, 387, *388*
Pelotas Basin *26,* 32, *328*
Perseus Graben 269
Pilbara Craton 206, *207*
plate cooling models 137, 152–153, 157
plate reconstruction, Equatorial Atlantic 102–105
plate tectonics, and bathymetry 135
Portimão Canyon 409, 412, 414
Portimão drift 413–414
Portugal Coastal Counter Current (PCCC) **407**
Portugal Coastal Current (PCC) *406,* 416
Portugal Current (PC) **407**, 416
proto-oceanic domain 24, 26
  SE Indian rifted margin 26, 30
    interpretation *29,* 30, *31*
    scenarios 31–32, 38–39
  Uruguay rifted margin *34,* 35
    interpretation 35
    scenarios 35–37, 38–39
proximal domain *4,* 26, 353, *354*
  Brazilian equatorial margin *168,* 171, 181
  Gulf of Mexico salt basin *346*
  SE Indian rifted margin 26, *29,* 30
  South Atlantic salt basin *346*
  Uruguay rifted margin *34,* 35
Punta del Este Basin *26,* 33, *34,* 35

Quseir Formation 51, *54, 58–59, 61, 64*

Rajmahal Traps 27
ramp synclines 49
  Dampier Sub-Basin 275
  Kwanza Basin 300, 311
  São Paulo Plateau *312,* 313

ramp-flat-ramp fault geometry 271, 279
Ranga Formation 51, *54*
Rankin Fault System *260*, *261*, 264, *266*, 269, 270, 271, *274*, 275, *276*, 278, 279, 281
Rankin Platform *260*, 262, *263*, 264, *265*, 269, 271, *274*, 275
Rawson Basin *26*
Red Sea, NW
  accommodation zone 50, 52
  corner folds 72–73, *75*, 76
  extensional fault-related folds 50, 56–77
    evolutionary model *74–75*, 76
    segmentation and linkage 71–72
  fault blocks *55*, 56
  fault displacement *55*, 56, 67, *70*, 71
  fault systems *55*, 56
  geological setting 50–51, *52*
  half-graben basins *52*, *55*, 56, 60, *65*, 66–67
  longitudinal synclines and transverse folds 72
  stratigraphy 51–52, *54*, *58–59*, *61*
    basement 51, *53*, *58–59*, *61*
    pre-rift 51, *53*, 56, *58–59*, *61*
    synrift 51–52, *53*, 56, *58–59*, *61*
  structure *52*, *53*, *54*, *55*, 56
Red Sea Rift Basin, opening 50
relay ramps
  Duwi half-graben 56, *57*, 60, *63*, *64*, *74–75*
  Exmouth Plateau 241, *248*
Rhaetian, Exmouth Plateau fault-scarp degradation *239*, 241, 242, *243*, *244*, *246*, 247, *248*, 249, *250*, *251*, 253
Rhaetian unconformity 212
ridges
  aseismic, South Atlantic 128, 150, *151*
  Côte d'Ivoire margin *115*, *116*, 119, *121–122*, *127*
rift-drift evolution, Côte d'Ivoire margin 127
rift-to-drift transition 24, 27
  outer margin collapse 320
rifted margins
  magma budget 23–24
  magma-rich, versus magma-poor 23, 31, 39–41
rifting
  Dampier Sub-Basin 271–272, 275–281
  North African margin *368*
  oblique 278
  Red Sea 50
  salt basins 335, 353, *354*
  and salt deposition 347–349
Rio Grande fracture zone 32, *136*
Rio Grande Rise *136*, 143, 147, 157
  see also Walvis-Rio Grande ridge
ripple marks 414
Roberts, Professor David Gwynn (1943-2013) 1, 11–20, *12*, *16*, *18*
  AAPG 18
  in the Alps 16, *18*
  at BP 16–19
  at IOS 12–16
  as consultant 19
  DSDP Leg 48 13–14, *15*
  on salt tectonics 18
Rockall, work of David Roberts 12–13
Roebuck Basin 206, *207*, 259, *260*
Roebuck-1 well 262, 267

Romanche fracture zone 5, *87*, 91, *92*, *93*, 94, 95, 101–102, *110*, 111, 128, *129*, 163, *165*
Rosemary Fault System *260*, *261*, 264, *266*, 269, 270, 275, *276*, 278, 279

sag-sequence
  deposition 350, 351
  generation of accommodation 349
  Gulf of Mexico 333, 336, 340–341, *342*, 346, 347
  NW Florida basin margin *319*, 320, *321*
  Santos Basin 324, *326*
  SE Indian rifted margin 28, 30
  South Atlantic 333, 339–340, *341*, 346, 347
Sagres Drift 414
St Helena Hotspot *136*, 148
St Paul Fracture Zone 5, *87*, 91, *92*, *93*, 94, 95, 102, *110*, 163, *165*, 166, 183–184
  seismic section *172*, *173*, *175*
  volcanic complex 178, *179*
Salado Basin *26*, 32, 33
salt basins 1, 6, 194, *196*, 333–357
  base-salt relief 287, *288*, 289–313
  basinward-dipping faults 340, *342*, 346, 347
  evolutionary model 353, *354*
  fault geometry *337*, *338*, *345*, 346
  formation models 333, 335
  magma-rich v. magma-poor margins 333, 347
  mega-salt basins
    salt deposition 317, *318*
      outer marginal collapse model 324–327, 352
      see also salt deposition, in mega-salt basins
  syn-exhumation 333, 353, *354*
  three-dimensional aspects 353, 355–357
salt canopies 291, *292*, 307, *308*, 309, *310*, 311, 340
salt deposition 351–353
  accommodation generation 349–350
  in mega-salt basins 317
    air-filled hole model 317, *318*, 329
    deep-water model 321
    hypersaline model 317
    lateral tectonic accommodation 327–329, 349
    OMC model 317, 325–329, 335, 352
      Campos Basin 322, 327
      NW Florida margin 317, 318–321, *318*
      Santos Basin 322–324, 325–327
      and South Atlantic break-up 327, *328*
  and rifting 347–349
  timing, and outer marginal trough 350–351
salt diapirs 291, *292*, 299, 300–302, *303*, *304*, 305, *308*, *309*
  Gulf of Mexico 336, 342, 346
  Kwanza Basin 342, *345*, 346
salt flow 287, 289, 352
  experiment set 1
    isolated blocks 289, *290*, **291**, 305, *307*
    multiple blocks 289, *290*, **291**, 298–305, *306*
    single block 289, *290*, **291**, 293–296, *297*
  experiment set 2, diapiric growth and salt sheets 289, 291, *292*, 305, 307–311
  Gulf of Mexico 320
  overburden deformation 295–296, *297*, *301*
  Santos Basin 325, *326*, 327
  streamlines, flux and basal drag 294–296, 311
salt nappes, Gulf of Mexico 356–357

salt sheets 291, *292*, 299, 300–301, *303*, *304*, 306
  suturing 291, *292*, 307, *308*, 309
salt streams 311
Saltpond and Central basins *110*
Samadi Formation 51, *54*
San Jorge Basin *26*
San Vicente Canyon 417
Santos Basin *328*
  salt deposition 335
    outer marginal collapse 6, 317, 322–324, 325, *326*, 327, 328, 329
Santos Drift 426–427
Santos-Kwanza conjugate basin 324
São Luis Craton 166
São Paulo Plateau 323–324, 325, *326*
scarp collapse 231
  see also fault-scarp degradation
SDRs (Seaward Dipping Reflectors) 25, *26*, 340
  Gulf of Mexico 340, 347
  Uruguay rifted margin 33, *34*, 35–37
SE Asia microplates, collision with Australian plate 233
SE Indian rifted margin
  accommodation space *29*, 30
  exhumation faults *29*, 30, 31
  geological setting 27
  magma-poor 24, 27, 32
  magmatism 28, *29*, 30, 32
    budget 31, 37–39
    emplacement *29*, 37
  proto-oceanic domain
    interpretation 30–31
    scenarios 31–32, 38–39
  reflective layer 28, *29*, 32
  rifting 27
  seismic Moho 28, *29*, 32
  seismic profiles 28, *29*
    interfaces 28, *29*
  stratigraphic and basin architecture 28
  tectono-magmatic evolution 27
SE Mediterranean passive margin 366
  basement 371
  faulting *366*, 370, 374–376, 377–378
  geological setting 369
  Gulf of Sirt 370–371
  Levant-north Egyptian margin 370
  megasequences 371–377
    Jurassic-Cretaceous SR2 *368*, 371, *372*, *373*, 374
    Late Cretaceous PR1 *368*, *372*, *373*, 374
    Miocene PR4 *368*, *372*, *373*, 376
    Oligocene-Miocene PR3 *368*, *372*, *373*, 375
    Paleocene-Eocene PR2 *368*, *372*, *373*, 375
    Pliocene-Recent post-salt *368*, *372*, *373*, 377
    Triassic-Jurassic SR1 *368*, 371, *372*
    unconformities 371, *372*, *373*, 374
  NE Egyptian-Cyrenaica margin 370
  Pan-African and Hercynian tectonic inheritance 392
  Pelagian Shelf-Malta Escarpment 371
  regional structure 377–387
  rifting and breakup, tectonic evolutionary model *367*, *390*, 392–394
  segmentation *366*, 387, 389–390
    rifted margin 387, 389
    transform-dominated margin 389–390
  seismic surveys 369–370
  tectono-stratigraphic evolution 6, 365–396
  uplift and inversion *391*, 394–396
seafloor spreading 27
  Côte d'Ivoire 111, *112*, 128
  Neotethys Ocean 365, *367*, *368*, 393–394
  and outer marginal collapse 320–321, 325, *326*
  subsalt 327
seamounts
  crustal thickness variation 147–150
  separated contourite drifts 409, **411**, *415*
sediment thickness
  data 84, 85
    Equatorial Atlantic *88*, 89
  effect on bathymetry 141, *144*
  prediction from residual bathymetry anomalies 147, *148*
  uncertainty in palaeobathymetric calculation 154, **155**
Setúbal Canyon 417
Shagara Formation 51, *54*
shale, delta systems 5
  Niger Delta mobile shale model 194, *195*, *196*, 197
Shona Ridge *136*, 143, 148, 150
Sierra Leona Rise *136*
Sierra Leone Basin *110*
Sigsbee Canopy *309*, 311
sills
  Côte d'Ivoire margin *112*, *115*, *117*, 122, *124–125*, *126*, 127
  in salt basin formation 317, *318*
  Uruguay rifted margin 33, *34*, 35
Sinemurian, Exmouth Plateau fault-scarp degradation 249, *250*, *251*
Sines Drift 417, *419*
Sirt Basin 365, *366*, 369, 370
  megasequence *368*, 371, *373*, 374, 375, 376, 377, 381, 384
  rifting 387, 393–394
  structure 381, 384–385, *386*
slide complexes, Brazilian equatorial margin 178, *180*, *181*, 184, 186
slipline theory of plasticity 205
slumps, cuspate 215, 249
South African rifting margin 32
South American equatorial margin
  crustal structure 90, *92*, 93–94
  gravity anomaly inversion 91, *92*, 96–100
South American Plate *136*
South Atlantic *136*
  aseismic ridges 150, *151*
  break-up, and salt deposition 327, *328*
  COTZ depth through time model 137–141
  depth-age curve 137, *138*
  margins 2, 3, 32
    onset of rifting 32, 105
    Uruguay segment *26*
  palaeobathymetry workflow 135–160
    present-day bathymetry *158*
    residual anomalies 147–150, *151*, *152*
    uncertainty in grid models 150–157
  plate reconstruction 102–105
  sag-sequence 339–340
  salt basins 333, *334*, 335, 337–357
    accommodation generation 349–350
    magma-rich v. magma-poor margin 347

Statfjord Field, fault-scarp degradation 249, *252*
Strait of Gibraltar
  contourite features 409, 412, *416*
  water masses *406*, 407
subsidence, thermal 137
  uncertainties in calculation 150, 152–153
Syrian Arc *366*, *368*, *370*, *378*

talus wedges, hanging-wall, Exmouth Plateau 243, *244*, 245, *247*, 249, *250*, *251*, 253, 275
Tano Basin *see* Ivorian Tano Basin
Tauride microplate *390*, 393
Tethyan margins, Western Alps, work of David Roberts 16
Thebes Formation 51, *54*, *58–59*, 60, *63*, *64*
thinned domain 26–27
  SE Indian rifted margin 26, *29*, 30
  Uruguay rifted margin *34*, 35
thinning, salt basins 353, *354*
Tocantins River *164*
toe folding 320
Top Cretaceous Unconformity 215
Top Messinian unconformity 376, 377
transform margins 3
  Côte d'Ivoire 128–129
  NE Brazil 163, *164*, 165
  SE Mediterranean 393
transgressions
  Beagle Sub-Basin 209, 215
  Exmouth Plateau 233, 237, 238, *239*
transitional domain, Brazilian equatorial margin *168*, *173*, 176–177, 181, 183
Trealla Limestone *263*
Tristán da Cunha Hotspot *26*, 32, *136*, 148
Tupi Field 323
turbidites 404, 405
  mixed drift deposits *see* contourites, turbidite mixed drift deposits
Turonian, Exmouth Plateau fault-scarp degradation *244*, *246*, *247*
Tyrrhenian Dense Water (TDW) *406*, 407, **407**

Um Mahara Formation 51, *54*
underplating 91
uplift *see* SE Mediterranean passive margin, uplift and inversion
Uruguay rifted margin
  magma-rich 24, 32–37, 40
    accommodation space 35
    geological setting 32
    magmatism 33, *34*, 35–37
      timing and budget 37–39
    proto-oceanic domain
      interpretation 35
      scenarios 35–37, 38–39
    SDRs 33, *34*, 35–37
    seismic interfaces 33, *34*
    seismic Moho 33, 35

  stratigraphy and basin architecture 33, *34*
  transitional crust 37
UruguaySPAN deep seismic profile 24, *26*
  PSDM *34*
  PSTM 33, *34*

Valanginian, Exmouth Plateau fault-scarp degradation 242, *244*, *246*, *247*, 249, *250*, *251*, 253
Valanginian unconformity 209, 215, 218, 233, 238, 242, 245, 249
Valdes Basin *26*
Valencia contourite features 408, **410**, *412*
Viking Graben, fault-scarp degradation 245, 249, *252*, 253
Vizco High 421
volcanoes
  Brazilian equatorial margin *168*, *172*, *173*, *174*, 177–178, *179*, *180*, *181*, 184, 185
  Côte d'Ivoire margin 119, *120*, 122, *124–125*, *126*, 127, 128

Walcott Formation *263*
Walvis Basin *26*, 32
Walvis Ridge *26*, 143, 147, 157
Walvis-Rio Grande ridge 150, 317
West African equatorial margin
  crustal structure *93*, 94–96
  gravity anomaly inversion 100–102
West Dalia South Fault, footwall degradation scarps 241, *242*, 243, 245, *248*
West Fault System 242
West Gondwana, breakup 32
West Indian rifted margin 32
Western Alboran Gyre (WAG) 407
Western Desert *366*, *368*, 370
western Iberian margin
  contourite features 407, **411**, 417–420, *419*
  oceanographic setting *416*
Western Intermediate Water (WIW) 407, **407**
western Ivory Coast transform segment *164*, 169, 185
Western Margin 261, *263*, 264, 271
Western Mediterranean Deep Water (WMDW) *406*, 407, **407**, 424
Westralian Super-basin 207, 209, 232, 259
Wilcox Formation *263*
Windalia Radiolarite *263*
Winter Intermediate Water (WIW) *406*
Withnell Formation *263*
Wombat Plateau *208*, *232*
Worm Channel *417*

Yucatan Platform 335, 340–341

Zohr hydrocarbon reservoir 374
Zug El Bahar fault segment 52, *53*, *54*, *55*, *56*, *67*, *68–69*
  displacement *70*, 71
Zug El Bahar half-graben 67, *68–69*
Zug El Bahar Sub-Basin 51